Choyke/Matsunami/Pensl (eds.)
Silicon Carbide
Volume I

Silicon Carbide

A Review of Fundamental Questions and Applications to Current Device Technology

edited by
Wolfgang J. Choyke
Hiroyuki Matsunami
Gerhard Pensl

Volume I

Reprinted from a special issue of the journal
physica status solidi (b) – basic research

Akademie Verlag

Editors:

Wolfgang J. Choyke, University of Pittsburgh, Pittsburgh/USA
Hiroyuki Matsunami, Kyoto University, Kyoto/Japan
Gerhard Pensl, Universität Erlangen-Nürnberg, Erlangen/Germany

Volume I with 419 figures and 75 tables

Special edition of the journal issue
physica status solidi (b) **202**, No. 1 (1997)

1st edition
Library of Congress Card Number pending

Die Deutsche Bibliothek – CIP-Einheitsaufnahme

Silicon carbide : a review of fundamental questions and applications
to current device technology / ed. by Wolfgang J. Choyke ... –
Berlin : Akad. Verl.
   ISBN 3-05-501792-7

Vol. 1. – Special ed., 1. ed. – 1997
   Aus: Physica status solidi (b) ; 202. 1997

Cover illustration:
Crystal structure of 4H-SiC, taken from the paper of U. Starke in this volume. The stacking direction is
depicted by the enhanced Si–C bond train parallel to the $(11\bar{2}0)$ direction.

Printing and Bookbinding: Druckhaus „Thomas Müntzer" GmbH, Bad Langensalza

Printed in the Federal Republic of Germany

Akademie Verlag GmbH
Mühlenstr. 33–34 · D-13187 Berlin
Federal Republic of Germany

Preface

SiC is not a stranger to our Universe, having been generated in the atmospheres of stars for billions of years. Indeed, only recently isotopic analysis has indicated that hexagonal shaped microcrystals of SiC found in meteorites have their origin outside of our Solar System and were formed before the gravitational coalescence of our Solar System. Nevertheless, the germ of the idea that there is a silicon–carbon bond had to await the work of the great Swedish chemist, Jöns Jakob Berzelius, who first published his results in 1824. Things then moved more quickly and by 1891 Eugene G. Acheson had established a commercial process for making SiC in large quantities for the abrasives and steel industry. Soon thereafter, polytypism as well as semiconducting electrical characteristics were discovered in Acheson type SiC crystals. A significant breakthrough came in the late forties when Jan Antony Lely, at Philips in Eindhoven, started to develop a new class of high temperature crystal furnaces. With the Lely type crystals, fundamental studies on the electronic and lattice vibrational properties of SiC became possible. In the late seventies and early eighties new developments in boule and epitaxial growth of SiC gave us the glimmer of hope that SiC could be grown in an industrially suitable fashion and with the purity and doping control required for modern semiconductor devices. This decade has brought spectacular progress in the growth and device development of SiC. We now look with confidence to the next decade where SiC will take an important role in electric power distribution systems, turbines, electric power controls and high frequency power devices and will be used in all sorts of applications in high temperature ambients as well as dangerous radiation environments.

We have selected relatively mature areas in SiC from which to solicit our contributions to this volume. The complete volume is subdivided into sections as follows: theory, crystal growth, surface properties, characterization, processing and devices. There may be a little overlap among the papers but in general each paper represents the individual tastes, perceptions and expertise of the authors. Papers were updated to the middle of April 1997.

Such new and important areas as radiation damage and radiation monitoring devices were left out because the available literature stems by and large from decades ago and new developments are just in the formative stage.

For practical reasons not all of the papers of this compendium can be published in one volume and hence they will first appear as Vol. 202, No. 1 of physica status solidi (b) (Part I) and Vol. 162, No. 1 of physica status solidi (a) (Part II), July 1997. Contrary to the normal practice of this journal the papers have not been divided into "fundamental" or "applied" categories since we strongly believe that this collection of papers represents a coherent body of work and should be viewed as such by the reader. After publication in physica status solidi, the compendium will be published as a book consisting of two volumes. The complete Table of Contents will appear in both Volume I and Volume II and the index for both volumes will be placed at the back of Volume II. These two volumes will only be available as one entity.

We wish to express our great debt to all the authors who not only worked hard and long on their own papers but also helped in reviewing the papers of their colleagues. Many improvements resulted from the review process and we feel sure that this volume will be of greater value to the reader due to these additional efforts. We also wish to thank Dr. H.-J. Hänsch and Mrs. K. Müller of physica status solidi who have been very flexible and have accomodated all of the wishes of the guest editors. Last but certainly not least, we wish to thank Prof. Martin Stutzmann who initially conceived this project and who has been most helpful and encouraging in seeing it through to a successful conclusion.

Finally, we hope this book will in some measure enlarge the general appreciation by engineers, materials scientists and physicists for the fascinating properties of the poly-types of SiC.

Pittsburgh, Kyoto, and Erlangen, May 1997   W. J. Choyke
                                                                       H. Matsunami
                                                                       G. Pensl

Contents

Theory

Crystal growth

Surface Properties

phys. stat. sol. (b) **202**, 5 (1997)

Subject classification: 71.20.Nr; 61.50.−f; 78.20.Ci; 79.60.Bm; S6

Electronic Band Structure of SiC Polytypes: A Discussion of Theory and Experiment

W. R. L. LAMBRECHT, S. LIMPIJUMNONG, S. N. RASHKEEV, and B. SEGALL

Department of Physics, Case Western Reserve University, Cleveland, OH 44106-7079, USA

(Received January 31, 1997)

After a brief discussion of the origin of polytypes and a few general remarks on the relationship between polytypism and properties of SiC, we focus on a comparative study of their electronic band structures. We first explain how the different band structures can be put on the same footing by examining Brillouin zone folding effects. Then we discuss the dependency of some of the important eigenvalues on hexagonality. Next, we discuss some of the available spectroscopic information on the band structures. Finally, we examine some of the details near the gaps in further detail such as the location of the conduction-band minima, the effective masses and the crystal field splittings and masses of the upper valence bands. Open questions and areas where experimental verification is needed are pointed out.

1. Introduction

Much of the recent work on SiC is driven by its technological potential as a semiconductor enabling operation of devices at high-temperatures, high-power, and high-frequencies. One of the aspects that makes SiC fascinating from a fundamental sciences point of view is that in some sense SiC is not a single semiconductor but a whole class of semiconductors because of its polytypism [1]. As is well known, the term polytypism refers to one-dimensional polymorphism, i.e. the existence of different stackings of the basic structural elements: in the present case, the {111} Si–C bilayers of the cubic (zincblende) structure, or {0001} layers of the hexagonal modifications. More than 200 polytypes have been reported to date. Luckily only a few of these have practical importance. These include the cubic form 3C and the 6H and 4H hexagonal forms. The rhombohedral 15R and 21R polytypes are also fairly common and substantial information about their properties is available. The 2H form (also known as wurtzite), although rare in SiC, is of interest as the extreme hexagonal case. Once we fix the orientations of the dangling bonds (which really implies that we are thinking of the third layer!), we can define the stacking between two successive layers to be either cubic (as in 3C, where the bonds are staggered) or hexagonal (as in 2H or as near a coherent twin boundary in a bicrystal of 3C) where the bonds are eclipsing (see Fig. 1). One can then define hexagonality H as the fraction of two layer stackings that are hexagonal out of all the possible ones (cubic + hexagonal ones): $H = h/(h + c)$ where h is the number of hexagonal stackings and c the number of cubic stackings. Most of the higher period polytypes are essentially combinations of the previous ones with periodic stacking errors linking the pieces. Some of the polytypes may be considered rather exotic. In fact, one may in some cases, where only a few long periods were observed, wonder if we should not think of these as

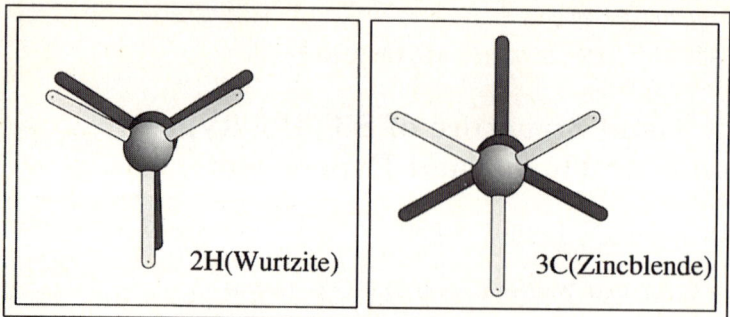

Fig. 1. Eclipsed and staggered bonds representative of hexagonal and cubic stacking

accidental (almost random but seemingly periodic) arrangements of stacking faults. On the other hand, some simpler conceivable polytypes such as $2nH$ with $n \geq 3$ occur very infrequently. Also, polytypes close to the fully hexagonal case have not been reported.

Several fundamental questions arise regarding the polytypism. Why do they occur? Why are certain polytypes more common than others? Which ones have the lowest energy, or, more generally what is the equilibrium diagram of SiC? More specifically: are the polytypes true thermodynamic phases, each with a specific domain of stability, or are they metastable states whose occurrence is kinetically determined? If the latter is true, which growth factors determine the polytype and can these factors be controlled? Finally, how do the physical properties depend on the polytype? In the second section of this paper, we make some comments on the state of knowledge of some of these questions.

Since our own work has mostly focused on the last question, that is, on how the band structures and closely related properties vary with polytype, the rest of the paper is restricted to it. We first briefly mention the main features of the computational method used in our work. Next, we examine theoretical aspects of the band structure, such as the relationship of band structures of different polytypes, the dependence on hexagonality of the minimum gap and some specific $\mathbf{k}$-point gaps. While some discussion of these issues has been given previously [2, 3], we here include rhombohedral polytypes and hypothetical high hexagonality ones. Next, we discuss the state of experimental knowledge of the band structures. First we examine the scale of several eV based on UV reflectivity and photoemission. Finally, we discuss some finer scale aspects of the band structure near the band edges, more precisely the effective mass tensors, the location of the conduction band minima and the corresponding parameters and crystal field splittings of the valence-band maximum. We present here for the first time our results on the valence bands and a new discussion of an old set of experiments probing optical transitions between the lowest few conduction bands.

2. General Discussion of Polytypism

2.1 Comments on the origin of polytypism

A very attractive theory in answer to the fundamental question "Why are there polytypes?" has in large part been given by the work of Heine and coworkers, summarized for example in [4]. On the basis of results of first-principles calculations of the total energy of the different polytypes, a model spin-Hamiltonian describing the total energy in terms of effective interlayer interactions was obtained by this group. The pseudo-spins in this model are associated with each double Si–C layer and parallel spins corresponds to cubic stacking while antiparallel spins corresponds to hexagonal stacking. First, second, and third neighbor pair interactions between the spins are included. At the simplest level of understanding, the reason for polytypism is, of course, that the difference in energy between the cubic and hexagonal stacking is small. However, this appears also to be the case in a number of other materials [5]. According to Heine et al.'s theory, the crucial point distinguishing SiC from other semiconductors is that the second nearest neighbor plane interaction is opposite in sign to the first nearest neighbor plane interaction and has about half its absolute value: $J_2 \approx -J_1/2$. This leads to the prediction that the twin boundary energy between a region of up spin and a region of down spin, which is $2(J_1 + 2J_2) \approx 0$. Hence twin boundaries are very likely to form. Since a twin boundary is nothing but a locally hexagonal stacking, this leads to a "frustration" between a preference for cubic and hexagonal stackings and hence a large number of different compromises. If the above relation between J_1 and J_2 held exactly and longer range interactions were negligible, all polytypes which have consecutive bands of two or three cubically stacked layers, alternated with hexagonal stackings would all have the same energy. These include the most common polytypes: 6H, 4H and 15R. Heine et al. [6] also offered an explanation of why the cubic stacking may occur frequently in growth: for a surface layer being deposited, only one of the second neighbor interactions is present. Hence it prefers cubic stacking. According to Heine et al.'s argument, if only surface equilibrium is achieved but not bulk equilibrium, i.e. no re-arrangements of the deeper layers take place, the resulting stacking would always be cubic!

Recent local density calculations, [7 to 9] presumably of higher numerical accuracy and convergence and which include individual atomic relaxations more completely than previously, no longer find this crucial relation between J_1 and J_2 which is at the heart of Heine et al.'s model. These calculations are further discussed by Bechstedt et al. [10] elsewhere in this volume. They find J_1 values smaller in magnitude than J_2 which predicts a definitely lower energy for 4H than 6H. The energy difference between these two and 3C is also increased relative to Heine et al.'s results. With the new values, the twin boundary energy in 3C, $2(J_1 + 2J_2)$ becomes negative. Thus, the occurrence of a large number of twins in an otherwise cubic material is even more favorable than in Heine's model. If the twins occur every two or three layers one essentially recovers the most common polytypes. So, the criterion for favoring the type of polytypism observed in SiC perhaps should be stated as $J_1 > 0$, $J_1 + 2J_2 \leq 0$.

Other fundamental problems concerning polytypes remain unanswered. Once one understands that many stackings should occur, the question arises, why are there not infinitely many and why are they not disordered? The question of what gives rise to long-range ordering in some of the more unusual polytypes, and what precisely leads to there being an advantage in having simple periodic arrangements of bands of 2 and 3 such as 6H (which is $\langle 3 \rangle$ in the Zhdanov notation [11]), 4H ($\langle 2 \rangle$) and 15R ($\langle 23 \rangle$), is still unanswered. Heine's group in their later papers [12] identified phonon contributions to the free energy as playing a role in the relative stability of 4H, 6H and 15R. For more recent work on the phonon contributions to the free energy, see Bechstedt et al. [10].

2.2 Growth issues

Heine's theory is a purely bulk thermodynamic theory. Clearly, the prediction that surface equilibrium only during growth should always lead to cubic is untenable. Heine et al. cautioned that this conclusion might be changed if the interaction parameters are different near the surface. Furthermore, if $J_1 < |J_2|$ Heine's arguments about preference for cubic growth does not hold at all because second layer interactions are dominating. However, there are other considerations to take into account. As pointed out by Matsunami and Kimoto [13], surface steps play a crucial role in maintaining the stacking of the substrate. Heine's argument hinges on the energetics of a complete layer coming down on a flat surface. Real growth, of course, proceeds by attachment of smaller fragments and the question arises how the energetics for cubic versus hexagonal deposition depends on the size of such molecular size deposits and their interaction with surface features such as steps. Experimentally, it appears that cubic attachment is indeed favorable in the middle of wide terraces. However, occasional errors occur in the very beginning of the deposition, and in those cases lead to double positioning boundaries. Near steps, the interaction with the step appears to favor the stacking already present in the substrate. This appears to be the result of a lateral interaction rather than an interplanar interaction with deeper layers, although a simple model Hamiltonian type description of such interactions has yet to be developed. Nevertheless, a crude argument can be given which indicates that indeed an elementary growth unit approaching a pre-existing island on an incomplete surface or a step, would rather conform to the step orientation. As in Heine's theory it is helpful to think of these orientations as pseudospins and we can think of the growth unit as an atom with a pseudospin. (Actually, it is useful to think of it as an atom with tetrahedrally arranged dangling bonds sticking out in well-defined directions, much as in the popular plastic molecular building-block models or Fig. 1).

A typical growth situation is illustrated in Fig. 2. The spin for an "atom" just landing on the surface is than essentially random because it can still easily reorient as it diffuses over the surface until it meets another island or step to attach to. This is essentially equivalent to saying that J_1 (of order 1 meV) for an "atom" is small compared to kT. In fact, considering a growth temperature of the order of 1000 K or about 100 meV, islands

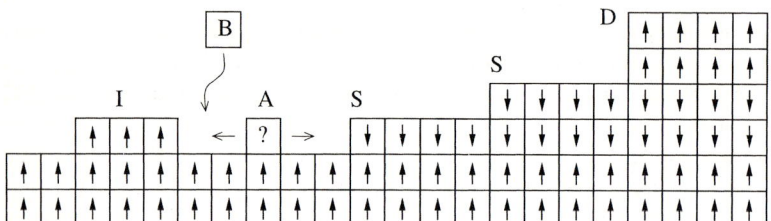

Fig. 2. Illustration of polytype growth issues. The picture shows a schematic model of six layers of a 4H polytype (4H because the spins flip every two layers) surface with (arbitrarily chosen) steps in a typical snapshot during the process of growth. On the far right is a double step D while just left of it are two monoatomic steps S with short terraces. On the far left is an island I which has grown with spin parallel to underlying substrate (according to "Heine's preference" for cubic). The "atom" A which just arrived at the surface is still undecided about its spin; can move left or right and attach either to the step or the island. As more atoms like B fill in the layer, a defect will unavoidably occur

of several tens of "atoms" still have a stacking energy difference with respect to the underlying layer which is lower than kT. So, islands of this size essentially do not care how they are stacked. What we are saying then is that once an island of a certain critical size of say ↑ exists and is approached by an undecided "spin-atom" the latter pays a relatively large energy to attach in a ↓ configuration to the underlying layer. The argument why this is so is that when we consider the result of this after a few layers, we have essentially developed an incoherent twin boundary normal to the surface. An examples of such a boundary is the $(\Sigma = 3)$ $\{211\}$ twin boundary. Calculations by Kohyama and Yamamoto [14] indate that the energies of such boundaries are of the order of 1 eV/atom as opposed to J_1 which is of the order of meV/atom. This means that during growth the pseudospin orientation is decided by the lateral interaction on the surface rather than by the interactions normal to the layers. Ultimately, however, this explains why indeed layers, or at least large islands, of a well defined spin are forming.

Beyond a certain critical size of the island, the stacking with respect to the underlying layers does regain importance. An important question is whether the islands that originally grew in an unfavorable position can overcome the barriers to move to the more energetically favorable stacking. Essentially, this requires a slip of the whole island which could be mediated by dislocations.

The other crucial ingredient in polytype selection clearly is the surface mobility which determines whether growth species move towards the steps or pre-existing islands or grow independently. One can easily see that if mobility is limited and there is a random incoming flux of particles, after a while some particles must get trapped in between pre-existing islands which may have opposite spin and hence defects are bound to occur as a result of the interplay of the stochastic nature of the processes involved.

The surface mobilities depend on the energy "landscape" of the surface, including the presence of surface impurities or protuberances, the size of terraces, the growth species and mechanisms of diffusion, and the substrate temperature. From the computational point of view, such questions are still difficult to handle but are coming into reach with recent advances in first-principles molecular dynamics. They could advantageously be combined with semi-emprical modelling such as kinetic Monte Carlo modelling of growth. The model sketched above is a straightforward generalization of the so-called cube-on-cube model analyzed and discussed for example by Zangwill [15]. In fact, Zangwill has already discussed an A/B alloy type of generalization of his model, which can be thought of as an antiferromagnetic spin model (if A wants to be coordinated with B atoms). Here we are essentially dealing with the ferromagnetic counterpart.

Evidence exists that polytypes can transform into each other under a stress driven partial dislocation motion [16]. A crucial point here is the relative energies of partial dislocations with C–C wrong bonds versus those with Si–Si wrong bonds, a problem which was recently addressed by first-principles calculations [17]. As stresses may build up during crystal growth, these processes may also play a role in determining the selection of polytypes during growth. Impurities in small amounts are also believed to play a role in stabilizing certain polytypes. Heine et al. [6] presented a tentative explanation for this in terms of the band offsets between polytypes. Since the shallow donor or acceptor levels are linked to the band edges, the total energy of the system would be changed from polytype to polytype by their relative differences in the energetic position of the band edges. This explanation, however, requires rather large impurity concentrations to have any substantial effect on the stability. However, small amounts of impurities may also influence surface energies and mobilities or, in other words, act as surfactants and hence affect growth properties. Thus, again, the question arises whether the impurity effects are kinetic or thermodynamic in origin. Clearly, understanding some of the kinetic effects alluded to above on the basis of first-principles calculations of the relevant processes is a significant challenge for future research.

2.3 Dependence of properties on polytypism

A more modest task than understanding the origin and selection of polytypism during growth is the study of the relationship between properties and polytype structure. First of all, intuition and simple insights immediately reveal that certain properties are essentially independent of polytype while other properties are quite sensitive to it. As a rule of thumb, any property which is derived from an average over the electronic properties should be fairly insensitive to polytypism. For example, the bonding is very similar and only shows minor differences and anisotropies. Relatedly, the elastic constants are quite similar, once this tensorial quantity is expressed in the same cartesian coordinates. For a discussion of the relations between cubic and hexagonal elastic constants, see e.g. Lambrecht et al. [18] and Kim et al. [19].

On the other hand, from the above we understand that polytypes are essentially regular arrangements of twin boundaries in an either cubic or hexagonal underlying arrangement. Electronic band structures and phonon dispersion relations in these systems essentially describe wave propagation through an assembly of such twin boundaries. The resulting formation of standing wave patterns clearly may lead to intricate dependencies on the polytype. This point of view on the polytype band structures emerges particularly clearly from the work of Backes et al. [20, 21]. A complicating factor is that evanescent waves at the interfaces must be included in this picture to be accurate. In Backes' theory this is accomplished using complex band structure technique. From this point of view, it is useful to think of the polytypism as a novel type of superlattices in which one does not vary the composition in the regions put together in a periodic fashion but their orientation or twist. General treatments of this problem have also been given by Ikonić et al. [22]. The band structure dependence on the polytypes forms the main subject of the rest of the paper. Before delving into the results, however, we need to mention briefly the computational method used in our work.

3. Computational Method

The general framework of our computational approach is the density functional method in the local density approximation [23]. It should be appreciated that the eigenvalues or band structures appearing in this theory are strictly speaking not true quasi-particle excitation energies as measured for example by photoemission or as involved in optical excitations. The LDA eigenvalues thus require self-energy corrections, an example of which is the correction of the gap, which is well-known to be necessary. These corrections will be extensively discussed in our comparisons to experiment.

The method used to calculate the band structures in our work is the linear muffin-tin orbital method [24]. While in our earlier work we found the so-called atomic sphere approximation (ASA) to the latter to be generally adequate for understanding band structures to a precision of a few 0.1 eV, we found that some important details depend rather sensitively on non-spherical corrections included in a so-called full-potential (FP) treatment. For that, we use Methfessel's approach [25]. A treatment of potentials, and more importantly, of charge densities of general shape is also crucial for accurate total energy and relaxation calculations. We found that the structural relaxations in the SiC polytypes from the ideal structures are very small and change the energies by less than 1 meV/atom in 2H SiC, where the relaxations are expected to be largest. This agrees with Cheng et al.'s [26] and Park et al.'s [9] conclusions but not with [7]. The FP-LMTO method should also enable us to calculate polytype energy differences. This requires extreme care in choice of basis sets, k-point sampling and other convergence parameters of the method. Preliminary results indicate fairly good agreement on the ordering of polytype energies and their respective energy differences with the recent calculations [7 to 9]. Needless to say, in the ASA one should not even attempt to address this problem. We thus caution the reader not to use the values obtained for the cohesive energies of zincblende and wurtzite given in [3] to extract these small energy differences. For the calculation of UV reflectivities, discussed below, we still use the ASA because it allows for a simpler procedure for including the dipole matrix elements using one-center partial wave expansions of the wave functions [27].

4. Theoretical Aspects of the Band Structures

In this section, we describe selected aspects of the band structures. The actual full band structures can be found elsewhere [2, 28]. The overall aspects of our band structures agree with those of other recent first-principles calculations, which can be found elsewhere in this volume and the references quoted below.

4.1 Brillouin zone relations

Since the early work on band structures of SiC [29, 30], it has been clear that the band structures of different polytypes can best be compared by examining the relations between their Brillouin zones (BZ). The band structures can then be thought of as being related to each other approximately by simple foldings. Of course, each polytype has a slightly different band structure but these differences only become apparent if we first display the band in the smallest common BZ of the two polytypes to be compared. For example, in comparing 2H to 4H SiC, we would use a non-primitive (along the c-axis doubled) unit cell for 2H and hence fold its bands so as to compare it directly to the bands of 4H in the same BZ. All the doubly degenerate bands along the top of the BZ face H–A–L of 2H will be folded on the basal plane K–Γ–M. One difference between 2H in a doubled unit cell and 4H is that in 4H these degeneracies will be lifted because these are related to the non-symmorphic nature of the space group and are only required at the BZ top and bottom face. By the very folding the bands in the mid-plane (i.e. the bisector plane of e.g. the Δ ≡ Γ–A line) of the 2H BZ will now become the top face of the 4H BZ and will obviously be doubly degenerate. More precisely, the degeneracies on the top face of the hexagonal BZ are due to the hexagonal screw axis. As the pitch of the screw axis becomes longer (in higher polytypes) the symmetry is lowered in some sense because there is a larger non-primitive translation vector involved. It is this breaking of the symmetry that lifts the degeneracies. Alternatively, we may fold the bands of 4H out into the larger Brillouin zone of 2H, except that this cannot be done completely unambiguously because there is no criterion to decide which part of the band to display in the first half and which in the second half of the BZ. Similarly, 6H is related to 2H by two foldings about the $1/3$ and $2/3$ points along the c-axis.

The relation between the band structures in 3C, or zincblende (ZB), and 2H, or wurtzite (WZ), is slightly more complicated because the symmetry is different. The relation is based on the fact that the wurtzite basal planes are essentially the same as the zincblende $\{111\}$ planes, but with a different relative stacking, ABC in cubic and AB in wurtzite. Thus, the first step to obtain a common framework to compare the band structures is to use a set of cartesian coordinates for cubic materials which is more closely related to the one used for the wurtzite. We can do this by choosing a z'-axis along the [111] direction of the cubic crystal. In the (111) plane of zincblende and the (0001) plane of wurtzite, the crystal structure consists of buckled hexagonal rings of alternating silicon and carbon atoms. Thus, by aligning these hexagons, we specify completely the relation between the new set of coordiate axes in zincblende and those in wurtzite.

Specifically, one has the rotation matrix

$$\begin{pmatrix} x' \\ y' \\ z' \end{pmatrix} = \begin{pmatrix} 1/\sqrt{2} & -1/\sqrt{2} & 0 \\ 1/\sqrt{6} & 1/\sqrt{6} & -2/\sqrt{6} \\ 1/\sqrt{3} & 1/\sqrt{3} & 1/\sqrt{3} \end{pmatrix} \begin{pmatrix} x \\ y \\ z \end{pmatrix}. \tag{1}$$

The above transformation can now be applied to any $\mathbf{k}$-point of interest to find equivalences between cubic BZ high-symmetry points and their location in the hexagonal Brillouin zone. These are summarized in Table 1. The relations are further clarified in Fig. 3. For the nomenclature of the high symmetry points, see e.g. Bradley and Cracknell [31].

In rhombohedral polytypes, only a threefold symmetry axis exists along the c-axis normal to the basal planes instead of a sixfold screw axis. The primitive unit cell in this case does not correspond to a prism normal to the hexagonal network basal plane. The simplest rhombohedral polytype in some sense is 3C when viewed as a system with only trigonal symmetry along a chosen [111] axis. Although it only repeats after three layers along the [111] axis, (ABC) (hence the number 3 in the notation 3C), it can, of course, be described by a unit cell containing only one SiC unit. Although this is not the commonly used unit cell for the cubic structure, we may choose the basis translation vectors as $\mathbf{a}_1 = a(1, 0, 0)$, $\mathbf{a}_2 = a(1/2, \sqrt{3}/2, 0)$, and $\mathbf{a}_3 = a(1/2, \sqrt{3}/6, \sqrt{2/3})$, where $a = a_c/\sqrt{2}$ is the distance between Si atoms in a $\{111\}$ plane and a_c is the cubic lattice constant and the cartesian coordinates are adapted to the $\{111\}$ planes as mentioned above. We will refer to this unit cell as that of 3R. The same description can be used for any rhombohedral lattice.

Table 1

Equivalence between wurtzite and zincblende symmetry $\mathbf{k}$-points, assuming the ideal c/a ratio for wurtzite

zincblende	wurtzite
$L_{\parallel}$	Γ
$L_{\perp}$	U at 2/3 of ML
X	U at 2/3 of ML
W	T at 3/4 of ΓK
U, K	Σ at 3/4 of ΓM

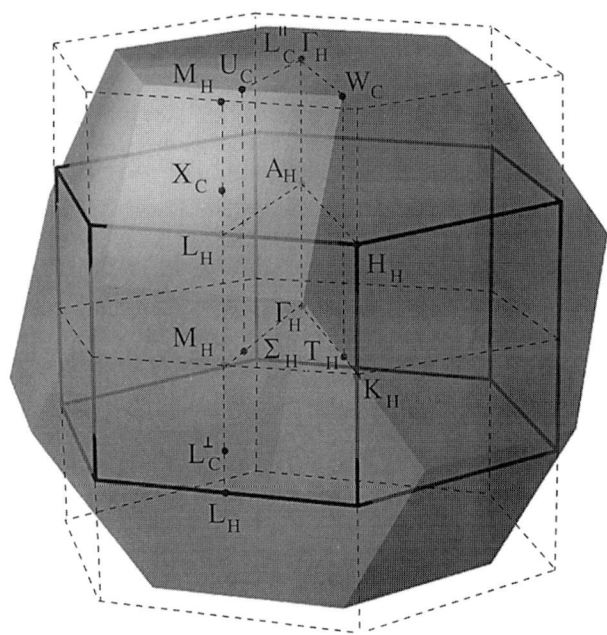

Fig. 3. Relation between the zincblende (ZB) and the wurtzite (WZ) Brillouin zones. Halves of the adjacent wurtzite Brillouin zones along the c-axis are shown in addition to the central one. In this way, one can see that both the X_C (center of the square face) and $L_C^\perp$ (center of one of the ZB hexagonal sides) points lie at 2/3 of the hexagonal U $\equiv$ ML$_H$ line. The $L_C^\parallel$ point on the top and bottom hexagonal faces correspond to the center of the hexagonal BZ, i.e. fold on to Γ_H. It can also be seen that the cubic LW$_C$ line will we folded on to the hexagonal ΓK$_H$ line

The next larger rhombohedral structure is 9R, which more precisely can be described as ↑↑↓ or the ABCBCACAB stacking of close packed {111} planes. In the c-direction, this only repeats periodically after nine layers. However, there is a smaller translation vector connecting the A layer with the second B layer which forms a unit cell one third the size. The 9R form has to our knowledge not been reported to exist naturally for SiC, but is nevertheless of some theoretical interest because it is a 2/3 or 66% hexagonal structure. This allows us to check the band gap behavior of polytypes with hexagonality between that of 4H (50%) and 2H (100%). Another rhombohedral polytype of interest is 15R. It corresponds to ↑↑↑↓↓ and can be described by a unit cell containing five chemical formula units of SiC, although its repeat period along the threefold rotation axis involves 15 layers. We also consider another hypothetical polytype, which we label 15R'. It corresponds to the stacking ↑↓↑↓↑ or ABABABCBCBCACAC which repeats after 15 layers and has rhombohedral symmetry but differs from the commonly known 15R. (This shows that the latter notation is incomplete.) The present 15R' polytype is 80% hexagonal.

Fig. 4 shows a cut through the BZ of the 9R polytype. Inside it, we show the BZ (one-third the size) that would be obtained if we described it with a non-primitive unit cell using the nine layer repeat period along the c-axis. For convenience, we denote the

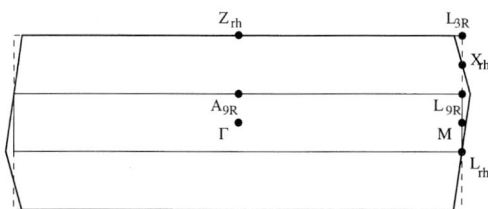

Fig. 4. A cut through the Brillouin zone of the 9R rhombohedral polytype (thick line) illustrating its relation to a pseudohexagonal zone corresponding to a nine layer unit cell (thin line) and a three layer cell (dashed line) of 3R (equivalent to 3C)

k-points in this small BZ with notations typical of the hexagonal structure. We chose this particular cut of the BZ because it contains the line ML along which we have previously found the locations of the conduction-band minima to occur in most polytypes (the exception being 2H). Also indicated is a similar cut for the 3R polytype (remember this is actually the cubic structure). The relation between the hexagonal notation of high-symmetry points (M–L–A) to the conventional rhombohedral and cubic notations is indicated. We then see that the L_{3R} point, which is equivalent to the cubic X point at which the minimum occurs in cubic SiC, will be folded onto the L_{9R} point of the small 9R zone.

4.2 Comparison of 3C and 2H band structures

Fig. 5 shows the band structures of 3C and 2H SiC both represented in the 2H BZ. The symmetry labeling of the 2H bands follows the notation of Rashba [32]. This presentation allows one to appreciate the similarities and the true differences brought about by the different structure without being confused by the conventional way of plotting these bands in different BZs. It should be understood that when we plot the zincblende bands in a wurtzite Brillouin zone, we plot the bands of a unit cell twice the size of the primitive unit cell of ZB. (It consists of two {111} layers.) Even so, the reciprocal lattice of this supercell does not correspond to that for WZ. We merely show the bands along the high-symmetry lines of WZ in **k**-space, which we can always do even if this is not a proper BZ. This implies among other things that the ZB bands do not have the full symmetry required of a hexagonal Brillouin zone. Thus, some of the eigenvalues at Γ_H are true Γ_C eigenvalues of ZB and some are folded $L_C^{\parallel}$ points. For example, the Γ_3 states are essentially folded L_{1C} states. Similarly, the bands at the point 2/3 of $U \equiv ML_H$ correspond to both the X_C and $L_C^{\perp}$ points of ZB. While in the hexagonal crystal some of the hidden cubic symmetries are broken, there are also new symmetries related to the

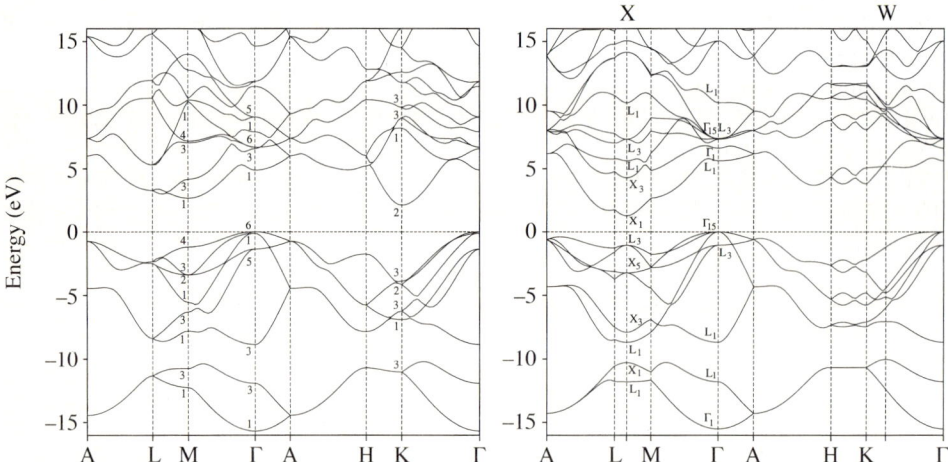

Fig. 5. Band structures of 2H (left) and 3C (right) SiC, both shown in the 2H Brillouin zone. In the right panel, the locations of some cubic high-symmetry points are indicated by additional vertical dashed lines, labeled at the top. The labeling of states is that for the cubic symmetry in the right panel and for hexagonal symmetry in the left

presence of a hexagonal axis instead of a trigonal axis. The wurtzite structure has point group C_{6v}. This is not a subgroup of the tetrahedral group T_d of zincblende. However, both have a common subgroup C_{3v}. In spite of these differences resulting from the different symmetry, the corresponding states are closely related in physical character, i.e. their predominant atomic orbital character. This is due to the underlying similarity in the bonding.

First we note that the lowest conduction band state at Γ actually corresponds to a folded L_C state. In any case, we see that this state is about 5 eV above the absolute minimum of the conduction band. When we proceed to higher period hexagonal polytypes, the bands at some points along the Δ axis, $\Gamma - A$, will be folded onto Γ leading to additional states at Γ (which will have lifted degeneracies if they were folded A-points from a lower period polytype). We note that the bands at A in both 3C and 2H are about equally high up in energy as the minimum at Γ. Hence, it is clear that no polytype will ever have a direct band gap.

The points "competing for the minimum" are K_H in 2H and a point along $U \equiv ML_H$ along which the cubic point X_C lies. Furthermore, we note that the minimum at K_H is lower in the 2H than in the 3C structure and that the minimum along ML is lowest in the 3C structure. Both of these conclusions also hold in a number of other materials (notably the group-III nitrides, diamond, BeO). This suggests that this is a geometric, or symmetry, effect rather than being specific to the potential of the particular atoms involved. The reason why the K_H state moves up when one goes from 2H to 3C and X_C moves up when one goes from 3C to 2H is that in their own structure these are genuine high-symmetry points while in the other structure the symmetry is broken. For example, consider the states of 2H-SiC at the cubic high symmetry $\mathbf{k}$-point X_C. Consider the $\mathbf{k}$-dependent Hamiltonian $H_k = \frac{1}{2}\left(-i\nabla + \mathbf{k}\right)^2 + v(r)$, acting on the periodic part of the Bloch function $u_k(\mathbf{r}) = \exp\left(-i\mathbf{k}\cdot\mathbf{r}\right)\psi_k(\mathbf{r})$. The crystal potential of wurtzite might be considered as that of 3C SiC plus a perturbation and we could, at least in principle, obtain the bands of 2H at this $\mathbf{k}$-point by using the bands of 3C as zeroth order approximation. The symmetry of the $H(\mathbf{k})$ at X_C in 3C-SiC is C_{4v} while that of wurtzite along the U-line is only C_{2v}. Hence, the perturbation will break the symmetry. This allows additional interactions between the states. Specifically, consider the X_{1c} state. In 3C-SiC, the only other state in the valence band with the same symmetry is in the C 2s band at about -20 eV. However, in 2H, there are two additional states of symmetry U_1. These derive from the X_{3v} and L_{1v} folded states in 3C-SiC, which both contain some hexagonal U_1 symmetry components and are thus allowed to interact with the conduction-band minimum of interest. This tends to push up the band in 2H. Of course, there are also interactions with higher conduction-band states which would tend to push the band down. In fact, there is a band of the same symmetry about at the same energy above the U_{1c} minimum as the band below it. However, these states have different orbital character and hence are likely to give smaller interaction matrix elements. The X_{3v} and X_{1c} states are known to be bonding and antibonding combinations of C 2p and Si 3s and thus have similar atomic orbital components. Hence, it is plausible that the net effect is an upward shift of the X_{1c} state. Relatedly, Yeh et al. [33] noted that in several semiconductors the U_{1c} state in wurtzite is about the average of the folded zincblende L_{1c} and X_{1c} states. Since both these states have U_1 symmetry components, this seems reasonable, although it is not clear why it should be fulfilled so accurately.

A similar analysis applies to the conduction band minimum state at K_H. At K_H the symmetry group of wurtzite is C_{3v}, while in 3C SiC, this point lies along the $\Sigma_C \equiv \Sigma_C$–K_C line which has C_{2v} symmetry. These two groups have no common subgroup besides the trivial group C_1 consisting of the identity operator. Hence, the perturbation will mix all states; as a result it is not simple to predict a-priori whether the K_{2c} conduction-band minimum will go up or down. We note that this state has a purely p_x, p_y character in 2H-SiC. The corresponding state in 3C, however, has p_z components mixed in. Furthermore, in 4H or 6H, it has no p_z components on the atoms nearest to the twin boundary in the structure (the h-layers), but has p_z components on the other layers. Thus, there is a gradual loss of symmetry and some trend can be expected in the K_C^{min} state with hexagonality. However, in the higher periodicity polytypes, the $P \equiv K$–H_H line is increasingly folded toward K_H. This complicates the analysis because the lowest state of the conduction band at K does not always turn out to be the K_2 state.

4.3 Trends of gaps as function of hexagonality

Fig. 6 shows the minimum gap and gaps at the relevant **k**-points according to the discussion given above. The minimum at K can indeed be seen to increase from 2H to 3C but shows some non-monotonic oscillations for the lower hexagonality polytypes. Secondly, we note that quite good agreement is obtained between the calculated values (shifted independent of polytype by a 1 eV self-energy correction) and the experimental values. The justification for the value of the gap correction will be discussed later. In this figure, we also have included results for the two hypothetical polytypes with high hexagonality mentioned earlier. The **k**-point location of the minimum gap is indicated for each polytype.

The dashed line in Fig. 6 indicates the well-known linear relation between minimum band gaps (along the U_H line) with hexagonality. We see that his extends all the way to 100%. Furthermore, this is the absolute minimum gap except very near to 100% hexagonality. All the minimum gaps along this line fall either at or close to the hexagonal M

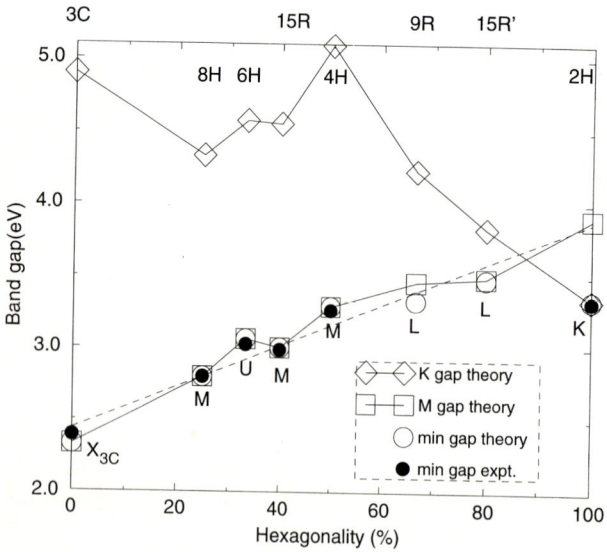

Fig. 6. Minimum and **k**-specific gaps in SiC polytypes as a function of hexagonality. A 1 eV self-energy correction was added to all LDA calculated gaps. The labels of the minimum gaps indicate their location in the BZ. The M, L, U, and K labels refer to the hexagonal BZ. For 15R, the M point corresponds to rhombohedral X_R, while for 15R' and 9R, the hexagonal L point corresponds to the L_R point

point. Of the polytypes studied, it only differs substantially in energy from the M point in the cubic polytype. The linear dependence and its small deviations from linearity very close to 3C (not shown here) are very nicely explained by Backes et al.'s work [20]. In view of this model, which is based on stacking cubic regions together, it is still somewhat surprising that it even holds for polytypes of very high hexagonality.

Since the minimum gaps occur near M in almost all polytypes, it is of interest to examine the evolution of these states with hexagonality in somewhat more detail in Fig. 7 in order to gain some insight in the location of the minima along the U line. The experimental evidence for the location of the minima in 4H and 6H will be discussed in a later section.

4.4 Minimum location of the gap

Fig. 7 illustrates the folding effects along the ML_H axis. The U_1 and U_3 non-interacting bands (because they are of different symmetry character in 2H) and crossing near 2/3 of the ML axis are subject to an increasingly stronger interaction along the series 2H, 4H,

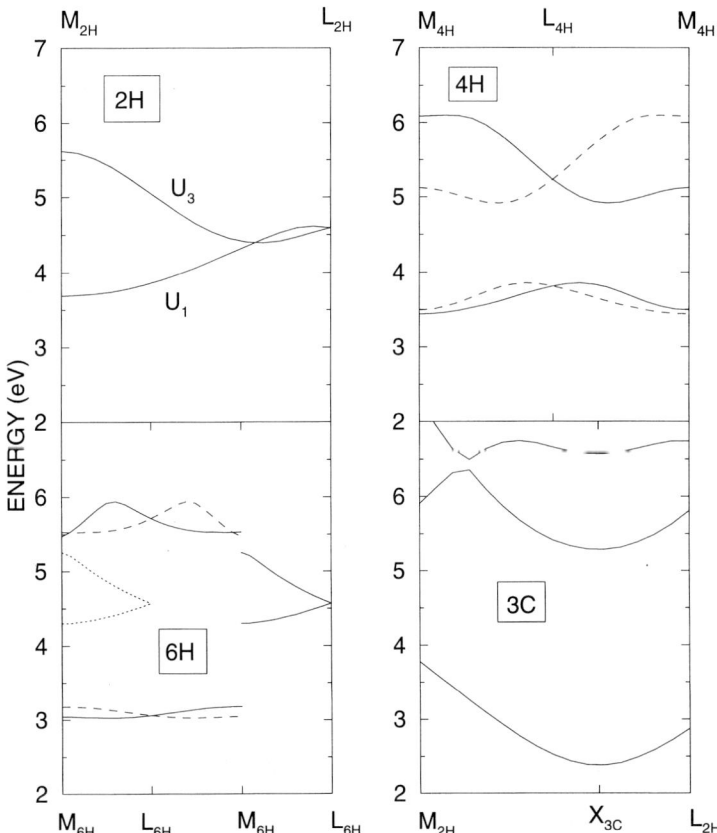

Fig. 7. Band structure near the conduction-band minimum along the ML line for 2H, 6H, 4H and 3C SiC all displayed in the Brillouin zone for 2H

6H and 3C. This interaction eventually leads to the minimum at the 2/3 point in 3C-SiC which corresponds to the X-point of the cubic BZ. In 4H, the coupling of the bands is still fairly weak and the minimum remains at M. In 6H-SiC, the degeneracy at L_{6H} imposed by time-reversal symmetry and the non-symmorphic nature of the space group (more precisely the occurrence of a sixfold screw axis along $\mathbf{c}$) leads to a degeneracy at the folded-out L_{2H} point while in 4H this folded out L_{2H} point is not degenerate. We can see that the maximum interaction between the bands in 6H (coinciding maximum in top band and minimum in lowest band) occurs either in the first or second third of the out-folded zone (there is an unresolvable ambiguity in how to fold the bands out into the extended zone). The reason why the minimum in 6H settles somewhere in between M and L_{6H} instead of at M which corresponds exactly to a folded 2/3 of ML_{2H} point is not easy to explain. This location is somewhat sensitive to details of the calculation (e.g. FP versus ASA, or precise c/a ratio and lattice constant). Also relevant is the fact that the energy difference between the two points is very small (about 10 meV in ASA and 5 meV in FP). To understand the origin of the increasing interaction we can think of these polytypes as stacking faults in the 2H structure. From this point of view 4H, or $\uparrow_t\uparrow\downarrow_t\downarrow$ has a stacking error (two consecutive $\uparrow$ or $\downarrow$'s) every other layer. We indicate the twin errors by a subscript t. On the other hand, 6H, or $\uparrow_t\uparrow_t\uparrow\downarrow_t\downarrow_t\downarrow$, has two consecutive twins. One can now think of these twin errors as introducing a coupling potential between the ideally non-interacting bands U_1 and U_3. It is clear that the larger the number of such interactions the stronger the symmetry breaking will be and hence the stronger the interaction of the bands. This point of view is in some sense complementary to that of Backes et al. [20] who consider interactions between twins of the cubic structure. This provides some insight into why the bands under discussion become more strongly interacting as the hexagonal symmetry is increasingly broken. It still does not explain why the minimim occurs where it does. The only thing we can say at this point is that the dispersion is clearly strongly influenced by whether n in the 2nH polytype is even or odd because this imposes a degeneracy at the L_{2H} point. For example, in 8H the minimum returns to M.

One final remark with respect to Fig. 7 must be made. The bands shown here were obtained in our earlier work which used the ASA. While this is generally adequate for the present discussion, we note that the lower two bands in 4H are crossing along ML in the FP results and lead to a reversal of the ordering of the solid line and dashed line band at M. This reversal is important to settle the symmetry of the lowest band as will be seen in our discussion of the band edge details.

5. Discussion of Experiments

5.1 UV optical properties

An important question is how well the LDA band structures describe the actual quasiparticle excitations, in other words, the experimental band structure. As already mentioned in the previous section, a self-energy correction of 1 eV for all the polytypes appears to bring the LDA values for the minimum gaps in good coincidence with experiment. This value is furthermore well justified by recent calculations of such corrections [34 to 36] using the GW approximation, (in which G stands for the one-electron Green's function and W for the dynamically screened Coulomb interaction, which are involved in the lowest energy diagram for the self-energy) and simple estimates of it [2]. Never-

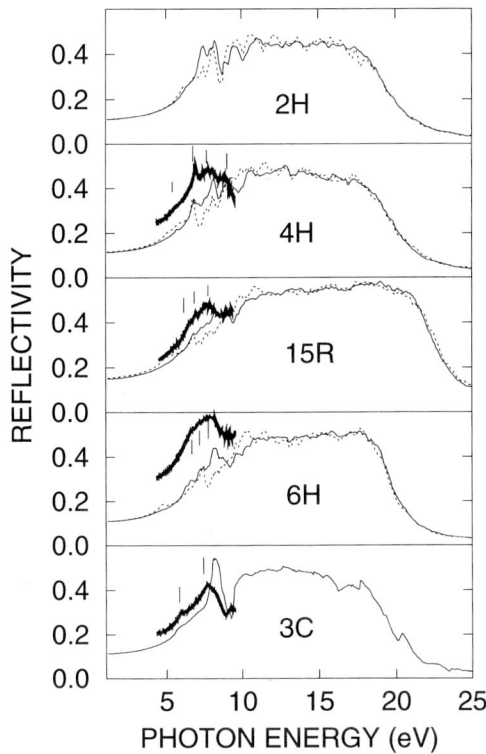

Fig. 8. Calculated UV reflectivity (in arb. units) of several SiC polytypes (arranged in order of increasing hexagonality from bottom up) for **E** ⊥ **c** (solid thin line) and **E** ∥ **c** (dotted line). The thick solid lines indicate the experimental spectra obtained by Choyke and coworkers [38]. Thin vertical lines indicate features discussed in the text

theless, the detailed calculations predict this correction to vary with **k**-point location and with specific eigenstate considered. In order to make a valid comparison with experiment, it is necessary to correctly identify the interband transitions with specific features in the data. Hence, we next discuss our results for the UV reflectivity and related optical response functions.

While previously we only showed the results for the energy range up to the 10 eV limit accessible to the present experiment and only for **E** ⊥ **c**, we here (Fig. 8) show the full computational results for both polarizations and over a larger energy region. The results were actually recalculated including higher precision (for example including f-partial waves which improves the convergence of the matrix element calculations and a more accurate treatment of symmetrization of the Brillouin zone integration). We also show here for the first time the calculated results for the 15R polytype. The discrepancies in absolute value should not be considered meaningful here because the data shown in this figure were not calibrated to the expected values of R at low frequency based on the known indexes of refraction. This allows one to more easily see the individual curves. This figure emphasizes the fact that we presently lack experimental confirmation of the higher energy part and of the polarization dependence. For a more detailed view of the low energy part (up to 10 eV) we refer the reader to the original papers [37, 38]. Fig. 9 shows the imaginary and real parts of the dielectric functions for **E** ⊥ **c** and **E** ∥ **c**. A constant shift of the $\varepsilon_2(\omega)$ spectra by 1 eV is included before calculating $\varepsilon_1(\omega)$ and the reflectivities. This value is the same as used above for minimum gaps, indicating overall consistency with the data. We note that our $\varepsilon_2(\omega)$ curves are in fairly good agreement with those by Bechstedt et al. [10] although there are some differences due to our omission of local field effects.

The reflectivity spectra for all polytypes show one or more sharp peaks in the region 5 to 10 eV followed by a large plateau stretching to about 15 eV or higher and then a decrease towards zero by about 25 eV. The present experimental data only see the start of this plateau. In the region below 10 eV, the features are sharpest in 3C, 2H and 4H while they are rather broad in 6H and 15R. In 3C, there is a single sharp peak at 7.7 eV

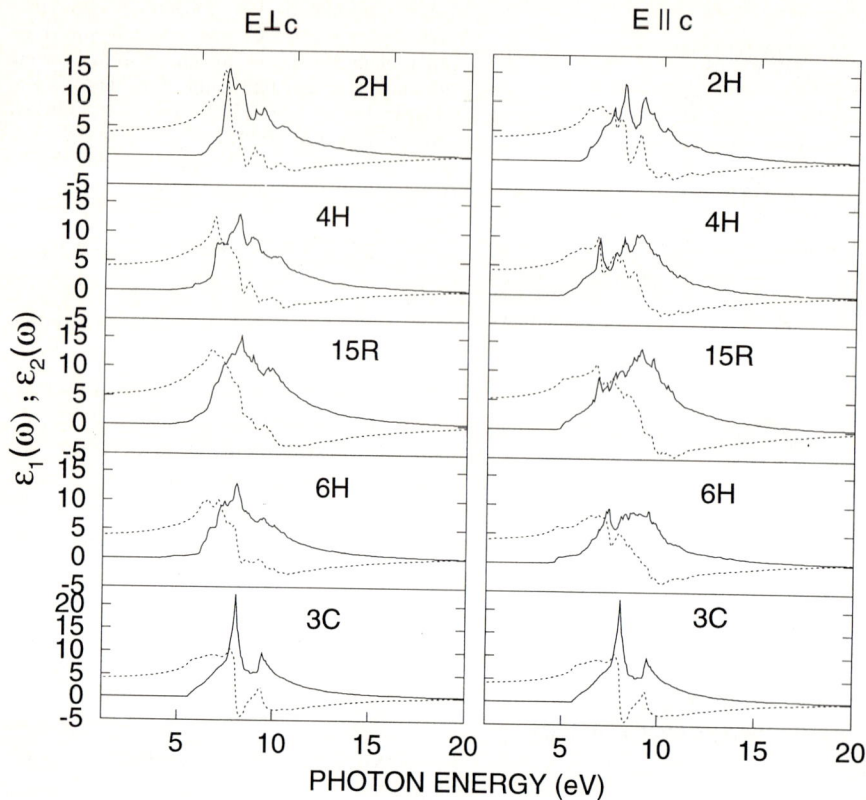

Fig. 9. Calculated optical response functions ε_1 (dotted lines) and ε_2 (solid lines) of several SiC polytypes (arranged in order of increasing hexagonality from bottom up) for $\mathbf{E} \perp \mathbf{c}$ (left), and $\mathbf{E} \parallel \mathbf{c}$ (right)

which is also present in the other polytypes. It corresponds to transitions along the cubic Γ–L line, well known in other semiconductors as the E_1, E_2 transitions. The theoretical spectra for all polytypes show a small bump at about 5 to 6 eV, indicating the lowest energy direct transitions. Experimentally, this is clearly resolved only in 3C and 4H. In the region 6.5 to 7 eV, features appear which are characteristic of all hexagonal and rhombohedral polytypes but absent in 3C. This consists of a single sharp peak in 4H at 7.0 eV, an even stronger peak at 7.4 eV in 2H, and a set of two shoulders at 6.5 and 7.0 eV in 6H and 15R. We also note that these features are predicted by the theory to stand out more sharply in $\mathbf{E} \parallel \mathbf{c}$ polarization because of the reduced intensity of the 7.7 eV peak common to all polytypes. These features can be traced back to direct transitions along the K–Γ line and along K–H where we can see nearly parallel top valence and lowest conduction bands in Fig. 5 for 2H SiC. Similar nearly parallel bands occur also in the other hexagonal polytypes (they are essentially the foldings of these two bands). In 3C, however, one can see in Fig. 5 that these bands are not nearly as nicely parallel. In fact, in 3C, the lowest conduction band at K is crossed by the second band along K–Γ. We may note that it is primarily the valence band that has a smaller slope because the K_3 state is raised in energy. This follows from the increased interactions

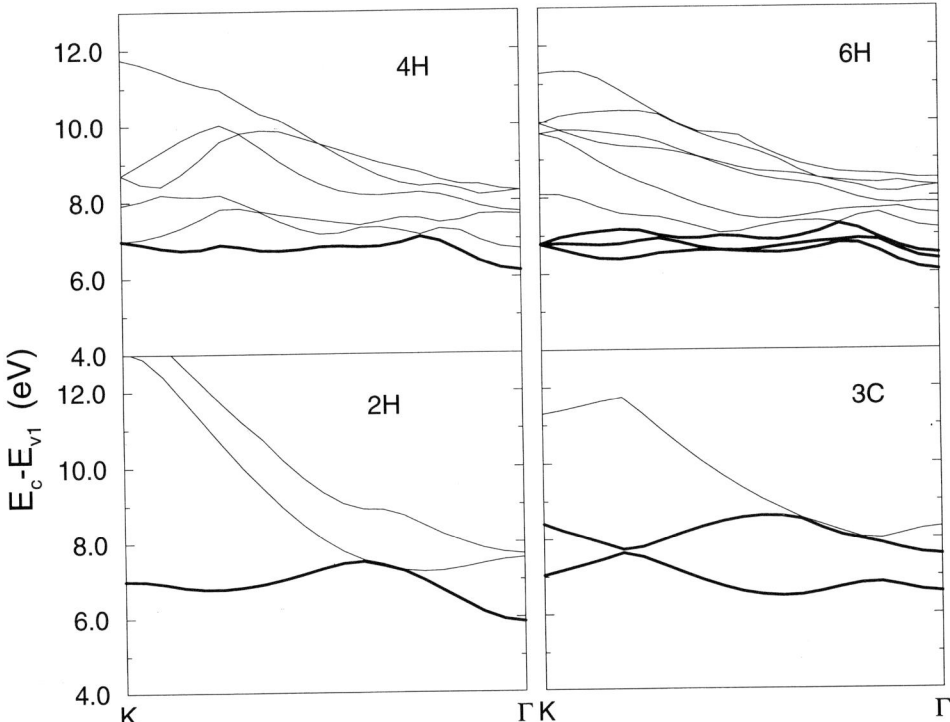

Fig. 10. Optical transition energies from the top valence band to the lowest few conduction bands along ΓK in several polytypes. The important transitions are indicated by thick lines

between valence bands due to the fact that K_{2H} is a point of lower symmetry for 3C SiC. Fig. 10 shows the difference in energy between the highest valence band and several conduction bands, i.e. the optical transition energies for transitions originating from the top valence band along the Γ–K line in several polytypes. Regions with flat dispersion correspond to nearly parallel bands and indicate a large joint density of states. This explains the occurrence of a single feature near 7 eV in 2H and 4H, two features in 6H and the absence of it in 3C. In 4H SiC, there is a third rather well defined peak at 9.2 eV before the plateau region sets in. This is in good agreement with the data with shows a decrease in reflectivity at the end of the observed spectrum while the other polytypes show increases.

Fig. 11 shows a partial decomposition of the $\varepsilon_2(\omega)$ functions of 2H SiC in band-to-band partial components. The spectra can roughly be divided into three energy regions: a region below 8 eV, one between 8 and 9 eV and one above 9 eV. The first region is seen to be due exclusively to transitions originating in the top three valence bands and terminating in the first two conduction bands. In cubic SiC this corresponds to transitions to the first conduction band, and in the other polytypes to the foldings of these bands. The region between 8 and 9 eV is dominated by transitions from the same set of three valence bands to the next higher two conduction bands. Again, this corresponds to the second conduction band in 3C and to the foldings of these bands in the higher polytypes. However, some of the intensity here also comes from transitions from deeper valence bands.

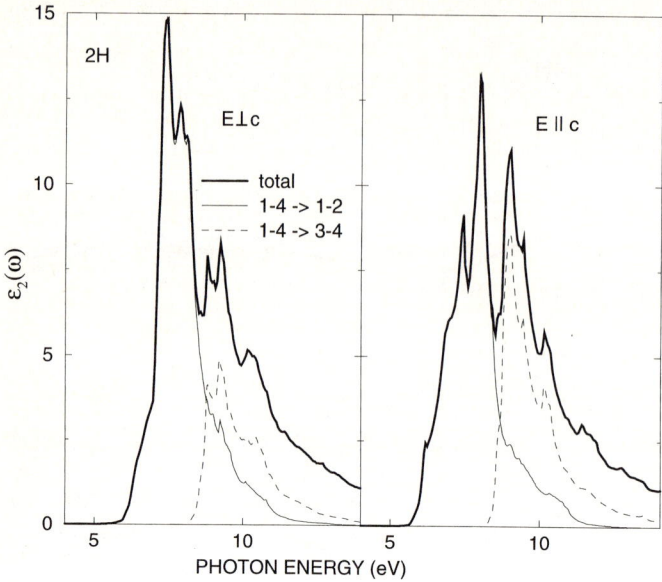

Fig. 11. Partial decomposition of $\varepsilon_2(\omega)$ into band-to-band contributions in 2H-SiC

While most peaks arise from rather extended regions of **k**-space, the calculated spectra and their decomposition in individual band-to-band contributions indicates certain fine features related to transitions at symmetry points. From this analysis, a few of these transitions can be associated with experimental features and the experimental values for them be determined. We have already mentioned the relation between the 7 eV peak and the K-point direct transition in 2H. So far, we have only carried out this critical point analysis in detail for 3C SiC. The results of these assignments are summarized in Table 2. The data for some of these, however, (e.g. Γ_{15}^v–Γ_{15}^c which we associate with the minimum in the reflectivity between the two main peaks) need to be sharpened up in order to be determined more accurately. To summarize the situation on the critical point transitions, we believe that since the overall reflectivity is well reproduced by our calculations based on ASA-LMTO band structure, the critical point transition energies obtained from the latter can probably be trusted to within a few 0.1 eV. To improve the accuracy, significant improvements in experimental resolution and associated detailed

Table 2

Experimental and theoretical critical point transition energies (in eV) in 3C-SiC

	exp.	LDA + 1 eV
Γ_{15}^v–Γ_1^c	7.4	7.56
L_3^v–L_1^c	7.5	7.71
X_5^v–X_1^c	5.8	5.6
X_5^v–X_3^c	8.3 ± 0.1	8.3
Γ_{15}^v–Γ_{15}^c	9.0 ± 0.2	8.7
L_3^v–L_3^c	9.4	9.4

analysis of the calculations would be needed. Since the **k**-dependence of the self-energy corrections for states near the band gap appears to be of only a few 0.1 eV, a detailed test of the GW calculations by experiment does not yet appear to be feasible.

5.2 Photoemission

Because only energy differences can be extracted from optical measurements the experimental knowledge of the band structure is still rather limited. Angular resolved photoemission and inverse photoemission could be of great assistance in mapping out the band structure. This information is now only available for the occupied states of 3C-SiC along the Γ–X direction [39].

Even in its integrated form, photoemission provides some information on the variation of the self-energy shifts within the occupied bands. Fig. 12 shows our calculated valence band density of states for 6H-SiC compared to an XPS measurement by King et al. [40]. In this figure, we have aligned the main features in the top valence band, which is dominated by C 2p states. More precisely, we have aligned the valence band maximum, determined experimentally by taking the tangent at the high energy of the distribution (shown in the figure as a dashed line), with the theoretical value. The sharp C 2p–Si 3s band is then located below the LDA peak by about 0.4 eV, in agreement with GW calculations. In fact, this peak corresponds to the X_{3v} state for which Rohlfing et al. [34] and Backes et al. [35] both obtain a self-energy shift of -0.4 eV with respect to aligned valence bands. Wenzien et al. [36] obtain -0.2 eV for this shift. A 0.2 eV shift of the valence band edge alignment is within the uncertainty range and hence, both of these calculations agree well with the data for the entire upper valence band. The experimental C 2s band is then found to exhibit a shift from the LDA theory by -1.0 eV. This is in good agreement with the former two GW calculations while Wenzien et al. [36] obtain a value of about -1.5 eV for this shift. This slightly less accurate value may probably be explained by Wenzien et al.'s use of a linearization in energy of the self-energy expecta-

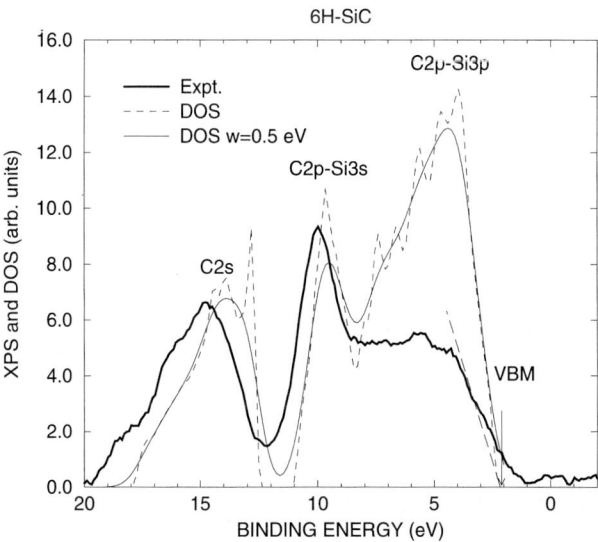

Fig. 12. Calculated density of states (DOS) (with (thin solid line) and without (dashed line) gaussian broadening) of 6H SiC compared to the XPS spectrum by King et al. [40] (thick solid line)

tion value. In any case, it is clear that all GW calculations correctly predict a discontinuity in self-energy corrections upon crossing the ionicity gap between C 2s and C 2p derived bands. It appears from this that the self-energy corrections are almost constant over a given band. This is consistent with the fact that they depend strongly on the localized or delocalized nature of the dominant kind of atomic orbitals involved. This suggests that some simple orbital dependent corrections would form a good approximation to the full dynamic energy and **k**-dependence of the self-energy.

5.3 Conduction-band edges

The locations of the conduction-band minima in the calculated bands have already been discussed extensively above from a theoretical point of view. Here we return to the question of how well they are established experimentally. First of all, we recall that we find that the minimum in 4H SiC lies at M while in 6H it lies along the ML line. In 15R, the minimum lies at X_R, which, as can be seen in Fig. 4 folds on M_H. For many years, there has been a controversy over the minimum location in 4H originating in the conjecture by Patrick et al. [30] that the minimum could not be at M, while most calculations found it there. Fortunately, this has recently been sorted out. Patrick et al.'s conjecture was based on observations of the phonon-replica of donor-bound excitons, which are due to the phonons at the **k**-point of the minimum location. They made assumptions about the phonon dispersions which appeared reasonable at the time but turned out to be inaccurate. Using recently calculated phonon spectra by Hofmann et al. [42], we showed in [43] that there is no discrepancy at all when assuming the minimum to be at M. Choyke et al. [44] redid the analysis on remeasured spectra which clearly resolve all phonon lines and came to the conclusion that the minimum in 4H is indeed at M. In 6H [41] the situation with the phonon replica spectra is not quite as clear because not all phonon lines have been resolved. Still comparison with the phonon calculations indicates a better match for a **k**-point between M and L than for a **k**-point at M.

Another set of data provides information on this question. In 1965, Biedermann [45] reported optical transitions from the lowest conduction band to the higher bands measured on moderately n-type doped samples. We have recently calculated the optical transition matrix elements for these transitions and find the assignments given in Table 3. For 4H, we note that the lowest conduction band at M has the same symmetry as the third conduction band only in the FP results, while in ASA the first and second

Table 3

Assignment of the Biedermann bands in 4H and 6H SiC. The notation $ci - cj$ indicates transitions from conduction band i to j (energies are in eV)

	polarization	ΔE (exp.)	ΔE (theory)	transition
4H	$\mathbf{E} \parallel \mathbf{c}$	2.0	2.04	c1–c3
	$\mathbf{E} \perp \mathbf{c}$	2.7	2.63	c1–c4
6H	$\mathbf{E} \parallel \mathbf{c}$	1.25	1.26	c1–c3 at M
	$\mathbf{E} \parallel \mathbf{c}$	1.6	1.51	c1–c3 at L
	$\mathbf{E} \perp \mathbf{c}$	2.0	2.21	c1–c4 at M
	$\mathbf{E} \parallel \mathbf{c}$	2.8	2.5	c1–c5 at ML min
	$\mathbf{E} \perp \mathbf{c}$	3.0	2.9	c1–c6 at ML min

bands are interchanged. Thus, the observation of the $\mathbf{E} \parallel \mathbf{c}$ peak at lowest energy confirms the correctness of the FP band structure. The transitions between the first and second conduction bands are allowed for $\mathbf{E} \perp \mathbf{c}$ but are overshadowed by the Drude term corresponding to the intraband transitions in the lower band. A remarkable fact about these spectra is their width, especially in 6H, where the $\mathbf{E} \parallel \mathbf{c}$ band has a two-peak nature. We presently do not fully understand the origin of widths or the line-shapes. Nevertheless, for 6H, we interpret the two-peak nature as evidence for the large effective mass in 6H along the c-direction. In other words, instead of seeing transitions very localized in $\mathbf{k}$-space near M as in 4H, we see transitions from a reasonable portion of the ML line and joint density of states effects come into play. It turns out that the bands in question are nearly parallel in the transverse directions. This implies a one-dimensional energy band difference is involved in the joint density of states calculation. At the M point an inverse square root singularity occurs (if no broadening is included). At the doping levels in this experiment and room temperature, the Fermi level can easily exceed the barrier height at the M point which our FP calculations indicate is possibly as low as 5 meV. In fact, we may even reach the L point which is at about 30 meV. Because of the almost purely one-dimensional nature of the van Hove singularity at M, a separate sharp peak can be expected. We associate this with the first peak in $\mathbf{E} \parallel \mathbf{c}$ at 1.25 eV. More detailed modeling of the joint density of states is in progress. If the present interpretation is corroborated by further experimental studies, it may constitute a definite proof of the double well nature of the minimum in 6H. An important prediction of the present theory is that this spectrum should show a strong dependence on doping. If one could precisely control the Fermi level, an accurate measurement of the barrier height at M may be obtained. Indeed, in such measurements, the first peak would be absent until the Fermi level exceeds the energy level of the M-point. Also, the second peak's width and high energy tail would be strongly reduced at lower Fermi level.

Further confirmation of the correct description of the conduction band minima follows from an analysis of the effective mass tensors. There has been considerable controversy over the anisotropy of the latter 6H and 4H SiC. The values extracted from an analysis of the IR absorption spectra of shallow donor states [46, 47] lead to rather small values for the masses, their anisotropies and difference between 6H and 4H. However, anisotropies in the electron mobility suggested a very large anisotropy in 6H and a smaller one of opposite sign in 4H [48, 49]. Optically detected cyclotron resonance (ODCR) [50 to 52] supported the large anisotropies and was shown by us to agree well with our calculated band structure results [53]. Since the point group symmetry at the M point and along the ML line is C_{2v}, one expects a fully anisotropic ellipsoid with axes aligned along the symmetry axes of the crystal. It is presently appreciated that the values obtained from the IR data are inaccurate because their analysis is based on a uniaxial model for the conduction-band minimum energy surface, and because the calculation of the Rydberg series of states in a simple effective mass model is not sufficiently accurate and/or sensitive to the mass anisotropy values [54]. Even in the ODCR measurements, the first reports suggested an axially isotropic model. For 4H, the fully anisotropic nature of the tensor has been subsequently confirmed by the experiment [52]. Excellent agreement is obtained for all three mass-tensor components between theory and experiment, as shown in Table 4. For 6H-SiC, the location of the minimum along the ML line leads to a very high mass in the c-direction and to considerable non-parabolicity. Our value for the

Table 4

Conduction band effective masses in 4H and 6H SiC in units of the free electron mass m_e

		theory	ODCR
4H	$m_{M\Gamma}$	0.58	0.58 [52]
	m_{MK}	0.28	0.31 [52]
	m_{ML}	0.31	0.33 [52]
	$m_\perp{}^{a)}$	0.40	0.42 [50]
	$m_\parallel{}^{b)}$	0.27	0.29 [50]
6H	$m_{M\Gamma}$	0.77	
	m_{MK}	0.24	
	m_{ML}	1.42	
	$m_\perp{}^{a)}$	0.43	0.42 [51]
	$m_\parallel{}^{b)}$	1.1	2.0 [51]

$^{a)}$ $m_\perp = \sqrt{m_{M\Gamma} m_{ML}}$
$^{b)}$ $m_\parallel m_\perp = \langle m_{ODCR}(\mathbf{B} \perp \mathbf{c})\rangle^2$ average in plane.

mass $m_\parallel$ near the minimum is 1.1 while the experimental value is 2.0. However, the effective mass is predicted to increase with filling of the band or carrier excitation and to reach a value close to 2.0 when the Fermi energy is just below the barrier height at M. Above the barrier height, a dramatic effect should happen since then the electrons start circling the double well centered about M. This is schematically indicated in Fig. 13, which uses our ASA calculations. As already mentioned, the barrier at M may be even lower. What this predicts for the ODCR signal in the plane is shown in Fig. 14. The signal for the magnetic field in the plane should gradually move up to higher effective mass values as we approach the barrier, should than make a jump to an even higher value and then gradually return to lower values. Unfortunately, it also broadens as we go to higher magnetic fields and this makes it rather difficult to see this effect. Further increases in lifetime and or the microwave frequency used may be required. Also, the calculations predict an anisotropy for fields in the plane in 6H. Again, because of the broad nature of the peak, this has not yet been confirmed experimentally. A series of

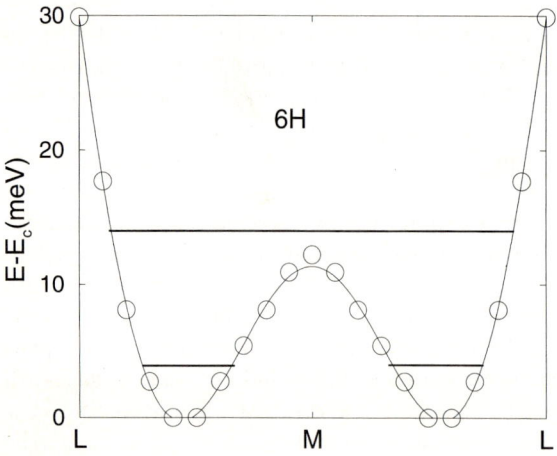

Fig. 13. Double well nature of the conduction band minima along ML in 6H SiC

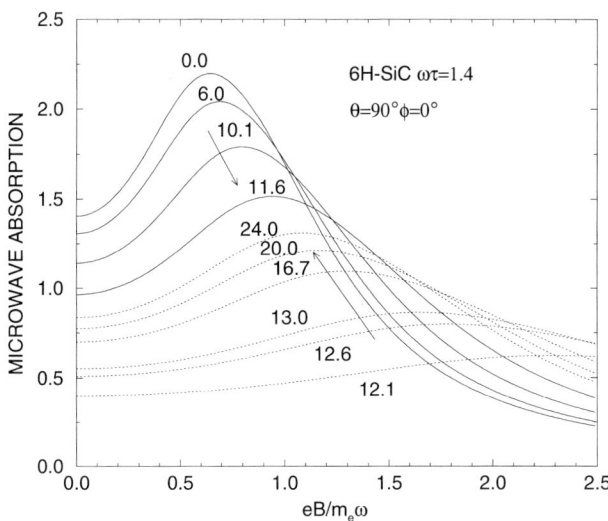

Fig. 14. Predicted changes of the ODCR signal with Fermi level position above the minimum. The latter is indicated on each curve in meV

experiments on samples with increasing level of doping and or with increasing level of optical pumping of carriers into the conduction band to shift the Fermi level would be very interesting because it may confirm the theoretical predictions based on there double well nature around M of the 6H conduction band minimum. The presently available measurements correspond to a rather small filling of the band and hence there remains a discrepancy on the magnitude of the effective mass. However, absorption of the microwave power may also have led to a non-equilibrium distribution of the carriers. A rather strong dependence of these ODCR signals on microwave power was observed by Son et al. [55] for 4H SiC.

For 15R SiC, our location of the conduction band minimum agrees with the calculations of Wellenhofer and Rössler [56]. Also, our values for the effective mass tensor, 0.653 along XΓ, 0.439 along XU and 0.22 along XK (using rhombohedral notations), agree well with their results, 0.67, 0.41 and 0.22. The experimental data on 15R were discussed by Wellenhofer and Rössler [56] and appear to be in good agreement with the most reliable set of data. Cyclotron resonance values are not presently available for 15R. For cubic SiC, the X minimum is well established and our values for the mass tensor $m_l = 0.645$ and $m_t = 0.226$ are in good agreement with the cyclotron resonance values 0.677 and 0.247 [57].

As a final observation, we note that the extremely good agreement for the conduction-band effective masses in 4H and 3C SiC achieved in the LDA calculations is actually somewhat surprising. For one thing, a straightforward increase in the underestimated LDA gap by a self-energy correction is expected to significantly increase the effective mass according to the usual $\mathbf{k} \cdot \mathbf{p}$ formula. This was considered for 3C SiC by Willatzen et al. [58]. However, straightforwardly replacing the LDA gaps by quasi-particle gaps is not justified within the framework of GW theory. As was shown by Zhu et al. [59] additional factors affecting the effective mass formulas appear in this theory. Furthermore, one would expect phonon renormalization effects to increase the experimental masses. Thus several issues remain to be resolved before we can claim a full understanding of the effective masses. But then, this is also the case for much more widely studied semiconductors such as Si and GaAs.

5.4 Valence band maximum

The valence band maximum in all the polytypes is located at Γ. As is well known, it is threefold degenerate and of symmetry Γ_{15} (without counting spin and without including spin–orbit coupling) in 3C and split into a doublet (of symmetry Γ_6 in hexagonal and Γ_3 in rhombohedral polytypes) and a singlet of symmetry Γ_1. This set of bands can be described by the Kohn-Luttinger [60] Hamiltonian for the cubic case and its equivalent for the wurtzite case given in Bir and Pikus [61] and referred to as the Rashba-Sheka-Pikus (RSP) Hamiltonian. The former is described in terms of three parameters A, B and C, which are essentially inverse masses, and a spin–orbit splitting Δ_{so}. The latter involves six inverse mass like parameters A_i, $i = 1$ to 6, a parameter A_7 describing a term linear in $\mathbf{k}$, the crystal field splitting $\Delta_{\mathrm{c}} = \Delta_1$, two spin–orbit coupling parameters Δ_2 and Δ_3 and a further set of linear in $\mathbf{k}$ parameters originating in relativistic effects. We will ignore the latter for the time being because they are expected to be quite small in SiC.

Since the hexagonal polytypes are very nearly ideal, one may expect that a "quasi-cubic" approximation would be reasonable. Within that approximation, the following relations hold between the A_i and the cubic parameters:

$$A_1 = (A + 2B + 2C)/3\,,$$
$$A_2 = (A + 2B - C)/3\,,$$
$$A_3 = -C\,,$$
$$A_4 = C/2\,,$$
$$A_5 = (B - A - 2C)/6\,,$$
$$A_6 = (2B - 2A - C)/3\sqrt{2}\,,$$
$$A_7 = 0\,. \tag{2}$$

Table 5 shows the values of these parameters obtained by direct fitting to our first-principles band structures. For 3C-SiC the values are obtained from the cubic parameters, $A = -5.156$, $B = -0.307$ and $C = -4.000$ (in units $\hbar^2/2m_{\mathrm{e}}$) by means of equations (2). These values are in reasonable agreement with those of Bimberg et al. [62] $A = -5.91$, $B = -1.95$ and $C = -6.42$ without phonon renormalization or $A = -4.85$, $B = -1.80$, $C = -4.84$ with phonon renormalization which were obtained from $\mathbf{k} \cdot \mathbf{p}$ theory using empirical data and matrix elements for Si. Our value of $|B|$ is notably smaller.

Table 5

Rashba-Sheka-Pikus valence band maximum effective Hamiltonian parameters. The A_i parameters are in units of $\hbar^2/2m_{\mathrm{e}}$ for $i = 1$ to 6 and $e^2/2$ for A_7

poly-type	hex. (%)	Δ_{cr} (meV)	A_1	A_2	A_3	A_4	A_5	A_6	A_7
3C	0	0	−4.59	−0.59	4.00	−2.00	1.34	−0.96	0
6H	33	46	−4.70	−0.54	4.10	−1.10	1.36	−0.94	0.009
15R	40	54	−4.70	−0.55	4.07	−1.12	1.35	−0.94	0.012
4H	50	66	−4.70	−0.56	4.07	−1.10	1.35	−0.94	0.013
2H	100	132	−4.66	−0.59	4.00	−1.13	1.30	−0.85	0.027

The term in the Hamiltonian involving A_7 is linear in $k_\perp$ and off-diagonal. It arises from interactions between the Γ_6 and Γ_1 states in nearly degenerate perturbation theory and is not zero in hexagonal polytypes because of the s–p mixing in Γ_1. As expected, its value is increasing with hexagonality. It leads to a lifting of the degeneracy at the crossing point of the light hole and split-off hole band for small $k_\perp$. Because of the linearly increasing crystal field splitting with hexagonality, the crossing point moves out further along the $k_\perp$-axis. It lies at $k_\perp = 0.034, 0.037, 0.041$ and 0.059 $2\pi/a$ respectively in 6H, 15R, 4H and 2H. The corresponding energies are 56, 66, 81, and 164 meV below the valence-band maximum. This interaction also affects the in-plane split-off and light- hole masses, which are respectively given by $m_{\mathrm{sh}}^\perp = -(A_2 - 2A_7^2/|\Delta_1|)^{-1}$, and $m_{\mathrm{lh}}^\perp = -(A_2 + A_4 - A_5 + 2A_7^2/|\Delta_1|)^{-1}$. The effect of the correction term $2A_7^2/|\Delta_1|$ which is increasing with hexagonality can be seen clearly in the trend of the split-off mass which is more strongly affected by it than the light-hole mass. With the exception of A_4, the quasi-cubic model is found to hold well since the other parameters can be seen to be nearly polytype independent and close to the quasi-cubic prediction. The uncertainty on these parameters is about 0.03. A_4 is found to be smaller than the quai-cubic prediction by almost a factor 2 for all non-cubic polytypes but otherwise almost independent from polytype to polytype. This can be traced back to the in-plane heavy hole mass $m_{\mathrm{hh}}^\perp = -(A_2 + A_4 + A_5)^{-1}$ which is indeed found to be significantly heavier in hexagonal (and rhombohedral) polytypes than in 3C-SiC. This, in turn, is related to the significant deviations from axial symmetry (about the [111] axis) of the heavy hole mass in cubic material.

For higher-order hexagonal polytypes $2n$H, the same effective Hamiltonian applies in principle for the highest valence states because the form of the latter is only based on symmetry arguments. Similarly, for rhombohedral polytypes, a general form involving a few extra parameters was derived in Bir and Pikus [61] and again the extra terms are related to the cubic ones,

$$A_8 = (B - A + C)\sqrt{2}/3\,,$$
$$A_9 = (B - A + C)/6\,. \tag{3}$$

It turns out that these two only affect terms of higher order than quadratic in the actual bands, so we have not attempted to determine them from the masses. The crystal field splitting is found to vary almost exactly linearly with hexagonality as shown in Fig. 15. The crystal field splitting in 2H-SiC is found to be 132 meV. This is somewhat larger than the value given by Käckel et al. [64].

As we go to higher period polytypes, more and more valence bands along the line Γ–A are folded onto the Γ-point. Thus, at some point, we might start worrying about these folded bands interacting too strongly with the levels deriving from the Γ_{15} manifold. At that point the three-band model (or six band when including spin–orbit splitting) would no longer be a good model. However, for the polytypes we have investigated, the folded states stay below the crystal field split state because the latter splitting becomes smaller and smaller as the hexagonality decreases.

We note that the RSP Hamiltonian without spin–orbit coupling leads to valence bands which are axially symmetric about the c-axis. This axial symmetry is not obeyed by the hole effective masses given by Käckel et al. [64]. These authors report a large anisotropy in the c-plane. We think that this derives from fitting the bands to parabola

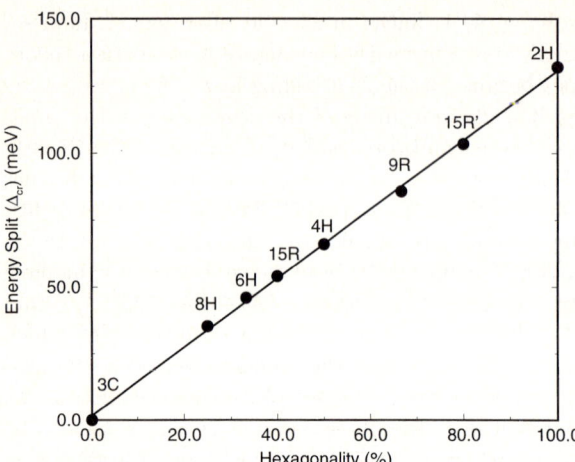

Fig. 15. The crystal field splitting $\varDelta_{cr}$ of the valence band maximum at Γ of SiC polytypes as a function of hexagonality

over a region of k-values beyond the point of validity of the RSP Hamiltonian. We indeed also find anisotropies at larger k-values. The average values in the plane, however, agree reasonably well with our values derived from the parameters A_i.

For the spin–orbit splitting, within the quasi-cubic approximation, one expects $\varDelta_2 = \varDelta_3 = \frac{1}{3} \varDelta_{so}$, where $\varDelta_{so}$ is the spin–orbit splitting in 3C-SiC which we calculated to be 8 meV, in good agreement with the experimental value of 10 meV [63]. When this is now added to the Hamiltonian, the bands become strongly non-parabolic and the effective masses change due to the interaction between the bands. The resulting masses are given in Table 6. In particular, the in-plane heavy hole effective mass is strongly reduced and the exact degeneracy of the heavy hole and light hole bands along the c-axis is lifted. Note that each of these masses, or rather the parabolic approximation to the

Table 6

Hole effective masses in SiC polytypes in units of the free electron mass m_e

polytype	hex. (%)	$m_{hh}^{\perp}$	$m_{lh}^{\perp}$	$m_{sh}^{\perp}$	$m_{hh}^{\parallel}$	$m_{lh}^{\parallel}$ a)	$m_{sh}^{\parallel}$
		without spin–orbit coupling					
3C	0	1.69	0.25	0.90	1.69	1.69	0.22
6H	33	3.57	0.34	1.70	1.67	1.67	0.21
15R	40	3.13	0.34	1.61	1.64	1.64	0.21
4H	50	3.23	0.34	1.59	1.60	1.60	0.21
2H	100	2.38	0.35	1.35	1.52	1.52	0.21
		with spin–orbit coupling					
3C	0	1.19	0.33	0.52	1.69	0.31	0.52
6H	33	0.62	0.62	1.68	1.67	1.56	0.21
15R	40	0.61	0.61	1.59	1.64	1.56	0.21
4H	50	0.62	0.62	1.58	1.60	1.55	0.21
2H	100	0.61	0.61	1.35	1.52	1.51	0.21

a) $m_{hh}^{\parallel} = m_{lh}^{\parallel}$ in the absence of spin–orbit coupling.

bands, is only valid for **k**-points close enough to Γ that the bands disperse by less than the spin–orbit and crystal field splittings. At larger energies, the masses become similar to those without spin–orbit coupling since at this scale of energies the spin–orbit splittings are negligible. In addition to the change in masses, the combination of the A_7 term with the spin–orbit coupling leads to a small linear splitting of the spin degeneracies of the light hole and crystal field split-off bands.

At present there is no experimental evidence for the linear relationship between crystal field splitting and hexagonality to the best of our knowledge. Acceptor bound excitons or acceptor excited states could contain fine structure relating to the split valence band edges. However, unraveling this information is complicated by the fact that central cell corrections due to different sites (cubic versus hexagonal) may also lead to splittings. The acceptor binding energies are of order 200 meV or larger and indeed exhibit splittings due to different sites [65]. They show a decreasing trend with hexagonality [65, 66]. One might a first think that this could be explained in terms of the decreasing trend of heavy hole masses (from 6H to 2H) obvious from the first half of Table 6. Additionally, the increasing crystal field splitting which should lower the number of states function at the relevant energy and hence the effective density of states mass. However, a closer inspection of this hypothesis shows that a simple effective mass theory is not tenable. A multiband effective mass calculation based on the RSP Hamiltonian would be of interest, although its applicability to the ground state of the acceptor is still doubtful.

Recently, an analysis of the valence band maximum closely related to and in good agreement with ours but using a slightly different set of parameters was presented by Persson and Lindefelt [67].

6. Conclusion

We have reviewed some general aspects of polytypism in SiC and described our current state of knowledge of their band structures both from a theoretical and experimental point of view. From the theoretical point of view, there is presently good agreement between various first-principles calculations of the band structures of the most important polytypes. There is also a good deal of understanding of their inter-relationships in terms of Brillouion zone folding and of the origin of the linear dependence of the gap on hexagonality. We caution though that not all properties vary monotonically with hexagonality. For example, the exact location of the minimum of the gap and the variation of other gaps at K and/or direct gaps as observed in the UV reflectivity depend on more subtle effects such as the occurrence of an even or odd number for n in the $2n$H polytypes.

Experimentally, the UV optical and photoemission spectra obtained so far are rather well explained by the band structures assuming a roughly constant gap correction. However, the polarization dependence remains to be confirmed and extension of the data to higher energy ranges is desirable. As for the details of the band edges, there is encouragingly good agreement between theory and ODCR spectra for the conduction-band minima. Nevertheless, further experimental work is desirable to confirm the theoretical prediction of a double well minimum in 6H-SiC. We have shown that the optical transitions between this band and the next few higher bands may contain interesting information regarding this question. As for the valence-band analysis, no direct experimental verification of the linear dependence of the crystal field splitting is available at this point.

Acknowledgements This work was supported by Wright-Laboratories and the National Science Foundation. We thank Dr. W. J. Choyke for suggesting some of the interesting challenges to the theory presented by the experiments.

References

[1] A. P. VERMA and P. KRISHNA, Polymorphism and Polytypism in Crystals, Wiley, New York 1966.

[2] W. R. L. LAMBRECHT, in: Diamond, SiC and Nitride Wide Bandgap Semiconductors, Ed. C. H. CARTER, JR., G. GILDENBLAT, S. NAKAMURA, and R. J. NEMANICH, Mater. Res. Soc. Symp. Proc. **339**, 565 (1994).

[3] W. R. L. LAMBRECHT and B. SEGALL, in: Wide Band Gap Semiconductors, Ed. T. D. MOUSTAKAS, J. I. PANKOVE, and Y. HAMAKAWA, Mater. Res. Soc. Symp. Proc. **242**, 367 (1992).

[4] V. HEINE, C. CHENG, G. E. ENGEL, and R. J. NEEDS, Mater. Res. Soc. Symp. Proc. **242**, 507 (1992).

[5] C. Y. YEH, Z. W. LU, S. FROYEN, and A. ZUNGER, Phys. Rev. B **46**, 10086 (1992).

[6] V. HEINE, C. CHENG, and R. J. NEEDS, J. Amer. Ceram. Soc. **74**, 2630 (1991).

[7] P. KÄCKEL, B. WENZIEN, and F. BECHSTEDT, Phys. Rev. B **50**, 17037 (1994).

[8] K. KARCH, G. WELLENHOFER, P. PAVONE, U. RÖSSLER, and D. STRAUCH, in: Proc. 22nd Internat. Conf. Phys. Semicond., Ed. D. LOCKWOOD, World Scientific, Singapur 1995 (p. 401).

[9] C. H. PARK, B.-H. CHEONG, K.-H. LEE, and K. J. CHANG, Phys. Rev. B **49**, 4485 (1994).

[10] F. BECHSTEDT, P. KÄCKELL, A. ZYWIETZ, K. KARCH, K. TENELSEN, and J. FURTHMÜLLER, phys. stat. sol. (b) **202**, 35 (1997).

[11] G. S. ZHDANOV, C.R. Academ. Sci. USSR **48**, 43 (1945).

[12] C. CHENG, V. HEINE, and I. L. JONES, J. Phys.: Condensed Matter, **2**, 5097 (1990).

[13] H. MATSUNAMI and T. KIMOTO, Mater. Res. Soc. Symp. Proc. **339**, 369 (1994).

[14] M. KOHYAMA and R. YAMAMOTO, Mater. Res. Soc. Symp. Proc. **339**, 9 (1994).

[15] A. ZANGWILL, J. Cryst. Growth **163**, 8 (1996).

[16] P. PIROUZ, Scripta Metall. **23**, 401 (1989).

[17] P. K. SITCH, R. JONES, S. OBERG, and M. I. HEGGIE, Phys. Rev. B **52**, 4951 (1995).

[18] W. R. L. LAMBRECHT, B. SEGALL, M. METHFESSEL, and M. VAN SCHILFGAARDE, Phys. Rev. B **44**, 3675 (1991).

[19] K. KIM, W. R. L. LAMBRECHT, and B. SEGALL, Phys. Rev. B **50**, 1502 (1994).
K. KIM, W. R. L. LAMBRECHT, and B. SEGALL, Phys. Rev. B **53**, 16310 (1996).

[20] W. H. BACKES, P. A. BOBBERT, and W. VAN HAERINGEN, Phys. Rev. B **49**, 7564 (1994).

[21] W. VAN HAERINGEN, P. A. BOBBERT, and W. H. BACKES, phys. stat. sol. (b) **202**, 63 (1997).

[22] Z. IKONIĆ, G. P. SRIVASTAVA, and J. C. INKSON, Phys. Rev. B **48**, 17191 (1993).

[23] W. KOHN and L. J. SHAM, Phys. Rev. **140**, A1133 (1965).

[24] O. K. ANDERSEN, O. JEPSEN, and M. ŠOB, in: Electronic Band Structure and Its Applications, Ed. M. YUSSOUF, Springer-Verlag, Heidelberg 1987 (p. 1).

[25] M. METHFESSEL, Phys. Rev. B **38**, 1537 (1988).

[26] C. CHENG, V. HEINE, and R. J. NEEDS, J. Phys.: Condensed Matter, **2**, 5115 (1990).

[27] M. ALOUANI, L. BREY, and N. E. CHRISTENSEN, Phys. Rev. B **37**, 1167 (1988).

[28] S. LIMPIJUMNONG and W. R. L. LAMBRECHT, unpublished.

[29] F. HERMAN, J. P. VAN DYKE, and R. L. KORTUM, Mater. Res. Bull. **4**, S167 (1968).

[30] L. PATRICK, W. J. CHOYKE, and D. R. HAMILTON, Phys. Rev. **137**, A1515 (1965).

[31] C. J. BRADLEY and A. P. CRACKNELL, The Mathematical Theory of Symmetry in Solids: Representation Theory for Point Groups and Space Groups, Clarendon Press, Oxford 1972.

[32] É. I. RASHBA, Fiz. Tverd. Tela **1**, 407 (1959) (Soviet Phys. – Solid State **1**, 368 (1959)).

[33] C.-Y. YEH, S.-H. WEI, and A. ZUNGER, Phys. Rev. B **50**, 2715 (1994).

[34] M. ROHLFING, P. KRÜGER, and J. POLLMANN, Phys. Rev. B **48**, 1791 (1993).

[35] W. H. BACKES, P. A. BOBBERT, and W. VAN HAERINGEN, Phys. Rev. B **51**, 4950 (1994).

[36] B. WENZIEN, P. KÄCKEL, F. BECHSTEDT, and G. CAPPELLINI, Phys. Rev. B **52**, 10897 (1995).

[37] W. R. L. LAMBRECHT, B. SEGALL, W. SUTTROP, M. YOGANATHAN, R. P. DEVATY, W. J. CHOYKE, J. A. EDMOND, J. A. POWELL, and M. ALOUANI, Appl. Phys. Lett. **63**, 2747 (1993).

[38] W. R. L. LAMBRECHT, B. SEGALL, M. YOGANATHAN, W. SUTTROP, R. P. DEVATY, W. J. CHOYKE, J. A. EDMOND, J. A. POWELL, and M. ALOUANI, Phys. Rev. B 50, 10722 (1994).

[39] H. HÖCHST and M. TANG, J. Vacuum Technol. A 5, 1640 (1987).

[40] S. KING, M. C. BENJAMIN, R. J. NEMANICH, R. F. DAVIS, and W. R. L. LAMBRECHT, in: Gallium Nitride and Related Materials, Ed. F. A. PONCE, R. D. DUPUIS, S. NAKAMURA, and J. A. EDMOND, Mater. Res. Soc. Symp. Proc. 395, 375 (1996).

[41] W. J. CHOYKE and L. PATRICK, Phys. Rev. 127, 1868 (1962).

[42] M. HOFMANN, A. ZYWIETZ, K. KARCH, and F. BECHSTEDT, Phys. Rev. B 50, 13401 (1994).

[43] W. R. L. LAMBRECHT, S. LIMPIJUMNONG, and B. SEGALL, Inst. Phys. Conf. Ser. No. 142, 263 (1996).

[44] W. J. CHOYKE, R. P. DEVATY, L. L. CLEMEN, M. F. MacMILLAN, and M. YOGANATHAN, Inst. Phys. Conf. Ser. No. 142, 257 (1996).

[45] E. BIEDERMANN, Solid State Commun. 3, 343 (1965).

[46] W. SUTTROP, G. PENSL, W. J. CHOKE, R. STEIN, and S. LEIBENZEDER, J. Appl. Phys. 73, 3708 (1992).

[47] W. GÖTZ, A. SCHÖNER, G. PENSL, W. SUTTROP, W. J. CHOYKE, R. STEIN, and S. LEIBEN-ZEDER, J. Appl. Phys. 73, 3332 (1993).

[48] W. J. SCHAFFER, G. H. NEGLEY, K. G. IRVINE, and J. W. PALMOUR, Mater. Res. Soc. Symp. Proc. 339, 595 (1994).

[49] M. SCHADT, G. PENSL, R. P. DEVATY, W. J. CHOYKE, R. STEIN, and D. STEPHANI, Appl. Phys. Lett. 65, 3120 (1994).

[50] N. T. SON, W. M. CHEN, O. KORDINA, A. O. KONSTANTINOV, B. MONEMAR, B. E. JANZÉN, D. M. HOFMANN, D. VOLM, M. DRECHSLER, and B. K. MEYER, Appl. Phys. Lett. 66, 1074 (1995).

[51] N. T. SON, O. KORDINA, A. O. KONSTANTINOV, W. M. CHEN, E. SÖRMAN, B. MONEMA, and E. JANZÉN, Appl. Phys. Lett. 65, 3209 (1994).

[52] D. VOLM, B. K. MEYER, D. M. HOFMANN, W. M. CHEN, N. T. SON, C. PERSSON, U. LINDE-FEL, O. KORDINA, E. SORMAN, A. O. KONSTANTINOV, B. MONEMAR, and E. JANZÉN, Phys. Rev. B 53, 15409 (1996).

[53] W. R. L. LAMBRECHT and B. SEGALL, Phys. Rev. B 52, R2249 (1995).

[54] P. SRICHAIKUL and A.-B. CHEN, Inst. Phys. Conf. Ser. No. 142, 285 (1996).
P. SRICHAIKUL, PhD Dissertation, Auburn University, 1995.

[55] N. T. SON, E. SÖRMAN, W. M. CHEN, J. P. BERGMAN, C. HALLIN, O. KORDINA, A. O. KONSTANTINOV, B. MONEMAR, D. M. HOFFMAN, D. VOLM, B. K. MEYER, and E. JANZÉN, in Silicon Carbide and Related Materials, edited by S. NAKASHIMA, H. MATSUNAMI, S. YOSHIDA, and H. HARIMA, Inst. Phys. Conf. Ser. No. 142, 353 (1996).

[56] G. WELLENHOFER and U. RÖSSLER, Solid State Commun. 96, 887 (1995), Inst. Phys. Conf. Ser. No. 142, 297 (1996); phys. stat. sol. (b) 202, 107 (1997).

[57] R. KAPLAN, R. J. WAGNER, H. J. KIM, and R. J. DAVIS, Solid State Commun. 55, 67 (1985).

[58] M. WILLATZEN, M. CARDONA, and N. E. CHRISTENSEN, Phys. Rev. B 51, 13150 (1996).
M. CARDONA, N. E. CHRISTENSEN, and G. FASOL, Phys. Rev. B 38, 1806 (1988).

[59] X. ZHU, M. S. HYBERTSEN, and S. G. LOUIE, Mater. Res. Soc. Symp. Proc. 193, 113 (1990).

[60] J. M. LUTTINGER, Phys. Rev. 102, 1030 (1956).
J. M. LUTTINGER and W. KOHN, Phys. Rev. 97, 869 (1955).

[61] G. L. BIR and G. E. PIKUS, Symmetry and Strain-Induced Effects in Semiconductors, John-Wiley & Sons, New York 1974.

[62] D. BIMBERG, M. ALTARELLI, and N. O. LIPARI, Solid State Commun. 40, 437 (1981).

[63] R. G. HUMPHREYS, D. BIMBERG, and W. J. CHOYKE, Solid State Commun. 39, 169 (1981).

[64] P. KÄCKEL, B. WENZIEN, and F. BECHSTEDT, Phys. Rev. B 50, 10761 (1994).

[65] M. IKEDA, H. MATSUNAMI, and T. TANAKA, Phys. Rev. B 22, 2842 (1980).

[66] W. J. CHOYKE, private communication.

[67] C. PERSSON and U. LINDEFELT, Phys. Rev. B 54, 10257 (1996).

phys. stat. sol. (b) **202**, 35 (1997)

Subject classification: 61.50.Ah; 63.20.Dj; 65.50.+m; 68.35.Md; 71.20.Nr; 78.40.Fy; S6

Polytypism and Properties of Silicon Carbide

F. BECHSTEDT, P. KÄCKELL, A. ZYWIETZ, K. KARCH, B. ADOLPH,
K. TENELSEN, and J. FURTHMÜLLER

*Institut für Festkörpertheorie und Theoretische Optik, Friedrich-Schiller-Universität,
Max-Wien-Platz 1, D-07743 Jena, Germany*

(Received January 31, 1997)

The relationship between crystal structure and related material properties is discussed for the common 3C, 6H, 4H, and 2H polytypes of SiC. The theoretical results are derived in the framework of well converged density-functional calculations within the local-density approximation and the pseudopotential-plane-wave approach. In the case of electronic excitations additionally quasiparticle corrections are included. The lattice-dynamical properties of the noncubic polytypes are described within a bond-charge model. We focus our attention on the actual atomic structures, the accompanying lattice vibrations, thermodynamical properties, properties of layered combinations of polytypes, optical spectra, and surface equilibrium structures. On the one hand, the influence of the polytype on the material properties is considered. On the other hand, indications for driving forces of the polytypism are extracted.

1. Introduction

The ability of compounds and elements to occur in more than one crystal structure is called poly*morphism*. Poly*typism* is a one-dimensional variant of this phenomenon. The so-called polytypes differ by the stacking sequence along one direction [1]. Silicon carbide (SiC) is one of the few compounds, which form such stable and long-range modifications. Moreover, it is the only known naturally stable group-IV compound. More than 200 SiC polytypes have been determined [2]. The most extreme polytypes are zincblende SiC (3C in the Ramsdell notation [3]) — with pure cubic stacking of Si–C double layers in the [111] direction — and wurtzite SiC (2H) — with pure hexagonal stacking in the [0001] direction. The other polytypes represent hexagonal (H) or rhombohedral (R) combinations of these stacking sequences with n Si–C bilayers in the primitive cell [1]. Important examples are 4H- and 6H-SiC with four (six) double layers and, hence, eight (twelve) atoms in the corresponding hexagonal unit cell.

Despite many years of research in this field the driving forces for the occurrence of so many polytypes are not very clear. Theories explaining the polytypism of the material may be roughly divided in two categories [4]. The first category deals with considerations about the thermodynamical stability of the common short-period polytypes. The relatively stable short-period structures are then believed to act as basic structural elements for the generation of long-period polytypes. The second category puts emphasis on the growth mechanism for long-period structures around screw dislocations. One argues that the corresponding spiral growth mechanism essentially provides the driving forces towards a long-range ordering. Nevertheless, there is no final picture of the polytypism. A satisfactory explanation of the physical mechanism should contain both types of considerations to a certain degree in dependence on the actual growth and preparation conditions of the SiC crystals and layers. Moreover, the picture of the SiC polytypism is complicated because of the occurring polytypic transformations [5]. A typical example concerns the solid–solid phase transition 3C → 6H at elevated temperatures.

The different stacking of the Si–C bilayers remarkably influences the properties of the various SiC polytypes. The most pronounced example concerns their electronic structure. The indirect band gaps appear to vary in a wide energy range from 2.4 to 3.3 eV going from the cubic polytypic 3C-SiC to the polytype with complete hexagonal stacking, 2H-SiC. A linear relation between gap and hexagonality of the polytype was experimentally established by Choyke et al. [6] for a couple of polytypes. The polytype also determines the position of the conduction band minimum in the k-space (cf. [7], and references therein). The components of the effective-mass tensor vary in dependence on the location at a high-symmetry point, line or plane in the Brillouin zone. Therefore, the k-space location has considerable consequences for the application of a certain polytype in microelectronic devices. There also seems to be an internal relationship between the stability of a certain SiC polytype and the exact atomic positions, which are displaced [8 to 10] with respect to that of the ideal tetrahedral structure realized in zinc-blende 3C-SiC. These atomic relaxations differ from polytype to polytype and may, therefore, tell us somewhat about the driving forces of polytypism.

In the present paper we discuss the relationship between SiC crystal structure and material properties. For that purpose we focus our attention on the most common polytypes 3C, 6H, 4H, and 2H. As examples we study the relation of atomic geometry and vibrational frequencies to the crystal structure and their feedback to the phenomenon of polytypism. Consequences of the accompanying electronic structures are derived for the optical properties and the character of heterocrystalline combinations of different polytypes. Finally, the interplay of surface structure and polytype growth is discussed. Starting points are theoretical considerations. They are mainly based upon ab initio calculations. The theoretical results are interpreted in the light of available experimental data.

2. Calculational Methods

The parameter-free total-energy, force, and electronic-structure calculations are performed within the density-functional theory (DFT) [11]. The exchange-correlation functional of the many-body electron–electron interaction is taken in the local-density approximation (LDA) [12]. Explicitly the electron gas data of Ceperley and Alder [13] as parametrized by Perdew and Zunger [14] are taken into account. The lattice-dynamical and dielectric properties are determined within the framework of the density-functional perturbation theory (DFPT) [15 to 17]. The DFPT allows the direct calculation of the high-frequency dielectric tensor and the effective charge for each atom in the unit cell [18]. Moreover, it is also able to determine the harmonic force constants [19]. Full calculations have been performed for the 3C, 2H, and 4H polytypes [19 to 21].

Within the applied self-consistent methods the electronic wave functions are expanded in terms of plane waves. The number of plane waves in such an expansion is determined by the kinetic energy cutoff. In order to limit this number we make use of the fact that all properties which should be discussed are governed by the valence electrons of SiC. The Si 1s, 2s, and 2p as well as the C 1s electrons contribute to the corresponding atomic cores which therefore have an effective static charge of +4. The interaction of these cores with the Si 3s and 3p as well as the C 2s and 2p electrons is treated by ab initio pseudopotentials. Generally, we use norm-conserving, fully separable pseudopotentials [22 to 24] in the Kleinman-Bylander form [25]. They are based on relativistic all-electron calculations for the free atoms by solving a Dirac-like equation self-consistently.

Due to the lack of core p states the carbon core is rather small. Many plane waves per atom are usually required. This holds especially for the Bachelet-Hamann-Schlüter (BHS) pseudopotentials which are used together with the computer code fhi93cp of Stumpf and Scheffler [26] to calculate the majority of structural, electronic, and optical properties. The use of the degrees of freedom left in the construction procedure for BHS-type pseudopotentials [22] allows a reduction of the energy cutoff to 34 Ry for SiC [27]. In the case of DFPT calculations [18 to 21] we work from the beginning with the scheme of Troullier and Martins [24] in which the potential softness is an important criterion. The cutoff is fixed at 40 Ry. In the case of the surface calculations the condition of the norm conservation is dropped and ultrasoft Vanderbilt pseudopotentials [28] are used to describe the interaction of electrons with carbon cores. In this particular case the plane-wave expansion can be restricted by a cutoff of 13 Ry. Explicitly this procedure is implemented in the Vienna Ab Initio Simulation Package (VASP) [29, 30].

In order to determine the equilibrium atomic positions in the zinc-blende case (space group T_d^2 (F$\bar{4}$3m)), only the cubic lattice constant a_0 has been varied. In the hexagonal cases with the C_{6v}^4 (P6$_3$mc) space-group symmetry, we varied the corresponding hexagonal lattice constants a (with $a_0 = \sqrt{2}a$) as well as the ratio of the hexagonal lattice constants c/a. For each considered pair, c and a, the positions of the atoms in the unit cell of the crystal are relaxed towards the total-energy minimum and vanishing atomic forces. In the calculations with the BHS pseudopotentials [26], we use a steepest-descent method for the atomic displacements together with a Car-Parrinello-like molecular-dynamics approach [31] for bringing the wave functions to self-consistency. In the case of the DFPT approach the Kohn-Sham equations [12] are directly diagonalized. To derive the equilibrium from the volume dependence of the total energy, the Vinet equation of state [32] is solved. In the case of the VASP code the ground state of the polytypes and surfaces is evaluated by means of a residuum-minimization or conjugate-gradient technique.

The Brillouin-zone (BZ) integrations over the electronic wave vectors have been replaced by sums over special points of Chadi-Cohen (CH) [33] or Monkhorst-Pack (MP) [34] type. Typical meshes use 6 CH points for 6H and 4H, 12 CH points for 2H, and 10 CH points for 3C in the irreducible wedge of the BZ. 8 **k**-points in the half of the surface BZ seem to be sufficient to describe 3C-SiC(111)/nH-SiC(0001) ($\sqrt{3} \times \sqrt{3}$) surfaces [35]. The equilibrium geometries are identified when all atomic forces are smaller than a certain limit. A typical value used is about 10^{-4} at. units, i.e., 5 meV/Å. This corresponds approximately to uncertainties in the energies (per pair) of about 1 meV and atomic positions of about 0.002 Å for given cutoff, **k**-point set, and pseudopotentials [36].

The calculations of the structural, electronic, and dynamical properties are performed at the theoretical lattice constants. The eigenvalues of the Kohn-Sham equations automatically give an electronic band structure. However, such an electronic structure does not account for the excitation aspect. Whereas the band dispersion is satisfactorily described, there is a remarkable underestimation of the energy gaps between empty and occupied states [37]. Quasiparticle corrections have to be taken into account. We have

developed an efficient approach to calculate such corrections also for noncubic crystals
by a reasonable approximation of the exchange-correlation self-energy [38]. As example
the quasiparticle band structures have been calculated for the 3C, 6H, 4H, and 2H polytypes [7] and used to explain optical transition energies [39].

Unfortunately, ab initio calculations of phonon frequencies for arbitrary wave vectors
consume a lot of computer time. Therefore, full calculations which are necessary to describe thermal properties [19, 40] are restricted to 3C-SiC. For 6H and 4H polytypes
with 12 or 8 atoms per unit cell only the determination of Γ-point phonons is possible.
In order to derive the complete lattice-dynamical properties of the hexagonal polytypes
we have developed a generalized adiabatic bond-charge (BC) model [41]. Massless bond
charges and ions interact via central and angular forces up to second-nearest neighbours.
The BC model contains 10 (16) free parameters for the cubic (hexagonal) SiC case.
These parameters have been fitted to reproduce first-order Raman frequencies.

3. Atomic Structures

3.1 Equilibrium positions

If we describe the Bravais lattice by $\mathbf{R} = \sum_{i=1}^{3} n_i \mathbf{a}_i$ $(n_i \in N_0)$, then the atomic positions of
the different polytypes are defined as $\mathbf{R}_s = \mathbf{R} + \mathbf{r}_s$ $(s = 1, \ldots, 2n)$. n denotes the number
of Si–C double layers per hexagonal unit cell (if the cubic 3C structure is represented in
a hexagonal 3H cell, n equals three). $\mathbf{r}_s$ defines the atomic basis.

The primitive vectors for the zinc-blende structure are well known. Each primitive cell
contains two atoms, Si at $a_0(0, 0, 0)$ and C at $a_0(1, 1, 1/4) = (\mathbf{a}_1, \mathbf{a}_2, \mathbf{a}_3/4)$, where a_0 is
the cubic lattice constant and the Cartesian coordinate system is defined by the cubic
axes. Considering the hexagonal polytypes, it is useful to transform the coordinate sys-
tem orthogonally to another Cartesian system with $x_{\mathrm{hex}} \parallel [1\bar{1}0]$, $y_{\mathrm{hex}} \parallel [11\bar{2}]$, and
$z_{\mathrm{hex}} \parallel [111]$. A possible choice for the primitive vectors then is

$$\mathbf{a}_1 = a(1, 0, 0), \qquad \mathbf{a}_2 = a\left(-\frac{1}{2}, \frac{\sqrt{3}}{2}, 0\right), \qquad \mathbf{a}_3 = c(0, 0, 1), \tag{1}$$

where a and c are the hexagonal lattice constants. In this coordinate system the vectors
of a characteristic tetrahedron around one atom take the form

$$\boldsymbol{\tau}_1 = \left(0, 0, \frac{3c}{4n}\right), \qquad \boldsymbol{\tau}_2 = \left(\frac{a}{2}, \frac{a}{2\sqrt{3}}, -\frac{c}{4n}\right),$$

$$\boldsymbol{\tau}_3 = \left(-\frac{a}{2}, \frac{a}{2\sqrt{3}}, -\frac{c}{4n}\right), \qquad \boldsymbol{\tau}_4 = \left(0, -\frac{a}{\sqrt{3}}, -\frac{c}{4n}\right). \tag{2}$$

In the case of ideal tetrahedra the relation between c and a is fixed by
$c/(na) = \sqrt{2/3}$. This is the case for zinc-blende 3C-SiC. The corresponding space group
is T_d^2 and it holds $c/a = \sqrt{6}$. This is no longer valid for hexagonal polytypes, where the
atomic positions obey operations of the space group C_{6v}^4: In the (x, y) plane, the positions are constrained to the trigonal axes, but perpendicular to that plane, they are in
principle allowed to move freely. The tetrahedra therefore can be pressed or stretched,
and c/a may differ from its ideal value.

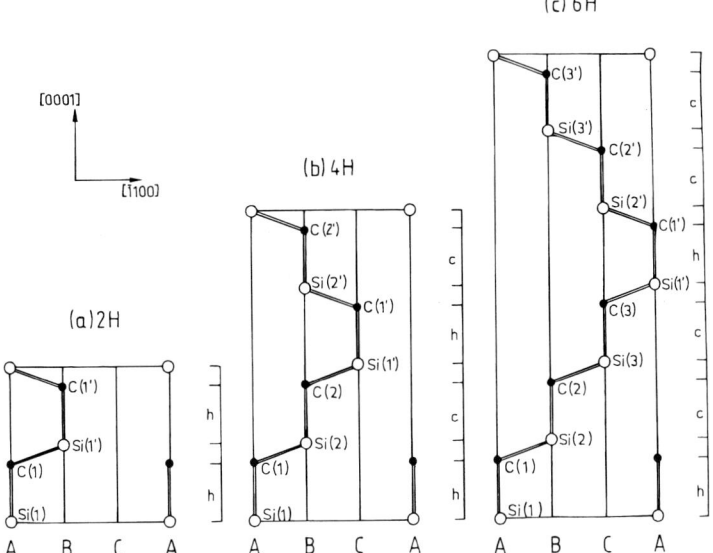

Fig. 1. The zig-zag chain structure of hexagonal SiC indicating the atomic positions. All atoms are located in the $(11\bar{2}0)$ plane. The different $(\bar{1}100)$ planes within the hexagonal unit cells are denoted by A, B, and C. The cubic (c) or hexagonal (h) character of Si–C bilayers in [0001] direction is given according to the parallel (c) or nonparallel (h) limiting bonds

The atomic positions of the hexagonal polytypes are indicated in more detail in Fig. 1. Here the characteristic zig-zag chains are plotted in the $(11\bar{2}0)$ plane within a rectangular coordinate system spanned by $-\mathbf{a}_1 + \mathbf{a}_2$ and $\mathbf{a}_3$. Half of the atoms are labeled by $X(i')$ ($X = \mathrm{Si}, \mathrm{C}; i' = 1, \ldots, n/2$), indicating the fact that their positions can be derived from those of the atoms $X(i)$ by a translation of $(n/6, n/6, 1/2)$ and a following rotation of $180°$ around the c-axis. Additionally the cubic (c) or hexagonal (h) character of the Si–C double layers is indicated. The ratio of the number of hexagonal to the total number of double layers gives the percentage of hexagonality h, which is 0% for 3C, 33% for 6H, 50% for 4H, and 100% for 2H.

The formation laws for the atomic positions can be found in [42]. Describing them in terms of the primitive vectors, the general formula for positions A, B and C in Fig. 1 are as follows:

$$(0, 0, u), \qquad (\tfrac{1}{3}, \tfrac{2}{3}, v), \qquad (\tfrac{2}{3}, \tfrac{1}{3}, w). \tag{3}$$

For the 2H polytype this leads to

$$u(\mathrm{Si}) = 0, \quad u(\mathrm{C}) = \tfrac{3}{8} + \varepsilon. \tag{4}$$

ε is a dimensionless parameter measuring the deviation of the ideal tetrahedral positions. In the case of 4H-SiC we obtain

$$u(\mathrm{Si}) = 0, \qquad u(\mathrm{C}) = \tfrac{3}{16} + \varepsilon(1),$$

$$v(\mathrm{Si}) = \tfrac{1}{4} + \delta(2), \qquad v(\mathrm{C}) = \tfrac{7}{16} + \varepsilon(2), \tag{5}$$

Table 1

Lattice constants of different SiC polytypes. Theoretical values are taken from [36] (a) or have been calculated recently (b, see text). Experimental values (c) have been taken from [45] (2H), [43] (3C), and [44] (4H and 6H)

polytype	ref.	a (Å)	c/n (Å)	$c/(na)$
3C	(a)	3.034	2.477	0.8165
	(b)	3.063	2.501	0.8165
	(c)	3.083	2.517	0.8165
6H	(a)	3.033	2.480	0.8177
	(b)	3.062	2.502	0.8172
	(c)	3.081	2.520	0.8178
4H	(a)	3.032	2.482	0.8185
	(b)	3.061	2.503	0.8179
	(c)	3.081	2.521	0.8184
2H	(a)	3.031	2.480	0.8185
	(b)	3.057	2.508	0.8201
	(c)	3.076	2.524	0.8205

where we set $\delta(1) = 0$ without loss of generality. Finally, for 6H-SiC, we arrive at

$$u(\text{Si}) = 0, \qquad u(\text{C}) = \tfrac{1}{8} + \varepsilon(1),$$
$$v(\text{Si}) = \tfrac{1}{6} + \delta(2), \qquad v(\text{C}) = \tfrac{7}{24} + \varepsilon(2),$$
$$w(\text{Si}) = \tfrac{1}{3} + \delta(3), \qquad w(\text{C}) = \tfrac{11}{24} + \varepsilon(3). \tag{6}$$

ε and δ characterize the additional degrees of freedom which have to be optimized during the calculation, i.e. one for 2 H, three for 4H, and five for 6H.

3.2 Atomic relaxations

Using a minimization technique as described above we finally arrive at the optimum structures. Before we go in for the relaxations, a short note concerning the other ground state properties, especially the lattice constants, seems to be useful. The lattice constants a and c per Si–C bilayer calculated with different pseudopotentials are listed in Table 1 in comparison with experimental data [43 to 45]. The values in (a) have been taken from [36] and were calculated with a modified carbon pseudopotential of BHS-type [22, 27]. Values (b) have been calculated by the authors using the ultrasoft Vanderbilt pseudopotential for carbon (see Section 2).

In general, theoretical and experimental data show the same trends. The absolute magnitude of the theoretical lattice constants is somewhat too small. The latter is a typical feature of well converged DFT-LDA calculations. Other authors find similar deviations [46] which become smaller [47] using the Wigner formula for exchange and correlation [48]. The lattice constant a decreases somewhat with percentage hexagonality, whereas c increases. The absolute failures in the theoretical values cancel each other in the ratio $c/(na)$. Indeed, a comparison for 6H and 4H shows almost perfect agreement for the results (a) with the findings from recent high-precision X-ray diffractometry [44]. Unfortunately, no such measurement exists yet for 2H-SiC. Nevertheless, the slight in-

Table 2

Internal geometrical parameters of hexagonal SiC polytypes. The interatomic distances $L(j)$ and $l(j)$ (in Å), and finally the deformation of the j-th Si–C bilayer $\varDelta(j) = 1 - L(j)/l(j)$ (in %) (cf. Fig. 2) are calculated from the absolute relaxations $\varepsilon(j)$ and $\delta(j)$

poly-type	j	with BHS pseudopotential					with Vanderbilt pseudopotential				
		$10^4 \times \varepsilon(j)$	$10^4 \times \delta(j)$	$L(j)$	$l(j)$	$\varDelta(j)$	$10^4 \times \varepsilon(j)$	$10^4 \times \delta(j)$	$L(j)$	$l(j)$	$\varDelta(j)$
2H	1	8.0	0	1.866	1.855	0.549	8.6	0	1.886	1.872	0.694
4H	1	4.8	0	1.866	1.855	0.625	6.0	0	1.884	1.873	0.584
	2	−2.3	−2.1	1.862	1.858	0.197	−0.8	−0.2	1.877	1.875	0.108
6H	1	3.7	0	1.865	1.856	0.506	5.1	0	1.884	1.873	0.578
	2	0.6	0.3	1.860	1.857	0.201	1.5	1.4	1.877	1.874	0.145
	3	−0.9	−1.2	1.860	1.858	0.127	−1.3	−0.9	1.876	1.876	0.017

crease of the ratio $c/(na)$ with rising hexagonality is clearly reproduced. It indicates a stretching of the tetrahedra parallel to the c-axis on the average and, hence, a certain biaxial strain in the hexagonal polytypes compared to 3C.

The atomic relaxations (cf. Table 2), i.e., the deviations from the ideal tetrahedron structure, are obtained after a minimization of the Hellmann-Feynman forces. They are not very large. Nevertheless, they are very important for the stabilization of the hexagonal polytypes. Furthermore, they have influence on the volume per Si–C pair. Experimentally a slight decrease with hexagonality is observed. Theoretically, this is only observed by inclusion of atomic relaxations [36, 47]. Neglecting the relaxations or taking them not fully into account, the trend observed [49 to 52] cannot be reproduced.

In Table 2 we compare the deviations from the ideal positions for two types of pseudopotentials. Qualitatively, we find an agreement with the calculation of Cheng et al. [51], although these authors let relax the structures only at the experimental value of the lattice constant for 3C-SiC, and only until the mechanical stresses equal those of the cubic polytype.

The changes in the bond lengths can be best demonstrated by calculating the deformation of the ideal tetrahedra, $\varDelta(j) = 1 - L(j)/l(j)$, around the carbon atoms. $L(j)$ denotes the bond length parallel to the c-axis, the three other bonds of a tetrahedron are labeled with $l(j)$ (cf. Fig. 2). The deformations $\varDelta(j)$ are positive altogether. This indicates an elongation of the polytypes along the c-axis, which goes together with a short-

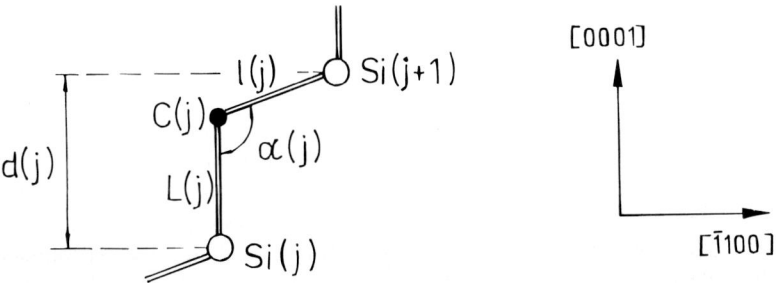

Fig. 2. Geometrical parameters for one Si–C bilayer in [0001] direction

ening of the other three bonds $l(j)$ — as can be seen from Table 2 — and is in agreement
with X-ray diffraction measurements for 2H-SiC [9] and 6H-SiC [8, 44]. The calculated
deviations from the ideal structures are, however, somewhat smaller than those reported
in [8] and those calculated in [51]. An analysis of weak "quasiforbidden" reflections, observed in recent X-ray diffraction measurements performed by Bauer et al. [44] gives
structure factors which are between those calculated from our data [36] and from [51],
but farer away from the ones extracted from the data in [8].

Similar values for the deformations $\Delta(j)$ have been calculated by Vignoles [10] using
the relaxed structures of Cheng et al. [51]. Following Vignoles, the deformations determine the epitaxial growth along the [0001] direction. Obviously they are more important
for hexagonal layers ($\Delta(j) > 0.5\%$) than for cubic ones ($\Delta(j) < 0.2\%$). Responsible for
these findings could be an interplay between electrostatic interactions: At short distances, steric repulsions between third neighbours in hexagonal bilayers disfavour a con-
formation, whereas the electrostatic attraction has the opposite effect at larger distances. The dynamic layer formation could therefore sensitively depend on the
interatomic distances and the microscopic, but long-range, stresses. Indeed, a calculation
of a silicon-terminated relaxed 3C(111) ($\sqrt{3} \times \sqrt{3}$) or 2H(0001) ($\sqrt{3} \times \sqrt{3}$) surface shows
a trend to rotate the Si trimers in an additional Si layer on top locally by about 18° [53].

The atomic relaxations also influence the cohesive energies. Without atomic relaxations the cubic structure is slightly favoured over the hexagonal polytypes. The latter
become the favoured structures only by taking into account the relaxations. They are
essential for a stabilization of the hexagonal crystal structure over the cubic polytype,
apart from the 2H polytype, which has in each case a lower cohesive energy than 3C.
However, the difference in energy between the two extreme polytypes is only 2.4 meV
per atom. The same energetical ordering is found by other authors [46, 47], whereas in
[51] 6H becomes the favourable (but not stress-free) polytype. Finally, the 3C polytype
is always favoured over 2H. This goes conform with calculations performed by Yeh et al.
[54] for several III–V compound semiconductors. They found that the wurzite structure
of a material is always favoured over its cubic structure if the wurzite c/a ratio is *smaller* than the ideal cubic value and vice versa.

4. Lattice Dynamics and Energetics

4.1 Phonon frequencies and consequences for the electronic structure

A generalization of the adiabatic bond-charge (BC) model is used to calculate the phonon frequencies as well as the corresponding phonon eigenvectors within the harmonic
approximation [41]. The BC model has been chosen, since it describes correctly not only
the phonon frequencies and the phonon eigenvectors [41] but also the thermal properties
of cubic and hexagonal SiC polytypes [55].

The equilibrium atomic positions are fixed by the experimental values of a and c from
[42]. The ten parameters of the model in the cubic case have been fitted to the ten
experimental frequencies at the high-symmetry points Γ, X, and L in the BZ [56]. The
frequencies calculated within the BC model are identical with the experimental ones and
given in Table 3. The comparison with results from our ab initio calculations within the
DFPT shows excellent agreement. The discrepancies in certain phonon frequencies range
between 0 and 1.6%. In order to describe the hexagonal polytypes correctly, the number

Table 3

Phonon frequencies (in units of cm^{-1}) of 3C-SiC at the high symmetry points Γ, X, and L

	Γ_{TO}	Γ_{LO}	X_{TA}	X_{LA}	X_{TO}	X_{LO}	L_{TA}	L_{LA}	L_{TO}	L_{TO}
ab initio BC model	783	956	366	629	755	829	261	610	766	838
experiment	795	972	372	639	760	829	261	610	765	837

of parameters of the BC model has been increased to 16. Thereby, the splitting of the model parameters has been adjusted to reproduce reasonably the Raman frequencies.

Results for zone-boundary phonons at the high-symmetry points L and M are summarized in Tables 4 and 5 for the polytypes 4H and 6H, respectively. In the 6H case also frequencies for the center of the LM line parallel to the c-axis, the so-called U point, are listed. Because of the combined time-reversal symmetry and the space-group symmetry (screw axis operation) the modes at the L point are twofold degenerate, whereas this degeneracy is lifted along the LM line. The maximum splitting is generally reached at the M point whereas it exhibits intermediate values at the U point. In Tables 4 and 5 the calculated frequencies are compared with phonon frequencies extracted from the shift of the phonon replica from the zero-phonon line in low-temperature photoluminescence spectra [57, 58].

From the symmetry of the L and M points as well as from the behaviour of the phonons at these points or along the LM line some conclusions for the electronic structure can be drawn. The little group of the vector $\mathbf{k}$ for points on the LM line is C_{2v}. Consequently, the electron effective-mass tensor has three independent components. This has been theoretically shown [7, 46, 59, 60] and experimentally confirmed [61, 62]. The magnitude of the independent components is governed by the position of the conduction-band minimum. In contrast to former predictions we have shown in our band-structure calculations [59, 63] that the conduction band minima in 4H are located at M and those in 6H at the LM line, although the exact position on LM depends on the numerical details [7]. However, in principle there is agreement between the majority of band-structure calculations now [7, 46, 47, 59, 60, 64].

Table 4

Phonon frequencies (in units of cm^{-1}) of 4H-SiC at the high-symmetry points L and M

L	M	ref. [57]	ref. [58]	L	M	ref. [57]	ref. [58]
249	246	266	267	655	743	775	762
	274	294	296		749	785	764
272	316	328	329	754	752	789	774
	319	331	331		758		779
362	320	336	338	758	760		
	370	372	372		760	799	797
387	396	410	409	766	768	809	835
	422	425	425		781	820	838
544	554	547	549	856	841	852	857
	559	554	554		845	857	861
631	611	618	616	921	857	873	
	619	633	630		865	877	879

Table 5

Phonon frequencies (in units of cm^{-1}) of 6H-SiC at the high-symmetry points L, U, and M

L	U	M	ref. [57]	L	U	M	ref. [57]
259	250	247		682	714	746	
	272	271	261		717	751	
288	291	287	285	726	744	752	
	304	288			747	755	
320	307	306	316	753	753	755	
	334	346	348		754	756	
330	338	347		759	757	757	760
	357	347	353		761	763	766
364	361	359	358	767	765	764	770
	368	369	366		768	768	778
375	371	402	402	792	786	771	786
	404	422	424		787	775	
553	558	554		831	835	843	836
	567	562			837	848	840
574	585	601	594	890	868	849	845
	588	606	604		868	852	850
636	616	606		910	883	863	857
	627	612	617		884	864	861

There are also strong experimental arguments for this **k**-space location of the conduction-band minima in 4H- and 6H-SiC. The momentum conservation rule for the
luminescence together with the location of the valence band maximum at the Γ points
yields that phonon replica of the zero-phonon line can only appear when phonons with
the wave vector of the conduction-band minimum are involved. From the rather good
agreement of theory and experiment in Table 4 with respect to the number of phonons
and their energetical positions the identification of the M point as the position of the
conduction-band minima in 4 H seems to be justified [57, 58]. In the case of 6H
(Table 5) the situation is less clear. Although not all 36 phonons are observed experimentally, their variety indicates at least that the point of location is on the LM line.
From the comparison of the frequency values we cannot conclude that this point is
identical with M. Rather, we believe that a point somewhere on LM has to be favoured.

4.2 Thermodynamics

The thermodynamics of the vibrating crystal lattices is governed by the Helmholtz free
energy $F_{vib}(T, V)$. In our calculations, we replace the electronic contribution $F_{el}(T, V)$ to
the total free energy, $F(T, V) = F_{el}(T, V) + F_{vib}(T, V)$, by its zero-temperature limit [36,
46], since the electronic entropy is negligibly small in the interesting temperature range.
The lattice contribution is described by the sum over a set of independent harmonic oscil-
lators. The entropy of the vibrating lattice follows from the free energy as
$S_{vib} = -(\partial F_{vib}/\partial T)_V$. Its knowledge allows the calculation of the internal energy according to $U_{vib} = F_{vib} + T S_{vib}$.

As shown in [55] the **k**-point sampling is crucial in achieving a reasonable accuracy in
the energy differences. Therefore, the special-point technique of Monkhorst-Pack with

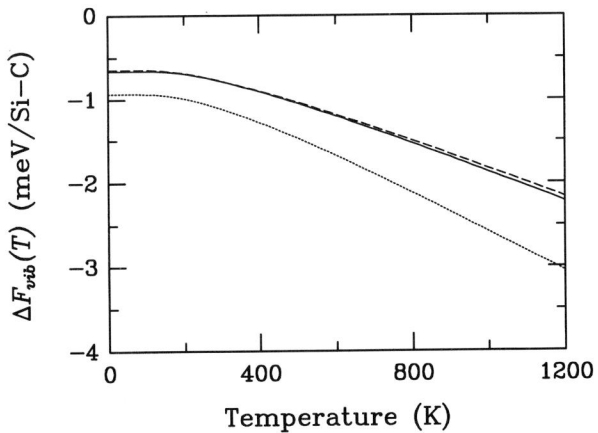

Fig. 3. Differences of the free energy of nH polytype with respect to the 3C value. $F_{\text{vib}}^{2H} - F_{\text{vib}}^{3C}$ (solid line), $F_{\text{vib}}^{4H} - F_{\text{vib}}^{3C}$ (dotted line), and $F_{\text{vib}}^{6H} - F_{\text{vib}}^{3C}$ (dashed line)

equivalent **k**-point meshs in all investigated structures is used to perform the integration over the irreducible part of the BZ [34]. In our computations the uncertainty for the energy differences using this finite **k**-point sampling is estimated to be below 0.01 meV per Si–C pair, whereas the differences of the free energy between the SiC structures are of the order of 0.1 meV.

For a discussion of the polytype stabilization by the vibrating lattice only the differences of the lattice contributions to the total energy are substantial. Such differences with respect to the free energy of the zinc-blende structure of SiC are plotted for the hexagonal polytypes 6H, 4H, and 2H with increasing percentage hexagonality h of 33, 50, and 100% in Fig. 3. The differences in the vibrational free energy per Si–C pair are of the same order of magnitude, i.e., several meV, as the differences of the static total internal energy [36, 46, 51]. In the low-temperature region the main contribution originates from the variation in the zero-point vibrational internal energy. For higher temperatures there is also a contribution from the entropy variation [55]. However, the latter one does not change the order of the static energies of the investigated polytypes, but rather enlarges the differences among them. In general, the (lattice) free energies are lower for the hexagonal lattices than for the cubic one. We explain this fact by the general tendency that the limiting frequencies slightly decrease with rising hexagonality. This has been clearly shown for the zone-center optical phonon frequencies [41] as well as for the Debye temperatures [55]. Therefore, the free energy decreases on the average. On the other hand, there is a nearly parabolic dependence of the free energy on the hexagonality for the polytypes 6H, 4H, and 2H, although the variations are smaller than the differences to the 3C values. The lowest free energy is observed for 4H ($h = 50\%$), whereas the values for 6H ($h = 33\%$) and 2H ($h = 100\%$) are nearly identical. Consequently, the vibrating lattice strengthens the tendency for the stabilization of the 4H polytype in the thermodynamic equilibrium. This happens surprisingly not only for low but also for high temperatures. Our results for the lattice free energy are somewhat in contrast to those of Cheng et al. [65]. These authors found differences of the phonon free energies that are one order of magnitude smaller. The reason may be related to the fact that the same valence-overlap shell model parameters derived for cubic SiC have been applied to the hexagonal polytypes.

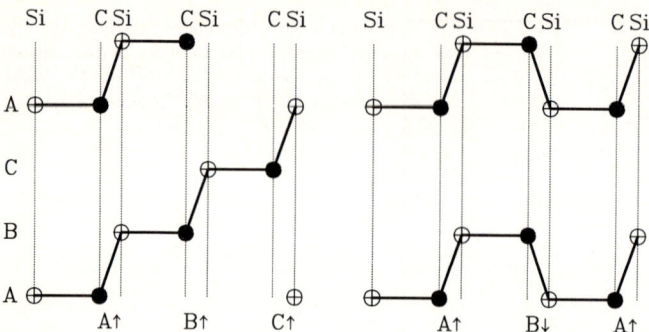

Fig. 4. Zig-zag chain structure of silicon and carbon atoms in the (11$\bar{2}$0) plane of 3C-(left panel) and 2H-SiC (right panel). The different relative lateral position A, B, and C of the Si–C layers as well as the pseudospins are indicated

4.3 Temperature-dependent ANNNI model

Geometrically the polytypes differ by the stacking along the cubic [111] or the equivalent hexagonal [0001] direction. As shown in Fig. 4 for the 3C and 2H structures, all polytypes may be imagined as different arrangements of cubic or hexagonal Si–C bilayers along this particular direction. The one-dimensional character of the stacking differences suggests the description of the polytype in terms of an axial next-nearest-neighbour Ising (ANNNI) model [66], where the i-th cubic (hexagonal) bilayer is represented by the pseudospin up $\sigma_i \equiv +1$ (down $\sigma_i \equiv -1$). This model has been already successfully applied to the discussion of polytypism of the material system silicon carbide neglecting the contributions of the vibrating lattice [49, 50, 67]. In a simplified version the total free energy of the system per Si–C pair may be represented by

$$F(T) = F_0(T) - \frac{1}{n} \sum_{i=1}^{n} \sum_{j=1}^{\infty} J_j \sigma_i \sigma_{i+j} \,, \tag{7}$$

where j runs over the interacting bilayers. The label i accounts the bilayers in the unit cell of the nH polytype. More complicated interactions, such as four-spin terms, are negligible and therefore have been left out. The parameters J_j are the interaction energies of two bilayers. The largest term $F_0(T)$ in Eq. (7) represents the energy of the crystal without interaction of the bilayers. Assuming that the long-range interactions are small we restrict the j-sum up to third neighbours ($j = 3$).

In order to derive the interaction parameters J_j we replace the total free energy $F(T)$ at given volume in Eq. (7) by the sum of the lattice contribution and the total internal energy of the electrons [36, 46, 51]. The entropy of the electronic system as well as the configurational contribution, which is related to the arrangement of boundaries between bilayers belonging to different spins, are not taken into account [65]. Explicitly, differences are considered, i.e., $F_{2H} - F_{3C} = 2J_1 + 2J_3$, $F_{4H} - F_{3C} = J_1 + 2J_2 + J_3$, and $F_{6H} - F_{3C} = \frac{2}{3} J_1 + \frac{4}{3} J_2 + 2J_3$. The results of such fitting procedure are given in Table 6.

The variation of the parameters starting from three different static total energy calculations [36, 46, 51] may be traced back to three facts: (i) The atomic relaxation taken into account by Käckell et al. [36] remarkably reduces the nearest-neighbour interaction J_1. (ii) The ab initio calculation of Cheng et al. [51] favours the 6H polytype instead of

Table 6

Interaction parameters J_i (meV per Si–C pair) of the ANNNI model for different temperatures and different electronic contributions to the total free energy from [36, 46, 51]

parameter	J_1			J_2			J_3		
reference	[51]	[36]	[46]	[51]	[36]	[46]	[51]	[36]	[46]
static value	4.80	1.08	2.33	−2.93	−2.45	−3.49	−0.45	−0.18	0.25
0 K	4.49	0.76	2.02	−3.23	−2.75	−3.79	−0.47	−0.19	0.24
400 K	4.38	0.66	1.91	−3.34	−2.86	−3.90	−0.48	−0.21	0.22
800 K	4.10	0.37	1.63	−3.60	−3.13	−4.16	−0.52	−0.24	0.18
1200 K	3.79	0.07	1.32	−3.90	−3.42	−4.46	−0.55	−0.28	0.15

the 4H one. (iii) A much larger energetical difference between 3C and 2H is calculated by Karch et al. [46] and Cheng et al. [51]. The introduction of the phonon free energy reduces all interaction parameters. We observe a tendency for a reduction of J_1, J_2, and J_3 and a shift towards more negative values. That means, the vibrating lattice reduces the attractive character of the nearest-neighbour interactions. The effect on the second-nearest-neighbour interaction is rather small. More important is the phonon influence on J_1 and J_3, especially starting from the ab initio results of [46]. In this particular case the third-nearest-neighbour interaction, J_3, will be repulsive for layers of equal pseudospin.

The phase diagram of the described ANNNI model including the phonon contributions is plotted in Fig. 5. We choose the ratios J_3/J_2 and J_1/J_2 as coordinates. In the selected parameter region two multi-phase degeneracies appear. For $J_3 = 0$ and $J_1 = -2J_2$ the phases of low hexagonality 3C, 6H, and 4H degenerate. For the more unrealistic parameter configuration $J_1 = J_2 = J_3$ a triple point of the hexagonal phases under consideration appears. When the lattice contributions to the total free energies are neglected, the results of the three ab initio calculations appear close to the first triple point. After inclusion of the lattice free energy the calculated phases appear much more in the stability region of the hexagonal phases. This tendency away from the cubic phase is increased with rising temperature, what confirms the paradox situation [68] that SiC appears to prefer to grow in cubic form, more than in any other, in spite of the fact that this is never the stable structure. Surprisingly, we observe a stabilization of the

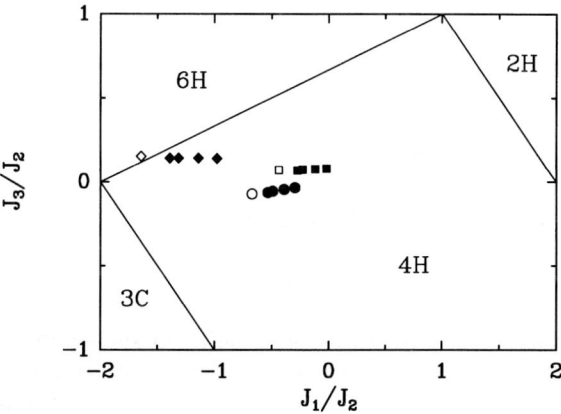

Fig. 5. The phase diagram of the ANNNI model. The stability regions of the four polytypes considered are shown. The phases obtained starting from the ab initio results of [36, 46, 51] are indicated by diamonds, circles, and squares, respectively. Open symbols: without the lattice contribution to the free energy, black symbols: with phonon free energy. From left to right increasing temperatures $T = 0$, 400, 800, and 1200 K are considered

4H polytype with increasing temperature. Although our estimations show also a tendency to increase the ratio J_3/J_2 and thereby to approach the 6H stability region, in the framework of the ANNNI model we cannot reproduce the results of the experiments [2, 69] which indicate that 6H is the stable form at high temperatures and probably 4H at low temperatures. However, as has been pointed out in [2] the experimental results are somehow uncertain.

5. Optical Properties and Many-Body Effects

5.1 Absorption

In order to examine optical properties of SiC polytypes we have first calculated the imaginary parts of the elements of the second-rank dielectric tensor $\varepsilon_{\alpha\beta}(\omega)$ in the optical limit [39, 70]

$$\operatorname{Im} \varepsilon_{\alpha\beta}(\omega) = \frac{8\pi^2 e^2 \hbar^2}{V} \sum_{c,v} \sum_{\mathbf{k}} \frac{\langle c\mathbf{k}| v_\alpha |v\mathbf{k}\rangle \langle c\mathbf{k}| v_\beta |v\mathbf{k}\rangle^*}{[\varepsilon_c(\mathbf{k}) - \varepsilon_v(\mathbf{k})]^2} \, \delta[\varepsilon_c(\mathbf{k}) - \varepsilon_v(\mathbf{k}) - \hbar\omega] . \quad (8)$$

The optical transition matrix elements, defined by the commutator of single-particle Hamiltonian and space operator, contain local as well as nonlocal contributions due to the nonlocal pseudopotentials in the Hamiltonian. The neglect of the nonlocality effects increases the average oscillator strength by about 15%. A similar overestimation arises for the dielectric constants and the low-frequency reflectivity spectra. The collective plasmon peak in the energy loss spectra is shifted to higher energies by about 2 to 3 eV [70]. For that reason all calculations are performed including nonlocal contributions. The real parts of the components of the dielectric tensor $\operatorname{Re} \varepsilon_{\alpha\beta}(\omega)$ are calculated from the imaginary parts by means of a Kramers-Kronig transformation. Expression (8) is taken within the independent-particle approximation. Local-field effects have been neglected. Their contributions vary in dependence on the photon energy [39, 71]. To investigate many-body effects the DFT-LDA eigenvalues $\varepsilon_n(\mathbf{k})$ are shifted by quasiparticle corrections calculated according to a scheme developed by Wenzien et al. [7]. The calculated quasiparticle shifts depend on the band index and the wave vector. In the 3C case the values are in satisfactory agreement with results of Rohlfing et al. [72] and Backes et al. [73]. The shifts are only included in the spectral properties but not in the matrix elements [39]. The effect of the energy dependence of the exchange-correlation self-energy on the spectral strengths is neglected.

Fig. 6 shows the frequency dependence of the imaginary parts of the dielectric function for the most important hexagonal polytypes 2H, 4H, 6H and the cubic one 3C. The anisotropy of the dielectric function for the hexagonal polytypes can be clearly seen by a comparison of the parallel and perpendicular components. The parallel components are more strongly influenced by the polytypism than the perpendicular components. These findings are related to the atomic structure which is in each bilayer essentially the same but differs due to the different stacking of the Si–C bilayers parallel to the c-axis. Comparing the spectra of the parallel tensor components the pronounced peaks of 3C split up, shift to lower energies and become broadened for the hexagonal polytypes. This holds especially for 6H, where practically no pronounced peak structure exists any more. Thus, the folding effect seems to induce an effective smearing out of the peak structures as follows from the comparison of the spectra from 3C and 2H over 4H to 6H. It is

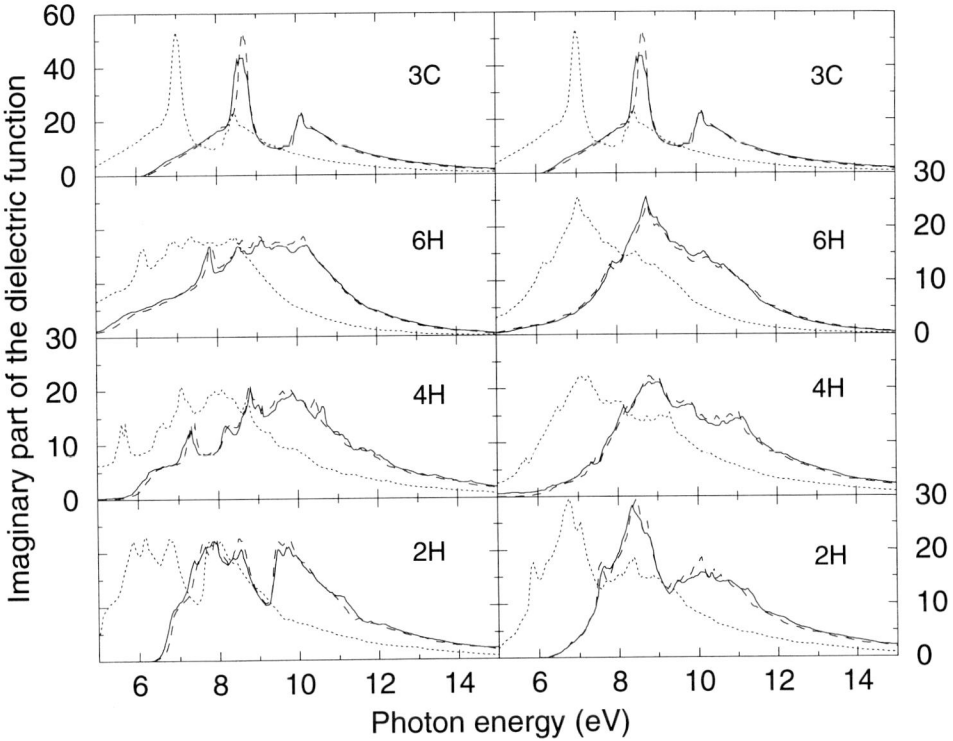

Fig. 6. Imaginary part of the dielectric tensor versus photon energy for the parallel (left) and perpendicular (right) component. Solid lines: DFT-LDA including quasiparticle corrections, dotted
lines: DFT-LDA, dashed lines: DFT-LDA with quasiparticle corrections within scissors operator
approximation

therefore mainly related to the period of the translational symmetry in c-axis direction,
but not to the relative number of twisted bonds, i.e., the hexagonality of the polytype.
The perpendicular tensor components behave in a different way. In general, for all polytypes two major peaks are observed with some fine-structure. However, the double structure is redshifted for 3C and 2H in comparison with its position for 4H and 6H. The
polytypes with more atoms in the unit cell, 6H and 4H, give again rise to more broadened features and are rather similar. Above photon energies of 10 eV the influence of
both the polytypism and the anisotropy almost vanishes.

Fig. 6 also includes results for taking into account quasiparticle corrections. Compared
to the curves within DFT-LDA the spectra are nearly rigidly shifted towards higher
energies. The absolute value of this shift seems to be independent of the polytype as well
as of anisotropy. Furthermore the peak intensities decrease. This effect is also independent of the polytype. For a more quantitative discussion we consider a scissors operator
approximation. Within this approximation the spectra are shifted by a scissors operator
Δ according to the replacement $\mathrm{Im}\,\varepsilon(\omega) \cong \mathrm{Im}\,\varepsilon^{\mathrm{DFT}}(\omega - \Delta/\hbar)$. The constant shifts Δ are
determined by variation until the spectra with band-index- and wave-vector-dependent
shifts and scissors operator show maximum agreement. In fact the absolute values of Δ
exhibit only small variations with the polytype and the tensor component, as can be

Table 7

Influence of many-body effects on the energetical positions of optical (Δ) and loss (ΔE) spectra expressed by relevant scissors operator for the polytypes and the two different polarizations

polytype	$2H_\parallel$	$2H_\perp$	$4H_\parallel$	$4H_\perp$	$6H_\parallel$	$6H_\perp$	$3C$
Δ (eV)	1.74	1.68	1.71	1.74	1.71	1.71	1.68
ΔE (eV)	1.43	1.30	1.38	1.06	1.55	1.33	1.20

seen in Table 7. Comparing the results within the scissors operator approximation with those including band-index- and wave-vector-dependent quasiparticle corrections one observes a tendency for broadening in the latter case. The intensity of the principal peaks decreases whereas the spectral weights of the low- and high-energy tails increase. In addition the peak positions in the quasiparticle spectra are slightly changed. This holds especially for those at the high-energy side. This behaviour is independent of the polytype as well as the tensor component under consideration.

5.2 Dielectric constant and energy loss spectrum

The trends with the crystal structure in the spectra of the real parts of the dielectric tensor as well as of the reflectivity follow the behaviour of the imaginary parts. For a polarization of the electric field in the plane perpendicular to the c-axis the polytype-induced changes are smaller than for the opposite polarization. In addition we conclude again a remarkable influence of the size of the unit cell and the accompanying folded band structure, whereas the hexagonality has less influence. Due to the folding of the bands in k-space parallel to the c-axis, there is a tendency for flattening the spectral features. Comparing our calculated reflectivity with experimental data [74, 75] we observe a slight overestimation of the absolute values in comparison to the experimental spectra. This discrepancy may be traced back to the overestimation of the high-frequency dielectric constant and the neglect of broadening effects, but also to experimental details as sample quality and surface treatment. The inclusion of many-body effects leads to results corresponding to the case of the imaginary parts of the dielectric function. The spectra are shifted towards higher photon energies and the peak intensities are somewhat lowered. The absolute values of the tensor of electronic dielectric constants $\varepsilon_\infty^{\parallel/\perp} = \operatorname{Re} \varepsilon_{\parallel/\perp}(0)$ are remarkably reduced by about 35 to 40%. Corresponding to the imaginary parts there are no mentionable differences according to the different polytypes or tensor components [70]. In comparison to the experiment the optimized shifts Δ are too large. Instead of values of 1.7 eV, one needs only one half to bring the main structures in the theoretical and experimental spectra to coincidence. These findings that the quasiparticle shifts of the main structures are overestimated seem to be a general result and are also valid for other semiconductors. We point out that the values for the fundamental indirect gaps are smaller. Typical values vary around 1.2 or 1.3 eV. Results [70] without quasiparticle shifts are listed in Table 8 and compared with values from the density-functional perturbation theory [18, 76], other calculations [77], and experimental values [43].

Considering the accompanying energy loss functions [70] for energies near the bulk plasma frequency we find the polytypism to be at least of the same importance as in the

Table 8
Tensors of dielectric constants for different polytypes

	2H		4H		6H		3C
	$\varepsilon_\infty^\parallel$	$\varepsilon_\infty^\perp$	$\varepsilon_\infty^\parallel$	$\varepsilon_\infty^\perp$	$\varepsilon_\infty^\parallel$	$\varepsilon_\infty^\perp$	ε_∞
ref. [70]	8.02	7.64	7.61	7.54	7.49	7.48	7.33
ref. [18, 76]	7.28	6.88	7.17	6.95	7.24	7.02	7.02
ref. [77]	7.32	6.91	7.20	6.96	7.24	7.00	6.95
ref. [43]	6.84	6.51	6.78	6.56	6.68	6.52	6.52

case of the optical properties. The intensities of the loss functions derived from the parallel components of the dielectric tensor are nearly independent of the polytype, whereas the energetical positions of the plasma peaks vary remarkably by about 1.5 eV. On the other hand, the positions of the main peaks of the loss functions derived from the perpendicular components coincide quite well. We suggest to use the position of the main plasmon loss for the polytype identification by means of the electron-energy loss spectroscopy in the parallel situation. Neglecting quasiparticle and local-field effects the absolute positions of the plasmon peaks are overestimated. We have to mention that the shape and the position of the plasmon peak depends dramatically on the numerical details. The inclusion of quasiparticle corrections again causes an energetical shift ΔE of the loss spectra towards higher energies by a value nearly independent of polytype and tensor component, as can be seen in Tab. 7. The reduction of the intensity of the main peak exhibits no clear tendency with respect to the polytype considered. Thus, quasiparticle corrections influence the loss spectra in a similar way as the optical properties discussed above.

6. Heterocrystalline Structures

6.1 Band line-ups

Thinking about possible applications, a combination of different polytypes could be of interest because of their different electronic structures. The combination of the cubic and the hexagonal polytype with a gap variation of about 1 eV should allow, for example, the generation of a two-dimensional electron gas in the cubic region near the interface after modulation doping of the hexagonal region. The mobility of the electrons should be rather high even at room temperature. Besides the inefficiency of the optical phonons at this temperature, the expectation is related to the perfect structure of the interface between two polytypes made from chemically identical materials.

Experimentally, a combination of polytypes to heterocrystalline structures should be possible since the lattice mismatch is negligible. Recently, large steps forwards in molecular-beam-epitaxy (MBE) growth of SiC have been reported [78 to 81] and a controlled growth of a certain polytype seems to be possible in the near future. An example for a heterocrystalline structure grown by means of solid-source MBE [81] is shown in Fig. 7. A rough estimation of the expected electronic properties follows already from the natural band discontinuities of the pure polytypes, which dominate the band line-ups at the interfaces. They can easily be determined by an alignment of the calculated bulk band structures [7, 59] at a certain internal reference energy. We use the vanishing of the

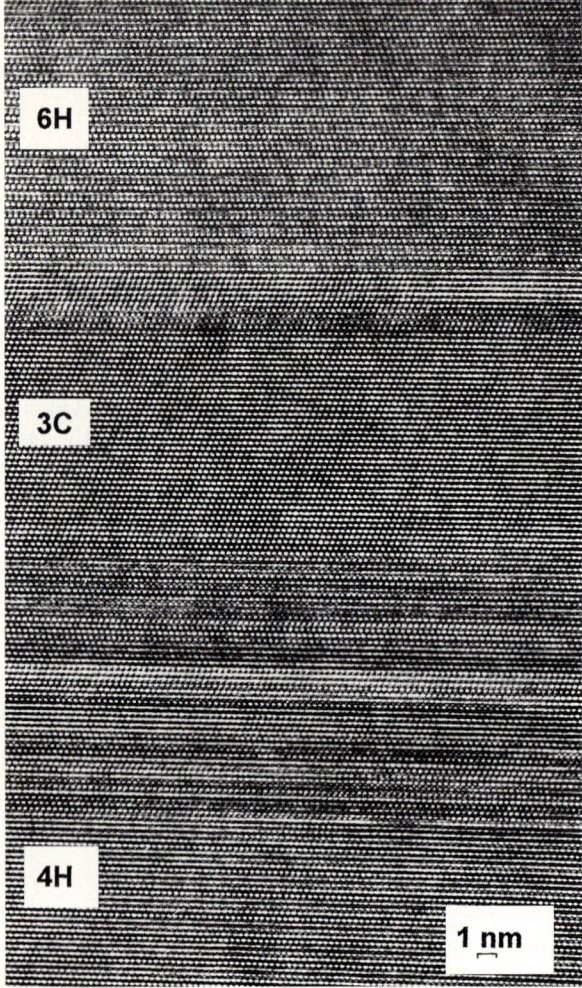

Fig. 7. Cross-sectional high resolution TEM image of SiC grown on a 4H-SiC(0001) substrate by means of MBE

pseudopotentials at infinite distances to determine this reference level. This is possible since all polytypes consist of the same chemical elements.

Table 9 show the valence and conduction band discontinuities relative to 3C-SiC. From these values one therefore expects a type-II-heterostructure behaviour for a combi-

Table 9

Natural band discontinuities for valence (ΔE_{v}) and conduction band (ΔE_{c}) (in eV) relative to 3C-SiC (from [59]). Band discontinuities including quasiparticle effects [7] are given in parentheses

polytype	ΔE_{v}	ΔE_{c}
6H	−0.02 (−0.00)	0.74 (0.70)
4H	−0.05 (−0.10)	0.99 (0.97)
2H	−0.13 (−0.24)	0.99 (1.10)

nation of a hexagonal with the cubic polytype. Consequently, electrons and holes should be localized in different regions in the real space as well as in the reciprocal space. More strictly, for a 3C/nH combination, electrons are expected to be localized in the cubic part and holes in the hexagonal region. In the literature, data for valence-band discontinuities are available for the extreme 3C/2H combination only. They agree very well with our findings. A tight-binding calculation by Murayama and Nakayama [82] gives $\Delta E_v = -0.146$ eV, and a calculation by Qteish et al. [83] using the geometry of Cheng et al. [51] gives a value of -0.13 eV.

Such natural discontinuities [84] should approach the real band offsets between two semiconductors if the interface dipoles vanish. However, for SiC the situation is somewhat more complicated than this simple estimation might entice to assume. The hexagonal polytypes are pyroelectric materials. That means they possess a macroscopic dipole moment per Si–C pair which is accompanied by a nonvanishing macroscopic polarization field parallel to the c-axis. We discuss the consequences in the case of two examples, $(3C)_4(2H)_3$ and $(3C)_6(2H)_6$, i.e. heterocrystalline superlattices made by the two polytypes 3C and 2H. In this notation, the product of index and polytype number n gives the number of Si–C bilayers of this polytype in the considered heterocrystalline supercell. The sum of both numbers, therefore, equals the overall number of double layers in the cell, that is 18 in the first and 30 in the second case (cf. Fig. 8 and 9; the left panel). Other structures are discussed in [85, 86].

In our calculations, we choose the weighted mean values of a and c (the differences are less than 0.2%) for the lattice constants. Again, the atomic positions are optimized until the Hellmann-Feynman forces vanish. It turns out that the structures are energetically stable and have cohesive energies of the same order as the pure polytypes. The atomic relaxations are different for both of the inequivalent interfaces. At the transition

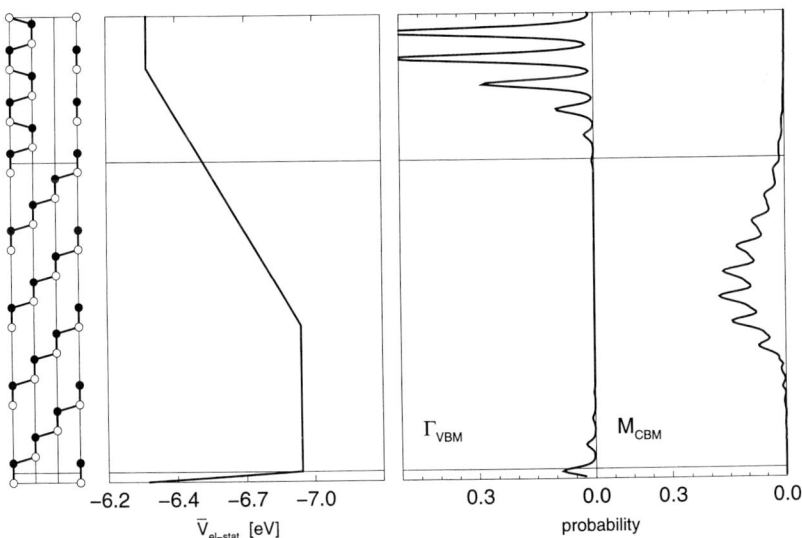

Fig. 8. Structure, mean electrostatic potential (schematically) and squared wave functions at VBM and CBM for a $(3C)_4(2H)_3$ structure. The horizontal lines separate cubic (lower) and hexagonal (upper) parts

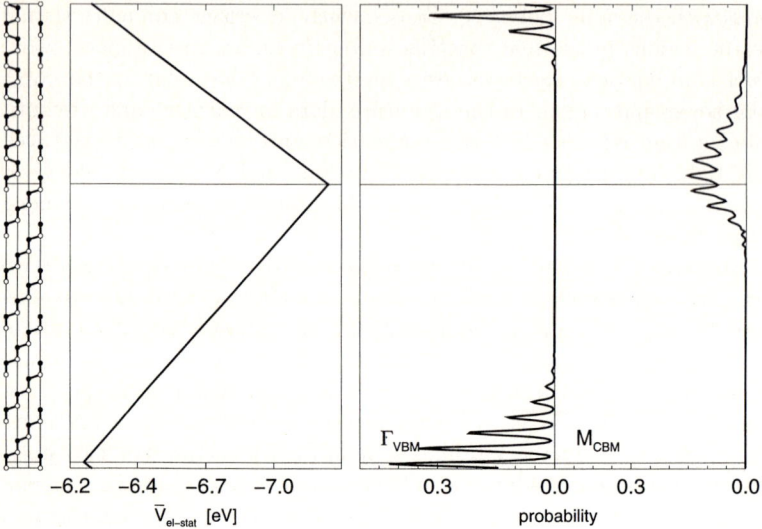

Fig. 9. Structure, mean electrostatic potential (schematically) and squared wave functions at VBM and CBM for a $(3C)_6(2H)_6$ structure. The horizontal lines separate cubic (lower) and hexagonal (upper) parts

from 2H to 3C (lower line in the figures) the bonds are stretched by about 0.01 to 0.02 Å. On the other hand, at the transition from 3C to 2H the changes are negligible, since the bond sequence in the $(11\bar{2}0)$ plane cannot be continued in the first case contrary to the second interface.

6.2 Electronic structure and pyroelectricity

The stacking arrangement in the superlattices of heterocrystalline combinations changes the electronic structure. From the electronic structure of the pure polytypes, one expects fundamental gaps similar to the one in the cubic parts of the superlattices. The 3C polytype has the lowest-lying conduction band minimum (CBM) of the considered structures. The valence-band maximum (VBM), on the other hand, should be determined by the hexagonal polytype, but with little influence on the gap due to the small valence-band discontinuities (cf. Table 9). For the combination of 2H and 3C we therefore expect a gap of about 1.14 eV, i.e., the energy gap of 3C diminished by the valence band discontinuity ΔE_v for 2H. Surprisingly at first sight, we find indirect fundamental gaps $\Gamma \to M$ of 0.75 eV for $(3C)_4(2H_3)$-SiC and 0.86 eV for $(3C)_6(2H)_6$-SiC. The calculated heterocrystalline superlattices therefore behave like type-II heterostructures. However, the actual band discontinuities are larger in the real systems than expected from the discussion of the natural discontinuities. Furthermore, we have to mention that the gaps have been calculated within DFT-LDA where the excitation aspect is not taken into account. Similarly to the bulk case [7] the superlattice gaps should be also opened by about 1.2 eV in order to compare with experimental values.

Obviously, the reason for the observed gap reduction with respect to the 3C polytype should be related to the change in the electrostatic potential. The averaged electrostatic potential is not only changed in the interface regions. Rather, compared to the static combination of two polytypes there is also a variation within the bulk polytype regions due to the self-consistency in the ab initio calculation. The mean electrostatic potential — integrated over planes vertical to the c-axis and averaged over Si–C bilayers parallel to it — is plotted along the c-axis in the middle panels of Figs. 8 and 9. The potential is modified by two different effects. However, in both of the structures considered, only one of them dominates. In the short-periodic structure (Fig. 8), a charge transfer from cubic to hexagonal material occurs at the interfaces. This results in an interface dipole moment which is accompanied by a step-like potential. For that reason, the ratio of valence to conduction band discontinuities changes whereas their sum remains nearly constant. The mean electrostatic potential is about 0.56 eV lower in the cubic regions compared to the hexagonal parts. This difference approximately matches the reduction of the energy gap compared to the value in 3C-SiC. Identifying the spatial extent of the dipole with a Si–C bond length it follows an electron transfer of about 0.08 electrons from the discontinuity in the potential. The second effect related to the pyroelectricity of the hexagonal material plays only a minor role in the $(3C)_4(2H)_3$ structure.

This is not valid for the longer-periodic structure (Fig. 9). Here the difference in the mean potentials of the cubic and hexagonal part amounts only to 0.03 eV. The effect of the interface dipole seems to be negligible. Actually, the potential behaves as one would expect from a more macroscopic picture, where the 2H crystal layers possess a macroscopic polarization due to the nonvanishing dipole moment of the unit cell parallel to the c-axis, whereas the dipole moment is zero for 3C because of symmetry arguments. Periodically repeated supercells with layers of alternating static polarization give rise to a sawtooth-like potential. Its maxima and minima are located in the two inequivalent interfaces, what has consequences on the localization of the wave functions.

We study the localization of the Bloch functions at the VBM and the CBM, i.e., the wave functions of the highest occupied and the lowest empty state. The squares of these states are shown in the right panels of Figs. 8 and 9, integrated over a plane vertical to the c-axis. In accordance with the simple estimation above, considering only the natural band discontinuities, in Fig. 8 the CBM wave function at the M point in $(3C)_4(2H)_3$-SiC is localized in the cubic region of the unit cell. The electrons should be confined there. The wave function of the VBM at Γ is mainly localized in the hexagonal part. Consequently, holes should be confined there. A larger supercell seems to give a more realistic description of a real system, as we have seen above. Again, in Fig. 9 we find a type-II heterostructure, where the Bloch states at the band extrema are localized in different spatial regions. Seemingly, they can be formally interpreted as interface states. However, one should keep in mind that the localization at the interfaces is caused by the pyroelectricity in the hexagonal parts. The periodic arrangement of cubic and hexagonal material layers results in the sawtooth-like potential. The state at the CBM is therefore localized at the minimum of the electrostatic potential, the state at the VBM at its maximum.

7. 3C-SiC(111) and nH-SiC(0001) Surfaces

7.1 Energetical stability and growth

During solid-source MBE growth a quite frequently observed surface reconstruction is related to the $(\sqrt{3} \times \sqrt{3})$ R30° translational symmetry. Fig. 10 shows a phase diagram for selected structure *proposals* for a Si-terminated 3C-SiC(111) $(\sqrt{3} \times \sqrt{3})$ R30° surface

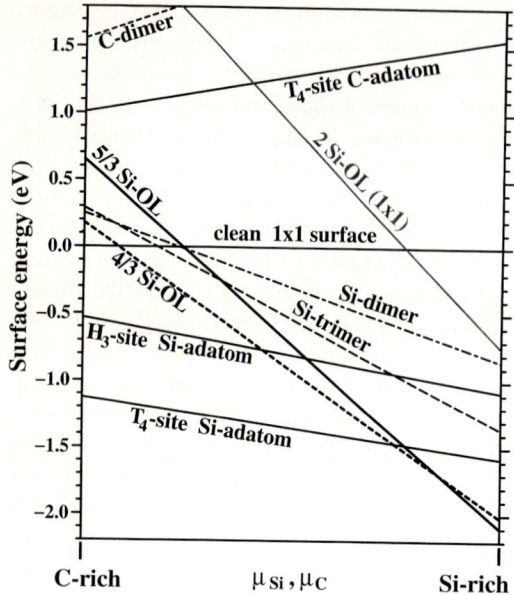

Fig. 10. Phase diagram of several structural models for the $(\sqrt{3} \times \sqrt{3})$ R30° surface. "4/3 (5/3, 2) Si-OL" corresponds to surfaces with 4/3 (5/3, 2) additional Si-overlayers on a Si-terminated face. The "2 Si-OL" surface is assumed to be unreconstructed. The other structural models contain adsorbed Si trimers and additionally one (two) Si adatom(s) between the Si trimers for the 4/3-OL (5/3-OL) geometry

reconstruction. The surface energy is plotted against the chemical potential of Si(C) over the whole range of relevant chemical potentials ("from Si or C bulk towards SiC"). The energy zero is defined by the clean *unreconstructed* (but relaxed) surface. The basic structural elements which give usually rise to a $(\sqrt{3} \times \sqrt{3})$ R30° reconstruction are either adatoms or trimers or may be also their combinations, e.g., trimers with adatoms on top or in between. For completeness we include also dimer structures and consider Si-covered as well as C-covered surfaces. The most important result is that adsorption of Si is always energetically preferable over adsorption of C on "nearly clean" surfaces — even under C-rich conditions. Over a wide range of chemical potentials (except for Si-rich environments) a T_4-site Si-adatom structure has the lowest energy and is expected to be observed under thermodynamic equilibrium conditions. This adatom structure has also been favoured in previous theoretical and experimental work [87 to 89]. However, under Si-rich conditions we find that surfaces with more than one additional Si layer adsorbed are even lower in energy. In this case the energy does not vary too much with the stoichiometry of the surface. The corresponding structures can be characterized as "trimerized Si overlayers with a varying amount of additional Si adatoms between the trimers". All other structures can only exist as *metastable* states. It is possible to prepare more-C-rich surfaces in experiment and to keep them stable, but they do not represent the ground-state structure in thermodynamic equilibrium.

These findings are also confirmed by experiment. Solid-source MBE growth experiments find that the best growth conditions can be achieved on Si-rich surfaces [81, 90 to 92]. Furthermore, one observes that the surface reconstruction changes from $(\sqrt{3} \times \sqrt{3})$ to "(1×1)" [81, 90 to 92] for rather Si-rich surfaces and finally to (3×3) [81, 90 to 93] for very Si-rich surfaces. Theoretical results for the (3×3) reconstruction are not yet available. A reasonable assumption is that a certain very Si-rich structure should exist which lies energetically even lower than the rather Si-rich ones, discussed in the phase diagram of a $(\sqrt{3} \times \sqrt{3})$ surface cell for very Si-rich environments.

The phase diagram (Fig. 10) explains that a "two-step-growth" method should work best where first Si-rich surfaces are exposed under Si-rich conditions in a first step and where the Si-rich surfaces are carbonized by adsorbing C under more-C-rich conditions in a second step [81, 90 to 92]. Si-rich surfaces are throughout the whole range of chemical potentials energetically preferred over C-rich surfaces and under rather Si-rich conditions one has to expect that more-Si-rich surfaces are most stable. From the phase diagram it becomes also clear that under C-rich conditions the very Si-rich surfaces have a higher surface energy than the only slightly Si-rich adatom surface. This means that, in contrast to clean or less-Si-rich surfaces, the adsorption of C atoms leads now to an energy gain, i.e., C can be easily adsorbed on Si-rich surfaces. By adsorption of C a Si-layer will be transformed into a Si-C bilayer, i.e., SiC grows. Although we can explain the growth scenario in principle, the real growth process itself is not yet described by the phase diagram in Fig. 10. The most important and also most complicated step is the C adsorption (second growth step). Here not only an atom has to be adsorbed somewhere *on* the surface, rather the C atom must afterwards also diffuse *into* the surface.

7.2 Surface energy and atomic structure

Supplementary to the 3C-SiC(111) surface structures we have also performed calculations for selected surface structures on a 2H-SiC(0001) "substrate". Table 10 lists the resulting surface energies, i.e., the slab energies minus the cohesive energies of the substrate atoms and the adsorbed Si atoms. The unrelaxed surface is taken as the energy zero. Energy values are given for seven models of the Si-rich ($\sqrt{3} \times \sqrt{3}$) surface and two different substrates. As the most extreme cases the pure cubic stacking (3C) as well as the pure hexagonal stacking (2H) of the Si–C bilayers are considered. The general trend for the energetics of the different structural models is independent of the polytype of the substrate. Apart from the case of the surface relaxation there is only a rigid shift of 0.1 to 0.2 eV between the results for 3C and 2H. The energy gains are somewhat larger in the hexagonal case. On the other hand, the hexagonal stacking seems to prevent that the atoms in the ideal Si-terminated surface are substantially displaced during relaxation. Moreover, the absolute surface energies are usually higher in the hexagonal case by 0.05 to 0.100 eV per ($\sqrt{3} \times \sqrt{3}$) unit cell. Exceptions are the most-Si-rich surfaces considered.

Besides the investigation of the surface energetics, the investigation of the atomic structure might also give hints towards "polytype selection rules". Although the energies

Table 10
Surface energies (in eV per ($\sqrt{3} \times \sqrt{3}$) cell) with respect to the unrelaxed, clean surface. To compare different surface stoichiometries bulk energies of adsorbed Si atoms are subtracted. In those cases where more than one (meta)stable structure exist we have only compiled the most stable one

	3C	2H
clean relaxed surface	−0.18	−0.01
T_4-site Si adatom	−1.76	−1.89
H_3-site Si adatom	−1.25	−1.37
Si dimer structure	−1.03	−1.19
Si trimer structure	−1.53	−1.73
"Si trimers + adatom"	−2.20	−2.39
"Si trimers + dimer"	−2.27	−2.47

Table 11

Characteristic geometrical data (as explained in the text) for the T_4-site adatom structure on different polytypes. All distances are given in Å. Positive (negative) signs for Δr denote outward (inward) deviations

	3C-3	3C-6	2H-3	2H-6	6H-c-6	6H-h-6
Δh_{ad}	1.746	1.742	1.744	1.744	1.746	1.741
d_{ad}	2.432	2.431	2.428	2.428	2.432	2.428
Δr_{Si1}	−0.075	−0.074	−0.077	−0.077	−0.075	−0.075
Δr_{C2}	0.020	0.019	0.021	0.021	0.023	0.018
Δz_{Si1-C1}	0.628	0.627	0.635	0.635	0.629	0.631
Δz_{C1-Si2}	1.891	1.894	1.900	1.901	1.893	1.902
Δz_{Si2-C2}	0.617	0.618	0.621	0.623	0.617	0.622

differ only slightly, one might expect significant differences in the atomic structure which might indicate a certain tendency towards a specific polytype. A critical examination of the structural data shows that it is not the case. As an example for a typical SiC surface reconstruction we study the T_4-site Si-adatom structure of the $(\sqrt{3} \times \sqrt{3})$ R30° surface. In Table 11 we have compiled characteristic geometrical parameters as the vertical distance Δh_{ad} of the adatom above the Si-atoms of the topmost "substrate" bilayer, the corresponding Si–Si bond length d_{ad}, and the lateral radial displacements Δr_{Si1} and Δr_{C2} (with respect to the ideal bulk positions of the atoms) for the Si atoms of the topmost Si–C bilayer and the C atoms of the second-topmost Si–C bilayer of the "substrate" (radially symmetric towards the adatom). The relative vertical distances between the Si and C atoms within the two topmost bilayers are characterized by the distances between the *centers of mass* of the corresponding atomic layers, namely Δz_{Si1-C1}, Δz_{C1-Si2}, and Δz_{Si2-C2} (Si1, Si2, C1, and C2 denote the first and second "substrate" Si or C layers). The data are listed for a 3C substrate with three bilayers relaxed (3C-3) and six bilayers relaxed (3C-6), for a 2H substrate with three bilayers relaxed (2H-3) and six bilayers relaxed (2H-6). Finally we also give data for a 6H substrate with six bilayers relaxed (6H-c-6 denotes cubic termination, 6H-h-6 hexagonal termination).

One clearly sees that the atomic geometries in Table 11 differ only marginally (by about 10^{-3} Å). That means the influence of the underlying substrate polytype is practically negligible. The bonding behaviour of the surfaces is essentially the same for all polytypes and surface terminations of the hexagonal substrates. This was already indicated by the small energy differences in Table 10. Consequently, the "selection" mechanism which could govern the polytype growth should be widely independent of the substrate polytype. The current suspicion is that the "polytype selection" during growth of ideal SiC crystals might be mainly driven by thermodynamics. Perhaps any polytype might grow initially but afterwards there is a phase transition in the bulk region which fixes the final polytype. Possibly also elastic stresses (due to the slightly different lattice constants) might result in a preference for the polytype with the closest "lattice constant match". This might play a role in homo- and heterocrystalline epitaxy.

8. Summary

We have reviewed many results of theoretical but also experimental studies that concern the occurrence of SiC in various polytypes. The theoretical results have been mainly obtained by several ab initio density-functional methods combined with the plane-wave-pseudopotential method. Phonon frequencies and free energies of the vibrating lattice of the hexagonal polytypes have been also calculated by means of a generalized adiabatic bond-charge model. Optical properties have been described within the random-phase approximation and neglecting local-field effects. Starting points are the band structures constructed from the solutions of the Kohn-Sham equations. However, they are corrected by quasiparticle self-energy effects to account for the excitation aspect. The theoretical data obtained have been discussed in the light of available experimental data.

We found that the atomic relaxations within unit cells contribute to the stabilization of a certain polytype. Another very important point for the epitaxial growth of a certain polytype seems to be the ratio $c/(na)$ of the hexagonal lattice constants normalized to the number of Si–C bilayers. The calculated structural parameters are in good agreement with X-ray diffraction measurements. Neglecting the vibrating lattice the 4H polytype is found to be the most stable one from the energetical point of view. Adding the free energy of the vibrating lattice we observe a tendency for the stronger energetical favourization of bond twisting in SiC and, hence, a more pronounced hexagonality of the polytype with rising temperature. Phonons play not only a role for the stabilization of a certain polytype. Rather, their degeneracy/splitting and energetical positions can give important hints for the identification of the polytype-dependent localization of the conduction-band minima in **k**-space.

By means of total-energy calculations we show that combinations of cubic and hexagonal polytypes may give rise to novel heterocrystalline SiC structures with interesting electronic properties. We predict a type-II character of such "heterostructures" made by chemically identical layers. However, the spatial localization of the wave functions of the band extrema depends on the pyroelectricity of the hexagonal layers and the strength of the "interface" dipole. Explaining the peak structure of the optical spectra we find two opposite tendencies related to the hexagonality of the polytype and the extent of its unit cell parallel to the c-axis.

Studying polar surfaces with a normal vector parallel to the c-axis we find results that are contradictory at first sight. For a given two-dimensional translational symmetry (e.g. $\sqrt{3} \times \sqrt{3}$), surface stoichiometry and reconstruction model the positions of the surface atoms do not depend on the polytype beneath the surface. On the other hand, it is known that details of the surface reconstruction may be used to obtain a reconstruction-stabilized epitaxial growth of a certain polytype.

Summarizing our results for the atomic geometry, the lattice dynamics, the electronic structure of heterocrystalline combinations, the optical properties, as well as surface structures, we find that the actual crystal structure of SiC polytypes, i.e., the stacking of Si–C bilayers, remarkably influence the resulting physical properties. On the other hand, from several of these properties ideas for the driving forces of the polytypism of SiC may be derived.

Acknowledgements The authors would like to thank their colleagues A. Fissel, K. Goetz, H. Hobert, S. Karmann, W. Richter, and B. Schröter from the Sonderforschungsbereich 196 for helpful discussions. This work has been supported by the Deutsche Forschungsgemeinschaft (Sonderforschungsbereich 196, project A8).

References

[1] A. R. VERMA and P. KRISHNA, Polymorphism and Polytypism in Crystals, Wiley New York 1966.

[2] N. W. JEPPS and T. F. PAGE, Progr. Cryst. Growth Charact. **7**, 259 (1983).

[3] R. S. RAMSDELL, Amer. Mineralogist **32**, 64 (1947).

[4] P. KRISHNA, Crystal Growth and Characterization of Polytype Structures, Pergamon Press, Oxford 1983.

[5] P. PIROUZ and J. W. YANG, Ultramicroscopy **51**, 189 (1993).

[6] W. J. CHOYKE, D. R. HAMILTON, and L. PATRICK, Phys. Rev. **133**, A 1163 (1964).

[7] B. WENZIEN, P. KÄCKELL, F. BECHSTEDT, and C. CAPPELLINI, Phys. Rev. B **52**, 10897 (1995).

[8] A. H. GOMESSS DE MESQUITA, Acta cryst. **23**, 610 (1967).

[9] H. SCHULZ and K. H. THIEMANN, Solid State Commun. **32**, 783 (1979).

[10] G. L. VIGNOLES, J. Crystal Growth **118**, 430 (1992).

[11] P. HOHENBERG and W. KOHN, Phys. Rev. **136**, B 864 (1964).

[12] W. KOHN and L. SHAM, Phys. Rev. **140**, A 1133 (1965).

[13] D. M. CEPERLEY and B. J. ALDER, Phys. Rev. Lett. **45**, 566 (1980).

[14] J. P. PERDEW and A. ZUNGER, Phys. Rev. B **23**, 5048 (1981).

[15] S. BARONI, P. GIANNOZZI, and A. TESTA, Phys. Rev. Lett. **58**, 1861 (1987).

[16] P. GIANNOZZI, S. DE GIRONCOLI, P. PAVONE, and S. BARONI, Phys. Rev. B **43**, 7231 (1991).

[17] X. GONZE and J. P. VIGNERON, Phys. Rev. B **39**, 13120 (1989).

[18] K. KARCH, F. BECHSTEDT, P. PAVONE, and D. STRAUCH, Phys. Rev. B **53**, 13400 (1996).

[19] K. KARCH, P. PAVONE, W. WINDL, D. STRAUCH, and F. BECHSTEDT, Internat. J. Quantum Chem. **56**, 801 (1995).

[20] K. KARCH and F. BECHSTEDT, Europhys. Lett. **35**, 195 (1996).

[21] K. KARCH, F. BECHSTEDT, P. PAVONE, and D. STRAUCH, J. Phys.: Condensed Matter **8**, 2945 (1996).

[22] G. B. BACHELET, D. R. HAMANN, and M. SCHLÜTER, Phys. Rev. B **26**, 4199 (1982).

[23] X. GONZE, R. STUMPF, and M. SCHEFFLER, Phys. Rev. B **44**, 8503 (1991).

[24] N. TROULLIER and J. L. MARTINS, Phys. Rev. B **43**, 1993 (1990).

[25] L. KLEINMAN and D. M. BYLANDER, Phys. Rev. Lett. **48**, 1425 (1982).

[26] R. STUMPF and M. SCHEFFLER, Comput. Phys. Commun. **79**, 447 (1994).

[27] B. WENZIEN, KÄCKELL, and F. BECHSTEDT, Surface Sci. **307/309**, 989 (1994).

[28] D. VANDERBILT, Phys. Rev. B **41**, 7892 (1990).

[29] G. KRESSE and J. HAFNER, Phys. Rev. B **49**, 14251 (1994).

[30] G. KRESSE and J. FURTHMÜLLER, Comput. Mat. Sci. **6**, 15 (1996).

[31] R. CAR and M. PARRINELLO, Phys. Rev. Lett. **55**, 2471 (1985).

[32] P. VINET, J. FERRANTE, J. R. SMITH, and J. H. ROSE, J. Phys. C **19**, L467 (1986).

[33] D. J. CHADI and M. L. COHEN, Phys. Rev. B **8**, 5747 (1973).

[34] H. J. MONKHORST and J. K. PACK, Phys. Rev. B **13**, 5188 (1976).

[35] J. FURTHMÜLLER, P. KÄCKELL, and F. BECHSTEDT, unpublished.

[36] P. KÄCKELL, B. WENZIEN, and F. BECHSTEDT, Phys. Rev. B **50**, 17037 (1994).

[37] F. BECHSTEDT, Adv. Solid State Phys. **32**, 161 (1992).

[38] B. WENZIEN, G. CAPPELLINI, and F. BECHSTEDT, Phys. Rev. B **51**, 14071 (1995).

[39] B. ADOLPH, V. GAVRILENKO, K. TENELSEN, F. BECHSTEDT, and R. DEL SOLE, Phys. Rev. B **53**, 9797 (1996).

[40] K. KARCH, P. PAVONE, A. P. MAYER, F. BECHSTEDT, and D. STRAUCH, Physica **219/220B**, 448 (1996).

[41] M. HOFMANN, A. ZYWIETZ, K. KARCH, and F. BECHSTEDT, Phys. Rev. B **50**, 13401 (1994).

[42] R. W. G. WYCKHOFF, Crystal Structures, Interscience Publishers, New York 1964.

[43] Landolt-Börnstein: Numerical Data and Functional Relationships in Science and Technology, Vol. 17a, new Series, Ed. K.-H. HELLWEGE and O. MADELUNG, Springer-Verlag, Berlin 1982.

[44] A. BAUER, S. KRÄUSSLICH, L. DRESSLER, P. KUSCHNERUS, S. WOLF, K. GOETZ, R. KÄCKELL, S. FURTHMÜLLER, and F. BECHSTEDT, Phys. Rev. B, submitted.

[45] R. F. ADAMSKY and K. M. MERZ, Z. Krist. **111**, 350 (1959).

[46] K. KARCH, G. WELLENHOFER, P. PAVONE, U. RÖSSLER, and D. STRAUCH, in: Proc. 22nd Internat. Conf. Phys. Semicond., Ed. D. LOCKWOOD, World Scientific Publ. Co., Singapore 1995 (p. 401).
[47] C. H. PARK, B.-H. CHEONG, K.-H. LEE, and K. J. CHANG, Phys. Rev. B **49**, 4485 (1994).
[48] E. WIGNER, Trans. Faraday Soc. **34**, 678 (1938).
[49] C. CHENG, R. J. NEEDS, V. HEINE, and N. CHURCHER, Europhys. Lett. **3**, 475 (1987).
[50] C. CHENG, R. J. NEEDS, and V. HEINE, J. Phys. C **21**, 1049 (1988).
[51] C. CHENG, V. HEINE, and R. J. NEEDS, J. Phys.: Condensed Matter **2**, 5115 (1990).
[52] W. R. L. LAMBRECHT, B. SEGALL, M. METHFESSEL, and M. V. SCHILFGAARDE, Phys. Rev. B **44**, 1685 (1991).
[53] J. FURTHMÜLLER, private communication.
[54] C.-Y. YEH, Z. W. LU, S. FROYEN, and A. ZUNGER, Phys. Rev. B **46**, 10086 (1992).
[55] A. ZYWIETZ, K. KARCH, and F. BECHSTEDT, Phys. Rev. B **54**, 1791 (1996).
[56] D. W. FELDMANN, J. PARKER, W. J. CHOYKE, and L. PATRICK, Phys. Rev. **173**, 787 (1968).
[57] W. J. CHOYKE, R. P. DEVATY, L. L. CLEMEN, M. F. MACMILLAN, and M. YOGANATHAN, in: Proc. 6th Internat. Conf. SiC and Related Materials, Kyoto 1995, Ed. S. NAKASHIMA, Inst. Phys. Conf. Ser. **142**, 257 (1996).
[58] I. IVANOV, A. HENRY, U. LINDEFELDT, C. PERSSON, T. EGILSSON, O. KORDINA, C. HALLIN, B. MONEMAR, and E. JANZEN, in: Proc. 23rd Internat. Conf. Phys. Semiconductors, World Scientific Publ. Co., Singapore 1996 (p. 233).
[59] P. KÄCKELL, B. WENZIEN, and F. BECHSTEDT, Phys. Rev. B **50**, 10761 (1994).
[60] C. PERSSON and U. LINDEFELT, Phys. Rev. B **54**, 10257 (1996).
[61] D. VOLM, B. K. MEYER, D. M. HOFMANN, W. M. CHEN, N. T. SON, C. PERSSON, U. LINDEFELT, O. KORDINA, E. SÖRMAN, A. O. KONSTANTINOV, B. MONEMAR, and E. JANZEN, Phys. Rev. B **53**, 15409 (1996).
[62] N. T. SON, O. KORDINA, A. O. KONSTANTINOV, W. M. CHEN, E. SÖRMAN, B. MONEMAR, and E. JANZEN, Appl. Phys. Lett. **65**, 3209 (1994).
[63] B. WENZIEN, P. KÄCKELL, and F. BECHSTEDT, in: Proc. 5th SiC Conf., Washington 1993, Inst. Phys. Conf. Ser. **137**, 223 (1994).
[64] W. R. LAMBRECHT and B. SEGALL, Phys. Rev. B **52**, R2249 (1995).
[65] C. CHENG, V. HEINE, and I. L. JONES, J. Phys.: Condensed Matter **2**, 5097 (1990).
[66] J. VON BOEHM and P. BAK, Phys. Rev. Lett. **42**, 122 (1979).
[67] J. J. A. SHAW and V. HEINE, Condensed Matter **2**, 4351 (1990).
[68] V. HEINE, C. CHENG, and R. J. NEEDS, J. Amer. Ceram. Soc. **74**, 2630 (1991).
[69] D. PANDEY, Phase Transitions **16/17**, 247 (1989).
[70] B. ADOLPH, K. TENELSEN, V. I. GAVRILENKO, and F. BECHSTEDT, Phys. Rev. B **55**, 1422 (1997).
[71] V. I. GAVRILENKO and F. BECHSTEDT, Phys. Rev. B **54**, 13416 (1996); B **55**, 4343 (1997).
[72] M. ROHLFING, P. KRÜGER, and J. POLLMANN, Phys. Rev. B **48**, 17791 (1993).
[73] W. H. BACKES, P. A. BOBBERT, and W. VAN HAERINGEN, Phys. Rev. B **51**, 4960 (1995).
[74] W. R. L. LAMBRECHT, B. SEGALL, M. YOGANATHAN, W. SUTTROP, R. P. DEVATY, W. J. CHOYKE, J. A. EDMOND, J. A. POWELL, and M. ALOUANI, Appl. Phys. Lett. **63**, 2747 (1993).
[75] W. R. L. LAMPRECHT, B. SEGALL, M. YOGANATHAN, W. SUTTROP, R. P. DEVATY, W. J. CHOYKE, J. A. EDMOND, J. A. POWELL, and M. ALOUANI, Phys. Rev. B **50**, 10722 (1994).
[76] G. WELLENHOFER, K. KARCH, P. PAVONE, U. RÖSSLER, and D. STRAUCH, Phys. Rev. B **53**, 6071 (1996).
[77] J. CHENG, Z. H. LEVINE, and J. W. WILKINS, Phys. Rev. B **50**, 11514 (1994).
[78] T. SUGII, T. AOYAMA, and T. ITO, J. Electrochem. Soc. **137**, 989 (1990).
[79] H. MATSUNAMI, Physica **185B**, 65 (1993).
[80] A. FISSEL, B. SCHRÖTER, and W. RICHTER, Appl. Phys. Lett. **66**, 3182 (1995).
[81] A. FISSEL, U. KAISER, K. PFENNIGHAUS, B. SCHRÖTER, and W. RICHTER, Appl. Phys. Lett. **68**, 1204 (1996).
[82] M. MURAYAMA and T. NAKAYAMA, Phys. Rev. B **49**, 4710 (1994).
[83] A. QTEISH, V. HEINE, and R. J. NEEDS, Physica **185B**, 366 (1993).
[84] F. BECHSTEDT and R. ENDERLEIN, Semiconductor Surfaces and Interfaces, Akademie Verlag, Berlin 1988.

[85] F. BECHSTEDT and P. KÄCKELL, Phys. Rev. Lett. **75**, 2180 (1995).

[86] P. KÄCKELL and F. BECHSTEDT, Mater. Sci. and Engng. B **37**, 224 (1996).

[87] J. E. NORTHRUP and J. NEUGEBAUER, Phys. Rev. B **52**, 17001 (1995).

[88] F. OWMAN and P. MARTENSSON, Surface Sci. **330**, L639 (1995).

[89] J. POLLMANN, private communication.

[90] A. FISSEL, U. KAISER, E. DUCKE, B. SCHRÖTER, and W. RICHTER, J. Crystal Growth **154**, 72 (1995).

[91] A. FISSEL, U. KAISER, K. PFENNIGHAUS, E. DUCKE, B. SCHRÖTER, and W. RICHTER, Proc. Conf. Silicon Carbide and Related Materials, Kyoto (Japan) 1995, Inst. Phys. Conf. Ser. **142**, 121 (1996).

[92] B. SCHRÖTER and A. FISSEL, private communication.

[93] R. KAPLAN, Surface Sci. **215**, 111 (1989).

phys. stat. sol. (b) **202**, 63 (1997)

Subject classification: 71.20.Nr; 61.50Ah; S6

On the Band Gap Variation in SiC Polytypes

W. van Haeringen, P. A. Bobbert, and W. H. Backes

Department of Physics, Eindhoven University of Technology, P.O. Box 513,
5600 MB Eindhoven, The Netherlands

(Received January 31, 1997)

Electronic band gaps of SiC polytypes are reproduced within an interface matching technique of electronic wave functions. Essential features resulting from this treatment are introduced in a one-dimensional model, leading to a transparent description of the electronic band gap variation among polytypes. It is discussed in what sense the polytypes of SiC are exceptional in showing a relatively strong band gap variation, contrary to e.g. polytypes of ZnS and hypothetical polytypes made up from Si, C or AlAs.

1. Introduction

Ground state properties of SiC have been studied extensively for polytypes with relatively small unit cells [1 to 6], specifically for the 3C, 2H, 4H and 6H modifications. Unfortunately, the employed ab-initio density functional schemes are not capable of producing reliable electronic band gaps, because of the involvement of excited states. It is generally accepted that this deficiency can be cured by incorporating dynamical nonlocal correlation effects, as is done in the celebrated GW approach [7, 8]. Indeed, ab-initio GW calculations of the band structure of 3C SiC gave perfect agreement with most of the experimentally obtained energy values [9, 10]. A simplified GW approach applied to 2H, 4H and 6H SiC yielded for the minimum energy gap good agreement with the experimental results [11]. If applied, the GW method would very likely also reproduce the experimentally obtained gap values for the remaining polytypes. The question can be raised, however, whether this would also contribute to our understanding of the very specific band gap variation among the various polytypes. In 1964 Choyke et al. [12] found an intriguing quasi-linear relation between the electronic gap and the degree of hexagonality percentage h up to $h = 50\%$. In our opinion, understanding of this phenomenon requires more than a mere reproduction of this relation by means of the application of GW.

In this article we try to explain the band gap variation in polytypes by first carrying through a wave function matching scheme. Polytypes are considered in this description as being built up from mutually twisted layers of varying size, consisting of 3C (cubic) SiC. The electronic wavefunctions of the polytypes are obtained by matching linear combinations of 3C SiC wavefunctions at the interfaces between these "building blocks". It is shown that in this way reliable electronic band structures (and thus band gaps) can be obtained. What is needed is knowledge of the basis material 3C SiC, together with the sizes of the 3C sublayers in the respective polytypes. The basis material 3C SiC can e.g. be characterized in terms of a band structure obtained with the empirical pseudopotential method (EPM) or with the local density approximation (LDA) of density functional theory. We emphasize that in this connection it is necessary to have the full complex band structure of 3C SiC at our disposal. In a second step we proceed by simplifying the description by reducing it to a one-dimensional model in which scattering characteristics at interfaces between the 3C layers form the essential ingredients. Finally, we succeed in understanding why SiC is so exceptional in showing its relatively strong variation in band gaps, contrary to, for example, ZnS polytypes, or hypothetical polytypes of Si.

In Section 2 we start by introducing the respective polytype structures. In doing so we slightly remodel the polytypes by allowing no other nearest-neighbour bond lengths and angles other than those present in 3C SiC. In Section 3 we introduce the idea of building up polytypes from blocks consisting of 3C SiC. It is shown that this simplification will have a very minor influence on the band gaps. In Section 4 we discuss the interface matching procedure, the results of which are given in Section 5. Section 6 is devoted to the description of the one-dimensional model. Section 7 deals with the question why SiC is so exceptional in showing a relatively strong band gap variation among polytypes. A summary and some concluding remarks are given in Section 8.

The material in this review covers part of a thesis work [13] and several papers of the authors [10, 14, 15].

2. Polytype Structures

A convenient classification scheme for polytypic structures can be obtained by first considering all close-packing possibilities of a collection of equally large spheres. To this end we start by considering Fig. 1, in which one layer of such close-packed spheres is displayed. Addition of a second layer on top of this layer can be achieved in one of two ways only, either by putting the spheres in the voids indicated by 1, 3 and 5, or in the voids indicated by 2, 4 and 6. If we mark the sphere positions in the ground layer by A, the possible set of sphere positions for the next layer is either B or C, as indicated in Fig. 1. Proceeding in this way, it will be clear that the third layer will be either A- or C-type (in case layer 2 is B-type) or of A- or B-type (in case layer 2 is C-type), etc. This procedure obviously leads to all conceivable close-packings of equally sized spheres. The piling direction of the spheres will be indicated as the c-, the z-, or the vertical direction.

An important subset of the above-mentioned stacking possibilities is obtained by demanding structures to be periodic in the piling direction. Simple examples of such periodic stackings are ABAB... and ABCABC..., with periodicity over two and three layers, respectively. The first stacking gives rise to a hexagonal structure, the second one

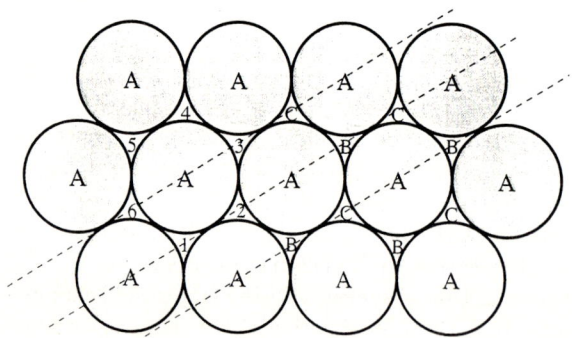

Fig. 1. Close packings of spheres. The displayed ground layer contains spheres in A-type positions, surrounded by six voids. Spheres of a second layer are situated either in the voids 1, 3 and 5 (B-type position) or 2, 4 and 6 (C-type). The spheres of any structure have their centres all located in parallel vertical planes, whose intersection with the ground plane is indicated by the dashed lines

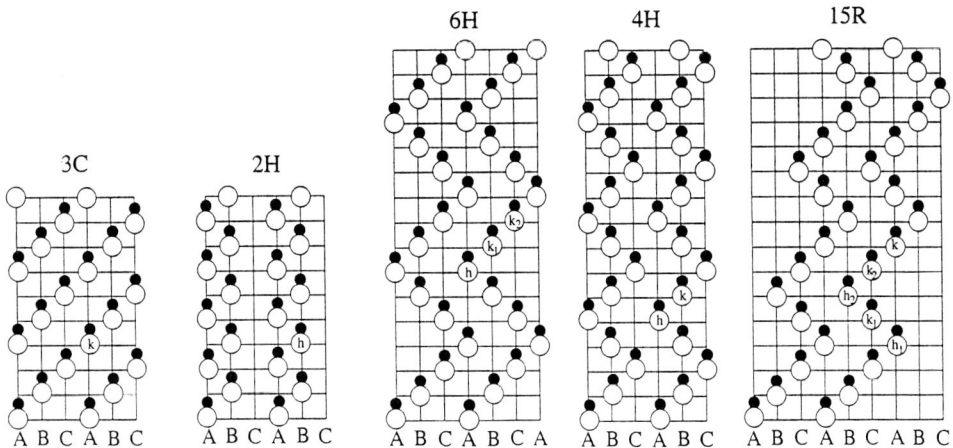

Fig. 2. Arrangement of Si (open circles) and C (filled circles) atoms in the polytypes 3C, 2H, 6H, 4H and 15R SiC. The plane in which the atom centers are situated is one of the vertical planes introduced in Fig. 1

to the face-centered cubic one. By exploring all possibilities, an infinite number of periodic stacking sequences results.

Let us now choose the diameters of the spheres such that the distance between two neighbouring planes of sphere centres coincides with the distance between neighbouring Si planes in cubic SiC. Let us furthermore place Si atoms in the center of each sphere, and C atoms precisely a bond length, as valid in cubic SiC, above them. Let us subsequently abandon the original A, B, and C spheres. The resulting periodic structures (almost perfectly) represent all conceivable SiC polytypes. The reason that this representation is not entirely perfect is the (very) small variation in bond lengths which may occur, both within one given polytype and also mutually between different polytypes.

In Fig. 2 five different polytypes, 3C, 2H, 6H, 4H, and 15R SiC are displayed by giving the positions of Si and C atoms in a plane going through the c-axis and one of the dashed lines in Fig. 1. Their periodicity is over 3, 2, 6, 4 and 15 bilayers of Si and C, respectively. The structure is either cubic (C), hexagonal (H) or rhombohedral (R). The layer sequences are ABC..., AB..., ABCACB..., ABCB... and ABCACBCABA-CABCB..., respectively.

It is important to note that not all atomic bilayers of Si and C in a given structure, and thus not all Si or C atoms, are equivalent. In Fig. 2 we denote inequivalent layers by the letters h and k, in some cases with additional subindices. The inequivalence is related to differences in the relative positioning of layers. The hexagonality of a polytype can be defined as the percentage of h-type layers. The existence of distinguishable atomic positions has been demonstrated experimentally for several SiC polytypes, by substituting nitrogen (N) atoms for C atoms situated at the different positions [16]. The implantation of such N donors at different locations gives rise to a number of different binding energies showing up in luminescence spectra, equal to the number of non-equivalent C atom layers in the given polytype. In this way it has been established that 3C, 2H, 4H, 6H and 15R SiC indeed contain 1, 1, 2, 3 and 5 kinds of C atom positions, respectively, in agreement with the number of layer kinds indicated in Fig. 2.

The existence of non-equivalent Si and C atoms in the various polytypes is very probably the reason why an earlier attempt [17] to give a universal description of their band
structures (and especially their electronic band gaps), in terms of transferable empirical
atomic pseudopotentials for the elements Si and C, was only partially successful. Indeed,
if the various Si and C atoms are shown to behave differently (see above), then their
effective atomic pseudopotentials will undoubtedly be different as well. In [17] it was in
fact attempted to describe the band structures of all polytypes by means of a single set of
"averaged" atomic pseudopotentials for Si and C. The largest difference between atomic
pseudopotentials is to be expected between Si and C atoms in 3C SiC on the one hand,
and those in 2H SiC on the other hand, the reason being that the difference in surround-
ings of the atoms in these two cases is largest. One can therefore speculate that exclusion
of the 2H-polytype would possibly have led to better results. Incidentally, it will be seen
furtheron in the description of our new approach to the band structures of polytypes,
that we are forced to exclude the polytype 2H SiC, the reason being again related to the
above-mentioned "large difference" in Si and C atoms in 3C SiC and 2H SiC. It is explicitly verified, however, that all other hitherto known polytypes perfectly fit within the
new description. The conclusion is clearly that 2H SiC has to be separately dealt with.

3. The Building-Block Scheme

The central idea in our approach comes from the insight that in all polytype structures
finite parts of the cubic structure can be recognized. We will use the 6H configuration,
with stacking sequence ABCACB, as an example to outline the idea in some detail. For
this structure the stacking sequence of the first three layers is ABC, which is the cubic
structure. The next three layers of the 6H structure correspond to ACB, which is, however, nothing but ABC rotated over 60° about the c-axis. This means that the 6H polytype can be represented as the periodic stacking of two equally thick, but mutually
twisted, parts (building blocks) of the cubic polytype. Any of the polytype structures
introduced in Section 2 can be subdivided in a similar way in cubic parts of various
sizes. This model representation will certainly be not precise enough to describe all
subtle details of polytypes, but it is completely adequate for our purpose of predicting
band gaps. We will in this connection critically have to consider the dependence of the
band gap on lattice constants, the positions of the atoms in the primitive cells, and the
charge density distributions. As for the lattice constants, the experimental values
[18, 19] deviate from the "constructed" values by less than 0.5%. Also, the actual bond
lengths in the short-period polytypes 2H, 4H and 6H SiC deviate by at most 0.5% from
their ideal tetrahedral values [20, 21]. A rough estimate of the influence of lattice constant variations and atomic relaxations on the band gap values can be obtained by considering the case of the (hydrostatic) pressure derivative of the band gap of 3C SiC.
Since the pressure derivative dE/dp of the band gap, as well as the equilibrium bulk
modulus $B_0 = -\Omega_c\, dp/d\Omega_c$, and the primitive cell volume Ω_c are known, we have, with
a_c the lattice constant,

$$\frac{dE}{da_c} = \frac{dE}{dp} \frac{dp}{d\Omega_c} \frac{d\Omega_c}{da_c} .$$

(1)

The primitive cell volume equals $\Omega_c = \frac{1}{4}\, a_c^3$, yielding $d\Omega_c/da_c = \frac{3}{4}\, a_c^2$. The experimental
values for dE/dp and B_0 in 3C SiC SiC are -0.34 meV/kbar and 2.2 Mbar, respectively.
Making use of these values, we find that the energy band gap for 3C SiC increases

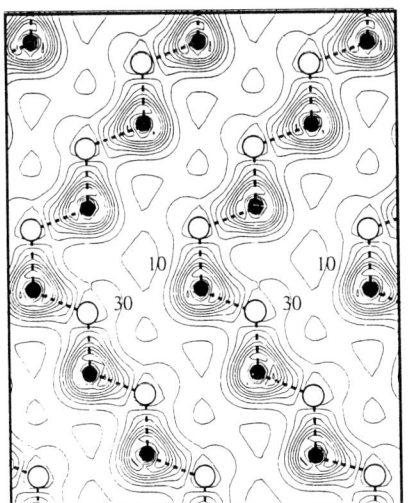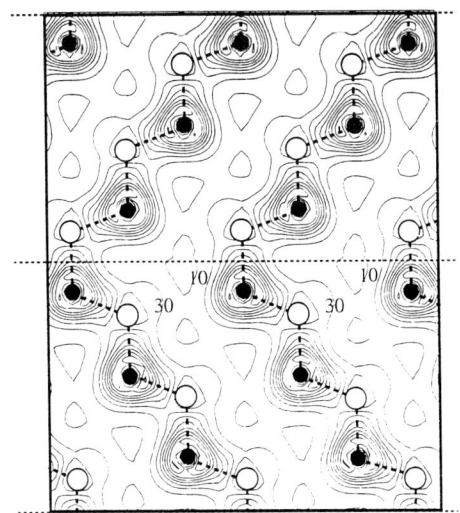

Fig. 3. Contour plots of the valence-charge density distribution of actual (unrelaxed) 6H SiC (left) and of the corresponding building-block system (right). The building blocks are indicated by the dashed lines. The plane of intersection is comparable to the one of Fig. 2 (the Si and the C atoms are interchanged). The contours indicate ten plus integral multiples of twenty valence electrons per 6H unit cell; taken from [13]

approximately 10 meV if the lattice constant is elongated 0.5% or by 0.02 Å. A more proper estimate would include the derivatives pertaining to the respective polytypes. However, according to recent pseudopotential calculations of Park et al. [22], the polytype dependent quantities in Eq. (1) are generally smaller. Only for the 2H polytype they may be a factor of two higher than the ones for the 3C structure, but the 2H structure is not covered by our approach anyhow. For the other polytype structures, our estimates reveal that the influence on the band gap of atomic relaxations is in the order of 10 meV or appreciably less, in any case small enough to justify the use of the building-block model.

It is also observed that the actual charge density distribution around the Si and the C atoms in a given polytype is surprisingly similar to that of the related building-block system. This was already noticed by Denteneer and coworkers [4, 23], who mapped the 3C density to a 2H structure. More specifically, when comparing the theoretically obtained valence-charge density of actual 6H SiC with that of the related building-block system (both structures with the same bond lengths) the differences turn out to be minor if the interface position in the building-block system is chosen precisely in between a Si and a C atom of an h-type bilayer. To this end we compared the respective LDA-obtained charge densities, contour plots of which are given in Fig. 3. The largest plane-wave Fourier components of the valence-charge density of the building-block structure and the actual one differ less than 2%. We may display the differences in charge density $\Delta\varrho(\mathbf{r})$ in another way by investigating it planewise as a function of z, and expanding it in two-dimensional Fourier components, using the set of two-dimensional reciprocal lattice vectors $\mathbf{Q}_\parallel$ parallel to the interface planes. The expansion reads

$$\Delta\varrho(\mathbf{r}) = \sum_{\mathbf{Q}_\parallel} e^{i\mathbf{Q}_\parallel \cdot \mathbf{r}_\parallel}\, \Delta\varrho_{\mathbf{Q}_\parallel}(z)\,. \tag{2}$$

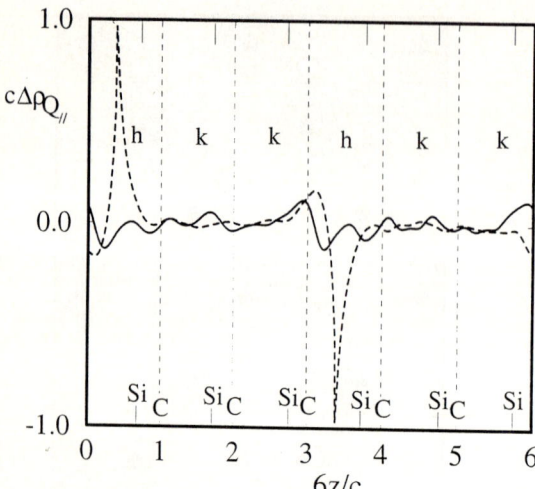

Fig. 4. Difference in valence-charge density $\Delta\varrho_{\mathbf{Q}_\parallel}(z)$ betwen the actual density and the building-block density of 6H SiC, for $\mathbf{Q}_\parallel = 0$ (solid line) and the smallest non-zero $\mathbf{Q}_\parallel$ (dashed line), as calculated within the LDA. The type, h or k, of the bilayers is also indicated; taken from [13]

Fig. 4 shows two planar Fourier components $\Delta\varrho_{\mathbf{Q}_\parallel}(z)$. It can be deduced from this figure that $\Delta\varrho_0(z)$, which is the z-dependent plane average of the valence-charge density difference, is at most 0.2% of the value $48/c$, where c is the 6H unit cell length and 48 the number of valence electrons per cell.

The important insight gained from Fig. 4 is that the Fourier components are clearly largest within a region of approximately one bilayer around the interface positions. This allows us to say that 6H SiC can to a large extent be considered to contain four out of six bilayers which to a high degree are not only structurally, but also electronically, cubic-like. After having constructed our matched wavefunctions for a 6H building-block structure in the next section, we are able to estimate the differences in band gap, brought about by replacing the actual 6H potential distribution by the building-block distribution. Application of first-order perturbation theory leads to a difference of 0.06 eV only, compared to a LDA gap value of 1.83 eV.

4. The Interface Matching Method

In this section the polytype 6H SiC will again be the example to demonstrate the determination of the characteristic equation from which the band structure results. The determination for the other polytypes runs analogously. Essential is that we have at our disposal the (complex) band structure of the basis material 3C SiC of the layers in our building-block system. We start by giving indices to the successive layers in the 6H polytype. In Fig. 5 three successive layers of the 6H superlattice are indicated by $j = 1$, 2 and 3. The layers $j = 1$ and $j = 3$ consist each of three bilayers of twisted 3C SiC, indicated by ACB. Layer $j = 2$ consists of three bilayers of untwisted 3C SiC, indicated by ABC. Note that $z_{j-1} < z < z_j$ defines the region of layer j, where $z_j = z_0 + 3jc_1$ gives the successive interface positions, with c_1 the width of one bilayer. We choose z_0 equal to $z_0 = -3c_1$. From the complex band structure analysis [24] (e.g. within the EPM) of 3C SiC, we know for each layer j in principle all possible independent wavefunction solutions at a given energy E and given wave vector $\mathbf{k}_\parallel$, parallel to the interface planes. As $\mathbf{k}_\parallel$ is a good quantum number (there is no loss of translational symmetry along the

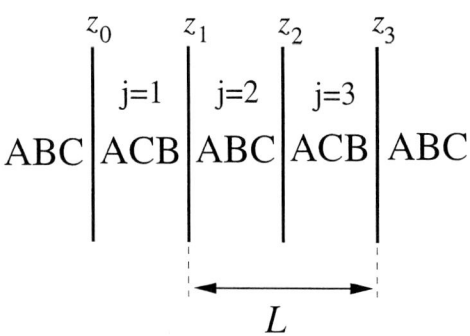

Fig. 5. The indexing of the layers and the interfaces for the 6H structure

interfaces), we may use these solutions to construct wavefunctions pertaining to the 6H structure. These wavefunctions can generally be written as

$$\psi_{l,\mathbf{k}_\parallel,K_z}(\mathbf{r}_\parallel,z) = e^{i(\mathbf{k}_\parallel \cdot \mathbf{r}_\parallel + K_z z)}\, u_{l,\mathbf{k}_\parallel,K_z}(\mathbf{r}_\parallel,z)\,, \tag{3}$$

with $\mathbf{k}_\parallel + K_z\mathbf{e}_z$ the wave vector of the 6H superlattice wavefunction. The functions $u_{l,\mathbf{k}_\parallel,K_z}(\mathbf{r}_\parallel,z)$ have the periodicity of the 6H lattice. On the other hand, in each layer j the 6H wavefunction at given E and $\mathbf{k}_\parallel$ can quite generally be written as a linear combination of the set of 3C (basis) wavefunctions $\varphi^j_{E,\mathbf{k}_\parallel,k_z^{j,s}}$. Using plane wave expansions for periodic functions, with N plane waves $\exp(i\mathbf{G}\cdot\mathbf{r})$, there are $2M$ of such basis functions, with $M \le N$ the number of different projections $\mathbf{G}_\parallel$ of the reciprocal lattice vectors $\mathbf{G}$ on the x–y plane. The linear combination now reads

$$\psi^j_{E,\mathbf{k}_\parallel}(\mathbf{r}_\parallel,z) = \sum_{s=1}^{2M} \alpha^j_s \varphi^j_{E,\mathbf{k}_\parallel,k_z^{j,s}}(\mathbf{r}_\parallel,z)\,, \tag{4}$$

where in all wavefunctions on the right-hand side we have replaced the band index by the corresponding energy E and wave vector $\mathbf{k} = (\mathbf{k}_\parallel, k_z^{j,s})$. Note that the $k_z^{j,s}$ may be either real or complex. The unknown variables in Eq. (4) are the coefficients α^j_s. The 3C wavefunctions on the right-hand side of Eq. (4) can be written as

$$\varphi^j_{E,\mathbf{k}_\parallel,k_z^{j,s}}(\mathbf{r}_\parallel,z) = \sum_{\mathbf{G}_\parallel}^{M} e^{i(\mathbf{k}_\parallel+\mathbf{G}_\parallel)\cdot\mathbf{r}_\parallel}\, v^{j,s}_{E,\mathbf{k}_\parallel,\mathbf{G}_\parallel}(z - z_{j-1})\,, \tag{5}$$

i.e. in terms of M z-dependent two dimensional Fourier components $v^{j,s}$, which are

$$v^{j,s}_{E,\mathbf{k}_\parallel,\mathbf{G}_\parallel}(z) = e^{ik_z^{j,s}z} \sum_{G_z(\mathbf{G}_\parallel)} d_{E,\mathbf{k}_\parallel+k_z^{j,s}+\mathbf{e}_z}(\mathbf{G}_\parallel + G_z\mathbf{e}_z)\, e^{iG_z z}\,. \tag{6}$$

In this expression the summation is meant to take place over all G_z pertaining to the indicated $\mathbf{G}_\parallel$. The coefficients d in Eq. (6) are coefficients in the plane wave expansion of the functions $\varphi^j_{E,\mathbf{k}}$. Note that in Eq. (5) we shifted the arguments of each $v^{j,s}$ towards its left interface position z_{j-1}. Such a shift in the [111] direction of 3C SiC is allowed if it is equal to an integral multiple of the 3C period length of ABC. We have applied these shifts to keep the values of the arguments of the $v^{j,s}$ functions relatively small, which otherwise might cause numerical trouble if the $k_z^{j,s}$ have large imaginary parts. Substituting Eq. (5) in to Eq. (4) yields for the 6H wavefunction in layer j

$$\psi^j_{E,\mathbf{k}_\parallel}(\mathbf{r}_\parallel,z) = \sum_{\mathbf{G}_\parallel}^{M} e^{i(\mathbf{k}_\parallel+\mathbf{G}_\parallel)\cdot\mathbf{r}_\parallel} \sum_{s=1}^{2M} \alpha^j_s v^{j,s}_{E,\mathbf{k}_\parallel,\mathbf{G}_\parallel}(z - z_{j-1})\,. \tag{7}$$

In order to properly connect the wavefunctions in the three layers $j = 1, 2$ and 3, we have to specify the matching prescriptions at the three interfaces $z = z_1 = 0$, $z = z_2 = 3c_1$ and $z = z_3 = 6c_1$, respectively. In the first place the matching prescriptions have to be such that the constructed 6H function is continuous and continuously differentiable at any position $\mathbf{r}$ located on the interface planes $z = z_1$ and $z = z_2$. To achieve this, we have to demand for each $\mathbf{G}_\parallel$

$$\sum_{s=1}^{2M} a_s^j v_{E, \mathbf{k}_\parallel, \mathbf{G}_\parallel}^{j, s} (3c_1) = \sum_{s=1}^{2M} a_s^{j+1} v_{E, \mathbf{k}_\parallel, \mathbf{G}_\parallel}^{j+1, s} (0) \,, \tag{8}$$

and

$$\sum_{s=1}^{2M} a_s^j \frac{\partial v_{E, \mathbf{k}_\parallel, \mathbf{G}_\parallel}^{j, s} (3c_1)}{\partial z} = \sum_{s=1}^{2M} a_s^{j+1} \frac{\partial v_{E, \mathbf{k}_\parallel, \mathbf{G}_\parallel}^{j+1, s} (0)}{\partial z} \,, \tag{9}$$

which we apply for $j = 1$ and $j = 2$. This gives $4M$ conditions for $6M$ unknown coefficients a_s^j. In the second place we have to demand that the wavefunction to be constructed satisfies the Bloch condition for 6H SiC, which reads

$$\psi_{E, \mathbf{k}_\parallel} (\mathbf{r}_\parallel, L) = \mathrm{e}^{iK_z L} \, \psi_{E, \mathbf{k}_\parallel} (\mathbf{r}_\parallel, 0) \,, \tag{10}$$

where $\mathbf{k}_\parallel + (\mathrm{Re} \, K_z) \, \mathbf{e}_z$ is a wave vector lying in the 1BZ of 6H SiC, and the length $L = 6c_1$ represents the period of the 6H supercell in the c-direction. Equation (10) gives the remaining $2M$ conditions for the $6M$ unknowns.

Solving the system of equations (8) to (10) for different combinations of energy E and parallel wave vector $\mathbf{k}_\parallel$, gives in principle the complete (complex) band structure of 6H SiC. To this end Eqs. (8) and (10) are written in matrix form,

$$D^j(3c_1) \, \underline{a}^j = D^{j+1}(0) \, \underline{a}^{j+1} \,, \tag{11}$$

in which the dimensions of D are $2M \times 2M$ and the subscripts E and $\mathbf{k}_\parallel$ have been omitted. Imposing the 6H Bloch condition between $z = z_1$ and $z = z_3$ yields the characteristic equation

$$[D^2(0)]^{-1} \, D^1(3c_1) \, \underline{a}^1 = \mathrm{e}^{iK_z L} \, [D^2(3c_1)]^{-1} \, D^3(0) \, \underline{a}^1 \,, \tag{12}$$

where $\mathrm{e}^{iK_z L}$ is the complex eigenvalue. The coefficients a_s^1, pertaining to layer $j = 1$, appear in Eq. (12) as the elements of the eigenvector columns $\underline{a}^1$. The coefficients for the next two layers, $j = 2$ and 3, can subsequently be obtained by substituting $\underline{a}^1$ in Eq. (11). In this way we can construct the superlattice wave function throughout the entire superlattice.

Similar techniques as described above have been applied to GaAs/AlAs superlattices [25 to 28] and twinned superlattices of Si and Ge [29].

5. Resulting Band Structures and Band Gaps

In Fig. 6 the first Brillouin zone (1BZ) of 6H SiC is displayed. In Fig. 7 the complex bandstructure for 6H SiC, resulting from the matching procedure, is given along the ML-axis of the 1BZ. For the calculation of this band structure we have chosen the component $\mathbf{k}_\parallel$ equal to $\mathbf{k}_M = (2\pi/a_h \sqrt{3}) \, \mathbf{e}_x$, with a_h the in-plane hexagonal lattice constant. The reason for this choice is that we are interested in finding the band edge of the low-

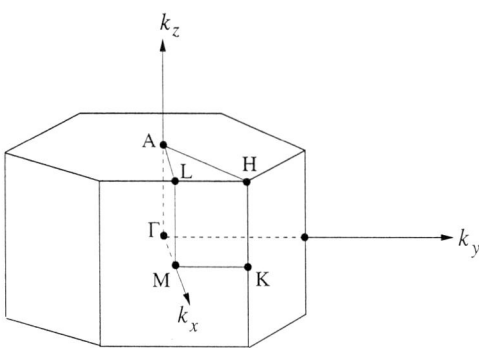

Fig. 6. First Brillouin zone of the hexagonal Bravais lattice, with symmetry points Γ, M, L, A, K and H. The cubic [111] direction is parallel to the ΓA-line and is chosen as the k_z-axis. The complex band structure is always calculated for a fixed $\mathbf{k}_\parallel$, along a path parallel to the k_z-axis

est conduction band in 6H SiC. This band edge is expected to lie in or near $\mathbf{k}_M$, as it is known that in 3C SiC it lies in $\mathbf{k}_X$. It is easily shown that if we consider 3C SiC in the larger 6H SiC cell, the point $\mathbf{k}_X$ (of the 1BZ of 3C SiC) reduces to $\mathbf{k}_M$ (of the 1BZ of 6H SiC). Indeed, by choosing other $\mathbf{k}_\parallel$-vectors, invariably higher conduction band states of 6H SiC are found. In calculating the band structure of Fig. 7 the number of 3C wavefunctions employed was equal to $2M = 16$.

By calculating the 6H SiC band structure along a variety of parallel directions we are able to find those points $\mathbf{k} = \mathbf{k}_\parallel + K_z \mathbf{e}_z$ at which the valence bands have their energy maximum and the conduction bands their energy minimum. We find that the top of the valence band for 6H SiC, as in fact for all other polytypes, is located at the Γ-point of the 1BZ, albeit with a very small energy shift with respect to the valence band top energy in 3C SiC. Fig. 7 shows a camel's back structure for the lowest conduction band along the ML-axis. Experiments [30, 31] confirm that the minimum is indeed located between $\mathbf{k}_M$ and $\mathbf{k}_L$. The value of $2M$, used in our calculation in order to obtain well-converged results, appears to be dependent on the magnitude and the direction of $\mathbf{k}_\parallel$ and generally satisfies $2M < 30$. The wavefunctions of the basis material 3C SiC have been expanded in 113 plane waves. Convergence of the indirect gap has been achieved

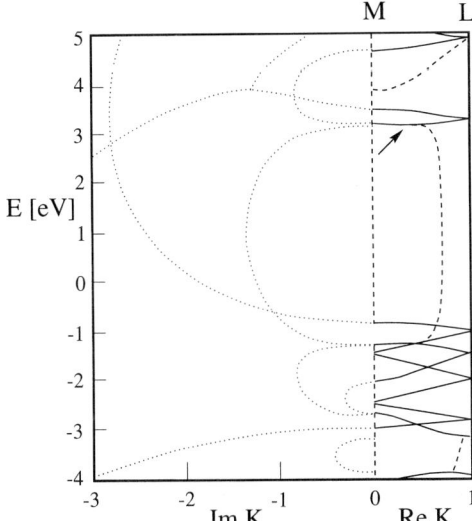

Fig. 7. Complex band structure for 6H SiC along the ML-axis of the 1BZ, as obtained by the interface matching method. K_z is in units of $\pi/6c_1$. The solid lines pertain to Bloch states, while the dashed and dotted lines describe the real and imaginary parts, respectively, of K_z of the evanescent states. A camel's back type of minimum is visible (arrow) in the lowest conduction band; taken from [14]

Table 1

Valence and conduction band edges and gaps (in eV) for ten different SiC polytypes, obtained with our interface matching method within the EPM and as found experimentally. The band edge energies are given relative to the valence band edge of 3C SiC. The corresponding hexagonality (in %) is also given. The valence band edges E_v are all at Γ. The calculated conduction band edges E_c are all at M (except for 6H, see Fig. 7), whereas the experimental conduction band edge for 2H SiC is found at K

polytype	h	E_v	E_c	calc. gap $E_c - E_v$	expt. gap [12, 32]
3C	0	0	2.40	2.40	2.390, 2.389, 2.416
24H	8.3	−0.04	2.41	2.45	
18H	11.1	0.00	2.47	2.47	
12H	16.7	0.03	2.60	2.57	
10H	20	0.04	2.70	2.66	
8H	25	0.04	2.80	2.76	2.80, 2.86
6H	33.3	0.04	3.09	3.05	3.023, 2.86
15R	40	0.05	3.08	3.03	2.986, 3.05
4H	50	0.06	3.34	3.28	3.265, 3.20
2H	100	0.10	4.14	4.04	3.330

within 0.02 eV. In Table 1 we have collected some of our results for a number of polytypes. They are compared with experimentally obtained gaps.

We have repeated the above analysis in terms of the LDA band structure and wavefunctions for the basis material 3C SiC. The results of both EPM and LDA band gaps are displayed in Fig. 8. The LDA results show the same almost linear relationship between (Kohn-Sham) gap and h. When raised by a common constant value of 1.2 eV, the EPM and experimental gaps are reproduced. This necessary energy shift of 1.2 eV, indicated in Fig. 8, and suggestively called the quasi-particle shift Δ_{QP} is to be identified with the self-energy correction obtained within the GW theory and has to be added due to the inherent inability of density functional theory to yield correct excited state ener-

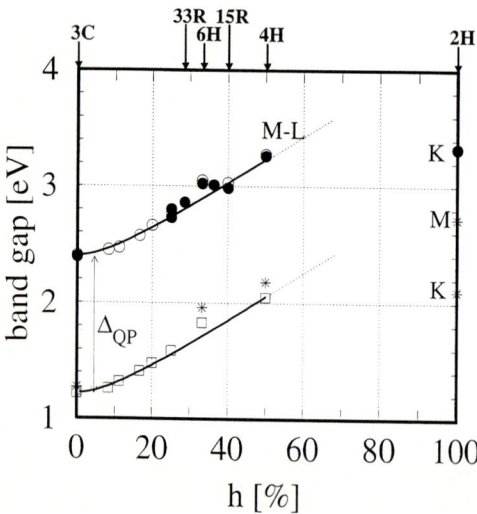

Fig. 8. The band gaps of SiC polytypes as a function of the hexagonality h, as obtained by experiment (filled circles), by EPM matching (open circles), by LDA matching (open squares) and by a self-consistent LDA calculation [33] (asterisks). An overall shift Δ_{QP} brings the LDA results in accordance with experiment and with the results obtained within the EPM. The drawn lines are the result of the analysis of Section 6 for hexagonal structures; taken from [13]

gies [8]. The fact that the size of the non-locality hole of the GW self-energy is of the order of a bond length [8] explains the polytype insensitivity of Δ_{QP}.

In comparing our results, obtained within the EPM, with the experimental gap values as obtained by Choyke and coworkers [12, 32], it should be kept in mind that the experimental values suffer from inaccuracies in the order of 0.015 eV. The reported values refer, moreover, to exciton gaps, which are typically up to 0.03 eV [18] lower than the electronic band gaps. In view of this, the agreement between our EPM results, as presented in Table 1 and Fig. 8, and the experimental data is very satisfactory indeed, except for 2H SiC. This agreement implies that we find an almost linear relationship between gap and h for $15\% \lesssim h \leq 50\%$. In Fig. 8 the relationship between band gap and hexagonality h is displayed. For very small h values $h \lesssim 15\%$ we predict deviations from the "linear" behaviour. This prediction can be considered to be reliable, since small h corresponds to relatively large building blocks, for which our method works best.

All considered polytypes happen to have indirect gaps. Except for the 2H and 6H case the band gap is between the Γ- and M-point. For 6H SiC the conduction band minimum lies on the ML-axis, as mentioned above, while it is lcoated in the K-point for 2H SiC. The deviation between our calculated 2H gap within the EPM and the experimental gap is considerable. This is not unexpected, as announced earlier. Indeed, the assumption that single bilayers in 2H SiC exhibit 3C SiC character is rather provocative. Because the 2H polytype itself is an exception, hypothetical SiC superlattices that are partly composed of this 2H polytype are exceptions as well. In this connection it is worth mentioning that Käckel and Bechstedt [34] have recently calculated LDA band gaps for the hypothetical SiC superlattices $(3C)_4(2H)_3$ and $(3C)_2(2H)_6$, but also for $(3C)_2(4H)_3$ and $(3C)_2(6H)_2$, which are smaller than the 3C SiC LDA band gap.

6. A One-Dimensional Model Description

The successful calculation of polytype band gaps, as described in the previous section, confirms our central assumption that the polytypes can be considered as building-block structures, in which the building blocks are mutualy twisted bulk-like fragments of 3C SiC. Two important elements obviously play a role in the determination of the band gap size, namely the transmission characteristics at interfaces (which are independent of the building-block size) and the respective building-block sizes themselves. In what follows we will construct a one-dimensional model, in which both features are present. The model will contain the essence of the formation of the lowest conduction band state in SiC polytypes. We will focus on electron waves in a 3C SiC layer with energy close to the minimum of the conduction band of 3C SiC, when such waves enter the twisted counterpart. The wave vector of the incoming wave is decomposed as $\mathbf{k} = \mathbf{k}_\parallel + k_z\mathbf{e}_z$. The lowest conduction band state in most of the polytypes appears at the parallel wave vector component $\mathbf{k}_\parallel = \mathbf{k}_M$. We therefore concentrate on the Bloch electrons in 3C SiC with $\mathbf{k} = \mathbf{k}_M + k_z\mathbf{e}_z$, and investigate how such electrons are scattered at an interface. In each layer of a given polytype, the complete wavefunction, at an energy slightly higher than the lowest conduction band energy in 3C SiC, is constructed as a linear combination of cubic wavefunctions, as described in detail in Section 4. Some of these wavefunctions have real k_z, some complex k_z. Both types of waves contribute considerably in each building block in making up the wavefunction of the polytype. Two oppositely travelling Bloch waves, pertaining to the lowest conduction band and present in each block, represent the main propagating part in the linear combination. It is found that the evanescent waves in the linear combination are only important close to the interfaces, because of the magnitude of the imaginary parts ($|\mathrm{Im}\, k_z| \approx 0.7/c_1$) of the corresponding wave vectors. These waves extinguish almost completely over distances larger than c_1.

In the one-dimensional model, only the effective coupling of the two participating Bloch waves in the adjacent layers with $\mathbf{k} = \mathbf{k}_M \pm k_z \mathbf{e}_z$ is accounted for. Though the evanescent waves are left out in this model, for the above-mentioned reason, they are extremely relevant for the matching of the wavefunctions at the interfaces. Essential in the model is that the proper matching is fully accounted for, by letting the Bloch waves (with real k_z) obey discontinuous boundary conditions. The Bloch waves of the superlattice are then subsequently described in terms of envelope functions $\mathscr{F}^j(z)$, obeying the following boundary conditions at the interface $z = z^j$:

$$
\begin{cases}
\mathscr{F}^{j+1}(z_j) = \beta \mathscr{F}^j(z_j) + \gamma c_1 \left. \dfrac{\mathrm{d}}{\mathrm{d}z}\, \mathscr{F}^j \right|_{z=z_j}, \\[2mm]
\mathscr{F}^j(z_j) = \beta \mathscr{F}^{j+1}(z_j) - \gamma c_1 \left. \dfrac{\mathrm{d}}{\mathrm{d}z}\, \mathscr{F}^{j+1} \right|_{z=z_j},
\end{cases}
\tag{13}
$$

where j again labels the particular layer. The envelopes $\mathscr{F}^j(z)$ in each layer are linear combinations of plane waves $\exp(\pm ik_z z)$. The particular form of the boundary conditions in Eq. (13) is determined by the invariance under interchange of layer indices and substitution of $-z$ for z. All effects of the evanescent waves are incorporated in the coupling parameters β and γ. These have to be real-valued in order to conserve the probability flow across the interfaces. The actual values of the parameters β and γ can be deduced from the energy dependent transmission properties of a single interface. The relation between β and γ on the one hand and the (complex) transmission coefficient T on the other hand can be determined as follows. Consider an incoming wave $\exp(ik_z z)$ in the untwisted layer j. It leads to a reflected wave $R \exp(-ik_z z)$, where R is the complex reflection coefficient. The envelope function in the layer j then reads $\mathscr{F}^j(z) = \exp(ik_z z) + R \exp(-ik_z z)$. For the outcoming wave in the twisted layer $j+1$ we find $\mathscr{F}^{j+1}(z) = T \exp(ik_z z)$. Substituting these expression for $\mathscr{F}^j(z)$ and $\mathscr{F}^{j+1}(z)$ in Eq. (13) at $z = z_j = 0$ yields

$$
\begin{cases}
T = \beta(1+R) + ik_z \gamma c_1 (1-R), \\[1mm]
(1+R) = \beta T - ik_z \gamma c_1 T,
\end{cases}
\tag{14}
$$

which gives

$$
T = \left(\beta - i\, \frac{1 - \beta^2 + (\gamma c_1 k_z)^2}{2\gamma c_1 k_z} \right)^{-1}.
\tag{15}
$$

It is now crucial to realize that T can be calculated from the complete matching analysis of Section 4. The actual values of β and γ can therefore be directly derived from the energy dependent transmission properties of a single interface by putting the right-hand side of Eq. (15) equal to the complex transmission amplitude, as obtained from our earlier analysis.

In the energy region of interest the conduction band in 3C SiC is nearly parabolic and can accurately be represented by $E = E_c + \hbar^2 k_z^2/2m^*$ with effective mass $m^* = 0.48m$

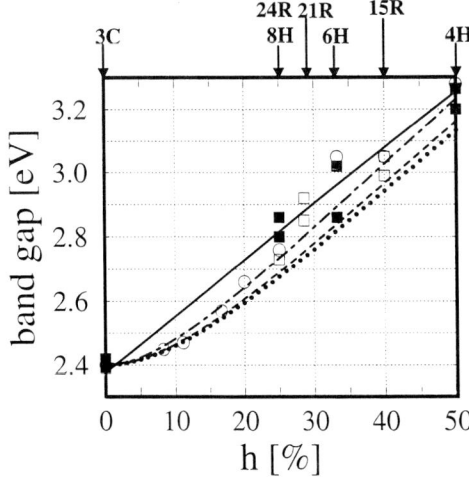

Fig. 9. Band gap values as a function of hexagonality as calculated in our one-dimensional model. Three ratios of layer sizes are indicated: $a = b$ (dashed-dotted line), $a = 1.3b$ (dashed line) and $a = 1.5b$ (dotted line). The empirical linear relation given by Choyke et al. [12] is also indicated (solid line). The available experimental gaps [18] for hexagonal (filled squares) and rhombohedral (open squares) structures are indicated, as well as the gaps of our matching procedure within the EPM (circles); taken from [15]

and E_c the conduction band minimum. The dependence of T on k_z in Eq. (15) can therefore easily be transformed into a dependence of T on E. A full calculation shows that $|T|^2$ increases from zero at the conduction band minimum to a maximum value of $1/\beta^2$ at some higher energy. Slightly above the conduction band minimum of 3C SiC the coupling parameters appear to be $\beta = 1.63$ and $\gamma = 0.88$. In the relevant region from 2.4 to 3.3 eV the parameters increase slightly, with $\beta = 1.76$ and $\gamma = 0.93$ as average values. The fact that these parameters vary only weakly with energy is not unexpected since the relevant k_z branches in the band structure of 3C SiC are quite steep and the periodic parts of the wavefunctions do not vary much with energy. The model would be of much less use if β and γ were strongly energy dependent.

Stacking cubic segments with different sizes a and b, and imposing the Bloch condition for the period $L = a + b$, yields the band structure equation

$$\cos K_z L = \frac{1}{(2\gamma c_1 k_z)^2} \{ (1 - \beta^2 - (\gamma c_1 k_z)^2)^2 \cos k_z(b - a) - [(1 - \beta^2 + (\gamma c_1 k_z)^2)^2$$
$$- (2\gamma c_1 k_z \beta)^2] \cos k_z L - 4\gamma c_1 k_z \beta (1 - \beta^2 + (\gamma c_1 k_z)^2) \sin k_z L \}, \quad (16)$$

with K_z the superlattice wave vector. For rhombohedral (R) structures the layers have different sizes $(a \neq b)$, but for hexagonal (H) structures the layers are equally sized $(a = b)$, simplifying the equation to

$$\cos \frac{K_z L}{2} = \beta \cos \frac{k_z L}{2} - \frac{[1 - \beta^2 + (\gamma c_1 k_z)^2] \sin \frac{k_z L}{2}}{2\gamma c_1 k_z}. \quad (17)$$

Fig. 9 displays the band gap predictions following from Eq. (16) for hexagonal and rhombohedral structures with $a = 1.3b$ and $a = 1.5b$, as a function of h. Note that the interpolating curves in Fig. 8 are also the result of the one-dimensional analysis. Note furthermore that the prediction is that R structures generally have smaller gaps. Unfortunately, the number of experiments on electronic energy gap values of rhombohedral polytypes is rather low. It is found that for not too small $h \gtrsim 15\%$ the calculated relation for H structures is linear and reproduces the experimental results found by Choyke et al. [12] within about 0.1 eV. The slight deviation from the nearly linear behaviour due to the camel's back in the lowest conduction band of 6H SiC is not included in our one-dimensional model.

The one-dimensional model focusses in fact on the conditions under which the formation of the lowest conduction states in a polytype takes place. It describes the periodic frustration of propagating waves in the respective cubic building blocks. The sizes of the respective building blocks are decisive in determining what the gap value of a given polytype will be.

7. Why is SiC so Exceptional?

Although we have given a successful description of the band gap variation among polytypes of SiC in terms of matching of wavefunctions, and also in terms of a one-dimensional frustration model of propagating waves in building blocks, it is of interest to identify the characteristic features responsible for the rather strong band gap variation among the polytypes of SiC, and to understand why other materials, such as ZnS (which is also a polytype material) and hypothetical polytypes of Si, C, and AlAs, do not show this rather strong variation.

ZnS is a polytype compound of which the band gaps are experimentally known for quite a number of different polytype structures. All these experimental band gap values appear to be (within 0.1 eV) the same, namely 3.7 eV. An important difference with SiC is that ZnS is a direct semiconductor with the band gap located at Γ. In order to make a comparison with the SiC case we have applied the matching method as described in Section 4 to the three ZnS polytypes 3C, 6H and 4H. We have chosen the 4H and the 6H structure because in SiC they have relatively small building-block sizes with band gaps that differ considerably from the 3C SiC band gap. In the calculation we have utilized the EPM form factors of [35]. We find the direct band gaps to be 3.64, 3.65 and 3.64 eV for 3C, 6H and 4H ZnS, respectively. Our calculations therefore confirm that the band gaps of ZnS polytypes are (almost) independent of the polytype structure. The extremely small energy variation of the conduction band edge in ZnS is clearly related to the fact that the two wave functions of the lowest conduction band state, pertaining to the two adjacent and mutually twisted layers, happen to be highly similar. For SiC this is not at all the case. As a measure of this similarity we define an "overlap" by the the inner product between one wavefunction and a twisted second wavefunction. In SiC, the overlap between the two lowest conduction band states at $\mathbf{k}_M$ is equal to 0.13, only. For ZnS, on the other hand, the corresponding overlap at $\mathbf{k}_\Gamma$ is equal to 0.95, implying that the wavefunctions almost perfectly match. Completely in line with this, we find (i) that the single-interface transmission analysis, as described in Section 6, yields a transmission probability of 98% for the lowest conduction band state in ZnS, in contrast to the lowest conduction band state of SiC, which is almost completely reflected, and (ii) that in the construction of the superlattice Bloch wavefunctions of 6H and 4H ZnS evanescent waves hardly contribute, in contrast to SiC where the contribution of evanescent waves close to the interfaces is considerable.

The conclusion therefore seems to be that in order to obtain a strong variation of the conduction band edge in building-block polytypes it is necessary that the related Bloch wavefunctions in the mutually twisted cubic basis materials are highly dissimilar. This leads to a strongly frustrated transmission, which makes it difficult to construct propagating waves in the polytype.

In order to investigate whether possibly other features can be held responsible for a strong variation of band gaps we have also investigated a number of hypothetical polytypes of silicon (Si), diamond (C) and AlAs. The reason behind these choices is that all these materials have indirect band gaps. For 3C AlAs the lowest conduction state occurs at X, like for 3C SiC. For C and Si the lowest conduction state is located at the ΓX-axis near the X-point. Because the lowest conduction band energies in Si and C near X do not differ much (less than 0.1 eV) from the values at X, we will focus in all three cases on the X-point. The three materials have in common with 3C SiC that the overlap between the Bloch wavefunctions related to the two conduction band minima (at X) in the twisted and untwisted material is quite small (in all cases even smaller than for 3C SiC). From the correspondence with the SiC case, one would therefore expect some band gap variation among different (hypothetical) polytypes of Si, C and AlAs. Nevertheless, we observe no appreciable band gap variation. In applying the interface matching method within the EPM, using the empirical pseudopotential form factors of [35, 36], to the 3C, 6H and 4H polytypes of Si, C and AlAs, we obtained band gaps at $\mathbf{k}_M$ for the polytypes 3C, 6H and 4H of 2.08, 2.14 and 2.11 eV for AlAs, 1.17, 1.03 and 1.00 eV for Si and 5.48, 5.53 and 5.38 eV for C, respectively. All these (rather) small band gap differences are mainly due to small energy shifts of the valence band tops. The energy levels of the lowest conduction band states at $\mathbf{k}_M$ vary at most 0.03 eV. The difference with the SiC case can be traced back to the difference in the (complex) cubic band structures of these substances compared to SiC. Because of the twofold degeneracy in Si and C of the lowest conduction band at X, in the matching procedure at $\mathbf{k}_\parallel = \mathbf{k}_M$ four Bloch waves participate (two pairs of oppositely travelling waves), in contrast to 3C SiC, where only two oppositely travelling Bloch waves participate. This larger freedom facilitates the construction of a transmitted wavefunction in the twisted material that matches the incoming Bloch wave in the untwisted material. As a result, the frustration occurring in SiC is no longer present. In AlAs, like in 3C SiC, the degeneracy occurring in Si and C is lifted. However, the splitting between the two lowest conduction states X1 and X3 at X is approximately 0.5 eV, which is quite small in comparison to 3C SiC, where it is approximately 3.5 eV. This means that in AlAs not only two oppositely travelling Bloch waves participate considerably in the matching procedure at $\mathbf{k}_\parallel = \mathbf{k}_M$, but also two evanescent waves with relatively small values of $|\mathrm{Im}\,k_z|$, contributing significantly not only close to the interfaces, but also well within the layers. This leads to a situation which is qualitatively the same as in Si and C. The antisymmetric parts of the lowest Fourier components $v_C(\mathbf{G}) - v_{Si}(\mathbf{G})$ with $|\mathbf{G}| > 0$ of the empirical pseudopotential are responsible for the energy splitting between the two lowest conduction bands at X. In Si and C these are exactly zero and in cubic AlAs they are much smaller than in 3C SiC.

For the valence band top in SiC polytypes the situation is in fact comparable to the situation for the conduction band edge in Si and C. The valence band top has a threefold degeneracy in 3C SiC. This means that in order to make up the wavefunctions of the valence band top of the superlattice we deal with, depending on the energy, either six propagating waves or six very weakly evanescing waves. The superlattice Bloch wave function of the valence band top is therefore quite easily formed by a linear combination of the six evanescent states, at an energy slightly above the energy of the valence band edge of the cubic material.

8. Summary and Conclusions

Although first-principles methods such as the GW method should be capable of reproducing the band gap variation among polytypes of SiC, they do not contribute to the understanding of this relatively strong variation. By representing polytypes of SiC as building-block structures, we have been able to approach the band structure problem in polytypes in a completely different way, by starting from the intrinsic electronic properties of 3C SiC. In our opinion this is an important first step in the elucidation of the band gap variation in polytypes. For the characterization of 3C SiC it is necessary to have the complex band structure and related Bloch or evanescent electron wavefunctions of this basis material, e.g. in the empirical pseudopotential scheme. By matching linear combinations of these wavefunctions at interfaces, surprisingly adequate band structures and band gaps for polytypes are obtained. Only 2H SiC is an exceptional case, which is easily understood by realizing that no layers can be identified in this case which are sufficiently 3C-like. The matching analysis in itself still does not give an answer to the question why the SiC polytypes show a relatively strong variation in band gaps. To this end, we proceeded by putting the essentials of the matching analysis in an even more simple one-dimensional model in which the focus is on the propagation through the interfaces of 3C-type wavefunctions with an energy just above the conduction band minimum. Finally, it was understood why polytypes of SiC lead to such large band gap variation, contrary to polytypes of e.g. ZnS, or hypothetical polytypes of Si, C or AlAs. We have argued in this connection that direct semiconductors are bad candidates for strong band gap variation because of the high similarity of Bloch waves in the two adjacent material layers, just above the conduction band minimum of the 3C material. A second important point is the size of the X1–X3 splitting (if any) of the lowest conduction bands, if the 3C band minimum is at or near X. The larger this splitting, the larger the frustration in the propagation through the interfaces and the stronger the band gap variation. For Si or C this splitting is absent, and for AlAs it is much smaller than for SiC.

In conclusion, we believe that our matching analysis is a transparent and relatively easy way of producing reliable band structures of SiC polytypes. We have shown that the method leads to an understanding of the rather exceptional, relatively strong band gap variation in SiC polytypes.

References

[1] C. Cheng, R. J. Needs, V. Heine, and N. Churcher, Europhys. Lett. **3**, 475 (1987).
[2] C. Cheng, R. J. Needs, and V. Heine, J. Phys. C **21**, 1049 (1988).
[3] P. J. H. Denteneer and W. van Haeringen, Phys. Rev. B **33**, 2831 (1986).
[4] P. J. H. Denteneer and W. van Haeringen, Solid State Commun. **65**, 115 (1988).
[5] G. E. Engel and R. J. Needs, J. Phys. C **2**, 367 (1990).
[6] W. R. L. Lambrecht, B. Segall, M. Yoganathan, W. Suttrop, R. P. Devaty, W. J. Choyke, J. A. Edmond, J. A. Powell, and M. Alouani, Phys. Rev. B **63**, 2747 (1993); B **50**, 10722 (1994).
[7] M. S. Hybertsen and S. G. Louie, Phys. Rev. Lett. **55**, 1418 (1985); Phys. Rev. B **34**, 5390 (1986).
[8] R. W. Godby, M. Schlüter, and L. J. Sham, Phys. Rev. Lett. **56**, 2415 (1986); Phys. Rev. B **37**, 10159 (1988).
[9] M. Rohlfing, P. Krüger, and J. Pollmann, Phys. Rev. B **48**, 17791 (1993).
[10] W. H. Backes, P. A. Bobbert, and W. van Haeringen, Phys. Rev. B **51**, 4950 (1995).

[11] B. WENZIEN, P. KÄCKELL, F. BECHSTEDT, and G. CAPPELLINI, Phys. Rev. B **52**, 10897 (1995).
[12] W. J. CHOYKE, D. R. HAMILTON, and L. PATRICK, Phys. Rev. **133**, 1163 (1964).
[13] W. H. BACKES, On the Band Gap Variation in SiC Polytypes, PhD Thesis, Eindhoven University of Technology, 1996.
[14] W. H. BACKES, P. A. BOBBERT, and W. VAN HAERINGEN, Phys. Rev. B **49**, 7564 (1994).
[15] W. H. BACKES, F. C. DE NOOIJ, P. A. BOBBERT, and W. VAN HAERINGEN, Physica **217B**, 207 (1996).
[16] Y. M. TAIROV and Y. A. VODAKOV, in: Topics in Applied Physics, Ed. J. I. PANKOVE, Vol. 17, Springer-Verlag, Berlin 1977.
[17] H.-G. JUNGINGER and W. VAN HAERINGEN, phys. stat. sol. **37**, 709 (1970).
[18] Numerical Data and Functional Relationships in Science and Technology, Ed. K.-H. HELLWEGE and O. MADELUNG, Landolt-Börnstein, New Series, Group III, Vols. 17a and 22c, Springer-Verlag, Berlin 1982 and 1986.
[19] R. W. G. WYCKOFF, Crystal Structures, Vol. 1, Interscience Publ. Co., New York 1963.
[20] A. QTEISH, V. HEINE, and R. J. NEEDS, Phys. Rev. B **45**, 6376 (1992).
[21] P. KÄCKELL, B. WENZIEN, and F. BECHSTEDT, Phys. Rev. B **50**, 17037 (1994).
[22] C. H. PARK, BYOUNG-HO CHEONG, KEUN-HO LEE, and K. J. CHANG, Phys. Rev. B **49**, 4485 (1994).
[23] P. J. H. DENTENEER, The Pseudopotential-Density-Functional Method Applied to Semiconducting Crystals, PhD Thesis, Eindhoven University of Technology, 1987.
[24] Y. CHANG and J. N. SCHULMAN, Phys. Rev. B **25**, 3975 (1982).
[25] S. BRAND and D. T. HUGHES, Semicond. Sci. Technol. **2**, 607 (1987).
[26] D. Y. K. KO and J. C. INKSON, Phys. Rev. B **38**, 9945 (1988).
[27] J. P. CUYPERS and W. VAN HAERINGEN, J. Phys. C **4**, 2587 (1992).
[28] Z. IKONIC, G. P. SRIVASTAVA, and J. C. INKSON, Phys. Rev. B **46**, 15150 (1992).
[29] Z. IKONIC, G. P. SRIVASTAVA, and J. C. INKSON, Solid State Commun. **12**, 799 (1993).
[30] W. J. CHOYKE and L. PATRICK, Phys. Rev. **127**, 1868 (1962).
[31] L. PATRICK, Phys. Rev. B **5**, 2198 (1972).
[32] L. PATRICK, D. R. HAMILTON, and W. J. CHOYKE, Phys. Rev. **143**, 526 (1966).
[33] P. KÄCKELL, B. WENZIEN, and F. BECHSTEDT, Phys. Rev. B **50**, 17037 (1994).
[34] P. KÄCKELL and F. BECHSTEDT, Phys. Rev. Lett. **75**, 2180 (1995).
[35] M. L. COHEN and T. K. BERGSTRASSER, Phys. Rev. **141**, 789 (1966).
[36] A. BALDERESCHI, E. HESS, K. MASCHKE, H. NEUMANN, K.-R. SCHULZE, and K. UNGER, J. Phys. C **10**, 4709 (1977).

phys. stat. sol. (b) **202**, 81 (1997)

Subject classification: 71.20.Nr; 71.55.Ht; S6

Shallow Donor Levels and the Conduction Band Edge Structures in Polytypes of SiC

AN-BAN CHEN (a) and P. SRICHAIKUL (b)

(a) *Department of Physics, Auburn University, Auburn, AL 36849, USA*

(b) *High Performance Computing Center, NECTEC, Bangkok, Thailand*

(Received January 31, 1997)

This study focuses on the calculation of conduction band edge structures and nitrogen (N) donor levels in four polytypes of SiC: 3C, 2H, 4H, and 6H. A band-structure-based theory and a model potential are developed for the donor level calculation. A hybrid pseudo-potential and tight-binding Hamiltonian is used to obtain a set of comprehensive band structures for these four polytypes by incorporating useful experimental information in the band-structure calculation. The conduction band edge wave functions derived from these band structures and several sets of theoretical and experimental effective masses are then used to study the various effects on the donor level, which include anisotropic masses, central cell potential correction, conduction band edge wave functions, and intervalley coupling. With suitable sets of effective masses, the calculated donor energies agree semi-quantitatively with those deduced from experiments. The site dependence of the donor energies in 4H and 6H is found to be dominated by the conduction band wave functions and the intervalley coupling.

1. Introduction

1.1 General consideration

Various imperfections can exist in a semiconductor, ranging from simple atomic vacancies and substitution impurities to complex defects such as dislocations and grain boundaries. These defects can greatly alter the host electronic structures and influence the electrical and optical properties. While most of these crystal imperfections are native defects, donor and acceptor impurities are deliberately added to control the carrier concentrations. A quantitative characterization of the electrical and optical properties of a semiconductor device requires a detailed knowledge of the donor and acceptor energy levels. This chapter is to review the theory of donor levels for several polytypes of SiC, with a focus on nitrogen (N) donors.

The donor energy levels are intimately related to the defect structure. For example, a N impurity in SiC may reside at a C or a Si site, or at one of the interstitial sites. These different defects will have different structural energies and donor energy levels. A rigorous theory of donor levels should include the impurity structure determination as a part of the calculation. This is not done in the present work. In this study, we will only consider the simplest case: a substitutional impurity, for example, a N atom replacing one of the C atoms in SiC. Whether or not this is the structure corresponding to the experimental situations requires a more careful study.

The donor energy levels, as well as the band structures of pure semiconductors, are many-body spectra. A rigorous treatment of this problem is still not available. However, for pure crystals, considerable progress has been made to reduce the many-body interactions to a manageable one-electron potential. For example, the so-called GW approximation, Hybertson and Louie [1], has produced reasonably accurate band gaps for semiconductors. Such technique has yet to be extended to the donor level calculation. On the other hand, many empirical models have successfully described both the qualitative and quantitative aspects of donor and acceptor levels, Chang et al. [2]. To develop the theory, we shall start with the simple hydrogenic model, examine the effect of the mass anisotropy, present a formal donor state theory based on band structures, devise a model potential, obtain the band structures, and calculate the N donor energies in 3C, 2H, 4H, and 6H SiC.

1.2 Special features in SiC

What is special about the N donor levels in SiC? First, as shown in Table 1, the experimental ionization energies ε_I vary considerably from one polytype to another, despite that the local tetrahedral structure and the total energies of different polytypes are very similar. Second, also seen from Table 1, there is a striking dependence on the symmetry of the impurity sites in the 4H and 6H polytypes. Finally, all these energy levels are not really shallow. They fall in the difficult intermediate range between the shallow (say, $\varepsilon_I \leq 50$ meV) and deep (say, $\varepsilon_I \geq 100$ meV) levels.

The inequivalent donor sites can be best defined by considering the stacking of the host carbon layers perpendicular to the c-axis. In the 4H stacking sequence for C, ...|abac|abac|..., all the a sites are the cubic (k) sites and all the b and c sites are the hexagonal (h) sites, according to Suttrop et al. [3]. Similarly in 6H, the stacking sequence of C is ...|abcacb|abcacb|.... While all the a sites in this stacking sequence are the hexagonal (h) sites, there are two kinds of inequivalent quasi cubic sites, denoted k_1 and k_2, respectively. If the Si sublattice is displaced a distance d from the C sublattice along the c-axis, where d is the first neighbor distance, then the corresponding symmetry designation in this sequence for the C sites used in Table 1 and Section 4 is ...|$hk_1k_2hk_1k_2$ | $hk_1k_2hk_1k_2$|....

Assuming that all these N donors are on the C sites, the impurity potential, i.e. the potential change caused by the substitution of a C atom by a N atom as viewed from

Table 1

Experimental values of the ionization energy (in meV) of nitrogen donor in 3C, 4H, and 6H SiC

SiC polytype	ionization energy
3C[a]	57.0
4H[b] h site	52.1
k site	91.8
6H[c] h site	81.0
k_1 site	137.6
k_2 site	142.4

[a] Madelung [9]; Kuwabara et al. [31].
[b] Götz et al. [21].
[c] Suttrop et al. [3].

the impurity site, should be very similar for all polytypes and for different donor sites. The difference in the donor energy levels in different polytypes must arise from different total potentials, which are the sum of the host and the impurity potentials. In terms of band-structure description in Section 2.4, these large differences in the donor ground state energies in different polytypes are due to substantially different band structures, particularly the location and the effective masses of the conduction band edges. The site dependence of ε_D in a given polytype (4H or 6H) can be attributed to the spatial variation of the conduction band bottom wave functions as viewed from inequivalent donor sites.

Accurate calculation of donor levels requires a reliable impurity potential and accurate host band structures. The most important band quantities needed for shallow donors are the locations, the effective masses, and wave functions of the conduction band minima. Unfortunately, these quantities have not been well established for all four polytypes of SiC. Furthermore, there is no band calculation that has produced the band gaps and the locations of E_c for all the four polytypes that are consistent with the experimental identifications. The present work will combine available information, such as the band gaps, the masses, and the location of E_c, derived from experiment and theory, to deduce a comprehensive set of band edge structures for four SiC polytypes, and use them in the donor level calculation.

2. Development of Donor Level Theory

In this section, we shall review the theoretical models for donor level calculation using Si as an example, and then devise a model potential that is applicable to SiC. This potential will be tested against the well-known results for Si before it is applied to SiC in Section 4.

2.1 Simple hydrogenic model

This simple model is useful for introducing the terminology of donor states. For example, consider an As impurity in Si. The As atom has one more valence electron than Si. Since it only takes four electrons per atom to saturate the tetrahedral bonds, the extra electron is loosely bound to the As atom and is easily ionized to become a free electron in the conduction band. This bonding is similar to a hydrogen atom, but the electrostatic potential is now screened by all the valence electrons and has the form

$$V(r) = \frac{e}{\kappa r}, \tag{2.1}$$

where κ is the static dielectric constant. Because the dielectric constant is large, for example, $\kappa = 11.9$ for Si, this potential is long-ranged but weak. The shallow donor energy levels in this model can be calculated from the Schrödinger equation

$$\left(\frac{-\hbar^2}{2m^*} \nabla^2 - \frac{e^2}{\kappa r} \right) \Psi(r) = E\Psi(r). \tag{2.2}$$

Here m^* is an effective mass at the conduction band minimum. The size of the ground state wave function is now characterized by an effective Bohr's radius a^* rather than the usual value a_0 for the hydrogen atom. a^* and a_0 are related to each other by

$$a^* = \frac{a_0 \kappa}{m^*/m_0}, \tag{2.3}$$

Table 2

Binding energies (in meV) of group-V donors in Si

designated state	hydrogenic model[a]	experiment[b]		
		P	As	Sb
$3p_{\pm}$	2.75	3.06	3.08	2.99
$2p_{\pm}$	6.19	6.33	6.33	6.33
$2p_0$	6.19	11.39	11.45	11.39
1s (E)	24.79	32.37	31.19	30.40
1s (T$_2$)	24.79	33.74	32.60	32.82
1s (A$_1$)	24.79	45.47	53.69	42.68

[a] Using $\dfrac{1}{m^*} = \dfrac{1}{3}\left(\dfrac{2}{m_{\perp}} + \dfrac{1}{m_{\parallel}}\right)$, $m_{\parallel} = 0.91$, and $m_{\perp} = 0.19$.

[b] Ning and Sah [32]; Morita and Nara [33].

where a_0 is 0.52917 Å and m_0 is the free electron mass. The energy levels E_n are also modified from the hydrogen atomic levels ε_n. They are related to each other by

$$E_n = \frac{m^*/m_0}{\kappa^2}\, \varepsilon_n\,, \tag{2.4}$$

where $\varepsilon_n = -13.606/n^2$ eV with n being non-zero integers. The effective mass m^* is typically 10 times smaller than m_0 and κ is about 10. Thus a^* in Eq. (2.3) is typically 100 times a_0, and E_n is only several tens of meV. As shown in Table 2, the comparison with experimental results suggests that the simple model is reasonable for higher excited states. However, results for the ground state do not agree well. This is understandable, since the 1s state occupies the smallest orbit of all, and the simple screened Coulomb potential is not accurate near the impurity. Besides, the simple hydrogenic model does not describe the dependence on donor species, whereas experiment shows considerable variation of the ionization energies with different donors.

2.2 Hydrogenic model with mass anisotropy

The single effective mass used in Eq. (2.2) is only valid for a direct-gap semiconductor in which the conduction band about its minimum is isotropic. The theory must be extended to deal with mass anisotropy. The hydrogenic model with two different masses has been studied by several authors including Faulkner [4]. Recently Pfeiffer et al. [5] have extended the calculation to include three masses. The equation modified for anisotropic masses becomes

$$\left(\frac{-\hbar^2}{2}\left(\frac{1}{m_1}\frac{\partial^2}{\partial x^2} + \frac{1}{m_2}\frac{\partial^2}{\partial y^2} + \frac{1}{m_3}\frac{\partial^2}{\partial z^2}\right) - \frac{e^2}{\kappa r}\right)\Psi(r) = E\Psi(r)\,. \tag{2.5}$$

In Rydberg units, it becomes

$$-\left(\frac{1}{m_1}\frac{\partial^2}{\partial x^2} + \frac{1}{m_2}\frac{\partial^2}{\partial y^2} + \frac{1}{m_3}\frac{\partial^2}{\partial z^2} - \frac{2}{\kappa r}\right)\psi = E\psi\,, \tag{2.6}$$

where free electron mass m_0 and the Rydberg are taken to be unity, m_1, m_2, and m_3 are the effective mass components expressed as multiples of the electron mass. Following

Table 3
Ionization energies (in meV) of shallow donors in Si from (a) hydrogenic model, (b) hydrogenic model with anisotropic masses, and (c) experimental data

impurity	(a)	(b)	(c)
P	24.8	29.0	45.5
As	24.8	29.0	53.7
Sb	24.8	29.0	42.7

Pfeiffer et al. [5], two anisotropy parameters are defined as

$$\alpha_1 = 1 - \frac{m}{m_3} , \tag{2.7a}$$

$$\alpha_2 = \frac{m_2 - m_1}{m_2 + m_1} , \tag{2.7b}$$

where $\dfrac{1}{m} = \dfrac{1}{2}\left(\dfrac{1}{m_1} + \dfrac{1}{m_2}\right)$ with $m_3 \geq m_2 \geq m_1$. The energy E is scaled by

$$E = \frac{m}{\kappa^2}\,\varepsilon . \tag{2.8}$$

With the transformation $x' = x\sqrt{m_1/m}$, $y' = y\sqrt{m_2/m}$ and $z' = z\sqrt{m_3/m}$, Eq. (2.6) becomes

$$-\nabla^2\psi - \frac{2}{\sqrt{(1+\alpha_2)\,x^2 + (1+\alpha_2)\,y^2 + (1+\alpha_1)\,z^2}}\,\psi = \varepsilon\psi . \tag{2.9}$$

For $m_3 = m_2 = m_1$, $\alpha_1 = \alpha_2 = 0$, the isotropic hydrogenic model is recovered. The twofold mass anisotropy, such as in Si and Ge, has $m_2 = m_1 = m$. This case corresponds to $\alpha_2 = 0$ and $0 < \alpha_1 < 1$. The ground state anisotropic factors and first few excited states of this hydrogenic model for a wide range of values of the α_1 and α_2 are tabulated by Pfeiffer et al. [5]. Table 3 shows the comparison of results from the hydrogenic model with experimental data for Si. The inclusion of mass anisotropy only gives a small improvement.

Although the simple hydrogenic potential including mass anisotropy provides a qualitative description of the shallow donors in semiconductors, it is obviously not an accurate model. Below we shall present a formal donor level theory based on the band structure, and then return to discuss the impurity potential in Section 2.5.

2.3 Band structure theory of donor states

We start with the band structure theory of a pure semiconductor. The energy eigenvalues $E_n(\mathbf{k})$ and eigenfunctions $|\phi_{n,\mathbf{k}}\rangle$ of an electron in a perfect crystal satisfy the Schrödinger equation

$$H_0\,|\phi_{n,\mathbf{k}}\rangle = E_n(\mathbf{k})\,|\phi_{n,\mathbf{k}}\rangle , \tag{2.10}$$

where n is the band index, $\mathbf{k}$ the crystal wave vector, and H_0 the pure crystal one-electron Hamiltonian. These eigenkets form a complete orthonormal basis set

$$\langle\phi_{n',\mathbf{k}'} | \phi_{n,\mathbf{k}}\rangle = \delta_{n,n'}\delta_{\mathbf{k},\mathbf{k}'} . \tag{2.11}$$

When a donor impurity is introduced, it produces an impurity potential U. The total Hamiltonian is

$$H = H_0 + U. \tag{2.12}$$

The donor energy levels E and the associate eigenkets $|\Psi\rangle$ must satisfy

$$H |\Psi\rangle = E |\Psi\rangle. \tag{2.13}$$

$|\Psi\rangle$ can be expanded in terms of $|\phi_{n,\mathbf{k}}\rangle$,

$$|\Psi\rangle = \sum_{n,\mathbf{k}} C_{n,\mathbf{k}} |\phi_{n,\mathbf{k}}\rangle. \tag{2.14}$$

Here $\mathbf{k}$ runs over all the first Brillouin zone and n over all the bands. The $C_{n,\mathbf{k}}$ are the expansion coefficients. Substituting Eqs. (2.12) and (2.14) into (2.13), we obtain

$$(H_0 + U) \sum_{n,\mathbf{k}} C_{n,\mathbf{k}} |\phi_{n,\mathbf{k}}\rangle = E \sum_{n,\mathbf{k}} C_{n,\mathbf{k}} |\phi_{n,\mathbf{k}}\rangle. \tag{2.15}$$

Using Eq. (2.10), Eq. (2.15) becomes

$$\sum_{n,\mathbf{k}} (E_{n,\mathbf{k}} - E) C_{n,\mathbf{k}} |\phi_{n,\mathbf{k}}\rangle + \sum_{n,\mathbf{k}} U C_{n,\mathbf{k}} |\phi_{n,\mathbf{k}}\rangle = 0. \tag{2.16}$$

Taking the scalar product of $\langle \phi_{n,\mathbf{k}}|$ on Eq. (2.16), and using the orthonormal condition in Eq. (2.11), we obtain for every n and $\mathbf{k}$

$$(E_{n,\mathbf{k}} - E) C_{n,\mathbf{k}} + \sum_{n',\mathbf{k}'} \langle \phi_{n,\mathbf{k}}| U |\phi_{n',\mathbf{k}'}\rangle C_{n',\mathbf{k}'} = 0. \tag{2.17}$$

Eq. (2.17) is an integral equation, because the summation over $\mathbf{k}'$ can be converted to integral inside the first Brillouin zone. A solution of this equation requires a knowledge of the wave functions and band structures throughout the Brillouin zone. Note that this equation is still quite general regardless of whether the impurity states are shallow or deep.

Eq. (2.17) can be rewritten as

$$C_{n,\mathbf{k}} = \frac{-1}{(E_{n,\mathbf{k}} - E)} \sum_{n',\mathbf{k}'} \langle \phi_{n,\mathbf{k}}| U |\phi_{n',\mathbf{k}'}\rangle C_{n',\mathbf{k}'}. \tag{2.18}$$

The magnitude of $C_{n,\mathbf{k}}$ is primarily governed by the factor $(E_{n,\mathbf{k}} - E)^{-1}$. For shallow donor levels, the energy E lies just below the conduction band edge, so the magnitude of $C_{n,\mathbf{k}}$ has the largest value for those $E_{n,\mathbf{k}}$ belonging to the lowest conduction bands. An approximation to Eq. (2.17) for these shallow donor levels is to drop all but the lowest conduction bands.

A way to solve Eq. (2.17) is to expand $C_{n,\mathbf{k}}$ in terms of a set of basis functions $g_{na}(\mathbf{k})$ that are localized in the $\mathbf{k}$ space and centered at the minimum of each band,

$$C_{n,\mathbf{k}} = \sum_{a} A_{na} g_{na}(\mathbf{k}), \tag{2.19}$$

where A_{na} is the expansion coefficient. The coefficients A_{na} can be obtained through variational calculation,

$$\frac{\partial \langle \Psi| H |\Psi\rangle}{\partial A^*_{n'a'}} - E \frac{\partial \langle \Psi | \Psi\rangle}{\partial A^*_{n'a'}} = 0. \tag{2.20}$$

Eq. (2.20) leads to the following eigenvalue problem:

$$(\mathbf{T} + \mathbf{V})\,\mathbf{a} = \mathbf{Ha} = E\mathbf{Sa}\,, \tag{2.21}$$

where $\mathbf{a}$ is a column matrix with A_{na} as its matrix elements. The elements of the square matrices $\mathbf{T}$, $\mathbf{V}$, and $\mathbf{S}$ are given by

$$T_{a'a}^{n'n} \equiv \frac{\Omega}{(2\pi)^3} \int \mathrm{d}^3k\, g_{na'}^*(\mathbf{k})\, E_n(\mathbf{k})\, g_{na}(\mathbf{k})\, \delta_{n'n}\,, \tag{2.22}$$

$$S_{a'a}^{n'n} \equiv \frac{\Omega}{(2\pi)^3} \int \mathrm{d}^3k\, g_{na'}^*(\mathbf{k})\, g_{na}(\mathbf{k})\, \delta_{n'n}\,, \tag{2.23}$$

$$V_{a'a}^{n'n} \equiv \left(\frac{\Omega}{(2\pi)^3}\right)^2 \int \mathrm{d}^3k' \int \mathrm{d}^3k\, g_{n'a'}^*(\mathbf{k}')\, \langle \phi_{n',\mathbf{k}'}|\, U\, |\phi_{n,\mathbf{k}}\rangle\, g_{na}(\mathbf{k})\,. \tag{2.24}$$

Here Ω is the crystal volume. Note that the kinetic energy $\mathbf{T}$ and the overlap matrices $\mathbf{S}$ do not couple different bands. This is true even when the basis functions $g_{na}(\mathbf{k})$ of two different conduction band minima overlap.

For the case that involves only one conduction band but with several minima, we can drop the band index n in Eq. (2.17) and add a pocket index j to specify different inequivalent minima. Let $\mathbf{k} = \mathbf{k}_{j0} + \boldsymbol{\eta}$, where $\mathbf{k}_{j0}$ is the location of a particular minimum and $\boldsymbol{\eta}$ is the relative wave vector. Eq. (2.17) can be written as

$$(E_j(\boldsymbol{\eta}) - E)\, C_{j,\boldsymbol{\eta}} + \sum_{j',\boldsymbol{\eta}'} \langle \phi_{j,\boldsymbol{\eta}}|\, U\, |\phi_{j',\boldsymbol{\eta}'}\rangle\, C_{j',\boldsymbol{\eta}'} = 0\,. \tag{2.25}$$

Furthermore, we expand $C_{j,\boldsymbol{\eta}}$ in terms of the multiple-packet basis functions $g_{ja}(\boldsymbol{\eta})$ that are localized about these minima, where a denotes the orbital type, i.e., s, p_x, p_y or p_z and j refers to inequivalent conduction band minima. Then Eq. (2.25) becomes

$$\sum_{ja} [T_{a''a}^{j''j} + V_{a''a}^{j''j} - E S_{a''a}^{j''j}]\, A_{ja} = 0\,, \tag{2.26}$$

where the matrix elements are similar to those given in Eqs. (2.24) through (2.26) with the band index n replaced by the pocket index j.

2.4 Effective mass approximation

To the extent that $C_{j,\boldsymbol{\eta}}$ is important only for small $\boldsymbol{\eta}$, we can make a further approximation for the band energy and wave function. Consider Eq. (2.25) near the conduction band minimum. $E_j(\boldsymbol{\eta})$ can be expanded about the band minimum $E_j(\mathbf{k}_{j0})$ in a Taylor's series

$$E_j(\boldsymbol{\eta}) = E_j(\mathbf{k}_{j0}) + \frac{\hbar^2}{2}\left(\frac{\eta_x^2}{m_x} + \frac{\eta_y^2}{m_y} + \frac{\eta_z^2}{m_z}\right) + \dots\,, \tag{2.27}$$

where the first-order term is zero, because $\nabla_{\mathbf{k}} E_j(\mathbf{k}_{j0}) = 0$ at the minimum. The cross terms such as $\eta_x \eta_y$ do not appear, because x, y, and z are chosen to coincide with the principle axes. Setting

$$E = E_j(\mathbf{k}_{j0}) + \varepsilon\,, \tag{2.28}$$

and keeping up to the quadratic term, we reduce Eq. (2.25) to

$$\left(\frac{\hbar^2}{2}\left(\frac{\eta_x^2}{m_x}+\frac{\eta_y^2}{m_y}+\frac{\eta_z^2}{m_z}\right)-\varepsilon\right)C_{j,\boldsymbol{\eta}}+\sum_{j',\boldsymbol{\eta}'}\langle\phi_{j,\boldsymbol{\eta}}\,|U|\,\phi_{j',\boldsymbol{\eta}'}\rangle\,C_{j',\boldsymbol{\eta}'}=0\,. \qquad (2.29)$$

This is known as an Effective Mass Approximation (EMA). Under EMA, the wave function $\phi_{j,\boldsymbol{\eta}}(\mathbf{r})$ can be further approximated as

$$\phi_{j,\boldsymbol{\eta}}(\mathbf{r}) \cong e^{i\boldsymbol{\eta}\cdot\mathbf{r}}\,\phi_{j0}(\mathbf{r})\,, \qquad (2.30)$$

where $\phi_{j0}(\mathbf{r})$ is the wave function at the bottom of the conduction band of the j-th pocket. The potential matrix elements in (2.29) can then be evaluated as

$$\langle\phi_{j,\boldsymbol{\eta}}|\,U\,|\phi_{j',\boldsymbol{\eta}'}\rangle = \int d^3r\,e^{-i\boldsymbol{\eta}\cdot\mathbf{r}}\,\phi_{j0}^*(\mathbf{r})\,U(\mathbf{r})\,\phi_{j'0}(\mathbf{r})\,e^{i\boldsymbol{\eta}'\cdot\mathbf{r}}\,. \qquad (2.31)$$

If we define an envelope function $u(\mathbf{r})$ as

$$u_j(\mathbf{r}) \equiv \sum_{\boldsymbol{\eta}} e^{i\boldsymbol{\eta}\cdot\mathbf{r}}\,C_j(\boldsymbol{\eta}) \equiv \frac{\Omega}{(2\pi)^3}\int d^3\eta\,e^{i\boldsymbol{\eta}\cdot\mathbf{r}}\,C_j(\boldsymbol{\eta})\,, \qquad (2.32)$$

Eq. (2.29) becomes

$$\left(\frac{\hbar^2}{2}\left(\frac{\eta_x^2}{m_x}+\frac{\eta_y^2}{m_y}+\frac{\eta_z^2}{m_z}\right)-\varepsilon\right)C_j(\boldsymbol{\eta})+\sum_{j'}\int d^3r'\,e^{-i\boldsymbol{\eta}\cdot\mathbf{r}'}\,\phi_{j0}^*(\mathbf{r}')\,U(\mathbf{r}')\,\phi_{j'0}(\mathbf{r}')\,u_{j'}(\mathbf{r}')=0\,. $$

$$(2.33)$$

Multiplying Eq. (2.33) by $e^{i\boldsymbol{\eta}\cdot\mathbf{r}}$, summing over $\boldsymbol{\eta}$, and using definition of (2.32), we obtain

$$\left(\frac{-\hbar^2}{2}\left(\frac{1}{m_x}\frac{\partial^2}{\partial x^2}+\frac{1}{m_y}\frac{\partial^2}{\partial y^2}+\frac{1}{m_z}\frac{\partial^2}{\partial z^2}\right)-\varepsilon\right)u_j(\mathbf{r})+\sum_{j'}U_{jj'}^{\text{eff}}(\mathbf{r})\,u_{j'}(\mathbf{r})=0\,, \quad (2.34)$$

where the effective potential $U_{jj'}^{\text{eff}}$ is defined as

$$U_{jj'}^{\text{eff}}(\mathbf{r}) = \Omega\phi_{j0}^*(\mathbf{r})\,U(\mathbf{r})\,\phi_{j'0}(\mathbf{r})\,. \qquad (2.35)$$

Now, if the intervalley coupling is not important, i.e. if $\langle\phi_{j,\boldsymbol{\eta}}|\,U\,|\phi_{j',\boldsymbol{\eta}'}\rangle \approx 0$, for $j\neq j'$, then Eq. (2.34) becomes the one-valley approximation,

$$\left(\frac{-\hbar^2}{2}\left(\frac{1}{m_x}\frac{\partial^2}{\partial x^2}+\frac{1}{m_y}\frac{\partial^2}{\partial y^2}+\frac{1}{m_z}\frac{\partial^2}{\partial z^2}\right)-\varepsilon\right)u_j(\mathbf{r})+U_{jj}^{\text{eff}}\,u_j(\mathbf{r})=0\,. \qquad (2.36)$$

Eq. (2.36) reduces to the case of a simple hydrogenic model with anisotropic masses in Eq. (2.5), if $\phi_{j0}=1$ and $U(r)$ is set to be the screened Coulomb potential in Eq. (2.35). However, we shall show that all three factors — the form of U, the pocket wave function ϕ_{j0}, and the intervalley coupling, will have important effects on the donor states.

Now in EMA, the eigenvalue Eq. (2.26) becomes $(\mathbf{T}+\mathbf{V})\,\mathbf{a}=\varepsilon\mathbf{Sa}$, where the matrix elements in Eq. (2.22) to (2.24) can be conveniently evaluated in r-space. Explicitly, the matrix elements of $\mathbf{T}$, $\mathbf{V}$, and $\mathbf{S}$ are given by

$$T_{\alpha''\alpha}^{j''j} \equiv -\int d^3r\,F_{j''\alpha''}^*(\mathbf{r})\left[\sum_{\lambda}\frac{\hbar^2}{2m_\lambda}\frac{\partial^2}{\partial x_\lambda^2}\right]F_{j\alpha}(\mathbf{r})\,, \qquad (2.37)$$

$$S_{\alpha''\alpha}^{j''j} = \int d^3r\,F_{j''\alpha''}^*(\mathbf{r})\,F_{j\alpha}(\mathbf{r})\,, \qquad (2.38)$$

and

$$V_{a''a}^{j''j} = \Omega \int F_{j''a''}^*(\mathbf{r})\, \phi_{j''0}^*(\mathbf{r})\, U(\mathbf{r})\, \phi_{j0}(\mathbf{r})\, F_{ja}(\mathbf{r})\, \mathrm{d}^3 r$$
$$= \int F_{j''a''}^*(\mathbf{r})\, U_{j''j}^{\mathrm{eff}}(\mathbf{r})\, F_{ja}(\mathbf{r})\, \mathrm{d}^3 r \, . \tag{2.39}$$

Here $F_{ja}(\mathbf{r})$ is an envelope function related to the basis function $g_{ja}(\boldsymbol{\eta})$ by the transformation

$$F_{ja}(\mathbf{r}) = \frac{\Omega}{(2\pi)^3} \int \mathrm{d}^3 \eta\, \mathrm{e}^{i\boldsymbol{\eta}\cdot\mathbf{r}} g_{ja}(\boldsymbol{\eta}) \, , \tag{2.40}$$

and $U_{j''j}^{\mathrm{eff}}(\mathbf{r})$ is defined in Eq. (2.35). This formalism is equivalent to Luttinger-Kohn's [6, 7] effective mass approximation.

2.5 Model donor potential

The results from Sections 2.1 and 2.2 and the theory of Sections 2.3 and 2.4 show that the anisotropic hydrogenic model with the screened Coulomb potential is not suitable for a quantitative calculation of donor energy levels. In this section, we describe a model potential which, when incorporated into the theory of Section 2.4, will generate accurate ionization energies for shallow donors in Si.

The correction to the screened of Coulomb potential should take into account the impurity as well as the host atom. The fact that the calculated excited states which occupy the larger orbits are accurate to some extent within the hydrogenic model suggests that this so-called "central cell correction" ΔU should be short-ranged. Since the substitutional donor replaces one of the host atoms, ΔU should depend on the atomic sizes and the potential difference between the impurity and the host atoms. In the spirit of pseudo-potential, we consider the following simple donor potential U to incorporate these effects:

$$U(r) = U_0(r) + U_\mathrm{s}(r) \, , \tag{2.41}$$

where $U_0(r)$ is a spherical square-well potential of radius r_im and depth V_0, i.e.,

$$U_0(r) = -V_0 \quad \text{for} \quad r < r_\mathrm{im} \, ;$$
$$= 0 \quad \text{for} \quad r > r_\mathrm{im} \, ; \tag{2.42a}$$

and $U_\mathrm{s}(r)$ is given by

$$U_\mathrm{s}(r) = 0 \quad \text{for} \quad r < r_0$$
$$= \frac{-e}{\kappa r}\left(1 + (\kappa - 1)\,\mathrm{e}^{-r/r_1}\right) \quad \text{for} \quad r \geq r_0 \, . \tag{2.42b}$$

Here r_1 is a screening length, r_0 a truncated core radius, and κ the static dielectric constant of the host crystal. Note that $U_\mathrm{s}(r) \to -e^2/\kappa r$ at large r and behaves like $-e^2/r$ at small r.

From the experimental donor levels in Table 1, we observe that the ordering of the ground state donor levels of different impurities in Si is similar to that for the atomic first ionization energies. Therefore, the square-well depth V_0 in Eq. (2.42) is taken to be the difference of the first ionization energies between the impurity and the host atoms,

$$V_0 = E_{\mathrm{imp}}^{(1\mathrm{st})} - E_{\mathrm{host}}^{(1\mathrm{st})} \, . \tag{2.43}$$

Table 4

Atomic first ionization energies (in eV) and covalent radii (in Å) for C, Si, Ge, N, P, As, and Sb

	first ionization energy[a]	Pauling's covalent radii
C	19.814	0.77
Si	15.027	1.17
Ge	16.390	1.22
N	26.081	0.70
P	19.620	1.10
As	20.015	1.18
Sb	17.560	1.36

[a] Chen and Sher [26].

The square-well width r_{im} and the truncated radius r_0 are taken to be the Pauling covalent radii [34] of the impurity and host atoms, respectively,

$$r_{\text{im}} = r_{\text{imp}}^{\text{cov}} ; \qquad r_0 = r_{\text{host}}^{\text{cov}} . \tag{2.44}$$

Table 4 lists the first ionization energies of free atoms and the covalent radii of Si, Ge and several group V atoms. Finally, the screening length r_1 only depends on the host crystal and is adjusted to produce the best ionization energies of all donors. The best screening length for Si in our model, as deduced from the calculation to be described in the next section, is $r_1 = 1.35$ Å. The values of V_0, r_{im}, r_0, and r_1 used in the calculation are given in Table 5.

2.6 Tested results for shallow donors in Si

The theory of Section 2.4 and the model potential of Section 2.5 are applied to the well-known case of shallow donors in Si to test their validity. For the ground state energy, an efficient way is the variational method. We use a trial envelope function $F_{j\alpha}(\mathbf{r})$ of the form of an anisotropic Slater orbital for each pocket,

$$F_j(\mathbf{r}) = B \exp\left[-\sqrt{\beta_1 x_1^2 + \beta_2 x_2^2 + \beta_3 x_3^2} \right] , \tag{2.45}$$

where B is a normalization factor, x_1, x_2, and x_3 are coordinates in the three principal axes in each pocket and β_1, β_2, and β_3 are variational parameters. We drop the notation α since we only use one type of trial orbital for each pocket. This distorted Slater-like

Table 5

The truncated radius r_0 (in Å), the screening length r_1 (in Å), the width r_{im} (in Å) and the depth V_0 (in eV) of the potential well used in the central cell correction to impurity potentials of N, As, P, and Sb in Si

impurity	r_0	r_1	r_{im}	V_0
N	1.176	1.35	0.70	11.054
As	1.176	1.35	1.18	4.988
P	1.176	1.35	1.10	4.593
Sb	1.176	1.35	1.36	2.533

orbital set is preferred because of its exact solution to the 1s ground state of the hydrogen atom. For a Coulomb potential with three masses, this distorted trial wave function also yields the ground state energy with the same accuracy as the best results of Pfeiffer et al. [5]. While other orbitals such as Gaussians are also valid, we found that 7 to 8 of these orbitals are required in order to have the same accuracy as a single Slater orbital.

For a given impurity potential $U(r)$ and with the known host effective masses of the system, the variational calculation is carried out to achieve the lowest energy and the associated β_1, β_2, and β_3. The band-structure wave functions at all of the inequivalent conduction band minima are needed to perform the calculation. The wave function is obtained from the band-structure calculation in the form of plane wave expansion,

$$\phi_j(\mathbf{r}) = \frac{1}{\sqrt{\Omega}} \sum_{\mathbf{G}} a_{j\mathbf{G}} \exp\left[i(\mathbf{k}_{j0} + \mathbf{G}) \cdot \mathbf{r}\right], \tag{2.46}$$

where $a_{j\mathbf{G}}$ are the expansion coefficients and the $\mathbf{G}$'s are the crystal reciprocal lattice vectors. Note that in this expression the origin of $\mathbf{r}$ is at the impurity nucleus. Therefore, the potential matrix element in Eq. (2.39) becomes

$$V_{\alpha''\alpha}^{j''j} = \Omega \int F_{j''\alpha''}^*(\mathbf{r})\, \phi_{j''0}^*(\mathbf{r})\, U(\mathbf{r})\, \phi_{j0}(\mathbf{r})\, F_{j\alpha}(\mathbf{r})\, \mathrm{d}^3r$$

$$= \sum_{\mathbf{G}'} \sum_{\mathbf{G}} a_{j''\mathbf{G}'}^* a_{j\mathbf{G}} \int e^{i\mathbf{q}\cdot\mathbf{r}} F_{j''\alpha''}^*(\mathbf{r})\, U(\mathbf{r})\, F_{j\alpha}(\mathbf{r})\, \mathrm{d}^3r, \tag{2.47}$$

where $\mathbf{q} = \mathbf{k}_{j0} + \mathbf{G} - \mathbf{k}_{j''0} - \mathbf{G}'$. The computation of $V_{\alpha''\alpha}^{j''j}$ is the most time consuming part of the calculation.

The band structure Si used in this calculation was taken from Krishnamurthy et al. [8]. The conduction band minima in the first Brillouin zone are found at $\mathbf{k}_{j0} = (\pm 0.8, 0, 0)$, $(0, \pm 0.8, 0)$, $(0, 0, \pm 0.8)$ all in $(2\pi/a)$ with a being the lattice constant. The effective masses used are the experimental values, Madelung [9], $m_l = 0.91 m_0$ and $m_t = 0.19 m_0$. The static dielectric constant used is 11.9, Madelung [9]. The comparisons among different cases are shown in Table 6.

It should be mentioned that the lowest conduction band energy at X is doubly degenerate (see Fig. 1). Because of the small energy separation of the band near the minimum, we need to include both bands in the calculation. However, if we use the extended zone shown in Fig. 1b and treat bands labeled 1 and 2 as two separate bands, the $\mathbf{k}$ integration in Eq. (2.22) through (2.24) can be made throughout the whole $\mathbf{k}$ space without making large errors. A similar situation will be encountered in Section 4 when we consider the N donor in 6H SiC.

Table 6

Calculated ground state donor energy levels (in meV) for N, P, As, and Sb in Si in several models: (a) anisotropic hydrogenic model Eq. (2.6), (b) anisotropic hydrogenic model with the potential given by Eq. (2.41), (c) EMA with one valley wave function, Eq. (2.36), (d) full EMA model, and (e) experimental values

impurity	(a)	(b)	(c)	(d)	(e)
N	−29	−41	−41	−48	−45
P	−29	−42	−41	−47	−46
As	−29	−44	−42	−54	−53
Sb	−29	−42	−41	−44	−43

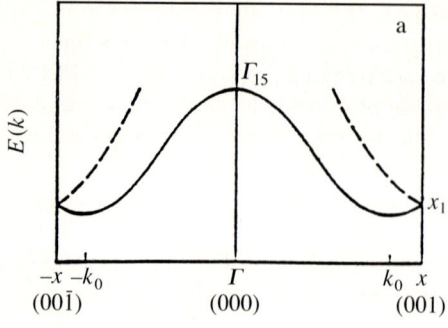

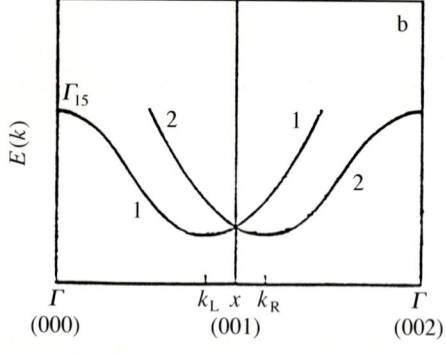

Fig. 1. a) Si conduction bands in the Brillouin zone. b) The same bands in an extended zone centered at X

Table 6 shows the ground state energies of donor levels in Si calculated in several intermediate steps and the final results. The potential models and the approximation made in these steps are:

a) the hydrogenic model with anisotropic masses and a screened Coulomb potential; a one-pocket calculation;

b) the central cell potential given by Eq. (2.41) and (2.42) with anisotropic masses; also a one-pocket calculation;

c) the effective mass approximation with both the central cell potential and the pocket wave function included for one valley only, i.e., Eq. (2.36);

d) the full model, i.e., extend c) to all inequivalent valleys including the intervalley interactions as given by Eq. (2.34).

Several observations can be made from Table 6: First, the central cell potential correction in b) introduces the impurity species dependence and lowers the donor levels quite significantly from those of the hydrogenic model. The single-valley wave function in c) has little effect on the levels. However, as will be shown later, the effect of the wave functions is important in SiC, particularly in their influence on the site dependence of donor levels in the 4H polytype. The intervalley coupling in d) further lowers the donor ground state energy. Note that in this model there is only one adjustable parameter in this model, the screening length r_1. The good agreement between columns (d) and (e) give a support of the applicability of this model potential. This approach will be extended to SiC in Section 4, after the discussion of SiC band structures in the next section.

3. Conduction Band Edge Structures of SiC

3.1 A brief review

In this section, we describe the general features of electronic band structures $\varepsilon_n(\mathbf{k})$ of 3C 2H, 4H, and 6H SiC, with an emphasis on the conduction band edge structures. The 3C SiC has the zincblende structure, 2H SiC has the wurtzite structure, while the 4H and 6H SiC have hexagonal structures. The discussion of the crystal structures can be found in other chapters of this volume. Table 7 lists the lattice parameters used in the band calculation. A plot of $\varepsilon_n(\mathbf{k})$ along several symmetry directions of the first Brillouin zone (BZ) is shown in Figs. 2a to d for 3C, 2H, 4H, and 6H SiC, respectively. Fig. 3 shows how the 3C BZ maps onto the wurtzite BZ.

Table 7

Lattice constants of four polytypes of SiC in the ascending order of hexagonal percentage (hp). The value of n is 1, 2, and 3 for 2H, 4H, and 6H, respectively

polytype	a_{zb} (Å)	a (Å)	c (Å)	$2c/na$	hp
3C	4.3585	3.0819	–	–	0
6H	–	3.0805	15.1145	1.6355	1/3
4H	–	3.0799	10.0814	1.6367	1/2
2H	–	3.0777	5.0486	1.6404	1

The most visible changes in the band structures between different polytypes in Fig. 2 are the "band folding" effects. The comparison between 3C and other polytypes is not simple because different BZs are involved. However, for 2H, 4H, and 6H the bands look similar along the same direction. The multiplication of bands can be seen when comparing Fig. 2c and d with b. Below we briefly comment on band gaps, the locations of conduction band minima, and the electron effective masses.

3.1.1 Band gaps

All four polytypes of SiC are known to have indirect gaps. The band gaps of all four polytypes have been deduced experimentally. Their values are 2.417 eV for 3C, Humphreys et al. [10]; 3.33 eV for 2H, Patrick et al. [11]; 3.265 eV for 4H and 3.023 eV for 6H, Choyke et al. [12].

3.1.2 Location of conduction band minimum

The conduction band minimum in 3C was found to be located at point X in f.c.c. BZ, Choyke and Patrick [13]. For 2H, it was determined to be at the point K on the edge of wurtzite BZ, Patrick et al. [11], instead of the mapped point of X in the hexagonal BZ, which lies between M–L (see Fig. 3). While the conduction band edge of 6H is believed to be on the M–L symmetry line excluding the end points, Patrick [14], that of 4H is still in debate. An experimental argument by Patrick et al. [15] pointed out that the conduction band minimum of 4H should be at the general F point in the large zone. Lambrecht and Segall [16], on the other hand, used their band-structure calculation results and a recent phonon calculation by Hofmann et al. [17] to show that the conduction band minimum of 4H is at the symmetry point M. This assignment is consistent with the conclusion from Choyke et al. [18] based on the analysis of the phonon replicas.

3.1.3 Electron effective masses

The effective masses of 3C were determined by Dean et al. [19] by analyzing the electron transitions in photoluminescence. The values obtained by Kaplan et al. [20] by far infrared cyclotron resonance agree with these earlier values, and are more precise. These masses have two-fold anisotropy with the corresponding values of $m_\parallel = 0.667 m_0$ and $m_\perp = 0.247 m_0$, with m_0 being free electron mass. Here, the subscript ($\parallel$) refers to the direction of Γ–X, and ($\perp$) for X–W and X–K directions. The effective masses of 2H SiC have not yet been determined by experiment. In other α-polytypes of SiC, the reported masses are still ambiguous. For 4H, Götz et al. [21] suggested values $m_\parallel = 0.22 m_0$ and

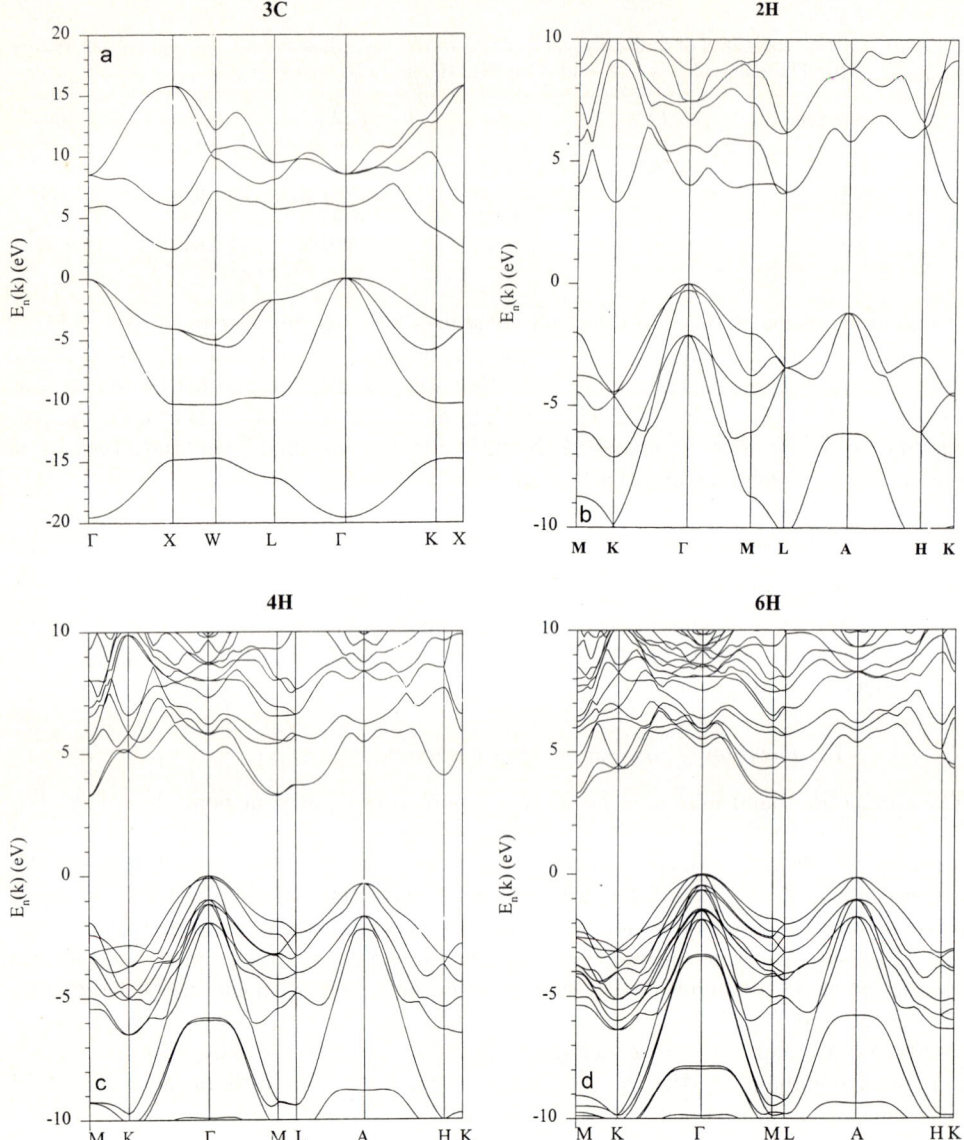

Fig. 2. Band structures of a) 3C SiC, b) 2H SiC, c) 4H SiC, and d) 6H SiC

$m_\perp = 0.18m_0$, which were deduced from the effective mass approximation (EMA) fitted to the excited states of the nitrogen donor. A set of larger masses for 4H SiC, $m_\parallel = 0.29m_0$ and $m_\perp = 0.42m_0$, were determined from optically detected cyclotron resonance by Son et al. [22]. For the conduction band effective masses in 6H SiC, a study of nitrogen donors in nitrogen-doped 6H by Suttrop et al. [3] suggested $m_\parallel = 0.34m_0$ and $m_\perp = 0.24m_0$, while a recent cyclotron resonance experiment, Son et al., [23] obtained $m_\parallel = 2.00m_0$ and $m_\perp = 0.42m_0$. The notation ($\parallel$) for α-polytypes refers to the direction

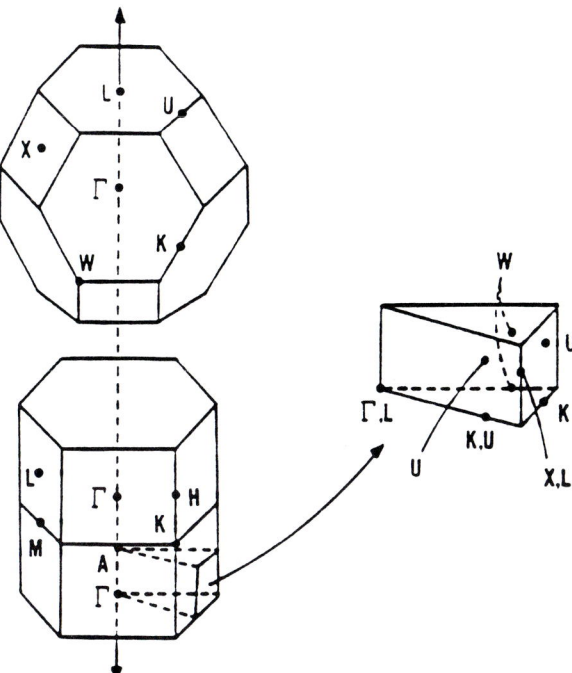

Fig. 3. Comparison of a) f.c.c. and b) wurtzite BZ in the same orientation; c) mapping of f.c.c. symmetry points onto the section of wurtzite BZ

of the c-axis and the ($\perp$) for the perpendicular plane. It should be noted that in all band-structure calculations, e.g. Srichaikul et al. [24], Lambrecht and Segall [16], and the present band structure calculation, the effective masses of 4H and 6H at the conduction band minimum are three-fold anisotropic. This full anisotropy of the masess in 4H SiC has been measured recently by Volm et al. [25]. These results will be discussed further in Section 4 in connection with the calculation of nitrogen ground state donor levels.

3.2 A hybrid pseudo-potential and tight-binding (HPT) approach

The band-structure method that we are interested in is one that can incorporate the known band-structure quantities, such as the one described above, in the calculation. A hybrid pseudo-potential tight-binding (HPT) method will be employed to achieve this goal. HPT encompasses the merits of both the empirical pseudo-potential method (EPM) and the tight-binding (TB) method. It has been used by Chen and Sher [26] to obtain accurate band structures for all common III–V and II–VI zincblende semiconductors and alloys. The application of HPT to SiC requires an extension from the zincblende to the wurtzite structure for 2H, 4H, and 6H SiC.

The key idea of HPT is to use the empirical pseudo-potentials to construct a long-range universal tight-binding Hamiltonian matrix $\mathbf{H}_0$, which generates a set of initial band structures. Then a short-range tight-binding perturbation Hamiltonian matrix $\mathbf{H}_1$ containing a handful of adjustable interaction parameters is added to $\mathbf{H}_0$ in order to produce the desired band gaps, location of the band gaps, and the effective masses etc.

Explicitly, an empirical pseudo-potential can be written as

$$U_0(\mathbf{r}) = \sum_{\mathbf{G}} e^{i\mathbf{G}\cdot\mathbf{r}} V(\mathbf{G}), \tag{3.1}$$

where $\mathbf{G}$ is a reciprocal lattice vector. $V(\mathbf{G})$ can then be written as the sum of the Fourier transforms of the atomic potential $V_j(\mathbf{G})$, the form factors, as

$$V(\mathbf{G}) = \sum_{j} e^{-i\mathbf{G}\cdot\boldsymbol{\tau}_j} V_j(\mathbf{G}), \tag{3.2}$$

where j runs over all the atoms in a unit cell and $\boldsymbol{\tau}_j$'s are the atomic positions relative to the lattice point. If the atomic pseudo-potentials are weak and are assumed to be spherical symmetric, then $V_j(\mathbf{G})$ is a function of only the magnitude of $\mathbf{G}$, and only several values of $V(G)$ at the first G's are needed in the band calculation.

To calculate $\mathbf{H}_0$, a set of local orbitals, denoted by the ket $|lj\alpha\rangle$, are chosen as the basis functions. Here l is a lattice vector, j again specifies the atoms in a unit cell, and α denotes the type of orbitals, i.e. s, p_x, p_y, and p_z for the minimum basis set used here. The local orbitals used in this calculation are Gaussian orbitals of the type $\phi_{lm}(r) = r^l \exp(-\lambda r^2)\, Y_{lm}$, where Y_{lm} is a real spherical harmonic. From these local orbitals, a set of Bloch basis of a wave vector $\mathbf{k}$ can be constructed,

$$|\mathbf{k}j\alpha\rangle = \frac{1}{\sqrt{N}} \sum_{l} e^{i\mathbf{k}\cdot(\mathbf{l}+\boldsymbol{\tau}_j)} |lj\alpha\rangle. \tag{3.3}$$

Given the pseudo-potential form factors, the matrix elements of $H_0 = p^2(2m) + U_0$ between these Bloch basis functions. i.e. $\langle \mathbf{k}j'\alpha'|\, H_0\, |\mathbf{k}j\alpha\rangle$, can be computed explicitly. Similarly, the overlap matrix $\mathbf{S}$ is defined as $\mathbf{S}_{j'\alpha',j\alpha}(\mathbf{k}) = \langle \mathbf{k}j'\alpha'\,|\,\mathbf{k}j\alpha\rangle$. The pertubation Hamiltonian $\mathbf{H}_1$ matrix is assumed to be a TB form with interactions up to second neighbors. Once the parameters of $\mathbf{H}_1$ are known, the band-structure energies are determined from the following eigenvalue equation:

$$(\mathbf{H}_0 + \mathbf{H}_1 - E\mathbf{S})\,\mathbf{b} = 0. \tag{3.4}$$

3.3 Pseudo-potential form factors and tight-binding parameters

For a zincblende semiconductor with a lattice constant a, the pseudo-potential form factors are required at only several values of the reciprocal lattice vectors with $G^2 = 3$, 4, 8, 11, 12, and 16 in units of $(2\pi/a)^2$. In many cases, $G^2 = 12$ and 16 are not needed. For wurtzite semiconductors, however, the values at several more G's are needed. In units of $(2\pi/a)^2$, where a is the cubic equivalent lattice constant and is related to the actual hexagonal lattice constant a_{wz} by the ideal condition $a = \sqrt{2}\,a_{\mathrm{wz}}$, the required G^2 are $\frac{3}{4}$, $2\frac{2}{3}$, 3, $3\frac{5}{12}$, $5\frac{2}{3}$, $6\frac{3}{4}$, 8, $8\frac{3}{4}$, $9\frac{5}{12}$, $10\frac{2}{3}$, 11, $11\frac{5}{12}$, 12, $13\frac{2}{3}$, $14\frac{2}{3}$. The number of G's required is more in 4H and even more in 6H polytypes. Hemstreet and Fong [27] obtained a set of empirical pseudo-potential form factors for 3C and another set for 2H. Since these two sets were determind separately for two different polytypes, there is no simple correlation between them. However, it is desirable to have a continuous $V_j(\mathbf{q})$ that can be applied to all polytypes. To this end we tried the following form for $V_j(q)$:

$$V_j(q) = \frac{(a + bq^2 + cq^4 + dq^6)\, e^{-aq^2}}{1 + e^{\gamma(q^2 - q_c^2)}}. \tag{3.5}$$

Table 8

Parameters for the atomic pseudo-potential form factors (in Ry) for silicon and diamond in silicon carbide according to Eq. (3.5), where q is in units of $(2\pi/a)$

parameter	Si	C
q_c^2	13.500	14.500
γ	3.000	3.000
α	0.050	0.445
a	-1.20	-0.753
b	0.233	0.373
c	0.005	-0.317
d	-0.001	0.033

The parameters a, b, c, d, α, and γ are treated as adjustable parameters and q_c is a cutoff such that $V(q) = 0$ for all $q > q_c$. These parameters are determined from a least square fit of $V(q)$ to Hemstreet and Fong's form factors. The fitted parameters are given in Table 8, and the plots of $V_j(q)$ for Si and C are shown in Fig. 4. Since 3C SiC has been studied more extensively, we require that the functional form produces the same band structures for 3C as those from a converged plane-wave calculation using the same pseudo-potential. By doing so, the quality of the 2H band structure is degraded a little. However, all we need to have is a $V(q)$ which produces a set of reasonable bands initially. The final bands will be refined later through the use of $\mathbf{H}_1$.

Given the pseudo-potentials, we still need to define the local orbitals and $\mathbf{H}_1$ in order to calculate the band structures. As mentioned, we used a minimum basis set of four Gaussian orbitals per atom. The Gaussian exponential constants λ_C and λ_{Si} for the C and Si atoms are optimized so that the band structures of the 3C polytype calculated from $\mathbf{H}_0$ agree well with those obtained from a converged plane wave calculation using the same pseudo-potentials. The optimized values for the two λ's in units of $(2\pi/a)^2$ are $\lambda_C = 0.36$ and $\lambda_{Si} = 0.34$, respectively. These values will be used for all four SiC polytypes throughout this chapter.

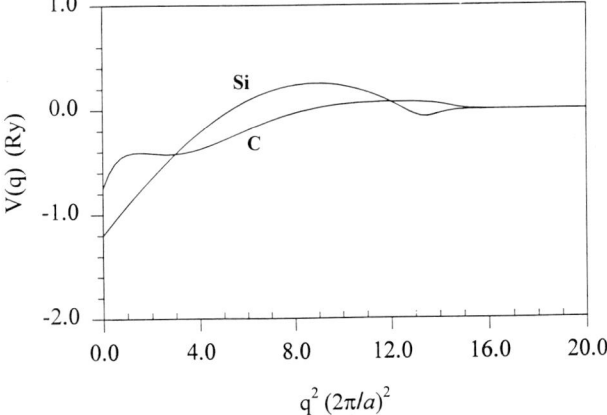

Fig. 4. Plot of $V(q) = (a + bq^2 + cq^4 + dq^6)\, e^{-\alpha q^2} / (1 + e^{\gamma(q^2 - q_c^2)})$ for silicon and diamond in SiC as a function of q^2 in units of $(2\pi/a)^2$ and $V(q)$ (in Ry)

Table 9

The tight-binding parameters (in eV) of $\mathbf{H}_1$ for SiC

TB parameter	values	TB parameter	values
ε_s^A	-0.0500	E_{ss}^A	0.1034
ε_p^A	-0.0914	E_{sp}^A	-0.0796
ε_s^C	0.6790	$E_{pp\sigma}^A$	0.1152
ε_p^C	0.3454	$E_{pp\pi}^A$	0.1350
V_{ss}	-0.0995	E_{ss}^C	-0.2200
$V_{sp\sigma}^{CA}$	0.0342	E_{sp}^C	0.1859
$V_{ps\sigma}^{CA}$	-0.2359	$E_{pp\sigma}^C$	0.0000
$V_{pp\sigma}$	0.1106	$E_{pp\pi}^C$	-0.0746
$V_{pp\pi}^{AC}$	-0.0742		

There are 19 TB parameters in $\mathbf{H}_1$, including the s and p term values for both Si and C and 5 first neighbor matrix elements $V_{aa'}$, and 10 second neighbor matrix elements $E_{aa'}$.

These parameters are determined by adjusting them in a heuristic manner to obtain a good fit for the band gaps, band edge locations, and effective masses. The final values that are used in our band calculations are listed in Table 9.

3.4 Calculated band structures

3.4.1 3C

The calculated band structure of 3C SiC is plotted along several directions of the f.c.c. BZ in Fig. 2a. The band gap is fitted exactly to the experimental value of 2.417 eV. The location of the band gap is at X in agreement with experiment. The onset of the direct transition in this calculation is at the Γ point. The top of valence band at Γ is triply degenerate since the calculation does not include the spin–orbit interaction. The aniso-tropy around the band gap is two-fold. The calculated effective masses are $0.678 m_0$ for longitudinal mass (along X–Γ) and $0.236 m_0$ for the transverse mass in comparison to experimental values of $0.667 m_0$ and $0.247 m_0$, respectively.

3.4.2 2H

A plot of the band structure of 2H along several BZ directions is shown in Fig. 2b. The top of valence band at the zone center Γ is doubly degenerate with a single state imme-diately below it. The principal band gap is fitted exactly to the experimental value of 3.33 eV at point K on the h.c.p. Brillouin zone edge.

There is a second minimum located between M–L symmetry line with an energy around 3.5 eV. One interesting point observed during the fitting procedure is that the conduction band minimum at K (K_{2c}) is a pure p-state in our sp model. In general, the effective masses at K have a two-fold anisotropy, with one mass $m_\parallel$ along the c-direction (the z-axis) and another $m_\perp$ in the xy-plane. However, the masses calculated from the present band structure are very close, with the values $m_\parallel = 0.311 m_0$ and $m_\perp = 0.310 m_0$. Experimental effective masses are still not available. Lambrecht and Segall [16] have obtained masses with a greater anisotropy, with $m_\parallel = 0.27 m_0$ (z-direction) and $m_\perp = 0.45 m_0$, based on a self-consistent local density functional band structure (LDA).

3.4.3 4H

A plot of the 4H band structure is shown in Fig. 2c following the same direction and notation as 2H. The general features of the 4H bands are similar to 2H with the width of the BZ in z-direction shrunk by half. The top of valence band is still at the zone center and has the characteristic 2−1 splitting similar to 2H. The conduction band minimum is located at 99% along the Γ−M direction with a calculated value of 3.27 eV, in excellent agreement with 3.265 eV from experiment. Analysis around the band edge shows that it is three-fold anisotropic anisotropy with a rather flat band along Γ−M. The calculated effective masses are $m_1 = 1.20 m_0$ (M–Γ direction), $m_2 = 0.19 m_0$ (M–K direction), and $m_3 = 0.33 m_0$ (M–L direction). These masses will be compared with the experimental and other theoretical values in Section 4.3, where the results for the N donor levels in 4H SiC are discussed.

3.4.4 6H

The calculated band structure of 6H is plotted in Fig. 2d. The conduction band minimum is calculated to be about 73% along the M–L direction and has the value of 3.07 eV. A more detailed study of the conduction band edges shows that the masses have three-fold anisotropy. The three calculated masses are $m_1 = 1.00 m_0$ (M–Γ direction), $m_2 = 0.19 m_0$ (M–K direction), and $m_3 = 0.54 m_0$ (M–L direction). These masses will be compared with other available values in Section 4.4, where they are used in the N donor calculation.

4. Calculation of Nitrogen Donor Levels in SiC

In this section, the donor theory and potential model of Section 2 and the band structure results of Section 3 are applied to the calculation of N donor levels in four polytypes of SiC. Since the conduction band edge structures are the most relevant band quantities for such studies, Fig. 5 shows the expanded plots of our band structures near the conduction band minima. Table 10 summarizes the important conduction band edge quantities obtained from our band calculation. These structures will be discussed along with the donor level calculation for each polytype.

Table 10

Calculated band gaps and effective masses for 3C, 6H, 4H, and 2H SiC and the location of the band minimum E_c. Energies are given in eV. Effective masses are in units of m_0

quantity	3C	6H	4H	2H
band gap	2.417	3.07	3.27	3.33
E_c location	X	73% along M–L	99% along Γ–M	K
expt. band gap	2.417	3.023	3.265	3.33
m_1^*	0.236 (XW)	1.00 (MΓ)	1.20 (MΓ)	0.310 (KΓ)
m_2^*	0.236 (XW)	0.19 (MK)	0.19 (MK)	0.310 (KΓ)
m_3^*	0.678 (XΓ)	0.54 (ML)	0.33 (ML)	0.311 (KH)

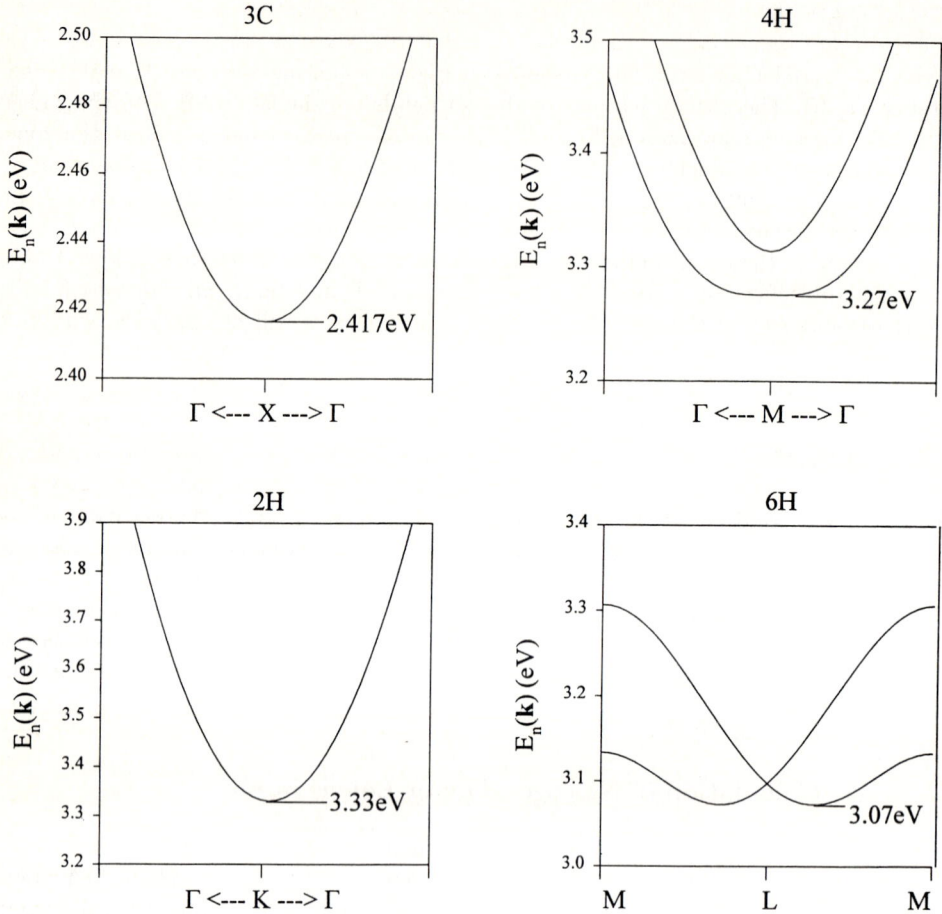

Fig. 5. Enlarged conduction band edge structures for four polytypes of SiC

4.1 N donor in 3C SiC

There are three inequivalent conduction band pockets with the minima located at the symmetry points X in the Brillouin zone, i.e. $(1, 0, 0)$ $2\pi/a$, $(0, 1, 0)$ $2\pi/a$, and $(0, 0, 1)$ $2\pi/a$, respectively. The effective masses at each minimum are two-fold anisotropic. The longitudinal direction is from the center of BZ, Γ, to X, and the transverse masses are along the other two perpendicular directions.

Table 11

The parameters for the N donor impurity potential in SiC given in Eq. (2.41) and (2.42). (r's in Å and V_0 in eV)

r_{im}	r_0	r_1	V_0	κ (3C)	κ (6H)
0.70	0.77	0.86	6.267	9.72	9.85

Table 12

Calculated ground state donor energy levels (in meV) for N donor in 3C SiC in several models: (a) anisotropic hydrogenic model, Eq. (2.6), (b) anisotrope hydrogenic model with the potential given by Eq. (2.41), (c) EMA with one-valley wave function, Eq. (2.36), (d) full EMA model, and (e) experimental values

masses	(a)	(b)	(c)	(d)	(e)
$m^*_\perp = 0.247$ $m^*_\parallel = 0.667$	−48	−57	−57	−57	−57
$m^*_\perp = 0.236$ $m^*_\parallel = 0.678$	−47	−55	−55	−55	−57

For the model potential defined in Eqs. (2.41) and (2.42), Table 11 lists all the parameters needed. The N impurities are assumed to reside on the C sites. The values of r_0 and r_{im} are the covalent radii of C and N with similar values of 0.77 and 0.70 Å, respectively. The well depth $V_0 = 6.267$ eV is the difference in first ionization energies between N and C. The static dielectric constant used for 3C SiC is 9.72, Patrick and Choyke [28]. Finally the screening length $r_1 = 0.86$ Å is obtained so that the model with the experimental masses produces the correct donor level.

Table 12 shows the N donor ground state energies ε_D calculated in several intermediate steps approximation leading to the final results. These steps and the calculational procedure are the same as those described in Section 2.6. Two sets of masses are used in the calculation. The first set, $m_\parallel = 0.667$ (along the c-axis) and $m_\perp = 0.247$, are those determined from the cyclotron resonance, Kaplan et al. [20]. The second set are our calculated masses, $m_\parallel = 0.678 m_0$ and $m_\perp = 0.236 m_0$. These results show that the only important correction in this polytype is the central cell potential.

4.2 N donor levels in 2H SiC

Experimentally, this polytype of SiC is probably the least understood and least investigated for its electronic properties. The band gap has been measured to be 3.33 eV at low temperature and determined to be at the zone boundary K. The calculated bands have incorporated these two aspects of the band structure. The calculated masses at the conduction band minimum K have two-fold anisotropy, with $m_\parallel = 0.310 m_0$ (along the c-axis), and $m_\perp = 0.311 m_0$. Experimental results for the effective masses and the donor

Table 13

Calculated ground state donor energy levels (in meV) for N donor in 2H SiC in several models: (a) hydrogenic model, Eq. (2.6), (b) hydrogenic model with central potential, Eq. (2.41), (c) EMA with one-valley wave function and central cell potential, Eq. (2.36), (d) full model, and (e) experimental values

masses	(a)	(b)	(c)	(d)	(e)
$m^*_\perp = 0.310$ $m^*_\parallel = 0.311$	−43	−50	−50	−50	−
$m^*_\perp = 0.45$ $m^*_\parallel = 0.27$	−52	−65	−65	−65	−

levels are not yet available. Using the conduction band edge wave function and taking the static dielectric constant to be 9.85, which is averaged from $\kappa_\perp = 9.66$ and $\kappa_\parallel = 10.03$ measured from 6H-SiC, Patrick and Choyke [28], we calculated the donor levels for two sets of theoretical effective masses: the first set from our calculation and the second set by Lambrecht and Segall [16]. The results are shown in Table 13. Note that the inclusion of the central cell potential correction has already produced the final donor energies. The pocket wave function and the intervalley coupling have no effect on the donor level at all in this polytype.

4.3 N donor levels in 4H SiC

As shown in Fig. 5, our calculated band structure for 4H has its conduction band minimum located near the symmetry point M of the BZ. Its actual location is at 99% from Γ to M. Thus, there are six inequivalent conduction band pockets about the minima. The masses along three principal axes are found to be different, with values given by $m_1 = 1.2m_0$ (M–Γ direction), $m_2 = 0.19m_0$ (M–K direction), and $m_3 = 0.33m_0$ (M–L direction). Due to the closeness of a pair of pockets around each inequivalent M point, we treat the case of 4H as if the conduction band minimum is located at the M point, which reduces the number of pockets to three. With the wave functions at the band edges, we calculated the ground state donor level ε_D using four sets of effective masses: the first set from our calculation, the second set are the calculated values of Lambrecht and Segall [16], the third and the fourth sets are the experimental sets by Son et al. [22] and Götz et al. [21], respectively. The results are shown in Table 14. Several qualitative observations can be made: 1. The first three sets of masses produce significantly different ε_D for two inequivalent impurity sites h and k, although the differences are not as large as that suggested by experiment. 2. The average magnitude of the two ε_D for the first three sets is about the same, because the average isotropic masses $m^* = (m_1 m_2 m_3)^{1/3}$ for the three sets are about the same. 3. The last set produces donor levels that are too

Table 14

Calculated ground state donor levels for N in 4H SiC at h and k site for four sets of effective masses in several models: (a) hydrogenic model, Eq. (2.6), (b) hydrogenic model with central cell potential, Eq. (2.41), (c) EMA with one-valley wave function and central cell potential, Eq. (2.36), (d) full model, and (e) experimental values

	(a)	(b)	(c)	(d)	(e)
$m_1 = 1.20$ h site	-55	-71	-65	-71	-52
$m_2 = 0.19$ k site			-80	-92	-92
$m_3 = 0.33$					
$m_1 = 0.58$ h site	-51	-63	-60	-65	
$m_2 = 0.28$ k site			-68	-78	
$m_3 = 0.31$					
$m_1 = 0.58$ h site	-54	-68	-64	-70	
$m_2 = 0.31$ k site			-77	-89	
$m_3 = 0.33$					
$m_1 = 0.18$ h site	-27	-28	-29	-31	
$m_2 = 0.18$ k site			-29	-29	
$m_3 = 0.22$					

shallow and essentially no site dependence as compared with experiment. We note that the inclusion of the pocket wave function produces the site dependence of ε_D even with a single valley. This site dependence is further enhanced by the intervally interaction.

4.4 N donor levels in 6H SiC

The calculated band gap for 6H is 3.07 eV, which is slightly larger than the experimental value 3.023 eV at low temperature. The conduction bands are located along the line M–L and at 70% from M to L. By including the second conduction band in the extended zone as shown in Fig. 5, there are a total of six inequivalent pockets for this case.

Table 15 shows the results of the calculated ε_D of N at three inequivalent sites using five sets of available masses: the first set is from our calculation, the second set uses the calculated values of Lambrecht and Segal [16], the third and fourth sets are the experimental sets by Son et al. [23] and Götz et al. [21] respectively, and the final set uses adjusted masses. The calculated donor levels from the last set compare reasonably well with experiment both in magnitudes and site dependence. With only slightly larger effective masses in the first three sets, the levels become unphysically deep. This sudden change indicates the breakdown of the effective-mass approximation used here, because the trial wave functions in the k-space start to spread into the whole Brillouin zone. In contrast, the fourth set has the smallest masses and yields levels that are too shallow as compared to experiment. Clearly, this table shows that intervally interaction has a very important effect on ε_D both on magnitude and on the site dependence in 6H SiC. It also indicates that the donor levels in this polytype are too deep to be treated with the effective-mass approximation.

Table 15

Calculated ground state donor levels for N in 64H SiC at h, k_1, and k_2, sites for five sets of effective masses in several models: (a) hydrogenic model, Eq. (2.6), (b) hydrogenic model with central cell pontential, Eq. (2.41), (c) EMA with one-valley wave function and central cell potential, Eq. (2.36), (d) full model, and (e) experimental values

	(a)	(b)	(c)	(d)	(e)
$m_1 = 1.00$ h site	−62	−87	− 74	− 190	− 81
$m_2 = 0.19$ k_1 site			− 85	− 797	−138
$m_3 = 0.54$ k_2 site			−110	−1089	−142
$m_1 = 0.77$ h site	−83	−169	−110		
$m_2 = 0.24$ k_1 site			−159		
$m_3 = 1.42$ k_2 site			−344		
$m_1 = 0.42$ h site	−92	−217	−125		
$m_2 = 0.42$ k_1 site			−201		
$m_3 = 2.0$ k_2 site			−460		
$m_1 = 0.24$ h site	−38	−42	− 42	− 55	
$m_2 = 0.24$ k_1 site			− 43	− 78	
$m_3 = 0.34$ k_2 site			− 45	− 88	
$m_1 = 0.22$ h site	−43	−50	− 48	− 79	
$m_2 = 0.17$ k_1 site			− 50	− 140	
$m_3 = 0.96$ k_2 site			− 54	− 168	

5. Summary and Discussion

One central theme of this chapter is to show, from the theoretical point of view, the close relationship between donor energy levels and the host band structures. Given an impurity potential U, the donor energies and wave functions can be calculated rigorously from Eq. (2.17) and (2.21) using many bands. For shallow donors, only the lowest conduction bands are needed and the effective-mass approximation (EMA) in Eq. (2.34) should be adequate, which has been used throughout this work.

An accurate derivation of donor impurity potentials has yet to be established from first principles. The model potential U in Eqs. (2.41) and (2.42) is shown to be a reasonable empirical model, as evidenced from the excellent ground state energies ε_D in Table 6 calculated for several shallow donors in Si. When applied to SiC, the screening length is adjusted to yield the correct ε_D in 3C SiC, and the same model potential is used for other polytypes.

Within the effective-mass approximation (EMA), the band-structure quantities needed for the donor level calculation are the locations, the effective masses, and the wave functions of the conduction band minima E_c. A hybrid pseudo-potential and tight-binding (HPT), Chen and Sher [26], band scheme is used to incorporate experimentally determined band quantities in the band structure calculation. The calculated conduction band edge structures and the model potential are then used in the calculation of ε_D of N donor in four polytypes of SiC. The results are summarized below.

For the cubic SiC, the combined results of the calculated band edge structures of Fig. 5, the band quantities in Table 10, and the associated N donor level shown in Table 12 indicate that we have a good model for treating the conduction band edges and the N donor energy in this polytype.

For the 2H SiC, the calculated conduction band minimum is at K, as shown in Figs. 2 and 5. The location of E_c and the band gap are the same as those identified experimentally, Patrick et al. [29].

These results cannot be deduced from a simple mapping of the 3C band edges from the f.c.c. the h.c.p. Brillouin zone. The wave function at K is a pure p state. There are no experimental results for the N donor level to compare with. However, with similar effective masses as those in 3C and well separated conduction band pockets, the predicted donor level should be accurate. When the epitaxial growth of 2H SiC starts to produce good and thick single crystal films, experimental measurement of the effective masses and the N donor levels should be made to compare with the calculated results.

Our 4H band structure (Figs. 2c and 5) shows that the minimum conduction band edges lies inside the Brillouin zone along the $\Gamma-M$ direction and near the M point, instead of the point F inside the large zone, Patrick et al. [15]. As a result, we have six minimum pockets in the first BZ instead of twelve. If E_c is exactly at the M point as was reported by Backes et al. [30] and Lambrecht and Segall [16], there will only three conduction band minima. Because of the closeness of the two conduction band minima along the $\Gamma-M$ line in BZ in our band structure, our donor calculation assumed M to be the minimum and treated the two pockets as one. The first three sets effective masses in Table 14 yield qualitatively correct ε_D for N donors in 4H as compared with experiment. All these three sets of effective masses produced a marked site dependence, which can be attributed to the conduction band wave function and intervalley coupling (see Table 14). However, one should not use the agreement between the calculated and experimental ε_D as a measure of the accuracy of the effective masses at this point, because two important effects have not been treated properly in our donor energy calculation. The first is the nearly degenerate band edge problem shown in Fig. 5 just mentioned, which was approximated as one minimum in the donor energy calculation. Secondly, there is a conduction band which is only 40 meV above E_c (see Fig. 5). This band should be included in an improved donor calculation for this polytype.

Finally for 6H SiC, our calculated band edge is along the M–L symmetry line (see Fig. 5) as was predicted by Patrick [14] and the recent analysis, Lambrecht and Segall [16]. For a set of moderate masses (the last set in Table 15), our calculation produced very nice site dependence of ε_D for N donors. The intervalley coupling is seen to be the dominant mechanism for producing the site dependence in this polytype. However, as clearly seen in Table 15, the EMA model breaks down for the cases with large values of effective masses. Along M–L in 6H, the band dispersion is very flat. The EMA does not accurately describe the band structure along this line.

In conclusion, the chapter has started the ground work for the calculation of the conduction band edge structures and donor energies in SiC. We have obtained a comprehensive set of conduction band edge structures, developed an applicable theory and model potential, and achieved a semi-quantitative description of the N donor ground state energies for the four polytypes of SiC studied here. However, there is still plenty of room for improvement in this area, which will require refined experiments and calculations plus more correlation between the two. With the flexibility of the present band scheme, we can gradually fine tune the band structure to incorporate more and better band information. For the case where EMA is not valid, the calculation of donor levels should use multiple bands and perform the Brillouin zone integration using the actual band structures as required in Eq. (2.17). The model impurity potential model should be further tested by extending the donor energy calculation to excited states and spectroscopic analysis. At the same time, the present calculation should be extended to acceptors. From the first principles point of view, a detailed band calculation using the GW approximation, Hybertson and Louie [1], will provide another useful input into the present theory. Future work should also include the impurity structural energies in the defect level calculation. These structural energies will be useful to correlate with experiments in the determination of the locations and distributions of impurity atoms in host crystals.

Acknowledgements   We would like to thank Professors Jim Choyke, Ben Segall, and Walter Lambrecht and Dr. Bob Pfeiffer for useful discussions. The work was supported in part by ONR, Cray Research, and NASA through the CCDS at Auburn University.

References

[1] M. S. Hybertson and S. G. Louie, Phys. Rev. Lett. **58**, 1551 (1987).

[2] Y. C. Chang, T. C. McGill, and D. L. Smith, Phys. Rev. B **23**, 4169 (1981).

[3] W. Suttrop, G. Pensl, W. J. Choyke, R. Stein, and S. Leibenzeder, J. Appl. Phys. **72**, 3708 (1992).

[4] R. A. Faulkner, Phys. Rev. **184**, 713 (1969).

[5] R. S. Pfeiffer, Y.-J. Huang, and A.-B. Chen, Phys. Rev. B **48**, 8541 (1993).

[6] J. M. Luttinger and W. Kohn, Phys. Rev. **97**, 869 (1954).

[7] W. Kohn, Shallow Impurity States in Silicon and Germanium, Solid State Physics, Vol. 5, Eds. F. Seitz and D. Turnbull, Academic Press, Inc., New York 1957.

[8] S. Krishnamurthy, A. Sher, and A.-B. Chen, Phys. Rev. B **33**, 1026 (1986).
[9] O. Madelung, Semiconductors Group IV Elements and III–V Compounds, Springer-Verlag, Berlin/Heidelberg/New York 1991.
[10] R. G. Humphreys, D. Bimberg, and W. J. Choyke, Solid State Commun. **39**, 163 (1981).
[11] L. Patrick, D. R. Hamilton, and W. J. Choyke, Phys. Rev. **143**, 526 (1966).
[12] W. J. Choyke, D. R. Hamilton, and L. Patrick, Phys. Rev. **133**, A1163 (1964).
[13] W. J. Choyke and L. Patrick, Phys. Rev. **127**, 1868 (1962).
[14] L. Patrick, Phys. Rev. B **5**, 2198 (1972).
[15] L. Patrick, W. J. Choyke, and D. R. Hamilton, Phys. Rev. **137**, A1515 (1965).
[16] W. R. Lambrecht and L. B. Segall, Phys. Rev. B **52**, R2249 (1995).
[17] M. Hofmann, A. Zywietz, K. Karch, and F. Bechstedt, Phys. Rev. B **50**, 13401 (1994).
[18] W. J. Choyke, R. P. Devaty, L. L. Clemem, M. F. MacMillan, and M. Yoganathan, Inst. Phys. Conf. Ser. No. 142, 257 (1996).
[19] P. J. Dean, W. J. Choyke, and L. Patrick, J. Lum. **15**, 299 (1977).
[20] R. Kaplan, R. J. Wagner, H. J. Kim, and R. F. Davis, Solid State Commun. **55**, 67 (1985).
[21] W. Götz, A. Schöner, G. Pensl, W. Suttrop, W. J. Choyke, R. Stein, and S. Leibenzeder, J. Appl. Phys. **73**, 3332 (1993).
[22] N. T. Son, W. M. Chen, O. Kordina, A. O. Konstantinov, B. Monemar, E. Janzen, D. M. Hofman, D. Volm, M. Drechsler, and B. K. Meyer, Appl. Phys. Lett. **66**, 1074 (1995).
[23] N. T. Son, O. Kordina, A. O. Konstantinov, W. M. Chen, E. Sörman, B. Monemar, and E. Janzen, Appl. Phys. Lett. **65**, 3209 (1994).
[24] P. Srichaikul, A.-B. Chen, and W. J. Choyke, in: Amorphous and Crystalline SiC IV, Springer-Verlag, Berlin/New York 1992 (p. 170).
[25] D. Volm, B. K. Meyer, D. M. Hofman, W. M. Chen, N. T. Son, C. Person, U. Lindefelt, O. Kordina, E. Sorman, A. O. Konstantinov, B. Monemar, and E. Janzen, Phys. Rev. B **53**, 15409 (1996).
[26] A.-B. Chen and A. Sher, Semiconductor Alloy: Physics and Materials Engineering, Plenum Press, New York/London 1995.
[27] L. A. Hemstreet, Jr. and C. Y. Fong, Solid State Commun. **9**, 643 (1971); Phys. Rev. B **6**, 1464 (1972); Silicon Carbide, 1973, Proc. Internat. Conf. Silicon Carbide, Eds. R. C. Marshall, J. W. Faust, Jr., and C. E. Ryan, Univ. of South Carolina Press, Columbia (SC) 1974 (p. 284).
[28] L. Patrick and W. J. Choyke, Phys. Rev. **186**, 775 (1969); Phys. Rev. B **2**, 2255 (1970).
[29] L. Patrick, D. R. Hamilton, and W. J. Choyke, Phys. Rev. **132**, 2023 (1963).
[30] W. H. Backes, P. A. Babbert, and W. van Haeringen, Phys. Rev. B **49**, 7564 (1993).
[31] H. Kuwabara, K. Yamanaka, and S. Yamada, phys. stat. sol. (a) **37**, K157 (1976).
[32] T. H. Ning and C. T. Sah, Phys. Rev. B **4**, 3468 (1971).
[33] A. Morita and H. Nara, J. Phys. Soc. Japan, Suppl. **21**, 234 (1966).
[34] L. Pauling, The Nature of the Chemical Bond, Cornell University Press, Ithaca, New York 1960.

phys. stat. sol. (b) **202**, 107 (1997)

Subject classification: 71.20. Nr; S6

Global Band Structure and Near-Band-Edge States

G. Wellenhofer and U. Rössler

*Institut für Theoretische Physik der Universität Regensburg,
D-93040 Regensburg, Germany*

(Received January 31, 1997)

The global band structure from ab-initio calculations using DFT-LDA concepts is presented and discussed for 2H, 3C, 4H, 6H, and 15R SiC in view of the different stacking sequences and point lattices of these polytypes. Details of the conduction and valence band structure close to the band edges are described by using **k · p** models and thermal density-of-states effective masses are calculated.

1. Introduction

SiC with its unique variety of polytypes would not have attracted so much interest recently were it not for its potential as device material, see e.g. [1]. Both aspects, the fundamental one of how to explain polytypism and the applied one of designing electronic devices have to do with the electronic band structure, though in different ways. The former is certainly related to the ground state properties of individual polytypes including the electronic states all over the Brillouin zone, while the latter makes use of electron states bordering the energy gap, which accommodate the free carriers of electron–hole pairs. We want to discuss both aspects of electronic structure by presenting results of the global band structure as well as of near band-edge states for the polytypes 2H, 3C, 4H, 6H, and 15R SiC.

A question of principal interest is the comparison of band structures for different polytypes. This can be achieved by looking at unit cells and Brillouin zones and how they change with the stacking sequence. Extending the lattice period in one direction reduces the size of the Brillouin zone in the corresponding direction of **k**-space with the consequence that by backfolding the number of energy bands is increased in the reduced Brillouin zone (some aspects are discussed in the contribution of van Haeringen et al. [47]). A generalized concept presented in Section 2 allows to make a more detailed assignment of special points in the different Brillouin zones of hexagonal, cubic and rhombohedral structures.

State-of-the-art calculations of ground-state properties of crystalline semiconductors are all based on the concepts of density functional theory (DFT) in the local density approximation (LDA) (for a more recent review see [2]). They are designed as ab-initio calculations, which means that given the sort of atoms to constitute a solid the stable crystal structure, its lattice parameters, and other ground states properties are determined from minimizing the total energy with respect to the atomic configuration. In reality this is done by minimizing the total energy with respect to the lattice parameters (including relaxation of the atoms in a unit cell) of a given crystal structure and comparing the results for different structures. Concepts of total energy calculations as well as results and consequences for polytypism in SiC are presented in the articles by Lambrecht et al. and by F. Bechstedt [49]. Within DFT-LDA several methods are in use and have been applied to SiC polytypes. The most common one is based on plane wave expansion and *first principles* pseudopotentials (see, e.g., [3 to 6]) which are tailored to achieve fast convergence with respect to the number of plane waves. Results of energy bands all over the Brillouin zone from such calculations are presented in Section 3. The method of choice is the linearized muffin tin orbital (LMTO) [7] or augmented plane wave (APW) [8] method if relativistic effects as the spin–orbit coupling are to be considered because they include an accurate calculation of the electronic structure close to the atomic sites. The effect of spin–orbit splitting, although not expected to be large for light atoms like Si and C will turn out to have some nonnegligible effect on the valence band dispersion close to its maximum (see [8] and Section 4).

With respect to the excited single-particle states the DFT-LDA concept suffers from the well-known *gap problem* which has been overcome by modifying the exchange-correlation energy, e.g., within the GW approximation in application to 3C SiC (see [9] and the contribution J. Pollmann [50]), using Gaussian instead of plane wave expansion. A simplified GW approximation has been used for 3C SiC and for some hexagonal polytypes [10]. These calculations result in larger separations between valence and conduction bands and yield gap energies in close agreement with experimental data. In addition the dispersion of energy bands is slightly changed with consequences for some details of near-band-edge states, e.g., the vanishing of the camel's back (obtained by van Haeringen's matching technique [11] or in DFT-LDA [12]).

The topmost valence bands close to its maximum (which for all polytypes is at or very close to the Γ-point) shows a rather complex dispersion. They can be described by using $\mathbf{k} \cdot \mathbf{p}$ theory or the invariant expansion [13] with parameters which can be determined by fitting to the ab-initio band structure. Using this parameterized valence band dispersion the density of states can be calculated. Results are presented together with those for the conduction band minimum in Section 4. The number of carriers available at a given temperature and doping is a key quantity in designing a device. It can be conveniently obtained using the thermal density-of-states (DOS) effective mass, which due to band nonparabolicity depends on temperature. Plots of thermal DOS effective masses of electrons and holes calculated from the near-band-edge states will also be presented in Section 4.

2. Unit Cells and Brillouin Zones

The polytypes of SiC (and likewise of other semiconductors with tetrahedral coordination) differ in the stacking sequence of double layers of Si and C atoms along the hexagonal c-axis. Depending on the projection of the atomic positions to the (xy)-plane (perpendicular to the c-axis) one distinguishes three different double layers denoted by A, B, and C. The situation is demonstrated in Fig. 1 with the hexagonal unit cell spanned by the indicated primitive translation vectors and the dash-dotted vertical lines on which the atomes of the A, B, and C layers are placed (see also Fig. 2 in the article by van Haeringen [47]). In the case of 2H SiC the unit cell contains two Si–C pairs, one on the A line and one on the B line. Their relative position is determined by the tetrahedral coordination leading to an ideal ratio of the lattice constants $c_{\text{hex}}/a_{\text{hex}} = \sqrt{8/3}$ (which in real crystals is slightly relaxed). 3C SiC can be described by a hexagonal unit

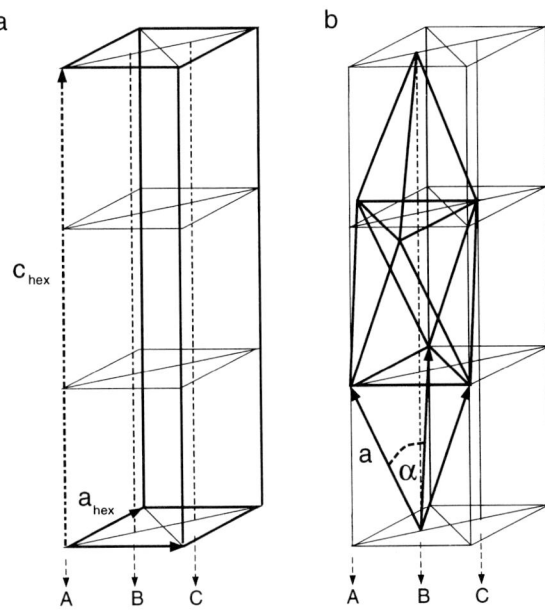

a

b

Fig. 1. Comparison of unit cells. a) Hexagonal unit cell; the dash-dotted vertical lines indicate where the Si–C pairs of the A, B, and C layers are placed. b) Rhombohedral unit cell aligned to the hexagonal cell

c_{hex}

a_{hex}

A   B   C

A   B   C

cell with three layers on A, B, and C and $c_{hex}/a_{hex} = \sqrt{6}$, which is 3/2 that of 2H SiC. Similarly 4H and 6H SiC are obtained with stacking sequences ABCB and ABCACB, respectively.

The most complex polytype considered in this article is 15R SiC with the sequence ABCACBCABACABCB with 30 atoms in the hexagonal unit cell of Fig. 1a and an ideal $c_{hex}/a_{hex} = \frac{15}{3}\sqrt{6}$. Exploiting the symmetry of the rhombohedral (or trigonal) lattice we can choose instead the Wigner-Seitz cell shown in Fig. 1b which is smaller by a factor of 3. It is spanned by the three primitive translations indicated by arrows of length a which include the angle α. Rhombohedral and hexagonal lattice parameters transform into each other according to

$$\frac{c_{hex}}{a_{hex}} = \frac{\sqrt{3(1 + 2\cos\alpha)}}{2\sin\alpha/2} \quad \text{and} \quad a_{hex} = 2a\sin\alpha/2.$$

For 15R the ideal α is 13°55′50″. If the intersection of the B line with the base of the hexagonal unit cell is taken as one atomic position, then no atom on the A or C line in the hexagonal cell belongs to the rhombohedral Wigner-Seitz cell. It accommodates ten atoms which are all placed on the B line. A unit cell of 3C SiC can be obtained for $\alpha = 60°$ which, however, is not the Wigner-Seitz cell as it does not show the symmetry of the face-centered cubic (f.c.c.) point group. The cubic and rhombohedral structures differ from the hexagonal ones by missing symmetry under nonsymmorphic operations, but instead their point groups have additional elements, which lead to the reduced unit cells.

Knowing the alignment of the unit cells (and thus of the point lattices) of the hexagonal and rhombohedral structures in real space (Fig. 1b) one can now look into the corresponding structures in reciprocal space. As a reminder the Brillouin zones and their irreducible sections with special points are shown in Fig. 2 for hexagonal and f.c.c. point lattices. The rhombohedral Brillouin zone aligned with the corresponding hexagonal

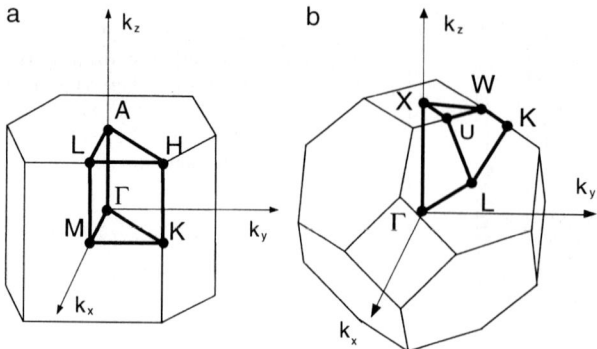

Fig. 2. Brillouin zones of a) hexagonal and b) face-centered cubic lattice

Brillouin zone is presented in Fig. 3. The former has three times the volume of the latter one as indicated by the dashed lines with three hexagonal cells put on top of each other along k_z to meet the extension of the rhombohedral zone (solid line). The heavy solid and dashed lines mark the irreducible sections: in the hexagonal zone it is twice as large as in Fig. 2a for the hexagonal lattice because of the reduced symmetry around the c-axis; in the rhombohedral zone it is four times as large as in the f.c.c. lattice whose point group contains 48 elements compared to only 12 in the rhombohedral lattice.

The aligned Brillouin zones in Fig. 3 allow to assign lines and points to each other. The correspondence between the hexagonal Γ–A line and the rhombohedral Γ–T line is obvious (it corresponds to the cubic Γ–L line) and leads to a three times backfolding of the energy bands along Γ–T onto the Γ–A line. A similar construction for the hexagonal and f.c.c. Brillouin zones leads to the well-known doubling of bands along Γ–L when backfolded onto the Γ–A line. As seen from Fig. 3 the hexagonal M-point can be identified with the rhombohedral (or cubic) X-point and the hexagonal K and H points are close to the rhombohedral K and W points, respectively. The hexagonal L point can be

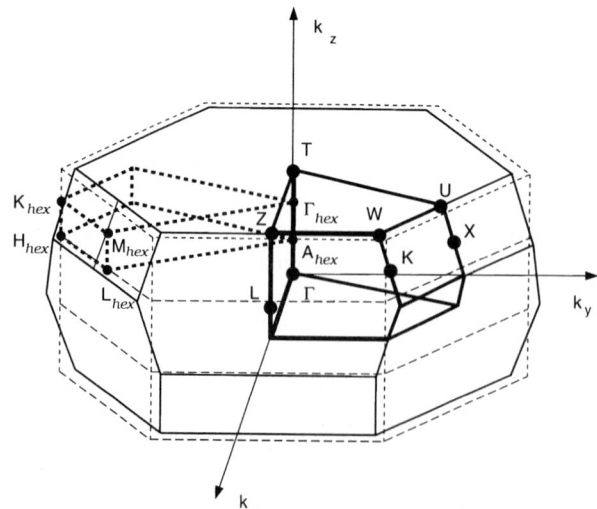

Fig. 3. Rhombohedral BZ embedded in three hexagonal BZs (dashed lines). Irreducible parts of the rhombohedral BZ (bold lines) with high symmetry points (big dots) and the hexagonal BZ (bold dashed lines) with high symmetry points (small dots)

related as well to the L as to the U or Z points of the rhombohedral zone. With this scheme at hand the global band structure of different structures and polytypes can be compared with each other.

3. Global Band Structure

Ab-initio band calculations of different authors are expected to give identical results. This is, however, strictly speaking not the case because even ab-initio concepts leave some freedom, e.g., in the choice of pseudopotentials and in the installed software. Nevertheless the differences are small and comparable concepts lead to almost identical data. Therefore, we show here band structures of 2H, 3C, 4H, 6H, and 15R SiC from our DFT-LDA calculations [12] based on plane wave expansion and pseudopotentials, which (except for 15R SiC [14, 15]) have not been published so far.

To characterize our concepts and numerics: we have used the parameterized exchange-correlation potential of Perdew and Zunger [16], the method of Martins and Troullier [17], for constructing norm-conserving pseudopotentials, and **k**-space integration with special points constructed according to Chadi and Cohen [18] for cubic and hexagonal polytypes and Hama and Watanabe [19] for 15R SiC. We used ten special points for 3C and 15R SiC and twelve ones for 2H, 4H, and 6H SiC. The ground state structural parameters have been determined by fitting calculated total energy values for different lattice parameters to Murnaghan's equation of state. Internal relaxations of the atoms with respect to their positions defined by the relaxed c/a ratio and quasi-particle corrections within the GW approximation have not been taken into account. Our calculated values for equilibrium lattice constants, the bulk modulus and its pressure derivative differ from the available experimental data and other theoretical results by less than 1% [12, 14]. The global band structures are shown in Figs. 4 to 8 along the main symmetry lines of the Brillouin zones in Fig. 2 and 3. The valence band maximum, which (when spin is neglected as in our calculations) for all polytypes is at the center of the

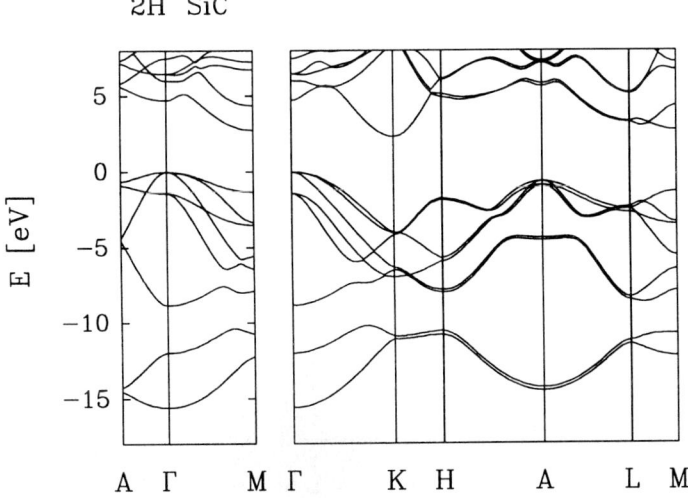

Fig. 4. Global band structure of 2H SiC

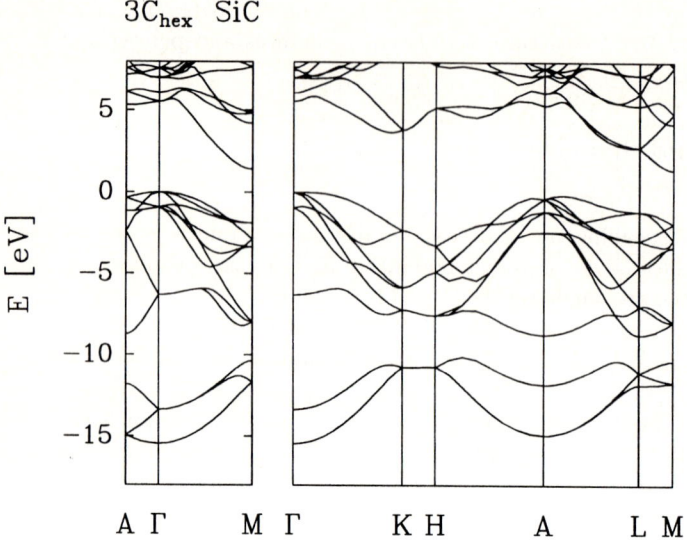

Fig. 5. Global band structure of 3C$_\text{hex}$ SiC

Brillouin zone at Γ, is taken as the zero of energy. The valence bands (down to -15 eV) derive from the bonding s- and p-orbitals of the outer shell electrons of Si and C. Their number is determined by the number of atoms in the unit cell each having 4 valence electrons. 2H SiC with 4 atoms in the unit cell has 16 valence electrons to fill 8 bands (Fig. 4); 3C SiC with 2 atoms in the cubic Wigner-Seitz cell has 4 valence bands which triplicate in the hexagonal symmetry (Fig. 5) where the unit cell contains 6 atoms. Simi-

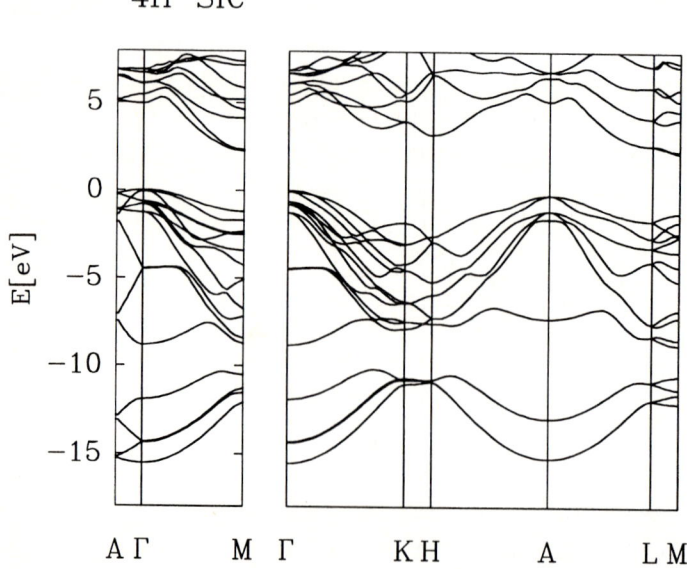

Fig. 6. Global band structure of 4H SiC

6H SiC

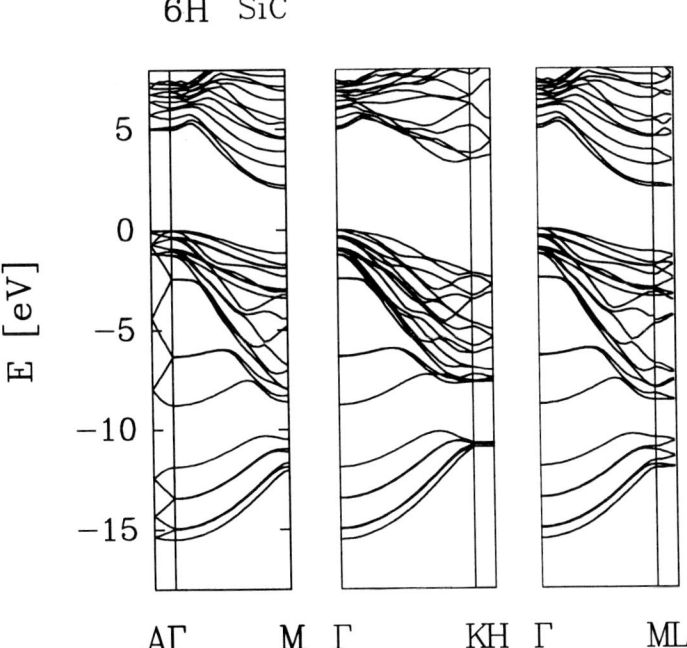

Fig. 7. Global band structure of 6H SiC

larly the number of valence bands increases to 16 for 4H SiC (Fig. 6) and 24 for 6H SiC (Fig. 7) but only 20 in the rhombohedral Brillouin zone of 15R SiC (Fig. 8). The conduction band minimum is at the K-point for 2H SiC but at the M-point for $3C_{hex}$ (note that the cubic X-point is mapped onto the hexagonal M-point, Fig. 3) and 4H SiC. For

15R SiC

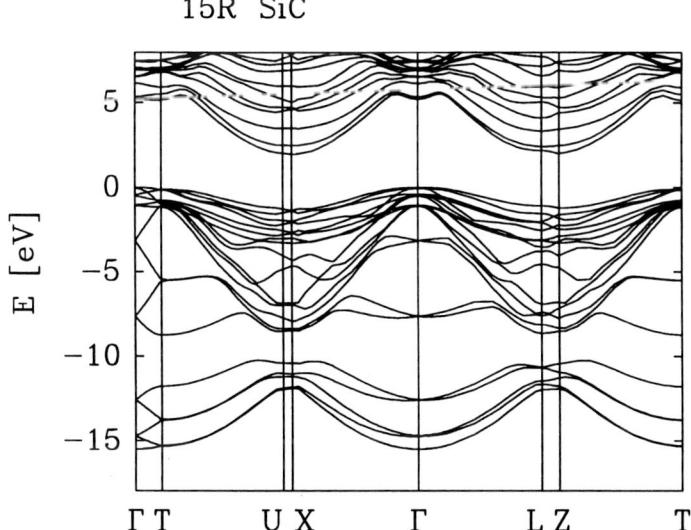

Fig. 8. Global band structure of 15R SiC

6H SiC it seems also to be at M (Fig. 7) but in a higher resolution it is shifted to the ML line (see Section 4). In 15R SiC the conduction band minimum is at X, which corresponds to the hexagonal M-point.

The details of the global band structure are almost the same as in other ab-initio calculations (see e.g., [20 to 22]) and differences are hard to detect on the scale shown in these figures. The most striking difference occurs in the separation between conduction and valence bands which in DFT-LDA calculations is in general too small by about 1 eV compared with concepts that consider quasi-particle corrections within the GW approximation [9, 10] and are in good agreement with experimental data. However, the internal valence and conduction band structue is not strongly affected by the quasi-particle corrections which essentially result in a more or less rigid shift of the conduction against the valence bands (frequently simulated by the *scissors operator* [23]). A small dependence of these corrections on the wave vector which takes influence on details of the near-band-edge dispersion will be discussed in the next section.

4. Near-Band-Edge States

Electronic device operation is based on carriers which by thermal excitation from impurities or by optical excitation across the band gap occupy states in the band structure close (within few 100 meV) to the conduction or valence band edge. The physical con-

Table 1

Effective conduction band masses of SiC polytypes

$m\ [m_0]$	theory					experiment			
2H SiC, minimum at K									
ref.	[12]	[25]	[26]	[8]					
$m_{\perp K-H}$	0.40	0.45	0.45	0.43					
$m_{\parallel K-H}$	0.26	0.26	0.27	0.26					
3C SiC, minimum at X									
ref.	[12]	[25]	[26]	[8]	[7]	[28]	[29]	[30]	
$m_{\parallel X-\Gamma}$	0.70	0.67	0.63	0.68	0.60	0.667	0.67	0.67	
$m_{\perp X-\Gamma}$	0.23	0.25	0.23	0.23	0.29	0.247	0.25	0.22	
4H SiC, minimum at M									
ref.	[12]	[25]	[26]	[8]	[31]	[32]	[35]		
$m_{M-\Gamma}$	0.66	0.57	0.58	0.57	0.58	$m_\perp$: 0.18	0.30		
m_{M-K}	0.31	0.32	0.28	0.28	0.31				
m_{M-L}	0.30	0.32	0.31	0.31	0.33	$m_\parallel$: 0.22	0.48		
6H SiC, minimum along M–L									
ref.	[12]	[25]	[26]	[8]	[35]	[33]	[36]	[37]	
$m_{\parallel M-\Gamma}$	0.78	0.75	0.77	0.75	$m_\perp$: 0.35	0.24	0.25	0.42	
$m_{\parallel M-K}$	0.23	0.27	0.24	0.24					
m_{M-L}	1.2–2.0	1.95	1.24	1.83	$m_\parallel$: 1.4	0.34	1.7	2.0	
15R SiC, minimum at X									
ref.	[14]				[38]	[34]			
$m_{\parallel X-\Gamma}$	0.67				$m_\perp$: 0.28	0.24			
$m_{\parallel X-K}$	0.22								
$m_{\parallel X-U}$	0.41				$m_\parallel$: 0.53	0.38			

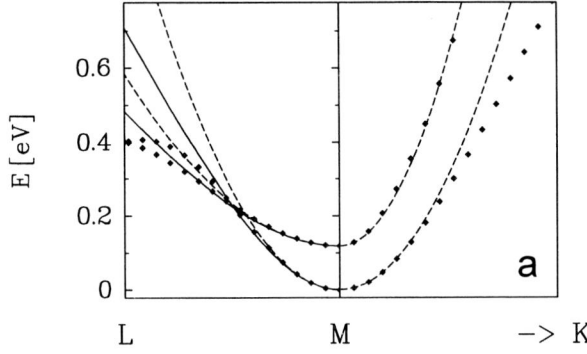

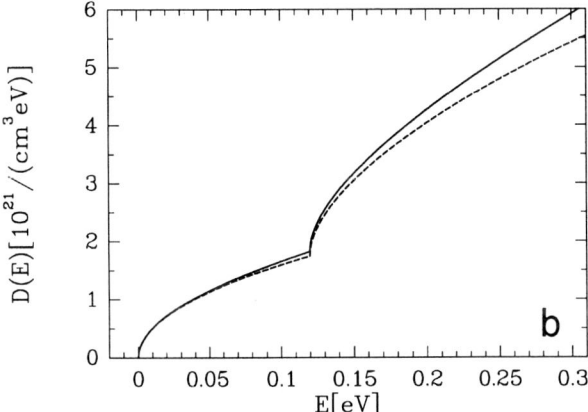

Fig. 9. a) Conduction band minimum of 4H SiC. Diamonds: DFT results, dashed lines: parabolic approximation, solid lines: hyperbolic approximation. b) DOS of 4H SiC from parabolic approximation (dashed line) and from parabolic approximation including hyperbolic approximation along M–L at the conduction band edge (solid line)

cepts used to describe operation treat these carriers as free particles, characterized by mass parameters and temperature dependent densities which can be derived from the band-edge dispersion by making use of the ab-initio results. The concept of an effective mass (or effective mass tensor) is restricted to near-band-edge states that can be properly described by a parabolic (in general anisotropic) dispersion with twofold (due to spin) degeneracy. This is the case for the lowest conduction band of 2H, 3C, 4H, and 15R SiC. For the conduction band minimum of 6H SiC and for the valence band maximum the situation is more complex as outlined below. In all cases the near-band-edge structure is modeled using $\mathbf{k} \cdot \mathbf{p}$ concepts or the invariant expansion [13] with parameters being determined by fitting to the ab-initio band structure close to the band edges.

4.1 Conduction band edge

The ab-initio conduction band of 2H SiC around its minimum at K can be fitted up to about 150 meV by parabolas along three independent directions, e.g., K–Γ, K–M, and K–H. The mass parameters obtained from this fit (see Table 1) correspond to an isotropic dispersion in the plane perpendicular to the K–H line which is invariant under sixfold rotations. Similarly, the conduction band of 3C SiC around its minimum at X is

Table 2

Effective conduction band masses and parameters of the hyperbolic approximation for 4H SiC. The index 2 refers to the second conduction band

c_z [eV]	$m^2_{M\Gamma}$ $[m_0]$	m^2_{MK} $[m_0]$	m^2_{ML} $[m_0]$, c^2_z [eV]
1.3	0.74	0.17	0.80, 1.2

characterized by two mass parameters (Table 1) due to fourfold symmetry around the Γ–X line. The DOS obtained for this parabolic dispersion is described by

$$D(E) = \frac{n_v}{\pi^2} \frac{\sqrt{2m^2_{\|}m_{\perp}}}{\hbar^3} \sqrt{E},$$ (1)

where n_v is the number of conduction band minima within the Brillouin zone.

The comparison of parabolic fits (dashed lines) to the ab-initio band structure (diamonds) for 4H SiC around the conduction band minimum at M is depicted in Fig. 9a along M–K and M–L. Again the fit is very accurate up to 200 meV for the lowest band but deviates and overlaps with a second band which starts about 120 meV above the bottom of the lowest band. For both bands a modified model (hyperbolic approximation) using

$$E(k_z) = E_M - c_z/2 + \sqrt{\frac{c_z^2}{4} + \frac{c_z \hbar^2 k_z^2}{2m_{M-L}}}$$

improves the fit for the M–L line ($k_z = 0$ corresponds to the M-point) up to about 300 meV above the minimum of the lowest band (solid lines in Fig. 9a). The effective masses of the second band and the parameter c_z for the hyperbolic approximation of both bands are presented in Table 2. The density of states in Fig. 9b (dashed line) shows the overlapping square root contributions of the parabolic (but anisotropic) bands. The corresponding densities of states calculated from both fits deviate from each other with increasing energy, as expected (Fig. 9b).

In 6H SiC the conduction band structure along M–L has a very low curvature corresponding to reported mass values up to about twice the free electron mass (Table 1 and dashed lines in Fig. 10a). Our ab-initio results (diamonds) for the lowest conduction band along M–L can be fitted nicely with a camel's back dispersion (Fig. 10a, solid lines) [11, 12, 24],

$$E(k_z) = E_M + \frac{\hbar^2 k_z^2}{2m^*_{M-L}} - \sqrt{\frac{\Delta^2}{4} + P^2 k_z^2},$$ (2)

with a minimum about halfway between M and L. As the second conduction band is not properly described as the backfolded continuation of the lowest conduction band using Eqn. (2), we use instead the hyperbolic approximation (dash-dotted line in Fig. 10a) with the parameters given in Table 3. With the height of only 6 meV the camel's back in the lowest conduction band is very sensitive with respect to changes in the band structure calculation. A vanishing camel's back and a shift of the conduction band minimum to the M-point is reported as a consequence of quasi-particle correction

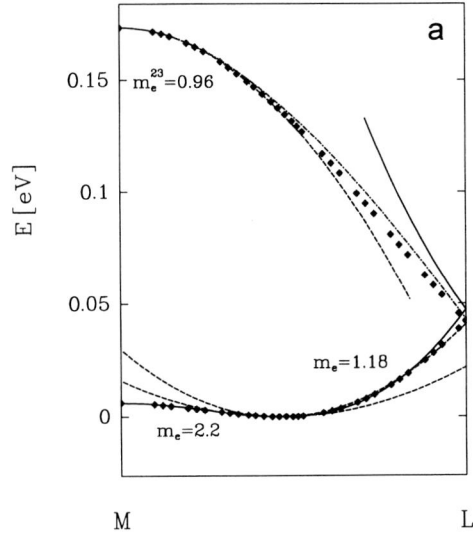

Fig. 10. a) Conduction band minimum of 6H SiC. Diamonds: DFT results, dashed lines: parabolic approximation, solid lines: camel's back relation, dash-dotted line: hyperbolic approximation. b) DOS of 6H SiC at the conduction band edge

[10]. Perpendicular to the M–L line the dispersion is parabolic yet with different curvatures along lines parallel to M–Γ and M–K (see Table 1). With the parameters from Tables 1 and 3 the density of states of Fig. 10b (solid line) is obtained.

Comparing these results with available effective masses for the lowest conduction band (Table 1) we find close agreement for all polytypes within the theoretical work. For the lowest conduction band along M–L in 6H SiC, however, the chamel's back relation allows a much better fit than a parabola with a single mass parameter m_{M-L}. For comparison with experimental data we are not aware of any results for 2H SiC and find good agreement for 3C SiC. Only in very recent experiments using optically detected cyclotron resonance (ODCR) [31] all threee components of the mass tensor in 4H SiC have been determined independently. All other experiments have been interpreted assuming axial symmetry of the mass tensor as in 3C or 2H SiC. This assumption is, especially, questionable for the masses derived from the FIR impurity data of [32 to 34]. Moreover, 15R SiC experiments have been interpreted assuming alignment of the energy surfaces with the k_z-axis (instead of the Γ–M line). For 6H SiC the data of [35 to 39][1] (the latter two references

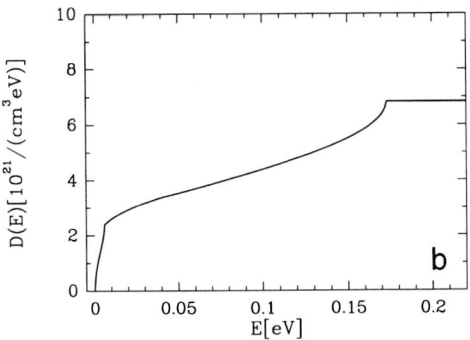

Table 3

Effective conduction band masses and parameters of the hyperbolic approximation and of the camel's back relation for 6H SiC. The index 2 refers to the second band

Δ [eV]	P [Å eV]	m^*_{ML} [m_0]	$m^2_{M\Gamma}$ [m_0]	m^2_{MK} [m_0]	m^2_{ML} [m_0], c^2_z [eV]
1.100	3.323	0.44	0.78	0.23	−0.96, 0.4

---

[1]) The authors of [38] assume the conduction band minimum along Γ–A.

are not quoted in Table 1) are consistent with the large mass along M–L from band calculations; the transverse mass $m_\perp$ of [37] can be related to $\sqrt{m_{\|\mathrm{M\Gamma}} m_{\|\mathrm{MK}}}$ which from our calculation is $0.42 m_0$.

4.2 The valence band edge

The dispersion of the valence bands close to its maximum at Γ is determined by the six-fold (with spin) degeneracy of the p-bonding states which is removed due to spin–orbit and crystal field splitting. A proper phenomenological model is provided by the $\mathbf{k} \cdot \mathbf{p}$ theory or an invariant expansion in terms of a 6×6 Hamiltonian [13, 40],

$$H_{6\times 6}(\mathbf{k}) = \begin{pmatrix} F & 0 & -H^* & 0 & K^* & 0 \\ 0 & G & \Delta & -H^* & 0 & K^* \\ -H & \Delta & \lambda & 0 & I^* & 0 \\ 0 & -H & 0 & \lambda & \Delta & I^* \\ K & 0 & I & \Delta & G & 0 \\ 0 & K & 0 & I & 0 & F \end{pmatrix} \tag{3}$$

with

$$F = \Delta_1 + \Delta_2 + \lambda + \theta, \qquad k_+ = k_x + ik_y, \qquad H = iA_6 k_z k_+ - A_7 k_+,$$
$$G = \Delta_1 - \Delta_2 + \lambda + \theta, \qquad \Delta = \sqrt{2}\,\Delta_3, \qquad \lambda = A_1 k_z^2 + A_2(k_x^2 + k_y^2),$$
$$I = iA_6 k_z k_+ + A_7 k_+, \qquad K = A_5 k_+^2, \quad \text{and} \quad \theta = A_3 k_z^2 + A_4(k_x^2 + k_y^2),$$

containing linear and bilinear expressions in the components of the wave vector $\mathbf{k}$, whose parameters can be determined (as for the conduction band) by fitting to the ab-initio valence bands. With the exception of [7] and [8] all ab-initio band calculations have been performed without including spin–orbit coupling. For this case ($\Delta_2 = \Delta_3 = 0$) the 6×6 Hamiltonian falls into two identical 3×3 Hamiltonians, which in the basis (where $|Z\rangle$ is oriented along the c-axis)

$$\frac{1}{\sqrt{2}}\,|X + iY\rangle, \qquad |Z\rangle \quad \text{and} \quad \frac{1}{\sqrt{2}}\,|X - iY\rangle$$

read

$$H_{3\times 3}^w(\mathbf{k}) = \begin{pmatrix} \Delta_1 + \lambda + \theta & -H^* & K^* \\ -H & \lambda & I^* \\ K & I & \Delta_1 + \lambda + \theta \end{pmatrix}. \tag{4}$$

It can be used to determine the parameters $A_1 \ldots A_7$ and Δ_1 by fitting to the ab-initio calculations without spin–orbit coupling (see Table 4). The result of the fitting procedure is exemplified for 6H SiC in Fig. 11a, by comparing the ab-initio data (diamonds) with the dispersion obtained from Eqn. (4) with the parameters of Table 4 (dashed lines). In order to include the effect of spin–orbit coupling we use the 6×6 model of Eqn. (3) with $\Delta_2 = \Delta_3$ and the spin–orbit splitting $3\Delta_2 = 7$ meV taken from experiment [41 to 43] (solid lines in Fig. 11a). Except for the small spin–orbit splitting of the top-most valence bands, the removal of the spin–splitting of the second and third valence bands, and the anticrossing along the Γ–M line Fig. 11a demonstrate the close agreement of the different models. This is also seen in the density of states (Fig. 11b). This

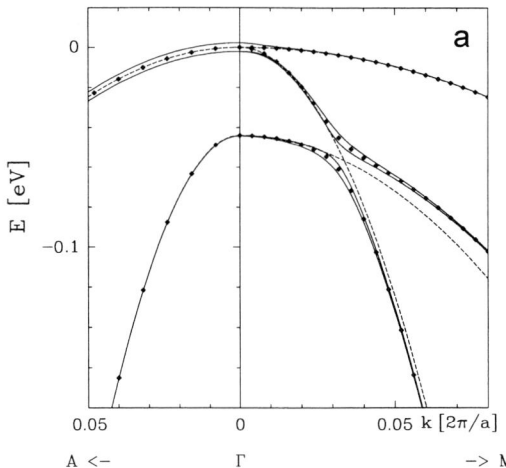

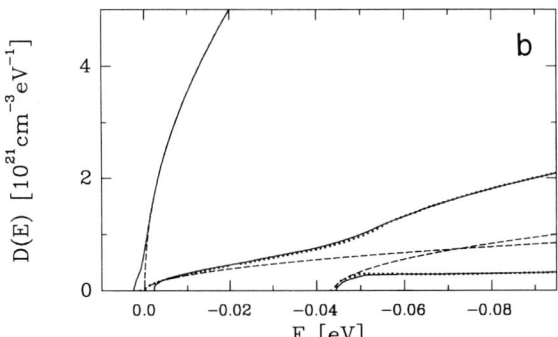

Fig. 11. a) Valence band maximum of 6H SiC. Diamonds: DFT results, solid lines: eigenvalues from Eqn. (3) including spin–orbit interaction, dashed lines: parabolic approximation. b) DOS of 6H SiC from Eqn. (3) (solid lines), DFT results (dotted lines), and parabolic approximation (dashed lines) at the valence band edge

$\mathbf{k} \cdot \mathbf{p}$ model has not been applied so far to SiC polytypes and, therefore, no parameter values are reported in the literature to compare with those given in Table 4. The hole mass values of [27] obtained by fitting parabolas to the ab-initio valence bands are questionable, because they are not consistent with the axial symmetry of the $\mathbf{k} \cdot \mathbf{p}$ model. The analytic expressions for the parabolic valence band dispersion given in [8] are derived under the assumption of a large crystal field splitting and apply only to the immediate vicinity of $\mathbf{k} = 0$.

Table 4

Mass parameters A_i (in at. units) and crystal field splitting (in meV) for 2H, 4H, and 6H SiC

	A_1	A_2	A_3	A_4	A_5	A_6	A_7	Δ_{cr}
2H SiC	-4.83	-0.576	4.18	-1.24	-1.41	-1.49	0.0232	144
4H SiC	-4.76	-0.628	4.14	-1.11	-1.45	-1.49	0.0127	74
6H SiC	-4.76	-0.590	4.14	-1.15	-1.50	-1.34	0.0119	44

The Hamiltonian of Eqn. (4) does not directly apply to 3C SiC, because the point group of the wurtzite-type structure (C_{6v}) is not a subgroup of the zinc-blende point group (T_d) [13]. The $3 \times 3\,\mathbf{k} \cdot \mathbf{p}$ Hamiltonian for the valence bands of zinc-blende is given by [44]

$$H_{3\times3}^{zb}(\mathbf{k}) = \begin{pmatrix} Lk_x^2 + M(k_y^2 + k_z^2) & Nk_xk_y & Nk_xk_z \\ Nk_xk_y & Lk_y^2 + M(k_x^2 + k_z^2) & Nk_yk_z \\ Nk_xk_z & Nk_yk_z & Lk_z^2 + M(k_x^2 + k_y^2) \end{pmatrix}. \quad (5)$$

Here, the basis $|X\rangle$, $|Y\rangle$ and $|Z\rangle$ is oriented along the cubic axes, and L, M, and N are the standard parameters of the cubic 3×3 model. The valence band parameters L, M, and N of 3C SiC can be related to those of the hexagoal polytypes if both are compared with the parameters corresponding to a Hamiltonian with the common subgroup C_{3v} (cubic approximation [13]). Including spin–orbit coupling the valence bands of 3C SiC close to $\mathbf{k} = 0$ for the fourfold (twofold) subspace corresponding to angular momentum $J = 3/2(1/2)$ are described by the analytic expressions [13, 44]

$$E_{1/2} = A\mathbf{k}^2 \pm \sqrt{B^2\mathbf{k}^2 + C^2(k_x^2k_y^2 + k_x^2k_z^2 + k_y^2k_z^2)} \quad \text{and} \quad E_3 = A\mathbf{k}^2 .$$

The parameters A, B, and C are related to L, M, and N by

$$A = \frac{L + 2M}{2} , \qquad B = \frac{L - M}{2} , \qquad \text{and} \quad C^2 = \frac{N^2 - (L - M)^2}{3} .$$

In Table 5 we compare our valence band parameters of 3C SiC with the values of A, B, and C of [8] and L, M, and N derived from [7] (considering the different notations used there). The larger difference to the values of [7] can be ascribed to the adjustment in [7] of the energy gap to the experimental value, which leads to reduced absolute values of L, M, and N.

4.3 Thermal DOS effective masses

With the parametrized near-band-edge dispersion and DOS of Sections 4.1 and 4.2 the temperature dependent carrier concentration can be computed according to

$$n(T) = \int\limits_{E_c}^{\infty} D(E)\, f_\mu(E, T)\, \mathrm{d}E$$

for electrons and

$$p(T) = \int\limits_{-\infty}^{E_v} D(E)\,(1 - f_\mu(E, T))\, \mathrm{d}E \quad (6)$$

Table 5
Valence band mass parameters of 3C SiC

	A [at. units]	B [at. units]	C [at. units]		L [at. units]	M [at. units]	N [at. units]
this work	−1.98	−0.28	2.38	this work	−2.54	−1.70	−4.20
[8]	−1.96	−0.30	2.27	[7]	−2.44	−1.42	−3.76

for holes, once the chemical potential (depending on impurity concentration and position of donor/acceptor ground states with respect to the band edge) is known. f_μ is the Fermi-Dirac distribution function and $E_c (E_v)$ the conduction (valence) band edge. Examples for $p(T)$ in 6H SiC are presented in [45]. A more compact information for easy use in device application is the *thermal DOS effective mass*, defined for holes and electrons [46] according to

$$ n(T) = \int\limits_{E_g}^{\infty} \frac{n_\nu (m_e^d(T))^{3/2}}{\pi^2 \hbar^3} \sqrt{2E}\, f_\mu(E,\, T)\, \mathrm{d}E \tag{7} $$

and

$$ p(T) = \int\limits_{-\infty}^{0} \frac{(m_h^d(T))^{3/2}}{\pi^2 \hbar^3} \sqrt{2E}\, (1 - f_\mu(E,\, T))\, \mathrm{d}E\, , $$

respectively, where $n(T)$ and $p(T)$ are the carrier densities calculated according to Eqn. (6) with the accurate $D(E)$. It turns out that for impurity binding energies typical for shallow donors and acceptors in SiC (which are larger than 50 meV) the thermal occupation is well described by the Boltzmann distribution.

For the lowest conduction band in 2H or 3C SiC with its parabolic dispersion up to 150 to 200 meV above the minimum, the DOS is a strict square root function of the energy and $m_e^d = \sqrt{m_\perp^2 m_\parallel}$ is independent of T, as long as no higher conduction band states are involved. The more complex dispersion of the lowest conduction band in 4H and 6H SiC (see Section 4.1) and of the topmost valence bands (see Section 4.2) leads to the thermal DOS effective masses shown in Fig. 12. For electrons in 4H SiC m_e^d increases slightly from the band-edge value $m_e^d(T = 0) = 0.394 m_0$ obtained for the masses of Table 1 with increasing occupation of states in the second conduction band. In 6H SiC $m_e^d(T)$ shows a peak due to the camel's back structure: at low temperature mainly states below the saddle point at M are occupied while with increasing temperature more carriers in higher states of the conduction band with smaller masses contribute to

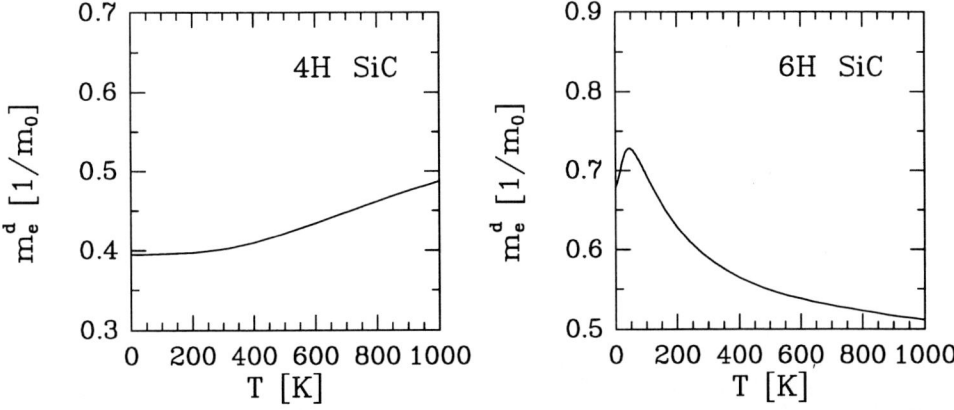

Fig. 12. Electron thermal DOS effective masses for 4H and 6H SiC

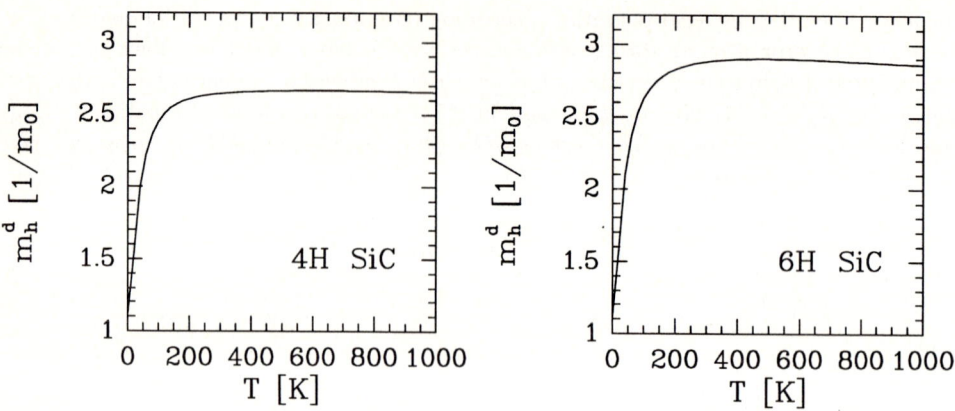

Fig. 13. Hole thermal DOS effective masses for 4H and 6H SiC

$m_e^d(T)$. In the valence bands (Fig. 13) the mass increases from a value defined by the immediate band-edge curvature of the topmost (or A) band at Γ and saturates at about room temperature at a value which is determined by the flat dispersion of the A-band as seen in Fig. 11a. This behavior is a consequence of the small spin–orbit splitting and can be derived also from the DOS in Fig. 11b, which for small energies is representative for all SiC polytypes.

5. Summary

The intimate relation between crystal and electronic structure is displayed in the modifications of the energy band structure with stacking sequences of increasing length when going from 2H to 15R SiC. Besides this more principal aspect we have presented detailed results for near-band-edge states including band parameters and thermal DOS effective masses as key information needed to design electronic devices.

Acknowledgement   The work has been supported by the BMBF within the project "SiC-Elektronik".

References

[1] G. PENSL and R. HELBIG in: Festkörperprobleme/Adv. Solid State Phys. **30**, 133 (1990).
[2] M. C. PAYNE, M. P. TETER, D. C. ALLAN, T. A. ARIAS, and J. D. JOANNOPOULOS, Rev. Mod. Phys. **64**, 1045 (1992).
[3] N. CHURCHER, K. KUNC, and V. HEINE, Solid State Commun. **65**, 177 (1985).
[4] P. J. H. DENTENEER and W. VAN HAERINGEN, Phys. Rev. B **33**, 2831 (1986).
[5] K. J. CHANG and M. L. COHEN, Phys. Rev. B **35**, 8196 (1987).
[6] P. E. VAN CAMP, V. E. VAN DOREN, and J. T. DEVREESE, phys. stat. sol. (b) **146**, 573 (1988).
[7] M. WILLATZEN, M. CARDONA, and N. E. CHRISTENSEN, Phys. Rev. B **51**, 13150 (1995).
[8] C. PERSSON and U. LINDEFELT, Phys. Rev. B **54**, 10257 (1996).
[9] M. ROHLFING, P. KRÜGER, and J. POLLMANN, Phys. Rev. B **48**, 17791 (1993).
[10] B. WENZIEN, P. KÄCKELL, F. BECHSTEDT, and G. CAPPELLINI, Phys. Rev. B **52**, 10897 (1995).
[11] W. H. BACKES, P. A. BOBBERT, and W. VAN HAERINGEN, Phys. Rev. B **49**, 7564 (1994).

[12] K. KARCH, G. WELLENHOFER, P. PAVONE, U. RÖSSLER, and D. STRAUCH, in: Proc. 22nd Internat. Conf. Physics of Semiconductors, Vancouver, Ed. D. J. LOCKWOOD, World Scientific Publ. Co., Singapore, 1995 (p. 401).

[13] G. L. BIR and G. E. PIKUS, Symmetry and Strain-Induced Effects in Semiconductors, John Wiley & Sons, New York/Toronto 1974.

[14] G. WELLENHOFER and U. RÖSSLER, Solid State Commun. **96**, 887 (1995).

[15] G. WELLENHOFER and U. RÖSSLER, Conf. Silicon Carbide and Related Materials, Kyoto, Inst. Phys. Conf. Series **142**, 297 (1996).

[16] J. P. PERDEW and A. ZUNGER, Phys. Rev. B **23**, 5048 (1981).

[17] N. MARTINS and J. L. TROULLIER, Phys. Rev. B **43**, 1993 (1991).

[18] D. J. CHADI and M. L. COHEN, Phys. Rev. B **8**, 5747 (1973).

[19] J. HAMA and M. WATANABE, J. Phys.: Condensed Matter **4**, 4583 (1992).

[20] C. H. PARK, B.-H. CHEONG, K.-H. LEE, and K. J. CHANG, Phys. Rev. B **49**, 4485 (1994).

[21] V. I. GAVRILENKO, A. V. POSTNIKOV, N. I. KLYUI, and V. G. LITOVCHENKO, phys. stat. sol. (b) **162**, 477 (1990).

[22] P. KÄCKELL, B. WENZIEN, and F. BECHSTEDT, Phys. Rev. B **50**, 17037 (1994).

[23] F. BECHSTEDT, in: Festkörperprobleme/Adv. Solid State Phys. **32**, 161 (1992).

[24] P. LAWAETZ, Solid State Commun. **16**, 65 (1975).

[25] P. KÄCKELL, PhD Thesis Jena 1996, unpublished.

[26] W. R. L. LAMBRECHT and B. SEGALL, Phys. Rev. B **52**, R2249 (1995).

[27] P. KÄCKELL, B. WENZIEN, and F. BECHSTEDT, Phys. Rev. B **50**, 10761 (1994).

[28] R. KAPLAN, R. J. WAGNER, H. J. KIM, and R. F. DAVIS, Solid State Commun. **55**, 67 (1985).

[29] J. KONO, S. TAKEYAMA, H. YOKOI, N. MIURA, M. YAMANAKA, M. SHINOHARA, and K. IKOMA, Phys. Rev. B **48**, 10909 (1993).

[30] G. PENSL, private communication.

[31] D. VOLM, B. K. MEYER, D. M. HOFMANN, W. M. CHEN, N. T. SON, C. PERSON, U. LINDEFELT, O. KORDINA, E. SÖRMAN, A. O. KONSTANTINOV, B. MONEMAR, and E. JANZÉN, Phys. Rev. B **53**, 15409 (1996).

[32] W. GÖTZ, A. SCHÖNER, G. PENSL, W. SUTTROP, W. J. CHOYKE, R. STEIN, and S. LEIBENZEDER, J. Appl. Phys. **73**, 3332 (1993).

[33] W. SUTTROP, G. PENSL, W. J. CHOYKE, R. STEIN, and S. LEIBENZEDER, J. Appl. Phys. **72**, 3708 (1992).

[34] TH. TROFFER, W. GÖTZ, A. SCHÖNER, W. SUTTROP, G. PENSL, R. P. DEVATY, and W. J. CHOYKE, Silicon Carbide and Related Materials, Inst. Phys. Conf. Ser. **137**, 173 (1994).

[35] H. HARIMA, S. NAKASHIMA, and T. UEMURA, J. Appl. Phys. **78**, 1996 (1995).

[36] A. V. MELNICHUK and Y. A. PASECHNIK, Soviet Physics − Solid State **34**, 227 (1992).

[37] N. T. SON, O. KORDINA, A. O. KONSTANTINOV, W. M. CHEN, E. SÖRMAN, B. MONEMAR, and E. JANZÉN, Appl. Phys. Lett. **65**, 3209 (1994).

[38] B. ELLIS and T. S. MOSS, Proc. Roy. Soc. A **299**, 383, 393 (1967).

[39] G. A. LOMAKINA and YU. A. VODAKOV, Soviet Phys. − Solid State **15**, 83 (1973).

[40] M. SUZUKI, T. UENOYAMA, and A. YANASE, Phys. Rev. B **52**, 8132 (1995).

[41] G. B. DUBROVSKII and V. I. SANKIN, Fiz. Tverd. Tela **14**, 1200 (1972) (Soviet Phys. − Solid State **14**, 1024 (1972)).

[42] R. G. HUMPHREYS, D. BIMBERG, and W. J. CHOYKE, Proc. 15th Internat. Conf. Physics of Semiconductors, Kyoto, 1980; J. Phys. Soc. Japan **49**, Suppl. A, 519 (1980).

[43] R. G. HUMPHREYS, D. BIMBERG, and W. J. CHOYKE, Solid State Commun. **39**, 163 (1981).

[44] G. DRESSELHAUS, A. F. KIP, and C. KITTEL, Phys. Rev. **98**, 368 (1955).

[45] G. WELLENHOFER and U. RÖSSLER, Diamond and Related Materials, in print.

[46] R. G. HUMPHREYS, J. Phys. C: Solid State Phys. **15**, 2935 (1981).

[47] W. VAN HAERINGEN, B. A. BOBERET, and W. H. BACHES, phys. stat. (b) **202**, 63 (1997).

[48] W. R. L. LAMBRECHT, S. LIMPIJUMNONG, S. N. RASHKEEV, and B. SEGALL, phys. stat. sol. (b) **202**, 5 (1997).

[49] F. BECHSTEDT, P. KÄCKEL, A. ZYWIETZ, K. KARCH, B. ADOLPH, K. TENELSEN, and J. FURTHMÜLLER, phys. stat. sol. (b) **202**, 35 (1997).

[50] J. POLLMANN, P. KRÜGER, and M. SABISCH, phys. stat. sol. (b) **202**, 421 (1997).

phys. stat. sol. (b) **202**, 125 (1997)

Subject classification: 71.55Ht; 71.15.Hx; S6

First-Principles Calculations of Impurity States in 3C-SiC

A. FUKUMOTO

Toyota Central Research and Development Laboratories, Inc., 41-1 Aza Yokomichi, Oaza Nagakute, Nagakute-cho, Aichi-gun, Aichi-ken 480-11, Japan

(Received January 31, 1997)

The electronic structures, atomic geometries, and formation energies of several substitutional impurities in 3C-SiC are calculated by the first-principles pseudopotential method. N is considered as the n-type dopant and Al and B as the p-type dopants. It is shown that N forms a shallow donor level on a C site and a localized state on a Si site, and both Al and B form a shallow acceptor level on a Si site and a deep level on a C site. For N, a C site has the lower formation energy than a Si site. For Al, a Si site has the lower formation energy than a C site. These features do not depend on the composition. For Al and N, the impurities on the favorable sites form shallow levels while for B, the lower formation energy site depends on the composition. A C site is favorable for B under Si-rich conditions and a Si site under C-rich conditions. These results suggest that the low-resistivity p-type SiC crystal can be fabricated under C-rich conditions when B is employed as a dopant.

1. Introduction

Silicon carbide has a technological potential for high-temperature, high-power, and high-frequency devices. However, there are several technological difficulties that must be overcome before realizing SiC electron devices. The development of appropriate doping processes is very important for fabricating low-resistivity crystal. Nitrogen ion implantation technique is one of the promising candidates for selective impurity doping for n-type crystal. However, it is observed that the carrier activation rate is very low although the defect density is sufficiently low [1] and N is known to be a good dopant which forms a shallow donor level [2]. On the other hand, in order to produce low-resistivity p-type crystal, it is essential to find a good dopant which forms a shallow acceptor level. Al and B are usually employed as p-type dopants. It is known from the experimental photoluminescence spectra [2, 3] that the acceptor level of Al substituted on a Si site is at 257 meV, and that of B substituted on a C site at 735 meV in 3C-SiC; 735 meV is too deep, and 257 meV is not sufficiently shallow for realizing low resistivity. Moreover, Al causes troubles in the process of producing devices because it is easily oxidized [4]. Therefore, another type of impurity which forms a shallower acceptor level is desirable.

Recently, two B-related levels, a shallow level and a deep level, have been observed by several researchers in hexagonal SiC [5]. Unlike the elemental semiconductors, the compound semiconductors such as SiC have two substitutional sites for impurities. Considering that B has middle values between Si and C both in atomic covalent radius and electronegativity, and also that both Al and B form a shallow level in Si crystal, it is expected that B substituted on a Si site in SiC will have low formation energy and will form a shallow level.

Table 1
The atomic radii and electronegativities

	N	C	B	Si	Al
atomic radius (Å) [7]	0.70	0.77	0.88	1.17	1.26
electronegativity [8]	3.0	2.5	2.0	1.8	1.5

In this paper, the substitutional impurity states in 3C-SiC are investigated by first-principles pseudopotential calculations. We employ N as n-type dopant [1] and Al and B as p-type dopants [6]. The atomic covalent radii and electronegativities of the five kinds of atoms considered here are sumarized in Table 1. The atomic geometries, electronic structures, and formation energies of six types of substitutional impurities, N on a Si site (N_{Si}), N on a C site (N_C), Al on a Si site (Al_{Si}), Al on a C site (Al_C), B on a Si site (B_{Si}), and B on a C site (B_C) are calculated and the possibility of realizing a low-resistivity crystal is discussed.

2. Method

The calculations are based on the local-density approximation [9] and norm-conserving pseudopotential method [10]. The exchange-correlation interactions are described with Wigner's formula [11]. The pseudopotentials are generated by the method proposed by Troullier and Martins [12]. A 64-atom cubic supercell is used, and one Si or C atom in the zincblende structure is replaced by N, Al or B atom in order to represent the isolated impurity. The lattice constant is fixed to the experimental value of 4.36 Å. Only the Γ point is used for the Brillouin-zone summations. Wave functions are expanded in a plane-wave basis with energy up to 21.9 Hr; that involves 21975 plane waves. Due to the softness of Troullier and Martins pseudopotentials, this cutoff energyy gives convergent total energy as well as 1 mHr/atom in a SiC perfect crystal of the zincblende structure. The nonlocality of the pseudopotentials is treated exactly [6].

In order to initiate the impurity calculations, the wave functions at fixed ionic configurations are obtained with the preconditioned conjugate gradient procedure [13]. After well-converged wave functions are obtained, the atomic degrees of freedom and the electronic degrees of freedom are allowed to relax and to be simultaneously optimized by the Car and Parrinello method [14]. The atoms in the unit cell are moved according to Newton's equations of motion. In order to find atomic coordinates corresponding to the local energy minima, the dissipative force for atomic motion is provided. The velocity of any atom is reduced by a constant factor when the inner product of force and velocity is negative [15]. The evolution of the wave functions is calculated according to the classical equations of motion under the constraint of orthonormality. The preconditioning for the electronic degrees of freedom is also applied at this stage since it makes a larger Δt acceptable and has the effect of introducing a dissipative force on the electronic degrees of freedom. After the forces on atoms have become negligible (less than 5×10^{-4} at. units), the wave functions are optimized with fixed atomic coordinates in order to confirm the system is at the ground state or not. If the forces appear as a result of the optimization with respect to the electronic degrees of freedom, the procedure should be repeated. However, in our experience, the forces do not appear since the wave functions do not deviate from the vicinity of the Born-Oppenheimer surfaces during the structural optimization procedure.

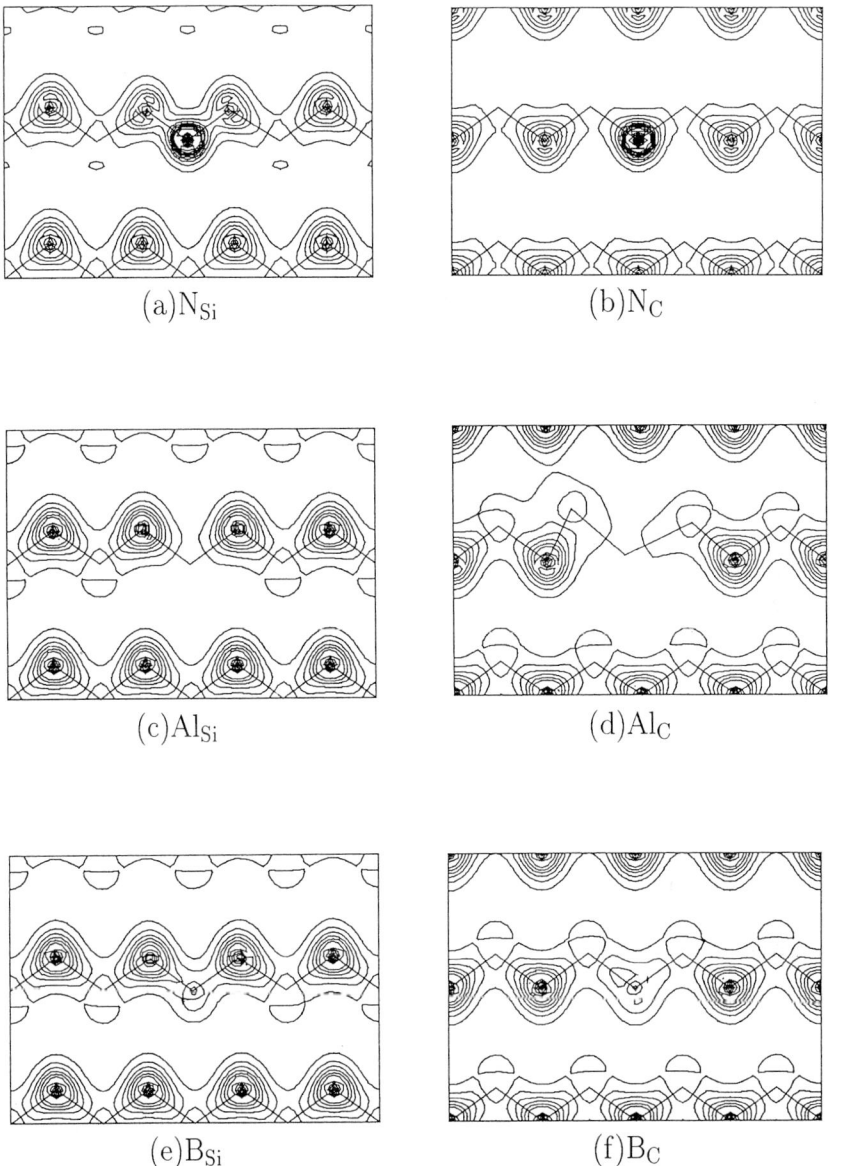

Fig. 1. The atomic configurations and the valence charge densities in the $(1\bar{1}0)$ plane for a) N_{Si}, b) N_C, c) Al_{Si}, d) Al_C, e) B_{Si}, and f) B_C. The contour spacing is 5×10^{-2} at. units for a) and b), and 4×10^{-2} at. units for c) to f). The central atom is the impurity

3. Results and Discussion

3.1 Atomic configurations and charge densities

The atomic configurations and the valence charge densities in the $(1\bar{1}0)$ plane are shown in Fig. 1. The atomic configuration around substitutional N holds the T_d symmetry of

Table 2

The bond lengths between the impurities and the neighbouring host atoms (Å). The numbers in parentheses are the deviations from the Si–C bond length in the perfect crystal

	d_{111}	$d_{\bar{1}\bar{1}1}$	$d_{1\bar{1}\bar{1}}, d_{\bar{1}1\bar{1}}$
N_{Si}	1.70 (-9.7%)	1.70 (-9.7%)	1.70 (-9.7%)
N_C	1.93 ($+2.4\%$)	1.93 ($+2.4\%$)	1.93 ($+2.4\%$)
Al_{Si}	1.99 ($+5.3\%$)	1.96 ($+3.7\%$)	1.96 ($+3.7\%$)
Al_C	2.43 ($+29\%$)	2.23 ($+18\%$)	2.30 ($+22\%$)
B_{Si}	1.86 (-1.6%)	1.76 (-7.0%)	1.76 (-7.0%)
B_C	1.97 ($+4.2\%$)	1.91 ($+1.1\%$)	1.91 ($+1.1\%$)

the host material. The p-type impurities give rise to a Jahn-Teller distortion because of the partially filled degenerate level in the neutral charge state. The charge density difference between the crystal with the impurity and the perfect crystal is small when N is substituted on a C site and when Al or B is substituted on a Si site. It is relatively large in the other cases, particularly for Al_C. The distortion of the atomic configuration is also very large for Al_C. This large distortion is due to the fact that a small C atom is replaced by a large Al atom. The symmetry observed for Al_{Si}, B_{Si}, and B_C is C_{3v}, lowered from the T_d symmetry of the host material. For Al_C, the Al atom moves toward the $[1\bar{1}0]$ direction. The bond lengths between the impurities and the nearest neighbouring host atoms are shown in Table 2. The Al impurity bond lengths expand in both cases. The N and B impurity bond lengths contract when impurities are substituted on a Si site, while they expand when substituted on a C site. These results agree with the relation of the atomic covalent radii shown in Table 1 except for the case of N_C. (The definition of the atomic radii is not so rigid to discuss the difference between N and C.) The relaxation of the atomic configuration at the second neighbour is very small (less than 0.025 Å) except for the case of Al_C.

The squared wave functions of the half-occupied highest state, which are equivalent to the donor (for N) or acceptor (for Al and B) densities, are shown in Fig. 2. The donor density of N_C is widely spread over the crystal. The density is high at the antibonding sites around Si atoms. This is the typical feature of the wave functions at the conduction band bottom in the perfect crystal. Therefore, N_C is expected to form a shallow donor level. The donor density of N_{Si} is very localized around N atom. This is the typical feature of a deep level. Al_{Si} and B_{Si} have very similar hole densities. The wave functions of the hole are composed of the p-orbitals of the C atoms. This feature is qualitatively the same as the wave functions at the valence band top of the SiC perfect crystal. B_C has also the same character, but the p-orbital of the B atom contributes largely to the hole density. In the case of Al_C, the wave function is highly localized around the Al atom.

3.2 Electronic structures

The character of the donor levels can be identified accurately from the dispersion of the band energy. Impurity levels which have no dispersion for an isolated impurity have dispersion when using finite-size supercells. Fig. 3 shows the band structure of $Si_{31}C_{32}N$ and $Si_{32}C_{31}N$ around the conduction band bottom. It is known that the local-density

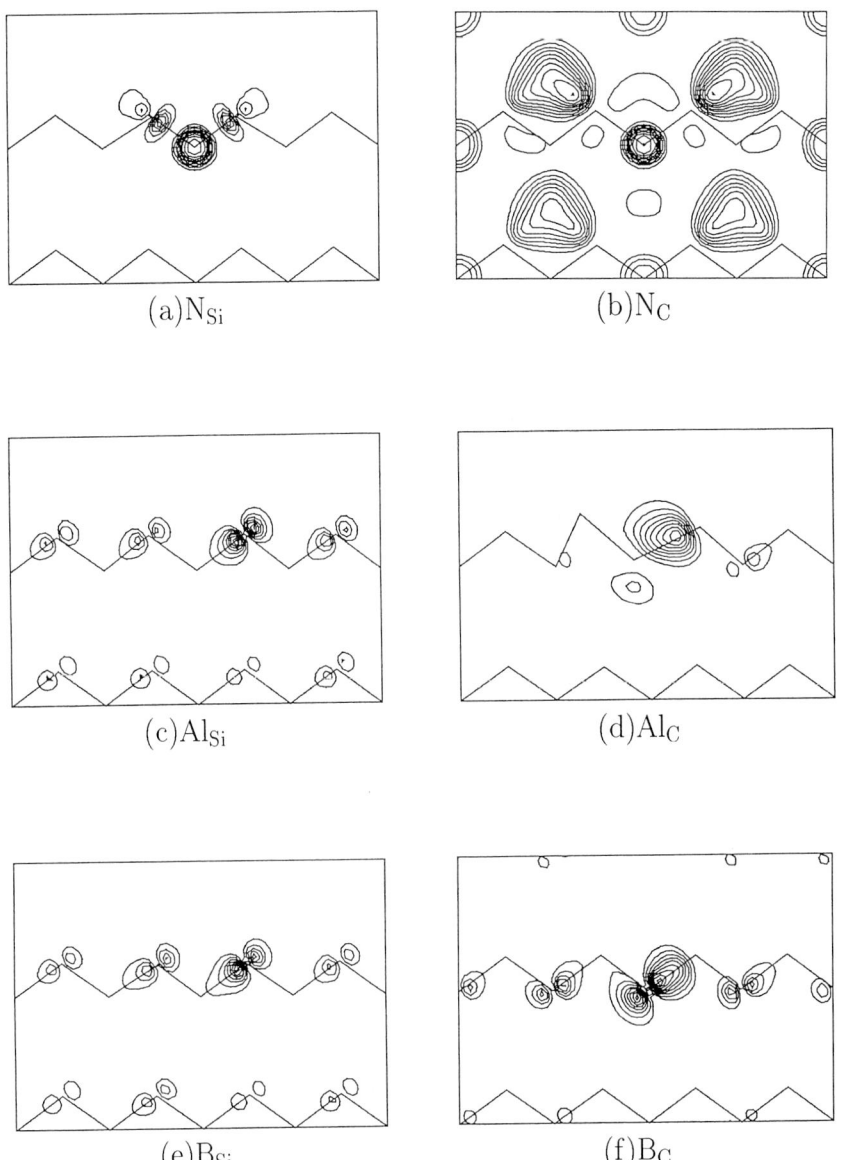

Fig. 2. The squared wave functions of the impurity states in the $(1\bar{1}0)$ plane for a) N_{Si}, b) N_C, c) Al_{Si}, d) Al_C, e) B_{Si}, and f) B_C. The contour spacing is 5×10^{-3} at. units for a), 5×10^{-4} at. units for b), and 2×10^{-3} at. units for c) to f). The central atom is the impurity

approximation gives rather accurate dispersion of the band energy although it underestimates the band gap. The dispersion of the $Si_{32}C_{31}N$ system is very similar to that of the perfect crystal. (The conduction band minimum in Fig. 3 is at the Γ point while it is well known to be located at the X point in the zincblende structure. This is a result of the band folding due to the 64-atom supercell geometry.) The t_2 state at the conduction

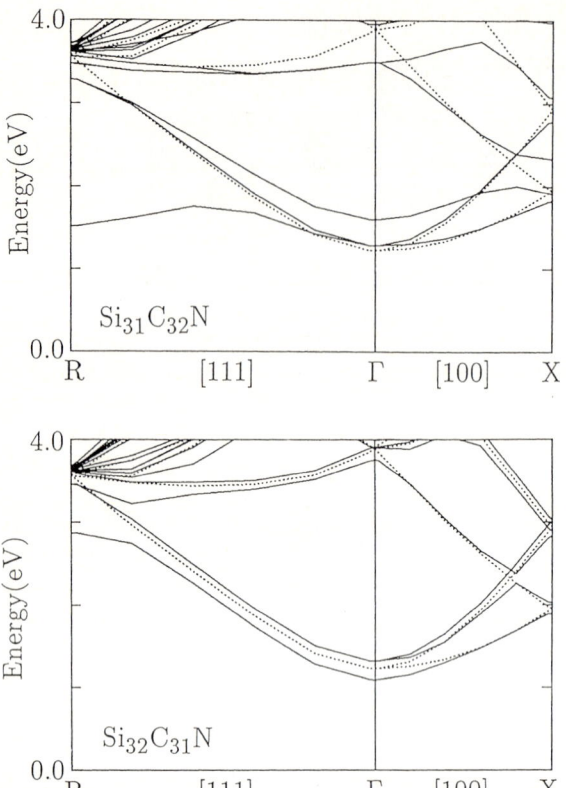

Fig. 3. Energy band structure of $Si_{31}C_{32}N$ (upper part) and $Si_{32}C_{31}N$ (lower part) around the bottom of the conduction band. The dotted lines denote the band structure of the perfect crystal. The origin of the energy is at the top of the valence band

band bottom in the perfect crystal splits into a_1 and e states and the a_1 state can be considered to be the shallow donor level. In order to study the donor level, the ionization energy of an impurity E_{ion} is calculated by

$$E_{ion}(q) = E_{tot}(q) - E_{tot}(0) - q\mu_F , \qquad (1)$$

with $q = 1$, where $E_{tot}(q)$ is the total energy of a charge state q, and μ_F is the Fermi energy which is assumed to be the energy of the conduction band bottom [16]. The total charge of the simulation supercell must be zero in order to compensate for the long-range Coulomb interactions. A uniform background charge is introduced in order to neutralize the unit cell. The atomic configurations are optimized again for the calculation of $E_{tot}(q)$. The calculated ionization energy is 152 meV for the $Si_{32}C_{31}N$ system. This is quite larger than the experimental value of 54 meV [3]. This is due to the fact that the size of the supercell is too small. We need to enlarge the size of the supercell until the dispersion of the impurity level becomes negligible in order to evaluate the ionization energy quantitatively. (The effect of the cell size will be discussed later for the case of the acceptor levels.) In the case of the $Si_{31}C_{32}N$ system, the dispersion differs very much from that of the perfect crystal. The t_2 state at the conduction band bottom holds and another branch with relatively weak dispersion appears around the conduction band bottom. This branch can be identified as a highly localized state. The wave function shown in Fig. 2a is the lowest a_1 state at the Γ point. If a larger supercell is used,

Table 3

The ionization energies of acceptors

	Al_{Si}	Al_C	B_{Si}	B_C
E_{ion} (meV)	50	1130	10	260

this localized state is expected to appear in the band gap. The method of evaluating an impurity level with Eq. (1) is not appropriate for the case of $Si_{31}C_{32}N$ since the donor level does not come out from the conduction band bottom states of the perfect crystal.

The ionization energies of the acceptors are also calculated by Eq. (1) with $q = -1$ and μ_F assumed to be the energy of the valence band top. In the -1 charge state, the T_d symmetry is recovered except for the case of Al_C, as expected for a closed-shell structure. The results are shown in Table 3. The calculations indicate that Al_{Si} and B_{Si} have shallow levels. B_C has a deeper level than those two impurities, and Al_C has a much deeper level.

It is known from the experimental photoluminescence spectra [2, 3] that Al_{Si} has an acceptor level at 257 meV and B_C has an acceptor level at 735 meV in the cubic SiC. The calculation and the experiment do not agree well quantitatively, even if we take into account that the ionization energy and the photoluminescence spectra are physically not equivalent. The author confirmed that in the 64-atom Si system, the calculated ionization energies of Al and B showed much better agreement with the experiment. We may need a larger unit cell in the case of compound semiconductors with ionicity. The screening property of the compound (polar) semiconductors is different from that of the elemental (homopolar) semiconductors since the compound semiconductors consist of alternately positively and negatively charged ions, and the electrons are nonuniformly distributed in the crystal [17]. However, calculations with a larger unit cell are impracticable because of their large computation cost. In order to examine the effect of the cell size, the same procedures were performed with a 32-atom b.c.c. supercell. The difference of the ionization energy between the 64-atom supercell and the 32-atom supercell was less than 20 meV for Al_{Si} and B_{Si}, and was about 100 meV for B_C. (The calculation of Al_C with the 32-atom supercell was not performed.) The order of the ionization energy of impurities was unchanged. From this examination and the dispersion of the band energy, the error in ionization energy due to the cell size can be estimated to be in the order of 100 meV. This estimation and the comparison of calculation and experiment indicate that Table 3 gives qualitatively meaningful results.

The most important result of Table 3 is that B_{Si} has a shallow acceptor level. The fact that B_{Si} forms a shallower level than B_C can be explained by the amount of contribution of the p-orbital of the B atom to the hole density. The p-orbital of B has a larger contribution in B_C than in B_{Si} as shown in Fig. 2e and f.

3.3 Formation energies

Now, we will consider the favorable sites of the impurities. In the system under consideration, the formation energy E_f of an impurity [18] is given by

$$E_f = E_{tot}(n_{Si}, n_C, n_X) - n_{Si}\mu_{Si} - n_C\mu_C - n_X\mu_X, \tag{2}$$

where E_{tot} is the total energy of a supercell, n is the number of atoms, X means N, Al or B, and μ is the chemical potential. The charge state is assumed to be neutral. The formation energy difference between the case of X on a Si site and on a C site can be written as

$$E_f(X_{Si}) - E_f(X_C) = E_{tot}(31_{Si}, 32_C, 1_X) - E_{tot}(32_{Si}, 31_C, 1_X) + \mu_{Si} - \mu_C. \qquad (3)$$

The chemical potentials depend on the experimental conditions under which the material is grown. The chemical potentials for Si and C are not independent, since both species are in equilibrium with SiC, but their sum is equal to the bulk energy per pair [19],

$$\mu_{Si} + \mu_C = \mu_{SiC(bulk)}, \qquad (4)$$

where $\mu_{SiC(bulk)}$ is the total energy of SiC in the zincblende structure. Thus, only one free parameter remains. In the following, $\mu_{Si(bulk)}$ and $\mu_{C(bulk)}$ are the total energy of Si and C in the diamond structure. The total energy difference between diamond and graphite is small enough [20] to prevent a serious error in the following discussion of the formation energy. Considering the relation

$$\mu_{SiC(bulk)} = \mu_{Si(bulk)} + \mu_{C(bulk)} - \Delta H_f, \qquad (5)$$

where ΔH_f is the heat of formation, we can define the lower and upper limits of the chemical potentials as

$$\mu_{Si(bulk)} - \Delta H_f \leq \mu_{Si} \leq \mu_{Si(bulk)}, \qquad (6)$$

$$\mu_{C(bulk)} - \Delta H_f \leq \mu_C \leq \mu_{C(bulk)}. \qquad (7)$$

ΔH_f is equal to 0.7 eV in the case of SiC. If these conditions are not satisfied, the system would be thermodynamically unstable and segregate into different components. Finally, defining $\Delta\mu$ by

$$\mu_{Si} = \mu_{Si(bulk)} + \Delta\mu, \qquad (8)$$

the formation energy differences are written as

$$E_f(N_{Si}) - E_f(N_C) = 8.0 + 2\Delta\mu \text{ (eV)}, \qquad (9)$$

$$E_f(Al_{Si}) - E_f(Al_C) = -4.3 + 2\Delta\mu \text{ (eV)}, \qquad (10)$$

$$E_f(B_{Si}) - E_f(B_C) = 1.0 + 2\Delta\mu \text{ (eV)}; \qquad (11)$$

$$-0.7 \text{ eV} \leq \Delta\mu \leq 0 \text{ eV}. \qquad (12)$$

$\Delta\mu = -0.7 \ (=0)$ means that the crystal is under the limiting case of C-rich (Si-rich) condition. The results are summarized in Table 4 where the charge states are assumed to be neutral.

Table 4
The formation energy difference (eV)

$\Delta\mu$	−0.7 (C-rich)	0 (Si-rich)
$E_f(N_{Si}) - E_f(N_C)$	6.6	8.0
$E_f(Al_{Si}) - E_f(Al_C)$	−5.7	−4.3
$E_f(B_{Si}) - E_f(B_C)$	−0.4	1.0

A C site is always more favorable than a Si site for N, and a Si site is always more favorable than a C site for Al regardless of the composition. However, the favorable site for B depends on the composition. Under the Si-rich condition, a C site is favorable, while a Si site is favorable under the C-rich condition. These results hold when the donors and acceptors are assumed to be positively and negatively charged, respectively.

3.4 Discussion

It is known that N is a good dopant which forms a shallow donor level [2]. However, the carrier activation rate is very low when the ion implantation technique is employed for doping [1]. The defect density is sufficiently low to be undetectable by Rutherford back-scattering spectrometry. One possible reason for the low activation rate is the presence of the N atoms substituted on a Si site which form a deep donor level. However, our calculation shows that the favorable site of N is a C site regardless of the composition. We should consider other microscopic structures such as the complex defect composed of interstitial carbon and substituted nitrogen proposed by Miyajima et al. [1] in order to explain the low activation rate. Si-rich condition is appropriate in order to prevent the presence of interstitial carbon atoms. In addition, the author has confirmed that a C vacancy, which is the dominant native defect in the Si-rich SiC when the Fermi level is close to the valence band top [21], forms a shallow donor level. When the Fermi level is close to the conduction band bottom, the dominant native defect is a Si_C antisite which does not have electronic states in the gap. Therefore, Si-rich condition is suitable for n-type doping.

When Al is employed as a dopant, only one acceptor level has been observed. According to the results of our calculations, this is due to the fact that a Si site is always favorable for Al regardless of the composition. For B, on the other hand, the two B-related acceptors have been observed in the hexagonal SiC [5]. Bernholc et al. [22] have pointed out by first-principles calculations of 3C-SiC that B on Si site and on C site have very similar formation energies in C-rich SiC. The results of 3C-SiC will be qualitatively the same for other polytypes because the polytype dependency of the acceptor levels is weak (less than 100 meV) [23]. The microscopic structure of the two B-related acceptors in hexagonal SiC can be explained from our results. The shallow and deep levels are due to B_{Si} and B_C, respectively. B atoms substituted on Si site and on C site have similar formation energies since B has middle values between Si and C both in atomic covalent radius and electronegativity. B_{Si} forms a shallower level than B_C since the p-orbital of B gives a smaller contribution to the hole density in B_{Si} than in B_C as shown in Fig. 2e and f. In addition, this hole density distribution of B_{Si} supports the electron paramagnetic resonance (EPR) results of Baranov and Mokhov [5] that the unpaired electron of the shallow acceptor is localized on the carbon atom. Of course, a more detailed investigation is needed to confirm that the two B-related acceptors are really the simple substitutional impurities.

The concentration ratio of B_{Si} and B_C depends on the crystal growth conditions since they have similar formation energies. It is very important to control the [Si]/[C] ratio during the process of crystal growth in order to make the concentration of B_{Si} higher than that of B_C. As a matter of fact, it is reported from the studies of lattice parameters and densities that the cubic SiC is nonstoichiometric and Si-rich with a deviation of less than 1% from stoichiometry [24, 25]. Si-rich SiC is not suitable for p-type doping because of the presence of the C vacancy as mentioned above. The C-rich SiC is suitable for p-type doping. The dominant native defect in the C-rich SiC is a C_{Si} antisite regardless of the position of the Fermi level [21]. This defect is electronically inactive, and does not give rise to compensation. The low-resistivity p-type SiC crystal can be obtained when B is employed as a dopant if C-rich conditions can be realized.

4. Summary

We have performed pseudopotential total-energy calculations of the substitutional impurity states in 3C-SiC with N, Al and B as dopants. Substitutional N forms a shallow donor level on a C site, and a deep donor level on a Si site. Substitutional Al and B form shallow acceptor levels on a Si site and deep acceptor levels on a C site. The favorable site is always a C site for N and a Si site for Al regardless of the composition. For B, the favorable site depends on the composition. B on Si site is favorable under C-rich conditions and forms a shallow acceptor level. On the other hand, B on C site is favorable under Si-rich conditions and forms a deep level. The difference of the acceptor levels between B_{Si} and B_C can be explained by the amount of contribution of the p-orbital of the B atom to the hole density. The suitable crystal condition is different between n-type and p-type. Si-rich SiC is suitable for n-type doping and C-rich SiC is suitable for p-type doping.

Acknowledgement   The author would like to thank K. Hara, N. Tokura, T. Miyajima, H. Fuma, H. Hayashi, and K. Miwa for helpful discussions.

References

[1] T. MIYAJIMA, N. TOKURA, A. FUKUMOTO, H. HAYASHI, and K. HARA, Jpn. J. Appl. Phys. **35**, 1231 (1996).
T. MIYAJIMA, N. TOKURA, A. FUKUMOTO, H. HAYASHI, and K. HARA, Extended Abstracts, Internat. Conf. Solid State Devices and Materials, Osaka 1995 (p. 758).
[2] J. A. FREITAS, JR., W. E. CARLOS, and S. G. BISHOP, Amorphous and Crystalline Silicon Carbide III, Ed. G. L. HARRIS, M. G. SPENCER, and C. Y.-W. YANG, Springer-Verlag, Berlin/Heidelberg/New York 1990 (p. 135).
[3] S. G. BISHOP, J. A. FREITAS, JR., T. A. KENNEDY, W. E. CARLOS, W. J. MOORE, P. E. R. NORDQUIST, and M. L. GIPE, in: Amorphous and Crystalline Silicon Carbide, Ed. G. L. HARRIS and C. Y.-W. YANG, Springer-Verlag, Berlin/Heidelberg/New York 1987 (p. 90).
[4] C.-M. ZETTERLING and M. ÖSTLING, in: Diamond, SiC and Nitride, Wide Bandgap Semiconductors, Ed. C. H. CARTER, JR., G. GILDENBLAT, S. NAKAMURA, and R. J. NEMANICH, MRS Symp. Proc. No. 339, 209 (1994).
[5] T. TROFFER, C. HÄSSLER, G. PENSL, K. HÖLZLEIN, H. MITLEHNER, and J. VÖLKL, Silicon Carbide and Related Materials, Ed. S. NAKASHIMA, H. MATSUNAMI, S. YOSHIDA, and H. HARIMA, Institute of Physics Publishing, Bristol/Philadelphia 1996 (p. 281).
P. G. BARANOV and E. N. MOKHOV, ibid. (p. 293).
R. A. STEIN, R. RUPP, K. O. DOHNKE, J. VÖLKL, D. STEPHANI, C. HÄSSLER, and G. PENSL, ibid. (p. 505).
[6] A. FUKUMOTO, Phys. Rev. B **53**, 4458 (1995).
[7] C. KITTEL, Introduction to Solid State Physics, 5th ed., Wiley, New York 1976.
[8] L. PAULING, The Nature of the Chemical Bond, 3rd ed., Cornell University Press, Ithaca (New York) 1960.
[9] P. HOHENBERG and W. KOHN, Phys. Rev. **136**, B864 (1964)
W. KOHN and L. J. SHAM, Phys. Rev. **140**, A1133 (1965).

[10] D. R. Hamann, M. Schlüter, and C. Chiang, Phys. Rev. Lett. **43**, 1494 (1979).
[11] E. Wigner, Phys. Rev. **46**, 1002 (1934).
[12] N. Troullier and J. L. Martins, Phys. Rev. B **43**, 1993 (1991).
[13] M. P. Teter, M. C. Payne, and D. G. Allan, Phys. Rev. B **40**, 12555 (1989).
     K. Miwa and A. Fukumoto, Phys. Rev. B **52**, 147748 (1995).
[14] R. Car and M. Parrinello, Phys. Rev. Lett. **55**, 2471 (1985).
[15] Q.-M. Zhang, J.-Y. Yi, and J. Bernholc, Phys. Rev. Lett. **66**, 2633 (1991).
[16] G. A. Baraff and M. Schlüter, Phys. Rev. B **30**, 1853 (1984).
[17] S. T. Pantelides, Rev. Mod. Phys. **50**, 797 (1978).
[18] C. G. Van de Walle, D. B. Laks, G. F. Neumark, and S. T. Pantelides, Phys. Rev. B **47**, 9425 (1993).
[19] G.-X. Qian, R. M. Martin, and D. J. Chadi, Phys. Rev. B **38**, 7649 (1993).
[20] S. Fahy, X. W. Wang, and S. G. Louie, Phys. Rev. B **42**, 3503 (1990).
[21] C. Wang, J. Bernholc, and R. F. Davis, Phys. Rev. B **38**, 12752 (1988).
[22] J. Bernholc, S. A. Kajiwara, C. Wang, A. Antonelli, and R. F. Davis, Mater. Sci. Engng. B **14**, 265 (1992).
[23] M. Ikeda, H. Matsunami, and T. Tanaka, Phys. Rev. B **22**, 2842 (1980).
[24] K. L. More, J. Ryu, C. H. Carter, Jr., J. Bentley, and R. F. Davis, Cryst. Lattice Defects Amorphous Mater. **12**, 243 (1985).
[25] D. P. Birnie III, J. Amer. Ceram. Soc. **69**, C-33 (1086).

phys. stat. sol. (b) **202**, 137 (1997)

Subject classification: 68.55.Jk; 68.55.Ln; 73.61.Le; S6

Physical Vapor Transport Growth and Properties of SiC Monocrystals of 4H Polytype

G. Augustine, H. McD. Hobgood, V. Balakrishna, G. Dunne, and R. H. Hopkins

Northrop Grumman Corporation, Electronic Sensors & Systems Division, Science and Technology Center, 1350 Beulah Road, Pittsburgh, PA 15235-5080, USA

(Received January 31, 1997)

The physical vapor transport technique can be employed to fabricate large diameter silicon carbide crystals (up to 50 mm diameter) exhibiting uniform 4H-polytype over the full crystal volume. Crystal growth rate is controlled to first order by temperature conditions and ambient pressure. 4H-polytype uniformity is controlled by polarity of the seed crystal and the growth temperature. 4H-SiC crystals exhibit crystalline defects mainly in the form of dislocations with densities in the 10^4 cm^{-2} range and micropipe defects, the latter having densities as low as 10 cm^{-2} in best crystals. Electrical conductivity in 4H-SiC bulk crystals ranges from $<10^{-2}\,\Omega$ cm, n-type, to insulating ($>10^{15}\,\Omega$ cm) at room temperature.

1. Introduction

For semiconductor devices operating under high temperature or high power conditions, silicon carbide (SiC) offers an advantageous combination of electronic and physical properties. The large bandgap of SiC (see Table 1) results in an increased maximum operating temperature. SiC diodes and FETs for example have been operated at temperatures greater than 200 °C with little loss in electrical properties. In addition, as shown in Table 1, compared to Si and GaAs, SiC exhibits a higher thermal conductivity $\varkappa_T$, a higher critical electric field (E_b) at which breakdown occurs, and a saturated carrier velocity (V_{sat}) equal to that of GaAs at the high fields desirable for high-power devices. Furthermore, SiC is physically rugged due to its high mechanical strength, which is an advantage in wafer handling and device fabrication.

Table 1

Comparison of fundamental properties of Si, GaAs, and SiC for device applications

	E_g (eV)	E_b (MV/cm)	V_{sat} (10^7 cm/s)	$\varkappa_T$ (W/cm K)	ε
Si	1.12	0.6	1.0	1.5	11.8
GaAs	1.42	0.6	2.0	0.5	12.8
6H-SiC	3.02	3.2	2.0	3.0[*]	10.0
4H-SiC	3.26	3.0	2.0	3.0[*]	9.7

[*] For substrate doping density of $\approx 5 \times 10^{18}$ cm^{-3}.

2. Advances in Crystal Growth Technology

Single-crystal SiC was formerly produced by the Lely [1] growth method and existed only as small platelets of varying size (up to 10 mm) as shown in Fig. 1. In the Lely technique, platelets are grown within a cavity formed in a charge of SiC by heating to approximately 2500 °C under essentially isothermal conditions in an argon atmosphere. Under these conditions, platelets nucleate randomly along the vapor transport flow paths into the growth cavity [2].

Although as-grown junction and diffused junction devices were fabricated using Lely platelets [3, 4], they generally displayed non-uniform physical and electrical properties, and the individual SiC platelets were difficult to handle. Consequently, work on SiC material and devices virtually stopped in the U.S. in the mid 1970s.

Fig. 1. a) Lely cavity and b) grown platelets

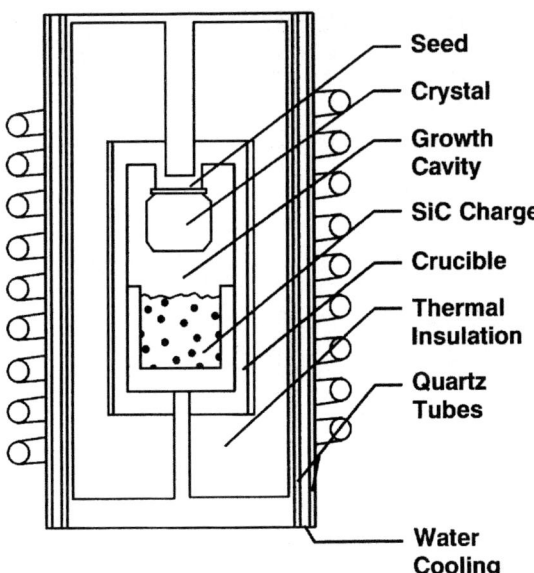

Fig. 2. Schematic of physical vapor transport growth cavity

Seed

Crystal

Growth Cavity

SiC Charge

Crucible

Thermal Insulation

Quartz Tubes

Water Cooling

Recent advances in the growth of large single-crystal boules and substrates of SiC [5 to 11] and the preparation of SiC epitaxial layers [12] have enabled the fabrication of practical SiC devices such as MESFETs [13], static induction transistors [14], MOSFETs [15], and GTOs [16].

At our laboratories, crystal growth of SiC has been developed with the goal of fabricating large-diameter crystals with uniform electrical properties and improved quality. 4H-SiC crystals are grown by the physical vapor transport (PVT) technique (see Fig. 2), in which growth proceeds by the sublimation of a SiC source and deposition of the vapor species upon a high-quality SiC monocrystalline seed wafer in an ultra-high purity

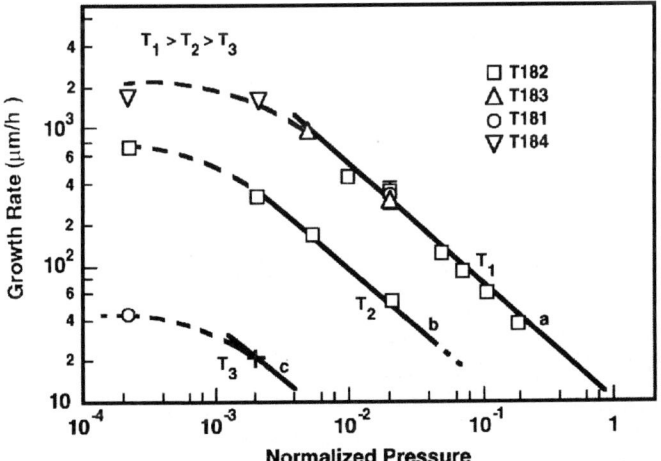

Fig. 3. Crystal growth rate as a function of ambient pressure for three different growth temperatures

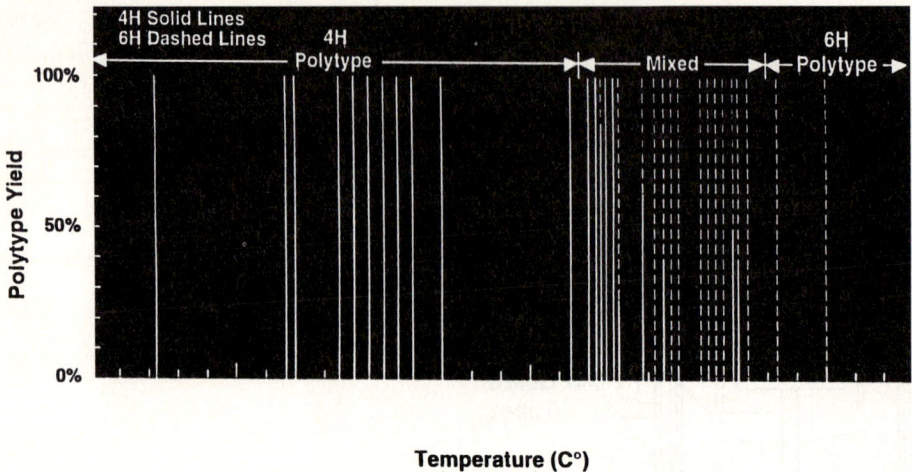

Fig. 4. Effect of crystal growth temperature on crystal polytype

inert ambient. The growth technique, already described elsewhere [8], utilizes an induction-heated, cold-wall system in which high-purity graphite materials constitute the hot-zone of the furnace. Undoped crystals are grown in high purity inert ambient provided by the boil-off of ultra-high-purity (UHP) liquid argon. N^+-doped crystals are prepared by adding controlled amounts of high-purity nitrogen to the argon ambient.

As-grown crystals are precisely oriented and machined to the required substrate dimensions. Wafers are sliced using conventional Si wafering technology modifed to accommodate the increased hardness of SiC. Substrates are prepared using a diamond-based abrasive polish technique [17]. Substrates fabricated from undoped high resistivity 4H-

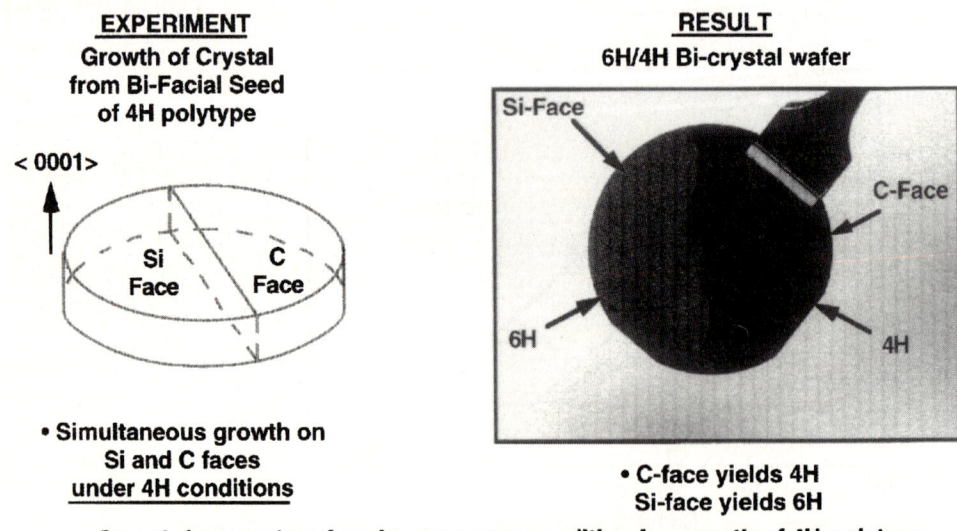

Fig. 5. Effect of seed surface polarity on crystal polytype

Fig. 6. 4H-SiC crystal grown by the physical vapor transport technique

SiC crystals are colorless and transparent in contrast to N^+ nitrogen-doped substrates which transmit mainly in the yellow-orange part of the visible spectrum due to the well-known Biedermann mechanism [18].

In general for PVT growth, the crystal growth rate will depend to first order on the parameters of: T_X, T_S, ΔT, L, and P_a where T_X is the crystal temperature, T_S is the source temeprature, $\Delta T = T_S - T_X$, L is the crystal to source spacing, and P_a is the ambient pressure. Fig. 3 shows the dependence of crystal growth rate on normalized ambient pressure as well as growth temperature, for our growth system. Clearly, the growth rate is dependent on crystal/source temperatures, falling dramatically as growth temperatures are reduced (compare rates for $T_1 > T_2 > T_3$). For a given crystal–source

	Current Wafer Characteristics		In Development	
dia:	1.188 in (30 mm)	1.375 in (35 mm)	1.500 in (38 mm)	2.000 in (50 mm)
rho:	10 to 15 mΩ cm	10 to 15 mΩ cm	10 to 15 mΩ cm	10 to 15 mΩ cm
pores:	150 cm^{-2} (avg)	200 cm^{-2} (avg)	375 cm^{-2} (avg)	500 cm^{-2} (avg)
	<25 cm^{-2} (best)	<25 cm^{-2} (best)	100 cm^{-2} (best)	100 cm^{-2} (best)
CEBs	no	no	yes (moderate)	yes

Fig. 7. Evolution of large diameter 4H-SiC wafers at Northrop Grumman

temperature condition, the growth rate is observed to increase as the ambient pressure is reduced. Over the pressure range studied, the growth rate is found to vary almost inversely as ambient pressure. This behavior is qualitatively consistent with a $1/P_a$ dependence on the molecular diffusion coefficient [19]. At very low pressures, the growth rate exhibits a saturation behavior.

Besides affecting the growth rate, the crystal growth temperature also has a profound effect on the crystal polytype (Fig. 4). 4H-SiC growth is favored at lower temperatures whereas the 6H polytype dominates at the higher temperatures [20]. An equally important factor affecting crystal polytype is the seed polarity [21, 22]. SiC crystals can be grown along the c-axis (along the $\langle 0001 \rangle$ direction) from either the silicon (0001) or the carbon (000$\bar{1}$) face of the seed crystal. To investigate the effect of seed crystal face on the crystal polytype the experiment, as illustrated in Fig. 5, was performed. A 4H-SiC seed crystal was split into two halves (left side of Fig. 5) so as to present simultaneously a silicon and a carbon terminated surface to the incident vapor flux during growth. The crystal was grown with nitrogen doping under temperature conditions favorable for 4H-SiC growth. The resulting nitrogen-doped crystal was a bicrystal of which one half was of the 6H-polytype while the other half consisted of the 4H-polytype. X-ray diffraction and optical absorption verified that growth on the Si-face of the seed produced the 6H-polytype (green in color), while growth on the C-face yielded the 4H-polytype (yellow-orange color). This result demonstrates that in the absence of impurity-related polytype stabilization effects [23], to achieve the 4H-polytype, growth must proceed from the carbon face of the seed crystal.

A typical 38 mm (1.5 inch) diameter SiC boule grown at our laboratories by the physical vapor transport technique is shown in Fig. 6. The as-grown surface of the boule exhibits a central (0001) facet and is shiny and smooth with minimum imperfections. Substrates with diameters varying from 30 to 50 mm have been produced to date (Fig. 7). The substrates exhibit uniform 4H-polytype over the entire crystal diameter.

3. Characteristics of 4H-SiC Crystals

3.1 Defect density

The distribution and density of crystalline defects in as-grown crystals have been determined by chemical etching, X-ray topography, and transmission electron microscopy (TEM). Dislocation etch pit densities ranging from 10^3 to 10^4 cm^{-2} are revealed by etching polished substrates in molten KOH at 450 to 500 °C for 20 to 30 min. Etching experiments and X-ray topographs can also reveal the presence of low-angle boundaries, crystallographically oriented edge boundaries (CEBs), and micropipes or pores within the substrate. Micropipes appear as dark, faceted pits corresponding to tube-like voids of μm size or less, which can extend the full substrate thickness. Micropipe defects tend to be replicated from the substrate into the epitaxial device layer, and are therefore considered to be particularly serious from the point of view of device fabrication and yield of working devices [24].

We have observed micropipes of differing sizes and characteristics, and have classified them into Type 1 and Type 2 micropipes, as shown in Fig. 8. Type 1 micropipes tend to exhibit faceted walls, usually have diameters >1 μm, lie approximately along the $\langle 0001 \rangle$ crystal axis, and can extend over long distances (up to several mm). Type

Type 1 Micropipes

Type 2 Micropipes

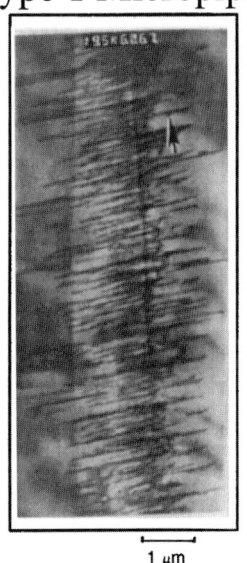

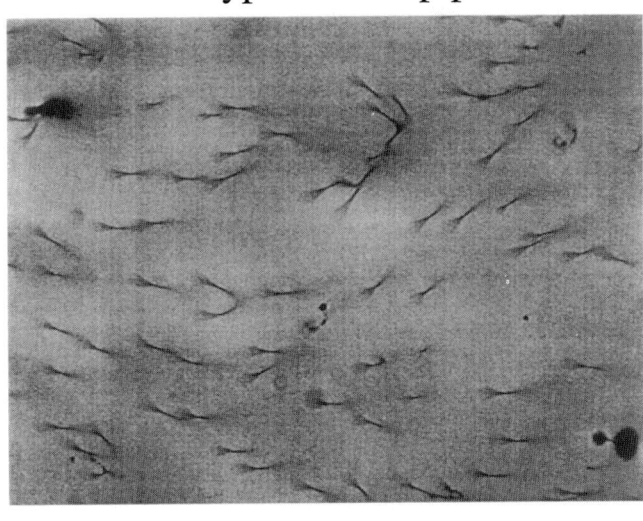

1 μm

•Directional
•Approx. <0001> axis
•Large Diameter
•Faceted

•Non-directional
•Oblique to <0001>
•Small diameter(<1μm)
•Short Range
•Unfaceted

Fig. 8. Classification of micropipes based on principal characteristics

Micropipes form as a result of........

| Reduction in strain energy along axis of screw dislocation | Silicon droplet or particle entrapment | Vacancy condensation at a helical dislocation |

ℓ

Hollow Core

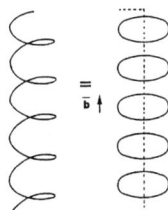

• Strain Energy Reduced
• Stable Hollow Core
• $r_{HC} \propto \dfrac{\mu b^2}{8 \pi b^2 K \vartheta}$

a   b   c

Fig. 9. Models of micropipe generation in SiC

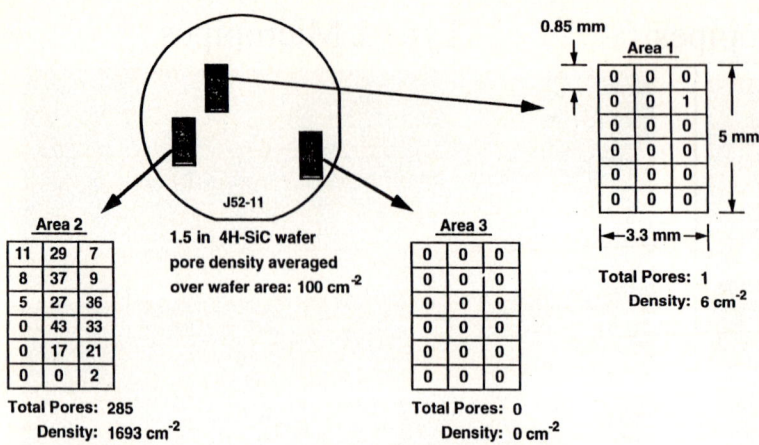

Fig. 10. Microscopic examination of micropipe distribution in 4H-SiC crystal

2 micropipes are generally sub-μm in size, unfaceted, have short ranges, starting and stopping abruptly, and can lie at highly oblique angles to the $\langle 0001 \rangle$ crystal growth axis.

Although the mechanism of micropipe formation is not fully understood in all details, various models have been suggested. Three of these are outlined in Fig. 9. Frank [25] has suggested (Fig. 9a) that in crystal structures, such as SiC, where the Burgers vector associated with a screw dislocation is sufficiently large, a stable hollow core will form along the dislocation axis as a result of the reduction in the local strain energy of the dislocation. This hypothesis is supported to a large extent by atomic force microscopy (AFM) [26] and X-ray topography [27] correlations between micropipe diameters and the magnitude of the associated Burgers vectors, as well as TEM studies [10] which confirm a tendency for micropipes to align along dislocation cores, where they are often bounded by periodic arrays of stacking faults (Fig. 8, left part). In addition, micropipes are often associated with low-angle grain boundaries in SiC crystals, suggesting an inter-action with the dislocations comprising the boundary. A second model [28, 29] (Fig. 9b) relates micropipe formation to impurities, particulates, or Si droplets entrapped at the crystal growth front, and may account for the high densities of Type 2 micropipes ob-served under certain growth conditions (Fig. 8, right part). Micropipe generation and propagation also appear to be sensitive to specific seeding techniques employed and to such growth parameters as pressure and temperature. These latter observations support a temperature activated mechanism involving segregation of point defects to dislocation cores [30] (Fig. 9c).

At our laboratories, studies taking these principal models into account have resulted in a reduction in the micropipe density observed in our 4H-SiC crystals. Current sub-strates exhibit micropipe densities of <100 cm^{-2} averaged across the full wafer area, and densities approaching 10 cm^{-2} in "best wafers" have been achieved. An important obser-vation regarding micropipes is that they are generally not homogeneously distributed throughout the crystal volume, as shown in Fig. 10. Detailed microscopic analysis of this substrate reveals that micropipes tend to occur in clusters while large areas of the wafer exhibit micropipe densities approaching zero.

3.2 Electrical characterization

3.2.1 N-type 4H-SiC crystals

Nitrogen is a shallow donor in SiC and easily compensates the residual acceptors resulting in n-type crystals. The range of resistivity achievable by nitrogen doping in 4H-SiC
and the doping density were studied as a function of the fraction of nitrogen in the
nitrogen–argon crystal growth ambient. The fraction of nitrogen in the argon ambient
$\{[N_2]/([N_2]+[Ar])\}$ was varied from zero to 100% as shown in Fig. 11. Calibrated secondary ion mass spectroscopy (SIMS) measurements reveal that the metallurgical content of nitrogen increases with increasing amounts of nitrogen in the ambient. The nitrogen content in the crystal is approximately proportional to the square root of the
nitrogen content in the gas phase, a result which is consistent with similar observations
on Lely-grown platelet crystals [1]. Four-point probe resistivity measurements show a
decrease in the bulk resistivity of the SiC as the nitrogen content in the solid increases
(Fig. 11). For crystal growth in 100% nitrogen ambient we observed a minimum resistivity of 0.007 Ω cm, corresponding to a nitrogen concentration of approximately
5×10^{19} cm^{-3} in the crystal as determined by calibrated SIMS analysis. Fig. 11 also
shows the resulting metallurgical nitrogen concentration in the crystal as a function of
the nitrogen content in the ambient. Under these growth conditions little is to be gained
in carrier density for ambient nitrogen concentration beyond 40%. By comparing the
carrier density with the metallurgical nitrogen content, it can be seen that virtually all
of the incorporated nitrogen appears to be ionized. This observation is consistent with
the onset of degeneracy at the high impurity concentrations investigated in this study.

3.2.2 Undoped and semi-insulating 4H-SiC

Glow discharge mass-spectroscopy (GDMS) and SIMS have been used to determine relative dopant concentrations in SiC sources and crystals. Under certain growth conditions,
boron is the dominant residual acceptor in unintentionally doped crystals (aluminum is
the next most abundant). Table 2 shows chemical analysis of a sublimation source and
of the resulting "undoped" crystal grown from that source. In addition to several metal-

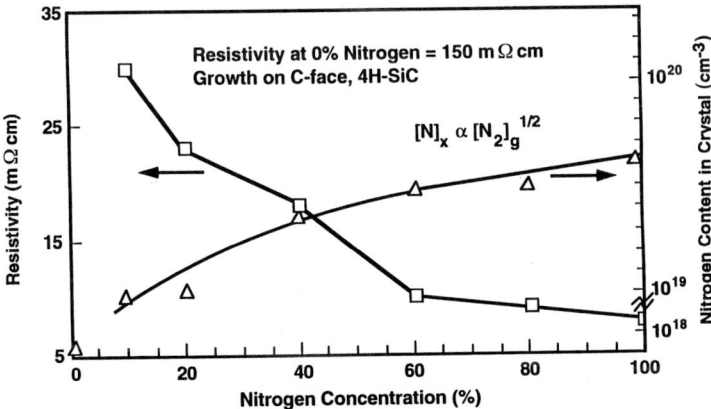

Fig. 11. High conductivity N$^+$ 4H-SiC growth: effect of nitrogen/argon ratio

Table 2

Chemical analysis of SiC sublimation source material and as-grown undoped 4H-SiC crystals

element	B	Na	Al	S	Ti	V	Cr	Fe	Co	Ni
source: GDMS (ppm wt)	5.0	0.6	9.0	0.3	75	130	0.1	2.0	0.02	0.08
crystal: GDMS (ppm wt)	3.0	0.2	0.3	0.1	0.5	1.0	0.2	0.5	0.05	0.02
SIMS (cm^{-3})	4×10^{17}		5×10^{16}		1×10^{17}	3×10^{16}				

lic impurities, boron and aluminum, which are known acceptor impurities in SiC, are also present in the starting source material. Compared to the starting SiC source material, the as-grown crystals exhibit improvements in purity by one to two orders of magnitude for most of the metallic elements. However, boron concentrations show almost no reduction after crystal growth. Both boron and aluminum remain as significant residual acceptor impurities in the as-grown crystal. This observation is supported by SIMS measurements of as-grown crystals for which B, Al, Ti, and V concentrations were determined using SIMS measurements calibrated against SiC samples implanted with known amounts of these elements. Boron and titanium are present in the 10^{17} cm^{-3} range while aluminum and vanadium are about an order of magnitude less.

The resistivity of the undoped 4H-SiC crystals was assessed by means of Lehighton contactless rf eddy-current measurements performed on substrates sliced from the crystals. Fig. 12 shows the room temperature resistivity of the undoped crystals as a function of crystal length expressed as increasing wafer number. The resistivity is observed to vary in the 10^2 to 10^3 Ω cm range. The spread in the observed resistivity is attributed to variability in the contamination level of boron as well as variability in compensation effects induced by residual shallow donors and deep-level impurities, such as vanadium.

To produce semi-insulating behavior, 4H-SiC crystals have been intentionally doped with deep-level donor impurities, vanadium in particular, which in 6H-SiC is known to introduce a deep-donor state lying near the middle of the band gap [31, 32]. A dramatic increase in the resistivity of these vanadium-doped crystals is observed, as shown in Fig. 13, where resistivity is plotted as a function of reciprocal

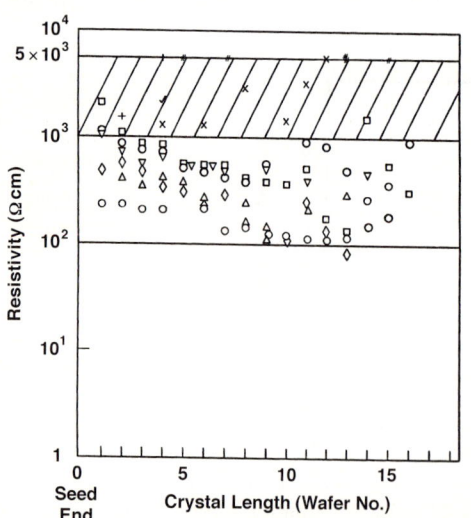

Fig. 12. Room temperature resistivity of undoped SiC crystals as a function of crystal length, expressed as increasing wafer number. Each symbol represents a different crystal

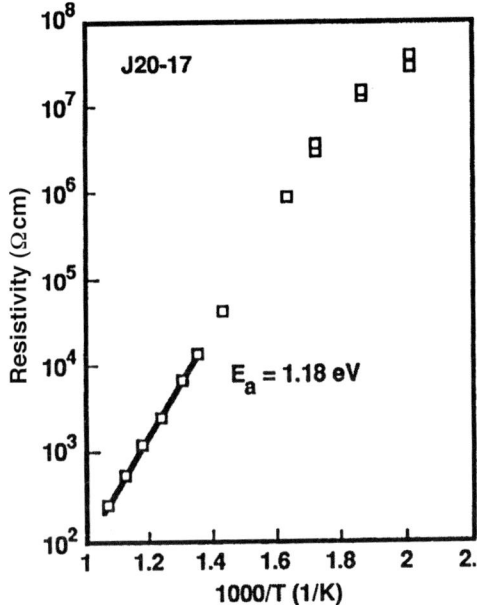

Fig. 13. Resistivity as a function of reciprocal temperature for vanadium-doped 4H-SiC grown by the physical vapor transport method

temperature. Since direct measurements of the resistivity near room temperature are complicated by the exceedingly high sample resistance ($>10^{15}\,\Omega$ cm) and spurious surface leakage effects, temperatures in the 250 to 650 °C range were used. A linear fit of the data over the higher temperature range yields an activation energy of approximately 1.18 eV, consistent with a deep-donor level in 4H-SiC attributed to a vanadium-related deep level [33].

4. Summary

Large diameter (up to 50 mm) 4H-SiC crystals have been grown by the PVT technique with micropipe densities approaching 10 cm^{-2} in the best wafers grown to date. Crystal growth rate is to first order predominantly a function of temperature and ambient pressure. 4H-polytype uniformity is controlled by polarity of the seed crystal and the growth temperature. Carbon-face seeding stabilizes the 4H-polytype. N$^+$ 4H-SiC crystals with resistivities $<0.01\,\Omega$ cm have been produced by controlled nitrogen doping. Undoped high resistivity 4H-SiC crystals exhibit resistivities in the 10^2 to $10^3\,\Omega$ cm range, are p-type, with electrical behavior determined principally by residual boron impurities. Resistivity levels have been extended into the semi-insulating range ($>10^8\,\Omega$ cm at 200 °C) by controlled addition of vanadium deep-level impurities.

Acknowledgements This work was supported in part by the Advanced Research Project Agency under contract No. F33615-95-C-5427 and by the U.S. Air Force under contract No. F33615-92-C-5912. The authors are grateful to P. G. Kennedy for his guidance and support. We would also wish to thank W. J. Choyke (University of Pittsburgh), G. A. Slack (Rennsalaer Polytechnic Institute), and M. Skowronski (Carnegie Mellon University) for their helpful discussions. Our sincere appreciation goes W. Gaida, R. R. Ronallo, J. Bogdon, D. P. Nebel, A. M. Stewart, and C. L. Chamberlain for their excellent technical assistance.

References

[1] J. A. LELY, Berichte Deutche Keramik Geselshaft **32**, 229 (1955).
[2] D. L. BARRETT, J. Electrochem. Soc. **113**, 1215 (1966).
[3] C. GOLDBERG and J. W. OSTROSKI, in: Silicon Carbide: A High Temperature Semiconductor, Eds. J. R. O'CONNOR and J. SMILTENS, Pergamon Press, Oxford 1960 (p. 453).

[4] H. C. Chang, C. Z. LeMay, and L. F. Wallace, ibid (p. 496).

[5] Y. M. Tairov and V. F. Tsvetkov, J. Cryst. Growth **43**, 209 (1978).

[6] G. Ziegler, P. Lanig, D. Theis, and C. Weyrich, IEEE Trans. Electron Devices **30**, 277 (1983).

[7] D. L. Barrett, R. G. Seidensticker, W. Gaida, R. H. Hopkins, and W. J. Choyke, Springer Proc. Phys. **56**, 33 (1990).

[8] H. McD. Hobgood, D. L. Barrett, J. P. McHugh, R. C. Clarke, S. Sriram, A. A. Burk, J. Greggi, C. D. Brandt, R. H. Hopkins, and W. J. Choyke, J. Cryst. Growth **137**, 181 (1994).

[9] S. Nishino, Y. Kojima, and J. Saraie, Springer Proc. Phys. **56**, 15 (1990).

[10] D. L. Barrett, J. P. McHugh, H. McD. Hobgood, R. H. Hopkins, P. G. McMullin, R. C. Clarke, and W. J. Choyke, J. Cryst. Growth **128**, 358 (1993).

[11] Y. M. Tairov, Technical Digest Internat. Conf. Silicon Carbide and Related Materials, Kyoto (Japan), 1995 (p. 11).

[12] A. A. Burk, Jr. and L. B. Rowland, J. Cryst. Growth **167**, 586 (1996).

[13] S. Sriram, G. Augustine, A. A. Burk, Jr., R. C. Glass, H. McD. Hobgood, P. A. Orphanos, L. B. Rowland, T. J. Smith, C. D. Brandt, M. C. Driver, and R. H. Hopkins, IEEE Electron Device Lett. **17**, 369 (1996).

[14] R. C. Clarke, R. R. Siergiej, A. K. Agarwal, C. D. Brandt, A. A. Burk, Jr., A. Morse, and P. A. Orphanos, Proc. IEEE/Cornell Conf. Advanced Concepts in High Speed Semiconductor Devices and Circuits, Ithaca (New York) 1995 (p. 47).

[15] J. W. Palmour, S. T. Allen, R. Singh, L. A. Lipkin, and D. G. Waltz, see [11] (p. 813).

[16] A. K. Agarwal, G. Augustine, V. Balakrishna, C. D. Brandt, A. A. Burk, Li-Shu Chen, R. C. Clarke, P. M. Esker, H. M. Hobgood, R. H. Hopkins, A. W. Morse, L. B. Rowland, S. Seshadri, R. R. Siergiej, T. J. Smith, Jr., and S. Sriram, IEDM Technical Digest, San Francisco, 9.1.1 (1996).

[17] W. Qian, M. Skowronski, G. Augustine, R. C. Glass, H. McD. Hobgood, and R. H. Hopkins, J. Electrochem. Soc. **142**, 4290 (1995).

[18] E. Biedermann, Solid State Commun. **3**, 343 (1965).

[19] E. Kaldis and N. Piechotka, in: Handbook Crystal Growth, Ed. D. T. J. Hurle, North-Holland, Publ. Co., Amsterdam, 1994 (p. 615).

[20] M. Kanaya, J. Takhashi, Y. Fujiwara, and A. Moritani, Appl. Phys. Lett **58**, 56 (1991).

[21] R. Stein and P. Lanig, J. Cryst. Growth **131**, 71 (1993).

[22] V. D. Heydemann, N. Schulze, D. L. Barrett, and G. Pensl, Appl. Phys. Lett. **69**, 3728 (1996).

[23] E. N. Mokhov, A. D. Roenkov, Y. A. Vodakov, G. V. Saparin, and S. K. Obyden, Techncial Digest Internat. Conf. Silicon Carbide and Related Materials, Kyoto (Japan), 1996 (p. 245).

[24] P. G. Neudeck and J. A. Powell, IEEE Electron Device Lett. **15**, 63 (1994).

[25] F. C. Frank, Acta Cryst. **4**, 497 (1951).

[26] J. Giocondi, G. S. Rohrer, M. Skowronski, V. Balakrishna, G. Augustine, H. McD. Hobgood, and R. H. Hopkins, J. Cryst. Growth, accepted for publication.

[27] S. Wang and M. Dudley, MRS Symp. Proc. **307**, 249 (1993).

[28] K. Koga, Y. Fujikawa, Y. Ueda, and T. Yamaguchi, Springer Proc. Phys. **71**, 96 (1992).

[29] J. Yang, in: PhD Thesis, SiC: Problems in Crystal Growth and Polytypic Transformation, Case Western Reserve University, Cleveland (Ohio) 1993.

[30] J. W. Mitchell, J. Appl. Phys. **33**, 406 (1962).

[31] J. Schneider, H. D. Muller, K. Maier, and F. Fuchs, Appl. Phys. Lett. **56**, 1184 (1990).

[32] H. McD. Hobgood, R. C. Glass, G. Augustine, R. H. Hopkins, J. Jenny, M. Skowronski, W. C. Mitchel, and M. Roth, Appl. Phys. Lett. **66**, 1364 (1995).

[33] J. R. Jenny, M. Skowronski, W. C. Mitchel, H. McD. Hobgood, R. C. Glass, G. Augustine, and R. H. Hopkins, Appl. Phys. Lett. **68**, 1963 (1996).

phys. stat. sol. (b) **202**, 149 (1997)

Subject classification: 61.50.−f; 61.72.Lk; 72.80.Jc; S6

SiC Seeded Crystal Growth

R. C. Glass, D. Henshall, V. F. Tsvetkov, and C. H. Carter, Jr.

Cree Research, Inc., 2810 Meridian Parkway, Suite 176, Durham, NC 27713, USA

(Received January 31, 1997)

The availability of relatively large (30 mm) SiC wafers has been a primary reason for the renewed high level of interest in SiC semiconductor technology. Projections that 75 mm SiC wafers will be available in 2 to 3 years have further peaked this interest. Now both 4H and 6H polytypes are available, however, the micropipe defects that occur to a varying extent in all wafers produced to date are seen by many as preventing the commercialization of many types of SiC devices, especially high current power devices. Most views on micropipe formation are based around Frank's theory of a micropipe being the hollow core of a screw dislocation with a huge Burgers vector (several times the unit cell) and with the diameter of the core having a direct relationship with the magnitude of the Burgers vector. Our results show that there are several mechanisms or combinations of these mechanisms which cause micropipes in SiC boules grown by the seeded sublimation method. Additional considerations such as polytype variations, dislocations and both impurity and diameter control add to the complexity of producing high quality wafers. Recent results at Cree Research, Inc., including wafers with micropipe densities of less than $1 \, \text{cm}^{-2}$ (with $1 \, \text{cm}^2$ areas void of micropipes), indicate that micropipes will be reduced to a level that makes high current devices viable and that they may be totally eliminated in the next few years. Additionally, efforts towards larger diameter high quality substrates have led to production of 50 mm diameter 4H and 6H wafers for fabrication of LEDs and the demonstration of 75 mm wafers. Low resistivity and semi-insulating electrical properties have also been attained through improved process and impurity control. Although challenges remain, the industry continues to make significant progress towards large volume SiC-based semiconductor fabrication.

1. Introduction

Production of semiconductor grade SiC ingots is one of the most challenging and exciting areas of the increasingly important SiC semiconductor industry. In order to take advantage of the superior properties of SiC devices we must be able to produce high quality SiC thin films, which in turn require high quality substrates of similar thermophysical properties. For these reasons, the development of low cost, high quality single crystal SiC substrates has been pursued since the first Lely platelets [1] were used to make devices. As exemplified in the accompanying papers within this publication, there are highly experienced industrial and academic research groups working toward SiC substrate fabrication. Their success and the successful development of the SiC bulk growth process translates to the future success of the whole SiC industry.

The availability of SiC wafers on a commercial basis has led to the demonstration of many types of electronic and optoelectronic devices which exploit the unique electrical and physical properties of this material. These devices include microwave MESFETs with $f_{\max}$ of 50 GHz [2], maximum power densities of 3.1 W/mm with PAE of 38.9% in Class A operation at 850 MHz (or 2.3 W/mm with PAE of 65.7% at 1.8 GHz in Class B) [3], 4500 V p−n junction diodes [4], >1 kV Schottky diodes [5], 175 V, 2 A power

MOSFETs [6], 700 V, 6 A thyristors which operate well to 500 °C [7], and high volume production of high brightness blue LEDs made from III–nitrides grown on SiC.

While the commercialization of small area devices has been realized based on currently available wafers, significant improvements in both wafer size and quality are necessary to allow fabrication of the large area devices required for SiC to fulfill its full potential for power conditioning and control. The primary obstacle to the production of large area SiC devices is the presence of micropipe defects which, to date, have occurred in virtually all seeded sublimation grown boules and wafers. As discussed below, these defects are caused by a number of mechanisms, but the result is a small diameter (≈ 0.1 to $5\,\mu m$) hole in the material which may extend through the entire length of a boule along the growth direction. While it is intuitive that these defects would be harmful for devices, especially at high voltage, Neudeck and Powell [8] very effectively showed the failure of high voltage diodes in reverse bias caused by microplasmas generated in micropipes. In addition to micropipes (which are typically reported in the 10 to $10^3\,cm^{-2}$ density range), dislocations in density ranges of 10^3 to $10^6\,cm^{-2}$, inclusions, striations (stress and impurity induced microcracks) and mosaic structural variations must be reduced in order to fully realize SiC devices.

As in the rest of the semiconductor industry, the desired size of SiC wafers available for device fabrication is primarily an economics issue. For most non-power applications, 50 to 75 mm wafers will be sufficient for economical commercialization while high current power device production will require 100 mm or larger diameter wafers. Cree is currently producing 50 mm diameter 4H and 6H wafers for internal use and 76 mm diameter capability has been demonstrated by Hobgood and coworkers [9].

The required ranges for controlled doping extend from the needs of power devices which have a vertical structure requiring low resistivity substrates through the applications of lateral structures such as power microwave devices, where very high resistivity substrates ($>2000\,\Omega\,cm$) are required. Previously, the lack of high resistivity (or semi-insulating) SiC substrates had severely limited the maximum frequency and efficiency of SiC microwave MESFETs and had been a roadblock for the development of SiC MMIC technology. In the past several years, however, both Cree (see below) and Northrop Grumman [10] have demonstrated advances in high resistivity 4H-SiC substrates and devices fabricated from them. With this substrate material, high frequency devices are now approaching the properties required for commercialization. In this brief article our goal is to introduce the reader to some issues in the field of SiC bulk growth.

After describing the typical vapor phase growth process, we will discuss key parameters involved in this process, the defects of primary interest, and finally some comments on the future direction of work. This paper is further development of the review paper presented at the 6th International Conference on Silicon Carbide and Related Material in Kyoto, Japan 1995 [11].

2. SiC Crystal Growth via Physical Vapor Deposition

While most single crystal semiconductor boules are grown either from a melt or a solution, the properties of SiC render either of these approaches impractical at present. The SiC phase diagram has a peritectic at $\approx 10^5\,Pa$ and our calculations indicate that stoichiometric melting would only occur for pressures $>100\,000\,atm$ and temperatures $>3200\,°C$ [11]. Although it is possible to create similar growth conditions such as those used for diamond crystal growth, it is currently not feasible to consider this process for commercial production of 50 to 100 mm diameter semiconductor grade SiC boules. Additionally, although the solubility of C in a Si melt is 0.01 to 19% in the temperature interval of 1412 to 2830 °C, at temperatures higher than 1700 to 1750 °C the evaporation of Si increases to the point that growth processes become unstable. The solubility of C can be increased to >50% by adding certain metals to the melt (e.g. Pr, Tb, Sc) which in principle would allow use of crystal pulling techniques. However, this technology has not been developed, because there are currently no crucible materials which would be stable with these melts and solvent evaporation is still a problem. Finally, the solubility of the metal additives in the growing SiC boule is probably too high to be acceptable for use as a semiconductor.

It is interesting to note that SiC boule growth using CVD (see accompanying paper Kordina et al. and [12]) is currently being researched and progress in this area is increasing. This is a non-trivial process, however, because there are serious problems with optimization of the thermal and thermodynamic conditions required for growth rates suitable for boule growth. Published reports and our experience indicate that the growth rate and quality of SiC crystals grown by CVD are limited by the desorption of the reaction products, the diffusion of the source atoms to the growth face and an increasing etch rate with increasing temperature. In light of the above difficulties, physical vapor deposition via seeded sublimation has been the most investigated and successful technique for growth of SiC boules to date.

3. Process Considerations

Initial techniques used to make SiC have been discussed in other articles [13 to 16]. In brief, the first commercial process for manufacture of low-grade SiC powder in large quantities was developed and patented by Acheson. From this process it was possible to obtain irregularly shaped pieces of single crystal SiC platelets in the by-product material. In 1955, however, Lely [1] developed a method for reproducing such platelets with much higher purity. In this initial design, SiC source powder was loaded between a dense graphite crucible and a porous graphite thin-walled inner cylinder. The assembly was heated to temperatures in excess of 2500 °C and after an amount of time the combination of surface and vapor transport produced a distribution of high-quality, irregularly shaped SiC platelets typically 0.5 cm diameter on the inner surface of the porous graphite thin-walled cylinder.

Although these Lely platelets can exhibit good quality with micropipe and dislocation densities of 1 to 3 cm^{-2} and 10^2 to 10^3 cm^{-2}, respectively, the poor ability to control growth surfaces, rates and direction hampered control of platelet thickness, doping and polytype. Hence the development of a seeded technique was necessary. The seeded, or modified Lely method was reported by Tairov and Tsvetkov [17] in 1978. As with the Lely method, this was a sublimation deposition process where the growth rate is proportional to, and the crystallization process is facilitated by, supersaturation of a vapor phase. This method, more generically termed physical vapor transport (PVT), has been further refined by Davis et al. [18], Stein and Lanig [19] and Barrett et al. [20] for producing larger SiC boules. The seeded methods utilize a SiC seed placed within a crucible containing a charge or source of SiC powder. Fig. 1 shows a common representation of this type of growth set-up where the crucible is heated to temperatures greater than

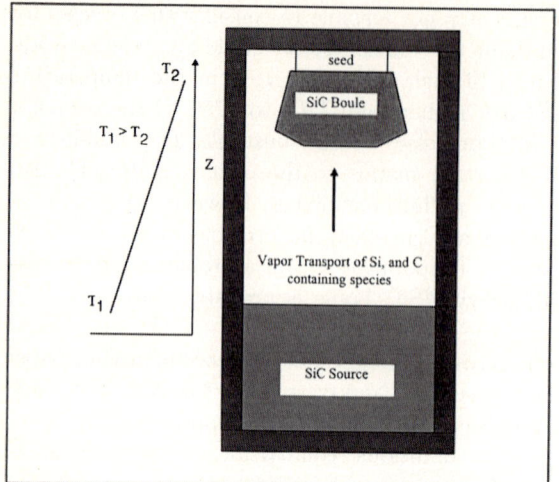

Fig. 1. Schematic representation of a seeded PVT sublimation growth system

2000 °C. A temperature gradient is created causing vapor phase transport of Si and C containing species between the source and the seed.

Due to the extreme temperatures, primarily carbon materials are used in these furnaces which makes qualifying materials as well as monitoring and controlling the process a challenge. A typical furnace will consist of two types of carbons: hard, dense iso-pressed electrically and thermally conducting material and less dense, thermally insulating material. Selection of this material is critical in order to reduce thermal stresses, vapor loss and impurities in the growing crystal. As an example consider material purity. Although purifying carbon/graphite material to below the parts per million (wt ppm) range is possible, it can be expensive and variable in efficiency resulting in the incorporation of these impurities (primarily S, B and metallic elements such as W, V, Ti, and Al) into the growing crystal at ppm levels. The transfer coefficients of these impurities vary by species, but they all can have a strong effect on material properties [21]. For example, uncontrolled impurities at levels in the range of 1 to 10 wt ppm in the source, will begin to rival in low-doped $(1 \text{ to } 5) \times 10^{17} \text{ cm}^{-3}$) n-type substrates. Although the transport inefficiency of most metallic impurity reduces their concentration in the crystal by a factor of 0.5 or 0.1 from that in the source, controlling substrate resistivity and quality with this level of interference is very difficult.

Process control is another area in which furnace design and material selection become important technological factors in producing high quality SiC. How this is accomplished greatly affects the growing crystal. The technology of controlling nucleation, composition and thermal geometry requires the ability to monitor pressure, temperature and growth rate. Temperature control alone is a major hurdle due to the high temperatures, the strong radiative component in the heat loss process and the large range over which control must be maintained. Minor variations in crucible design coupled with improper attention to heat transfer mechanisms can drastically affect crystal shape. Defects can result from these variations and from the environment of strong thermal radiation in which the SiC boule growth process occurs. The SiC crystal is semitransparent to this radiation which is absorbed at the growth interface and within the bulk of the growing boule. The formation of a thermo-elastic field inside the boule is a result of radiation absorption, thermal conduction and thermal boundary conditions. Our calculations [11] show that the thermal conductivity of SiC at the growth temperature is ≈20 times less than at room temperature and that radiation absorption strongly depends on absorption by free carriers, impurity atoms and the lattice. Our modeling of the thermo-elastic field shows a higher gradient at the growth interface than inside the crystal. This difference depends on the growth temperature, doping level and thermal boundary conditions. By varying these conditions, the elastic field in the growing crystal may have a V or W shape. The modeling results are in good agreement with our experimental results on increasing boule diameter and for the formation of micropipes and dislocations in the growing crystal.

Seeding technology is critical and the crystal quality is greatly affected by seed doping, thickness, orientation and surface quality. The latter is difficult to maintain due to premature breakdown of the seed surface at growth temperatures, since the vapor pressure of Si is much greater than that of C. This loss of Si from the seed surface can result in a C-rich and poor-quality surface. Thus, maintaining Si in the vapor above the seed must be accomplished by careful manipulation of the temperature differential and the source material qualities. One advantage to these considerations, however, is the ability to back-etch the seed with a reverse temperature gradient. In this process the Si over pressure can be manipulated to remove the first several layers of the seed thereby giving a pristine starting surface.

The SiC source technology and material properties are very important for controlling the vapor phase concentrations of Si and C containing molecules. A pure form of Si and C is obviously required (as shown above and in [21]), but as important is the source structure, particle strain and polytype. Commercial grades of SiC powder are the most common source, however, these are often contaminated by impurities during handling and refining. Lely platelets or other forms of SiC which have been transported at least once provide a very clean, although rather inefficient, source. Finally, silicon and C precursors can be reacted to synthesize SiC powder.

Understanding the effect of any source material on the vapor state requires an analytical look at the dissociation process. As mentioned above, in the seeded vapor deposition process Si and C containing molecules are driven from the source hot zone area to condense on the cooler seed. The vapor phase over SiC contains many different compound molecules with the primary ones being Si, Si_2C, and SiC_2. There have been several reported attempts to determine the temperature dependence of the partial pressures of these components. However, most of the results were from temperatures which were too low for useful modeling of bulk growth processes or intrinsic point defect formation. In addition, most of these data have a large amount of scatter and do not all agree with the phase diagram or all crystal growth experimental results. Therefore, we have calculated the temperature dependencies of the partial pressure of the main components for the two corners of the phase diagram (SiC–C and SiC–Si) using the most reliable thermodynamic data available [11]. We also determined the relationships between the partial pressure of Si and those of Si_2C and SiC_2 (in Pa) as

$$P_{Si_2C} = 2.85 \times 10^2 \exp\left(-1.79 \times 10^4/T\right) \times P_{Si}$$

and

$$P_{SiC_2} = 9.41 \times 10^{28} \exp\left(-14.35 \times 10^4/T\right)/P_{Si}.$$

Based on these results, one can calculate the temperature dependence of the ratio of Si to C atoms in the vapor phase. Used in a proper model, the sublimation growth process can be optimized and will (as in our own work) show satisfactory agreement with experiments.

In addition to the thermodynamic considerations mentioned above, an understanding of kinetic limitations to the growth process is required. The background gas generally used in SiC bulk growth is Ar with process pressures ranging from ≈ 1 Pa down to below $\approx 10^{-2}$ Pa. In a diffusion limited processes such as this there is a direct correlation between growth rate and the background pressure at a given temperature. As recently published by Tuominen et al. [22], but also studied by ourselves and others, the activation energy for the dissociative process provides only part of the necessary information. One must also understand the growth rates and effective dissociative energies for various growth pressures in order to accurately predict the growth process. With this information, one of the most common methods for controlling the growth process experimentally is through background pressure control. The recipes for these processes probably vary by group, but typically as the pressure is reduced the vapor diffusion rates increase and constituents move more rapidly down the concentration gradient from the source toward the cold seed.

Concomitantly, this concentration gradient will be a direct result of the relative source and seed temperatures. As mentioned above, the temperature of the seed must be controlled to ensure that any loss of Si from the surface is minimized by maintaining the over-pressure of free Si in the vapor near the seed. As importantly, the source temperature must be controlled in order to develop the proper temperature gradient (and thus the concentration gradient) between the source and seed. Thus, one can see that the pressure and temperature profiles are important in the growth process. Controlling these two key parameters is not trivial, and the fluctuations which do occur will be a primary cause of defect formation. For example, local fluctuations in temperature over the seed can cause constitutional super-cooling and the creation of micropipes, precipitates, or inclusions. Axial temperature variations can cause edge striations and cracking. This discussion leads to the next section and the topic of defects.

4. Polytype Determination, Growth Mechanisms and Defects

One of the most unique and traditionally confounding properties of SiC is its ability to exist in over 120 different polytypes. The term *polytype* denotes a unique class of materials which possess the same chemical make-up, but whose structure varies perpendicular to the close packed direction by a fundamental unit. For SiC this unit is a covalently bonded tetrahedron of C atoms with each C atom tetrahedrally surrounded by Si atoms (for details please see review by Davis and Glass [23]). The manner with which these units are stacked and rotated on top of one another determines the polytype. This sequential stacking of tetrahedral units results in a polar structure of repeating Si and C layers where all tetrahedra have one apex terminating the outer plane. Thus the {0001} surfaces have either a C or a Si termination layer. This polarity becomes important during growth when discussing polytype stability, doping processes and defect formation.

Polytypes currently produced routinely are 4H, 6H, and 3C. The nomenclature used here stems from the simple hexagonal and cubic stacking which intermix to form the wide variety of larger period structures. The number in the nomenclature depicts the number of unit layers in the repeating sequence along the c-direction. The letters C, H, and R refer to which basic crystallographic structure: cubic, hexagonal, and rhombohedral, respectively, is formed by the sequence. For example, 4H indicates the stacking repeat sequence is four units with strongly hexagonal character. The only purely cubic structure is denoted as 3C.

Silicon carbide has a very low stacking fault energy which can allow solid-state transformations and formation of these many different polytypes during crystal growth. This fact greatly increases the demands on the thermodynamic, kinetic and thermal conditions of the process and design of the growth cell. If the conditions are not optimized, inclusions of 3C, 4H or rhombohedral polytypes may occur in 6H crystals and 3C, 6H or rhombohedral polytypes in 4H. These factors become increasingly important for larger crystal diameters.

The kinetic and thermodynamic requirements for polytype stability are not well understood, although Knippenberg [13] first determined an empirical observation of relative stability. His work investigated the characteristics and number of various polytypes existing along isothermal zones within a growth cell. From this he determined a thermally based ranking. Unfortunately this data is not easily translated to seeded sublimation where growth rates exceed those of the Lely cell, but Hopkins [24] has shown that under certain circumstances in seeded sublimation similar relationships can be used to predict polytype stability. His work also indicated that the frequency of 4H and 6H polytypes was in some part related to the growth temperature. With the physical vapor technique, however, where growth typically is occurring far from equilibrium and under vapor and surface diffusion limited conditions, one must consider surface and vapor kinetic and impurity effects.

Stein et al. [25] were one of the first to discuss the effect of the C versus Si surface on the stability of 4H and 6H polytypes. It was determined that 4H is more stable when growing on the C face. This stability has been explained by some [26] as resulting from sp^2-hybridized bonds of the C-terminated face reducing the effect of the substrate. Additionally, impurity additives (CeO_x, Sc) have been used which possibly enrich the vapor with C via carbides.

A unique feature observed on PVT grown SiC surfaces is the existence of clearly discernible growth spirals. The density of these can vary, but the PVT growth rates (approximately 0.5 to 2.0 mm/h) require a relatively high supersaturation of the growth constituents which can result in a high density of these spirals. As mentioned above, both surface and vapor phase diffusion kinetics affect the growth rate, and if local instabilities occur in temperature or pressure and surface diffusion limitations prevail, the potential for secondary nuclei formation and three-dimensional nucleation increases. The existence of these growth centers and the possible misorientation between them has been discussed previously [27] where results were presented showing the severely mosaic structure of seeded sublimation material as compared to Lely wafers. A possible relation between the mosaicity, pinholes, dislocations and growth spirals (screw dislocations) was based on high resolution X-ray diffraction (HRXRD) topography, etching studies and the theoretical work by Sunagawa and Bennema [28]. In these works the Lely wafer HRXRD showed typical rocking curves across the (0006) reciprocal lattice point with FWHMs of $\approx 0.0041°$ and $0.0028°$ in the $2\theta/\omega$ and ω directions, respectively. The same point from a seeded sublimation grown wafer was typically asymmetrically broadened in

Fig. 2. Optical photograph of crystal surface showing several growth centers apparently formed via a spiral growth mechanism around screw dislocations

ω with multiple peaks separated at ≈ 0.01 to $0.05°$ displaying individual FWHMs of 0.004 to 0.007. This indicated that these wafers had a more domain-type structure. Under this type of growth regime the opportunity for defect formation is high as steps move across one another, islands contact each other at slightly different angles and localized thermodynamic and kinetic variations are easily created. Fig. 2 shows a growth

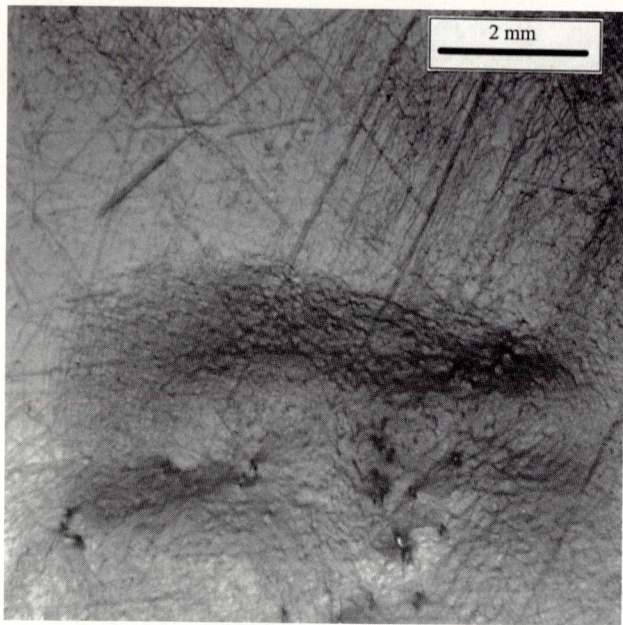

Fig. 3. Transmission SWBXT image of a 4H-SiC wafer showing dislocation tangles, and holes where pinholes are located. The large straight defects are polishing scratches

surface exhibiting a high density of spiral sources. Within these spirals can be seen holes or micropipes.

Monitoring micropipe densities is an important aspect of developing processes to reduce/eliminate them. In our work, wafers are lapped, polished and etched in molten KOH at a temperature of 650 °C for 20 to 30 min prior to examination with transmitted light microscopy. This is one of the most straightforward detection methods to use. Although somewhat destructive, (wafers can be reclaimed), this technique preferentially removes material around the micropipe center, which delineates the pipe from the surface for either manual or automated counting. Another highly useful method to analyze micropipes in wafers is synchrotron white beam X-ray topography (SWBXT) which has been used extensively by Wang et al. [29]. We have used this technique to develop our understanding of etching data for micropipes. In SWBXT the strain fields around the pipe disrupt the diffraction process to the extent that actual holes appear in the topographs (Fig. 3).

Micropipes are seen by many as the single largest threat to the large scale commercialization of SiC technology. There are many conflicting views on the source and formation processes of micropipes in SiC. Most views are based around Frank's theory of a micropipe being the hollow core of a screw dislocation with a huge Burgers vector (several times the unit cell) and with the diameter of the core having a direct relationship with the magnitude of the Burgers vector. Our results show that there are several mechanisms or combinations of these mechanisms which cause micropipes in SiC boules grown by the seeded sublimation method. The mechanisms can be separated into two classifications [11]; those which are fundamental to the growth process and those which depend on the technological process conditions. The mechanisms are listed in Table 1 according to these categories.

Stein [30] discussed several possible reasons for the occurrence of the hollow cores: 1. kinetic: too high growth rate, 2. thermodynamic: variations in constituent supersaturation, 3. dislocation centers, and 4. defect formation after growth. In all these cases one must consider the seed surface quality, the growth process stability and cleanliness, as well as the specific parameters controlling nucleation density and growth rate. Stein showed that in his specific case the hollow cores developed after boule formation as a result of secondary evaporation from defects or variations in the boule and not as a result of instabilities at the growth surface.

Table 1

Primary mechanisms of micropipe formation in SiC sublimation growth

fundamental	
1. thermodynamic	2. kinetic
a) thermal field uniformity	a) nucleation processes
b) dislocation formation	b) inhomogeneous supersaturation
c) solid-state transformation	c) constitutional supercooling
d) vapor phase composition	d) growth face morphology
e) vacancy supersaturation	e) capture of gas phase bubbles

technological		
1. process instabilities	2. seed preparation	3. contamination

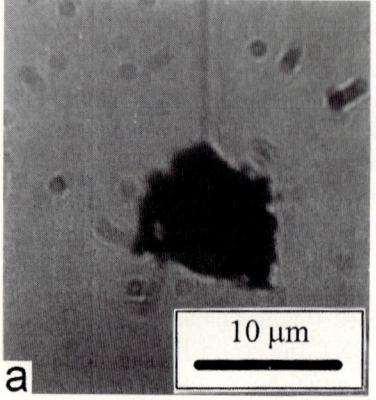

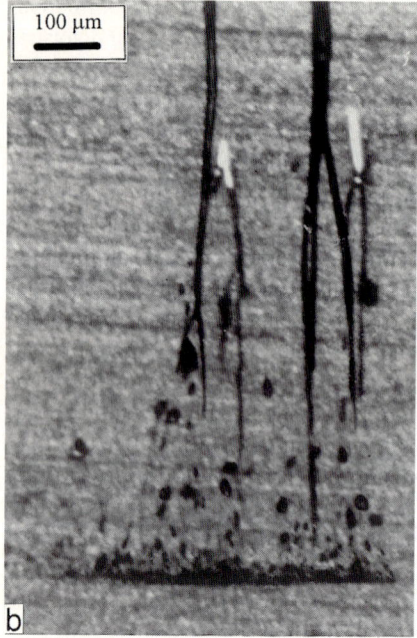

Fig. 4. a) Graphite particle with resulting micropipe (technological issue) and b) results of a process instability issue where constitutional supercooling has produced Si droplets on the growth surface leading to micropipe streaming

Others have reported their findings on micropipe formation as well. Most recently Takanaka et al. [31] used etching to identify micropipes and define vertical and planar defects in their material. They described how some micropipes terminated at planar defects. The exact method of the planar defect formation was not discussed. Anikin et al. [32] have discussed the initiation of growth with *in situ* etching. In their work the effect of excess Si on the growth surface was believed to cause disruptions to the growth process. The effect of this material impeding the surface movement of steps or the accumulation of free C or impurities in clusters on the surface have all been cited as possible modes of micropipe initiation. Fig. 4 shows an example of a) a technological issue where a graphite particle has caused a micropipe and b) a process-instability issue where constitutional supercooling has produced Si droplets on the growth surface which have led to micropipe streaming.

In our work we have seen that while micropipes are forming during growth, simultaneous processes occur which reduce their concentration such as micropipe dissolving, moving, transformation and recombination. Having made continuous reductions in micropipe density over the past three years, we continue to have success reducing the density of this defect. Fig. 5 shows a digitized image of a KOH etched 35 mm 4H-SiC wafer cut from a low micropipe density 4H-SiC boule. This wafer contains a total of eight micropipes, yielding a density of 0.83 cm^{-2}.

The interaction between two growth spirals can be assumed to result in low angle twist boundaries due to the misorientation between each spiral. It is expected that these low angle boundaries will be relaxed interfaces. The HRXRD results presented above showed little or no strain in the domains of the PVT grown wafers. Such a relaxed state of stress will be facilitated via an increase in the dislocation density between them, which may produce, for example, an array of dislocations. Further arguments for dislocation formation follow those above for micropipe formation. It is encouraging, however, that additional examination of 4H-SiC boules using the SWBXT technique show some

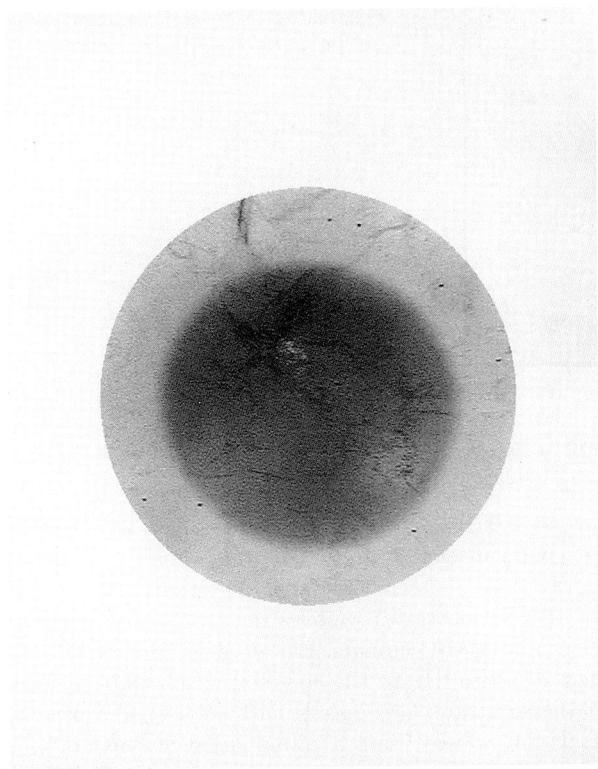

Fig. 5. Digitized image of a KOH etched 35 mm 4H-SiC wafer with a total of eight micropipes which is 0.83 cm^{-2}. The micropipes appear as dark points across the wafer

wafers have areas $\geq 0.5\,\text{cm}^2$ with a total line defect density of about $1000\,\text{cm}^{-2}$. Although current dislocation densities are typically in the 10^4 to $10^6\,\text{cm}^{-2}$, these results show that it will be possible to produce very low defect density 4H-SiC wafers in the future.

Nitrogen and Al are the main n- and p-type dopants, respectively, used for doping SiC boules because they create relatively shallow donor and acceptor levels in the SiC bandgap. Our results do not show any major distinctions for doping processes between 6H and 4H crystals. However, there is significant anisotropy for N and Al incorporation in sublimation boule growth for the crystal faces (0001)Si, (000$\bar{1}$)C, (1$\bar{1}$00) and (11$\bar{2}$0). For nitrogen there is an increase in incorporation from (0001)Si to (1$\bar{1}$00) to (000$\bar{1}$)C, while for Al it is the reverse. Nitrogen incorporation is approximately a function of the square root of the N_2 partial pressure in the growth cell and for Al the dependence is roughly linear with Al concentration in the source for concentrations up to the $10^{20}\,\text{cm}^{-3}$ level. Nitrogen incorporation decreases with growth temperature while Al increases. Calculation of the bond energies with N and Al in the SiC lattice indicates that the main reason for the abnormal dependence of N incorporation with temperature is the large difference in s–p splitting when N replaces C atoms versus Al replacing Si [11].

Data for resistivity as a function of N concentration is shown in Fig. 6 for both 6H and 4H boules. This data is important because low resistivity substrates are required for vertical power device structures which is a major application area for SiC, especially for 4H. As shown, we have produced 4H-SiC boules with resistivities as low as $0.0028\,\Omega\,\text{cm}$

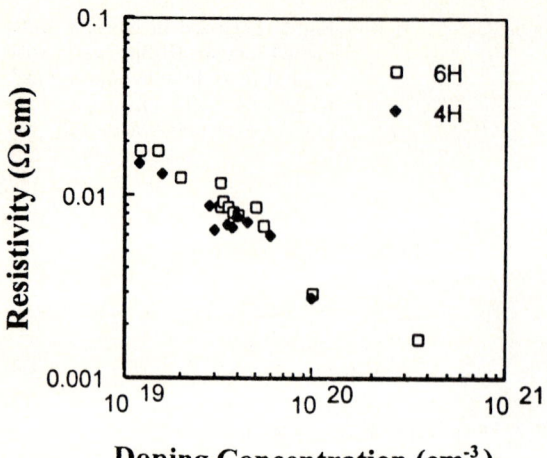

Fig. 6. Resistivity vs. N doping concentration for 6H and 4H-SiC boules

while 6H has been produced as low as 0.0016 Ω cm. With increased level of effort in developing low resistivity 4H wafers, one can expect to see 4H wafers with even lower resistivities in the near future.

The availability of semi-insulating 4H-SiC substrates removes the final impediment for development of SiC microwave devices and MMIC circuits. Earlier generations of Cree's SiC MESFETs had limited frequency response due to the conductive substrate, just as Si bipolar transistors are limited by the substrate. Cree has recently developed a process for producing very high resistivity 4H-SiC wafers using a combination of dopants and intrinsic point defects [11]. The substrates are too resistive to be measured with a conventional Hall effect technique so two alternate types of measurements were made to characterize the substrates. The first used a network analyzer to make microwave measurements of transmission lines on the substrates. The data was then used to extract the

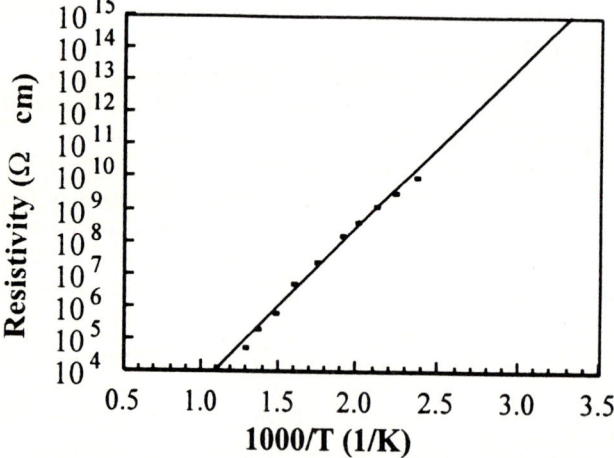

Fig. 7. Resistivity as a function of temperature for a semi-insulating SiC substrate as determined with I–V measurements

four parameters that completely define a transmission line: L, C, R, and G. The measured loss of 10 mm long transmission lines were compared for several substrates: 1. a lightly doped p-type 4H-SiC buffer layer which had been previously used to separate the conductive substrate from the device, 2. a semi-insulating GaAs substrate and 3. a high resistivity 4H-SiC substrate. The loss from the high resistivity 4H-SiC substrate matched that of the semi-insulating GaAs substrate throughout the measured frequency range of 1 to 20 GHz. The loss from the lightly doped p-type 4H-SiC buffer layer was 2.8 dB greater at 1 GHz and 3.7 dB greater at 20 GHz than the other two. The second method was to measure that I–V characteristics using a picoammeter with the substrates at high temperatures. A plot of resistivity versus temperature is shown in Fig. 7. At 100 °C, which was the lowest temperature for which we could measure a current, the resistivity was 10^{10} Ω cm and at 350 °C the resistivity was still $\approx 10^5$ Ω cm. Extrapolating to room temperature gives a resistivity of 10^{15} Ω cm. Initial MESFETs fabricated on these substrates had an increase in f_{max} to 30 GHz from 16 GHz for previous devices, with recent results showing 4H-SiC MESFETs with an f_{max} of 50 GHz on these substrates.

5. Summary

Reductions in substrate resistivity as required for power devices, the development of semi-insulating substrates for high frequency devices and most importantly the increasingly large areas of micropipe free substrates are leading the SiC industry toward successful application of this wide bandgap material. Therefore, future focus must be directed toward: 1. empirically supported modeling of the growth process, stress states, defect formation mechanisms and material properties, 2. fundamental issues such as impurity control and doping mechanisms and 3. optimization of growth processes. Success in these areas will allow acceleration of improvements in substrate quality and size which will continue to drive the SiC industry.

Acknowledgements   The authors would like to thank Prof. Mike Dudley and Shaoping Wang of SUNY-Stony Brook for the SWBXT image. This research was partially funded by the Advanced Research Projects Agency through a contract administered by the Air Force Wright Laboratories.

References

[1] J. A. LELY, Ber. Dtsch. Keram. Ges. **32**, 229 (1955).
[2] S. ALLEN, presented at WOXAMAT, San Antonio (TX) 1997.
[3] C. E. WEITZEL, J. W. PALMOUR, C. H. CARTER, JR., K. MOORE, K. J. NORDQUIST, S. T. ALLEN, C. THERO, and M. BAHATNAGAR, IEEE Trans. Electron Devices **43**, 1732 (1996).
[4] O. KORDINA, J. P. BERGMAN, A. HENRY, E. JANZÉN, S. SAVAGE, J. ANDRE, L. P. RAMBERG, U. LINDERFELT, W. HERMANSSON, and K. BERGMAN, Appl. Phys. Lett. **67**, 1561 (1995).
[5] T. KIMOTO, T. URUSHIDANI, S. KOBAYASHI, and H. MATSUNAMI, IEEE Electron Device Lett. **14**, 548 (1993).
[6] J. W. PALMOUR, S. T. ALLEN, R. SINGH, L. A. LIPKIN, and D. G. WALTZ, Inst. Phys. Conf. Ser. No. 142, 813 (1996).
[7] J. W. PALMOUR, R. SINGH, and D. G. WALTZ, Abstract 54th Ann. Device Research Conf. Digest, IEEE Cat. No. 96TH8193, IEEE, Piscataway (NJ) 1996 (p. 54).
[8] P. G. NEUDECK and J. A. POWELL, IEEE Electron Device Lett. **15**, 63 (1994).

[9] D. L. BARRETT, J. P. McHUGH, H. M. HOBGOOD, R. H. HOPKINS, P. G. McMULLIN, and R. C. CLARKE, J. Cryst. Growth **128**, 358 (1993).

[10] H. M. HOBGOOD, R. C. GLASS, G. AUGUSTINE, R. H. HOPKINS, J. JENNY, M. SKOWRONSKI, W. C. MITCHEL, and M. ROTH, Appl. Phys. Lett. **66**, 1364 (1995).

[11] V. F. TSVETKOV, S. T. ALLEN, H. S. KONG, and C. H. CARTER, JR., Inst. Phys. Conf. Ser. No. 142, 17 (1996).

[12] O. KORDINA, C. HALLIN, A. ELLISON, A. S. BAKIN, I. G. IVANOV, A. HENRY, R. YAKIMOVA, M. TUOMINEN, A. VEHANEN, and E. JANZÉN, Appl. Phys. Lett. **69**, 1456 (1996).

[13] F. KNIPPENBERG, Philips Res. Rep. **18**, 161 (1963).

[14] YU. M. TAIROV and V. F. TSVETKOV, J. Cryst. Growth **52**, 146 (1981).

[15] D. L. BARRETT, R. G. SEIDENSTICKER, W. GAIDA, R. H. HOPKINS, and W. J. CHOYKE, J. Cryst. Growth **109**, 17 (1991).

[16] J. A. POWELL, P. PIROUZ, and W. J. CHOYKE, Growth and Characterization of Silicon Carbide Polytypes for Electronic Applications, in: Semiconductor Interfaces, Microstructures and Devices: Properties and Applications, Ed. ZHE CHUAN FENG, Institute of Physics, London 1993 (pp. 257 to 293).

[17] YU. M. TAIROV and V. F. TSVETKOV, J. Cryst. Growth **43**, 209 (1978).

[18] R. F. DAVIS, C. H. CARTER, JR., and C. E. HUNTER, US Patent No. Re 34,861 (February 14, 1995).

[19] R. A. STEIN and P. LANIG, Mater. Sci. Engng. B **11**, 69 (1992).

[20] D. L. BARRETT, J. P. McHUGH, H. M. HOBGOOD, R. H. HOPKINS, P. G. McMULLIN, and R. C. CLARKE, J. Cryst. Growth **128**, 358 (1993).

[21] R. C. GLASS, G. AUGUSTINE, V. BALAKRISHNA, H. McD. HOBGOOD, R. H. HOPKINS, J. JENNY, M. SKOWRONSKI, and W. J. CHOYKE, Inst. Phys. Conf. Ser. No. 142, 37 (1996).

[22] M. TUOMINEN, R. YAKIMOVA, A. S. BAKIN, I. G. IVANOV, A. HENRY, A. VEHANEN, and E. JANZÉN, Inst. Phys. Conf. Ser. No. 142, 45 (1996).

[23] R. F. DAVIS and J. T. GLASS, in: Advances in Solid State Chemistry, Vol. 2, JAI Press, Ltd., London 1991 (p. 1).

[24] R. H. HOPKINS, presented at the 1996 MRS Spring Meeting, San Francisco 1996, unpublished.

[25] R. A. STEIN, P. LANIG, and S. LEIBENZEDER, Mater. Sci. Engng. B **11**, 69 (1992).

[26] A. A. MALTSEV, A. YU. MAKSIMOV, and N. K. YUSHIN, Inst. Phys. Conf. Ser. No. 142, 41 (1996).

[27] R. C. GLASS, L.-O. KJELLBERG, V. F. TSVETKOV, J.-E. SUNDGREN, and E. JANZÉN, J. Cryst. Growth **132**, 504 (1993).

[28] I. SUNAGAWA and P. BENNEMA, J. Cryst. Growth **53**, 490 (1981).

[29] S. WANG, M. DUDLEY, C. H. CARTER, JR., V. F. TSVETKOV, and C. FAZI, Mater. Res. Soc. Symp. Proc. **375**, 281 (1995).

[30] R. A. STEIN, Physica **185 B**, 211 (1993).

[31] N. TAKANAKA, S. NISHINO, and J. SARAIE, Inst. Phys. Conf. Ser. No. 142, 813 (1996).

[32] M. M. ANIKIN, R. MADAR, A. ROUALT, I. GARCON, L. DI CIOCCIO, J. L. ROBERT, J. CAMASSEL, and J. M. BLUET, Inst. Phys. Conf. Ser. No. 142, 33 (1996).

phys. stat. sol. (b) **202**, 163 (1997)

Subject classification: 61.72.Nn; 61.72.Ff; S6

Modified-Lely SiC Crystals Grown in [1$\bar{1}$00] and [11$\bar{2}$0] Directions

J. Takahashi[1]) and N. Ohtani

Advanced Technology Research Laboratories, Nippon Steel Corporation,
5-10-1 Fuchinobe, Sagamihara, Kanagawa 229, Japan

(Received January 31, 1997)

This study has concentrated on the sublimation growth of SiC in directions perpendicular to the *c*-axis. 6H- and 4H-SiC crystals are grown in the [1$\bar{1}$00] and [11$\bar{2}$0] directions by the modified-Lely growth method. The crystals are different in many aspects from those grown in the conventional $\langle 0001 \rangle$ directions. The polytypic structure of crystals grown in the [1$\bar{1}$00] and [11$\bar{2}$0] directions perfectly succeeds to that of the seed, and thus polytype mixing never occurs. The crystals contain no hollow core dislocations (micropipes) and exhibit a characteristic strain relaxation. A number of stacking faults in the basal plane are introduced during growth, and the density of the stacking faults strongly depends on the crystal growth direction and polytype. We discuss the defect formation process during the [1$\bar{1}$00] and the [11$\bar{2}$0] growth and present an atomistic model for the stacking fault generation.

1. Introduction

A new degree of freedom in the seeded sublimation growth of α-SiC is the growth of crystals in directions perpendicular to the *c*-axis, e.g. [1$\bar{1}$00] and [11$\bar{2}$0]. Some very interesting aspects have emerged about defect formation [1 to 3], impurity incorporation [4, 5] and electrical properties [5] in sublimation growth in the [1$\bar{1}$00] and [11$\bar{2}$0] directions. In addition, there is also the possibility of the developing devices on {1$\bar{1}$00} and {11$\bar{2}$0} surfaces, and thin film epitaxial growth and device fabrication on {1$\bar{1}$00} and {11$\bar{2}$0} have been eagerly pursued [6, 7].

In general, for seeded SiC sublimation growth, SiC platelets or wafers with {0001} faces have been used as the seed crystal [8 to 12]. However, this brings two basic problems. Firstly, SiC polytypes tend to mix during growth; and secondly, the crystals contain a number of micropipes. In particular, micropipes, which penetrate the crystals as hollow tubes, are defects of the most serious concern since they cause serious damage to SiC devices [13].

One approach to grow SiC crystals free of micropipes is to grow crystals in directions perpendicular to the *c*-axis. We have recently proposed SiC crystal growth in the [1$\bar{1}$00] and [11$\bar{2}$0] directions using {1$\bar{1}$00} and {11$\bar{2}$0} wafers as the seed, and proved that micropipes can be eliminated in SiC crystals grown in these growth directions [14], which are naturally favored in the Acheson and Lely growth processes [15].

In this paper, we show the crystallographic properties of 6H- and 4H-SiC crystals grown in the [1$\bar{1}$00] and [11$\bar{2}$0] directions, in comparison with those grown in the

---

[1]) Fax: +81-427-68-5973, Tel.: +81-427-68-6176, e-mail: takahasi@adv.erl.nsc.co.jp

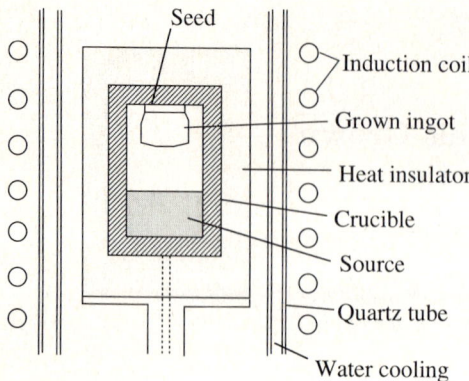

Fig. 1. Schematic drawing of a cross-sectional view of the SiC growth system used in this study

conventional ⟨0001⟩ directions. The results clearly demonstrate that there are major differences between the crystals grown along the *c*-axis and perpendicular to the *c*-axis.

2. Experimental

2.1 Crystal growth

Fig. 1 shows a schematic drawing of the growth system used in this study [1]. The crucible assembly consisted of a graphite cover to which the seed crystal was attached and a cylindrical graphite crucible containing the source powder. Mirror-polished SiC wafers were used as the seed crystals, and industrial-grade abrasive SiC powder was used as the source. The crucible assembly was covered with heat insulators made of graphite felt and placed in the reaction chamber, equipped with an rf induction heating system. Prior to growth, the crucible assembly was prebaked to reduce residual impurities.

Typical growth conditions were the seed temperature of 2200 to 2300 °C, source temperature of 2300 to 2400 °C, and thermal gradient of 10 to 20 K/cm. The Ar pressure in the reaction chamber was kept at 10 to 20 Torr. Crystal growth of 6H- and 4H-SiC was carried out using 6H- and 4H-SiC $\{1\bar{1}00\}$ and $\{11\bar{2}0\}$ wafers as the seed crystals under the same growth conditions, where the crystals grew along the $[1\bar{1}00]$ and $[11\bar{2}0]$ directions, respectively. In addition, growth in the ⟨0001⟩ directions was also conducted for comparison. The seed wafers were prepared from ingots grown in the ⟨0001⟩ directions by the modified-Lely method. They were sliced with a diamond blade, and then lapped and polished with diamond slurry.

2.2 Wafer preparation

We prepared wafers of several orientations for characterizations of grown SiC crystals. Well-oriented $\{0001\}$, $\{1\bar{1}00\}$ and $\{11\bar{2}0\}$ wafers were prepared for optical microscopy, defect preferential etching, and X-ray diffraction and topography. The $\{1\bar{1}00\}$ and $\{11\bar{2}0\}$ wafers were sliced transversely from $[1\bar{1}00]$ and $[11\bar{2}0]$ grown crystals, while the $\{0001\}$ wafers were sliced along the growth direction from both crystals. In addition, 5° off-oriented $\{0001\}$ wafers towards $[11\bar{2}0]$ were prepared, which reveal the basal plane dislocations [1].

a

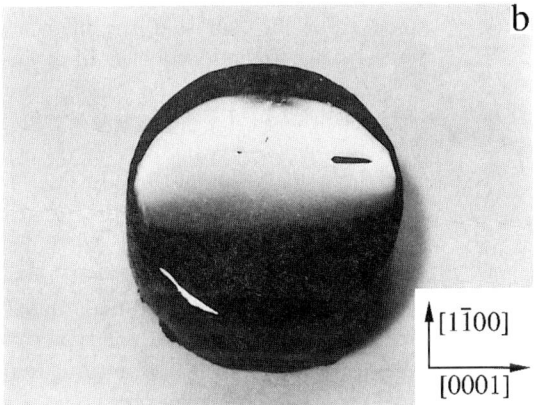

b

Fig. 2. Photographs of 6H-SiC ingots grown in a) $[1\bar{1}00]$ and b) $[11\bar{2}0]$ directions

3. Morphology and Polytype of Grown Crystals

3.1 Morphology

Fig. 2 shows the photographs of 6H $[1\bar{1}00]$ and 6H $[11\bar{2}0]$ grown ingots of 27 mm diameter. The growth surface of the 6H $[1\bar{1}00]$ grown crystal is covered with a large $(1\bar{1}00)$ facet, accompanied with $(1\bar{1}01)$ and $(1\bar{1}0\bar{1})$ facets (Fig. 2a). A grooved structure appears on the $(1\bar{1}00)$ facet of the 6H $[1\bar{1}00]$ grown crystal, which consists of a number of small $(h\bar{h}0l)$ facets (h and l are integers) extending along the $\langle11\bar{2}0\rangle$ direction. By contrast, the $[11\bar{2}0]$ grown crystal exhibits no characteristic facets in the $[11\bar{2}0]$ direction, and small $(01\bar{1}0)$ and $(10\bar{1}0)$ facets are observed at the shoulders of the ingot (Fig. 2b). Both the $[1\bar{1}00]$ and $[11\bar{2}0]$ grown crystals show no spiral step patterns, which are quite often observed on the $\{0001\}$ facets of $\langle0001\rangle$ grown crystals [16].

3.2 Polytype

Among more than 250 different polytypes of SiC reported, the 6H, 4H and 15H modifications are commonly observed in the sublimation growth process [15, 17]. They have very similar formation energies [18] and can simultaneously occur in a single crystal ingot, which has largely hindered the crystal growth of high-quality bulk SiC for years. The advantage of polytype control for the growth in the $[1\bar{1}00]$ and $[11\bar{2}0]$ directions has been recently experimentally demonstrated [1]. Fig. 3a shows the luminescence at 77 K under ultraviolet irradiation from an ingot grown in the $[1\bar{1}00]$ direction on a seed crystal of mixed polytypes. Due to the different band structures of the polytypes, each polytype shows a specific luminescence color: reddish purple for 6H, yellowish green for 4H, and orange for 15R. Fig. 3b shows a schematic drawing of the polytypic structure identified by the luminescence experiment. It was found that the polytypic structure of the grown crystal perfectly succeeds to that of the seed crystal in the $[1\bar{1}00]$ growth: 6H always grows on the 6H part of the seed, and 4H always grows on the 4H part of the seed crystal. The polytypic structure of the seed crystal was also perfectly inherited in

Fig. 3. a) Luminescence at 77 K under ultraviolet irradiation from the ingot grown in [1$\bar{1}$00] direction on a seed crystal of mixed polytypes and b) schematic drawing of the polytypic structure identified by the luminescence experiment

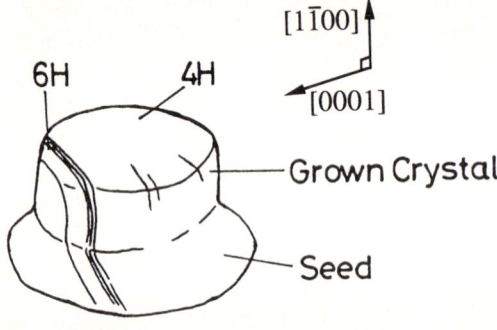

the [11$\bar{2}$0] growth, independently of the growth conditions such as the seed temperature and the Ar pressure. These results are reasonable understood from the fact that the information on the stacking sequence of atomic layers appears on the (1$\bar{1}$00) and (11$\bar{2}$0) surfaces. By contrast, for the growth in the ⟨0001⟩ directions, though the polytype of the grown crystal is mainly determined by the growth temperature, together with the polytype and face polarity of the seed crystal [19 to 21], growth conditions under which crystal of a specific polytype can be grown have yet to be formulated. The advantage of polytype control over the conventional ⟨0001⟩ growth should be stressed in the [1$\bar{1}$00] and [11$\bar{2}$0] growth.

4. Characterization of Structural Defects

4.1 Optical microscopy

Fig. 4 shows optical micrographs of {0001} wafers fabricated from [000$\bar{1}$] (Fig. 4a) and [1$\bar{1}$00] (Fig. 4b) grown ingots. The micrographs were taken under crossed polarizers and the interference contrasts in the figure reveal the strains accommodated in the wafers by a photoelastic effect. It is clear from the figure that there exist large strains associated with micropipes in the [000$\bar{1}$] grown crystal (Fig. 4a) [14]. Micropipes are hollow core screw dislocations with a large Burgers vector as predicted by Frank [22 to 25] and severely deteriorate devices, especially at high voltages [13]. On the other hand, no interference contrasts due to micropipes are observed for the [1$\bar{1}$00] grown crystal (Fig. 4b).

4.2 Defect preferential etching

Fig. 5 shows the optical micrographs of the molten KOH etched surface of 5° off-oriented {0001} wafers prepared from [000$\bar{1}$] (Fig. 5a) and [1$\bar{1}$00] (Fig. 5b) grown crystals. The crossed lines in the micrographs are surface scratches due to poor polishing. As seen in Fig. 5a, large dark hexagonal etch pits appear on the etched surface for the [000$\bar{1}$]

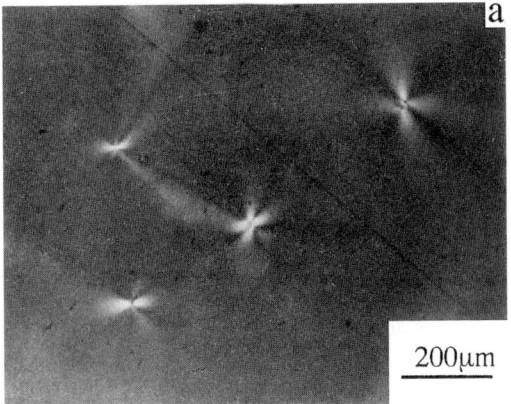

200μm

200μm

Fig. 4. Optical micrograph under crossed polarizers of {0001} wafers fabricated from a) [0001] and b) [1$\bar{1}$00] grown crystals

grown crystal. They were often observed at the center of growth spirals and formed at the outcrops of micropipes [1]. Medium and small roundish etch pits are also observed in Fig. 5a, which reveal dislocations extending along $\langle 0001 \rangle$ since they have pointed bottoms and retained the same shape as the etching proceeded. The medium etch pits correspond to $\langle 0001 \rangle$ screw dislocations with a unit c Burgers vector, while the small pits correspond to $\frac{1}{3}$ [11$\bar{2}$0] type edge dislocations [1, 3]. It was often observed that the small pits were aligned along $\langle 1\bar{1}00 \rangle$, indicating the existence of low-tilt-angle boundaries in the crystal.

On the other hand, no large hexagonal etch pit or medium etch pits were observed for the [1$\bar{1}$00] grown crystal (Fig. 5b), indicating that the crystal contained neither micropipes nor unit c screw dislocations. In both Figs. 5a and b, seashell shaped etch pits (called "shell" pits) are revealed, which correspond to dislocations lying in the basal plane and also often observed in Acheson and Lely platelets [1, 25 to 27]. The density of the "shell" pits in the [1$\bar{1}$00] grown crystal was about one order of magnitude larger than that in the [000$\bar{1}$] grown crystal. A number of basal plane dislocations are introduced upon the growth in the [1$\bar{1}$00] direction [3].

Defects lying in the basal plane can also be detected on the etched {1$\bar{1}$00} surfaces. Fig. 6 shows a Nomarski micrograph of the etched surface of a {1$\bar{1}$00} wafer prepared from a 6H [1$\bar{1}$00] grown crystal, where linear etch pits (labeled L) along $\langle 11\bar{2}0 \rangle$ observed. By a direct comparison with X-ray topographs, it was concluded that the etch pits were caused by the stacking faults in the basal plane [3]. The density of the stacking faults is in the range 10^2 to 10^4 cm^{-1} and was different between the [1$\bar{1}$00] and [11$\bar{2}$0] grown crystals [5]. They were hardly observed in $\langle 0001 \rangle$ grown crystals, which were used as the seed crystal for the [1$\bar{1}$00] and [11$\bar{2}$0] growth, thus implying that the stacking faults were generated during growth. Polytype mixing was not a cause for the stacking fault generation since the polytype of the grown crystal perfectly inherits that of the seed crystal for the [1$\bar{1}$00] and [11$\bar{2}$0] growth [1].

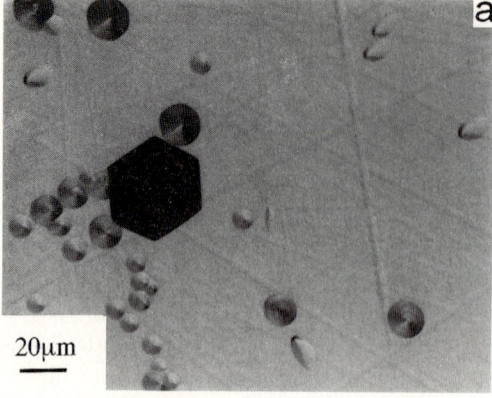

20μm

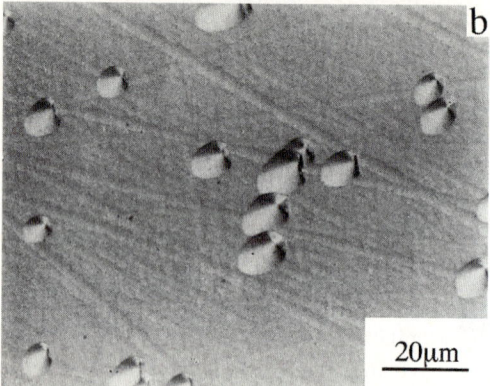

20μm

Fig. 5. Micrographs of the etched vicinal (0001)Si surface of 5° off-oriented {0001} wafers prepared from a) [000$\bar{1}$] and b) [1$\bar{1}$00] grown crystals

4.3 X-ray topography and HRXRD

The mosaic structure of SiC crystals grown in the $\langle 0001 \rangle$ directions was reported by several authors [3, 28 to 31] using X-ray topography and high resolution X-ray diffraction (HRXRD). The rocking curves of the crystals generally show asymmetrically broadened multiple peaks, indicating that the crystals consist of domains (subgrains) which are slightly misoriented to each other. The subgrains are bordered by a dislocation structure which forms a low angle boundary having tilt and twist components [3].

Fig. 7a and b show (positive) transmission topographs of a 6H-SiC {0001} wafer cut longitudinally from a [1$\bar{1}$00] grown ingot, with $\bar{1}\bar{1}$20 (Fig. 7a) and 3$\bar{3}$00 (Fig. 7b) diffraction vectors which are perpendicular and parallel to the growth direction, respectively. Intense band images (F) extend along the growth direction and perpendicularly intersect the top (1$\bar{1}$00) plane (Fig. 7a). On the other hand, no intense image is observed in the 3$\bar{3}$00 topograph (Fig. 7b), and the topograph shows an ideal uniform contrast, indicative of no defect having displacements along the [1$\bar{1}$00] growth

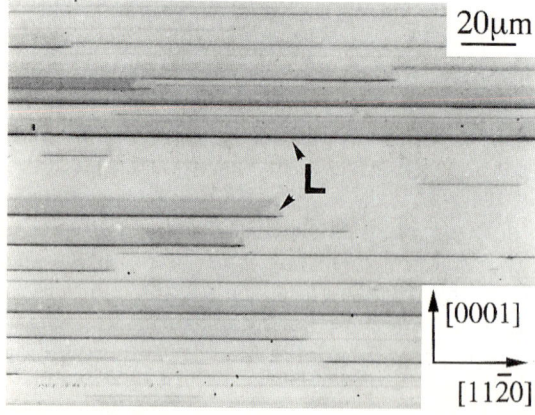

20μm

L

[0001]

[11$\bar{2}$0]

Fig. 6. Nomarski micrographs of the etched (1$\bar{1}$00) surface for the {1$\bar{1}$00} grown crystal

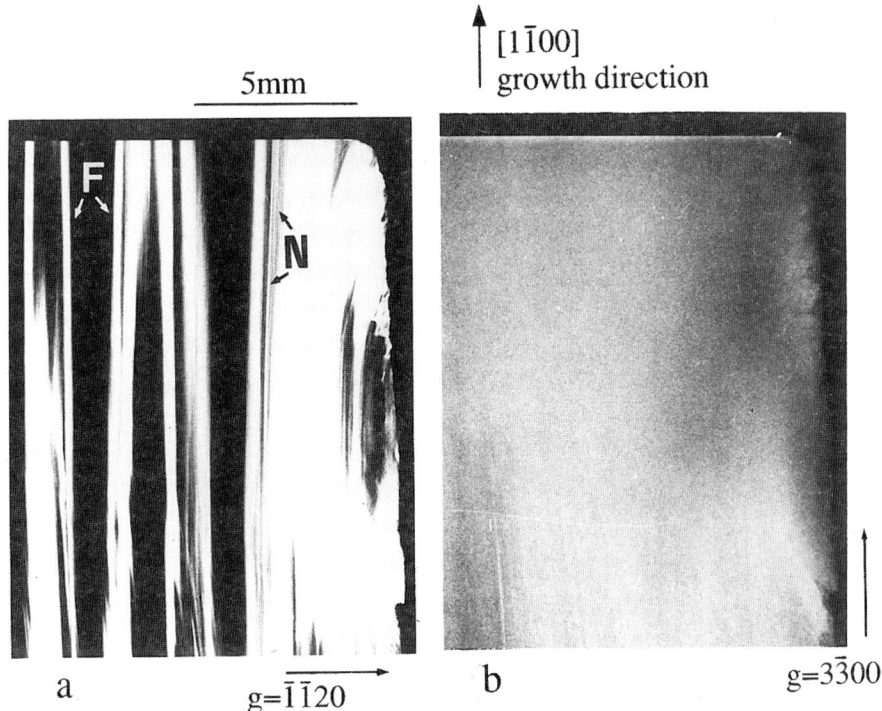

Fig. 7. Transmission topographs of a {0001} wafer prepared from a [1$\bar{1}$00] grown crystal with a) 1$\bar{1}$20 and b) 3$\bar{3}$00 reflections

direction, such as micropipes and screw dislocations. This is consistent with the results of defect preferential etching by molten KOH. Line images (N) are also observed in Fig. 7a, though their observations are largely hindered by the intense band images. The line images reveal edge dislocations parallel to the growth direction with a $\frac{1}{3}$ ⟨1$\bar{1}$20⟩ Burgers vector, which can be observed as "shell" etch pits on a vicinal (0001)Si surface by molten KOH etching [1].

Fig. 8 shows 0006 rocking curves of the {0001} wafer prepared from a [1$\bar{1}$00] grown crystal. The rocking curve in Fig. 8a was taken with the X-ray incident plane parallel to the [1$\bar{1}$00] growth direction, while the curve in Fig. 8b was taken with the incident plane perpendicular to the growth direction. The rocking curve in Fig. 8a shows a single symmetrical peak with a full width at half maximum (FWHM) of 20 arcs, which is as good as that for high quality Lely platelets [28]. In addition, as the wafer was scanned along the growth direction, the rocking curve hardly changed for the peak position and FWHM. On the other hand, the rocking curve showed peaks split by about 200 arcs when the X-ray incident plane was perpendicular to the growth direction (Fig. 8b). Furthermore the position and width significantly varied for scans across the wafer. The results suggest that the crystal contained subgrains separated by low-tilt-angle boundaries extending in the [1$\bar{1}$00] growth direction, which gave rise to the intense band images in Fig. 7a.

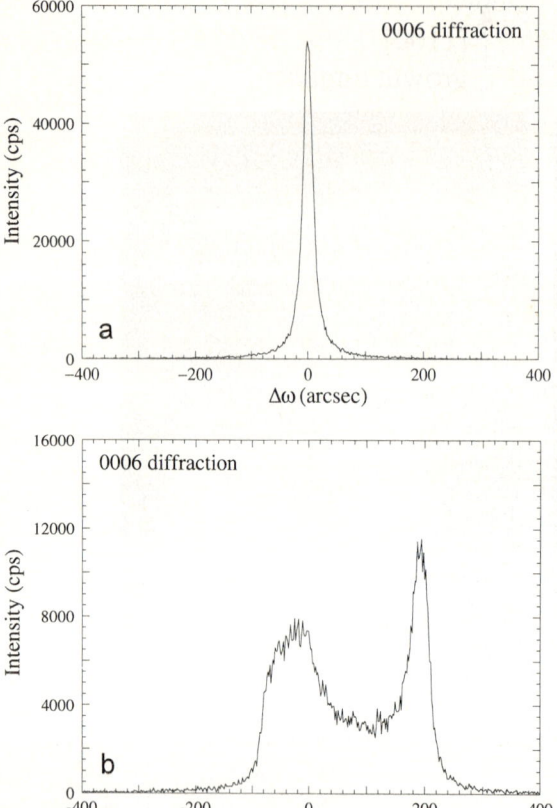

Fig. 8. Rocking curves of a {0001} wafer prepared from a [1$\bar{1}$00] grown crystal with the 0006 diffraction, where the X-ray incident plane was a) parallel and b) perpendicular to the [1$\bar{1}$00] growth direction

5. Discussions

5.1 Growth mechanism

For the $\langle 0001 \rangle$ growth, growth spirals and associated macrosteps are quite frequently observed on the {0001} facet of the crystals, implying that the spiral growth mechanism dominates the $\langle 0001 \rangle$ growth [1, 16, 23 to 25]. On the other hand, the [1$\bar{1}$00] and [11$\bar{2}$0] grown crystals contain no micropipes or screw dislocations extending along the growth direction, and thus spiral growth never occurs in these growth directions. The (1$\bar{1}$00) and (11$\bar{2}$0) surfaces have a higher surface energy and lower crystal symmetry compared to the {0001} surfaces, so that an "adhesive"-type growth mode is likely to be dominant for the [1$\bar{1}$00] and [11$\bar{2}$0] growth rather than a lateral growth (step flow growth) mode.

The growth rate and surface morphology for the [1$\bar{1}$00] and [11$\bar{2}$0] growth lend support to the above conclusion. The growth rate difference was measured by simultaneous growth on two or three seed crystals of different orientations. It was found that the growth rate in the [1$\bar{1}$00] direction was 10 to 20% higher than that in the [000$\bar{1}$] direction, and the growth rate in the [11$\bar{2}$0] direction was even higher than that in the [1$\bar{1}$00] direction. In addition, the [11$\bar{2}$0] growth showed no characteristic facet formation at the growth top, which indicates that an adhesive-type growth is dominant in the [11$\bar{2}$0] growth.

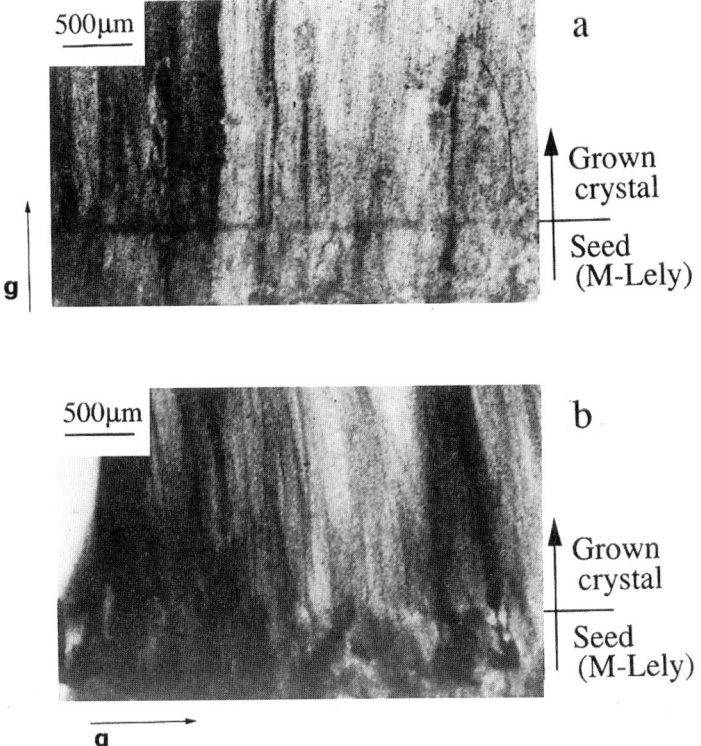

Fig. 9. Magnified transmission topographs of the interface region between the seed and the grown crystals: a) [000$\bar{1}$] growth on a {0001} seed wafer; $g = 0006$ and b) [1$\bar{1}$00] growth on a {1$\bar{1}$00} seed wafer; $g = 11\bar{2}0$

5.2 Defect formation mechanism

It was revealed that the crystals grown along the c-axis and perpendicular to the c-axis show a significant difference in both types and densities of structural defects. The crystals contain defects peculiar to each growth direction. The $\langle 0001 \rangle$ grown crystal contains a number of micropipes and screw dislocations, which are not observed in the [1$\bar{1}$00] and [11$\bar{2}$0] grown crystals and thus are supposed to be defects peculiar to the $\langle 0001 \rangle$ growth. On the other hand, the crystal grown in the [1$\bar{1}$00] and [11$\bar{2}$0] directions accommodate stacking faults and edge dislocations in the basal plane. All the defects extend along each growth direction, so that they are all grown-in-type defects and not introduced by plastic deformation during cooling or machining.

Fig. 9 shows magnified (negative) transmission topographs of the seed–crystal interface region of the [000$\bar{1}$] and [1$\bar{1}$00] grown crystals, where the seed crystals were {0001} and {1$\bar{1}$00} wafers prepared from modified-Lely [000$\bar{1}$] grown crystals. The seed crystals contained defects peculiar to the $\langle 0001 \rangle$ growth, i.e., micropipes, screw dislocations and subgrain boundaries. As seen in the figure, for the [000$\bar{1}$] growth (Fig. 9a), most of the contrast originates in the seed, implying that the grown crystal inherited the defects of the seed crystal. In particular, micropipes in the seed were perfectly inherited. The growth also introduced additional micropipes.

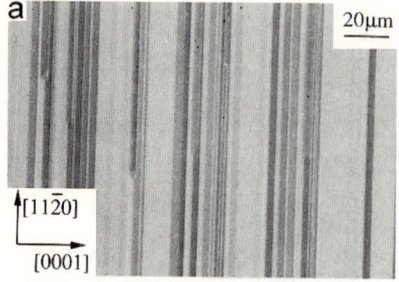

[112̄0]

[0001]

Fig. 10. Etched surfaces of (101̄0) wafers sliced from a 6H [11̄00] grown crystal. The etched surfaces correspond to a) the top and b) the middle parts of the crystal, and c) the interface region between the seed and the grown crystals. The interface is indicated by an arrow

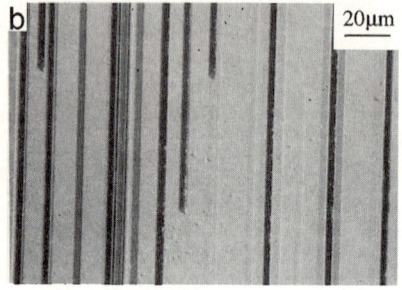

On the other hand, for the [11̄00] growth (Fig. 9b), discontinuities of contrast are observed at the interface between the seed and the grown crystals, and most of the contrast images are generated at the interface. The contrast, however, hardly changes from the bottom to the top portion of the crystal (see Fig. 7a), indicating that the defect formation occurred at the very initial stage of growth. Moreover, intense contrast images often originate at the distorted portions of the seed crystal.

The [11̄00] growth did not inherit the defects in the seed, which had been brought about by the ⟨0001⟩ growth. During the growth, they were converted to defects peculiar to the [11̄00] growth. This result suggests that strain relaxation (defect formation) processes depend on the crystal growth direction and are largely different between the crystals grown along the c-axis and perpendicular to the c-axis.

We believe that large thermal strains due to temperature variations in the crystals as well as the strains existing in the seed cause defect generation at the initial stage of growth. The strains have two components parallel and perpendicular to the seed surface. In general, large strains are relaxed by dislocations during growth. For the ⟨0001⟩ growth, the strains parallel to the seed surface bring about edge dislocations, while the ones perpendicular to the surface are relaxed by screw dislocations and micropipes. On the other hand, for the [11̄00] growth, the strains parallel to the seed surface are relaxed by edge-type dislocations running in the growth direction, while the ones perpendicular to the surface may be relaxed by the stacking faults in the basal plane, or possibly defects which are yet to be identified.

Fig. 10 shows micrographs of the etched surface of the (101̄0) wafer 60° off from the [11̄00] growth direction prepared from a 6H [11̄00] grown ingot: each micrograph corresponds to a) the top and b) the middle parts of the crystal, and c) the interface region between the seed and the grown crystals. The stacking faults appear as linear etch pits extending along the growth direction. As seen in the figures, the etch pit density in-

Table 1

Densities of stacking faults (in cm^{-1}) for 6H- and 4H-SiC crystals grown in various directions

growth direction	density of stacking faults	
	6H	4H
[1$\bar{1}$00]	10^3 to 10^4	10^2 to 10^3
[11$\bar{2}$0]	10^2 to 10^3	10^2 to 10^3
[000$\bar{1}$]	$\leq$10	$\leq$10

creases as the growth proceeds. Once the stacking faults are generated, they proceed to grow and are never terminated. Stacking faults are hardly observed in the seed crystal, and there are also very few stacking faults introduced at the interface between the seed and the grown crystals (Fig. 10c), implying that the stacking fault generation occurs throughout the entire growth process rather than at the initial stage.

It was found that the stacking fault generation strongly depends on the crystal growth direction and the grown polytype [5]. Table 1 summarizes the densities of stacking faults for the [1$\bar{1}$00], [11$\bar{2}$0] and [000$\bar{1}$] grown crystals of the 6H and 4H polytypes. In particular, the 6H [1$\bar{1}$00] grown crystal contains a large number of stacking faults, whose density is at least ten times higher than that for the 6H [11$\bar{2}$0] grown crystal and 10^3 times higher than that for the 6H [000$\bar{1}$] grown crystal. By contrast, for 4H-SiC, the [1$\bar{1}$00] and [11$\bar{2}$0] grown crystals have similar densities of stacking faults, though both still contain a higher density of stacking faults than the 4H [000$\bar{1}$] grown crystal.

We have ascribed a major cause for the stacking fault generation to the SiC growth kinetics on the (1$\bar{1}$00) surface [5]. Fig. 11 illustrates the configuration of Si and C atoms of the 6H (1$\bar{1}$00) surface seen from the [11$\bar{2}$0] direction. The 6H (1$\bar{1}$00) surface is assumed to comprise (1$\bar{1}$02) and (1$\bar{1}$0$\bar{2}$) subsurfaces of three Si–C double layers, which are alternately arranged in the $\langle$0001$\rangle$ direction. As seen in the figure, the (1$\bar{1}$02) and (1$\bar{1}$0$\bar{2}$) subsurfaces have bond configurations identical to (0001)Si and (000$\bar{1}$)C, respectively.

When the nucleation occurs on the subsurface, there are two possible bonding configurations p and p$'$, as depicted in Fig. 11. The two configurations have a small energy

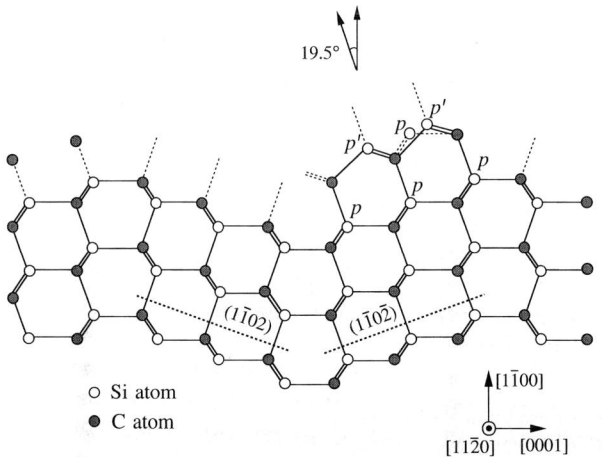

o Si atom

● C atom

Fig. 11. Atomistic surface model of 6H (1$\bar{1}$00)

difference [18], and thus while configuration p is most favorable in terms of the bulk total energy, p′ is also kinetically established during growth. Stacking faults are caused by this kinetically-induced misarrangement of surface adatoms. When a bilayer island nucleates on the subsurface, it brings about regions of disregistry at the subsurface boundaries. Stacking fault generation may relax this disregistry and relieve the associated large localized strains at the boundaries. The difference between 6H ($1\bar{1}00$) and 4H ($1\bar{1}00$) is the width of subsurfaces; narrower subsurfaces are on 4H ($1\bar{1}00$). This causes a large difference in the nucleation behavior of surface adatoms, i.e., they more frequently nucleate at the subsurface boundaries. When a bilayer island nucleates over the subsurface boundary, the bond configuration is uniquely determined, and regions of disregistry never happen.

For the ($11\bar{2}0$) surface, the bonding configuration is always uniquely determined and the kinetically-induced misarrangement of adatoms is prevented. This surface kinetics model well accounts for the observed difference in stacking fault density between the [$1\bar{1}00$] and [$11\bar{2}0$] grown 6H-SiC crystals. Stacking fault density, however, is still much higher even for the [$11\bar{2}0$] grown crystal compared to the [$000\bar{1}$] grown crystal, which implies that growth in directions perpendicular to the c-axis is inherently amenable to stacking fault generation.

6. Conclusion

Many aspects are different between the crystals grown along the c-axis and perpendicular to the c-axis. It has been shown that the crystals grown in the [$1\bar{1}00$] and [$11\bar{2}0$] directions are different from those grown in the ⟨0001⟩ directions in both types and densities of crystallographic defects. The results indicate the importance of the crystal growth direction to obtain SiC crystals of high crystallographic perfection.

Acknowledgements We are pleased to thank T. Ohsawa and O. Hashimoto for their technical help.

References

[1] J. TAKAHASHI, M. KANAYA, and Y. FUJIWARA, J. Cryst. Growth **135**, 61 (1994).
[2] H. M. HOBGOOD, J. P. MCHUGH, J. GREGGI, and R. H. HOPKINS, Inst. Phys. Conf. Ser. No. 137, 7 (1994).
[3] J. TAKAHASHI, N. OHTANI, and M. KANAYA, J. Cryst. Growth **167**, 596 (1996).
[4] Y. M. TAIROV and V. P. RASTEGAEV, Trans. 2nd Internat. High Temperature Electronics Conf. (HiTEC), North Carolina (USA), 1994 (p. 159).
[5] J. TAKAHASHI, N. OHTANI, M. KATSUNO, and S. SHINOYAMA, to be published in J. Cryst. Growth (1997).
[6] A. HENRY, I. G. IVANOV, T. EGILSSON, C. HALLIN, O. KORDINA, U. LINDEFELT, and E. JANZÉN, Book of Abstracts of 1st Europ. Conf. Silicon Carbide and Related Materials, Crete (Greece), 1996 (p. 44).
[7] K. HORINO, A. KURAMATA, K. DOMEN, R. SOEJIMA, and T. TANAHASHI, Proc. Internat. Symp. Blue Laser and Light Emitting Diodes, Chiba (Japan), 1996 (p. 530).
[8] YU. M. TAIROV and V. F. TSVETKOV, J. Cryst. Growth **43**, 209 (1978).
[9] YU. M. TAIROV and V. F. TSVETKOV, J. Cryst. Growth **52**, 146 (1981).
[10] G. ZIEGLER, P. LANIG, D. THEIS, and C. WEYRICH, IEEE Trans. Electron Devices **30**, 277 (1983).
[11] K. KOGA, T. NAKATA, and T. NIINA, Extended Abstracts 17th Conf. Solid State Devices and Materials, Japan Society of Applied Physics, Tokyo (Japan) 1985 (p. 249).

[12] D. L. BARRETT, J. P. McHUGH, H. M. HOBGOOD, R. H. HOPKINS, P. G. McMULLIN, R. C. CLARKE, and W. J. CHOYKE, J. Cryst. Growth **128**, 358 (1993).
[13] P. G. NEUDECK and J. A. POWELL, IEEE Electron Device Lett. **15**, 63 (1995).
[14] J. TAKAHASHI, M. KANAYA, and T. HOSHINO, Inst. Phys. Conf. Ser. No. 137, 13 (1994).
[15] W. F. KNIPPENBERG, Philips Res. Rep. **18**, 161 (1963).
[16] J. TAKAHASHI, N. OHTANI, and M. KANAYA, Jpn. J. Appl. Phys. **34**, 4694 (1995).
[17] P. T. B. SHAFFER, Acta Cryst. **B25**, 477 (1969).
[18] V. HEINE, C. CHANG, and R. J. NEEDS, J. Amer. Ceram. Soc. **74**, 2630 (1991).
[19] K. KOGA, T. NAKATA, Y. UEDA, Y. MATSUSHITA, Y. FUJIKAWA, T. UETANI, and T. NIINA, Extended Abstracts of Electrochemical Society Fall Meeting, Hollywood, Florida (USA), 1989 (p. 689).
[20] M. KANAYA, J. TAKAHASHI, Y. FUJIWARA, and A. MORITANI, Appl. Phys. Lett. **58**, 56 (1991).
[21] R. A. STEIN, P. LANIG, and S. LEIBENZEDER, Mater. Sci. and Engng. B **11**, 69 (1992).
[22] F. C. FRANK, Acta Cryst. **4**, 497 (1951).
[23] Y. INOMATA, H. KOMATSU, M. MITOMO, and Z. INOUE, J. Cryst. Growth **2**, 322 (1968).
[24] I. SUNAGAWA and P. BENNEMA, J. Cryst. Growth **53**, 630 (1981).
[25] P. KRISHNA, S.-S. JIANG, and A. R. LANG, J. Cryst. Growth **71**, 41 (1985).
[26] H. POSEN and J. A. BRUCE, Silicon Carbide-1973, Univ. of South Carolina Press, Columbia, South Carolina (USA) 1974 (p. 238).
[27] R. S. RAI, G. SINGH, and S. R. SE GUPTA, J. Cryst. Growth **60**, 170 (1983).
[28] R. C. GLASS, L. O. KJELLBERG, V. F. TSVETKOV, J. E. SUNDGREN, and E. JANZÉN, J. Cryst. Growth **132**, 504 (1993).
[29] S. WANG, M. DUDLEY, C. CARTER, JR., D. SBURY, and C. FAZI, Mater. Res. Soc. Symp. Proc. **307**, 249 (1993).
[30] M. TUOMINEN, R. YAKIMOVA, R. C. GLASS, T. TUOMI, and E. JANZÉN, J. Cryst. Growth **144**, 267 (1994).
[31] M. DUDLEY, S. WANG, W. HUANG, C. CARTER, JR., V. TSVETKOV, and C. FAZI, J. Phys. D **28**, A63 (1995).

phys. stat. sol. (b) **202**, 177 (1997)

Subject classification: 61.50.−f; 61.72.Lk; 68.55Ln; S6

Use of Ta-Container for Sublimation Growth and Doping of SiC Bulk Crystals and Epitaxial Layers

Yu. A. Vodakov (a), A. D. Roenkov (a), M. G. Ramm (a), E. N. Mokhov (a), and Yu. N. Makarov (b)

(a) A. F. Ioffe Physico-Technical Institute, Russian Academy of Sciences, Polytekhnicheskaya 26, 194021 St. Petersburg, Russia

(b) Fluid Mechanics Department, University of Erlangen-Nürnberg, Cauerstraße 4, D-91058 Erlangen, Germany

(Received January 31, 1997)

Analysis of specific features of sublimation growth of bulk SiC crystals in presence of Ta is performed. Control of doping and formation of different SiC polytypes is discussed. Description of mechanisms responsible for generation of micropipes during sublimation growth of bulk crystals is given. It is shown that use of Ta is promising for growth of bulk SiC crystals.

1. Introduction

The paper is devoted to the description of research works performed at A. F. Ioffe Physical Technical Institute and related institutions on sublimation growth and doping of SiC in the presence of Ta. It was a priority research and, therefore, main part of the references given in this paper is related to the studies of these groups.

At the late 70s and beginning of 80s there was a task to grow epitaxial layers of SiC, highly pure or doped by different dopants, with thickness up to 1 mm using the sublimation sandwich method (SSM) [1 to 4]. At this time graphite was considered as the only material which can be used to build systems for sublimation growth of SiC because it is stable in the presence of Si and SiC in the whole range of temperatures of practical interest. However, the graphite available at this time contained a significant amount of impurities like boron and nitrogen. Another feature was porosity of the graphite. All these factors stimulated attempts to find another material for the construction of growth systems, which is free of these disadvantages.

We started to work with Ta because TaC is stable at temperatures up to 3000 °C and silicides of Ta are stable at temperatures significantly higher than 2200 °C. We succeeded to pretreat Ta in a way that it becomes inert with respect to SiC up to 2600 °C [5]. Simultaneously, we tested Hf, Nb, and Zr. Hf is too expensive to be used extensively, and Nb and Zr are less stable at high temperatures necessary for sublimation growth of SiC compared with Ta. These studies resulted in a patent issued in 1980 [5].

During further research work on the use of Ta in systems for sublimation growth of SiC it was shown that it is possible to grow highly pure as well as doped epitaxial layers of SiC [6, 7]. Another important effect of the presence of Ta also in a very small amount inside the growth cell was that the growth process becomes much easier and controllable [5]. First of all, graphitization in the source was suppressed which allows to increase the growth temperature and, therefore, the rate of species transport and the growth rate of SiC. In addition the control of the SiC formation of the designed polytype is increased [3, 8 to 10]. The last experimental studies on the effect of the presence of Ta on the SiC growth process and the theoretical diagram obtained allowed to explain the observed behavior and to improve the process [11 to 15].

In the present paper the role of Ta in sublimation growth of SiC in sandwich system is considered in detail including the new results. Three major questions are addressed. First, specific features of the Ta–Si system. Second, doping of SiC by donor and acceptor impurities during the sublimation growth of crystals of different orientations and polytypes. This is especially critical for the growth of uniformly doped bulk crystals. Third, use of the Ta–SiC system for the growth of bulk SiC crystals.

2. Specific Features of the Ta–SiC System

In this section the effect of the presence of Ta inside the growth cell is discussed on the basis of specially performed studies. The major part of the growth experiments was done using the sublimation sandwich method. In Fig. 1 the geometry of the growth cell used in the present work for sublimation growth of epitaxial layers (Fig. 1a) and bulk crystals (Fig. 1b) is shown. The cell includes a graphite heater, tantalum container, source and seed. In some cases the container is made of massive Ta piece. The container is usually well closed to decrease vapor losses of the Si–C species. The efficiency of the sandwich cell under well-defined thermal conditions is dependent on the value of the geometrical factor $\theta = (1 - l/R)$. For the case of the closed container the vapor flux E from the vapor environment, which is created by solid SiC and is determined by the highest tem-

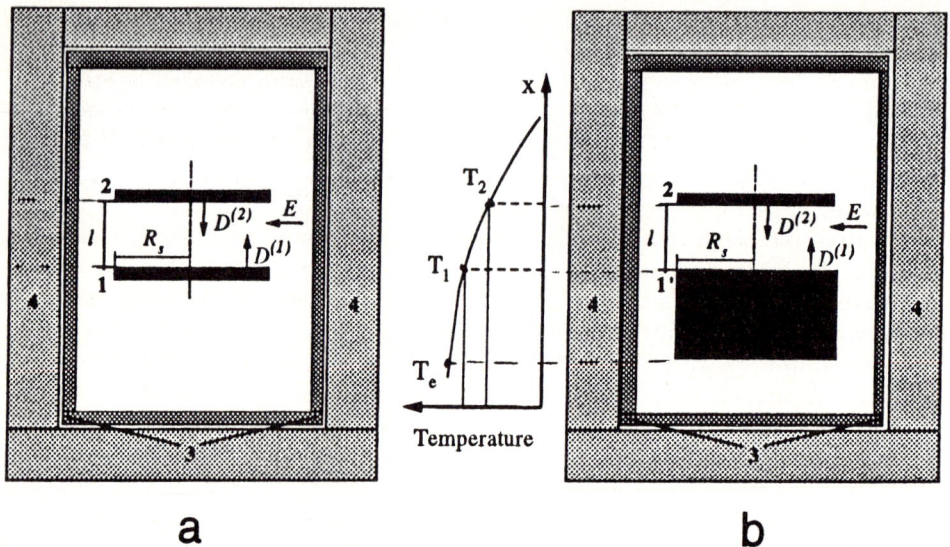

a   b

Fig. 1. Schematic view of the sublimation sandwich system used for growth of SiC a) epitaxial layers and b) bulk crystals. 1 Mono- and poly-crystalline SiC source, 2 seed crystal, 3 tantalum container, 4 graphite heater. $D^{(1)}$ and $D^{(2)}$ are the vapor fluxes of Si, Si$_2$C, SiC$_2$ from substrate and source and from external environment (E) with temperature T_e, l distance between source and seed, R_s radius of the source. The temperature distribution is also shown

perature in the solid SiC, T_e, influences the growth rate. T_e is one more process parameter of the sublimation growth process. The geometry shown in Fig. 1 and the set of process parameters was used also in theoretical studies of material transport in the systems SiC–C, SiC–Si and SiC–Ta [11 to 17].

Comparison of the SiC–Ta system with the well-known SiC–C and SiC–Si systems is very important for understanding of the specific features of the growth process in the presence of Ta. The key results are the following.

1. In papers [2 to 4, 12, 18] it was shown that during the sublimation growth in the SiC–C system the partial pressures of the species Si, Si_2C, SiC_2 are not determined strictly by temperature due to the fact that graphitization of the source occurs and is not constant during the time of the process. As a result the SiC growth rate also changes during the process, and it is difficult, in the range of temperatures appropriate for the sublimation growth, to obtain growth rates higher than 0.5 mm/h. Due to the graphitization of the source, in the sublimation sandwich system there is a high probability of graphitization of the growing crystal, and, as a consequence, uncontrolled formation of other SiC polytypes, dislocations and micropipes. Increase of temperature gradients in the growth cell decreases the degree of graphitization of the growing crystal during a certain time of the process, but, usually, the morphology of the growing crystal surface becomes worse.

2. It was shown in [11, 19, 20] that during the sublimation sandwich growth in the SiC–Si system the partial pressures of the vapor species are under better control, graphitization is suppressed and density of dislocations and microinclusions of graphite is much lower than in the SiC–C system. However, the SiC–Si system has also serious disadvantages in comparison to SiC–C: Decreased rate of material transport, possibility of formation of Si-droplets on the growing surface, formation of twin nuclei of cubic SiC which are very critical for high-quality growth of monocrystalline SiC, formation of shallow donor non-stoichiometrical defects due to excess of Si [20], and difficulties to establish long-time stability of growth conditions in the SiC–Si system which is critical for growth of bulk SiC crystals.

3. The SiC–Ta system has advantages compared to the SiC–C and SiC–Si systems. In the SiC–Ta system graphitization is significantly suppressed and this is especially critical during the beginning of the growth process [18]. It is possible to perform sublimation growth during a long time under very low pressures of the inert gas (up to $\approx 10^{-4}$ Pa). Therefore growth temperatures between 1600 and 2300 °C can be applied keeping the temperature difference between source and seed small (in practice, from 1 to 5 K/cm) [11, 12]. Growth of epitaxial layers and bulk crystals up to 20 mm in diameter without intermediate layer and with low micropipe densities (less than 10 cm^{-2}) was shown in the SiC–Ta system [11, 12, 21]. It is possible to decrease the background impurity concentration of N down to 10^{16} and B to 10^{15} cm^{-3} [6] using a Ta container made of non-porous material. As a result doping of SiC by different impurities including Cr, Mo, Mn, rare earth elements, Li, Be, Ge and V [7, 22 to 25] could be studied. Due to suppressed graphitization in the SiC–Ta system it is easier to perform controlled sublimation growth of different SiC polytypes like 4H, 15R, 21R and 3C [3, 8 to 10, 12, 21, 26 to 29].

One of the disadvantages of the SiC–Ta system is the following. When a new Ta container is used for the sublimation growth, Ta incorporates into the growing SiC especially during the initial stage of the growth process and the nucleation of polycrystal-

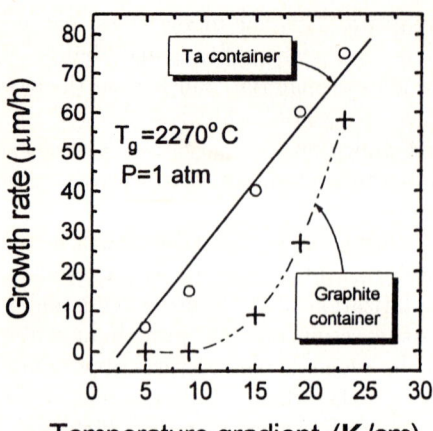

Fig. 2. Dependence of SiC growth rate on temperature gradient in SiC–Ta (solid line) and SiC–C
systems (dashed line). The processes were performed under atmospheric pressure of Ar and at
seed temperature 2270 °C

line 3C-SiC occurs similar to the SiC–Si system. A special pretreatment of Ta was
developed which allows to solve the problem [6].

In Fig. 2 the SiC growth rates in SiC–C and SiC–Ta systems are shown in dependence on the temperature gradient. The growth was performed in Ar at atmospheric pressure to suppress graphitization. To obtain a sufficient thickness of the grown layer for
the analysis of the growth rate at low temperature gradients the experiments were performed at 2270 °C. The advantages of the SiC–Ta system are seen especially for small
temperature gradients. However, noticeable growth in the SiC–C system under the considered conditions occurs only at sufficiently high temperature gradients.

The sublimation sandwich method allows to establish vapor conditions close to equilibrium inside the growth cell and to suppress graphitization of source and seed. This feature is used to compare the SiC–Ta and SiC–C systems in a wide range of temperatures
from 1650 to 2000 °C. The distance between source and seed is 0.5 mm and the temperature difference about 1 K which corresponds to a temperature gradient of about 20 K/
cm. In Fig. 3 the dependences of the SiC growth rate on temperature for both systems
are shown. It can be seen that the growth rate in the SiC–C system is slightly higher

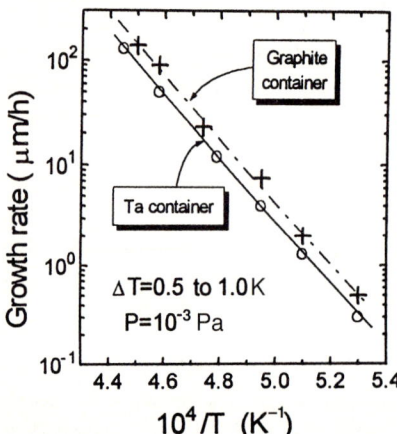

Fig. 3. Dependence of SiC growth rate on temperature
obtained during sublimation sandwich growth in Ta
and graphite containers

Fig. 4. Photo in transmitted light of a cross-section of the SiC crystal in a plane perpendicular to the growing surface (× 7)

than in the SiC–Ta system. This is the case if graphitization can be suppressed in the SiC–C system (Fig. 2 shows the results of experiments with graphitization). Taking into account that in the Ta–SiC system the formation of 3C-SiC is easier and that the growth rate is lower (this was shown experimentally and theoretically in [19]), we can conclude that the SiC–Ta system is closer to SiC–Si than to SiC–C.

In Fig. 4 a photo of a cross-section of the SiC crystal grown in SSM system is shown. One can see the absence of an intermediate layer between the Lely crystal used as a seed and the thick epitaxial layer.

A study of Ta annealed at high temperature close to the temperatures used for sublimation growth in the presence of graphite and/or SiC was performed in [12 to 14]. For characterization Auger analysis, phase analysis using X-ray diffractometry and precise weighing of the annealed Ta samples were used. It was obtained that annealing of Ta in the presence of graphite at temperatures which correspond to the growth temperature results in transformation of Ta into TaC. This occurs in a layer of 0.2 to 0.4 mm thickness within a few hours. In Fig. 5 a photo of a Ta container of 0.5 mm thickness annealed in presence of SiC is shown. It is clear that TaC is formed over the whole thickness of the container due to the fast diffusion of carbon through the grain boundaries. However, diffusion inside the Ta grains is much slower, and in a container of a few millimeter thickness Ta can be present after many hours of annealing at the temperatures used for the sublimation growth. Therefore the Ta container can work as a getter of carbon during many hours of the sublimation growth process.

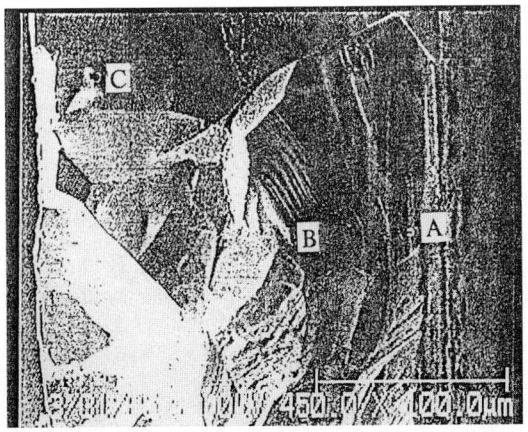

Fig. 5. Cross-section of the wall of the Ta container annealed in presence of SiC. A, B and C on the photo show the points where Auger measurements were performed and the following results are given: point A: 51.6 at% C, 45.6 at% Ta, 2.8 at% oxygen; point B: 53.8 at% C; 44.6 at% Ta, 1.6 at% oxygen; point C: 54.9 at% C; 43.4 at% Ta, 1.7 at% oxygen

In [12] it was proposed and later in [14] confirmed that during the annealing of Ta in the presence of SiC the surface layer of the container consists not only of TaC but includes also $TaSi_2$ and SiC. The obtained data show that the solid Ta–Si–C phase is present in a thin surface layer of the Ta container, the next layers in depth consist of TaC, and further ones of a mixture of TaC and Ta.

On the surface of SiC crystals present at high temperatures inside the Ta container some traces of solid $TaSi_2$ are also found. It was shown in [6, 18] that a special pretreatment of the Ta container results in a drastic decrease of the amount of Ta on the surface of growing SiC and allows to grow high-quality SiC crystals with concentration of Ta less than 10^{16} cm^{-3}.

The obtained experimental data were used for a theoretical analysis of the SiC–Ta system in [12, 16, 17, 30, 31]. An approach proposed earlier in [30] for phase analysis in SiC–C and SiC–Si systems was used also for the SiC–Ta system. A phase diagram of sublimation sandwich growth is shown in Fig. 6 for different external vapor environments of the SiC–C, SiC–Si and SiC–Ta systems. The temperature of the seed crystal and the difference between the Si and C fluxes arriving at the growing SiC surface are used as coordinates of the diagram. A point in the diagram corresponds to a growth process with a defined set of process parameters — seed temperature, temperature difference between source and seed, geometrical factor θ, and the external vapor environment of temperature T_e. The right part of the diagram corresponds to the conditions of stable growth of SiC without formation of additional condensed carbon or silicon phases. Two narrow lines show the area of formation of an extra liquid Si phase, and the so called "kinetically clean area" corresponds to the metastable growth of SiC where the formation of excess Si might be possible due to fluctuations. One can see that under the considered growth conditions the sublimation sandwich growth processes in SiC–Si and SiC–Ta are similar concerning the formation of additional phases in contrast to the SiC–C system where growth occurs close to the boundary of the graphitization areas.

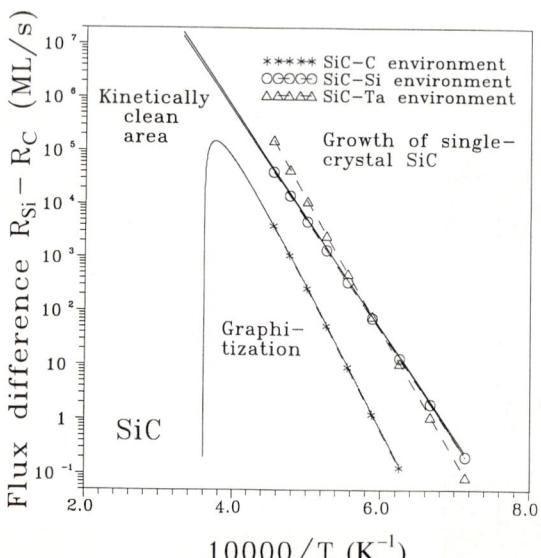

Fig. 6. Theoretical diagram of the formation of secondary phases during sublimation sandwich growth of SiC with different external vapor environments of the SiC–C, SiC–Si and SiC–Ta systems

Also at lower growth temperatures the SiC–Ta system shows advantages in comparison with the SiC–Si system with respect to the formation of a liquid Si phase.

We think that the presence of a small amount of Ta on the surface of the growing SiC prevents the formation of a graphite phase and improves the growth process. This can be considered as an additional specific feature of the SiC–Ta system.

For the growth of bulk SiC crystals in a Ta container using the sublimation sandwich method it is necessary to take into account material supply from the external vapor environment with temperature T_e which is dependent on the geometrical factor θ. The differences of fluxes of Si and Ca atoms arriving at the growing surface can be calculated in dependence on θ, and a phase diagram in coordinates θ and growth temperature can be obtained which is specific for SSM. Such diagrams were obtained for the systems SiC–C and SiC–Si in [31]. One of the key parameters of these diagrams is the maximal temperature of the external environment T_e which for the growth of bulk SiC crystals in SSM (see Fig. 1b) is the temperature of the lower part of the source. The phase diagrams for SSM growth of SiC in SiC–C and SiC–Si systems are shown in Fig. 7. The horizontal line with the value of the geometrical factor $\theta = 0.5$ corresponds to the value used in our experiments for the growth of bulk SiC crystals. Two vertical dashed lines show the range of growth temperatures used. The corresponding diagram for the SiC–Ta system is similar to that for SiC–Si except for the fact that the area of graphitization disappears totally. From this consideration we can conclude that T_e should be in the range 2000 to 2200 °C, to keep the temperature drop across the volume of the source and to use small temperature difference between the surfaces of source and seed. Such conditions should result in the suppression of graphitization inside the growth cell and provide stable growth of crystalline SiC. We will show in Section 4 where results of the growth of bulk SiC are given that such growth conditions are advantageous.

From the performed experimental and theoretical studies it can be concluded that it is promising to grow SiC crystals by the sublimation sandwich method in the presence of Ta. The proposed explanation of the role of Ta during the sublimation sandwich growth is valid and the shown advantages to use a Ta container are retained during the time when on the surface of the container a $TaSi_2$ film is available. For Ta containers of a few millimeter thickness this time might be around 100 h at growth temperatures close to 2000 °C.

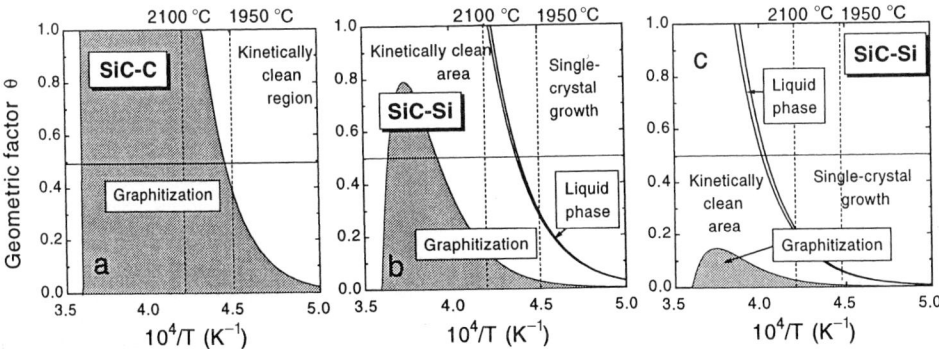

Fig. 7. Diagram of the formation of secondary phases during SSM growth of SiC with external environment of SiC–C and SiC–Si systems and different values of the environment temperature T_e: a) SiC–C and $T_e = 2027$ °C; b) SiC–Si and $T_e = 2027$ °C; c) SiC–Si and $T_e = 2327$ °C. $\theta = (1 - l/R_s)$

3. Doping of SiC by Donors and Acceptors; Polytype Formation

In this section doping of SiC crystals using the main donor and acceptor impurities N, P, Al, B and Ga during the sublimation sandwich growth in the presence of Ta is considered. Also the effect of doping on polytype formation and morphology of the growing crystals is discussed using the results of [3, 8 to 10, 26 to 29].

In our group studies of doping of different SiC polytypes using a wide range of impurities were performed since a long time [6, 7, 10, 22 to 24]. Some results which are important for the present papere are the following.

Doping of SiC crystals was peformed in two ways:

— from the vapor phase the source of the dopant being a gas (for example, nitrogen) or a solid at a given temperature;

— from a solid SiC source specially doped by a certain impurity and up to the necessary level at high temperature during the preparation procedure [7, 24].

The level of doping by the considered impurities depends on the partial vapor pressure of the dopant inside the growth cell, on temperature, as well as on the orientation of the growing SiC surface. In Fig. 8 and 9 the basic dependences of the doping level on the process parameters are shown. The doping level can be dependent also on the SiC growth rate, this is clearly seen in Fig. 10 for the case of doping by nitrogen.

It is important to know that the dependence of the doping level on the orientation of the growing SiC surface becomes weaker with increasing growth temperature. This dependence is also weaker when Si-rich growth conditions are available near the growing surface. But, for growth temperatures up to 2300 °C the effect of orientation is still visible. For example the concentration of N during SiC growth on the C-face is 2 to 3 times higher than on the Si-face. In contrary the Al and Ga incorporation is 5 to 10 times lower on the C-face than on the Si-face. It is quite general that the incorporation of almost all impurities except donors is higher on the Si-face during the sublimation growth of SiC. Incorporation of impurities is sensitive to small variations of the crystallographic orientation. Misorientation of a few degree from a singular face changes the amount of the incorporated dopant.

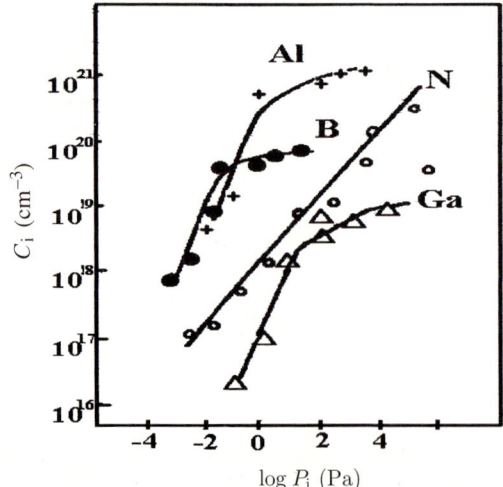

Fig. 8. Dependence of the impurity concentration in the growing SiC on the dopant partial pressure inside the growth cell for the cases of doping by N, B, Al and Ga. $T = 2000$ °C. Doping with N is performed on $(000\bar{1})$C SiC; with B, Al and Ga on (0001)Si SiC

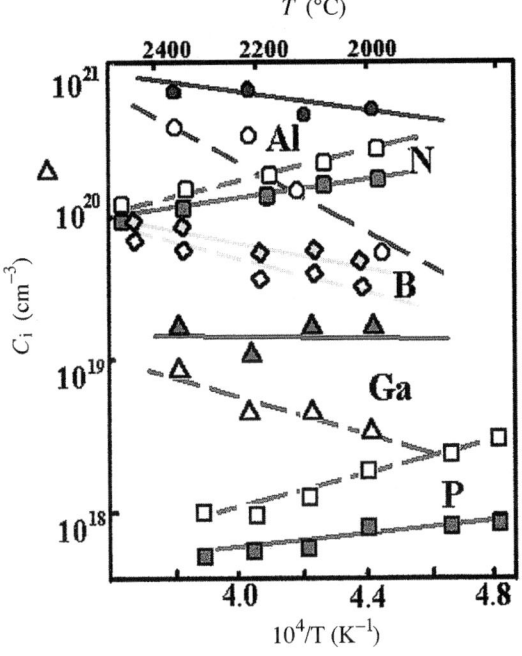

Fig. 9. Dependence of the concentrations of N, B, Ga, Al, and P in SiC on the growth temperature. Orientation of the growing SiC: (0001)Si solid lines; (000$\bar{1}$)C dashed lines

The discussed features of the incorporation of the main impurities make it difficult to obtain bulk SiC crystals which are uniformly doped in radial and axial directions. A crystallization front which is not plane results in non-uniform dopant incorporation. This is especially critical for the case of a convex shape of the growing SiC crystal due to the increased probability of formation of a singular face. Significantly non-uniform doping is observed when the crystal growth occurs simultaneously in two directions — along the C-axis and in perpendicular direction. For example, if the concentration of nitrogen is higher than 10^{18} cm^{-3}, the areas of crystal orientation (000$\bar{1}$)C parallel to the C-axis are much more colored in transparent light (4H — brown, 6H — green, 15R — yellow, 27R — red, etc.) than the areas grown in the perpendicular direction. In the case of doping by Al and Ga when the doping level is higher than 5×10^{18} cm^{-3} it is opposite — side walls are blue. For the case of (0001)Si orientation of the growing surface the change of color in dependence on the direction of growth is opposite. If simultaneous doping by donor and acceptor impurities occurs, the effect of the crystal orientation on the non-uniformity of doping becomes even stronger

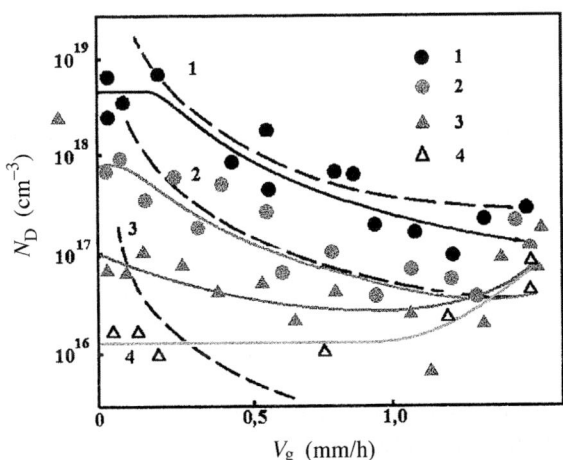

Fig. 10. Dependence of the donor concentration (N) in SiC epitaxial layer on growth rate. Concentration of nitrogen in the source: (1) 2×10^{19}, (2) 3×10^{18}, (3) 3×10^{17}, (4) 7×10^{16} cm^{-3}. The growth temperature is 1850 °C. Dashed lines are theoretical results

to the variation of the level of compensation. For example, transfer from C- to S-side can result in a change of electrical conductivity from n- to p-type.

The discussed effect of orientation on the doping level can be used for an analysis of the growth process. At sufficiently high doping levels the growing crystal is colored. Non-uniformity of the color over the crystal area shows the character of the variation of the crystallization front along the length of the crystal, specific features of the initial stage of growth and lateral growth of the crystal, and the formation of polytypes different from that of the seed.

For some applications it is necessary to obtain not only a uniform but also a highly doped SiC crystals (more than 10^{19} cm^{-3}). This is difficult due to a few reasons. First, there is the effect of doping on polytype formation [3, 9, 10, 26, 29]. For example, addition of a significant amount of N favors the formation and stabilization of "more cubic" polytypes like 3C, 21R and 15R, and makes difficult the stable growth of high-quality SiC of more "hexagonal" polytypes, first of all of 4H. An the contrary, addition of rare earth metals, Ge, Sn, Pb and less Al leads usually to easier transformation into the SiC polytypes with higher hexagonality (first of all into 4H) and their stabilization [9, 10]. It was shown for the first time in [3, 8, 26 to 28] that controlled transformation of the growing SiC polytype or uncontrolled change of the polytype is much easier on singular faces like (0001)Si and especially (000$\bar{1}$)C. On the (000$\bar{1}$)C face formation of the more hexagonal polytypes is advantageous (first of all 4H) and on the Si singular face, of the more cubic polytype like 3C, 15R and 21R.

Analysis of the available experimental data resulted in the conclusion that a deviation from stoichiometry at the phase boundary of the growing crystal determines formation and stabilization of a definite polytype [10]. Excess of active carbon on the growing surface significantly suppresses the formation of carbon vacancies in the growing SiC crystal and on the surface, and, therefore, stimulates formation of the more hexagonal SiC polytypes [9, 10]. Here we consider carbon not in the form of a secondary graphite phase but in atomic or molecular form in the adsorbed layer on the SiC surface. Excess of active silicon, however, stimulates generation of carbon vacancies and interstitial silicon on the phase boundary and close to it, which results in preferable growth of "more cubic" SiC polytypes [9, 10]. We can assume that significant concentrations of vacancies and interstitial atoms which are substantially different in theses two cases [10, 34, 35] should change the elastic energy in SiC in the vicinity of the phase boundary due to existence of local strain [10, 34 to 37] and, therefore, the elastic energy on the growing surface. This quantity can determine what kind of SiC polytype is more favourable and more stable.

It is known that growth temperature [32] and growth rate [33] also influence the stabilization of the growing SiC polytype.

In [34] results of studies on the evolution of agglomerations of point defects in SiC in a wide range of temperatures 450 to 2700 K are summarized. It was shown that there exist quite narrow temperature intervals where a significant change of the type of point defect clusters occurs. It was demonstrated in [38] that clusters of interstitial atoms are located close to the (0001) plane, but vacancies and clusters of vacancies are concentrated in perpendicular direction. In accordance with our assumption, the elastic energy of the nucleus and the polytype to be formed should depend on the character of the point defects, and, therefore, on the growth temperature. However, these effects should influence first of all the nuclei on the growing surface and be more visible when the nucleation growth mechanism prevails. During the layer by layer growth mechanism, when usually the polytype of the seed is retained, these effects should not be visible. Exactly this was shown for the first time for the case of SiC growth in [8]. High concentration of impurities in the vapor phase and, consequently, in the adsorbed layer, stimulates the change of the growth mechanism from the layer by layer to the nucleation mode.

This understanding was used in [11 to 13, 21, 26 to 28] for the control of polytype formation and also for the growth of bulk SiC which is discussed in the next section.

4. Specific Features of Sublimation Sandwich Growth of Bulk SiC Crystals of 6H, 4H and 15R Polytypes with Use of Ta

Growth of bulk SiC crystals is performed in the growth system shown in Fig. 1b in a closed Ta container with use of inductive heating. During the growth of SiC crystals of diameter up to 25 mm a Ta container with wall thickness up to a few millimeters is used. A specially developed technique to close the container allows to perform the growth process under external pressures up to 10^{-3} to 10^{-4} Pa with losses not exceeding a few percent of the mass of the grown crystal during 10 to 20 h growth. A different approach is used when it is necessary to grow doped SiC crystals, for example by nitrogen. In this case it is necessary to control the closeness of the container in order to provide uniform doping along the length of the crystal. In the latter case the yield of the source is usually between 70 and 90%. The reason of this variation is not clear up to now.

In Fig. 11 a schematic view of the closed Ta container is shown. The SiC seed is placed on the top and the polycrystalline SiC source on the bottom of the container. The distance between source and seed is usually between 5 and 10 mm depending on the size of the crystal. Before closing the container is pumped down to a pressure of 10^{-4} Pa at temperatures between 1300 and 1500 °C. All the operations are performed inside the inductively heated furnace during the growth cycle.

During the process development stage a controlled growth of SiC bulk crystals of 6H, 15R and 4H polytypes was achieved. 6H and 15R SiC crystals were grown on 6H SiC of orientation (0001)Si. The key task was to grow 4H–SiC bulk crystals which were grown

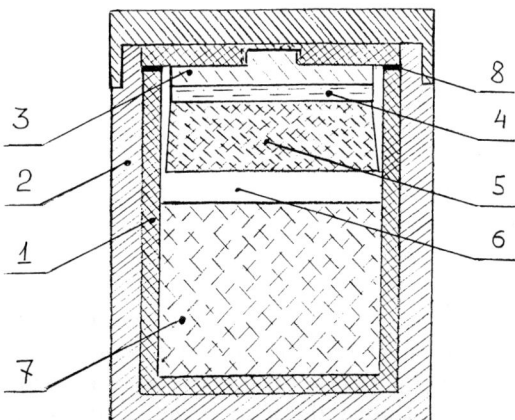

Fig. 11. Scheme of the Ta container for sublimation sandwich growth of bulk SiC crystals. 1 Ta container, 2 graphite container, 3 substrate holder, 4 substrate (SiC single crystal), 5 bulk SiC single crystal, 6 clearance between substrate and source, 7 Si vapour source, 8 vacuum sealing

3

2

1

7

8

4

5

6

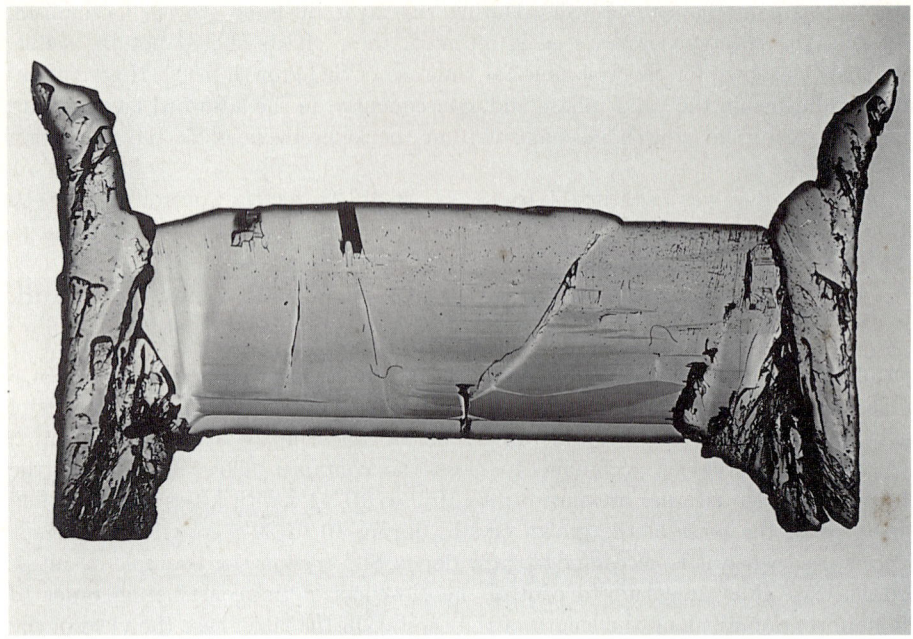

Fig. 12. Photo in transmitted polarized light of a cut in a plane parallel to the growth direction of two 4H SiC crystals grown on a seed which consists of two 4H SiC crystals

in presence of Sn [21, 26] using different SiC polytypes [8, 9] with $(000\bar{1})$C orientation as seeds. Special attention was given to crystals grown under non-optimal conditions and, therefore, including specific defects. This is important in order to better understand the mechanisms of defect formation.

For optimization of the growth processes with respect to the use of seeds with $(000\bar{1})$C and (0001)Si orientations the following procedure was used [12, 21]. Two symmetrically placed crystals of different orientation and polytype are used as seed. If the two seed crystals are of the same orientation, usually an increase of the size perpendicular to the c-axes of one of the crystals is observed, also when the seed crystals are of the same polytype (see Fig. 12). This effect is not due to some asymmetry of the temperature field, special effort is taken to make the thermal field symmetrical (rotation of the crucible, etc.). Probably, one of the two crystals has more efficient nucleation centers like screw dislocations or slightly different orientation. The normal growth rate is very similar for both crystals. Usually the laterally extended crystal (left part of Fig. 12) is faster grown than the laterally smaller one (right part of Fig. 12), and at a total length of the crystal of around 10 to 20 mm there exists a step height of 100 to 2000 μm between the two crystals. If the orientation of the two seed crystals is different, the behavior is similar with the height of the step close to the upper limit.

It is obtained that the doping level in the grown crystals is very different for different orientations. In Fig. 13 (see page 193) a cut of two crystals grown on two 6H seeds with orientations $(000\bar{1})$C (left) and (0001)Si (right) is shown. On the left side a 4H SiC crystal with n-type conductivity is grown, and on the right a 6H p-type one significantly doped by Al during the initial stage of growth. During the later stages of growth it becomes compensated and, therefore, colourless. The n-doped sample is blue in polarized light, and brown in non-polarized light. The color of 6H SiC is weakly dependent on polarization and is determined as in all other SiC crystals of different polytype and n-type conductivity by the nitrogen doping level, and is green–blue. Under optimal growth conditions and doping level lower than 10^{18} cm^{-3} in the absence of Sn vapor, both growing crystals are of the same polytype as the corresponding seed. If Sn is added, on the seed of (0001)C orientation quick transformation into 4H–SiC occurs independently of its polytype (see Fig. 13a), followed usually by the spreading of the growing crystal. It is possible to state that 4H SiC grows definitely faster than crystals of any other polytype under the same growth conditions.

Growth of the crystals shown in Fig. 12 and especialy in Fig. 13 is performed under less optimal conditions as compared to the crystal shown in Fig. 4. In contrast to the crystals in Fig. 4 and 12 which have seed crystals and intermediate layers of good quality, it is different in the crystal shown in Fig. 13. Inclusions of a secondary phase (dark points) are seen in crystals from Figs. 12 and 13, the amount of the inclusions increasing along the growth direction. One can see also visible morphological defects and vertical micropipes (dark lines). The mechanisms of formation of these defects will be discussed further in this paper.

In our experiments it was shown that one of the important factors determining the quality of the growing crystal is the quality of the back side of the seed and how well it is fixed on the holder [21]. This is especially important for the case of growth from top to bottom as in our system (see Fig. 1b). This direction of growth is more appropriate from our point of view because of a better possibility to insert the source material and to minimize the deposition of any kind of dust and particles onto the seed due to gravitation. If there exists an empty space or metallic droplets or small SiC crystals between the seed and the holder, there are conditions for the recrystallization of a part of the back side of the seed onto the holder. The layer which is growing in this case on the holder can pass through the growing bulk SiC crystal and form a secondary crystallization front (see defect region D in Fig. 13). Most probably such a recrystallization occurs through an empty space between seed and holder by sublimation, but we observed also recrystallization through a plane droplet via liquid phase epitaxy similar to the travelling solvent method. On the holder usually polycrystalline material is deposited, sometimes 3C SiC which is easier to be formed in presence of Ta. This results in the formation of the secondary crystallization front. Sometimes also small monocrystals are formed on the holder with orientation close (0001) but turned on an angle around the c-axis from the seed orientation. In the latter case the secondary crystallization front formed in the defect regions "D" moves through the growing crystal and behind this front the quality of the crystal becomes worse. Rate and direction of movement of the secondary crystallization front, and, therefore, thickness and location of the areas in the crystal with modified properties depend on the temperature gradient inside the crystal, on the thickness of the empty space between seed and holder, on the misorientation of the crystal of the secondary front, etc. Thus, different areas of recrystallization form secondary crystallization fronts moving with different velocities through the crystal. Therefore they stay at the end of the process at different heights inside the crystal. Such a case is seen in Fig. 14 (see page 193). A similar situation is shown in Fig. 15 where a cut perpendicular to the growth direction is shown. One can see that separate plane crystals of (0001) orientation formed on the back side of the seed pass through the earlier grown bulk SiC crystal. This effect was described morre detailed in [21].

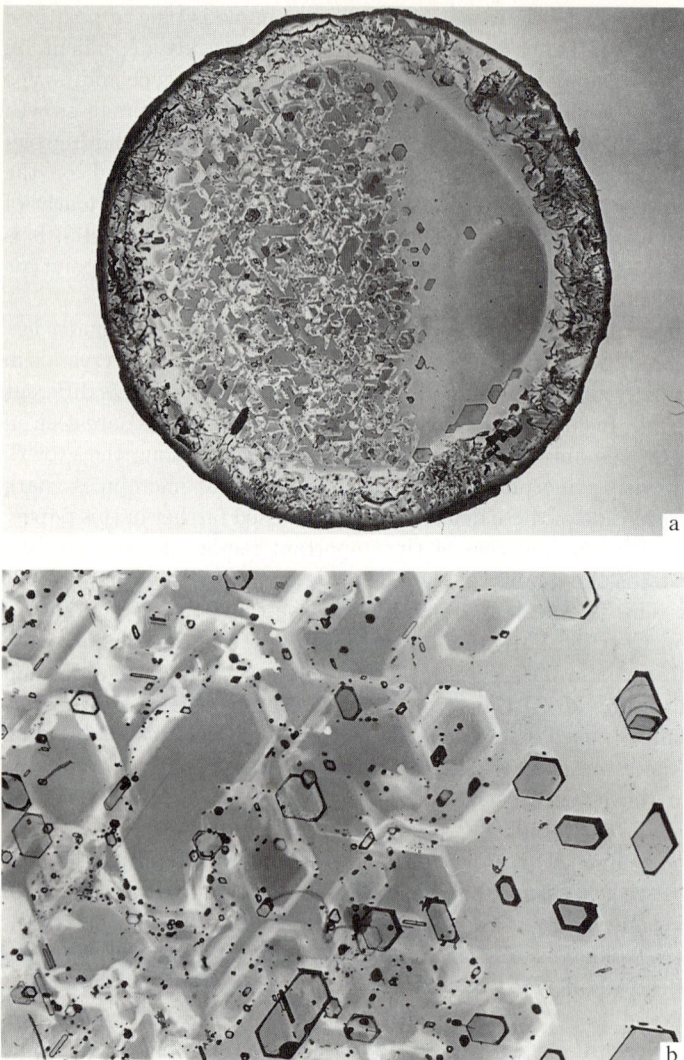

Fig. 15. Photo of a cut of 4H SiC bulk crystal similar to that shown in Fig. 14 in the plane perpendicular to the growth direction made in transmitted light. a) On the left side a recrystallized polycrystalline area is seen. The horizontal planes are of orientation close to $(000\bar{1})$C (magnification $2.5 \times$). b) One can see hexagonal structures in the grown crystal (basal plane tubes) (magnification $15 \times$)

In Fig. 13 on the back side of the seed two areas "D" moving very slowly are seen, and, what is quite typical, a few millimeters above these areas in the growth direction highly defective areas are formed in the crystal. On the crystal shown in Fig. 16a similar behavior can be seen in the left part. On the back side of the seed a polycrystalline 3C SiC layer is deposited, and inside the grown 4H SiC crystal (grown on 6H seed) highly defective areas are formed passing through the whole crystal. Careful analysis of all the cuts of the crystals presented in Figs. 13 to 16. shows that the sources of the vertical

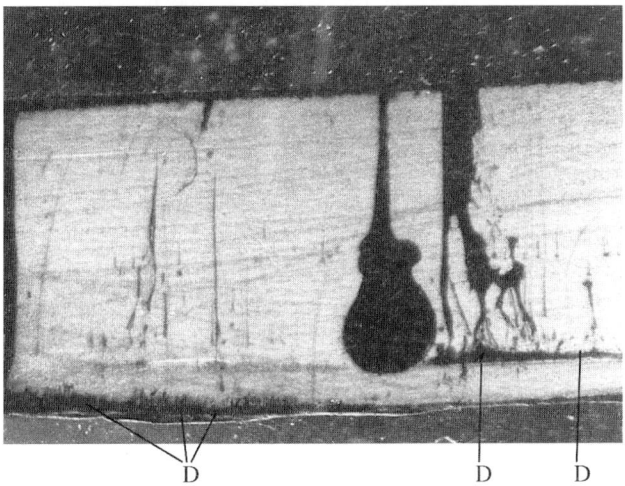

Fig. 16. Longitudinal cross section of SiC crystal. Photo in transmitted light. Micropipes are generated over the defect areas "D" on the back side of the seed. The micropipes formed over the secondary phase inclusions are only 0.5 to 1 mm in length and are further overgrown

micropipes passing through the whole or main part of the crystal are the polycrystalline defect areas. Some of them are related to the back side of the seed as discussed earlier and these areas are formed at the initial stage of growth. Sometimes they can be overgrown by the bulk crystal, but the micropipes do not disappear (see Fig. 16 on the right). In contrast, micropipes are not generated on inclusions of the secondary phases, which can be seen in Figs. 13 and 16 as dark or light points, or, if generated, they are overgrown after 0.5 to 1 mm.

Fig. 17. Photo of the cut of 4H SiC bulk crystal made in transmitted light. Thick layers of other SiC polytypes (6H and 15R) are seen from which micropipes start to grow. The secondary phase inclusions do not result in formation of micropipes (magnification 2.5 ×)

14*

All these data allow to conclude about the dominating mechanism of micropipe formation. We think that the micropipes are formed in SiC due to the existence of domains
with different degrees of misorientation from the growing bulk crystal. If the misorientation is small, micropipes can be generated in form of clusters. In the case of significant
misorientation locally strained areas can be available in the growing crystal. This could
be due to the well-known long range effect of strain, which may generate micropipes or
other macroscopic defects in the crystal far from the strained area.

In this paper it is mentioned that on (0001) planes the formation of other polytypes
and of nuclei is easier, and, therefore, the appearance of slightly misoriented domains is
more probable. The nuclei formed on surfaces usually are of (0001) orientation, but they
can be misoriented due to a rotation around the main symmetry axis. For example, we
often see that the 4H SiC formed on 6H is usually turned by 60° around the c-axis. Such
misorientations are easily formed in SiC because of the small energy of stacking faults in
this direction [4].

If our understanding of and assumption on the role of the misoriented domains in the
formation of micropipes are correct, layers of another polytype in the grown SiC crystal
should act as areas of generation and disappearance of micropipes. Such micropipes
should be generated on the singular surface of the layer which is formed later during the
growth process, and can end on a earlier formed one. In the growing bulk crystal such
layers of another polytype should be placed in the planes perpendicular to the growth
direction and definitely on singular surfaces. In the growing crystal the formation of
singular surfaces of significant large area is more probable in the case of plane or convex
phase boundaries. In Fig. 17 such a typical situation is shown. It is seen that layers of
another polytype and the micropipes generated on them are available only in the

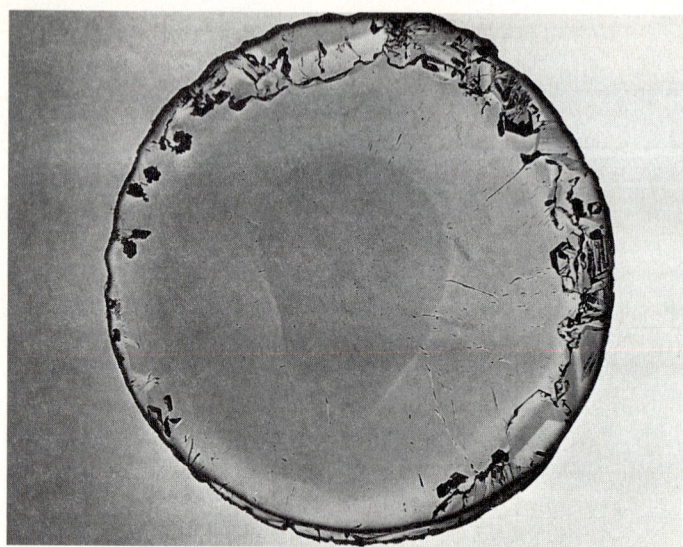

Fig. 19. Photo made in transmitted light of a cut in a plane perpendicular to the growth direction
of 4H bulk SiC crystal similar to that shown in Fig. 18. Significantly convex crystallization front is
seen during the initial stage of growth. The radial cracks are seen on the slope of the crystallization
front only. They are absent usually on a surface close to a (0001) plane

a

Fig. 13. Longitudinal cross section of bulk crystal grown on a seed which consists of two 6H SiC crystals. The photo in transmitted a) polarized and b) non-polarized light. The crystal grown on the left part is of $(000\bar{1})$C orientation, n-type conductivity and 4H polytype. The crystal on the right is of (0001)Si orientation, p-type conductivity and 6H polytype. The part b) is three times enlarged in comparison to part a). Areas "D" are defective areas of the crystal. More intensive color corresponds to areas of the crystal with higher doping level

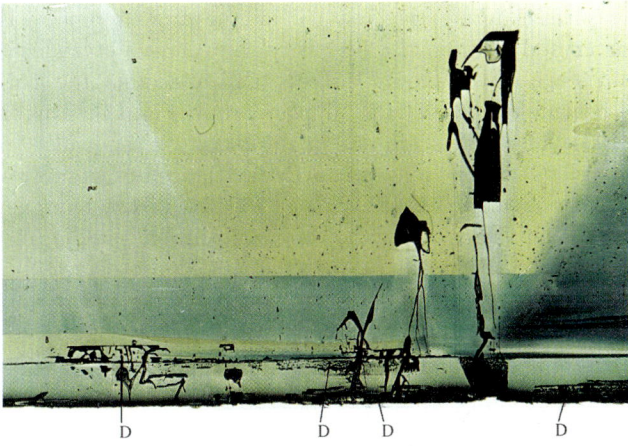

b

Fig. 14. Photo in transmitted polarized light of a SiC plate which is made by cutting the 4H SiC crystal parallel to the growth direction where recrystallization of seed and grown crystal took place (magnification $18 \times$)

areas where the crystallization front is slightly convex. In the areas where the phase boundary is slightly concave, no micropipes are generated in spite of the fact that there are secondary phase inclusions in a large amount. However, it is necessary to keep in mind that formation of secondary phases during the growth of bulk crystals results in significant degradation of the morphology of the growing surface as well as they stimulate the formation of layers of other polytypes on the singular surfaces with all the discussed negative consequences. This can be clearly seen in Fig. 17. In the earlier grown part of the bulk crystal the concentration of secondary phase inclusions is small and the growth of a high-quality crystal occurs. Only nonuniformity of doping can be seen. On the 6H SiC seed the 4H polytype is formed and in the beginning of the process it is growing normally to the seed on the whole surface. But, very soon the colder central area of the crystal which is growing faster starts to leave behind the periphery part of the crystal. The radial growth results in significant decrease of the nitrogen doping level. Approximately after half the time of the growth process the radial expansion of the crystal is changed to a growth with decreasing diameter of the crystal (see Fig. 17). This means that the slightly convex crystallization front which is available on the whole growing surface during the first half of the growth process is changed later. In this particular crystal only on the periphery the concave crystallization front is formed at a certain stage of the growth process, and after that the generation of polycrystalline areas becomes possible (right upper part in Fig. 17). This is also seen in Fig. 18 (see page 196). The wafers are cut in the plane perpendicular to the growth direction. When the crystallization front is convex (at the center the nitrogen doping level is higher − darker area in Fig. 18a), the growth is possible on the whole area of the crystal surface. When the crystallization front is concave (at the center the nitrogen doping level is lower, see Figs. 18b, c), the crystal starts to decrease in size and at the periphery monocrystals of different orientation and polytype appear.

We see that the growth should not be performed with significantly convex crystallization front to suppress the formation of micropipes. However, to keep the size of the crystal one has to establish a slightly convex phase boundary which contradicts the previous requirement.

There exists one more negative effect of a convex crystallization front. In crystals more than 30 mm in diameter we have found radial cracks which are clearly seen on the wafers obtained from the part of the crystal close to the seed (see Fig. 19, page 192). During the growth of the bulk crystal the cracks can extend more and more. The character of doping allows to conclude that at the beginning of the growth process there exists a significantly convex crystallization front with a plane central part of the crystal close to the singular surface. Cracks of different sizes are observed reproducibly only in the area of the maximal slope of the crystallization front. The following explanation can be given: If at a certain stage of the groth process, for example, at the initial stage of growth, a significantly convex crystallization front is formed, monocrystal growth is possible only from the center of the seed. For this, there might be several possibilities. The first and the most advantageous one is the radial growth of the crystal uniformly without significant faceting of the growing crystal. In this case the quality of the growing crystal should be the best but it is difficult to realize such growth process. Another possibility is nonuniform growth from the center to the periphery in the form of radial rays. If this happens, the side surfaces of the rays cannot be coupled well and form low-quality areas in the crystal. This problem is also related to the specific feature of monocrystal growth when the side surfaces of the radial rays have different orientations and facets and cannot coalesce. In this case formation of radial cracks occur which later can be kept or overgrown in dependence on the growth conditions. In the best case in the area where the surfaces of the rays occur, locally strained regions occure, but also the formation of clusters of micropipes and areas with high density of dislocations is possible. High doping level nonstoichiometry of the vapor phase with excess of Si or C, formation of secondary phases on the growing surface as well as dust or other particles stimulate this negative effect. An intermediate case, when the radial growth of the crystal is everywhere on the surface convex, but the side surfaces of the rays contain facets, results in formation of a hexagonal crystallization front and also in generation of strain and cracks in the vicinity of the corners of the hexagons.

One more effect which result in defects similar to that discussed above is the spontaneous formation of small monocrystals on the growing surface with orientations different from the orientation of the growing crystal. This process is also stimulated for example by the presence of inclusions of secondary phases, dust, not appropriate fixing of the seed, etc. [21]. Such crystals with significant misorientation cannot be overgrown and can pass through the whole length of the grown bulk crystal. In Fig. 20, a typical case of passing of such a small misoriented crystal through the grown bulk crystal is shown. Some cases of secondary growth are given in [21] and shown also in Figs. 13, 14 and 15. It has to be mentioned again that the use of Ta allows a better control of the formation of different SiC polytypes and doping of the crystals, and results in the understanding of the mechanisms of formation of micropipes and other defects discussed in the present paper.

In the grown SiC crystals we see inclusions of secondary phases which consist of particles or small SiC crystals (see Figs. 12, 13 and 14). In earlier experiments on the sublimation sandwich growth of epitaxial layers of SiC of a thickness up to a few millimeters, we did not observe the inclusions which typically occur in the present bulk crystals. During the sublimation growth of epilayers we used a small distance between the seed and the monocrystalline source wafers. We concluded that the inclusions are generated from the source material. For the growth of bulk crystals we use dense polycrystalline SiC sources. In the shown figures it can be seen that during one third or half the time of the process the amount of the inclusions is really minimal with further significant increase. This indicates degradation of the source with respect to the formation of particles and dust. We see also that the inclusions in the grown bulk SiC are mainly microscopic SiC crystals. It was shown in [39] that a protective semitransparent screen between source and seed significantly decreased the amount of inclusions in the grown crystals. All these facts convince us that there exists transport of solid particles from the source to the seed by the SiC vapor species against the gravitational force. In spite of serious doubts that the growth proceeds under very low external pressures around 10^{-3} Pa and total pressure of the Si-C species inside the container not higher than 10 Pa, we could not find any other mechanism to explain the observations than the gas transport of particles. Theoretical calculations performed recently in [31] confirmed this conclusion. It was shown that inside the growth cell under the growth conditions used for growth of bulk SiC crystals with growth rates around 1 mm/h transport of small solid particles is possible. This is due to the high rate of material transport which results in high velocities of macroscopic flow from source to seed, and, therefore, in a hydrody-

Fig. 18. Photo made in transmitted light of a cut in a plane perpendicular to the growth direction of 4H bulk SiC crystal. a) The cut is at the position where the crystallization front is convex. Dark region corresponds to higher doping level where orientation of the surface is close to $(000\bar{1})$C; b) and c) similar cuts but the crystallization front is concave. The central part has a lower doping level (magnification $3\times$)

a

b

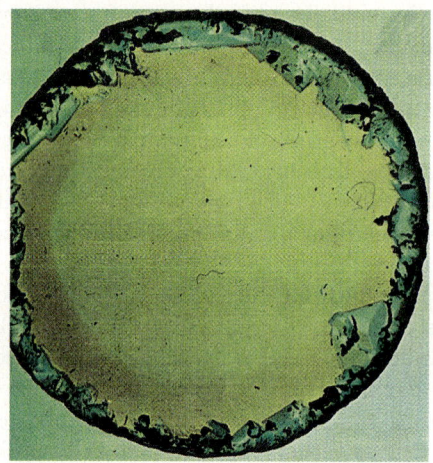

c

Fig. 20. Photo made in transmitted light of a cut in a plane perpendicular to the growth direction of 4H bulk SiC crystal similar to that shown in Fig. 18. On the singular part of the growing surface close to $(000\bar{1})$C a monocrystal of 6H polytype is seen

namic force affecting small solid particles and bringing them to the seed surface. From this consideration we can conclude that it is necessary to prepare a special source and to improve the design of the source part of the growth cell to minimize transport of SiC crystals from the source to the seed.

The obtained results show that control and variation of the shape of the crystallization front during the sublimation growth of bulk SiC crystals is a crucial point to grow high-quality crystals of considerable size. The shape of the phase boundary is determined by the radial distribution of the growth rate which depends on the rate of material transport and local values of equilibrium partial pressures of the species on the growing surface. The temperature distribution on the surface of the growing crystal is one of the key parameters determining the growth rate distribution, and therefore, the shape of the crystallization front. If the material transport is uniform over the surface of the seed, the lower surface temperature results in higher growth rate. The most appropriate way to optimize the temperature distribution on the surface of the crystal is to control the cooling regime of the crystal. One should keep in mind that the heat of crystallization released on the surface of the growing crystal is proportional to the growth rate and influences the surface temperature. Increasing length of the crystal changes the temperature distribution in it and results in significant variation of the growth conditions.

Material transport to the surface of the seed is determined by a number of factors. It is presently understood that the mechanism of the material transport in the growth cell made of Ta under the conditions used in the present work is not determined by diffusive or molecular transport mechanism as it was thought earlier. The material transport is provided by macroscopic flow of the gas which is generated on the surface of the source due to intensive sublimation of SiC and consists of Si–C vapor species. The flow is very intensive, at growth rates around 1 mm/h and the flow velocities are in the range of 10 m/s depending on the growth temperature. The local distribution of the rate of sublimation of the source determines the local flow velocities, which determine the distribution of the rate of material transport to the seed. The radial distribution of the rate of sublimation of the source is influenced by the temperature distribution in the source. It was shown in [40, 41] that the heat conductivity of SiC powder is significantly lower than the heat conductivities of graphite and crystalline SiC. This results in a significantly nonuniform temperature distribution in the source and, therefore, significant temperature gradients on the surface of the powder. As a result, material transport from the source is nonuniform and to control the shape of the crystal it is necessary to optimize also the temperature distribution in the source. Another factor which influences the whole growth process and the radial distribution of the material transport to the growing surface is the effect of the Ta container. A specific and very important feature of the SiC–Ta system is the suppression of graphitization in the source during the whole growth process. The explanation is the following. The mechanism to suppress graphitization and to keep stable growth conditions is related to the tightness of the container and to the effect of gettering of carbon inside the container. It is discussed earlier in this paper that inside the Ta container Ta coexists with TaC during many hours of the growth process due to slow diffusion of carbon inside the Ta grains. This results in continuous gettering of a small amount of carbon inside the container. In the beginning of the process, inside the growth cell one has only the stoichiometric SiC source and the SiC seed because the system is pumped down to a pressure of 10^{-4} Pa. After heating the system sublimation of the source starts. However, the gettering of carbon starts simultaneously which results in a very specific vapor composition inside the growth cell. The vapor consists of Si, Si_2C and SiC_2 species with atomic ratio between Si and C in the vapor close to 1 but shifted towards Si-rich vapor due to the effect of carbon gettering. It was shown in [30, 42] that such vapor composition prevents formation of a secondary graphite phase and, therefore, only stoichiometric evaporation of the source is possible. If the container is open, gettering of carbon inside the container is not strong enough to support the Si-rich vapor composition and graphitization of the source starts (this is confirmed in experiments). One has to take into account that the surface layer of the Ta container consists of a mixture of SiC, TaC and $TaSi_2$ solid phases, which results in chemical interaction of the wall of the container with the vapor inside the growth cell. The temperature distribution on the container wall influences the material transport to the growing surface especially in the vicinity of the crystal.

The described mechanisms which influence the shape of the crystallization front are complicated and significantly coupled. For example, due to intensive radiative and conductive heat transport a change of temperature in part of the growth cell results in temperature variations everywhere. For the optimization of the growth process modelling is the most appropriate tool for analysis of complex systems like the sublimation growth system. Presently available mathematical models and software allow to perform a qualitatively accurate analysis of tendencies of the sublimation growth process and to predict direction and effect of changes of process parameters or geometry of the system before starting the growth run [13].

5. Conclusion

In the present paper results on the use of Ta for sublimation growth of doped monocrystalline SiC are summarized. It is shown that the presence of Ta in the growth cell shifts the system close to SiC–Si, but the growth conditions are in fact free from the disadvantages of this system. The key advantage of using Ta is reliable and reproducible suppression of graphitization inside the growth cell during tens of hours of the growth process. Using this feature of the growth process it becomes possible:

– to obtain fundamental knowledge on doping of SiC;
– to explain the mechanisms of material transport during sublimation growth and to find conditions of stable growth of SiC;
– to study and explain mechanisms responsible for the formation of defects which are critical for the growth of high-quality crystals.

A conclusion on the most important mechanisms determining the formation of micropipes during the sublimation growth of SiC is made.

As a result of the studies it was possible during a quite short time – two years – to utilize the advantages of using Ta and to develop a growth process of highly doped bulk SiC crystals. Presently we grow SiC crystals of more than 30 mm in diameter and 20 mm length with uniform polytypes 6H, 15R and 4H, doped by nitrogen up to 5×10^{19} cm^{-3}. Crystallinity of the grown crystals is analyzed using X-ray rocking curves and for 4H as well as for other SiC polytypes it is in the range 16 to 24 arcs. Between crystal and seed there is no intermediate layer. The amount of micropipes in 4H crystals is not more than 50 cm^{-2}, but in the major part of wafers areas are available with only a few micropipes on an area of 0.5 to 1 cm^2. Dislocation density is usually in the range 5×10^3 to 5×10^4 cm^{-2}. The crystals are grown at temperatures between 1950 and

2100 °C under an external pressure around 10^1 Pa and with growth rates 0.8 to 1.2 mm/h. The yield of the source is usually around 90%.

At present, the key problem is the formation of radial cracks on the periphery of the crystal. To avoid this it is necessary to improve the control of the temperature distribution inside the growth cell and to vary it during the growth process as well as to use more perfect seeds.

Acknowledgements The authors are thankful to Dr. Temkin for continuous interest and support of the research work, to Dr. D. Hofmann from University of Erlangen-Nürnberg for systematic support and cooperation, and to S. Yu. Karpov, M. S. Ramm and R. A. Talalaev for assistance and continuous discussions. We acknowledge support of G. Saparin, S. Obydin, A. Bogomolov and A. Lobus from Moscow State University in characterization of the crystals using CREM and making of color photos, as well as M. Boiko for analysis of crystallinity. We would like to thank also M. Muchlinov, A. Ostroumov and E. Pokornyi for carrying out growth experiments and assistance.

References

[1] YU. A. VODAKOV and E. N. MOKHOV, USSR Patent 403275 (1970); Germany Patent DT 2409005 B 2 (1977); UK Patent 1458445 (1977); USA Patent 4,147,572 (1979).
[2] YU. A. VODAKOV, E. N. MOKHOV, M. G. RAMM, and A. D. ROENKOV, Kristall und Technik **14**, 729 (1979).
[3] YU. A. VODAKOV, E. N. MOKHOV, M. G. RAMM, and A. D. ROENKOV, Springer Proc. Phys. **56**, 323 (1992).
[4] E. N. MOKHOV, I. L. SHULPINA, A. S. TREGUBOVA, and YU. A. VODAKOV, Cryst. Res. Technol. **16**, 879 (1981).
[5] YU. A. VODAKOV, E. N. MOKHOV, M. G. RAMM, and A. D. ROENKOV, USSR Patent 882247 (1980).
[6] E. N. MOKHOV, M. G. RAMM, and YU. A. VODAKOV, Vysokochistie Veshestva, **3**, 98 (1992).
[7] YU. A. VODAKOV, E. N. MOKHOV, M. G. RAMM, and A. D. ROENKOV, Springer Proc. Phys. **56**, 329 (1992).
[8] YU. A. VODAKOV, E. N. MOKHOV, A. D. ROENKOV, and D. T. SAIDBEKOV, phys. stat. sol. (a) **51**, 209 (1979).
[9] YU. A. VODAKOV, E. N. MOKHOV, A. D. ROENKOV, and M. M. ANIKIN, Soviet Phys. − J. Tech. Phys. Lett. **5**, 147 (1979).
[10] YU. A. VODAKOV, G. A. LOMAKINA, and E. N. MOKHOV, Soviet Phys. − Solid State **24**, 1377 (1982).
[11] D. HOFMANN, S. Y. KARPOV, Y. N. MAKAROV, E. N. MOKHOV, M. G. RAMM, M. S. RAMM, A. D. ROENKOV, YU. A. VODAKOV, Inst. Phys. Conf. Ser. No. 142, 29 (1996).
[12] D. HOFMANN, S. Y. KARPOV, Y. N. MAKAROV, E. N. MOKHOV, M. G. RAMM, M. S. RAMM, A. D. ROENKOV, and YU. A. VODAKOV, presentation at ICSCRM-VII, Kyoto, 1995.
[13] E. N. MOKHOV, M. G. RAMM, A. D. ROENKOV, and YU. A. VODAKOV, J. Cryst. Growth (1997), in press.
[14] D. HOFMANN, S. Y. KARPOV, Y. N. MAKAROV, E. N. MOKHOV, M. G. RAMM, M. S. RAMM, A. D. ROENKOV, and YU. A. VODAKOV, EMRS-96, Spring Meeting, June 4 to 7, 1996, Strassburg (France) (p. A-16).
[15] P. G. BARANOV, YU. A. VODAKOV, E. N. MOKHOV, M. G. RAMM, M. S. RAMM, and A. D. ROENKOV, Ioffe Inst. Prize Winners (1996), 25−29.
[16] S. Y. KARPOV, Y. N. MAKAROV, E. N. MOKHOV, M. G. RAMM, M. S. RAMM, A. D. ROENKOV, R. A. TALALAEV, and YU. A. VODAKOV, see [14] (p. A-27).
[17] S. Y. KARPOV, Y. N. MAKAROV, E. N. MOKHOV, M. G. RAMM, M. S. RAMM, A. D. ROENKOV, R. A. TALALAEV, and YU. A. VODAKOV, Internat. Conf. Compound Semicond., Sept. 23 to 27, 1996, St.-Petersburg (p. 71).

[18] E. N. Mokhov, M. G. Ramm, A. D. Roenkov, A. A., Wolfson, A. S. Tregubova, and I. L. Shulpina, Soviet Phys. − J. Tech. Phys. Lett. **13**, 265 (1988).

[19] A. O. Konstantinov and E. N. Mokhov, Soviet Phys. − J. Tech. Phys. Lett. **7**, 247 (1981).

[20] A. A. Maltsev and E. N. Mokhov, Soviet Phys. − J. Tech. Phys. Lett. **18**, 453 (1992).

[21] E. N. Mokhov, M. G. Ramm, A. D. Roenkov, and Yu. A. Vodakov, see [14] (p. A-14).

[22] Yu. A. Vodakov and E. N. Mokhov, Intern. Conf. on Silicon Carbide, Miami Beach, Florida, Sept. 17 to 20, 1973, South Carolina University Press (p. 609).

[23] Y. M. Tairov and Yu. A. Vodakov, Topics Appl. Phys. **17**, 31 (1977).

[24] Yu. A. Vodakov, G. A. Lomakina, and E. N. Mokhov, in: Doped Semiconductor, Ed. N. Kh. Abrikosov, Nauka, Moscow 1982 (p. 230).

[25] St. G. Muller, D. Hofmann, A. Winnacher, E. N. Mokhov, and Yu. A. Vodakov, Inst. Phys. Conf. Ser. No. 142, 333 (1996).

[26] E. N. Mokhov, A. D. Roenkov, Yu. A. Vodakov, G. V. Saparin, and S. K. Obydin, Inst. Phys. Conf. Ser. No. 142, 245 (1996).

[27] G. V. Saparin, S. K. Obydin, E. N. Mokhov, A. D. Roenkov, and B. A. Akhmedov, Scanning **16**, 21 (1994).

[28] E. N. Mokhov, A. D. Roenkov, G. V. Saparin, and S. K. Obydin, Scanning **18**, 67 (1996).

[29] G. V. Saparin, E. N. Mokhov, S. K. Obydin, and A. D. Roenkov, Scanning **18**, 25 (1996).

[30] S. Y. Karpov, Y. N. Makarov, M. S. Ramm, and R. A. Talalaev, Inst. Phys. Conf. Ser. No. 142, 69 (1996).

[31] S. Yu. Karpov, Yu. N. Makarov, and M. S. Ramm, phys. stat. sol. (b) **202**, 201 (1997) (this issue).

[32] W. F. Knippenberg, Philips Res. Rep. **18**, 161 (1963).

[33] Yu. M. Tairov and V. F. Tsvetkov, J. Cryst. Growth **52**, 146 (1981).

[34] Yu. A. Vodakov and E. N. Mokhov, Inst. Phys. Conf. Ser. No. 137, 197 (1993).

[35] E. N. Mokhov and Yu. A. Vodakov, see [17] (p. 40).

[36] Yu. M. Tairov and V. F. Tsvetkov, Progress in Crystals Growth and Characterization, Vol. 7, Ed. P. Krishna, Pergamon, Oxford 1983 (pp. 111 to 162).

[37] A. P. Garshin, T. A. Lavrenova, Yu. A. Vodakov, and E. N. Mokhov, Inst. Phys. Conf. Ser. No. 142, 413 (1996).

[38] A. A. Sitnikova, E. N. Mokhov, and E. I. Radovanova, phys. stat. sol. (a) **135**, K45 (1993).

[39] A. Y. Maksimov, A. A. Maltsev, N. K. Yushin, and I. S. Barash, Soviet Phys. − J. Tech. Phys. Lett. **21**, 20 (1995).

[40] R. Eckstein, D. Hofmann, Yu. N. Makarov, St. G. Müller, G. Pensl, E. Schmitt, and A. Winnacker, Mater. Res. Soc. Symp. Proc. **423**, 215 (1996).

[41] Yu. E. Egorov, E. L. Kitanin, M. S. Ramm, V. V. Ris, and A. A. Schmidt, submitted to Mater. Sci. Engng. B (1997).

[42] S. Yu. Karpov, Yu. N. Makarov, M. S. Ramm, and R. A. Talalev, J. Appl. Phys. (1997), accepted for publication.

phys. stat. sol. (b) **202**, 201 (1997)

Subject classification: 61.50.−f; 64.75.+g; S6

Simulation of Sublimation Growth of SiC Single Crystals

S. Yu. Karpov (a), Yu. N. Makarov[1]) (b), and M. S. Ramm (c)

(a) Advanced Technology Center, P.O. Box 29, 194156 St. Petersburg, Russia

(b) Fluid Mechanics Department, University of Erlangen-Nürnberg, Cauerstrasse 4, D-91058 Erlangen, Germany

(c) A. F. Ioffe Physical-Technical Institute, Russian Academy of Sciences, Polytekhnicheskaya 26, 194 021 St. Petersburg, Russia

(Received January 31, 1997)

Modelling of sublimation growth of SiC is discussed with the goal to describe the mathematical models necessary to optimize the process and design of the growth system. An analysis of the mechanisms of growth of bulk silicon carbide crystals is performed. Growth conditions which provide stable growth of single SiC crystals without formation of secondary phases are considered. The phase diagram of the formation of extra phases during the sublimation growth of SiC is presented. Modelling of the growth of bulk SiC crystals is considered. Results of modelling the temperature distribution inside the inductively heated system for the growth of bulk SiC crystals are shown. A mechanism of material transport inside the closed Ta container in the absence of an inert gas atmosphere is proposed which is different from that of diffusive or free-molecular transport. First results of the model analysis of chemical processes inside the volume of SiC powder during the sublimation growth are demonstrated. It is shown that the sublimation and re-crystallization of the SiC source is sensitive to the temperature distribution in the source.

1. Introduction

Silicon carbide is considered presently as one of the most promising materials for high-power and high-temperature electronics due to the wide band gap, extremely high breakdown voltage, high thermal conductivity, chemical and radiation stability, and ease of p- and n-type doping of this semiconductor [1]. All these applications require high-quality SiC substrates. The deficiency of such substrates inhibits presently large-scale industrial production of SiC-based devices. The main problem which has to be overcome to obtain such substrates is the insufficient quality of bulk SiC crystals grown with the necessary dimensions.

The technique widely used to grow SiC bulk crystals is the modified Lely method. It is based on sublimation of the source (specially prepared material which contains the growth components − silicon and carbon) and transport of the vaporized species in the gas phase to the substrate where growth of single SiC crystals takes place. A specific feature of these techniques is that the growth of SiC occurs via transport of Si as well as the mixed Si_2C and SiC_2 vapor species, so that the carbon and silicon supply cannot be controlled independently. In particular, this makes the precise control of the tempera-

---

[1]) Corresponding author.

ture field in the growth chamber necessary. Another peculiarity of the sublimation method is that secondary graphite phases can easily appear on the substrate while growing
the SiC crystal. To prevent the graphitization a certain excess of silicon over carbon has
to be present in the gas phase. However, maintaining the gas phase composition often
faces difficulties due to possible chemical reactions of the gaseous species with the walls
of the growth container. The temperature field and vapor composition in the growth
chamber cannot be controlled independently of each other.

Even these two examples show the extreme importance and difficulties of the control
of the process of SiC growth using the sublimation technique. Detailed understanding of
the process is necessary to optimize the growth and to obtain high-quality crystals. The
experimental study of the sublimation growth is difficult due to extremely high growth
temperatures. On the other hand, an effective way for studying specific features of SiC
growth is the numerical simulation of the growth process. The success of modelling is
obviously dependent on an adequate choice of the physical model of the growth process.

In the present paper the physical background and mechanisms of sublimation growth
of SiC bulk crystals are considered. Before analysing modern growth systems we start
from a simple version of the sublimation technique — the so-called sublimation sandwich
method proposed in [2]. This method is very attractive for experimental and theoretical
analysis of the sublimation growth of SiC due to its simplicity and possibility to control
easily the process parameters. Modelling of sublimation growth of bulk SiC crystals is
analysed in Section 4.

2. Sublimation Growth in the Sandwich System

The sublimation sandwich method (SSM) was proposed in [2] for growing epitaxial SiC
films with low density of dislocations and micropipes. Sufficiently thick (up to 1 to
2 mm) epitaxial layers of SiC free from micropores and micropipes were obtained using
this method [3]. SiC epitaxial layers usually are grown in the sandwich system at temperatures of 1600 to 1900 °C, which are remarkably lower than the temperatures usually
used in the modified Lely process (2200 to 2400 °C). In addition, the formation of different SiC polytypes can be easier controlled in this technique as compared to other methods [4, 5]. The method is also applicable for growth of bulk SiC crystals [6], however, it
requires some modifications to provide high growth rates and stable growth conditions
to grow economically crystals of considerable size.

2.1 Configuration of the growth cell. Reactive environment

Mass transport in the sandwich system is arranged between two SiC wafers being maintained at different temperatures T_p and T_s. These wafers placed face to face form the
sandwich cell (see Fig. 1). The source wafer is at a higher temperature than the seed one
which results in sublimation of the source. The evaporated species are transported
through the space between the two wafers and condense on the seed. Both wafers are
placed inside a closed container. A specific feature of the sandwich system is that it is
essentially open — the species exchange between the growth cell and the vapor phase
inside the container plays an important role in the process of crystals growth.

Growth of high-quality SiC epitaxial layers in the sandwich system is usually performed with additional Si–C vapor environment enriched by silicon and formed in the

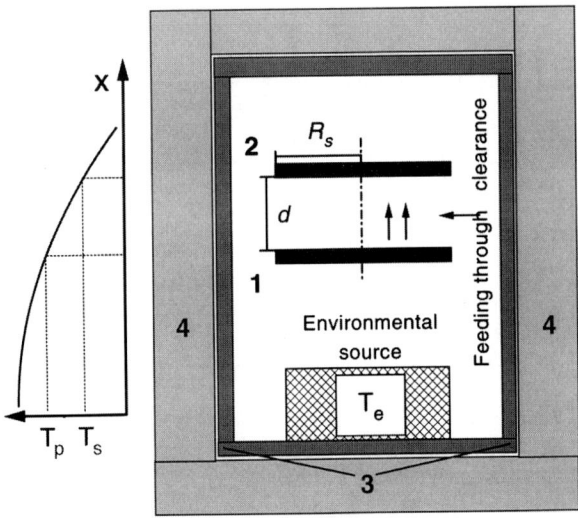

Fig. 1. Schematic view of the sublimation sandwich system: 1 source, 2 substrate, 3 graphite container, 4 thermal insulation. The temperature distribution is shown on the left

container surrounding the sandwich cell. This allows to suppress graphitizaton which starts usually first on the source and after some time occurs also on the substrate. Two different types of environment can be created by suitable choice of the environmental source (see Fig. 1). If polycrystalline SiC is placed into the hot zone of the graphite container touching directly the graphite walls, the composition of the vapor phase becomes close to that specific for the three-phase "vapor–solid–solid" SiC–C equilibrium. In this case both solid SiC and graphite are in equilibrium with the vapor phase providing the following relationships between the partial pressures of the main gaseous components, Si, Si_2C and SiC_2:

formation of crystalline SiC from the gaseous components

$$P_{Si_2C}P_{SiC_2} = K_1 \qquad [SiC_2(g) + Si_2C(g) \rightleftharpoons 3\,SiC(s)], \tag{1}$$

transformation of SiC_2 into Si_2C in the vapor phase

$$P_{SiC_2}P_{Si}^3 = K_2 P_{Si_2C}^2 \qquad [SiC_2(g) + 3\,Si(g) \rightleftharpoons 2\,Si_2C(g)], \tag{2}$$

formation of solid graphite from the gaseous species

$$P_{Si}^2 = K_3 P_{Si_2C} \qquad [Si_2C(g) \rightleftharpoons C(s) + 2\,Si(g)] \tag{3}$$

(the bracketed expressions in Eqs. (1) to (3) indicate the corresponding heterogeneous reactions occurring in the environmental source).

To grow SiC crystals under the vapor atmosphere enriched by silicon, some amount of liquid silicon is introduced into the hot zone of the container in addition to the polycrystalline SiC. As a result, the vapor composition near the environmental source is determined predominantly by the three-phase "vapor–solid–liquid" SiC–Si equilibrium. The relationships between the partial pressures of the main components corresponding to this type of equilibrium are determined by Eqs. (1) and (2) with the additional relation:

condensation of liquid silicon from the vapor phase on the surface

$$P_{Si} = P_{Si}^0 \qquad [Si(g) \rightleftharpoons Si(l)]. \tag{4}$$

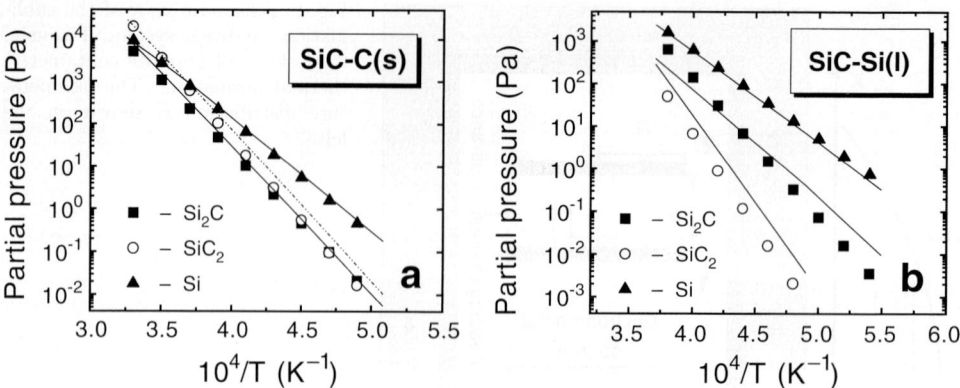

Fig. 2. Partial pressures of Si, Si$_2$C and SiC$_2$ vs. temperature calculated for three-phase equilibria:
a) SiC–C(s) and b) SiC–Si(l). Symbols are experimental points measured for the Si–C system under C-rich and Si-rich conditions [8]

Here P_{Si}^0 is the saturated vapor pressure of silicon over the liquid phase. The values of
P_{Si}^0, K_1, K_2 and K_3 in (1) to (3) can be calculated using standard thermodynamic properties of the relevant components [7].

By solving the system of equations (1) to (4) for the SiC–C(s) or SiC–Si(l) equilibrium one can find the pressures of the main gaseous components in dependence on
temperature. Fig. 2 shows the calculated partial pressures of Si, Si$_2$C and SiC$_2$ corresponding to SiC–C and SiC–Si three-phase equilibria. The points in these figures indicate results of measurements performed in [8] for the Si–C system under C-rich and Si-
rich conditions. Agreement between the calculated and measured data shows that the
assumption of three-phase equilibria of different types properly describes the experimental situation.

The environmental source placed in the hot zone of the container at temperature T_e
creates around the growth cell a vapor atmosphere with composition depending on the
type of the environment. Along with the substrate temperature T_s and the temperature
difference $\Delta T = T_p - T_s$, the value of T_e is an additional parameter of the growth process in the sandwich system.

2.2 Growth of SiC under free-molecular transport of components

At vapor pressure $P_{tot} \leq 10^{-1}$ to 1 Pa the free path length of the gaseous molecules
becomes greater than the clearance d typically used between substrate and source. In
this case the species evaporated from the source reach the substrate without collisions
between each other.

This experimental situation was studied in early papers [9 to 11]. The approaches
used in all these works have two main weak points. Firstly, the sandwich cell was considered without accounting for the material exchange with the external environment.
Secondly, a certain type of equilibrium (SiC–C or SiC–Si type) was assumed a priori to
exist on the surfaces of source and substrate. This assumption immediately resulted in
the conclusion that excess of silicon or carbon atoms is accumulated on the surfaces.
The only way to release the excess of atoms was continuous formation of an extra condensed – liquid silicon or solid graphite – phase (in particular, on the basis of such

consideration a "vapor–liquid–solid" VLS mechanism for SiC growth in the temperature range of 1800 to 2100 °C was proposed in [11, 12]. Meanwhile numerous experiments reveal stable growth of SiC in the temperature interval of practical importance without any extra phase formation (see, for example [5, 6]).

In our opinion, a correct consideration of SiC growth should include — (A) analysis of all possible pathways of the growth process corresponding to the different heterogeneous reactions shown in the bracketed expressions in Eqs. (1) to (4), and (B) determination of the reaction prevailing under the growth conditions chosen.

An alternative mathematical model of SiC growth was developed in [13, 14]. It was assumed that both evaporation of the source and growth of SiC on the substrate occur under heterogeneous two-phase "vapor–solid" equilibrium. Along with the conservation of silicon and carbon atoms on the reactive surfaces, the losses of the components through the clearance of the sandwich cell as well as the feeding from the environmental atmosphere were taken into account in the model. It was assumed that the efficiency of the material exchange between the sandwich cell and the environment is determined by the value of the so-called geometrical factor $\theta = 1 - \omega d/R_s$ (d is the clearance between source and substrate, R_s is the source/substrate radius; $\omega = 1$ for cosine and 3.7 for Maxwellian velocity distribution of the gas molecules [15]). The closer θ approaches one, the less significant are the losses of material through the clearance and the feeding from the environment.

A comparison of the model predictions with experimental data obtained for the case of growth of 6H-SiC in the sandwich system is shown in Fig. 3. In this figure the calculated growth rate V_g (solid line) is plotted versus the substrate temperature T_s (a) and the temperature difference ΔT (b) at fixed difference between environment temperature T_e and the substrate temperature T_s along with the experimental points. Reasonable agreement between the theoretical and experimental data is achieved without any fitting parameters.

Analytical formulae obtained in [13, 14] allow to make the following conclusions on the growth mechanism:

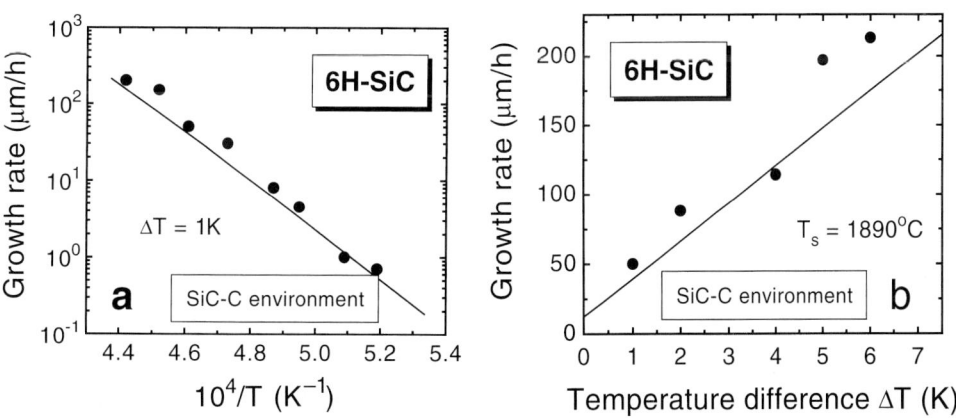

Fig. 3. Calculated (lines) and experimentally measured [9] (circles) dependence of the growth rate versus a) substrate temperature and b) temperature difference between source and substrate. Growth conditions: free-molecular mass transport in the sandwich cell; SiC–C(s) environment; $T_e - T_s = 10$ K; $\Delta T = 1$ K; $R_s = 5$ mm; $d = 0.5$ mm

(i) The partial pressures of the main gaseous components (Si, Si_2C and SiC_2) are determined entirely by the corresponding partial pressures of the species generated in the environmental source. They practically do not depend on the d/R_s ratio as well as on T_s and ΔT.

(ii) The growth rate under typical growth conditions depends primarily on the substrate temperature T_s and the temperature difference ΔT between source and substrate. The external environment provides only a minor contribution to the growth rate which is proportional to d/R_s and strongly depends on the excess of the environment temperature T_e over the substrate one.

(iii) The vapor composition (partial pressures of the components) in the sandwich cell can be modified by the proper choice of the type of the environment. For example, change of the SiC–C to the SiC–Si environment results in a remarkable increase of the Si/C ratio in the vapor phase. It will be shown in Section 3 that an increase of the Si/C ratio is essential for suppressing graphitization during the growth of SiC.

(iv) The role of the environment diminishes only if the temperature of the environment becomes less than those of substrate and source. In this case the vapor phase inside the sandwich cell is found to consist entirely of a $1:1$ mixture of SiC_2 and Si.

The described trends in the growth behavior are valid also in case of diffusive transport of the gaseous components from source to substrate.

2.3 Growth of SiC under diffusive transport of species

In some cases the sublimation growth of SiC in the sandwich system is carried out in the atmosphere of inert gas maintained at externally controlled pressure P_{tot}. At $P_{tot} \gtrsim 10^{-1}$ to 1 Pa the transport of the gaseous species from source to substrate is no longer collisionless. If the temperature in the system is not too high, the partial pressures of the Si–C species evaporated from the source (Si, Si_2C and SiC_2) are much lower than the background pressure of inert gas. In this case, neglecting effects of convection, one can reduce the analysis of the mass transfer to the solution of the steady-state diffusion equation [16]. Essential parameters of the mass transfer process become in this case the diffusion resistances between source and substrate, environmental source and substrate, and environmental source and source wafer.

One of the surprising results following from the consideration of SiC growth in the sandwich system using the diffusion model is that the partial pressures of the main gaseous components inside the sandwich cell are not dependent on diffusion resistances and, therefore, on the total pressure and geometrical configuration of the system. As in the case of the free-molecular transport of components, these pressures are practically equal to the partial pressures of the corresponding species formed in the environmental source.

In contrast, the growth rate of the crystal is found to depend on the diffusion resistances. Under high pressure, due to the change of the transport mechanism from free-molecular to diffusive, the growth rate becomes much less than that in vacuum. Fig. 4 shows the growth rate versus the substrate temperature T_s (a) and the temperature difference ΔT (b) at fixed difference between the environment temperature T_e and the substrate temperature T_s. Three types of environment are analyzed – SiC–C type, SiC–Si type, and "cold" environment corresponding to T_e essentially lower than T_s

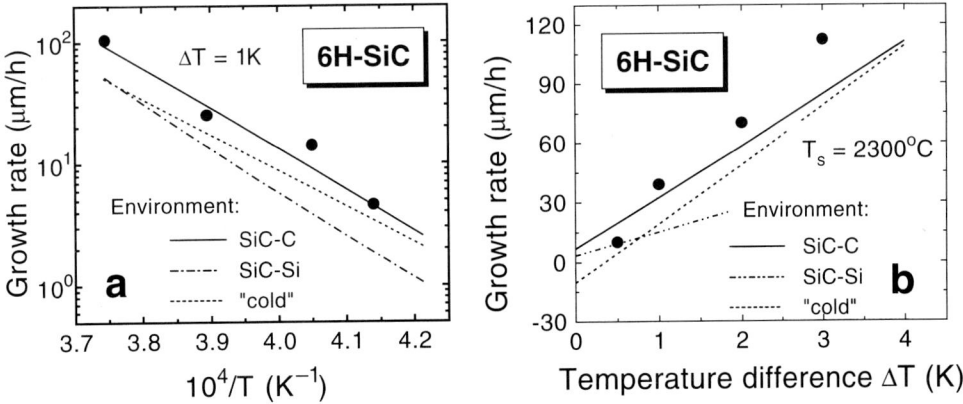

Fig. 4. Calculated (lines) and experimentally measured [16] (circles) growth rate versus a) substrate temperature and b) temperature difference between source and substrate. Growth conditions: diffusive mass transport in the sandwich cell; $T_e - T_s = 50$ K; $\Delta T = 1$ K; $R_s = 5$ mm; $d = 0.5$ mm; $P_{tot} = 10^5$ Pa. Theoretical curves are calculated for three types of the external environment. The experiments are carried out using SiC–C(s) environment

and T_p. In the latter case etching of the substrate at low temperature difference $\Delta T = T_p - T_s$ is predicted because of material losses through the clearance of the sandwich cell. The calculated dependence for the case of the SiC–C environment is in reasonable agreement with the experimental values obtained in an atmosphere of argon at $P_{tot} = 10^5$ Pa [16].

It should be noticed that the model of diffusive mass transfer in the sandwich cell can be extended to the case of free-molecular transport of components by introducing the diffusion resistance of the Knudsen layer associated with every surface and independent of P_{tot}, into the model. This generalization allows to build up a universal dependence of the growth rate on the total pressure in the system. A comparison of this curve with experimental data of [16] is shown in Fig. 5.

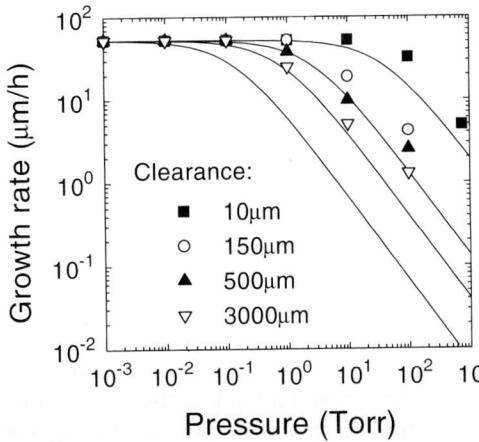

Fig. 5. Calculated (lines) and experimentally measured [16] (symbols) growth rate vs. total pressure in the sandwich system. Growth conditions: $T_s = 2300\,°C$; $T_e = 2350\,°C$; $\Delta T = 1$ K; $R_s = 5$ mm; $d = 0.5$ mm. The curves are calculated for different values of the clearance corresponding to those used in the experiments

3. Formation of Secondary Condensed Phases during Growth of SiC in the Sandwich System

Heterogeneous two-phase "vapor–solid" equilibrium provides preferable conditions for sublimation growth of SiC. However, some secondary condensed phases can be formed during growth either on the substrate or on the source. This effect is undesirable and should be minimized. Two secondary phases have to be considered in the Si–C system – solid graphite with small amount of solved silicon, and a liquid phase which consists primarily of silicon with small addition of diluted carbon.

It is important to note that standard methods of phase analysis based on minimization of the Gibbs free energy (see, for example, [17, 18]) are not applicable to the sandwich and other open systems because of the fact that crystal growth occurs. Firstly, when growth of the crystal takes place, the state of the system is shifted from the pure thermodynamic equilibrium. Secondly, it is difficult in principle to define formally the thermodynamic system to be analyzed. For example the composition (content) of the vapor species arriving at the surface very often differs from that corresponding to the outgoing components. Obviously this is impossible in the case of thermodynamic equilibrium.

A different approach to the phase analysis was proposed in [19] for growth of III–V compounds and applied in more elaborated form to the Si–C system in [20 to 22]. The main idea of the approach was to consider the early stage of the process of secondary phase formation along with the crystal growth occurring under the same conditions (temperature and component fluxes arriving at the subtrate). Then the chemical potentials of the obtained extra phase and the crystal had to be compared. The phase with the minimum chemical potential was assumed to appear on the surface.

3.1 Extra phase diagram for SiC growth under free-molecular transport of species

Using the approach [19, 22] a general diagram for the formation of secondary phases during the growth of SiC can be calculated (Fig. 6). In this diagram the areas of appearance of secondary phases are plotted in terms of the reciprocal temperature–silicon flux (F_{Si}) arriving at the substrate (the diagram is obtained for a fixed value of the carbon flux, but the boundaries depend only on the difference between silicon and carbon fluxes). Any particular growth process which occurs under the free-molecular mechanism of material transport can be indicated in this diagram by a point.

A major area on the diagram is the area of stable crystal growth where the appearance of crystalline SiC is thermodynamically preferable. The shadowed area in the diagram corresponds to graphitization of the surface. It is seen that a significant excess of silicon over carbon is required to prevent graphitization of the SiC surface during the sublimation growth. This conclusion is, however, valid only for temperatures $T_s < 2800\,^\circ C$. At $T_s > 2800\,^\circ C$ graphitization does not occur because of drastic changes of the Si/C ratio in the vapor phase. The dash-dotted line in Fig. 6 separates the areas where steady-state (above the line) and non-steady-state (below the line) graphitization occurs. Since the resulting effect in both cases is the same we do not differentiate these two regions hereafter.

More complicated is the situation with the area corresponding to liquid phase formation. A detailed consideration [22] shows that appearance of the liquid phase is thermodynamically advantageous if the value of the incident silicon flux is less than some cricital value F_{TD} which depends on temperature (this unexpected effect is due to the

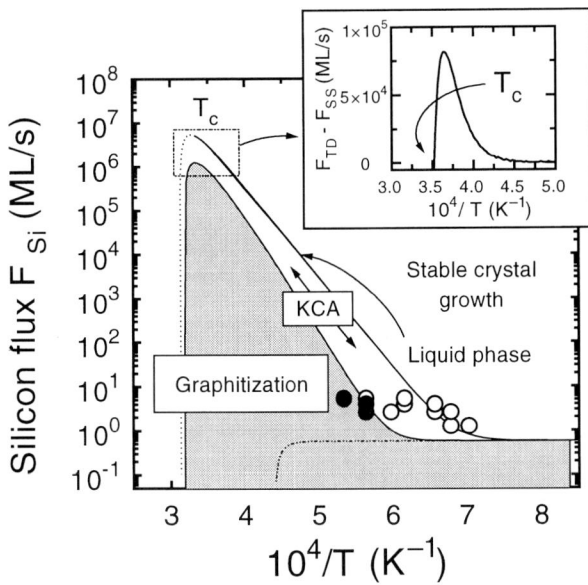

Fig. 6. Diagram of secondary phase formation during the sublimation growth of SiC, $F_C = 7 \times 10^{14} \, cm^{-2} \, s^{-1}$. Experimental data obtained during growth of 6H-SiC by MBE are taken from [23]. Open circles correspond to single-crystal growth. Solid circles indicate graphitization of the surface

interplay of the mixed Si_2C and SiC_2 species desorbing from the surface of the crystal). At the same time, steady-state formation of the liquid phase is possible only if the incident silicon flux exceeds another critical value F_{SS} (if opposite, the appeared liquid droplets rather evaporate from the surface than grow). Both critical fluxes lie very close to each other in the phase diagram, so that they look like one united line. Detailed examination of the behavior of these fluxes shows (see inset in Fig. 6) that the curves related to F_{TD} and F_{SS} intersect at the temperature $T_c \approx 2550\,°C$. This means that at higher temperatures appearance of the liquid phase becomes impossible.

Summarizing the results related to the formation of the liquid phase we can conclude:

(i) The liquid phase can appear only at $T_s < 2550\,°C$ and F_{Si} values from the narrow range between F_{TD} and F_{SS}. This means that formation of an extra liquid phase and, therefore, realization of the VLS mechanism of growth is very unlikely.

(ii) Between the lower boundary of the field related to liquid phase formation (corresponding to F_{SS}) and the boundary of the graphitization area the so-called "kinetically clean area" (KCA in Fig. 6) is placed where metastable growth of SiC crystal is predicted. Here the liquid phase is thermodynamically preferable but it cannot be formed on the surface due to the kinetic limitation caused by the rapid evaporation of liquid silicon.

In Fig. 6 the experimental data obtained while growing SiC by molecular beam epitaxy [23] are compared with the predictions. Open circles correspond to single-crystal growth of SiC, solid circles are related to growth regimes where graphitization of the surface was observed. One can see that the calculated boundary of graphitization agrees well with the experimental data.

3.2 Formation of secondary phases in the sandwich system

As was shown in [22] by using a linearization with respect to the content of components solved in the extra condensed phase, three factors govern the process of secondary phase formation in the sublimation sandwich system — the substrate temperature T_s, the tem-

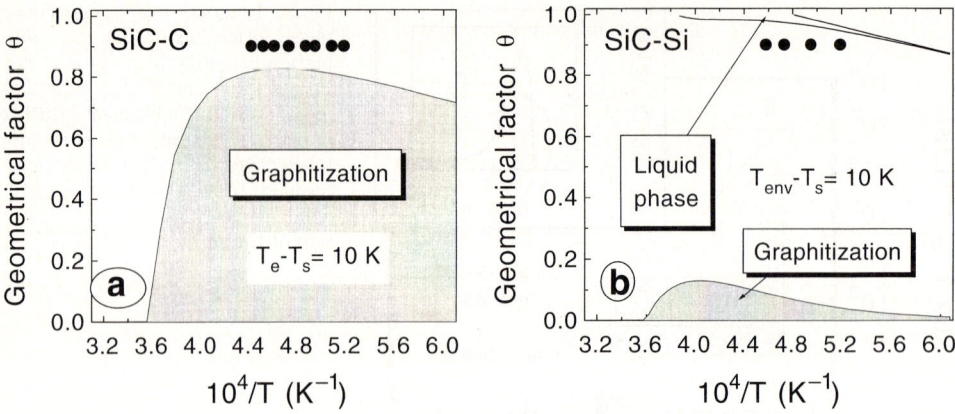

Fig. 7. Diagram of secondary phase formation for SiC growth in the sandwich system with a) SiC–C(s) and b) SiC–Si(l) environment; $T_e - T_s = 10$ K. Circles indicate growth conditions [6] where graphitization of the surface was not observed

perature of environment T_e and the value of the geometrical factor θ (there is no dependence on ΔT in the linearized approach). This fact allows determination of the geometrical configuration of the sandwich cell (via coefficient θ) in dependence on T_s and T_e providing growth of SiC without secondary phase formation.

Fig. 7 shows the extra phase diagram obtained for the sandwich system for two types of environment and fixed difference between T_e and T_s. The circles in the figure indicate the process conditions used in the experiments of [9, 6, 14] where growth of SiC without graphitization was observed.

In the case of SiC–C environment the range of allowable values of θ is limited. Growth of the crystal without graphitization is possible only at high θ, i.e. in a relatively closed sandwich cell where the material losses through the clearance are small (with increasing T_e the graphitization area becomes more narrow). In the opposite case of SiC–Si environment graphitization can occur only at low θ, i.e. in the cell exhibiting significant exchange of components between cell and environment. This comparison shows that enrichment of the vapor phase by silicon inside the sandwich cell is a crucial factor to avoid graphitization during growth of SiC crystals.

It should be noted that the experimental growth conditions used in [9, 6, 14] are located entirely in the kinetically clean area. Formation of the thermodynamically preferable liquid phase is forbidden in this area due to kinetic limitations. However, liquid droplets can appear here via fluctuations. This may be a factor which favors the formation of some types of defects, for instance, micropipes and micropores. To move the working point (corresponding to a particular growth process) into the area of stable crystal growth one has either to increase the growth temperature above 2550 °C or to increase significantly T_e over T_s. An alternative way is the further enrichment of the vapor phase with silicon by a proper choice of the external environment.

3.3 Influence of the tantalum container on secondary phase formation (modification of the environment)

In this section we discuss the use of a tantalum container for the growth of SiC by the sublimation sandwich method. The application of a tantalum container is found to be advantageous to obtain high-quality SiC crystals with a remarkable growth rate and without appearance of secondary phases during the growth [24 to 26]. The effect of tantalum on the growth process is not completely understood. The present consideration is based on the assumption that metastable tantalum disilicide is formed on the wall of the tantalum container during the sublimation growth [25, 26]. This mechanism can serve as an example of intentional modification of the environment to improve the growth conditions.

The main idea underlying the treatment of the effects of the tantalum container is that the partial pressures of the gas species in the growth chamber correspond to heterogeneous four-phase chemical equilibrium established on the wall of the container and involving solid SiC, TaC, TaSi$_2$ and the vapor components. Under such equilibrium conditions

formation of tantalum carbide,

$$P_{\mathrm{Ta}}P_{\mathrm{Si}_2\mathrm{C}} = K_4 P_{\mathrm{Si}}^2 \qquad [\mathrm{Ta(g)} + \mathrm{Si}_2\mathrm{C(g)} \rightleftharpoons \mathrm{TaC(s)} + 2\,\mathrm{Si(g)}]\,, \qquad (5)$$

formation of tantalum disilicide,

$$P_{\mathrm{Ta}}P_{\mathrm{Si}_2\mathrm{C}}^2 = K_5 \qquad [\mathrm{Ta(g)} + 2\,\mathrm{Si(g)} \rightleftharpoons \mathrm{TaSi}_2\mathrm{(s)}] \qquad (6)$$

are going on along with reactions (1) and (2). The simultaneous solution of Eqs. (1), (2), (5) and (6) determines the partial pressures of the main gas components (including Ta) as a function of temperature. Using these values we can analyse whether extra phases are formed during the growth of SiC in the sandwich system.

Fig. 8 shows the extra phase diagram plotted for the sandwich system with environment formed by assuming the four-phase heterogeneous equilibrium. One can see that the presence of Ta in the growth cell shifts the growth process into the area of stable SiC growth on the phase diagram.

We should notice that up to now the question remains unclear whether TaSi$_2$ exists at high temperatures where this material is metastable and can decompose into solid tantalum and liquid silicon. At the same time the evidence of formation of TaSi$_2$ on the wall of the container used for sublimation growth was shown in [25].

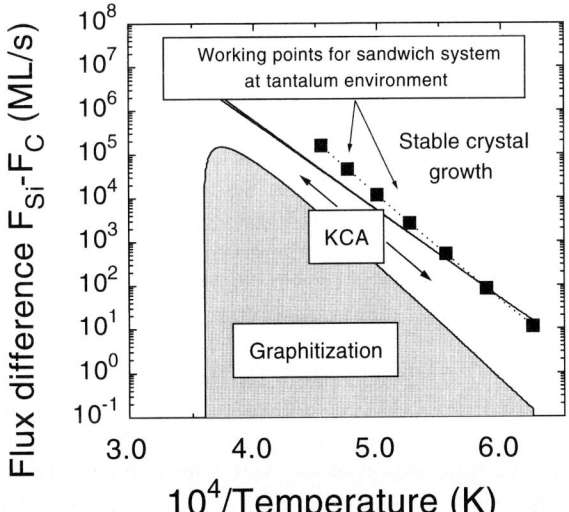

Fig. 8. Diagram of secondary phase formation for SiC growth in the sandwich system with the tantalum container; $T_{\mathrm{e}} - T_{\mathrm{s}} = 10\,\mathrm{K}$

In the present consideration we assume that the four-phase equilibrium determines the vapor composition in the environment surrounding the sandwich cell. This is valid if a significant amount of material is available in the environmental source. The model should be modified for the case of sublimation growth of bulk SiC crystals in a closed Ta container [26]. In this case accurate consideration of the material transport inside the closed container is necessary, taking into account the chemical interactions of the vapor with the Ta container, the SiC source and the seed.

In [26] experimental data on sublimation growth of SiC in the presence of Ta are summarized and a hypothesis on the effect of tantalum on the growth process and graphitization is given. The observations are in accordance with the proposed theoretical model.

4. Modelling Analysis of Sublimation Growth of Bulk SiC Crystals

Industrial use of SiC requires efficient production of high-quality SiC wafers. For this purpose the sublimation growth of bulk SiC crystals is usually used. Optimization of this process is of high practical importance. Modelling is actively used for the analysis and optimization of bulk growth of SiC crystals [27 to 29, 26]. A consistent model of the sublimation growth process which would allow to predict growth rate, vapor composition inside the growth cell, shape of the crystallization front, formation of deposits on the crucible wall, formation of secondary phases in source and seed and other important features of the growth process is still not demonstrated. It is still a long way to come to such a model which can be used for accurate predictions. In the present paper some advances in the development of a model of the sublimation growth of SiC crystals are shown.

4.1 Simulation of heat transfer in the growth chamber

Growth of high-quality SiC crystals by sublimation is only possible if the temperature distribution inside the growth system is well-controlled. First, it is necessary to establish a temperature field inside the growth cell which allows to obtain the necessary growth rate. The value of the growth rate is determined by the growth temperature, the total pressure in the system and the temperature difference between the surfaces of source and seed. Secondly, the radial distribution of the growth rate determines the shape of the growing crystal. In paper [26] it was shown that the shape of the crystallization front during the sublimation growth of bulk SiC crystals influences the formation of different macroscopic defects like micropipes, inclusions of different SiC polytypes, pores and cracks. Thirdly, it is advantageous for the growth of bulk SiC crystals to prevent the formation of SiC deposits on the crucible wall and around the seed. To provide this it is necessary to establish a temperature on the surface of the crucible which is everywhere higher than the temperature of the surface of the growing crystal. This results in suppression of deposit formation and provides material transport predominantly to the seed.

During the sublimation growth of bulk SiC crystals of significant length the temperature distribution inside the cell changes due to increase of the crystal size, formation of SiC deposits on the crucible wall, and changes in volume and porosity of the SiC powder due to sublimation and loss of material. Graphitization of the source also changes thermal conductivity and radiative properties of the powder.

A general description of the thermal processes in the system for sublimation growth of SiC requires the global solution of the Navier-Stokes equations for the flow of the gas mixture coupled with the equations for heat transfer and material transport. In [27] it was shown that the natural convection due to the temperature gradients as well as the solution convection due to the concentration gradients inside the growth cell are very weak and can be neglected. However, to control the temperature distribution on the crystallization front the simulation of the material transport inside the cell has to account for the heats of sublimation and crystallization. Simulation of only global heat transfer in the growth system does not allow to predict directly the shape of the crystallization front, but, it is very useful for the optimization of the temperature distribution inside the growth cell.

Modelling of the global heat transport was performed for the cases of resistively [27, 29] and inductively heated [28] systems for sublimation growth of bulk SiC crystals in graphite crucibles. In the present paper the simulation of global heat transfer is performed for the inductively heated sublimation growth system with Ta crucible. Specific features of this system are lower growth temperatures, smaller temperature gradients inside the growth cell and absence of inert gas atmosphere in comparison to the typical conditions used for the modified Lely growth in graphite crucibles.

Modelling of the heat source due to inductive heating is performed by solving the quasi-stationary Maxwell equations in a large system which includes the growth system, coil and surrounding hardware. Especially critical for an accurate modelling of the RF heating are highly conductive metallic parts surrounding the growth chamber. The heat source distribution in the whole growth system is used further for the simulation of the global heat transfer by conduction and radiation. For the modelling of the radiative heat transport the grey-diffusive model of radiation is used, and the heat exchange between the solid surfaces is modelled using the view-points of [30]. Details of the algorithm are given in [31].

The typical geometry of the growth cell used in [26] and chosen for the calculations is shown in Fig. 9. The growth cell consists of two containers. A pyrometer window with a

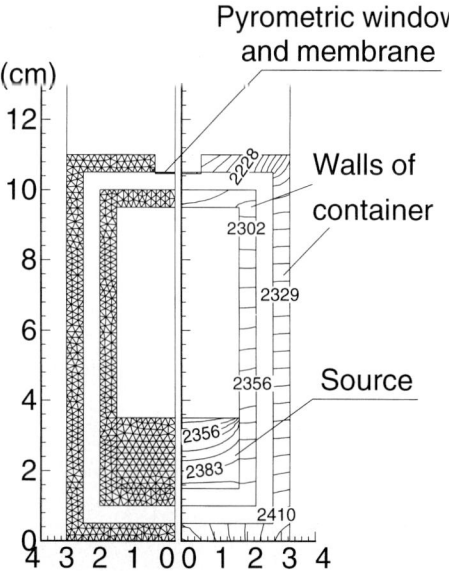

Fig. 9. Schematic configuration of the growth chamber used for growth of SiC bulk crystals by the sublimation technique (left part of the figure), and the temperature distributions under the typically used growth conditions (right part of the figure)

thin membrane is seen on the top of the external container. On the bottom of the inner container the SiC source and on its lid the seed are placed. The system is pumped during the whole process to a pressure less than $\approx 10^{-3}$ Pa.

The unstructured grid generated in the solid blocks as well as the temperature distribution inside the growth cell are shown in Fig. 9. The calculations are performed for the empty cell when the source and seed are not yet introduced. One can see significant radial temperature gradients inside the inner container.

Results of modelling the temperature distribution in the growth cell with source and seed are shown in Fig. 10. The axial temperature distribution is presented in Fig. 10a. One can see that the axial temperature profile inside the growth cell is far from linear which is usually assumed. The pyrometer window with the membrane provides efficient cooling of the top of the inner container which results in cooling of the seed.

In the experiments usually two types of the SiC source are used [26] — polycrystalline material or powder. It was shown in [36, 32] that the heat conductivity of SiC powder is almost two orders of magnitude lower than that of crystalline SiC or graphite. In the present paper a model proposed in [32] is used for calculating the heat conductivity of the SiC powder in dependence on its porosity, size of the granules, total pressure inside the growth cell and other factors. In Fig. 10b and c the effect of using two SiC sources on the temperature distribution inside the growth cell is demonstrated. The radial temperature distributions on the surfaces of seed, top and bottom of the source are shown.

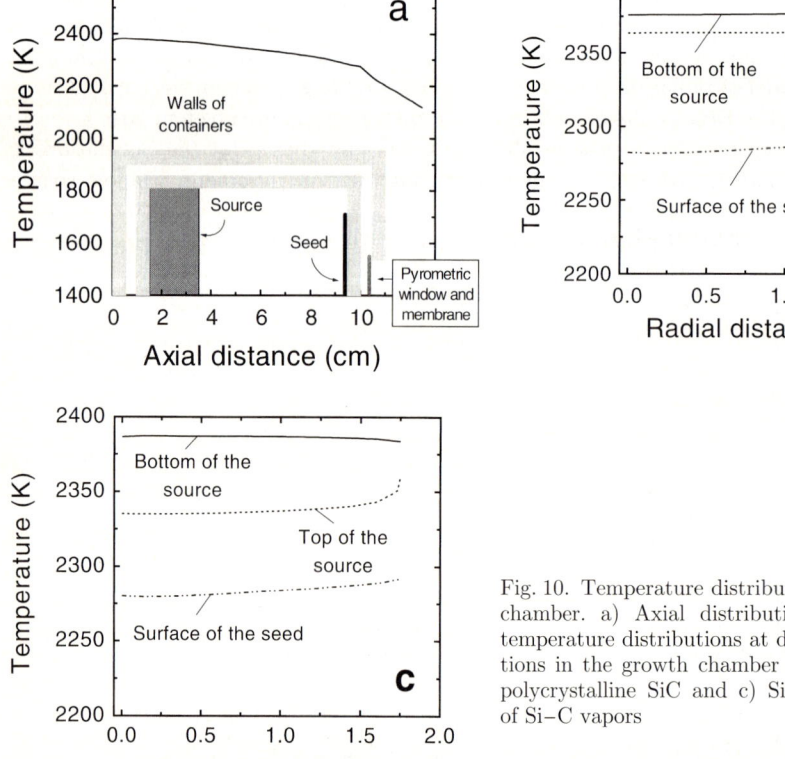

Fig. 10. Temperature distribution in the growth chamber. a) Axial distribution, b), c) radial temperature distributions at different axial positions in the growth chamber for the case of b) polycrystalline SiC and c) SiC powder sources of Si–C vapors

Fig. 10b corresponds to the use of the polycrystalline source and Fig. 10c to the use of SiC powder. It is important to note that the temperature of the surface of the seed is not influenced by the change of the typeof the source. However, the temperature distribution in the source including its surface and the bottom of the container is significantly different. In the case of the powder the significantly lower heat conductivity of the source results in a temperature on its surface which is almost 30 K lower than in the case of the use of the polycrystalline source. In practice, this changes the temperature difference between the surfaces of source and seed and results in totally different growth conditions. Inside the powder source there exists a high temperature drop which might result in a non-uniform sublimation rate across the source volume. The temperature distribution on the surface of the powder source is significantly non-uniform with a minimum in the center which provides conditions for radially non-uniform material supply from the source. Temperature non-uniformities in the powder might also stimulate graphitization of the source.

The temperature distributions on the surfaces of seed and crucible wall can be optimized by using an appropriate design of the growth cell. The best way is to combine conductive and radiative cooling of the seed. The modelling approach was used in [26] for optimizing the design of the system and the sublimation growth process.

4.2 Model of material transport from source to seed

In the present paper the material transport from source to seed during the sublimation growth of bulk SiC crystals in a Ta container is considered. The growth is usually performed in absence of an inert gas atmosphere [26] in a closed container. From experiments some specific features of the growth process were found [33]. Among them are a weak dependence of the growth rate on temperature, transport of small particles from source to seed opposite to the direction of gravitation as well as the necessity to introduce an excess external pressure of nitrogen to dope the growing SiC crystal. All these facts convince us that the material transport inside the container is totally different from the diffusive or free-molecular mechanisms. In [33] an alternative mechanism of material transport was proposed which is based on the accurate consideration of convective-diffusive transport of the Si–C vapor species from source to seed. Intensive sublimation of material in the source generates a macroscopic flow from source to seed which is modelled using the Navier-Stokes equations coupled with the equations for material transport with taking into account the Stephan flow on the surface. Relations describing the chemical processes on the surfaces of source and seed are coupled with those of material transport inside the container. An important feature of the model is the accurate consideration of the pressure inside the closed growth cell. It is shown that the pressure is highly uniform inside the container with only small variations due to the macroscopic flow from source to seed. The level of the pressure is determined by a mean value of the temperatures of source and seed. This explains why it is necessary to introduce somes overpressure of nitrogen to dope the growing SiC — one needs to exceed this inner pressure to pump nitrogen inside the growth cell.

The developed model is used for the simulation of sublimation growth of SiC under the conditions of growth of bulk SiC in a Ta container [26]. In Fig. 11 the temperature dependence of the growth rate of SiC in the absence of the inert gas atmosphere is shown. An important feature is that the activation energy of the growth process is tem-

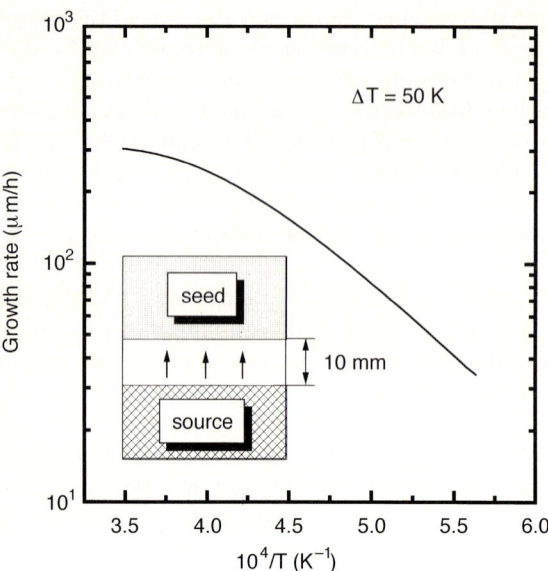

Fig. 11. Dependence of SiC growth rate on temperature for the case of sublimation growth of bulk SiC crystals in a Ta container in absence of an inert gas

perature dependent. The velocities of the macroscopic flow from source to seed are in the range of 10 m/s which is definitely sufficient to transfer particles of µm size from source to seed. Small SiC crystals of µm size were found in the bulk crystals grown under the considered conditions [26]. Another specific feature of the growth process is that the total flux of Si vapor is directed from seed to source opposite to the flow direction. Intensive diffusion of Si to the source compensates the transport due to the flow. To our understanding this effect is specific for the growth of SiC in the absence of an inert gas atmosphere and suppressed graphitization in the source in the presence of Ta.

From this analysis we can conclude that the mechanism of material transport inside the closed growth cell made of Ta in absence of an inert gas atmosphere is totally different from the diffusive or free-molecular ones and can be described by a complicated mathematical model which includes consideration of macroscopic flow coupled with material transport. It has to be noted that the proposed mechanism is valid also for the case of sublimation growth of bulk SiC crystals in presence of an inert gas and using a graphite crucible if the growth rate is high enough that the velocity of the Stephan flow becomes comparable to the rate of the diffusive transport.

4.3 Material transport in SiC powder

SiC powder is widely used for the growth of bulk crystals by sublimation. It is known that the chemical processes in the powder are of complicated character and include sublimation, graphitization, re-crystallization and others [34 to 37]. Analysis of the powder after the growth process showed that in some areas of the powder congruent sublimation occurs which results in the appearance of empty space in the source. Formation of a yellow deposit on the SiC granules in the source is observed in some cases which indicates the formation of cubic SiC due to re-crystallization of SiC in the source. Graphitization of the source can be significantly non-uniform in axial and radial directions. It is also known that the chemical processes in the SiC powder are sensitive to temperature variations inside the growth cell.

Material transport in the powder is still an open question in the modelling of the sublimation growth. In the present paper an attempt is made to simulate the chemical processes in the volume of the SiC source which consists of powder. The powder is considered as a porous medium. The mathematical model includes the interactions between the small gas volume inside the pores and the surface of the granules as well as the diffusion of species in the porous material. Mean values of the concentrations of species inside the pore and on the surface of the pore are introduced.

The equations for material transport by diffusion through the porous material take into account chemical interactions with the surface of the pore due to sublimation and re-crystallization. On the surface of the crucible chemical interactions with the crucible wall are also taken into account. Therefore, at every point of the powder volume a set of non-linear equations for the mean values of the concentrations of species in the volume of the pore and on the surface of the granules are solved. If the vapor composition inside the pore is supersaturated in comparison to the equilibrium composition at the local temperature at this point, re-crystallization of the powder takes place. Sublimation occurs if the vapor composition inside the pore is undersaturated with respect to the equilibrium with the surface of the pore. The model includes also an analysis whether forma-

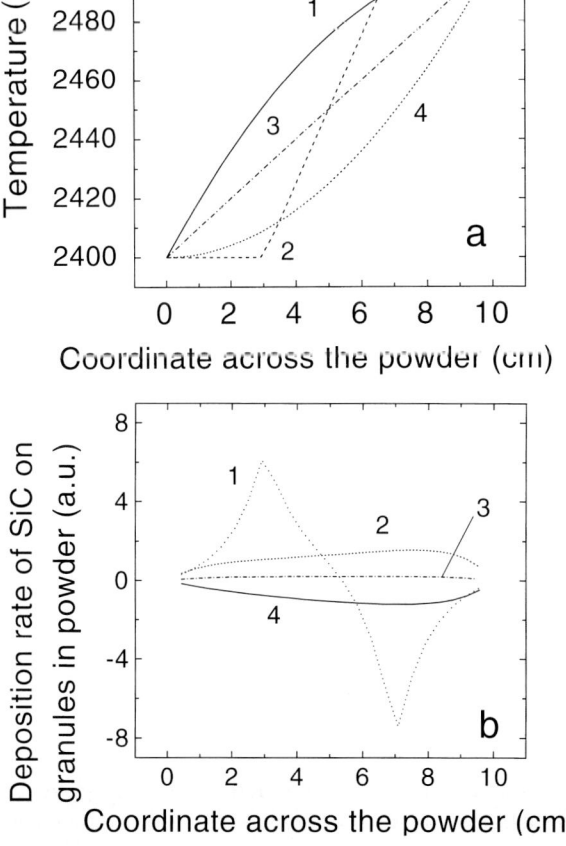

Fig. 12. a) Axial temperature profiles and b) rates of etching of SiC on the surfaces of the SiC granules in the powder

tion of secondary phases occurs on the surface of the granules using the approach described in [22]. If, for example, graphitization is predicted at a certain point in the powder, this is taken into account by calculating the chemical interactions between the volume of the pore and the surface of the granule which further is assumed to be a SiC–C solid phase.

Modelling of chemical processes inside the powder is performed for conditions similar to those used for the sublimation growth of cubic SiC in [34]. In Fig. 12a four different axial temperature profiles across the volume of the powder are shown which are selected artificially to demonstrate the features of material transport inside the powder. Among them there are linear, concave, convex and step-like temperature profiles. For simplicity, the case of congruent sublimation of SiC powder inside a SiC-coated crucible is shown in Fig. 12b. The rates of etching of SiC on the surface of the granules are presented in Fig. 12b for the temperature profiles shown in Fig. 12a. One can see that the linear temperature profile inside the powder results in dominating etching of the bottom of the crucible and re-deposition of SiC on the surfaces of the granules everywhere in the volume of the powder. This is more visible for the case of the concave temperature profile. It is opposite for the convex temperature profile – sublimation of SiC granules takes place everywhere in the volume of the powder. A step-like temperature profile results in intensive sublimation near the bottom of the powder and intensive re-deposition near the top.

From this simple consideration it is possible to conclude that stable material supply from the SiC powder to the seed requires accurate control of the temperature distribution inside the powder volume.

5. Conclusion

In the present paper modelling of sublimation growth of SiC is considered. The goal of the simulation has been an improvement of the basic understanding of the growth process, optimization of the design of the system and the process. It is demonstrated that a consistent model of the process which would allow to predict important features of the growth is not available presently. It is shown that coupled consideration of global heat transfer, material transport with chemical reactions taking into account interactions with the crucible wall and formation of secondary phases and deposits is necessary to predict growth rate and shape of the crystallization front. Modelling of global heat transport is discussed with the goal to optimize the temperature distribution inside the growth cell. The effect of the low thermal conductivity of SiC powder on the temperature distribution in the growth chamber is studied.

An advanced model of material transport during sublimation growth of bulk SiC crystals with high growth rates is proposed. The model is applied to the sublimation growth of bulk SiC in a Ta container in absence of an inert gas atmosphere. It is shown that some specific features of the growth process observed in the experiments are reproduced in the calculations. The model is free of any adjustable parameters. First results of modelling the chemical processes which occur in the volume of SiC powder during the sublimation growth are demonstrated.

Acknowledgements   The authors are thankful to Yu. A. Vodakov, M. G. Ramm, A. D. Roenkov and E. N. Mokhov for useful discussions and supply of experimental data. Support of A. S. Segal, A. N. Vorobjev, Yu. E. Egorov and R. A. Talalaev in studies of the sublimation growth is gratefully acknowledged. We would like also to thank D. Hofmann for interesting discussions on the growth mechanisms as well as M. G. Spencer for useful discussions and supply of experimental information.

References

[1] H. MORKOÇ, S. STRITE, G. B. GAO, M. E. LIN, B. SVERDLOV, and M. BURNS, J. Appl. Phys. **76**, 1363 (1994).
[2] YU. A. VODAKOV and E. N. MOKHOV, Pat. USSR N403275, 1970; Pat. G.B. N1458445 (21. 02. 74); Pat. USA N4147575 (03. 04. 79).
[3] E. N. MOKHOV, I. L. SHULPINA, A. S. TREGUBOVA, and YU. A. VODAKOV, Cryst. Res. Technol. **16**, 879 (1981).
[4] YU. A. VODAKOV, G. A. LOMAKINA, and E. N. MOKHOV, Soviet Phys. – Solid State **24**, 1377 (1982).
[5] YU. A. VODAKOV, E. N. MOKHOV, A. D. ROENKOV, and D. T. SAIDBEKOV, phys. stat. sol. (a) **51**, 209 (1979).
[6] YU. A. VODAKOV, A. D. ROENKOV, M. G. RAMM and E. N. MOKHOV, Mater. Sci. Engng. B (1997), accepted for publication.
[7] L. V. GURVICH, I. V. VEYTS, and CH. B. ALCOCK (Ed.), Thermodynamic Properties of Individual Substances, 4th ed., Hemisphere Publ. Co., New York 1989.
[8] J. DROWART and G. DE MARIA, J. Chem. Phys. **29**, 1015 (1958).
[9] YU. A. VODAKOV, E. N. MOKHOV, M. G. RAMM, and A. D. ROENKOV, Kristall und Technik **14**, 729 (1979).
[10] A. O. KONSTANTINOV and E. N. MOKHOV, Zh. Tekh. Fiz., Pisma **7**, 247 (1981).
[11] S. K. LILOV, Cryst. Res. Technol. **28**, 299 (1993).
[12] E. G. IVANOV, YU. M. TAIROV, and V. F. TSVETKOV, Univ. Rep., Non-Organic Mater. **21**, 588 (1985) (in Russian).
YU. M. TAIROV, Mater. Sci. Engng. B **29**, 83 (1995).
[13] S. YU. KARPOV, YU. N. MAKAROV, and M. S. RAMM, J. Cryst. Growth **169**, 491 (1996).
[14] S. YU. KARPOV, YU. N. MAKAROV, E. N. MOKHOV, M. G. RAMM, M. S. RAMM, A. D. ROENKOV, R. A. TALALAEV, and YU. A. VODAKOV, submitted to J. Cryst. Growth (1996).
[15] L. N. ALEKSANDROV, S. V. LOZOVSKII, and S. Y. KNYAZEV, phys. stat. sol. (a) **107**, 213 (1988).
[16] S. YU. KARPOV, YU. N. MAKAROV, E. N. MOKHOV, A. S. SEGAL, M. G. RAMM, M. S. RAMM, A. D. ROENKOV, YU. A. VODAKOV, and A. N. VOROBJEV, submitted to J. Cryst. Growth (1997).
[17] A. I. KINGON, L. J. LUTZ, P. LIAW, and R. F. DAVIS, J. Amer. Ceram. Soc. **66**, 558 (1983).
[18] G. S. FISCHMAN and W. T. PETUSKEY, J. Amer. Ceram. Soc. **68**, 185 (1985).
[19] S. YU. KARPOV, YU. V. KOVALCHUK, V. E. MYACHIN, and YU. V. POGORELSKY, J. Cryst. Growth **129**, 563 (1993).
[20] S. YU. KARPOV, YU. N. MAKAROV, and M. S. RAMM, Inst. Phys. Conf. Ser. No. 142, 177 (1996).
[21] S. YU. KARPOV, YU. N. MAKAROV, and M. S. RAMM, Inst. Phys. Conf. Ser. No. 142, 69 (1996).
[22] S. YU. KARPOV, YU. N. MAKAROV, M. S. RAMM, and R. A. TALALAEV, J. Appl. Phys. (1997), accepted for publication.
[23] S. KANEDA, Y. SAKAMOTO, T. MIHARA, and T. TANAKA, J. Cryst. Growth **81**, 536 (1987).
[24] YU. A. VODAKOV, E. N. MOKHOV, M. G. RAMM, and A. D. ROENKOV, USSR Patent 882247, 1980.
[25] D. HOFMANN, S. YU. KARPOV, YU. N. MAKAROV, E. N. MOKHOV, M. G. RAMM, M. S. RAMM, A. D. ROENKOV, YU. A. VODAKOV, Inst. Phys. Conf. Ser. No. 142, 29 (1996).
[26] YU. A. VODAKOV, A. D. ROENKOV, M. G. RAMM, E. N. MOKHOV, and YU. N. MAKAROV, phys. stat. sol. (b) **202**, 177 (1997).
[27] D. HOFMANN, M. HEINZE, A. WINNACKER, F. DURST, L. KADINSKI, P. KAUFMANN, Y. MAKAROV, and M. SCHÄFER, J. Cryst. Growth **146**, 214 (1995).

[28] M. Pons, E. Blanquet, J. M. Dedulle, I. Garcon, R. Madar, and C. Bernard, J. Electrochem. Soc. (1996), accepted for publication.

[29] R. Eckstein, D. Hofmann, Y. Makarov, St. G. Müller, G. Pensl, E. Schmitt, and A. Winnacker, Mater. Res. Soc. Symp. Proc. **423**, 215 (1996).

[30] F. Dupret, P. Nicodeme, Y. Ryckmans, P. Wouters, and M. J. Crochet, J. Heat Mass Transfer **33**, 1849 (1990).

[31] Yu. E. Egorov, A. O. Galjukov, L. A. Kadinski, Yu. N. Makarov, and A. I. Zhmakin, Proc. 3rd ECCOMAS CFD Conf., Wiley, New York 1996 (p. 704).

[32] Yu. E. Egorov, E. L. Kitanin, M. S. Ramm, V. V. Ris, and A. A. Schmidt, submitted to Mater. Sci. Engng. B (1997).

[33] S. Yu. Karpov, Yu. N. Makarov, E. N. Mokhov, A. S. Segal, M. G. Ramm, M. S. Ramm, A. D. Roenkov, Yu. A. Vodakov, and A. N. Vorobjev, submitted to ICSCIII-N '97, Stockholm.

[34] H. N. Jayatirtha and M. G. Spencer, Inst. Phys. Conf. Ser. No. 142, 61 (1996).

[35] M. G. Spencer, private communication (1996).

[36] D. Hofmann, private communication (1996).

[37] Yu. A. Vodakov, private communication (1996).

phys. stat. sol. (b) **202**, 221 (1997)

Subject classification: 61.50.−f; 61.72.Mm; 64.75.+g; S6

Cubic Silicon Carbide (3C-SiC): Structure and Properties of Single Crystals Grown by Thermal Decomposition of Methyl Trichlorosilane in Hydrogen

S. N. Gorin[1]) and L. M. Ivanova

Baikov Institute of Metallurgy, Russian Academy of Sciences,
Leninskii prospekt 49, Moscow 117334, Russia
Fax: (095)135-8680; tel.: (095)135-8630

(Received January 31, 1997)

This overview, based on earlier published papers, concerns the growth and some properties of single and polycrystalline cubic silicon carbide (3C-SiC) prepared by thermal decomposition of methyl trichlorosilane in hydrogen on resistively heated graphite substrates in a temperature range of 1500 to 2100 K. The morphology of faceted crystals and their specific twin structure, as well as the effects of crystallographic polarity of the 3C-SiC structure (sphalerite type) on impurity segregation and etching are considered in some detail.

1. Introduction

Silicon carbide is one of the most promising wide-bandgap semiconductors because, apart from a wide energy gap, it has many other valuable physicochemical properties such as a very high chemical stability, unique thermal and radiation resistance, high mechanical strength, etc. Semiconductor devices made of silicon carbide (light-emitting diodes, photodetectors, microwave devices, etc.) remain stable for periods as long as tens of thousands or even hundreds of thousands of hours, have working temperatures as high as 700 to 1000 K, withstand high irradiation doses, may operate in aggressive environments, in strong magnetic fields, and under other extreme conditions. However, no large-scale production of silicon-carbide devices exists at present. This may be explained by difficulties encountered in growing structurally perfect homogeneous pure and doped single crystals and epitaxial layers of silicon carbide. These difficulties are mainly due to some material properties of the compound such as very high growth temperatures (1773 to 2773 K), the absence of the liquid phase preventing melt growth, polytypism.

It is known that silicon carbide can form numerous modifications (the so-called polytypes). The most common polytypes are hexagonal modifications 6H, 4H and 8H, a rhombohedral polytype 15R, and a cubic polytype 3C. The last one occupies a special place, because it is the only polytype possessing a cubic structure (sphalerite type). It can be prepared at much lower temperatures (1473 to 2273 K) than hexagonal polytypes (2473 to 2773 K); it has a lower energy gap (2.39 eV at 0 K) and a higher electron

---

[1]) e-mail: gorins@lesr.imet.ac.ru

mobility (up to 1000 cm^2/(V s)). Unique semiconductor devices can be made of cubic silicon carbide, such as green light-emitting diodes with a very high brightness (greater than 10^6 cd/m^2), a high stability (operation lifetimes of 10^4 to 10^5 h), and working temperatures of up to 700 K; thermal indicators for measuring temperatures in a range of 360 to 1600 K in not easily accessible places, e.g. at objects of atomic industry and space technology; high-temperature photodetectors for recording power laser radiation.

High-purity polycrystalline layers of cubic silicon carbide may be used as protective coatings on items of graphite or commercial (Acheson's) silicon carbide as well as for making various structural elements such as chemically resistant reactors, containers and mirrors in microelectronics and power-device building.

This overview describes the morphology, internal structure, and some properties of cubic silicon carbide (3C-SiC, or beta-SiC) crystals grown from the vapor phase by thermal decomposition of methyl trichlorosilane (MTCS) in hydrogen on a resistively heated graphite substrate. The method was developed at the Baikov Institute of Metallurgy, Russian Academy of Sciences [1]. It ensures the preparation of single crystals (of up to 3 × 5 mm^2 in size) and thick polycrystalline layers (1 mm and thicker) that can be used for various semiconductor devices and passive elements [2].

The method is essentially a development of the so-called Van Arkel process [3] which was first used for silicon carbide growth by Pring and Fielding [4], Van Arkel and De Boer [3], Moers [5], and Kendall and Yeo [6]. In all these works, however, only small crystals (less than 1 mm in size) were obtained, and in those cases where the crystals were larger, they are considered as very imperfect from photographs and micrographs presented. This overview is based on the earlier published papers [7 to 9, 11 to 15, 19, 21 to 31].

2. Thermodynamics and Kinetics [7 to 9]

As starting compound for vapor deposition, methyl trichlorosilane CH$_3$SiCl$_3$ was selected, whose molecule contains Si and C in the stoichiometric ratio of SiC. The use of other organosilicon compounds, which have an excess content of carbon with respect to the 1 : 1 stoichiometry, is disadvantageous because a considerable amount of free carbon is formed during the decomposition. Thus, it was shown [7] that upon the decomposition of dichlorodimethyl silane (CH$_3$)$_2$SiCl$_2$ in a temperature range of 1173 to 1973 K, the solid products contain 7.5 to 14.5% C in addition to SiC. Depositing on the surface of crystals, this free carbon prevents the growth of low defect single crystals.

Three main reactions resulting in the formation of silicon carbide were analyzed:

(1) CH$_3$SiCl$_3$ ⇔ SiC + 3HCl;
(2) CH$_3$SiHCl$_2$ ⇔ SiC + 2HCl + H$_2$;
(3) (CH$_3$)$_2$SiCl$_2$ ⇔ SiC + C + 2HCl +2H$_2$.

In order to estimate the probability of the occurrence of these reactions, changes in the Gibbs thermodynamic potentials were determined at various temperatures using tabulated values of reduced thermodynamic potentials. These can be calculated with a high accuracy from spectroscopic data [32]. It was shown that the reactions can occur at temperatures as low as 850 K. At temperatures above 1000 K, the reactions are shifted completely toward the formation of SiC.

The composition of the solid, liquid, and gaseous products that formed upon the decomposition of MTCS in hydrogen was studied at $T = 1523$ to 1873 K [8]. The main reactions were found to occur with a high yield (60 to 80%) of SiC and be accompanied by the formation of an appropriate amount of HCl. At the same time, reactions leading to the formation of some simple chlorosilanes and hydrocarbons occur.

Chromatography and chemical analysis showed that elevated temperatures favor a more complete transformation of MTCS to SiC, whereas reduced temperatures lead to side reactions resulting in the formation of liquid and gaseous products. The main liquid by-products of MTCS decomposition are $SiHCl_3$ and $SiCl_4$. The main gaseous products are methane (1.18%), ethane and ethylene (0.31%), and acetylene (0.35%) [8].

The presence of these compounds can be explained by the occurrence of intermediate and side reactions. For example, upon the interaction of MTCS with hydrogen, trichlorosilane and methane can be formed. Upon the interaction of MTCS with HCl (the second main reaction product), silicon tetrachloride and methane can be formed.

Although hydrogen does not participate in the main reactions, it appears to take part in intermediate stages and leads to the formation of $SiHCl_3$, $SiCl_4$, and CH_4. In addition, hydrogen sharply decreases the rate of methane pyrolysis leading to the formation of free carbon (and hydrogen). This is confirmed in [7], where the use of helium instead of hydrogen was shown to markedly increase the amount of free carbon in solid products. For example, the decomposition of methyl dichlorosilane in helium at 1673 K yields 2 wt% free carbon, whereas in hydrogen its content is negligible at this temperature (less than 0.5%). A similar picture is observed upon the decomposition of MTCS in nitrogen. In the case of dichlorodimethylsilane, the amount of free carbon increases from 8.3 to 14.4% upon the substitution of helium for hydrogen.

In order to determine the limiting stages of SiC formation upon thermal decomposition of methyl trichlorosilane in hydrogen, the kinetics of the process was studied in a temperature range of 1523 to 1873 K [9]. From the dependence of the SiC yield on the rate of the MTCS supply into the reaction zone, we determined the value of the apparent activation energy E_a for the reaction [10]. This proved to retain a constant maximum value in the temperature range 1523 to 1623 K; at higher temperatures, the value of E_a began to decrease.

Based on the temperature dependence of the activation energy, we were able to select the following ranges:

(1) the kinetic range, occurring at $T = 1523$ to 1723 K. The activation energy is greatest in this range (95 to 98 kcal/mol).

(2) an intermediate range (1723 to 1873 K). In this range, the activation energy decreases to 30 kcal/mol.

(3) the diffusion range (above 1873 K). The activation energy decreases to about 12 kcal/mol at $T = 1873$ K.

Thus, in the range of relatively low temperatures (up to 1723 K), where polycrystalline layers of 3C SiC form, the rate of crystallization is reaction-controlled. At higher temperatures (above 1873 K), where separate crystals form, the limiting stage of SiC growth is diffusion, which controls the rate of the supply of the reacting agents to the surface of the growing crystals and the rate of the removal of the reaction products.

3. Growth, Morphology and Twin Structure of Beta Silicon Carbide Single Crystals Produced by Thermal Decomposition of Methyl Trichlorosilane [11 to 14, 19, 20]

Vapor growth of crystals in a flow system is controlled by many parameters such as the temperature of the graphite substrate (rod of 4 to 6 mm in diameter), composition of the reactive gas mixture, flow rate of the reaction mixture, temperature gradient, the shape of the heated surface, and some other factors such as the effects of impurities, thermal and concentration shocks, etc. In order to determine the optimal conditions for growth of single crystals and polycrystalline films, large systematic work was performed [15].

At temperatures from 1523 to 1723 K, a continuous dense polycrystalline film is formed. At temperatures below 1923 K, numerous small crystals grown together are deposited at the substrate. At 1923 to 2023 K, separate euhedral crystals grow, which frequently have the shape of regular hexagonal well-faceted platelets 3 to 5 mm in length and 1 to 1.5 mm in thickness (Fig. 1).

Above 2073 K, separate larger crystals form the basis of deposits consisting of thin intergrown "scales" of SiC that are coated with a thin film of free carbon. The scales consist of thin hexagonal or strongly disordered cubic SiC platelets (20 to 30 μm thick) forming a honeycombed structure. This structure has a poor thermal conductivity. Therefore, the temperature at the substrate under the deposit was found to increase to 2173 to 2273 K, whereas the temperature at the surface of crystals decreases by 200 to 300 K inhibiting the growth of crystals. Thus, the optimum temperature for growing single crystals lies in the range between 1923 and 2023 K.

During long-term runs, the temperature of the surface of growing crystals decreases gradually to 1773 to 1873 K (as measured with an optical pyrometer without any corrections for the degree of blackness). Nevertheless, the main body of crystals continues to grow normally. This indicates that, in the presence of a crystal surface that serves as a

Fig. 1. Photograph of three plate-like crystals of 3C-SiC without {100} faces. The trapezoid faces are Si{111}; the bootlike faces are Si{211}

0   1   2   3 mm

substrate, the range for growth of single crystals broadens. This strongly facilitates the problem of growing single crystals. However, even a short-time increase of the surface temperature to above 2073 K degrades the single-crystalline growth and leads to the formation of polycrystalline overgrowths.

In the general case, crystals are faceted and have various shapes (which frequently do not exhibit visible symmetry) and various degrees of perfection (which depends both on the growth conditions and the type of the crystal). However, it was found that the majority of crystals obtained under any conditions could be classified to three main types, which were arbitrarily called "platelets", "prisms", and "skeletal crystals" [13].

Crystals of the first type (Figs. 1 and 2) are flattened in one of the $\langle 111 \rangle$ directions and frequently have a characteristic shape of regular hexagonal platelets with one most developed mirror-like face [12, 13]. In all cases without exception, this face belongs to the positive tetrahedron (carbon $\{111\}$) faces. These crystals usually constitute 3 to 10% of the total number of crystals formed. Crystals of the second type, constituting 10 to 40% of the total number of crystals are extended along a direction $\langle 110 \rangle$ and have 6 to 10 faces of this crystallographic zone (pseudoprism faces) (Fig. 3). Crystals of the third type (Fig. 4) have at least two reentrant dihedral angles [see Section 5(ii)], which gives the crystals a characteristic skeletal appearance. These crystals may constitute 50 to 95% of the total number of crystals; they are, as a rule, very imperfect.

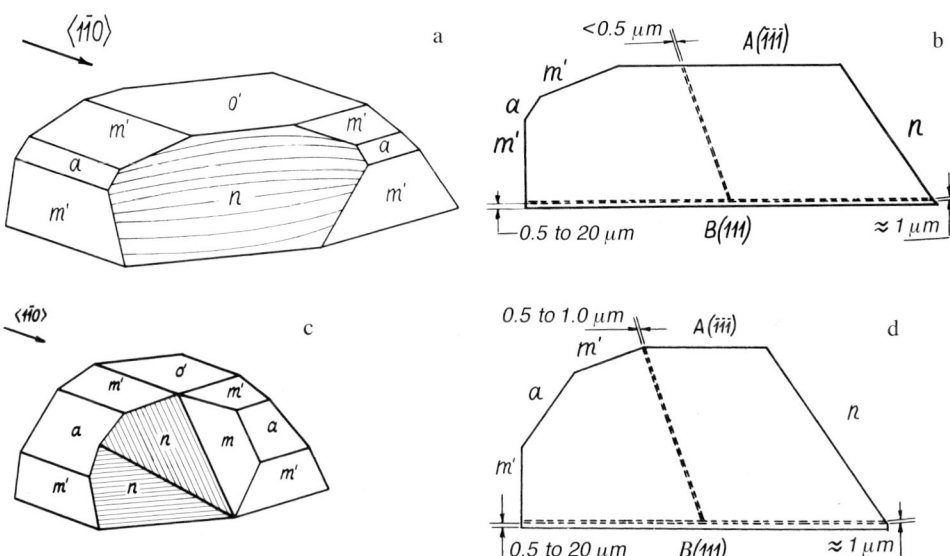

Fig. 2. (a, c) Morphology and (b, d) twin structure (in projection onto the $\{110\}$ plane perpendicular to the growth direction) of some types of platelike crystals (schematic; typical crystal size 2 to 5 mm): (a, b) platelike crystals without a reentrant dihedral angle; (c, d) crystals with a characteristic reentrant dihedral angle formed by vicinal faces n (see text below). Designations: $o' = A\{111\} = Si\{111\}$; $m' = A\{211\} = Si\{211\}$; $a = \{100\}$; $n = B\{hhl\} = C\{hhl\}$ (close to $\{552\}$, as a rule); $m = B\{hkk\} = C\{hkk\}$ (close to $\{211\}$, as a rule); $o = B\{111\} = C\{111\}$. Double dashed lines show microtwin lamellae (MTLs) that pass along the $\langle 110 \rangle$ direction parallel to the growth direction of the crystal. The horizontal MTLs lie a distance of 5 to 20 μm below the surface of the C$\{111\}$ face

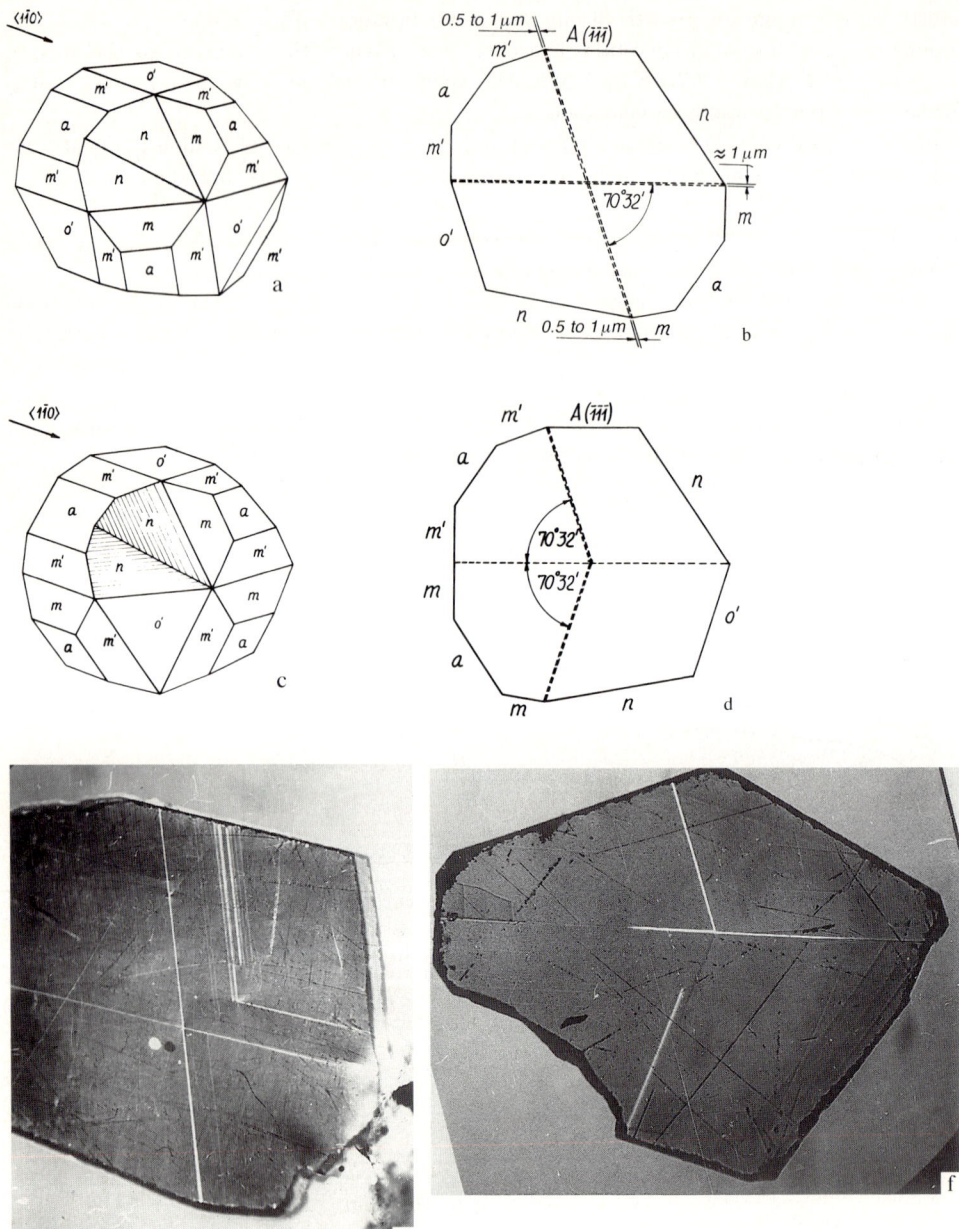

Fig. 3. (a, c) Morphology and (b, d) twin structures (in projection onto the {110} plane perpendicular to the growth direction) of some types of prismatic crystals (schematic; typical crystal size 2 to 5 mm): (a, b) morphologically single-crystalline prismatic crystal (both horizontal and inclined MTLs contain even numbers of twin planes); (c, d) morphologically twinned prismatic crystal (the horizontal MTL contains an odd number of twin planes); (e, f) twin structure of prismatic crystals viewed (along the ⟨110⟩ direction parallel to the growth direction) in polarized light with crossed Nicol prisms: (e) a single crystal; (f) a twin (note that the structure of the lower inclined MTL changes during growth). For other designations, see Fig. 2

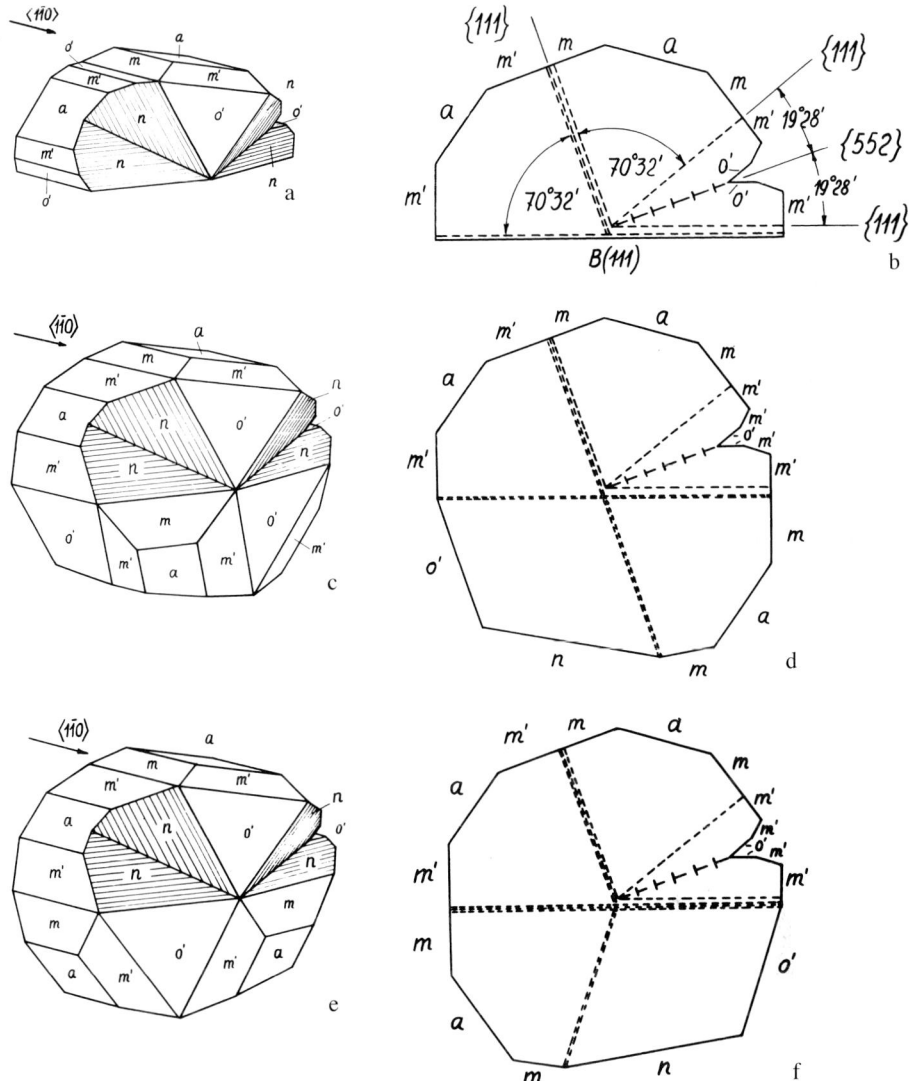

Fig. 4a–f (Legend on the next page)

Apart from these main types, crystals of other types occur, e.g., such as shown in Fig. 5. Those displayed in Figs. 5a and 5b are the simplest possible types of crystals. They are single crystals that only contain thin microtwin lamellae immediately under the surface of mirror C{111} faces intersecting one another at an angle of 70°32′. The difference in their shape is a very significant feature that will be discussed below. These crystals occur quite rarely; this refers also to triple twins (Fig. 5e, f).

The facets were identified by X-ray diffraction, etching, and goniometry (performed on a Goldschmidt two-circle optical goniometer with an accuracy of 1′). The investigation of the external and internal morphology (sectorial structure [16], i.e., the

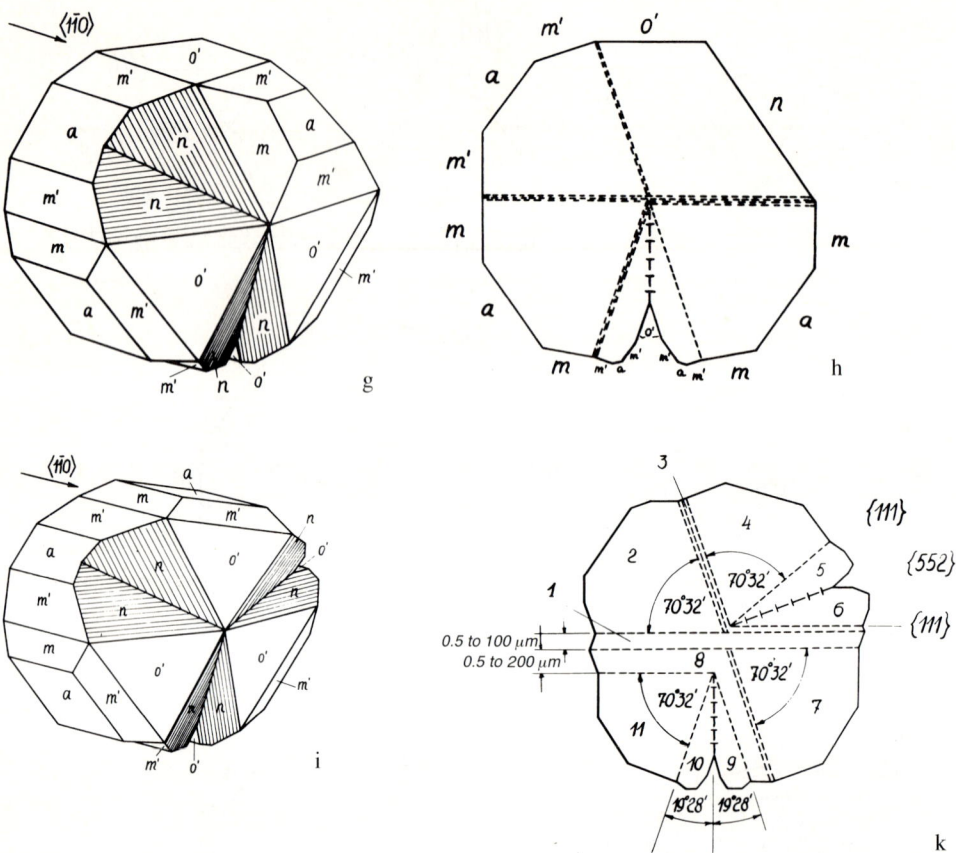

Fig. 4. (a, c, e, g, i) Morphology and (b, d, f, h, k) twin structures (in projection onto the {110} plane perpendicular to the growth direction) of some types of "skeletal" crystals (schematic; typical crystal size 2 to 5 mm): (a, b) skeletal platelike crystal; such crystals contain a reentrant dihedral angle related to a higher-order twin boundary; this boundary causes strong mechanical stresses in the crystal, for which reason its mirror-like C{111} face ceases to be flat and becomes distorted (areas deviating an angle of 3 to 7° from the flat portion can arise [12, 13]); (c to h) skeletal prisms, i.e., crystals that are, basically, of prismatic types but contain third-order {552}–{552}-type twin boundaries (cf. Figs. 3a to 3d). Twin boundaries shown by odd numbers of lines contain odd numbers of twin planes; boundaries composed from "dislocation symbols" are third-order {552}– {552}-type twin boundaries (see text). Crystals with larger numbers of reentrant angles can also occur, but they are imperfect (primarily, due to the presence of numerous MTLs and complex intergrowths) to an extent that makes no sense in their morphological analysis

shape of "growth cones" (Anwachskegel) [17], "growth pyramids" [16], or "face loci" [18]) in pure and doped crystals of 3C-SiC in various sections (primarily, in sections passing through the starting growth point of the crystal) showed the following results [13]:

  (1) the growth direction of the majority of crystals is ⟨110⟩;

  (2) the following faces are formed on most crystals: C{111} (positive tetrahedron), Si{111} (negative tetrahedron), Si{211} (negative trigon-tritetrahedron), and {100} (cube), all of which belong to the ⟨110⟩ zone;

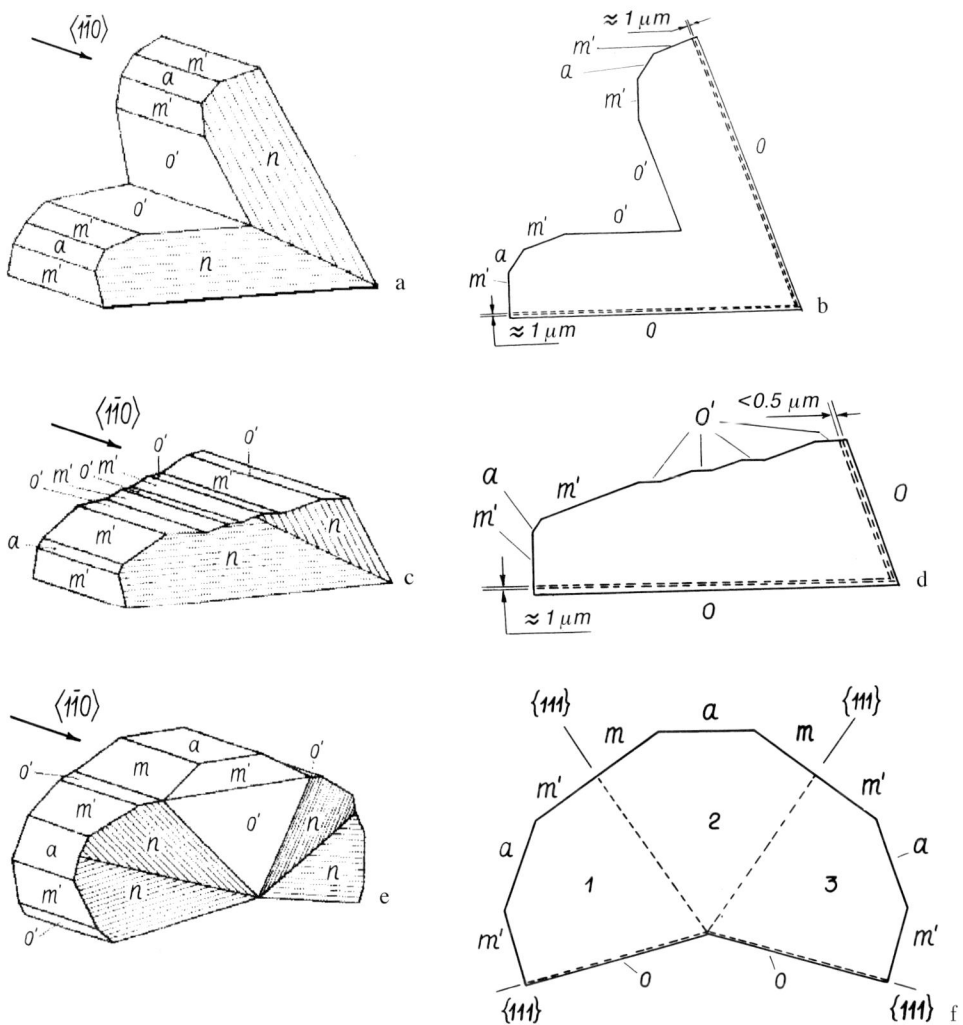

Fig. 5. (a, c, e) Morphology and (b, d, f) twin structures (in projection onto the {110} plane perpendicular to the growth direction) of some rarely occurring types of 3C-SiC crystals (schematic; typical crystal size 2 to 5 mm): (a to d) spikelike crystals (elongated tetrahedra); and (e, f) a triple cyclic twin (twinning on nonparallel {111} twin planes). For designations, see Fig. 2

(3) the C{111} faces have a zero growth rate [13, 14]; this can be deduced from the shape of growth pyramids in crystals doped with boron (see, e.g., Figs. 9a and 9b and Fig. 17). As a result, only one C{111} face is formed on most crystals instead of four faces characteristic of this simple form.

(4) Imperfect vicinal faces of the positive tetragon- and trigon-tritetrahedra appear in place of the other C{111} faces. Their tilts with respect to the parent C{111} faces and neighboring {100} and Si{111} faces differ in different crystals (and sometimes even in the same crystal). For this reason, they were called merely C{hhl} and C{hkk}, repectively. The number of these faces never exceeds four, although the corresponding hkl

forms consist of 12 faces each. Moreover, the {*hhl*} and {*hkk*} faces are located asymmetrically with respect to the symmetry elements of the crystal.

(5) The platelet-type crystals are morphological "monsters" [13]: On these crystals, faces of the hemisphere adjacent to the C(111) pole (which corresponds to the single mirror-like face of the positive tetrahedron) are absent, i.e., these crystals have only one face Si{111} instead of four, three {100} faces instead of six, and six Si{211} faces instead of 12.

(6) Morphologically, some of prismatic crystals look like single crystals (Figs. 3a, b), the other, like twins (Figs. 3c, d).

(7) It follows from (4) that the {*hhl*} and {*hkk*} faces have no independent morphological significance; taking into account that the single face of the C{111} type is the greatest face and is very perfect (it was recently shown [20] that this face is extraordinarily flat and uniform, with a main surface roughness of 1 to 2 Å), we assumed [11, 13] that the ideal form of beta SiC single crystals obtained by thermal decomposition of MTCS in hydrogen may be represented as a combination of the positive and negative tetrahedra, negative trigon-tritetrahedron, and cube (C{111}, Si{111}, Si{211}, and {100}, respectively), see Fig. 6a.

Investigations of 3C-SiC crystals by etching [19] and optical microscopy (in transmitted polarized light) showed that they contain stacking faults and microtwin lamellae (MTL), i.e., thin planar defects consisting of some number (even or odd) closely located twin planes and stacking faults that are revealed as a single defect upon etching and frequently exhibit strong birefringence when inspected in polarized light. It was shown that even the best crystals contain a characteristic structure (that was called twin structure) in the form of a few MTL passing through the crystal vertex looking in the growth direction, see Figs. 2 (b and d), 3 (b, d, f), 4 (b, d, f, h, k), and 5 (b, d, and f).

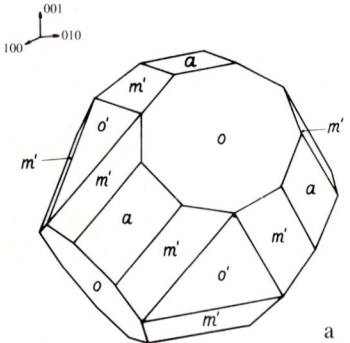 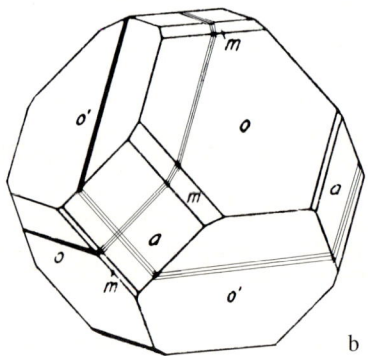

Fig. 6. (a) An "ideal" form of 3C-SiC crystals grown by thermal decomposition of methyl trichlorosilane in hydrogen on a resistively heated graphite substrate in the presence of a temperature gradient of 400 to 500 K/cm [13, 14]. The choice of the C{111} face as belonging to the positive tetrahedron is arbitrary and was made in [11] for considerations that appear to be important at those times (1964). If the Si{111} face were chosen as the positive tetrahedron, we would obtain an ideal form almost identical to that shown in Fig. 6b, which is a drawing of a 3C-SiC crystal (less than 0.5 mm in size) found by Thibault [33] in the cubic fraction ("amorph") of the silicon carbide crystal obtained by the Acheson method (note that Thibault's faces *m* were {311}, whereas on our crystals they are Si{211} without exception)

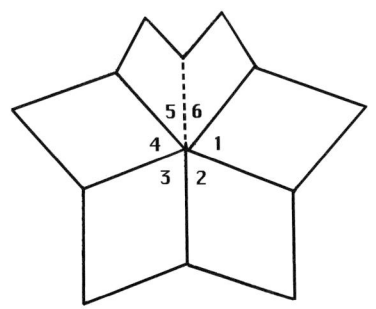

Fig. 7. A quasi-fivefold polysynthetic cyclic twin (schematic) characteristic of diamonds [35 to 37]. The dashed line shows a mismatch boundary (higher order twin boundary); numerals in the figure denote individual twins; the paper plane is a {110} plane

Skeletal crystals represent a special case. They contain, in addition to coherent twin boundaries on {111} planes, twin boundaries of higher order (symmetric third order tilt boundaries of the (110)[110] type with a tilt angle of 31°35′, composition twin plane {552}). These boundaries appear as a result of multiple twinning on nonparallel {111} planes. It should be emphasized that from four possible types of the mutual arrangements of such twin boundaries, only one type occurs in all crystals without exception. Etching and goniometry indicate that the boundaries are of the Si{552}–Si{552} type, i.e., they contain dislocations with "wrong" bonds of the Si–Si type or, which is far less probable, "broken" C bonds (beta-dislocations). This is a direct consequence of the crystallographic polarity of the sphalerite structure. Another related effect of the polarity is the absence of multiple twins with the apparent fivefold symmetry that are quite typical for diamonds (Fig. 7). Multiple twins in 3C-SiC always involve the above-described {552}-type boundaries (Fig. 4) and these are arranged in all cases as in Fig. 4 (b, d, f, h, and k).

4. Effects of Crystallographic Polarity: Impurity Segregation and Etching [11, 12]

Pure beta silicon carbide single crystals are transparent and have an amber yellow color. Some impurities make them dark (black or green). This enables to easily trace their behavior in crystals. It was found by visual inspection that Al, B, and N are distributed inhomogeneously in crystals. The crystals have a "sectorial" structure [11, 12] caused by preferential absorption of different impurities by different faces (Fig. 8, 9). The sectorial distribution of impurities was studied in sections cut along the growth pyramids of crystal faces in such a way that the section passed through the initial point of growth (the "root" of the crystal by which the crystal is attached to the graphite substrate, see Fig. 10). The distribution of nitrogen and boron was studied in detail. Because both these impurities are electrically active in SiC, their distribution found visually was confirmed by electrical measurements.

It was found that nitrogen is strongly "absorbed" by C{hhl} and C{hkk} faces and weakly absorbed by Si{111}, Si{211}, and {100} faces, whereas boron is absorbed most actively by Si{111} faces, less strongly by Si{211} faces, and only weakly by C{hhl}, C{hkk}, and {100} faces, i.e., preferential absorption of nitrogen is observed in growth pyramids of "carbon" faces (positive hkl forms) and is absent in "silicon" faces (negative hkl forms). For boron, the picture is opposite. This impurity is preferentially absorbed by negative hkl forms. Cube faces {100}, which are nonpolar in the sphalerite structure, exhibit no preferential absorption for either nitrogen or boron. The distribution of nitro-

a

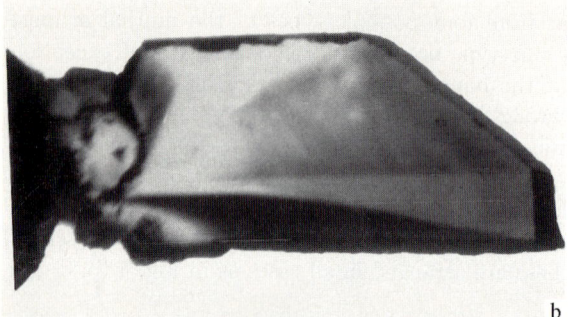

b

c

Fig. 8. Inhomogeneous (sectorial) distribution of nitrogen in 3C-SiC crystals (crystal size 3 to 5 mm): (a, c) plate-like crystals (a view in plan); (b) A–A section (see Fig. 10) of a prismatic crystal without {100} faces (cf. Fig. 11b). It follows from the darker color of the C{$h'h'l'$} growth pyramid in this crystal that the tilt of this face is less than 20° (i.e., this is a {332} or {553} rather than {552} face

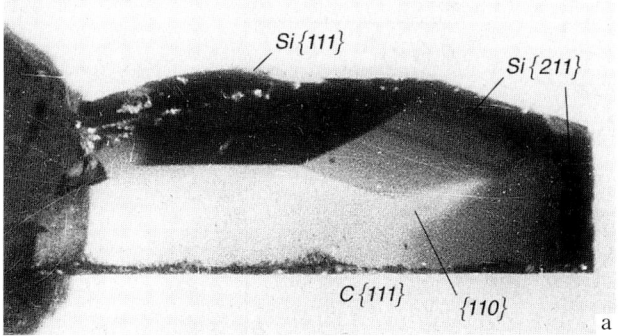

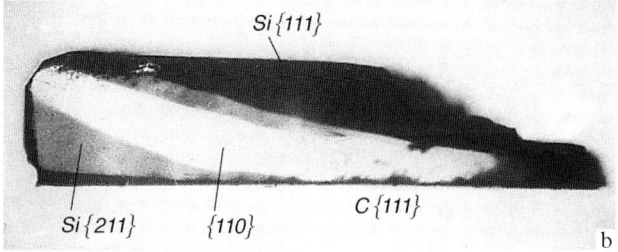

Fig. 9. Inhomogeneous (sectorial) distribution of boron in 3C-SiC crystals (length about 4 mm). (a) crystal in which the {100} face is eliminated during growth; (b) crystal in which the {100} face is retained for the whole growth period. It can easily be seen that the C{111} face has no growth pyramid, which unambiguously indicates that its growth rate is zero

gen and boron over growth pyramids of various faces in crystals of platelet and prismatic types is shown schematically in Fig. 11. Aluminum is distributed similar to boron.

Resistivity measurements show that the "facet effect" is very strong in 3C-SiC. In crystals heavily doped with nitrogen, the resistivity of neighboring growth pyramids

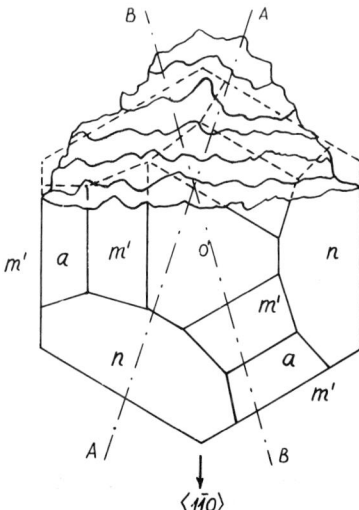

Fig. 10. The two most important sections of 3C-SiC crystals passing through the root (stepped upper portion)

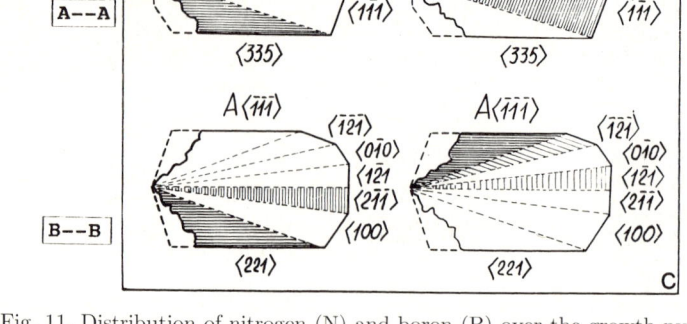

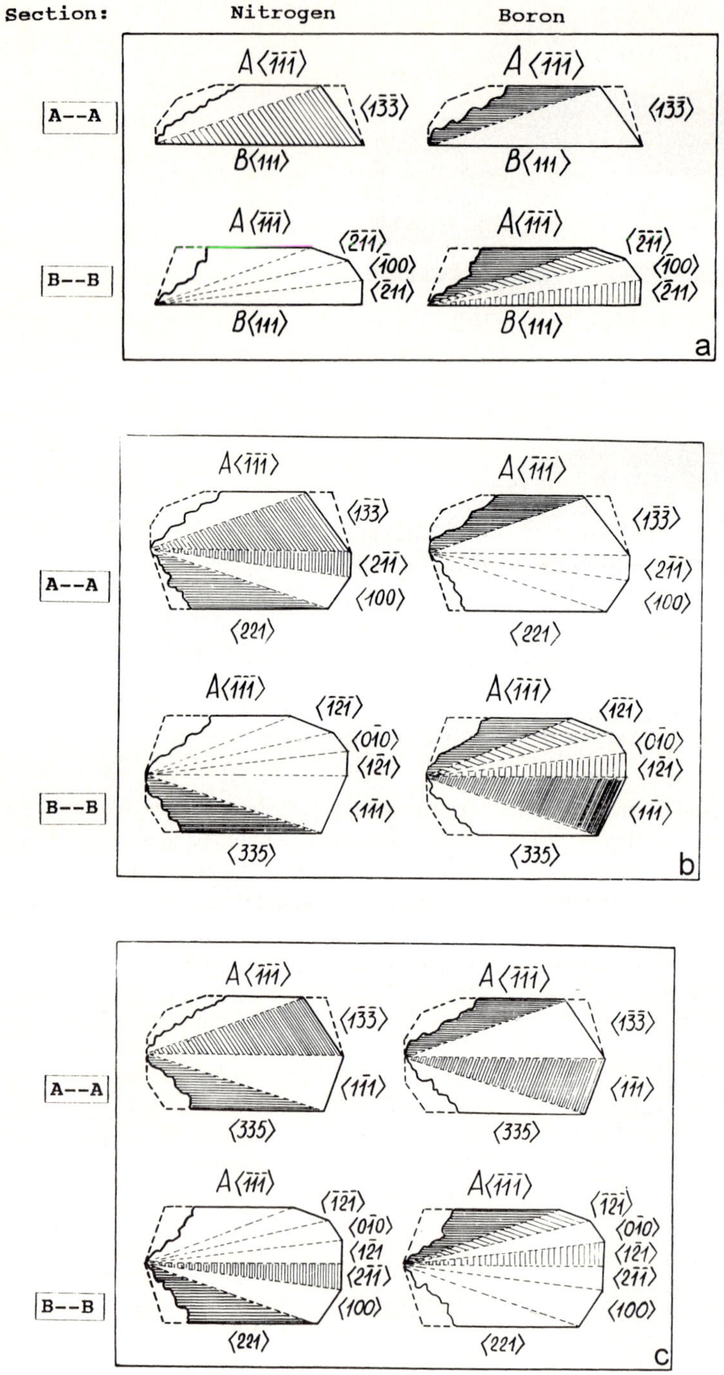

Fig. 11. Distribution of nitrogen (N) and boron (B) over the growth pyramids of 3C-SiC crystals in sections A–A and B–B: (a) platelike crystals; (b) prismatic crystals of single-crystal type (cf. Fig. 3a); and (c) prismatic crystals of twin type (cf. Fig. 3c). For designations, see Fig. 2

corresponding to faces of opposite polarity may differ by more than two orders of magnitude. The impurity content appears to differ slightly less, because the mobility at room temperature in beta SiC is primarily controlled by impurity scattering and, consequently, is smaller in heavily doped regions.

Crystallographic polarity of the beta SiC structure also strongly manifests itself during etching and abrasion. For example, when polishing crystals with a fine aluminum oxide powder, the polishing rate (as characterized by the inverse of the time needed for obtaining mirror finish) changes in the order $C\{111\} > C\{211\} > \{100\} > Si\{211\} > Si\{111\}$. The time required to obtain a good polish on $C\{111\}$ faces is smaller by a factor of ten or more than on $Si\{211\}$ faces, and $Si\{111\}$ faces could not at all be polished reasonably using Al_2O_3 as abrasive. This effect enables one to reveal twin structures of crystals even without etching. For example, Fig. 12 shows photographs of transverse cross sections of prismatic and skeletal crystals after polishing with a powder of Al_2O_3 with grain size of about 1 µm. The effect is absent when polishing with diamond powders as well as when lapping with coarse abrasives (e.g., with a powder of boron carbide with grain size of 10 to 30 µm).

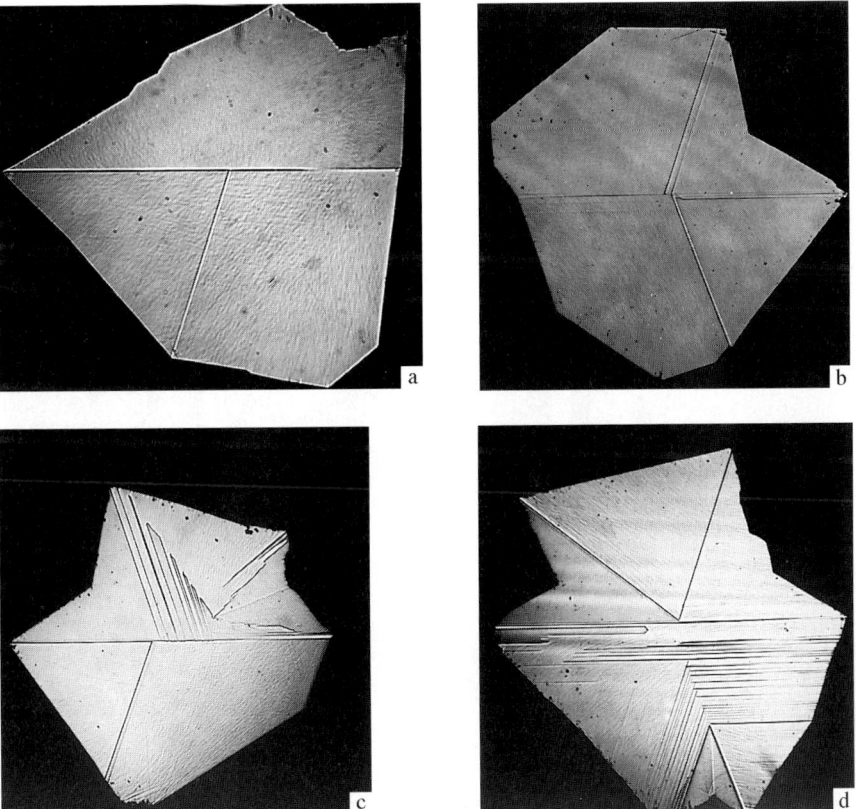

Fig. 12. Twin structures of prismatic (a, b) and skeletal (c, d) crystals (size 2 to 5 mm) as revealed in sections perpendicular to the growth direction by merely polishing the sections on a "mild" abrasive (Al_2O_3 powder with grain size of about 1 µm) without subsequent etching. Note a "perfect" structure of prismatic crystals and a far less perfect (and nevertheless, quite distinct) structure of skeletal crystals without any signs of polycrystalline overgrowths

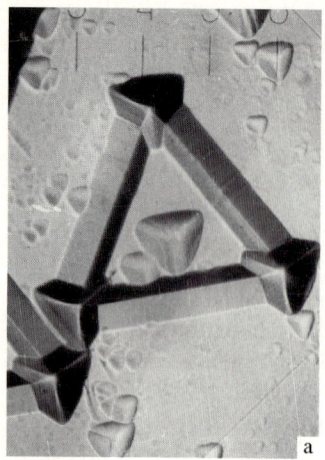

Fig. 13. Etch pits and stacking faults (original size about 50 to 75 μm): (a) Si{111} face etched in molten KOH at 673 K for 15 min and (b) Si{211} face etched in molten PbO (plus 5% sodium tetraborate) at 983 K for 45 s (500×)

The polarity effects upon etching are as follows [19]. Etching in most reagents (such as molten NaOH, KOH, PbO, Na_2O_2, sodium tetraborate, KNO_3 and its mixtures with KOH, as well as in gaseous chlorine) gives distinct triangular etching figures at the faces of negative hkl forms (silicon faces, see Fig. 13), whereas no such figures are formed at the faces of positive hkl forms (carbon faces). Moreover, the etching figures are depressions on silicon faces and hills on carbon faces. Etching pits on cube faces are lens-shaped (Fig. 14a) with lens axes aligned in only one of the two possible ⟨110⟩ directions.

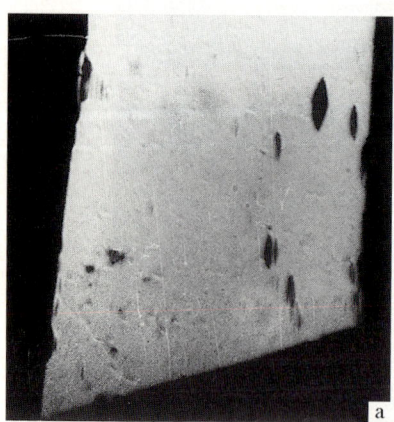

Fig. 14. Etch pits on {100} and {110} faces: (a) {100} face etched in molten KOH at 633 K for 1 h (magn. 60×); (b) section approximately perpendicular to the growth direction (close to the {110} plane) etched in a mixture of PbO + 46 mol % PbF_2 at 853 K for 1 h (magn. 40×). It should be noted the different orientation of the etch pits on the opposite sides of a twin lamella in the {110} section; the difference in the form of triangular etch pits on the Si{111}, Si{211} (Fig. 13), and {110} faces and the different shape of stacking faults on the Si{111} and Si{211} faces

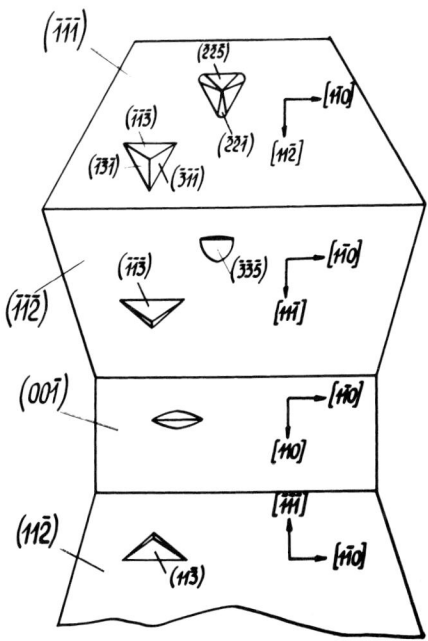

Fig. 15. Shape and orientation of etch pits on the Si{111}, Si{211}, and {100} faces of 3C-SiC crystals

Etching pits at the opposite {100} faces are mutually perpendicular. In some etchants (molten PbO, gaseous chlorine) the etching rates of carbon faces are many times greater than those of silicon faces. Upon oxidation in wet oxygen (at 1473 K), the oxide film on carbon faces is approximately twice as thick as that on silicon faces. The shapes of etch pits and stacking faults of some faces are shown in photographs and schematically in Figs. 13 to 15.

5. Growth Rates of Various Faces of Beta Silicon Carbide Crystals

A comparison of the faces occurring on real crystals with those characteristic of the ideal" form (Fig. 16) shows that microtwin lamellae (MTLs) can locally increase the growth rate of 3C-SiC crystals. Estimating real growth rates of various faces by studying sectorial structures of crystals (Fig. 17) reveals [13, 14] the following:

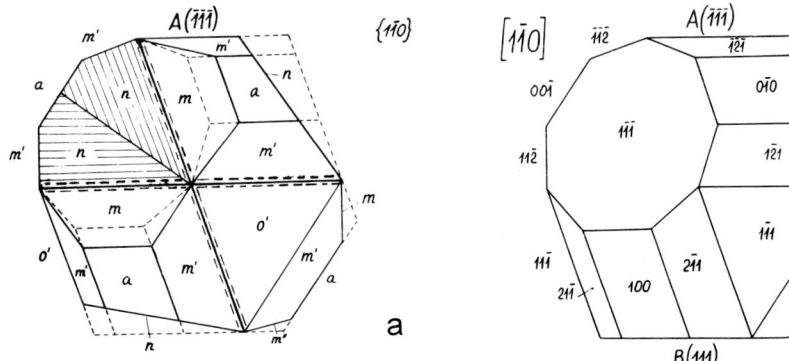

Fig. 16. Comparison of a real (a) and ideal (b) form of cubic silicon carbide crystals ({110} projections). The double dashed lines intersecting in the center of Fig. 16a show the MTLs that lead the crystal growth; single dashed lines outline the position of the "ideal" faces (cf. Fig. 16b). For other designations, see Fig. 2

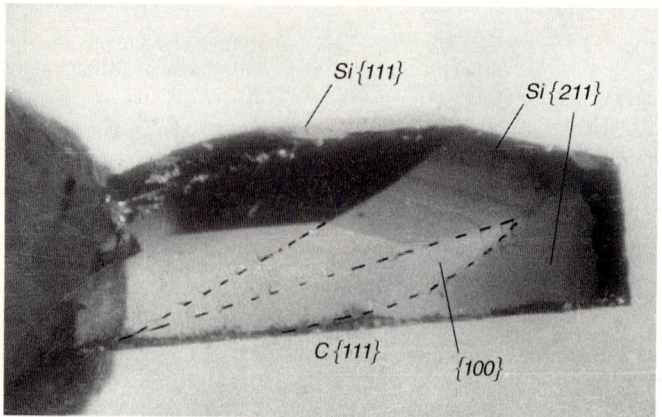

Fig. 17. Shape of the growth pyramids in a section of the B–B type (see Fig. 10) in a crystal doped with boron starting from a certain moment (corresponding to the boundary between the dark and light portions of the section)

(i) MTLs appear to have specific growth rates V^{MTL}. When an MTL intersects a face whose growth rate is smaller than $V^0 = V^{\mathrm{MTL}}\cos \alpha$, where α is the smaller angle between the MTL and the face, the face undergoes a transformation. It becomes replaced by two induced vicinal faces (Fig. 16a, 18a) whose growth rates are related to V^0 as $V = V^0\cos \beta$, where β is the angle between the induced and vicinal faces. The site where the MTL emerges onto the surface of the face coincides with the edge between the induced faces (which form a ridge). For the Si{111} faces, whose growth rate is about $V^{\mathrm{MTL}}/3$, the tilt of the vicinal faces {1, 1, 1} relative to Si{111} is 0.5 to 1.0°. For the C{111} faces, whose growth rate is zero, the tilt of the induced faces can reach 20°, i.e., these can be, e.g., faces $\{kkh\} = \{112\}$ and $\{hhl\} = \{552\}$.

(ii) On those C{111} faces that are intersected by two nonparallel MTL, a reentrant dihedral angle appears: because each MTL leads to the formation of a ridge, the intersection of two $\{hhl\}$ faces that are intersecting sides of these ridges results in the formation of a reentrant angle which does not relate to any internal boundaries in the crystal and represents a suppression in the single-crystal portion of the crystal; this reentrant angle is the consequence of the accelerated crystal growth near nonparallel edges caused by MTLs.

(iii) By no means all of MTLs exhibit an acceleration effect; rather, these are the MTLs that have a thickness of 1 to 2 μm. The thickness of MTLs seldom exceeds 2 to 3 μm and can change during growth, which usually is accompanied by changes in the habit and shape of the crystals, e.g., the appearance of new facets and vertices that are atypical for the ideal form (Fig. 19);

(iv) The induced $\{hhl\}$ and $\{hkk\}$ faces are not flat; their tilt relative to the parent {111} face may vary in different crystals and even in the same crystal, depending on the type of the MTLs, from 0.5 to 1.0° to about 20°.

An analysis of the morphology permits to understand the "disinclination" of the 3C-SiC crystals to grow. In the cubic system, all hkl forms are closed (i.e., enclosing space). It is known that a crystal, when growing, is finally faceted by the most slowly growing faces (this is essentially a geometric effect, with some restrictions that can be

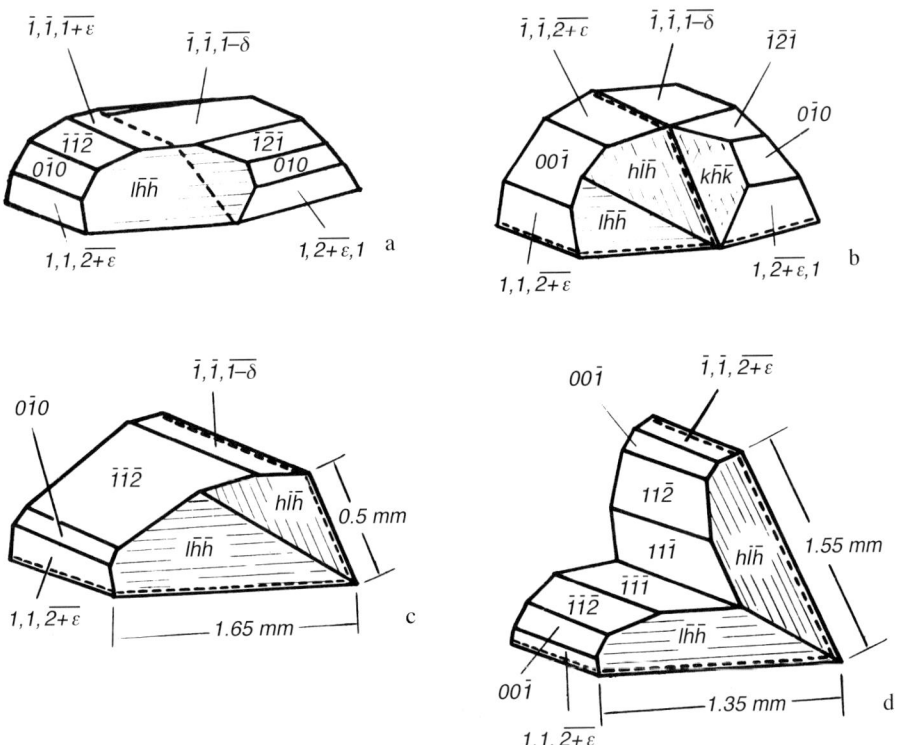

Fig. 18. Effects of "slow" ("thin") and "fast" ("thick") MTLs on the morphology of 3C-SiC crystals. (a, c) The horizontal MTLs are fast, the inclined MTLs are slow; (b, d) both the horizontal and the inclined MTLs are fast

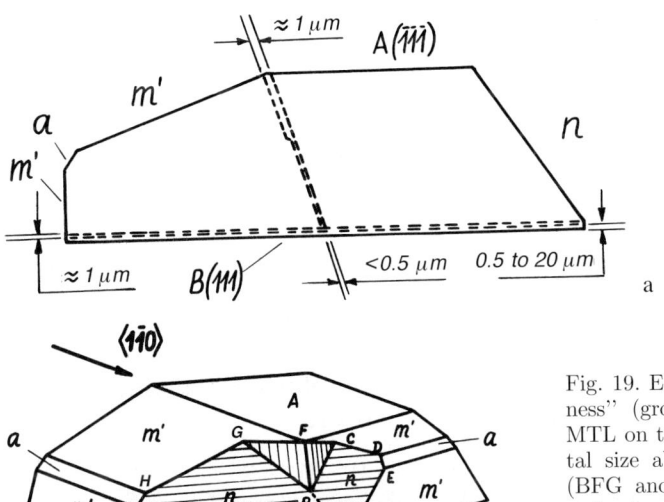

Fig. 19. Effect of a change in the "thickness" (growth activity) of the inclined MTL on the faceting of the crystal (crystal size about 2 to 3 mm): new facets (BFG and BFC) appear on the crystal and two reentrant dihedral angles (ABGH^BFG and ABCD^BFC) arise (cf. Fig. 2a)

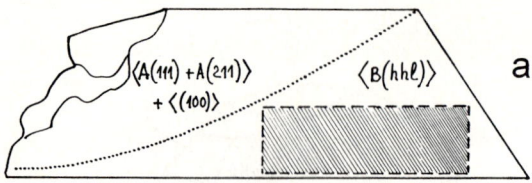

Fig. 20. Cutting of samples from individual growth pyramids. The hatched areas are used for the fabrication of devices (see Section 8)

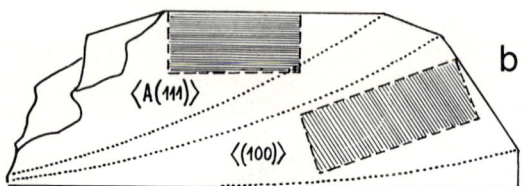

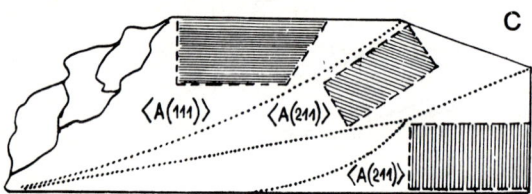

taken into account in the Borgstroem scheme) [34]. This means that a perfect 3C-SiC crystal must stop growing after taking the form of a tetrahedron bounded by C{111} faces with a "zero" growth rate. In reality such a crystal reaches only a microscopic size. It becomes "locked in the shell of the zero growth rate" of these faces. Microtwin lamellae disclose this shell, and the habit and faceting of all types of 3C-SiC crystals are unambiguously controlled by their twin structure, i.e., by the number, position, and thickness of the MTLs present and their type (i.e., by whether the number of twin planes is even or odd in each of them). As to the very fact of arising MTLs in 3C-SiC crystals, it may be supposed to be related to the rather high temperature gradient (400 to 500 K/cm) in the growth reactor used in this method: The temperature at the surface of the graphite rod (6 mm in diameter) which served as the substrate for the crystal growth, is 1800 to 2000 K, whereas the temperature at the surface of the quartz reactor (80 mm in diameter, coaxial with the graphite rod) was 300 to 320 K at most, because it was cooled by running tap water.

6. Pure Polycrystalline Layers of Cubic Silicon Carbide: Preparation and Properties [15, 21 to 24]

The yield, deposition rate and structure of polycrystalline layers of 3C-SiC (PCLs) that are obtained by thermal decomposition of MTCS at low temperatures depend on a number of parameters. The most important of these parameters are (a) the temperature at which the deposition is performed, the concentration of MTCS in the gas phase, the flow rate of hydrogen in the reaction chamber, and the area of the deposition surface. When studying the growth of PCL in our work, these parameters were varied in the following limits: the deposition temperature, 1523 to 1873 K; the MTCS concentration in hydrogen, 0.2 to 1.7 g/l; and the gas flow rate, 3 to 30 l/h.

Our investigations [15, 21] indicate that the following conditions are the optimum conditions for the preparation of pure dense polycrystalline layers of 3C-SiC: the deposition temperature, $T = 1673$ to 1773 K; the concentration of MTCS in hydrogen, $C = 0.5$ to 1.5 g/l; and the flow rate of the gaseous mixture, $l = 6$ to 10 l/h. Under these conditions, the 3C-SiC PCLs have a smooth surface and fine microstructure. The yield of SiC varies from 65 to 70%. The duration of the process is virtually unlimited; in a cycle of 3 to 6 h, layers of 1 to 3 mm thickness can be obtained.

The polycrystalline layers obtained under the above conditions have physicochemical and mechanical properties that are superior to those of the conventional carborundum materials obtained by powder metallurgy. Their density as determined by the hydrostatic weighing technique is 3.216 to 3.220 g/cm^3, which agrees within the experimental error with the literature data for the theoretical density of beta SiC.

Polycrystalline layers of 3C-SiC obtained by thermal decomposition of MTCS in hydrogen have a high purity. The impurity content is 0.5 to 3 ppm, which is 2 to 3 orders of magnitude lower than that in the conventional carborundum material obtained by recrystallization or hot pressing. Spectroscopic analysis indicates that the concentration of the most common impurities is as follows: Mg, 0.18; Mn, 0.5; Ti, 3; Al, 1.8; B, 1; and Fe, 8.8 ppm. These results agree with the data of activation analysis. The content of some impurities is lower than the sensitivity of the method: P, $<2 \times 10^{14}$; Cu, $<3 \times 10^{13}$; W, $<3.8 \times 10^{13}$; and Sb, $<5 \times 10^{13}$ cm^{-3}.

Pure polycrystalline layers of 3C-SiC have high resistance to high-temperature oxidation. They oxidize in air slower by a factor of 10 to 30 than other carborundum materials [22].

The value of the ultimate strength in compression for pure polycrystalline layers obtained by thermal decomposition of MTCS in hydrogen exceeds that of the material prepared by hot pressing by a factor of 4 to 40 and reaches 3500 to 5340 MPa at room temperature [23].

Studying infrared spectra of pure 3C-SiC shows that PCLs possess high reflectance in the range of 10 to 14 m: $R = 90\%$ [24].

Measurements of the electrical properties show that pure 3C-SiC PCLs have electronic conductivity; their resistivity varies from 0.1 to 1000 Ω cm; the charge carrier concentration is $n = 1.4 \times 10^{16}$ to 5×10^{17} cm^{-3}.

Pure polycrystalline beta silicon carbide has a high thermal conductivity. The maximum value is 5 W/(K cm) at 110 K. With increasing temperature, thermal conductivity decreases, but still remains sufficiently high: 2.5 W/(K cm) at room temperature.

The thermopower of pure polycrystalline beta silicon carbide with a charge carrier concentration $n = 5 \times 10^{17}$ to 1.8×10^{16} cm^{-3} ranges from -500 to -600 μV/K.

7. Doping Polycrystalline Cubic Silicon Carbide [25 to 28]

Doping of polycrystalline beta silicon carbide with nitrogen and boron was performed directly during growth by thermal decomposition of methyl trichlorosilane.

The most suitable compounds for doping with nitrogen are nitriles, which are the derivatives of aliphatic and aromatic acids. They only contain carbon, hydrogen, and nitrogen, i.e., there is no danger of contamination by foreign impurities when using nitriles. The nitriles are stable compounds that can be purified by distillation or rectification in air under atmospheric pressure.

From the variety of nitriles available, we selected acetonitrile as the simplest one [25]. In order to introduce it into the reaction chamber, a separate evaporator was used, which permitted us to easily vary its concentration in the gas phase.

Acetonitrile as a doping agent has some advantages over the nitrogen gas. First, the use of the nitrile allows one to easily obtain samples with a high conductivity. Second, in order to obtain a certain level of conductivity, one can use much smaller concentrations of acetonitrile in comparison with free nitrogen. Thus, to obtain beta SiC with a conductivity of 140 S/cm, the concentration of free nitrogen should be kept at a level of 240 mg/l, whereas the corresponding concentration of acetonitrile should be as small as 0.13 mg/l. An increase in the acetonitrile concentration from 0.1 to 15.1 mg/l in the gaseous mixture decreases the resistivity of beta SiC by a factor of more than two orders of magnitude (from 0.026 to 0.0006 Ω cm. The carrier concentration increases from 1.5×10^{18} to 1.2×10^{19} cm^{-3}. Similar efficiency is also exhibited by other nitriles such as propionitrile and capronitrile.

As an acceptor impurity, we used boron, which was introduced into the gas phase in the form of boron tribromide vapors. The investigations performed [26, 27] permit us to arrive at the following conclusions:

(a) An increase in the boron concentration in the reaction chamber from 0.28 to 4.7 mg/l increases its concentration in beta silicon carbide to a maximum value of 0.68 wt%. The further increase in its concentration to about 12 mg/l does not change its concentration in SiC, indicating that the solubility limit of boron corresponding to the temperatures used in our experiments appears to be reached. It should be noted that only one tenth of boron amount introduced into the gas phase was found to be present in the crystals .

(b) Doping with boron first markedly increases the resistivity of 3C-SiC. An increase of the boron concentration from 0.28 to 1.5 mg/l increases the resistivity from 40 to 380 Ω cm.

(c) All samples grown at boron concentrations lower than 1.5 mg/l retained electronic conductivity. Distinct inversion of the conductivity type from n to p type occurs at a boron concentration of 2.5 mg/l. The samples had in this case a resistivity of 2600 Ω cm. According to activation analysis, the maximum content of boron in beta SiC is about 4×10^{19} cm^{-3}.

(d) The introduction of excess silicon (in the form of silicon tetrachloride or trichlorosilane) into the gas phase strongly affects the process of doping beta SiC by both donor and acceptor impurities, decreasing the degree of doping by an order of magnitude or even greater.

Doping with boron substantially improves photoelectric properties of polycrystalline samples of 3C-SiC [28]. It increases the impurity conductivity in comparison with that of pure beta SiC and the exponent in the expression for the lux−ampere characteristics for both the intrinsic and extrinsic conductivities. The temperature dependence of the photocurrent becomes activative (the photoresponse increases with increasing temperature). The activation energy increases with increasing concentration of boron. The photoresponse times are much shorter in these samples (of about few seconds).

The photoelectric properties of polycrystalline cubic silicon carbide may be of interest for the production of photodetectors that can operate under extreme conditions such as high temperatures, high pumping levels, reactive media and severe irradiation.

8. Some Parameters and Applications of Pure 3C SiC Single Crystals [13, 29 to 31]

Single crystals of pure cubic silicon carbide prepared by thermal decomposition of methyl trichlorosilane in hydrogen have the following parameters. Color: yellow; size: up to $4 \times 5 \times 1.5$ mm^3; energy gap: 2.39 eV at 0 K; conductivity: n type; charge carrier concentration: 10^{14} to 10^{16} cm^{-3}; electron mobility: >1000 cm^2 V^{-1} s^{-1}; total content of impurities: 5×10^{-5} to 10^{-6} wt%; lattice parameter $a = 0.43602 \pm 0.00002$ nm.

On the basis of single crystals of pure and doped 3C-SiC that were grown at the Baikov Institute of Metallurgy (S.N. Gorin, L.M. Ivanova, A.A. Pletyushkin et al.), various semiconductor devices were fabricated primarily at the Kiev Polytechnic Institute (Yu.M. Altaiskii, V.S. Kiselev, L.S. Aivazova, V.P. Rodionov et al.) such as green light-emitting diodes of high brightness, high-temperature photodetectors, gamma detectors, high-temperature detectors of high-power radiation flows and violet emitters.

It is important to note that the good characteristics (high electron mobility, low concentrations of charge carriers in uncompensated material, etc.) were measured and many devices were prepared using samples that were cut from separate growth pyramids of proper faces rather than of a crystal as a whole. One example of cutting such samples is shown in Fig. 20. The thus-cut samples have sizes of about $1 \times 1 \times 0.2$ to 0.5 mm^3.

9. Conclusions

A thermodynamic analysis of the reaction of formation of SiC upon thermal decomposition of various methyl chlorosilanes indicates that above 1000 K the reactions are shifted completely to the formation of SiC. The use of dichlorodimethyl silane results in the formation of free carbon preventing the growth of single crystals. The study of the kinetics of the reactions shows that at 1523 to 1723 K the reaction is controlled by kinetics; above 1923 K, the reaction is controlled by diffusion.

The crystals formed in the diffusion region are faceted single crystals and may contain more or less complex twins. The study of their morphology shows that crystals of three main types are formed. Arbitrarily, they were called platelets, prisms, and skeletal crystals. All of them contain characteristic twin structures formed by intersecting microtwin lamellae parallel to the growth direction. This twin structure, along with the effects of crystallographic polarity of the sphalerite structure, unambiguously determines the growth behavior and structure of the crystals.

Platelet-type crystals are found to be the most perfect ones. They may be used for fabrication of semiconductor devices. The prismatic crystals are less perfect. They can be used for the preparation of devices such as thermal indicators for measuring maximal temperatures under extreme conditions. The skeletal crystals can be used as an ultra-pure starting material for growing SiC crystals e.g. by the Lely technique. The crystallographic polarity and the presence of natural facets lead to an inhomogeneous distribution of impurities in crystals (sectorial structure). These effects must carefully be taken into account when choosing crystal regions suitable for fabricating concrete semiconductor devices. In most cases, separate growth pyramids should be used rather than a whole crystal. The strong orientation dependence of impurity absorption in cubic silicon carbide crystals should also be borne in mind when growing thin films on substrates of various orientations.

On the basis of the platelet-type crystals of 3C-SiC grown at the Baikov Institute of Metallurgy, a number of devices were designed and fabricated at the Kiev Polytechnical Institute, such as light-emitting diodes of high brightness, high-temperature photodetectors, gamma-radiation detectors and violet emitters [29 to 31].

References

[1] A.A. Pletyushkin, S.N. Gorin, L.M. Ivanova, and N.G. Slavina, A Method for Producing Silicon Carbide Crystals, Inventor's Certificate No. 327779, Nov. 3, 1971 (with priority of invention from Nov. 6, 1959).

[2] Yu.M. Altaiskii, S.N. Gorin, L.M. Ivanova, N.P. Kalabukhov, and A.A. Pletyushkin, III Vsesoyuznaya Konferentsiya po Poluprovodnikovomu Karbidu Kremniya (III All-Union Conf. on Semiconductor Silicon Carbide), GIREDMET, Moscow 1970 (pp. 318 to 328).

[3] A.E. Van Arkel and J.H. De Boer, Z. anorg. Chem. **148**, 345 (1925).

[4] J.N. Pring and W.J. Fielding, J. Chem. Soc. **95**, 1497 (1909).

[5] K. Moers, Z. angew. allgem. Chem. **198**, 233 to 275 (1931).

[6] J.T. Kendall and D. Yeo, Proc. XI. Internat. Congr. on Pure and Applied Chemistry, Vol. 1, 171 (1947).
J.T. Kendall, J. Chem. Phys. **21**, 821 (1953).
J.T. Kendall, in: Silicon Carbide–A High-Temperature Semiconductor, Ed. J.R. O'Connor and J. Smiltens, Pergamon Press, New York 1960 (pp. 67 to 72).

[7] L.M. Ivanova, G.A. Kazaryan, and A.A. Pletyushkin, Izv. Akad. Nauk SSSR, Ser. Neorg. Mater. **2**, 223 (1966).

[8] L.M. Ivanova and A.A. Pletyushkin, Izv. Akad. Nauk SSSR, Ser. Neorg. Mater. **4**, 1089 (1968).

[9] L.M. Ivanova and A.A. Pletyushkin, Izv. Akad. Nauk SSSR, Ser. Neorg. Mater. **3**, 1817 (1967).

[10] G.M. Panchenkov and Yu.M. Zhorov, Neftekhimiya **1**, 172 (1961).

[11] S.N. Gorin and A.A. Pletyushkin, Dokl. Akad. Nauk SSSR **154**, 333 (1964).

[12] S.N. Gorin and A.A. Pletyushkin, Izv. Akad. Nauk SSSR, Ser. Fiz. **28**, 1310 (1964).

[13] S.N. Gorin and A.A. Pletyushkin, Rost Kristallov (Growth of Crystals), Vol. VI, Nauka, Moscow 1965 (pp. 210 to 219).

[14] S.N. Gorin and A.A. Pletyushkin, Protsessy Rosta Kristallov i Plenok Poluprovodnikov (Processes of Growth of Semiconductor Crystals and Films), Nauka, Sib. Otd., Novosibirsk 1970 (pp. 306 to 315).

[15] L.M. Ivanova and A.A. Pletyshkin, Karbid Kremniya: Stroenie, Svoistva i Oblasti Primeneniya (Silicon Carbide: Structure, Properties, and Application), Naukova Dumka, Kiev 1966 (pp. 151 to 158).

[16] G.G. Lemmlein, Sektorialnoe Stroenie Kristallov (Sectorial Structure of Crystals), Izd. Akad. Nauk SSSR, Moscow 1948

[17] F. Becke, Jahrb. Naturwiss. "Lotos" **14** (42), 1 (1894).

[18] C. Frondel, W.H. Newhouse, and R.F. Jarrel, Amer. Mineralogist **27**, 726 (1942).

[19] S.N. Gorin, M.D. Korsakova, Z.I. Palaguta, and A.A. Pletyushkin, Karbid Kremniya (Silicon Carbide), Naukova Dumka, Kiev 1966 (pp. 247 to 264).

[20] A.J. Steckl, J. Devrajan, S.N. Gorin, and L.M. Ivanova, in: Electronics Materials Conf. on Wide Band Gap Materials, Santa Barbara (CA), June 1996 (Abstracts).

[21] N.G. Slavina, A.A. Pletyushkin, S.N. Gorin, and L.M. Ivanova, A Method for Applying Silicon Carbide Coatings, Inventor's Certificate No. 145106, May 27, 1961

[22] V.K. Zakharenkov, A.A. Pletyushkin, L.M. Ivanova, V.T. Novikov, L.V. Miroshnichenko, and A.Kh. Kharbash, Izv. Akad. Nauk SSSR, Ser. Neorg. Mater. **12**, 1573 (1976).

[23] L.M. Ivanova, I.M. Kopev, and L.R. Protvina, Izv. Akad. Nauk SSSR, Ser. Neorg. Mater. **31**, 1204 (1995).

[24] A.V. Laptev and L.M. Ivanova, Izv. Akad. Nauk SSSR, Ser. Neorg. Mater. **30**, 1040 (1994).

[25] A.A. Pletyushkin, L.M. Ivanova, and M.S. Zhuravel, Elektron. Tekh.: Ser. Mater., No. 2, p. 48 (1970).

[26] A.A. Pletushkin, L.M. Ivanova, B.P. Zverev, and L.E. Krasivina, Svoistva Legirovannykh Poluprovodnikov (Properties of Doped Semiconductors), Nauka, Moscow 1977 (pp. 58 to 61).

[27] A.A. Pletyushkin, L.M. Ivanova, A.V. Laptev, and T.N. Sultanova, Legirovanie Poluprovodnikov (Doping Semiconductors), Nauka, Moscow 1982 (pp. 85 to 92).

[28] V.N. Rodionov, L.M. Ivanova, and A.A. Pletyushkin, Legirovanie Poluprovodnikovykh Materialov (Doping Semiconductor Materials), Nauka, Moscow 1985 (pp. 25 to 27).

[29] Vysokochistye Monokristally Kubicheskogo Karbida Kremniya dlya Mikroelektroniki (High-Purity Single Crystals of Cubic Silicon Carbide for Microelectronics), Inst. Metallurgii im. A.A. Baikova AN SSSR.

[30] Yu.M. Altaiskii, S.F. Avramenko, S.N. Gorin, V.S. Kiselev, E.N. Mineev, and V.P. Nikitina, Optiko-Mekhan. Promyshl., No. 4, p. 41 (1981).

[31] Yu.M. Altaiskii, S.F. Avramenko, V.N. Brukin, S.N. Gorin, V.S. Kiselev, and G.N. Polisskii, Tezisy Dokl. na VI. Vses. Konf. po Elektroluminestsentsii (VI All-Union Conf. on Electroluminescence, Abstracts), Dnepropetrovsk 1977 (p. 79).

[32] V.P. Glushko (Ed.),Termodinamicheskie Svoistva Individualnykh Veshchestv (Thermodynamic Properties of Individual Substances), Akad. Nauk SSSR, Moscow 1962, Vol. 1 and 2.

[33] N.W. Thibault, Amer. Mineralogist **29**, 249, 327 (1944).

[34] L.H. Borgstroem, Z. Krist. **62**, 1 (1925).

[35] R.H. Wentorf, Jr., in: Art and Science of Growing Crystals, Wiley, New York 1963.

[36] J.W. Faust and H.F. John, J. Phys. Chem. Solids **25**, 1407 (1964).

[37] G.G. Lemmlein, M.O. Kliya, and A.A. Chernov, Kristallografiya **9**, 231 (1964).

phys. stat. sol. (b) **202**, 247 (1997)

Subject classification: 68.55.Jk; 68.55.Ln; S6

Step-Controlled Epitaxial Growth of High-Quality SiC Layers

T. Kimoto, A. Itoh, and H. Matsunami

Department of Electronic Science and Engineering, Kyoto University,
Yoshidahonmachi, Sakyo, Kyoto 606-01, Japan

(Received January 31, 1997)

The growth mechanism in chemical vapor deposition (CVD) of silicon carbide (SiC) on off-oriented SiC{0001} substrates (step-controlled epitaxy) is reviewed. In step-controlled epitaxy, SiC growth is controlled by the diffusion of reactants in a stagnant layer. Critical growth conditions where the growth mode changes from step-flow to two-dimensional nucleation are predicted as a function of growth conditions using a model describing SiC growth on vicinal {0001} substrates. Step bunching on the surfaces of SiC epilayers is also investigated. Dominant step heights correspond to the half or full unit cell of SiC polytypes. The high quality of the SiC epilayers has been elucidated through Hall effect and deep level measurements. Excellent doping controllability in a wide range has been obtained by in-situ doping of a nitrogen donor and an aluminum acceptor.

1. Introduction

Silicon carbide (SiC) has received increasing attention as a wide bandgap semiconductor material for high-power, high-frequency, high-temperature and radiation-resistant devices. In particular, theoretical simulation has predicted that SiC power switching devices can replace the present-day Si power devices on account of much lower dissipation and reduced chip sizes [1].

Although chemical vapor deposition (CVD) has advantages in the precise control and uniformity of epilayer thickness and impurity doping, there had been a serious problem of polytype mixing in CVD growth of α-SiC [2 to 4]. In 1986 to 1987, the authors' group found that single crystalline 6H-SiC can be homoepitaxially grown on off-oriented 6H-SiC{0001} at low temperatures of 1400 to 1500 °C [5, 6]. Independently, Kong et al. [7] also succeeded in homoepitaxial growth of 6H-SiC using off-oriented substrates. This technique was named *"step-controlled epitaxy"*, since the polytype of epilayers can be controlled by surface steps existing on the off-oriented substrates. This technique was an epoch-making breakthrough in two senses that (i) growth temperature can be reduced more than 300 °C and (ii) epilayers have very high quality enough for device applications. Today, device-quality α-SiC epilayers have been produced by this technique through the world [8 to 13], supporting recent progress of SiC device fabrication [14].

In this paper, the mechanism of step-controlled epitaxial growth of SiC is reviewed. High-quality of SiC epilayers is presented through optical and electrical characterization.

2. Experiments

Fig. 1 shows a schematic diagram of the CVD growth system used at the authors'
group. Crystal growth was carried out by atmospheric pressure CVD in a horizontal
reaction tube. SiH_4 (1% in H_2) and C_3H_8 (1% in H_2) were used as source gases. The
carrier gas was H_2 purified with an Ag–Pd purifier. The flow rates of SiH_4 and C_3H_8
were 0.10 to 0.60 sccm (typically 0.30 sccm) and 0.10 to 0.80 sccm (typically 0.20 sccm),
respectively. The H_2 flow rate was fixed at 3.0 slm, which provides a linear gas velocity
of 6 to 10 cm/s above the substrates. N_2 was used for n-type doping, and trimethylaluminum (TMA : $Al(CH_3)_3$) and B_2H_6 for p-type doping. Hydrogen chloride (HCl) gas was
used for etching of a substrate surface before CVD growth.

Two kinds of α-SiC crystals were used as substrates, crystals grown by the Acheson method or a modified Lely (sublimation) method. Since the basal plane of the Acheson crystals is
{0001} face, the off-oriented substrates were prepared by angle-lapping of the basal plane.
As for crystals grown by a modified Lely method, both commercially available and home-
made wafers were used. The substrate off-angle was 0 to 10° (typically 5 to 6°) toward
⟨11$\bar{2}$0⟩. SiC{0001} is a polar face, being either (0001) Si or (000$\bar{1}$) C. Both Si and C faces
were used to investigate the substrate polarity effects. The polarity was identified by thermal oxidation at 1000 °C for 5 h utilizing the difference in oxidation rates between both
faces (oxidation is faster on (000$\bar{1}$) C faces) [15, 16]. The polytypes of substrates were identified by the absorption edges in ultraviolet-visible light transmission spectra and photoluminescence, and were confirmed by X-ray diffraction and Raman scattering.

Substrates were set on a SiC-coated graphite susceptor, and heated by radio frequency (rf) induction. Before the CVD growth, in-situ HCl etching was performed to
remove surface damage introduced by the polishing process. The growth temperature
was varied in the range of 1100 to 1600 °C (typically 1500 °C).

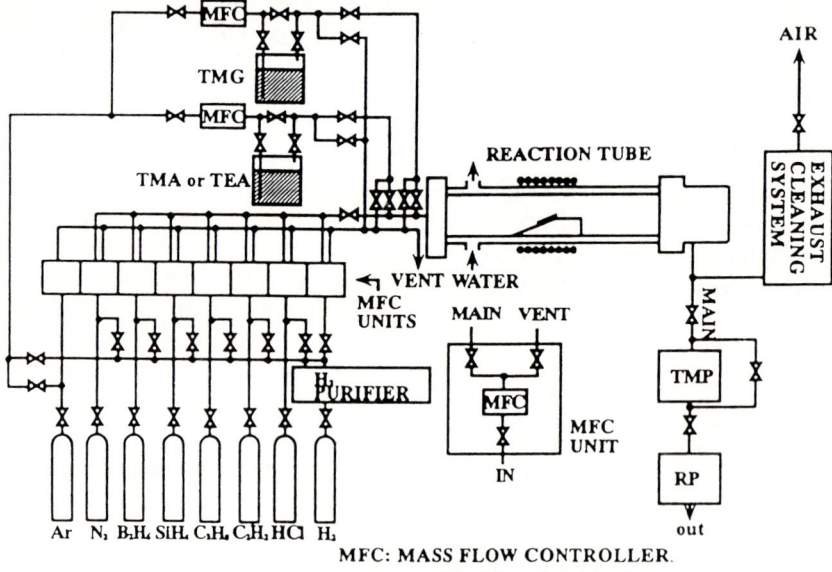

MFC: MASS FLOW CONTROLLER.
TMP: TURBO MOLECULER PUMP
RP : ROTARY PUMP

Fig. 1. Schematic diagram of the CVD system used in the authors' group

3. Epitaxial Growth

3.1 Growth mode

Fig. 2 shows Nomarski microphotographs of 5 μm thick SiC layers grown on a) well-oriented and b) 6° off-oriented 6H-SiC(0001) substrates at 1500 °C under a typical gas flow condition (SiH$_4$: 0.30 sccm, C$_3$H$_8$: 0.20 sccm). On a well-oriented (0001) face, the epilayer shows a mosaic pattern, and smooth domains are separated by step- or groove-like boundaries. From the reflection high-energy electron diffraction (RHEED) analysis, the grown layer was identified as 3C-SiC(111) with double positioning twinning [17]. On a well-oriented (000$\bar{1}$) C face, the grown surface is rough, and island-like growth is observed (not shown). The RHEED analysis revealed that the grown layer is also twinned crystalline 3C-SiC.

In contrast, epilayers on off-oriented substrates exhibit specular, smooth surfaces. The grown layer was identified as 6H-SiC(0001) by RHEED and transmission electron microscope (TEM) observation. Homoepitaxial growth with excellent surface morphology can also be achieved on an off-oriented (000$\bar{1}$) C face. Surface morphology depends on growth rate and especially temperature, of which details are discussed later.

On well-oriented {0001} faces, the step density is very low and vast terraces exist. Then, crystal growth may initially occur on terraces through two-dimensional nucleation due to the high supersaturation on the surface. The polytype of grown layers is determined by growth conditions, mainly growth temperature. This leads to the growth of 3C-SiC, which is stable at low temperatures. This phenomenon has been predicted by theoretical studies using a quantum-mechanical energy calculation [18] and an electrostatic model [19]. As the stacking order of 6H-SiC is ABCACB ... in the ABC notation [20], the growing 3C-SiC can take two possible stacking orders of ABCABC ... and ACBACB ..., as shown in Fig. 3a, leading to double positioning twins. On off-oriented substrates, the step density is high, and the terrace width is narrow enough for adsorbed species to reach steps. At a step, the incorporation site is uniquely determined by bonds from the step, as shown in Fig. 3b. Hence,

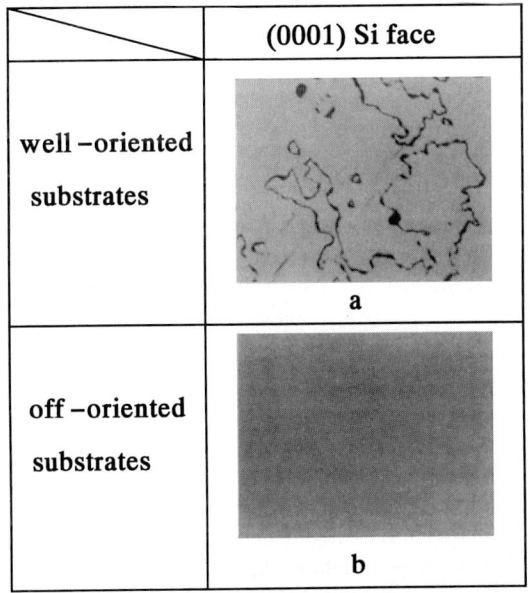

a

b

Fig. 2. Surface morphology of SiC layers grown on a) well-oriented and b) 6° off-oriented 6H-SiC(0001) Si substrates. Growth temperature and growth rate are 1500 °C and about 2 μm/h, respectively

—— 100 μm

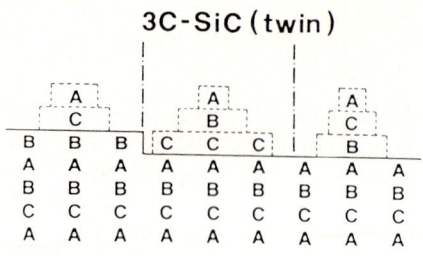

3C-SiC (twin)

a

Fig. 3. Schematic images of the relationship between growth modes and polytypes of layers grown on 6H-SiC{0001}. a) 3C-SiC is grown through two-dimensional nucleation, and b) homoepitaxy of 6H-SiC is achieved owing to step-flow growth

6H-SiC (single)

b

homoepitaxy can be achieved through the lateral growth from steps (step-flow growth), inheriting the stacking order of substrates, i.e. surface steps serve as a template which forces the replication of the substrate polytype in the epilayer. This growth technique is applicable to homoepitaxy of any other polytypes such as 4H-SiC [21], 15R-SiC, and 21R-SiC [22]. Tairov et al. [23] investigated the effects of substrate off-angle in epitaxial growth of SiC at 1600 to 2200 °C by a sandwich growth method. They observed the stable homoepitaxy without 3C-SiC inclusions on off-oriented substrates.

The cause of 3C-SiC nucleation on off-oriented substrates has been investigated by Powell et al. [24]. They have investigated the effcts of off-angle and surface treatment on the polytypes of grown layers, and found that homoepitaxy of 6H-SiC is possible even on substrates with a low-tilt angle of 0.2° and 3C-SiC nucleation takes place at defect sites on the surface. The defects can be screw dislocations or surface damages introduced by polishing. More recently, Hallin et al. [25] reported that substrate imperfection and surface defects induce the development of large (0001) facets on which 3C-SiC inclusions are created via spontaneous nucleation, which is much more pronounced in 4H-SiC growth than 6H-SiC. All these facts indicate that it is essential to keep supersaturation low enough to ensure "perfect step-flow growth", thus enabling homoepitaxy of α-SiC. The improvement of 6H-SiC epilayer quality by utilizing off-oriented substrates has also been reported in LPE [26]. Thus, the use of off-oriented substrates may be a key technique in various SiC growth methods.

The off-direction dependence in CVD growth of 6H-SiC has been reported by Kong et al. [7] and Ueda et al. [27]. On a 6H-SiC(0001) Si face inclined toward $\langle 1\bar{1}00 \rangle$, stripe-like morphology, which is caused by pronounced step bunching, appeared and the inclusion of 3C-SiC domains was observed by long-time growth. Based on these results, the off-direction toward $\langle 11\bar{2}0 \rangle$ has mainly been employed in almost all the groups.

3.2 Growth mechanism

At a C/Si ratio (the ratio between the number of C and Si atoms in supplied gases) greater than 1.4, where good morphology without Si droplets can be obtained in the present growth system, the growth rate increases proportionally with the flow rate of SiH_4, indicating that Si species limit SiC growth. Karmann et al. [28] also investigated the growth on off-oriented 6H-SiC(0001) faces using SiH_4 and C_3H_8 at 1500 to 1600 °C, and observed that the supply of SiH_4 controlled the rate of SiC growth.

Allendorf and Kee [29] have analyzed gas-phase and surface reactions at 1200 to 1600 °C in a SiH_4–C_3H_8–H_2 system. Stinespring and Wohmhoudt [30] also reported a similar analysis on gas-phase kinetics. Their analyses have shown that the dominant species which contribute to SiC growth may be Si, SiH_2, Si_2H_2 species from SiH_4, and CH_4, C_2H_2, C_2H_4 molecules from C_3H_8. This simulation suggests that Si (or SiH_2) may be preferentially adsorbed and migrate on the surface. In fact, almost no deposition occurs in the present CVD system without SiH_4 supply. Recently, the effective C/Si ratio in vapor in the vicinity of a substrate surface has been estimated to be about 100 or higher at an "input" C/Si ratio of 2 [31]. This fact means that growth environment is extremely C-rich, making the supply of Si species be a major limiting factor of growth rate.

The authors systematically investigated the effects of C/Si ratio on surface morphology. At 1500 °C, epilayers on Si faces showed specular, smooth surfaces for the C/Si ratios between 2 and 6. On C faces, the optimum C/Si ratio window is relatively narrow, from 2 to 3. This tendency is independent of substrate polytypes, 6H or 4H. However, epitaxial growth 4H-SiC on Si faces is very sensitive to polishing damage and/or substrate defects. Triangular pits and macrosteps are easily formed at the defect sites. A recent work revealed that 3C-SiC nucleation takes place on most triangular pits (triangular stacking faults) [25]. Though the appearance of these surface defects can be suppressed by reducing surface damage and improving CVD procedures [32], it is noteworthy that these defects appear only on Si faces. Epilayer surfaces on C faces are very flat even for 4H-SiC as far as the C/Si ratio is in the range from 2 to 3.

Fig. 4 shows the growth rates for various off-angles of the substrate at 1500 °C [33]. The flow rates of SiH_4 and C_3H_8 are 0.30 and 0.20 sccm, respectively. In this figure, open (on a Si face) and closed (on a C face) triangles mean the growth of 3C-SiC, and open and closed circles mean that of 6H-SiC. 6H-SiC can be homoepitaxially grown on off-oriented substrates with more than 1°. On well-oriented substrates (off-

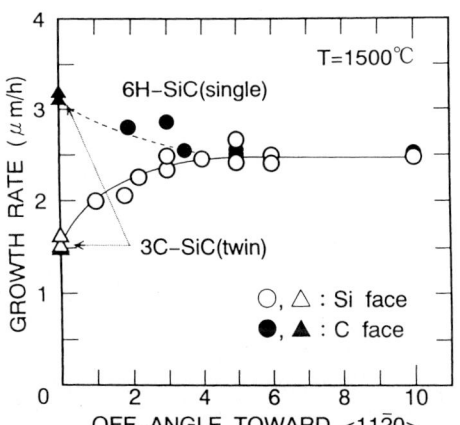

Fig. 4. Dependence of growth rate at 1500 °C on off-angle of the substrate. The flow rates of SiH_4 and C_3H_8 are 0.30 and 0.20 sccm, respectively. Open and closed triangles show growth of 3C-SiC, and circles denote that of 6H-SiC

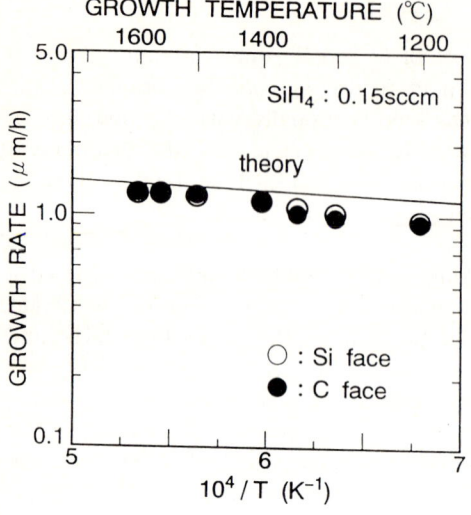

Fig. 5. Temperature dependence of growth rate on 6° off-oriented 6H-SiC(0001) Si and (000$\bar{1}$) C faces. The flow rates of SiH$_4$ and C$_3$H$_8$ are 0.15 and 0.10 to 0.14 sccm, respectively. The calculated result based on a stagnant layer model is shown by a solid curve

angle = 0°) higher growth rates are obtained on C faces, as has been reported [34]. This might be ascribed to the higher nucleation rate on (000$\bar{1}$) C terraces [35]. With increasing off-angle, the growth rates on both faces approach each other and become almost the same value of 2.5 µm/h for off-angles from 4° to 10°.

Fig. 5 shows the temperature dependence of the growth rate in the range between 1200 and 1600 °C, in which homoepitaxial growth of 6H-SiC occurs [33, 36]. The flow rates of SiH$_4$ and C$_3$H$_8$ are 0.15 and 0.10 to 0.14 sccm, respectively, and the off-angle is 5 to 6°. The temperature dependence of the growth rate yields a very small activation energy of 12 kJ/mol. For CVD growth on well-oriented {0001} faces, activation energies of 50 [34], 84 [37] or 92 kJ/mol [2] were reported. There is little difference between the growth rates on Si and C faces even at low temperatures. Karmann et al. [28] also reported similar insensitivity of growth rate to temperature in CVD on 6H-SiC substrates with a 2° off-angle toward $\langle 1\bar{1}00 \rangle$.

The solid curve in Fig. 5 denotes the theoretical growth rate calculated with a stagnant layer model, which has been developed in Si CVD [38]. The absolute values of growth rate calculated from the simple model show surprisingly good agreement with

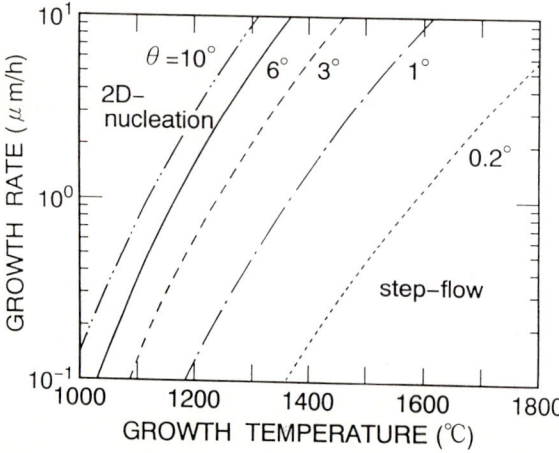

Fig. 6. Critical growth conditions as a function of growth temperature, growth rate, and off-angles of the substrate ($\theta = 0.2°$, 1°, 3°, 6°, and 10°). The top-left and bottom-right regions from the curves correspond to two-dimensional nucleation and step-flow growth conditions, respectively

experimental values. The predicted growth rate increases gradually with temperature increase, owing to the enhancement of diffusion in a stagnant layer. The calculated curve in the range of 1200 to 1600 °C yields an apparent activation energy of 10 kJ/mol using the method of least squares, in very good agreement with the experimental result. Therefore, the growth would be limited by mass transport in step-controlled epitaxy. This fact gives good explanation for little difference in the growth rates on Si and C faces, because no polarity dependence should be observed in the growth controlled by mass transport.

The "step-flow growth window", where homoepitaxy of SiC is realized through step-flow is of great interest. For example, the authors found that 6H-SiC can be homoepitaxially grown on a 6° off-oriented substrate at 1200 °C, but not at 1100 °C [33]. Using a simple surface diffusion model based on the BCF (Burton, Cabrera, and Frank) theory [39] and some experimental data, we determined the critical growth conditions where growth mode changes from step-flow (6H- or 4H-SiC growth) to two-dimensional nucleation (3C-SiC growth) [40]. Critical growth conditions are shown by the curves in Fig. 6 for substrate off-angles of 0.2°, 1°, 3°, 6°, and 10°. In the figure, the top-left and bottom-right regions separated by the curves correspond to the two-dimensional nucleation and step-flow growth conditions, respectively. Almost no difference in the critical conditions was obtained on Si and C faces. The higher growth rate and lower off-angle are available for step-flow growth at higher growth temperatures. At 1800 °C, a very small off-angle of 0.2°, which yields almost "well-oriented" faces, is enough to achieve step-flow growth with a moderate growth rate of 6 μm/h. This may be one of the reasons why 6H-SiC can be homoepitaxially grown on well-oriented faces if the growth temperature is raised up to 1700 to 1800 °C [2 to 4]. The role of defects for 3C-SiC nucleation may become important on substrates with small off-angles [24]. On the contrary, large off-angles more than 5° are needed to realize homoepitaxy of 6H-SiC at a low temperature of 1200 °C with a growth rate of 1 μm/h.

3.3 Step bunching

In SiC growth, quite a few studies have been reported about step structure on high-quality CVD-grown α-SiC surfaces [41 to 44]. In the present study, as-grown samples without any surface treatments were examined using atomic force microscopy (AFM) and TEM observations. Fig. 7a, b show the height profiles of 6H-SiC epilayers grown on (0001) Si

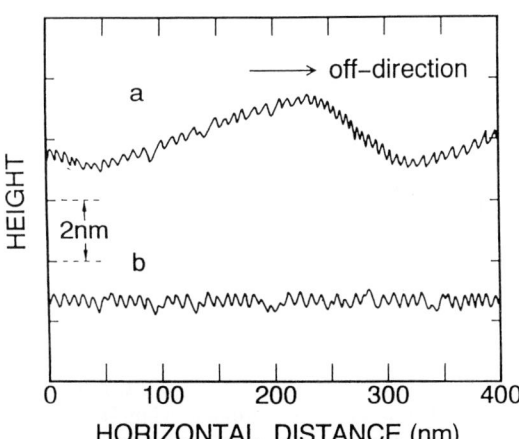

Fig. 7. Height profiles of 6H-SiC epilayers grown on 5° off-oriented (a) 6H-SiC(0001) Si and (b) (000$\bar{1}$) C faces. The surface steps go down from the left to the right

and $(000\bar{1})$ C faces, respectively, obtained from AFM data [42]. The off-angle of the substrates is 5°, and the steps go down from the left to the right. A distinctive difference in the surface structure between both the faces can be observed. Epitaxial growth on a (0001) Si face yields "apparent macrosteps" with a terrace width of 220 to 280 nm and a step height of 3 to 6 nm. Each macrostep is not a single multiple-height step but composed of a number of "microsteps" with different terrace widths as well as different step heights. Powell et al. [41] also reported macrostep formation on 6H-SiC(0001) Si faces [41]. On a $(000\bar{1})$ C face, the surface is rather flat and no macrosteps are observed. Although 4H-SiC epilayers had basically similar step structures, the 4H-SiC(0001) Si faces exhibited real macrosteps with 110 to 160 nm width and 10 to 15 nm height in some regions.

The mechanism of "apparent macrostep" formation on 6H- and 4H-SiC(0001) Si faces is not clear at present. However, the surface is quite similar to the so-called "hill-and-

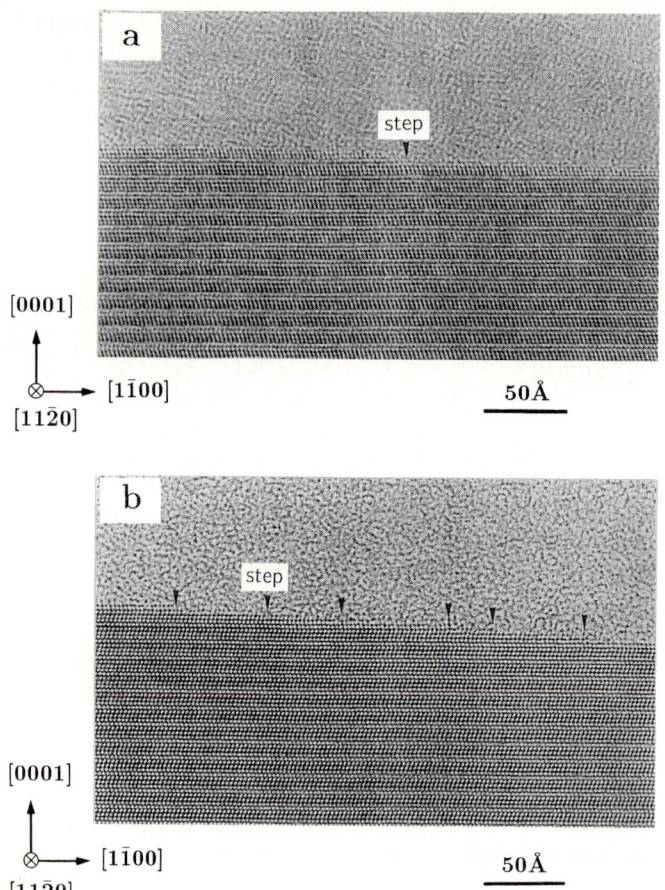

Fig. 8. Typical cross-sectional TEM images for 4H-SiC surfaces grown on a) Si and b) C faces. Substrates are (0001) Si 3.5° off-oriented toward $\langle 11\bar{2}0 \rangle$. The samples are examined along the $\langle \bar{1}2\bar{1}0 \rangle$ zone axis

valley (or faceted)" structure, which often appears on grown surfaces with off-orientation from a low-index plane [45, 46]. The off-oriented surfaces will spontaneously rearrange to minimize their total surface energies, even if this involves an increase in surface area. The surface free energies of SiC were calculated to be 1767×10^{-7} J/cm^2 for the Si face and 718×10^{-7} J/cm^2 for the C face [47]. Thus, the surface energy may be reduced by the formation of hill-and-valley structure on off-oriented (0001) Si, which has much higher surface energy.

Fig. 8a, b show typical cross-sectional TEM images for 4H-SiC surfaces grown on Si and C faces, respectively, with 3.5° off-angle. The epilayers were produced at a C/Si ratio of 2, and have a thickness of 10 μm. The samples were examined along the $\langle 11\bar{2}0 \rangle$ zone axis to obtain clear lattice images. No island growth on the {0001} terraces are observed, indicating step-flow growth. On a Si face (Fig. 8a), the number of Si–C bilayer at bunched steps is four. It should be noted that the bunched steps correspond to exactly the unit cell of 4H-SiC: ABCB steps in the ABC notation. On a C face (Fig. 8b), however, single bilayer-height steps dominate, and bunched steps are relatively few.

The authors examined more than 200 steps from at least two samples for each condition, and made histograms of step height and terrace width. Fig. 9 and 10 show the histograms of step height for the surfaces of 6H-SiC and 4H-SiC epilayers, respectively. On a 6H-SiC Si face, 88% of steps are composed of three Si–C bilayers (half unit cell), and 7% of steps have six Si–C bilayer height (unit cell). In contrast, single Si–C bilayer-height steps are dominant on C faces, showing a probability of 68%. On the other hand, four-bilayer-height (unit cell) steps are the most dominant (66%) and two-bilayer-height steps show the second highest probability (19%) on a 4H-SiC Si face. On a 4H-SiC C face, however, most (80%) steps have single Si–C bilayer height. It is also noteworthy that even on C faces, small amount of bunched steps have, again, three- or six-bilayer-height in 6H-SiC, and two- or four-bilayer-height in 4H-SiC. The origin of this striking polarity dependence is not known. The migrating species, surface coverage, and exact bond configuration at step edges should be analyzed to reveal the mechanism.

As shown in Figs. 9 and 10, the formation of half-unit-cell or unit-cell height steps seems to be inherent in α-SiC growth. Similar observation has been reported on 6H-SiC

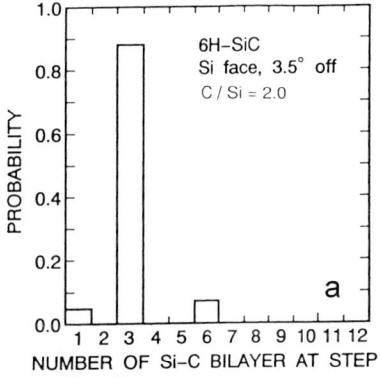
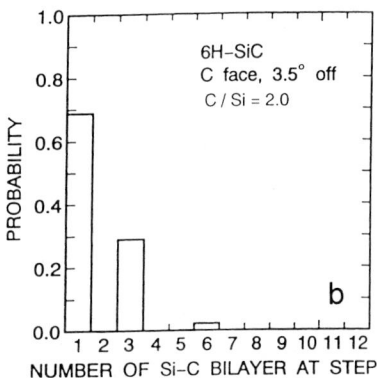

Fig. 9. Histograms of step height for surfaces of 6H-SiC epilayers grown on a) Si and b) C faces. The substrate off-angle is 3.5° toward $\langle 11\bar{2}0 \rangle$. The epilayers were produced with a C/Si ratio of 2

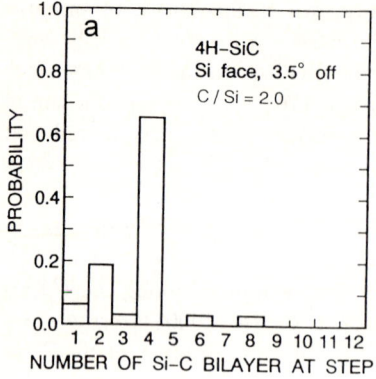
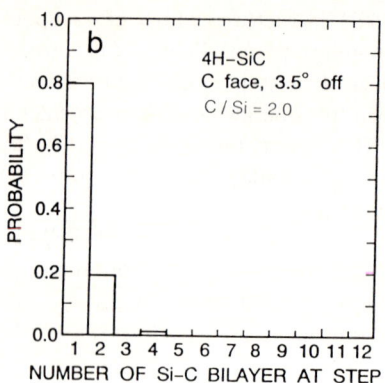

Fig. 10. Histograms of step height for surfaces of 4H-SiC epilayers grown on a) Si and b) C faces. The substrate off-angle is 3.5° toward ⟨11$\bar{2}$0⟩. The epilayers were produced with a C/Si ratio of 2

surfaces grown by the Lely method [48] and MBE [49]. Thus, the origin of step bunching in SiC may be correlated with the surface equilibrium process. Heine et al. [18] suggested that surface energies are different for each SiC bilayer plane owing to the peculiar stacking sequence. Different surface energy may lead to different step velocity among different Si–C bilayers, and thereby causes "structurally-induced macrostep formation", of which details will be discussed elsewhere [50].

On a Si face with 3.5° off-angle, the average terrace widths experimentally obtained were 12.4 nm for 6H-SiC and 16.8 nm for 4H-SiC. The different average terrace width between 6H-SiC and 4H-SiC, in spite of the identical off-angle, originates from the different height of multiple steps as described above. From a viewpoint of epitaxial growth, narrow terraces are preferable to achieve step-flow growth. This is crucial in SiC growth, because supersaturation increases on larger terraces, leading to 3C-SiC nucleation. In this sense, 4H-SiC, which shows larger terrace widths, may have the disadvantage of relatively higher probability for nucleation on terraces. To overcome this problem, a slightly higher growth temperature would be helpful, since the longer surface diffusion length of adsorbed species and lower supersaturation on terraces are expected at higher temperatures. Larger off-angles of substrates might be also effective in 4H-SiC growth [51]. On the other hand, C faces showed much smaller average terrace widths (4 to 5 nm), owing to fewer bunched steps. This might be one reason why epilayers grown on C faces exhibit a very flat surface even for 4H-SiC.

4. Impurity Doping and Characterization

4.1 Characterization of unintentionally doped n-type epilayers

For SiC epilayers grown under optimum condition, very smooth surfaces can be obtained on both Si and C faces (especially on a C face), and almost all the surface pits originate from so-called "micropipes" in the substrates, except for 4H-SiC epilayers on a Si face, on which a small density of triangular pits is still existing even on 8° off-oriented substrates. Although it is difficult to detect small micropipes on as-polished surfaces (before growth), triangular pits are formed at the micropipe positions after epitaxial growth. These pits are accompanied with "shadows" due to the impedance of step-advance.

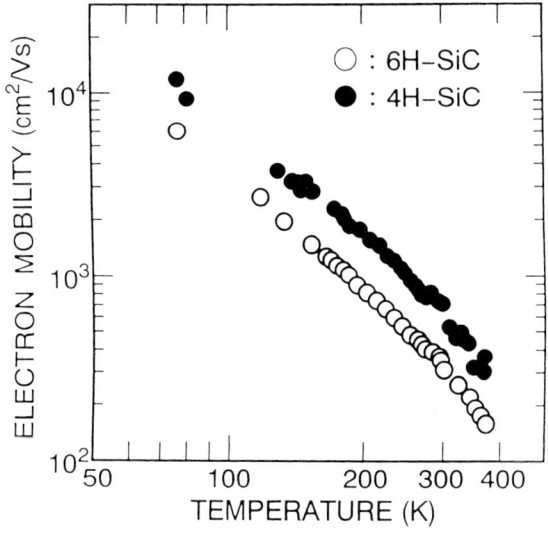

Fig. 11. Temperature dependence of electron mobility of 6H- and 4H-SiC epilayers with a net donor concentration of 4×10^{16} cm^{-3}

In photoluminescence (PL) measurements at 2 to 15 K, PL spectra are governed by the lines due to the recombination of an exciton bound to neutral nitrogen and free excitons. The N donor–Al acceptor pair band is very weak, whereas the donor–acceptor pair luminescence is dominant in substrates. PL spectra indicated very little contamination of Al acceptor, which would normally show the Al bound exciton peaks [52].

Fig. 11 shows the temperature dependence of electron mobility of 6H- and 4H-SiC with a net donor concentration of 4×10^{16} cm^{-3}. The mobility is 351 cm^2/Vs for 6H-SiC and 724 cm^2/Vs for 4H-SiC at room temperature [21]. The mobility increases with lowering temperature, and reaches up to 6050 cm^2/Vs for 6H-SiC and 11000 cm^2/Vs for 4H-SiC at 77 K. The increasing mobility at low temperature reflects the low impurity compensation in the epilayers.

Isothermal capacitance transient spectroscopy (ICTS) measurements on Schottky structures of n-type 6H-SiC epilayers have shown very small concentrations (below the detection limit) of deep traps [53]. Recently, the authors' group succeeded to reveal deep electron traps in both 6H- and 4H-SiC epilayers by high-resolution deep level transient spectroscopy (DLTS). The samples used in this study were 8 to 10 µm thick epilayers with a net donor concentration of 2×10^{15} to 1×10^{16} cm^{-3}. Fig. 12a, b show the DLTS

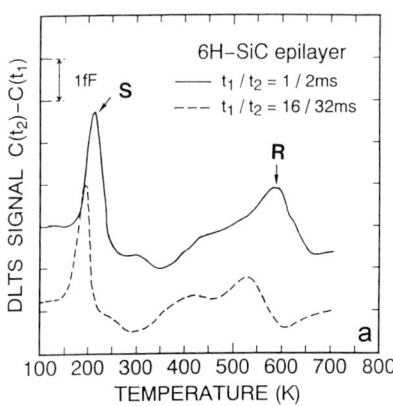

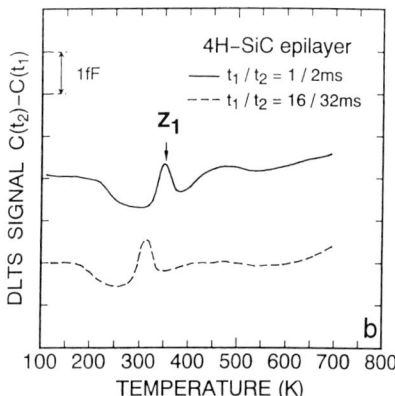

Fig. 12. DLTS spectra obtained from as-grown a) 6H-SiC and b) 4H-SiC epilayers with a net donor concentration of 2×10^{15} cm^{-3}

spectra obtained from as-grown 6H-SiC and 4H-SiC epilayers, respectively. The spectra of 6H-SiC exhibit two peaks both at concentrations of 4×10^{12} cm^{-3}. An Arrhenius plot evaluation of these two peaks revealed the activation energies of 0.39 to 0.43 and 1.17 to 1.27 eV with respect to the conduction band, respectively. These peaks are termed S and R, based on a previous report [55]. On the other hand, only one peak is observed in 4H-SiC epilayers. The trap concentration and activation energy are estimated to be 4×10^{12} cm^{-3} and 0.63 to 0.68 eV, respectively. This trap can be attributed to the Z_1-center, an acceptor-like complex containing intrinsic defects [56]. Thus, both 6H- and 4H-SiC epilayers have low concentration of deep levels, indicating high quality sufficient for device applications.

4.2 In-situ doping of impurities

In-situ n-type doping can easily be achieved by the introduction of N_2 during epitaxial growth. The donor concentration estimated from capacitance–voltage (C–V) characteristics was proportional to the N_2 flow rate in the wide range on both Si and C faces, in agreement with the results by Wang et al. [57] and Karmann et al. [9], though these previous studies employed only 6H-SiC(0001) Si substrates.

Recently, Larkin et al. [58] have found that the doping efficiency of impurities strongly depends on the C/Si ratio (or Si/C ratio) during CVD growth (site-competition epitaxy). The growth under a higher C/Si ratio leads to the lower N concentration in the epilayers. This phenomenon can be explained by the fact that the higher C atom coverage on a growing surface prevents the incorporation of N atoms, which substitute at the C site, into crystals. Fig. 13 shows our result on the C/Si ratio dependence of background doping level of unintentionally doped 4H-SiC epilayers. In the case of C/Si ratio of 2, no significant difference was observed between epilayers on Si and C faces. On a Si face, the donor concentration can drastically be reduced by increasing the C/Si ratio. The lowest value in our system is in the range of 5×10^{13} to 1×10^{14} cm^{-3}. On a C face, however, the donor concentration is not sensitive to the C/Si ratio. We observed a similar C/Si ratio dependence in intentional N doping [59]. On the other hand, the incorporation of aluminum (Al) and boron (B) atoms, which substitute at the Si site, is enhanced under C-rich conditions on a Si face.

Fig. 14 shows the electron mobility at room temperature in the basal plane versus carrier concentration of

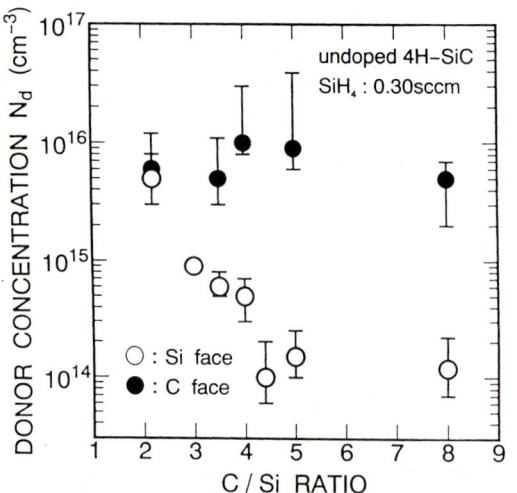

Fig. 13. C/Si ratio dependence of donor concentration for unintentionally doped 4H-SiC epilayers

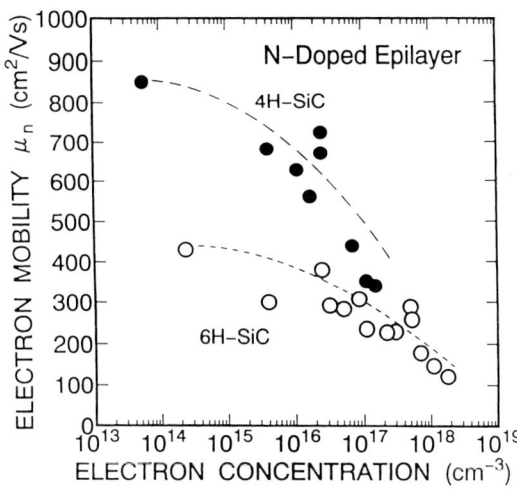

Fig. 14. Electron mobility vs. carrier concentration of n-type 6H- and 4H-SiC epilayers at room temperature

6H- and 4H-SiC epilayers. As is well-known, 4H-SiC exhibits a two times higher electron mobility than 6H-SiC. However, the difference seems to be small for heavily doped layers, in agreement with a previous report [60]. As shown in Fig. 13, very low-doped epilayers can be produced on Si faces by growing under C-rich conditions. For very low-doped epilayers, which were produced with a C/Si ratio of 4 to 5, high electron mobilities of 431 cm^2/Vs ($n = 2 \times 10^{14}$ cm^{-3}) for 6H-SiC and 851 cm^2/Vs ($n = 6 \times 10^{13}$ cm^{-3}) for 4H-SiC were obtained at room temperature. For device applications, 4H-SiC is much more attractive owing to its higher electron mobility and smaller anisotropy [60, 61].

The addition of a small amount of TMA is effective for in-situ p-type doping. Although most Al-doped epilayers showed very smooth surfaces, surface pits and hillocks were observed in heavily doped (Al concentration $>3 \times 10^{18}$ cm^{-3}) samples grown on C faces. The supply of TMA causes the shift of growth conditions toward C-rich ambience due to the release of CH$_3$ species from TMA molecules. The surface migration is suppressed and the nucleation is promoted under C-rich growth conditions [62]. Besides, C faces easily suffer from two-dimensional nucleation, due to its low critical supersaturation ratio [35, 40]. This may be the reason for the surface roughening of heavily Al-doped epilayers grown on C faces. The Al acceptor concentration versus TMA flow rate is shown in Fig. 15. The flow rates of SiH$_4$ and C$_3$H$_8$ are 0.30 and 0.20 sccm (C/Si ratio = 2.0), respectively. The acceptor concentration estimated from C–V measurements agreed well with the Al concentration determined by secondary ion mass spectro-

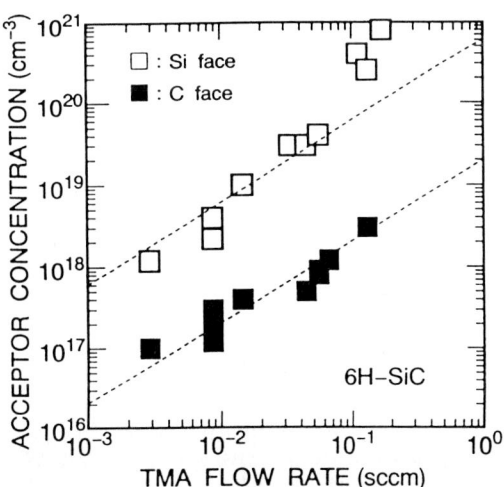

Fig. 15. Al acceptor concentration vs. TMA flow rate in epitaxial growth of 6H-SiC. The growth was performed at 1500 °C with a C/Si ratio of 2

scopy (SIMS) measurements. The doping efficiency is much higher on Si faces than on C faces by a factor of 10 to 80. On a Si face, the acceptor concentration increases super-linearly with the TMA supply. This superlinearity may be caused by the increased effective C/Si ratio under high TMA flow conditions, enhancing the Al incorporation, mentioned above. It should be noted that heavily doped p-type layers can be grown only on a Si face.

Because of the high ionization energy Al acceptors (242 meV in 6H-SiC) [63], the activation ratio p/N_a (p hole concentration, N_a acceptor concentration) was as low as 0.01 to 0.1 at room temperature. However, a very high hole concentration of 4 to 6×10^{19} cm^{-3} could be achieved for heavily doped epilayers (Al concentration is in the mid 10^{20} cm^{-3} range). This result might arise from the decreased ionization energy in heavily doped samples, or from the formation of impurity band caused by the impurity–impurity interaction [64]. The lowest p-type resistivity was 0.042 Ω cm for 6H-SiC and 0.025 Ω cm for 4H-SiC, which were obtained on Si faces. These p$^+$ epilayers can be successfully used for contact layers to reduce contact resistances. Thus, each surface (Si, C face) possesses its inherent properties, and the substrate polarity should be selected, depending on the device structure, to achieve the full potential of SiC [65].

5. Summary

Step-controlled epitaxial growth of SiC on off-oriented SiC{0001} substrates was reviewed. Step-flow growth is essential to realize polytype replication in epilayers without 3C-SiC inclusions through two-dimensional nucleation on terraces. The introduction of a substrate off-angle induces the change of rate-determining step from surface-reaction control to diffusion control. Critical growth conditions where growth mode changes from step-flow to two-dimensional nucleation were predicted as a function of growth temperature, growth rate, and substrate off-angle, by using a model describing SiC growth on vicinal SiC{0001}. Step structures of epilayer surfaces depended on the substrate polarity as well as polytypes. Dominant step heights corresponded to the half or full unit cell of SiC polytypes.

The background doping level of epilayers could be reduced to less than 1×10^{14} cm^{-3} by the growth under C-rich conditions, by which very high electron mobilities of 431 cm^2/Vs for 6H-SiC and 851 cm^2/Vs for 4H-SiC were obtained. Deep level analyses revealed that the trap concentration was in the 10^{12} cm^{-3} range, indicating very high quality of the epilayers. Excellent doping control has been obtained by in-situ doping of a nitrogen donor and an aluminum acceptor.

Acknowledgements The authors wish to express their gratitude to Prof. W. J. Choyke of University of Pittsburgh, Dr. G. Pensl and Mr. T. Dalibor of University of Erlangen-Nürnberg for collaborative works on characterization of epilayers. They also would like to thank Prof. T. Ohachi of Doshisha University for the use of RHEED equipment, and Mr. T. Nakata of Ion Engineering Research Institute and Mr. T. Okano of Matsushita Technoresearch, Inc. for TEM analyses.

References

[1] M. BHATNAGAR and B. J. BALIGA, IEEE Trans. Electron Devices **40**, 645 (1993).
[2] V. J. JENNINGS, A. SOMMER, and H. CHANG, J. Electrochem. Soc. **113**, 728 (1966).
[3] W. VON MUENCH and I. PHAFFENEDER, Thin Solid Films **31**, 39 (1976).

[4] S. Yoshida, E. Sakuma, H. Okumura, S. Misawa, and K. Endo, J. Appl. Phys. **62**, 303 (1987).
[5] N. Kuroda, K. Shibahara, W. S. Yoo, S. Nishino, and H. Matsunami, Extended Abstracts 34th Spring Meeting of the Japan Society of Applied Physics and Related Societies, Tokyo, 1987 (p. 135) (in Japanese).
[6] N. Kuroda, K. Shibahara, W. S. Yoo, S. Nishino, and H. Matsunami, Extended Abstracts 19th Conf. Solid State Devices and Materials, Tokyo, 1987 (p. 227).
[7] H. S. Kong, J. T. Glass, and R. F. Davis, J. Appl. Phys. **64**, 2672 (1988).
[8] J. A. Powell, D. J. Larkin, L. G. Matus, W. J. Choyke, J. L. Bradshaw, L. Henderson, M. Yoganathan, J. Yang, and P. Pirouz, Appl. Phys. Lett. **56**, 1442 (1990).
[9] S. Karmann, W. Suttrop, A. Schöner, M. Schadt, C. Haberstroh, F. Engelbrecht, R. Helbig, and G. Pensl, J. Appl. Phys. **72**, 5437 (1992).
[10] O. Kordina, A. Henry, C. Hallin, R. C. Glass, A. O. Konstantinov, C. Hemmingsson, N. T. Son, and E. Janzén, Mater. Res. Soc. Symp. Proc. **339**, 405 (1994).
[11] A. A. Burk, Jr., D. L. Barrett, H. M. Hobgood, R. R. Siergiej, T. T. Braggins, R. C. Clarke, G. W. Eldridge, C. D. Brandt, D. J. Larkin, J. A. Powell, and W. J. Choyke, Silicon Carbide and Related Materials, Eds. M. G. Spencer, R. P. Devaty, J. A. Edmond, M. A. Khan, R. Kaplan, and M. M. Rahman, Institute of Physics, Bristol, 1994 (p. 29).
[12] R. Rupp, P. Lanig, J. Volkel, and D. Stephani, J. Cryst. Growth **146**, 37 (1995).
[13] N. Nordell, S. G. Andersson, and A. Schöner, Silicon Carbide and Related Materials, 1995, Eds. S. Nakashima, H. Matsunami, S. Yoshida, H. Harima, Institute of Physics, Bristol, 1996 (p. 81).
[14] S. Nakashima, H. Matsunami, S. Yoshida, H. Harima (Eds.), Silicon Carbide and Related Materials, 1995, Institute of Physics, Bristol, 1996 (Chapter 4).
[15] W. von Muench and I. Pfaffeneder, J. Electrochem. Soc. **122**, 642 (1975).
[16] A. Suzuki, H. Ashida, N. Furui, K. Mameno, and H. Matsunami, Jpn. J. Appl. Phys. **21**, 579 (1982).
[17] H. Matsunami, T. Ueda, and H. Nishino, Mater. Res. Soc. Symp. Proc. **162**, 397 (1990).
[18] V. Heine, C. Cheng, and R. J. Needs, J. Amer. Ceram. Soc. **74**, 2630 (1991).
[19] W. S. Yoo and H. Matsunami, Amorphous and Crystalline Silicon Carbide IV, Eds. C. Y. Yang, M. M. Rahman, and G. L. Harris, Springer-Verlag, Berlin 1992 (p. 66).
[20] A. R. Verma and P. Krishna (Eds.), Polymorphism and Polytypism in Crystals, John Wiley & Sons, Inc., New York 1966.
[21] A. Itoh, H. Akita, T. Kimoto, and H. Matsunami, Appl. Phys. Lett. **65**, 1400 (1994).
[22] T. Kimoto, Doctoral Thesis, Kyoto University, 1995.
[23] Yu. M. Tairov, V. F. Tsvetkov, S. K. Lilov, and G. K. Safaraliev, J. Cryst. Growth **36**, 147 (1976).
[24] J. A. Powell, J. B. Petit, J. H. Edgar, I. G. Jenkins, L. G. Matus, J. W. Yang, P. Pirouz, W. J. Choyke, L. Clemen, and M. Yoganathan, Appl. Phys. Lett. **59**, 333 (1991).
[25] C. Hallin, A. O. Konstantinov, O. Kordina, and E. Janzén, see [13] (p. 85).
[26] Y. Matsushita, T. Nakata, T. Uetani, T. Yamaguchi, and T. Niina, Jpn. J. Appl. Phys. **29**, L343 (1990).
[27] T. Ueda, H. Nishino, and H. Matsunami, J. Cryst. Growth **104**, 695 (1990).
[28] S. Karmann, C. Haberstroh, F. Engelbrecht, W. Suttrop, A. Schöner, M. Schadt, R. Helbig, G. Pensl, R. A. Stein, and S. Leibenzeder, Physica **185B**, 75 (1993).
[29] M. D. Allendorf and R. J. Kee, J. Electrochem. Soc. **138**, 841 (1991).
[30] C. D. Stinespring and J. C. Wohmhoudt, J. Cryst. Growth **87**, 481 (1988).
[31] A. O. Konstantinov, C. Hallin, O. Kordina, and E. Janzén, J. Appl. Phys. **80**, 5704 (1996).
[32] A. A. Burk, Jr. and L. B. Rowland, J. Cryst. Growth **167**, 586 (1996).
[33] T. Kimoto, H. Nishino, W. S. Yoo, and H. Matsunami, J. Appl. Phys. **73**, 726 (1993).
[34] H. S. Kong, J. T. Glass, and R. F. Davis, J. Mater. Res. **4**, 204 (1989).
[35] T. Kimoto and H. Matsunami, J. Appl. Phys. **76**, 7322 (1994).
[36] H. Matsunami and T. Kimoto, Mater. Res. Soc. Symp. Proc. **339**, 369 (1994).
[37] B. Wessels, H. C. Gatos, and A. F. Witt, Sillicon Carbide, 1973, Eds. R. C. Marshall, J. W. Faust, Jr., and C. E. Ryan, University of South Carolina Press, Columbia, 1974 (p. 25).

[38] F. C. EVERSTEYN, P. J. W. SEVERIN, C. H. J. V. D. BREKEL, and H. L. PEEK, J. Electrochem. Soc. **117**, 925 (1970).
[39] W. K. BURTON, N. CABRERA, and F. C. FRANK, Phil. Trans. Roy. Soc. **A243**, 299 (1951).
[40] T. KIMOTO and H. MATSUNAMI, J. Appl. Phys. **75**, 850 (1994).
[41] J. A. POWELL, D. J. LARKIN, and P. B. ABEL, J. Electronic Mater. **24**, 295 (1995).
[42] T. KIMOTO, A. ITOH, and H. MATSUNAMI, Appl. Phys. Lett. **66**, 3645 (1995).
[43] S. TANAKA, R. C. KERN, R. F. DAVIS, J. F. WENDELKEN, and J. WU, Surf. Sci. **350**, 247 (1996).
[44] J. A. POWELL, D. J. LARKIN, P. B. ABEL, L. ZHOU, and P. PIROUZ, see [13] (p. 77).
[45] C. HERRING, Phys. Rev. **82**, 87 (1951).
[46] W. A. TILLER, The Science of Crystallization: Microscopic Interfacial Phenomena, Chap. 2, Cambridge University Press, Cambridge 1991.
[47] T. TAKAI, T. HALICIOGLU, and W. A. TILLER, Surf. Sci. **164**, 341 (1985).
[48] S. TYC, see [11] (p. 333).
[49] S. TANAKA, R. S. KERN, and R. F. DAVIS, Appl. Phys. Lett. **65**, 2851 (1994).
[50] T. KIMOTO, A. ITOH, H. MATSUNAMI, and T. OKANO, J. Appl. Phys. **81**, 3494 (1997).
[51] V. F. TSVETKOV, S. T. ALLEN, H. S. KONG, and C. H. CARTER, JR., see [13] (p. 17).
[52] L. L. CLEMEN, R. P. DEVATY, M. F. MACMILLAN, M. YOGANATHAN, W. J. CHOYKE, D. J. LARKIN, J. A. POWELL, J. A. EDMOND, and H. S. KONG, Appl. Phys. Lett. **62**, 2953 (1993).
[53] S. JANG, T. KIMOTO, and H. MATSUNAMI, Appl. Phys. Lett. **65**, 581 (1994).
[54] T. DALIBOR, G. PENSL, T. KIMOTO, H. MATSUNAMI, S. SRIDHARA, R. P. DEVATY, and W. J. CHOYKE, presented at the 1st Europ. Conf. Silicon Carbide and Related Materials, Crete, 1996.
[55] M. M. ANIKIN, A. N. ANDREEV, A. A. LEBEDEV, S. N. PYATKO, M. G. RASTEGAEVA, N. S. SAVKINA, A. M. STREL'CHUK, A. L. SYRKIN, and V. E. CHELNOKOV, Soviet Phys. — Semicond. **25**, 198 (1991).
[56] T. DALIBOR, C. PEPPERMÜLLER, G. PENSL, S. SRIDHARA, R. P. DEVATY, W. J. CHOYKE, A. ITOH, T. KIMOTO, and H. MATSUNAMI, see [13] (p. 517).
[57] Y. C. WANG, R. F. DAVIS, and J. A. EDMOND, J. Electronic Mater. **20**, 289 (1991).
[58] D. J. LARKIN, P. G. NEUDECK, J. A. POWELL, and L. G. MATUS, Appl. Phys. Lett. **65**, 1659 (1994).
[59] T. KIMOTO, A. ITOH, and H. MATSUNAMI, Appl. Phys. Lett. **67**, 2385 (1995).
[60] 85 (1995).
[61] W. J. SCHAFFER, G. H. NEGLEY, K. G. IRVINE, and J. W. PALMOUR, Mater. Res. Soc. Symp. Proc. **339**, 595 (1994).
[62] M. SCHADT, G. PENSL, R. P. DEVATY, W. J. CHOYKE, R. STEIN, and D. STEPHANI, Appl. Phys. Lett. **65**, 3120 (1994).
[63] T. KIMOTO and H. MATSUNAMI, J. Appl. Phys. **78**, 3132 (1995).
[64] A. SCHÖNER, N. NORDELL, K. ROTTNER, R. HELBIG, and G. PENSL, see [13] (p. 493).
[65] V. I. FISTUL, Heavily Doped Semiconductors, Plenum Press, New York 1969.
[66] T. KIMOTO, A. ITOH, O. TAKEMURA, S. KOBAYASHI, and H. MATSUNAMI, Extended Abstracts 38th Electron. Mater. Conf., Santa Barbara, 1996 (p. 19).

phys. stat. sol. (b) **202**, 263 (1997)

Subject classification: 68.55.Jk; 68.55.Ln; 73.61.Le; S6

Homoepitaxial VPE Growth of SiC Active Layers

A. A. BURK, JR. (a) and L. B. ROWLAND (b)

(a) Northrop Grumman Electronic Sensors and Systems Division,
Baltimore, MD-21203, USA

(b) Northrop Grumman Science and Technology Center, Pittsburgh, PA-15235, USA

(Received January 31, 1997)

SiC active layers of tailored thickness and doping form the heart of all SiC electronic devices. These layers are most conveniently formed by vapor phase epitaxy (VPE). Exacting requirements are placed upon the SiC-VPE layers' material properties by both semiconductor device physics and available methods of device processing. In this paper, the current ability of the SiC-VPE process to meet these requirements is described along with continuing improvements in SiC epitaxial reactors, processes and materials.

1. Introduction

Silicon carbide is a very promising semiconductor material for high-power and high-temperature microwave devices because of its superior thermal and electrical properties [1]. For example, SiC has an order of magnitude greater thermal conductivity and breakdown field strength, and a higher saturated electron drift velocity than GaAs. However, the lack of large-area SiC substrates, combined with the difficulties in processing this material due to its chemical, mechanical, and thermal stability, have hampered the development of SiC-based devices until recently. Although work has been done in molecular beam epitaxy [2, 3], as well as ion implantation [4] for device active layer formation, the most successful method for SiC device active layer fabrication has been vapor phase epitaxy (VPE) [5].

6H and 4H-SiC homoepitaxial layers have been grown by VPE exhibiting specular morphologies, net background doping densities of less than 1×10^{14} cm^{-3}, and controlled n- and p-type doping from 5×10^{15} to above 1×10^{19} cm^{-3} thereby enabling the development of SiC-based power microwave devices such as metal–semiconductor field effect transistors (MESFETs) [6] and static induction transistors (SITs) [7]. Higher-power SiC devices such as diodes, vertical MOSFETs, and GTOs have also been demonstrated [8 to 10].

A partial list of the requirements placed upon SiC active layers is summarized in Table 1. They are prioritized from the material properties needed for basic device function, such as good crystal quality and purity, to attributes that are required for production of devices, such as uniformity and throughput. Each of the requirements becomes more stringent upon progressing from "proof-of-concept" device work to production. The requirements are also necessarily device-type dependent. Those shown in the table are based on experience gained in GaAs monolithic-microwave integrated circuit (MMIC) production. The current progress of SiC VPE in satisfying these diverse requirements is summarized in the results section using our internal capabilities by way of example.

Table 1

Goals of epitaxy

active layer attribute	device proof-of-concept requirement	MMIC production requirement
good crystal quality/ morphology	specular, "10% good area"	<100 morphological defects/cm^2, 90% area
high purity/low background doping	$<1 \times 10^{16}$ cm^{-3} background doping	$<1 \times 10^{14}$ cm^{-3} background doping
wide intentional doping range	1×10^{16} to 1×10^{18} cm^{-3}	1×10^{14} to 1×10^{19} cm^{-3}
intrawafer thickness and doping uniformity	$< \pm 20\%$ total variation	$< \pm 5\%$ total variation
thickness and doping reproducibility	$< \pm 20\%$ with frequent calibration runs	$< \pm 5\%$ process capability
high throughput/low cost	single wafer satisfactory	multiwafer

2. Background

A resurgence of interest in SiC-based semiconductor devices over the last decade has occurred primarily because of the development of substrates of 6H and 4H-SiC upon which device-quality layers can be homoepitaxial grown. Bulk substrates of both n- and p-type SiC, formed using the modified Lely technique [11 to 14], have been demonstrated at up to two-inch diameter [15]. They are now available with both low ($<0.01\ \Omega$ cm) and high ($>10^7\ \Omega$ cm) resistivities [16] suitable for both vertical and horizontal devices. These wafers still exhibit defects which adversely impact device performance and yield. Micropipe defects, which are present in SiC bulk wafers in densities from 10 to 100 cm^{-2}, are particularly deleterious to device performance because they extend from the substrate through any epitaxial layers [17]. In addition to micropipe defects, polishing scratches, low-angle grain boundaries and inclusions of other polytypes can adversely impact the epitaxial layer quality. These unwanted features are currently the subject of intense investigations and are being significantly reduced as shown in the accompanying articles and the provided references.

Significant breakthroughs in the quality of SiC epitaxial layers have also recently been accomplished. In 1986 to 1988 Kuroda et al. [18] and Kong et al. [19] obtained specular homoepitaxial growth at reduced temperature (e.g., $\approx 1600\ ^{\circ}$C) by using slightly off-oriented c-axis substrates, thus enabling step-flow growth. While the two-dimensional step-flow growth model of Frank and van der Merwe, see [20], is well known in crystal growth, in SiC growth the use of off-axis substrates has the added benefit of exposing the stacking sequence of the substrate thereby facilitating effective polytype replication. In 1993 Larkin et al. [21] demonstrated a dependence of dopant incorporation efficiency on input Si/C ratio (analogous to the impact of III/V ratio in that semiconductor family [22]) helping to both reduce background doping density and extend the range of intentionally doped layers.

Several reactor configurations have been successfully developed for the growth of SiC epitaxial layers. Most will be described in greater detail in this volume so they are only outlined briefly here. All are constructed with high temperature tolerant materials such as quartz, graphite, and SiC. Most require active water cooling. Hydrogen carrier gas

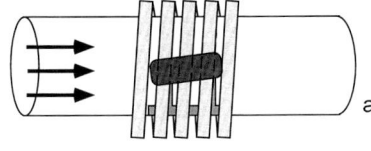

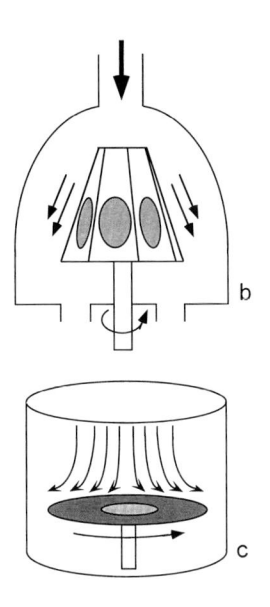

Fig. 1. Schematic drawings of a) cylindrical horizontal, b) barrel, and c) rapidly rotating vertical SiC-VPE reactors. The arrows indicate the direction of gas flow and component motion

along with silane and propane reagents are typically employed at atmospheric and reduced pressure and temperatures ranging from 1450 to 1600 °C. To reach these high temperatures inductively heated SiC-coated graphite susceptors are typically employed. The most basic configuration utilized by among others, Powell et al. [23], Karmann et al. [24], and ourselves [25] is a water cooled (cold wall) cylindrical quartz horizontal reactor shown in Fig. 1a. The group of Davis [26], have developed the multiwafer barrel reactor shown in Fig. 1b. Kordina et al. [27] have developed a hot-wall reactor for the purpose of reducing the large (20 to 40 kW) power requirements of the other reactors by using a hollow susceptor in an otherwise basic horizontal reactor. Most recently Rupp et al. [28] developed the single wafer rapidly rotating vertical reactor shown in Fig. 1c.

With the exception of the barrel reactor, all of the above SiC reactors were designed for single wafer use. Historically, however, it has been difficult to achieve the epitaxial layer uniformities required for production of microwave devices in scaled-up cylindrical horizontal reactors or even barrel reactors when using large diameter wafers. In III/V epitaxial growth two basic types of multiwafer MOCVD reactors have met with the greatest success for 3 and 4-inch diameter wafers, the rapidly rotating vertical reactor [29] and the horizontal planetary reactor [30]. Now that we have achieved device quality SiC horizontal epitaxial layers with our single wafer horizontal reactor, we are currently developing a SiC planetary reactor to improve uniformity, reproducibility, throughput and cost. Fig. 2 contains a rough sketch of the SiC planetary reactor that will be discussed in Section 4.5.

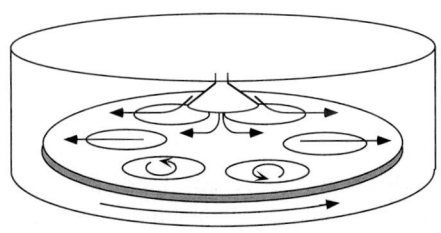

Fig. 2. Schematic drawing of the planetary SiC-VPE reactor

3. Experimental Procedures

Unless specified otherwise, the epitaxial growths and pretreatments conducted in this study were performed in the single wafer horizontal SiC VPE reactor. The hydrogen carrier gas was purified by diffusion through a Pd-cell membrane. The susceptor temperature was measured with a pyrometer by way of a sapphire light pipe placed in the growth chamber approximately 2 cm from the inductively heated SiC-coated susceptor. Temperature measurements were calibrated by observing the indicated melting point of silicon samples. All growths occurred at atmospheric pressure using resin-purified silane and propane (both 7% in H_2) in 12.0 slpm H_2 carrier gas at $\approx$1520 °C. Substrates of 6H-SiC cut and polished on the a-axis face $(11\bar{2}0)$, prismatic face $(\bar{1}010)$ and 3.5° off c-axis (0001) (Si-face) towards the $\langle 11\bar{2}0 \rangle$ direction, as well as in situ 4H-SiC oriented 3.5° and 8° off c-axis towards $\langle 11\bar{2}0 \rangle$ were used. Prior to growth, samples were exposed to a 1350 °C etch in 3% HCl (in H_2) for 2 min. Samples were grown using SiH_4 mole fractions of 3.0 to 5.2×10^{-4} and C_3H_8 mole fractions from 0.2 to 1.1×10^{-3}, resulting in a Si/C source gas phase ratio of 0.16 to 0.80. Intentional doping was accomplished using various flow rates of 0.5% and 10% N_2 in H_2, resulting in N_2 mole fractions of 1.3×10^{-5} to 1.5×10^{-3}.

Epitaxial layer morphology was measured by optical interference microscopy. Capacitance–voltage (C–V) measurements were used to determine carrier concentration as a function of depth, with gold electron beam evaporated Schottky diodes of 200 μm diameter defined by a shadow mask. Hall electron mobilities were obtained on clover leaf samples using p-type 4H-SiC substrates, a 50 nm thick n-type contact epilayer (10^{19} cm^{-3}) and Ti/Ni contact pads.

4. Results and Discussions

4.1 Crystal quality

Examining epitaxial layer morphology is a useful way of assessing crystal quality and obtaining insights to the underlying growth mechanism. Obtaining specular active layers is one of the most important requirements of epitaxy for device active layers. No matter how perfect an active layer is by all other measures it is unlikely to be processed if it does not look specular to the process engineer. While active layers with very poor morphology can supply critical "proof-of-concept" device demonstrations an excessive density of morphological defects interferes with key device fabrication steps such as lithography, etching, and metallization resulting in low yields particularly for devices with large footprints such as MMICs and power devices. More importantly, however, poor epitaxial layer morphology usually signals non-optimum growth conditions which can also adversely impact active layer purity and electrical transport properties. Detailed descriptions of SiC epitaxial layer morphology have been presented by several groups [31 to 36]. Following is a summary of some of these results.

4.1.1 Crystal face dependence

A comparison of the epitaxial layer morphology obtained on three 6H-SiC substrates having c-axis (3.5° off (0001) towards the $\langle 1\bar{2}10 \rangle$), a-axis $(\bar{1}2\bar{1}0)$, and prismatic $(10\bar{1}0)$ orientations is shown in Fig. 3. These photographs show nominally specular

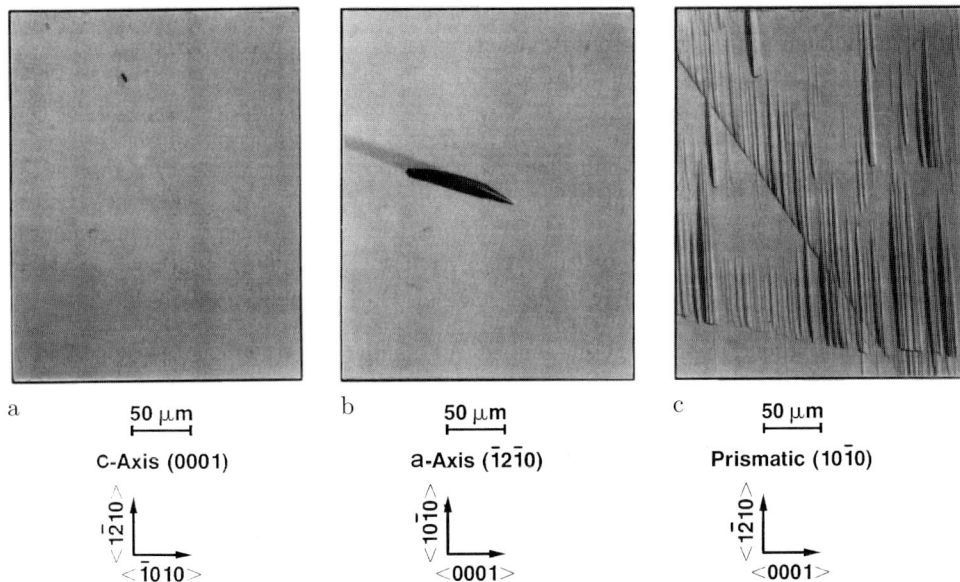

a   50 μm   b   50 μm   c   50 μm

C-Axis (0001)   a-Axis ($\bar{1}2\bar{1}0$)   Prismatic ($10\bar{1}0$)

Fig. 3. A comparison of 6H-SiC epitaxial layer morphology on a) 3.5° off c-axis, b) a-axis and c) prismatic oriented substrates

epitaxial layers on the c-axis and a-axis substrates whereas the prismatic substrate exhibits a hypersensitivity to surface imperfections resulting in elongated morphological defects along an a-axis direction. The c-axis sample also exhibits shallow tiny "amphitheater" shaped depressions with vertices pointing in the up-step a-axis direction. These observations and the studies of the growth on circular mesas by Kimoto and Matsunami [37] clearly indicate that the SiC crystal grows fastest in the a-axis directions. Most recently these orientation effects have also been observed by Hallin [36]. These observations are also consistent with earlier observations of Kuroda et al. [18] that step flow growth results in improved epitaxial layer morphology when using substrates oriented a few degrees toward the a-axis direction in comparison to exactly oriented c-axis samples [18] or samples oriented a few degrees in the prismatic direction [19].

4.1.2 Impact of Si droplets at growth initiation

Table 2 contains a summary of the conditions of SiC surfaces after exposure to a variety of pregrowth treatments. Si droplets were observed to develop on SiC surfaces when exposed to hydrogen at temperatures above the melting point of silicon (1410 °C). These droplets, which interfere with specular epitaxial growth, were effectively inhibited by the presence of propane or HCl overpressures prior to growth and by minimizing heat-up time. These results are presented in greater detail elsewhere [31]. Despite this significant reduction of droplets afforded by optimization of pregrowth conditions, varying numbers of morphological defects were still observed in subsequent epitaxial growths. These are described in the following section.

Table 2

Causes of silicon-droplet formation in SiC epitaxy

case No.	experimental condition	silicon droplets formed?	explanation
1	30 min @ 1450 to ≈1520 °C under 1 atm H_2 on 4H, 6H, a- and vicinal c-axis PVT SiC and 6H Lely SiC	yes on all types	1) $SiC + H_2 \rightarrow Si(l) + $ hydrocarbons 2) $Si(l)$ accumulates on surface
2	30 min @ 1450 to ≈1520 °C under 1 atm Ar	no	1) SiC is stable under Ar and 2) vapor pressure of $Si \gg C$
3	30 min @ 1450 to ≈1520 °C under 1 atm H_2 and 140 ppm C_3H_8	no	$SiC + H_2 \rightarrow Si(l) + $ hydrocarbons (C_3H_8 inhibits H_2 etching of SiC)
4	30 min @ 1450 to ≈1520 °C under 1 atm H_2 with exposed carbon	no	1) $C + H_2 \rightarrow $ hydrocarbons 2) hydrocarbons inhibit SiC etching
5	30 min @ 1450 to ≈1520 °C under 1 atm H_2 and 3% HCl	no, but etch pits	1) $SiC + H_2 \rightarrow Si(l) + $ hydrocarbons 2) $Si(l) + HCl \rightarrow$ chlorosilane gas
6	30 min @ 1450 to ≈1520 °C under 0.1 atm H_2	edges only	$Si(l)$ evaporation rate increased by reduced diffusion layer

4.1.3 Impact of substrate quality on 3.5° off c-axis 4H-SiC morphology

The following experiment was performed to determine the cause of the morphological defects that are still present in SiC epitaxial layers when using optimized start-growth conditions. Samples of 4H and 6H-SiC were subjected to an in situ etch and growth nucleation procedure that effectively minimized Si droplets and etch pits but then the growth was suddenly halted within a few seconds. Photomicrographs of the resulting surfaces were obtained. They still reveled a few barely visible submicron-sized Si drop-

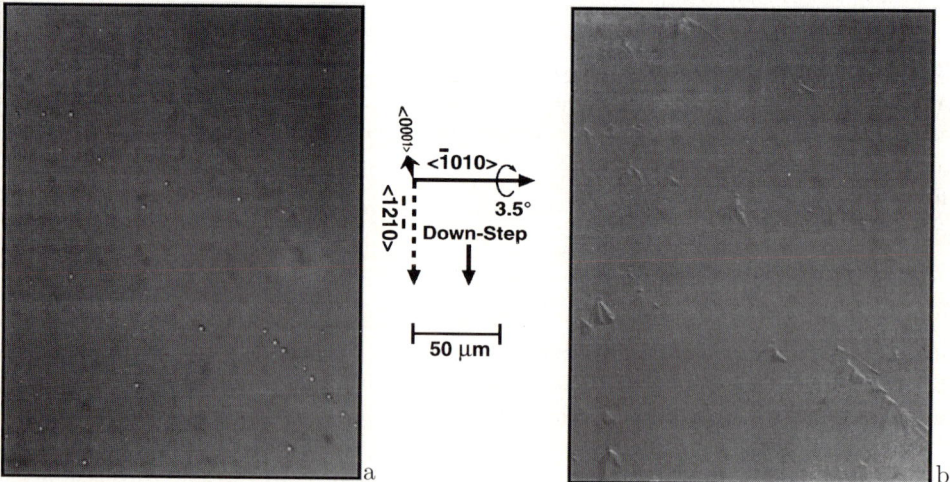

Fig. 4. Interference contrast photomicrographs of a 4H-SiC sample, a) before and b) after epitaxial growth. Every morphological defect in the epitaxial layer can be attributed to a substrate surface imperfection revealed by a previous growth nucleation

lets, impressions where Si droplets had rested and then evaporated, micropores and other unresolvable "point defects". Before this treatment (not shown) only the pores and some very faint polishing scratches had been discernible. Subsequently the samples were placed back into the growth chamber followed by another identical in situ etch and growth nucleation. This time however, the growth was allowed to continue to a thickness of 1 μm, after which the growth was ended with propane overpressure during cooldown to suppress end-growth Si droplet formation. Photomicrographs of the resulting epitaxial layer were taken and compared to those taken of the same location after the first nucleation.

On the 1 μm thick 4H-SiC sample shown in Fig. 4 all the features identified after the first nucleation resulted in morphological defects in the epitaxial layer. While most were "amphitheater"-shaped shallow depressions, some were larger and deeper tetrahedral-shaped pits. There was a rough correlation between tetrahedral pits and the sites that had Si droplets evident after the first nucleation step (as opposed to those that only had

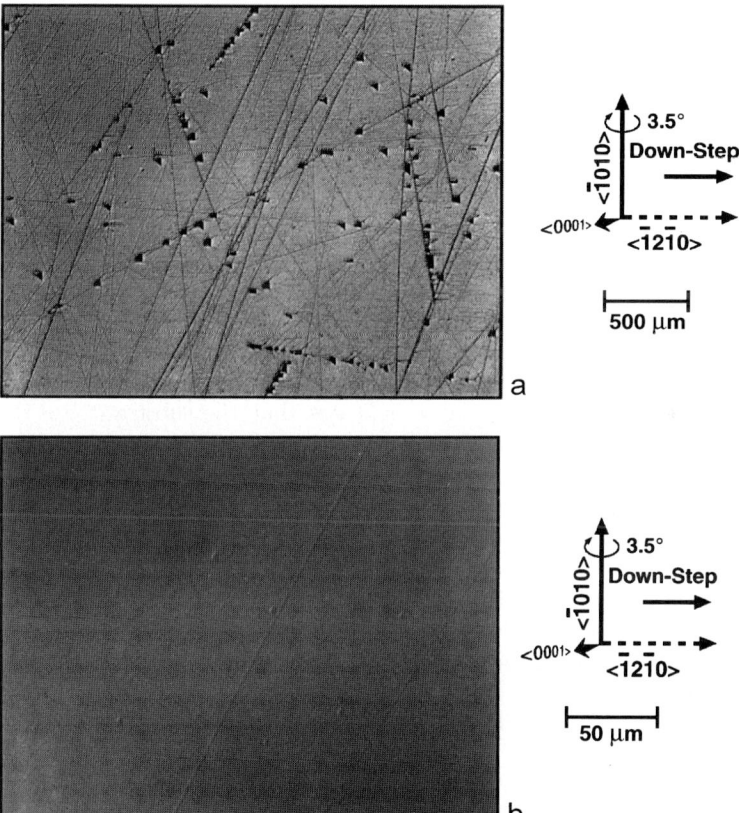

Fig. 5. Interference contrast photomicrographs of two different 4H-SiC epitaxial samples after 5 μm of epitaxial growth, both grown under conditions that inhibit Si droplet formation and etch pit formation. Note in sample a) the decoration of many of the residual polishing scratches with large faceted tetrahedral pits. Sample b) in contrast is specular showing only shallow features even at higher magnification (at low magnification the sample is featureless)

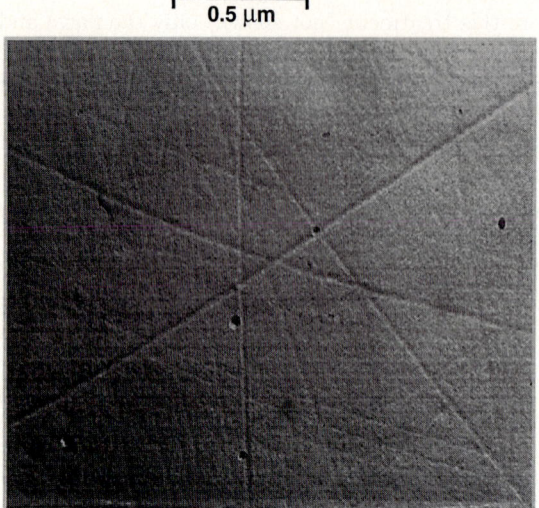

0.5 µm

Fig. 6. A replica-TEM photomicrograph of a 4H-SiC substrate surface before epitaxial growth reveals fine polishing scratches

impressions of Si droplets). In addition, large micropores were observed to initiate tetrahedral pit formation. The dominant source of tetrahedral pits, however, were residual substrate polishing scratches. This is clearly shown in Fig. 5a. It is noteworthy that many of the scratches are decorated with numerous tetrahedral pits while others have none. This distribution is consistent with that observed for Si droplets on polishing scratches. Identical experiments performed on 6H-SiC substrates resulted in only the small "amphitheater"-type defect even though the residual polishing scratches appeared to be comparable.

An important result of the above nucleation experiments is that as the first and second etching and nucleation procedures were identical, the lack of a second population of morphological defects in the final epitaxial layer indicates that the substrate was the source for each of the morphological defects in the epitaxial layer.

While the morphology of 3.5° off 4H epitaxial layers was typically observed to be inferior to that of 6H (particularly on small, 5 mm square, samples), equivalent results were observed on a small percentage of as-received whole 4H-SiC wafers as shown in Fig. 5b. Obvious polishing scratches were observable by interference contrast microscopy on some 4H substrates that yielded highly defective epitaxial layers, while others that showed no prior evidence of significant residual polishing problems also resulted in poor morphology. Closer inspection of one such substrate by replica TEM, however, revealed a field of very fine polishing scratches (Fig. 6). Another substrate (not shown) which resulted in excellent morphology exhibited no discernible polishing scratches by this method. This anecdotal evidence suggests that hidden polishing damage is the source of many of the morphological defects in 3.5° off 4H-SiC epitaxial layers but does not explain the marked polytype dependence.

4.1.4 8° versus 3.5° off 4H-SiC

As shown in the last section specular epitaxial growth can be achieved on 3.5° off 4H-SiC, although a superior polish is required than on the 6H polytype. Another way to

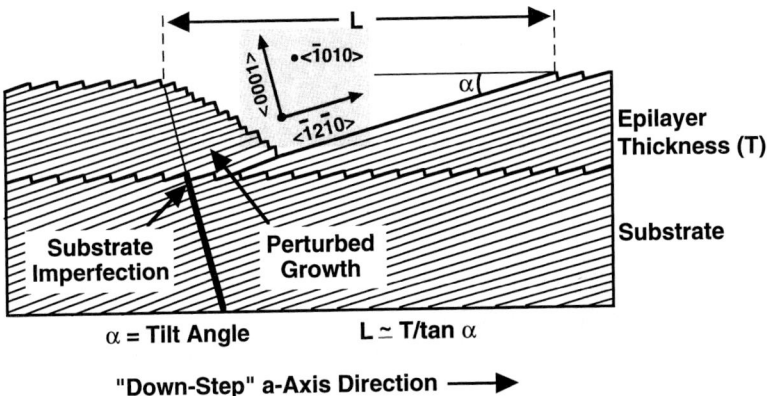

α = Tilt Angle $L \simeq T/\tan \alpha$

"Down-Step" a-Axis Direction ➔

Fig. 7. Schematic of the proposed mechanism for tetrahedral pit formation on off-axis 4H-SiC (3.5° from the $\langle 0001 \rangle$ direction towards the $\langle 1\overline{2}10 \rangle$ a-axis direction). A surface imperfection such as a pore, screw dislocation, scratch or Si droplet exposes a basal plane facet which then propagates during epitaxial growth. Shorter terrace widths increase the probability of adatoms diffusing to and incorporating in a step edge

improve 4H-SiC epitaxial layer morphology is to use higher off-orientations as first demonstrated by Carter and Tsvetkov [38] or higher growth temperatures as shown by Hallin [39]. These improvements are all consistent with increasing the ideality of the operative step-flow growth mechanism by either improving the quality of the start-growth surface, reducing surface terrace widths or increasing surface adatom mobility, respectively. These improvements all increase the probability of a surface adatom reaching a step edge or kink before being incorporated in the crystal. Fig. 7 contains a schematic of the step-flow growth mechanism and how interference with the steps by a scratch pore or Si droplet can cause a morphological defect. In the case of the relatively benign shallow "amphitheater"-shaped defect, the step flow effectively wraps around the defect nucleus. In the case of the large tetrahedral pit, an exposed basal plane facet grows laterally in the down-step direction 16.4 µm for every µm of vertical growth on a 3.5° off 4H-SiC surface. One likely reason for 4H-SiC's greater tendency to form these large defects can be explained by the observation of Kimoto et al. [40] that given the same off-axis orientation 4H-SiC exhibits larger average terrace widths than does 6H-SiC (16.8 versus 12.4 nm, for 3.5° off orientation). An additional explanation offered by Hallin [36] is that 4H-SiC may have a lower stacking fault energy than does 6H-SiC.

4.1.5 Statistical morphological data

Fig. 8 shows a compilation of morphological data that we have obtained confirming the benefit of using improved polishing, higher off-orientations and optimized growth conditions. Despite these improvements SiC epitaxial layers still typically exhibit 1000 micron-sized morphological defects/cm^2. The remaining defects have been shown to originate at polishing and other imperfections still present at the substrate surface. While these need to be further reduced it is important to note that a similar density of oval defects were common in "device quality" GaAs MBE layers only ten years ago.

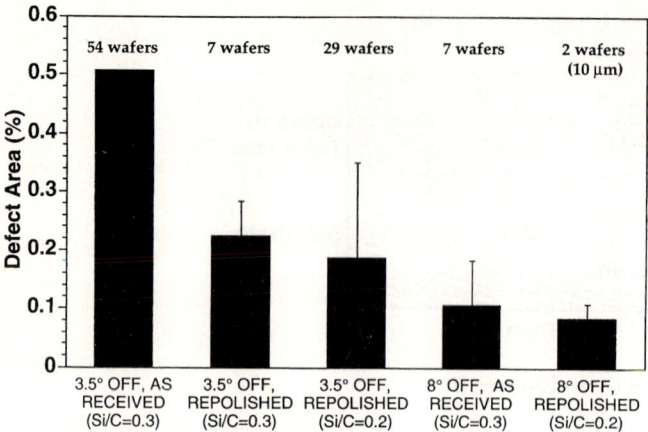

Fig. 8. Effective defect area of a series of ≈5 μm thick 4H-SiC epitaxial layers as a function of polishing, orientation and input Si/C ratio

4.2 Purity

Purity can be assessed by a number of techniques including CV, Hall, PL, and lifetime measurements. As shown in Fig. 9, background doping density as measured by CV can be reduced to $< 1 \times 10^{14}$ cm^{-3} by reducing the input Si/C ratio. Hall, PL, and lifetime measurements are good measures of both purity and crystal quality. As shown in Fig. 10a to c, our 4H epitaxial layers have demonstrated room temperature mobilities of over 950 cm^2/Vs, sharp excitonic spectra, and lifetimes as high as 3.1 μs. The Hall mobility sample indicated a peak mobility of 17666 cm^2/Vs at 47 K. When the temperature

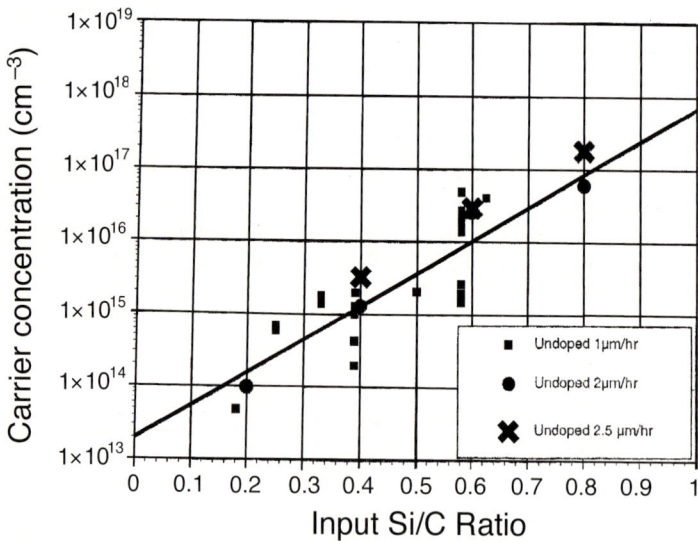

Fig. 9. Background n-type doping density of SiC-VPE epitaxial layers as a function of input Si/C ratio at 1 μm/h, 2 μm/h and 2.5 μm/h

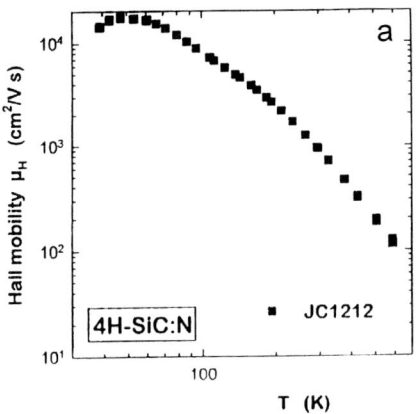

Fig. 10. Hall mobility a) as a function of temperature for a 4H-SiC epitaxial layer shows room temperature and peak mobilities (47 K) of 950 and
17666 cm^2/Vs. b) The low temperature photoluminescence spectrum shows sharp nitrogen bound exciton lines (P_0, Q_0) and little sign of aluminum compensation. c) Band edge photoluminescence decay
measurements, reveal long, 3.1 μs minority carrier
lifetimes. These results are all characteristic of high-
quality, high-purity 4H-SiC epitaxial layers

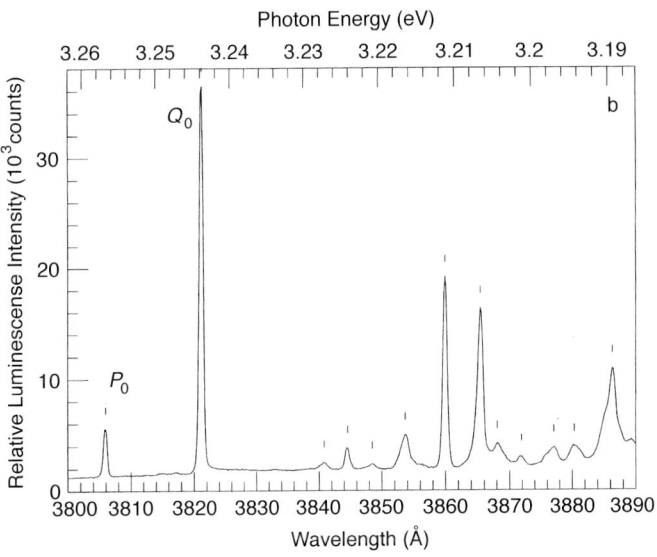

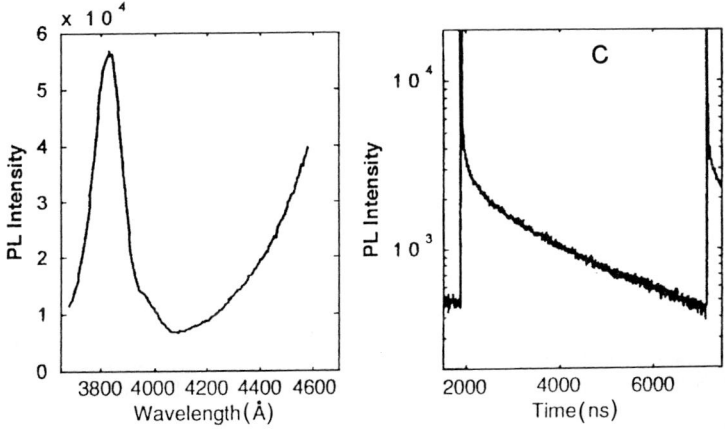

dependent Hall mobility was fit it indicated a room temperature free electron concentration of 2.97×10^{15} cm^{-3}, a nitrogen dopant concentration of 3.3×10^{15} cm^{-3}, and a compensation of only 1.8×10^{14} cm^{-3}. The sharp P_0 and Q_0 lines and absence of large Al or B lines in the PL spectrum are also consistent with low levels of compensation. The long lifetime (3.1 µs) was observed near the center of the 30 mm diameter, 10 µm thick sample, shown in Fig. 10. The lifetime dropped however only to ≈ 2.5 µs within 3 to 4 mm of the wafer's edge. As a point of reference, the next longest lifetime reported in the literature [41] is ≈ 2 µs and for a much thicker layer. Taken all together, these materials properties are all consistent with epitaxial layers having the highest purity currently available in SiC. Moreover, the background doping density compares very favorably with what can be achieved currently in GaAs MOCVD.

4.3 Intentional doping

Changes in the doping concentration of viscinal 4H-SiC(0001) were investigated as a function of variations in N_2 and propane mole fraction. Fig. 11 illustrates the effect of doping density on nitrogen mole fraction in the range of 1.3×10^{-5} to 1.5×10^{-3} for four different mole fractions of C_3H_8 at a constant SiH_4 mole fraction, corresponding to Si/C gas source flow ratios of 0.20, 0.34, 0.42, and 0.80. A linear relationship between nitrogen mole fraction and doping density with slope of 1.0 was observed for all four Si/C gas source ratios. With independent control and variation of Si/C ratio and nitrogen mole fraction, doping densities from lower than 1×10^{15} cm^{-3} to greater than 1×10^{19} cm^{-3} have been demonstrated.

Nitrogen is a very inefficient dopant species in SiC. The distribution coefficient of nitrogen in SiC is defined as

$$k = x_\mathrm{s}/x_\mathrm{v}, \tag{1}$$

where x_s is the doping density/concentration of C sites and x_v the concentration of N in the gas phase/concentration of C in the gas phase. The distribution coefficients deter-

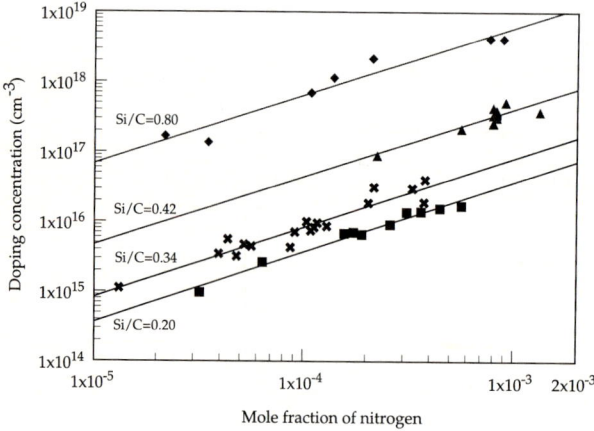

Fig. 11. Intentional n-type doping as determined by CV measurement for a series of input Si/C ratios as a function of nitrogen mole fraction. By varying these parameters a wide range of carrier concentrations can be obtained

Table 3

Distribution coefficients for nitrogen doping of SiC

polytype	Si/C	growth T (°C)	[Si]	[C]	k	ref.
4H-SiC	0.80	1540	5.2×10^{-4}	6.5×10^{-4}	6.4×10^{-5}	this work
4H-SiC	0.34	1540	5.2×10^{-4}	1.5×10^{-3}	2.5×10^{-6}	this work
4H-SiC	0.20	1540	5.2×10^{-4}	2.6×10^{-3}	2.0×10^{-6}	this work
4H-SiC	0.41	1540	3.0×10^{-4}	7.5×10^{-4}	5.2×10^{-6}	this work
6H-SiC	1.00	1500	2.2×10^{-4}	2.2×10^{-4}	1.7×10^{-5}	[46]
6H-SiC	0.83	1600	1.7×10^{-4}	2.0×10^{-4}	7.3×10^{-6}	[24]
6H-SiC	0.29	1800	1.1×10^{-4}	3.9×10^{-4}	3.7×10^{-7}	[47]
6H-SiC	0.50	1500	5.0×10^{-5}	1.0×10^{-4}	1.5×10^{-5}	[48]

mined from previous work in 6H-SiC, as well as in this work are shown in Table 3. Depending upon the Si/C ratio and the growth temperature, the distribution coefficient can vary from 10^{-7} to 10^{-5}. This is many orders of magnitude lower than typically seen for dopants in III–V semiconductors [42], but is not surprising considering the exceedingly high bond strength of N_2 (945 kJ/mol at room temperature) which must be overcome as part of the dopant incorporation process.

While not shown, we have also demonstrated intentional p-type doping ranging from 5×10^{15} to over 10^{20} cm^{-3} by varying the input Si/C ratio and admitting either trimethyl- or triethyl-aluminum dopant.

4.4 Device profiles

Beyond the ability to demonstrate high purity and a wide range of intentional doping, most device structures require that two or more layers of tailored carrier type and concentration be grown on top of one another with well defined, often abrupt junctions. Fig. 12a, b contain SIMS doping profiles for two such device structures, the metal–semiconductor field-effect transistor (MESFET) and the vertical metal–oxide–semiconductor (VMOS) transistor grown by SiC-VPE. These demanding profiles require undoped, p- and n-type doped layers ranging from 1×10^{14} to over 1×10^{20} cm^{-3}. The ability of SiC-VPE to demonstrate this level of control at this early stage of development is encouraging regarding the potential of producing SiC-based devices in the future.

4.5 Uniformity, reproducibility, and throughput

While impressive results have been achieved by us and others with regard to SiC materials properties, yielding key device demonstrations, SiC epitaxial layer production is still very immature in comparison to the more conventional semiconductors. Wafer diameters are still only between one and two inches, substrate and epitaxial layer quality, reproducibility, and uniformity are highly variable and the cost and availability of materials is not sufficient for large-scale device production. While recently excellent intra-wafer thickness uniformities (±2.5% total variation) have been reported by Rupp et al. [28] for their rapidly rotating single wafer reactor, no one has yet reported the simultaneous achievement of good materials quality, uniformity and reproducibility in a multiwafer SiC epitaxial reactor.

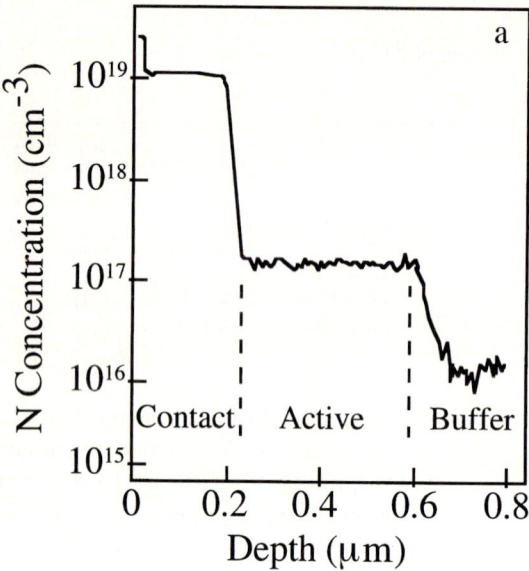

Fig. 12. Examples of device doping profiles grown by SiC-VPE, exhibiting a wide range of n- and p-type doping and abrupt layer interfaces. a) A MESFET nitrogen doping profile and b) vertical MOSFET nitrogen and aluminum profiles

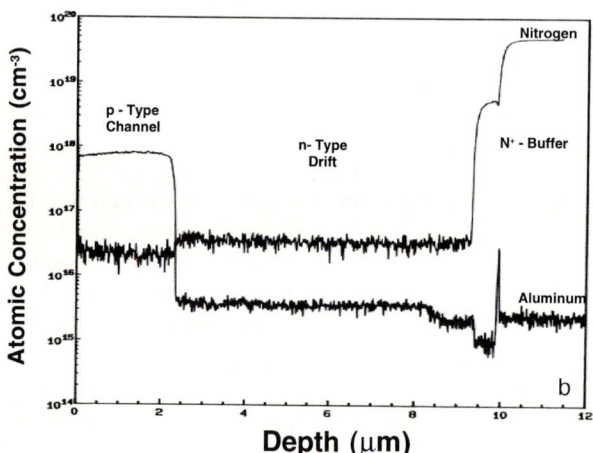

We have been attempting to develop this capability using the planetary SiC-VPE reactor shown in Fig. 2. As currently configured, the reactor is capable of growth on seven 2-inch diameter substrates at a time, greatly increasing layer throughput. Moreover the growth chamber is sized to allow future expansion to a 5×3-inch configuration.

The planetary reactor concept was originated by Frijlink [43] and refined and commercialized by Aixtron [44] for the growth of highly uniform III–V compound semiconductors. In this design the gases enter from the center of the reactor and flow outward radially. Because of the increase in cross-sectional area, a drop in gas velocity occurs along the flow direction, which results in an increased but more linear depletion of reagents (in comparison to a conventional horizontal cylindrical geometry). For diffusion limited growth, this yields a linear drop in growth rate along the flow direction. The

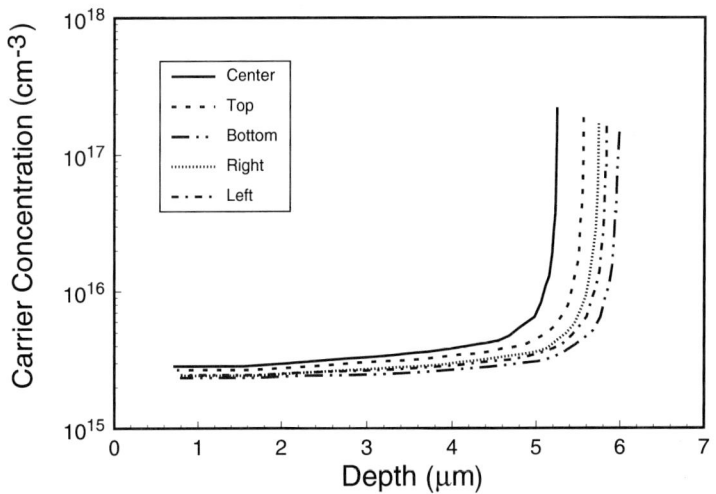

Fig. 13. A superposition of doping profiles obtained at the center and periphery (bottom refers to the side of the wafer with the flat) from an unintentionally doped buffer layer grown on a 30 mm diameter 4H-SiC substrate in the multi-wafer reactor. C–V measurements were performed on five $\approx 200\,\mu m$ diameter gold dots evaporated through a shadow mask. Indicated thickness and doping uniformity (standard deviation/mean) are 4% and 7%, respectively

rotation of the individual wafers about their central axis effectively averages this trend, yielding highly uniform layers. Additionally, the entire susceptor rotates about its center in order to further average any asymmetry in susceptor temperature or reagent flows. The means of rotating the susceptor and wafer holders is via a gas foil levitation. The end result of the planetary motion in the III–V case is reported to be ±1% thickness and doping uniformity both intra-wafer and wafer-to-wafer [45].

In the case of the SiC planetary reactor we are using roughly the same reactor design but extending it from the typical $< 800\,^\circ C$ III–V growth temperatures to the $\approx 1600\,^\circ C$ temperatures required for SiC epitaxial growth. These extreme temperatures have required the use of RF induction heating, custom designed high-temperature materials and construction.

After more than 300 developmental growth runs and extensive chamber optimization we can now present preliminary SiC epitaxial layer growth results from the planetary reactor. Currently we have achieved specular, reasonably pure ($< 1 \times 10^{15}\,cm^{-3}$ n-type background) epitaxial layers of up to 10 μm thickness. Growth rates from 2 to 5 μm/h have been observed. While our results are too preliminary to yet address questions of wafer-to-wafer and run-to-run reproducibility at this time, a demonstration of the initial intra-wafer thickness (4%) and doping (7%) uniformity obtained from the multiwafer reactor is contained in Fig. 13.

5. Conclusions

Device quality SiC active layers have been grown by vapor phase epitaxy. We have demonstrated specular morphology, net background doping density less than $1 \times 10^{14}\,cm^{-3}$, room temperature Hall mobilities of 950 cm^2/Vs, 3.1 μs minority carrier lifetimes, and intentional n-type doping from 1×10^{15} to above $1 \times 10^{19}\,cm^{-3}$. This has been achieved using optimized growth conditions in a basic single wafer horizontal SiC-VPE reactor. Similarly impressive results have also been reported in the literature using vertical, hot-wall, and barrel SiC reactors. These device quality SiC active layers have enabled the demonstration of SiC power microwave and higher-power dc devices with performance exceeding those possible with conventional semiconductor materials.

We are currently transferring our growth processes to a multiwafer planetary SiC-VPE reactor to improve layer uniformity, reproducibility and throughput. Initial results from this reactor already show reasonable epitaxial layer quality and purity and significant improvements in epitaxial layer uniformity.

Acknowledgements   We thank Lin Thomas, Jim Chance, and Vince Toth for performing epitaxial growths and characterization. Drs. Jim Choyke of the University of Pittsburgh, Gerhard Pensl of the University of Erlangen-Nürnberg and Peder Bergman of Linköping University are thanked for the PL, mobility and lifetime results. Al Burk thanks Dr. Egbert Woelk, Gert Strauch, Dietmar Schmitz and the rest of the engineering staff of Aixtron, Dr. Yuri Makarov, University of Erlangen−Nürnberg and Dr. Mike O'Loughlin of Northrop Grumman for their collaborative efforts in the development of the planetary SiC reactor. This work was supported in part by the Department of the Air Force under contract F33615-92-C-5912 and F33615-95-C-5427 (Tom Kensky contract monitor).

References

[1] H. MORKOÇ, S. STRITE, G. B. GAO, M. E. LIN, B. SVERDLOV, and M. BURNS, J. Appl. Phys. **76**, 1363 (1994).
[2] T. YOSHINOBU, M. NAKAYAMA, H. SHIOMI, T. FUYUKI, and H. MATSUNAMI, J. Crystal Growth **99**, 520 (1990).
[3] L. B. ROWLAND, R. S. KERN, S. TANAKA, and R. F. DAVIS, J. Mater. Res. **8**, 2753 (1993).
[4] M. GHEZZO, D. M. BROWN, E. D. DOWNEY, J. KRETCHMER, W. HENNESSY, D. L. POLLA, and H. BAKHRU, IEEE Electron Device Lett. **10**, 639 (1992).
[5] A. A. BURK, JR., Workshop on Compound Semiconductor Materials and Devices, New Orleans, Louisiana, February 19 to 22, 1995.
[6] S. SRIRAM, G. AUGUSTINE, A. A. BURK, JR., R. C. GLASS, H. M. HOBGOOD, P. A. ORPHANOS, L. B. ROWLAND, R. R. SIERGIEJ, T. J. SMITH, C. D. BRANDT, M. C. DRIVER, and R. H. HOPKINS, IEEE Electron Device Lett. **17**, 369 (1996).
[7] R. R. SIERGIEJ, S. SRIRAM, R. C. CLARKE, A. K. AGARWAL, C. D. BRANDT, A. A. BURK, JR., T. J. SMITH, A. MORSE, and P. A. ORPHANOS, Tech. Digest Internat. Conf. SiC and Related Materials, Kyoto (Japan), 1995 (p. 321).
[8] O. KÖRDINA, J. P. BERGMAN, A. HENRY, E. JANZÉN, S. SAVAGE, J. ANDRÉ, L. P. RAMBERG, U. LINDEFELT, W. HERMANSSON, and K. BERGMAN, Appl. Phys. Lett. **67**, 1561 (1995).
[9] J. N. SHENOY, J. A. COOPER, JR., and M. R. MELLOCH, IEEE Electron Device Lett. **18**, 93 (1997).
[10] A. K. AGARWAL, J. B. CASADY, L. B. ROWLAND, S. SESHADRI, W. F. VALEK, and C. D. BRANDT, submitted to IEEE Electron Device Lett.
[11] Y. M. TAIROV and V. F. TSVETKOV, J. Crystal Growth **52**, 146 (1981).
[12] G. ZIEGLER, P. LIANG, D. THEIS, and C. WEYRICH, IEEE Trans. Electron Devices **30**, 277 (1984).
[13] C. H. CARTER, JR., L. TANG, and R. F. DAVIS, 4th National Review Meeting on Growth and Characterization of SiC, Raleigh (NC), 1987.
[14] H. M. HOBGOOD, D. L. BARRETT, J. P. MCHUGH, R. C. CLARKE, S. SRIRAM, A. A. BURK, JR. GREGGI, C. D. BRANDT, R. H. HOPKINS, and W. J. CHOYKE, J. Crystal. Growth **137**, 181 (1994).

[15] H. M. HOBGOOD, J. P. McHUGH, J. GREGGI, R. H. HOPKINS, and M. SKOWRONSKI, Inst. Phys. Conf. Ser. No. 137, 7 (1994).
[16] H. M. HOBGOOD, R. C. GLASS, G. AUGUSTINE, R. H. HOPKINS, J. JENNY, M. SKOWRONSKI, W. C. MITCHEL, and M. ROTH, Appl. Phys. Lett. **66**, 1364 (1995).
[17] J. A. POWELL, Westinghouse SiC Technical Advisory Committee Meeting, December 9, 1994, Pittsburgh (PA).
[18] N. KURODA, K. SHIBAHARA, W. S. YOO, S. NISHINO, and H. MATSUNAMI, Extended Abstracts 19th Conf. Solid State Devices and Materials, Tokyo, 1987 (p. 227).
[19] H. S. KONG, H. J. KIM, J. A. EDMOND, J. W. PALMOUR, J. RYU, C. H. CARTER, JR., J. T. GLASS, and R. F. DAVIS, Mater. Res. Soc. Symp. Proc. **97**, 233 (1987).
[20] W. A. TILLER, The Science of Crystallization: Microscopic Interfacial Phenomena, Cambridge University Press, Cambridge 1991.
[21] D. J. LARKIN, P. G. NEUDECK, J. A. POWELL, and L. G. MATUS, Appl. Phys. Lett. **65**, 1659 (1994).
[22] S. TAKAGISHI and H. MORI, Jpn. J. Appl. Phys. **22**, L100 (1984).
[23] J. A. POWELL, D. J. LARKIN, L. G. MATUS, W. J. CHOYKE, J. L. BRADSHAW, L. HENDERSON, M. YOGANATHAN, J. YANG, and P. PIROUZ, Appl. Phys. Lett. **56**, 1442 (1990).
[24] S. KARMANN, W. SUTTROP, A. SCHONER, M. SCHADT, C. HABERSTROH, F. ENGELBRECHT, R. HELBIG, G. PENSL, R. A. STEIN, and S. LEIBENZEDER, J. Appl. Phys. **72**, 5437 (1992).
[25] A. A. BURK, JR., D. L. BARRETT, H. M. HOBGOOD, R. R. SIERGIEJ, T. T. BRAGGINS, R. C. CLARKE, G. W. ELDRIDGE, C. D. BRANDT, D. J. LARKIN, J. A. POWELL, and W. J. CHOYKE, Inst. Phys. Conf. Ser. No. 137, 29 (1994).
[26] H. S. KONG, J. T. GLASS, and R. F. DAVIS, U.S. Patent No. 5011549 (April 30, 1991).
[27] O. KORDINA, C. HALLIN, R. C. GLASS, and E. JANZÉN, Inst. Phys. Conf. Ser. No. 137, 305 (1994).
[28] R. RUPP, P. LANIG, J. VÖLKL, and D. STEPHANI, J. Crystal Growth **146**, 37 (1995).
[29] G. S. TOMPA, M. A. McKEE, C. BECKMAN, P. A. ZAWADZKI, J. M. COLABELLA, P. D. REINERT, K. CAPUDER, R. A. STALL, and P. E. NORRIS, J. Crystal Growth **93**, 220 (1988).
[30] H. JÜRGENSEN, Microelectronic Engng. **18**, 119 (1992).
[31] A. A. BURK, JR. and L. B. ROWLAND, J. Crystal Growth **167**, 586 (1996).
[32] J. A. POWELL, D. J. LARKIN, and P. B. ABEL, J. Electronic Mater. **24**, 295 (1995).
[33] T. KIMOTO, A. ITOH, and H. MATSUNAMI, Appl. Phys. Lett. **66**, 3645 (1995).
[34] Y. C. WANG and R. F. DAVIS, J. Electronic Mater. **20**, 869 (1991).
[35] H. S. KONG, J. T. GLASS, and R. F. DAVIS, J. Appl. Phys. **64**, 2672 (1988).
[36] C. HALLIN, PhD Thesis, Linköping University, Sweden, 1996.
[37] T. KIMOTO and H. MATSUNAMI, J. Appl. Phys. **76**, 7322 (1994).
[38] C. H. CARTER, JR. and V. F. TSVETKOV, see [7] (p. 11).
[39] C. HALLIN, to be published in J. Crystal Growth.
[40] T. KIMOTO, A. ITOH, and H. MATSUNAMI, Appl. Phys. Lett. **66**, 26 (1995).
[41] O. KORDINA, J. P. BERGMAN, C. HALLIN, and E. JANZÉN, Appl. Phys. Lett. **69**, 1 (1996).
[42] G. B. STRINGFELLOW, J. Crystal Growth **75**, 91 (1986).
[43] P. M. FRIJLINK, J. Crystal Growth **93**, 207 (1988).
[44] Aixtron Inc., Kackerstr. 15–17, D-52072 Aachen, Germany.
[45] P. M. FRIJLINK, J. L. NICOLAS, and P. SUCHET, J. Crystal Growth **107**, 166 (1991).
[46] Y. C. WANG, R. F. DAVIS, and J. A. EDMOND, J. Electronic. Mater. **20**, 289 (1991).
[47] W. V. MUENCH and I. PFAFFENEDER, Thin Solid Films **31**, 39 (1976).
[48] T. KIMOTO, A. ITOH, and H. MATSUNAMI, Appl. Phys. Lett. **67**, 2385 (1995).

phys. stat. sol. (b) **202**, 281 (1997)

Subject classification: 68.55.Jk; 68.55.Ln; S6

Silicon Carbide Epitaxy in a Vertical CVD Reactor: Experimental Results and Numerical Process Simulation

R. Rupp (a), Yu. N. Makarov (c), H. Behner (b), and A. Wiedenhofer (a)

(a) Siemens AG, Corporate Research and Development, Department ZT EN 6,
P.O. Box 3220, D-91050 Erlangen, Germany

(b) Siemens AG, Medical Engineering, P.O. Box 3220, D-91050 Erlangen, Germany

(c) Institute of Fluid Mechanics, University Erlangen-Nürnberg, D-91058 Erlangen,
Germany

(Received January 31, 1997)

In this paper an overview is given on the epitaxial growth of SiC in a vertical CVD reactor. Results concerning impurity incorporation and ways to achieve background doping levels as low as 10^{14} cm^{-3} are discussed. Precise control of the C/Si ratio in the gas phase, which is easily achieved in the described reactor, and the use of reduced pressure, lead to good control of dopant incorporation over more than three orders of magnitude, and smooth surface morphology at growth rates higher than 5 μm/h. Doping variations $< \pm 12\%$ across 35 mm wafers can routinely be obtained. The quality of the epilayers is proven by electrical brakdown fields as high as 2×10^6 V/cm at $N_A - N_D = 5 \times 10^{-15}$ cm^{-3} achieved in both pn and Schottky diodes and an electron mobility higher than 700 cm^2/Vs at 300 K (4H-SiC) estimated from the on-resistance of these test devices. Another important experimental boundary condition, the influence of the gas composition at the end of the epitaxial growth process on the surface properties of the epitaxial layer, is described. It will be shown that surfaces nearly resistant against oxidation can be generated in a hydrogen free atmosphere. As a second main topic of this paper, results of an elaborate numerical process simulation will be described including both fluid mechanical and chemical behavior. The influence of the main process parameters like total flow, chamber pressure, and rotation speed on the stability of the flow was investigated. The results achieved are compared with experimental observations showing excellent agreement. The experimental observation of an irradiant layer in the gas phase in front of the wafer under typical process conditions is explained with the help of the numerical model. The usefulness of this specific feature for the optimization of process conditions is discussed.

1. Introduction

The development of SiC substrates with low defect density and diameters up to 2″ within the last few years offers the prospect for the realization of a wide variety of SiC demonstrator devices. But to gain significant commerical interest for such devices it is necessary to obtain a reasonable yield in conjunction with a small variation of their properties. In the field of power electronics these demands are especially harsh, because these devices must be able to handle 1000 V or more in reverse mode and at least several amperes in forward direction. This leads to active device areas on the order of mm². In addition to the substrate defect density the quality and homogeneity of the epitaxial layer is the dominant issue for fulfilling these economic requirements. In the last ten years significant progress can be observed in the understanding and technical realization

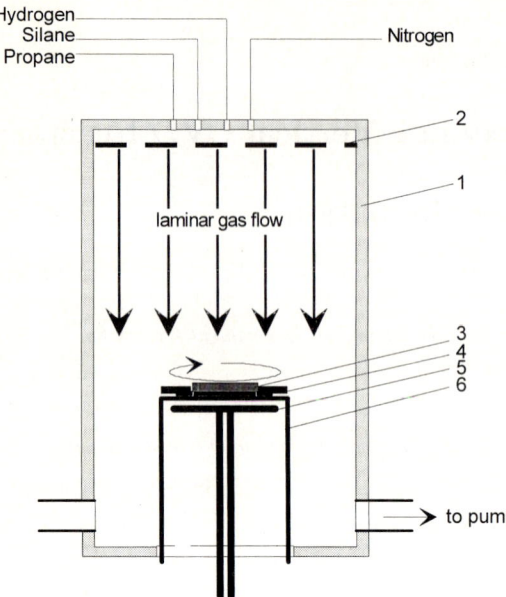

Hydrogen
Silane
Propane
Nitrogen

laminar gas flow

2

1

3
4
5
6

to pump

Fig. 1. Sketch of the principal design of the CVD reactor chamber. 1 Double-walled water cooled reactor chamber, 2 gas diffuser, 3 SiC wafer, 4 substrate holder, 5 rf pancake coil, 6 bell-jar-like graphite susceptor

of the SiC-VPE process. One very important milestone was the introduction of the step controlled epitaxial growth technique by Matsunami and coworkers [1, 2] in 1986. This was the key to grow epilayers of the same polytype as the substrate by using slightly off-oriented surfaces. Nevertheless, the reproducibility of dopant incorporation in the growing layers was a problem until Rottner and Helbig [3] showed the importance of excess carbon coming from un-coated graphite surfaces and Larkin [4, 5] reported the influence of the carbon-to-silicon ratio in the source gas on the incorporation of both acceptors and donors in SiC three years ago.

But even with all this new knowledge, still a lot of work has to be done to reach the point where SiC epitaxy can be called *production suited.*

This paper will give an overview about the state of the development and understanding of the vertical reduced pressure VPE process. The authors will report the achieved growth results. Process modeling results together with experimental observations will be described to explain special features of this process type. In an outlook, comments are made regarding the transfer of this single wafer process to a production suited multi-wafer reactor.

2. Experimental

2.1 Epitaxial growth

The CVD system used for this study was built by EMCORE Corp. (New Jersey). A palladium cell is used to provide high-purity hydrogen to the process and a loadlock equipped with a turbo pump is attached to the process chamber for batch processing. A pressure control system allows accurate pressure adjustment in the growth chamber, even at high flow rates. The process control and data acquisition for all relevant parameters is computer based. The special needs of SiC-CVD, e.g. very high temperature capability, are taken into account in the construction of the growth chamber which is displayed schematically in Fig. 1. The reactor allows processing of single wafers with diameters up to $2^1/_4''$. The rf-heated bell-jar-like susceptor and the substrate holder (aid for transferring the wafer from the loadlock into the reactor) are made of high-purity graphite. No SiC coating of the susceptor was used, in order to avoid unintentional SiC

Table 1
Typical growth conditions

silane flow 2% diluted in H_2	propane flow 5% diluted in H_2	C/Si ratio	temperature	rotation speed	hydrogen shroud flow	growth chamber pressure
500 to 950 sccm	75 to 325 sccm	0.5 to 2	1400 to 1550 °C	800 rpm	15 to 30 slm	50 to 300 Torr ($\approx 66.6 \times 10^2$ to 400×10^2 Pa)

deposition on the back of the wafer during growth. This is not only an advantage for the setup of vertical devices but also allows an easy determination of the thickness of the epilayers by measuring the weight difference before and after growth. The pancake-like rf coil is stationary and the susceptor is mounted on a rotary ferrofluidic feed-through sealing, both against the coil and the outer reactor wall. Rotation speeds up to 1500 rpm are possible with this setup at a leakage rate $< 10^{-5}$ Pa l/s. A two-color pyrometer is used for temperature control. The point of measurement is on the wafer surface, because the surrounding graphite surfaces change their optical properties upon being coated during the run. However, the absolute reliability of the temperature readings is still a weak point. It seems that large differences in the doping concentration of the wafer can influence the temperature measurement.

As a dopant source nitrogen is used, a hydrogen dilution stage is employed to allow an accurate control of the nitrogen flow in the reaction chamber over nearly three orders of magnitude.

4H-SiC wafers with a diameter of 30 and 35 mm purchased from Cree Res. Inc. (Durham NC) were used as substrates for most of our growth experiments. These wafers are oriented in the (0001) direction (Si face) with an off-angle of 3.5° and 8° towards (11$\bar{2}$0). Measurements of the lateral temperature distribution with the pyrometer showed that ΔT is less than 10 K across a 35 mm wafer.

Typical ranges of the most important parameters for our CVD growth runs are given in Table 1.

The reliability of a system is a very important industrial issue. With our epi-equipment we have made more than 1000 growth runs with an average duration of about 4 to 5 h within a time frame of $3^1/_4$ years. The average maintenance time in that period, including the periodical exchange of the graphite parts, was less than 20% of the process time.

2.2 Characterization of the epitaxial layers

The surface quality of the epitaxial layers was examined by means of optical and electron microscopy and with electron diffraction (LEED) combined with X-ray photoelectron spectroscopy (XPS) [6]. Secondary ion mass spectroscopy (SIMS) is used to evaluate the thickness homogeneity of the layers.

For the *CV* measurement we use Schottky contacts patterned by means of photolithography and etching, to get a precise contact area for calculating the doping concentration, a significant advantage over the widely used Hg probers. As Schottky metal we employ Ti or Ni vapor deposited on the as-grown surface. In order to determine the lateral and vertical doping distribution, we have installed an automatic wafer mapping system for current–voltage and capacitance–voltage measurements. Typically, we apply at least a 10×10 matrix of contacts. The reverse and forward properties of these contacts are also used to determine the electrical properties of the epitaxial layer.

3. Visualization of the Flow in the Reactor and Optimization of Flow Conditions

A special feature is observable in our reactor. Due to the reactor-specific temperature and concentration fields an irradiant layer in the gas phase close to the surface of the wafer can be observed under growth conditions without having a negative influence on the morphology of the growing layer. This observation is possible by naked eye via a radial viewport. In certain cases the layer is stable and we assume that the flow in the reactor is stationary and stable, too. In other cases time-dependent patterns are observed which look like swirls moving upward from the susceptor [7]. The behavior of this layer and its dependence on the process parameters (total flow rate, rotation speed and pressure) are used to optimize the flow conditions and to select the parameters which provide stable flow in the reactor. Fig. 2 shows a typical photography of the irradiant layer taken through the viewport at a chamber pressure of 200 Torr ($\approx 266 \times 10^2$ Pa). The shape of the irradiant layer is concave, showing the smallest distance to the susceptor at its center. The thickness of the layer is largest also at the center of the susceptor and disappears towards the periphery. For comparison, a similar picture, obtained under

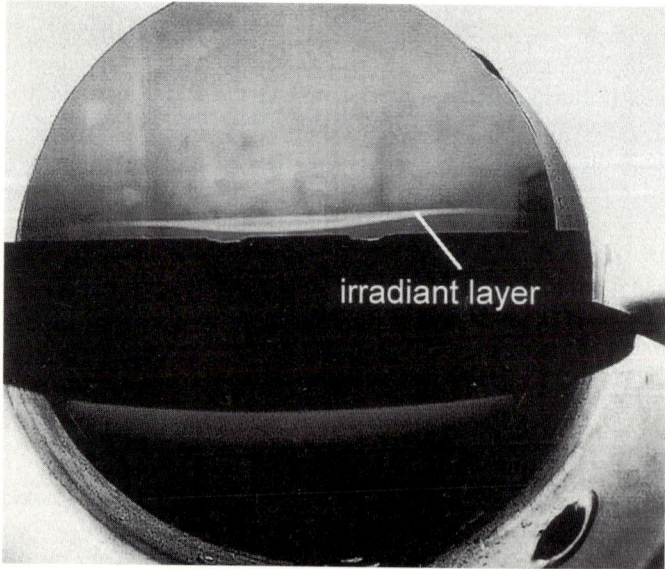

Fig. 2. Photography of the irradiant layer taken via the peripheral viewport. The black part in the center of the picture is the shutter, which is usually closed to prevent the window from heat radiation. The hot susceptor is hidden behind this item. Parameters: rotation speed 750 rpm, pressure 200 Torr ($\approx 266 \times 10^2$ Pa), total flow 36 slm hydrogen + 35 sccm silane, wafer temperature 1550 °C

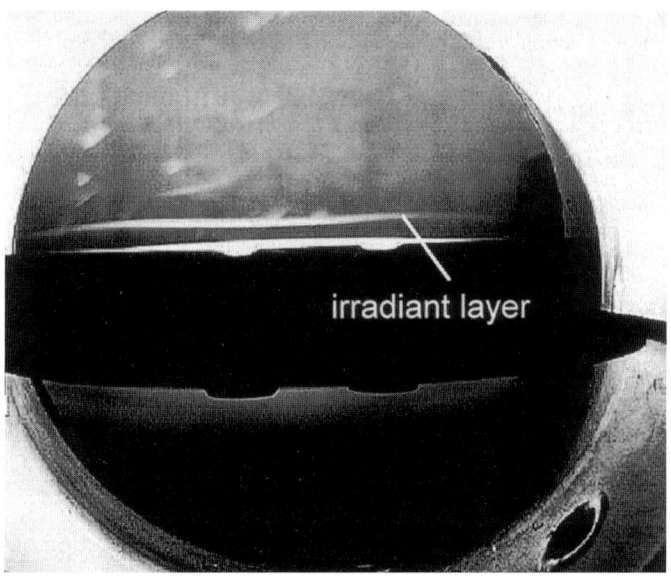

Fig. 3. Photography of the irradiant layer taken via the peripheral viewport, see Fig. 2. A small part of the hot susceptor is visible above the shutter. Parameters: rotation speed 750 rpm, pressure 350 Torr ($\approx 465 \times 10^2$ Pa), total flow 36 slm hydrogen + 35 sccm silane, wafer temperature 1550 °C

increased pressure (350 Torr $\approx 465 \times 10^2$ Pa), is shown in Fig. 3. The irradiant layer becomes more intense, its thickness is now less and it seems to be more concentrated. The distance between the susceptor and the layer is reduced.

This is a quite typical behavior of the irradiant layer for increasing total pressure. Further increase of pressure results in a time-dependent unstable feature characterized by many swirls observable in the chamber. Decrease of total pressure below 200 Torr leads to a less intense irradiant layer which moves away from the susceptor. At a pressure between 50 and 100 Torr (66.5×10^2 and 133×10^2 Pa) it vanishes, i.e. it is no longer observable by the naked eye. Overall, one gets the impression that the irradiant layer becomes more and more diluted with decrease of pressure.

The effect of variation of the total flow rate on the behavior of the irradiant layer is similar to the effect of pressure change. Decrease of the total flow rate usually results in an upward shift of the layer and a decrease of its intensity, as long as the flow in the reactor is stable.

The appearance of the irradiant layer is extremely helpful for a fast adjustment of stable flow conditions in our reactor, but the mechanisms of formation of this layer were not fully understood earlier [7]. In the present paper an attempt is made to explain this observations by using the additional experimental data and results of a numerical simulation of the flow and the chemical processes in the reactor during the CVD process.

4. Numerical Modelling
of the Chemical Vapor Deposition Process

4.1 Model description

For the present investigation we have employed numerical modelling of the deposition process for better understanding of the growth mechanisms and to optimize the deposi-

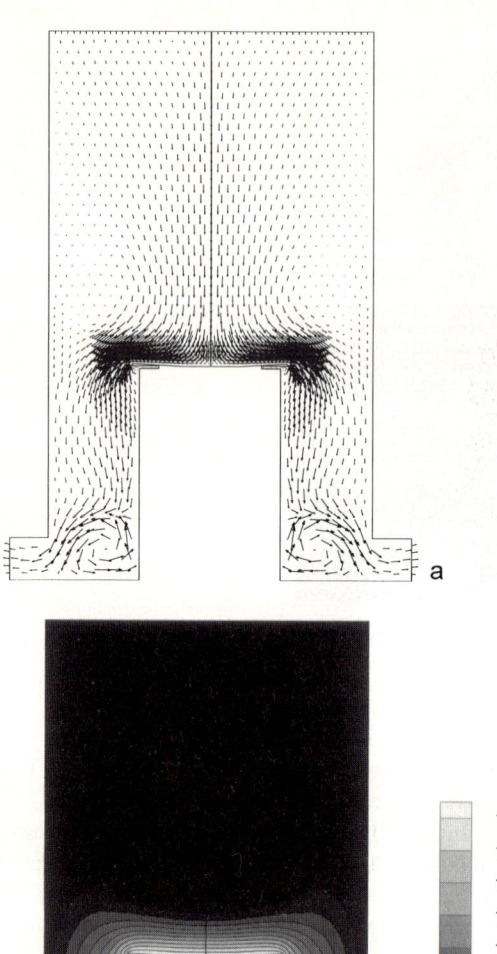

Fig. 4. Calculated gas flow velocity and temperature field in our reactor chamber. Parameters: rotation speed 750 rpm, pressure 200 Torr ($\approx 266 \times 10^2$ Pa), total flow 30 slm hydrogen, maximum temperature 1600 °C. Weak recirculation occurs near the wall. a) Velocity field, the length of the velocity vectors is proportional to the velocity. b) Temperature field, the temperatures are given in °C

T

1594.4
1464.9
1335.5
1206.1
1076.6
947.2
817.7
688.3
558.9
429.4

tion. The mathematical model used in the calculations is similar to that proposed in [8, 9]. It includes Navier-Stokes equations for flow, heat transfer and species transport with taking into account kinetics of chemical reactions in the gas phase, and a set of boundary relations to describe the growth and the chemical processes on the surface. The set of the homogeneous chemical reactions used in this work is obtained by reduction of the chemical model proposed in [10]. The key feature of the present approach is an assumption that a local equilibrium is established in the gas phase in the vicinity of the growing layer. This assumption is valid if the growth is limited not by kinetic pro-

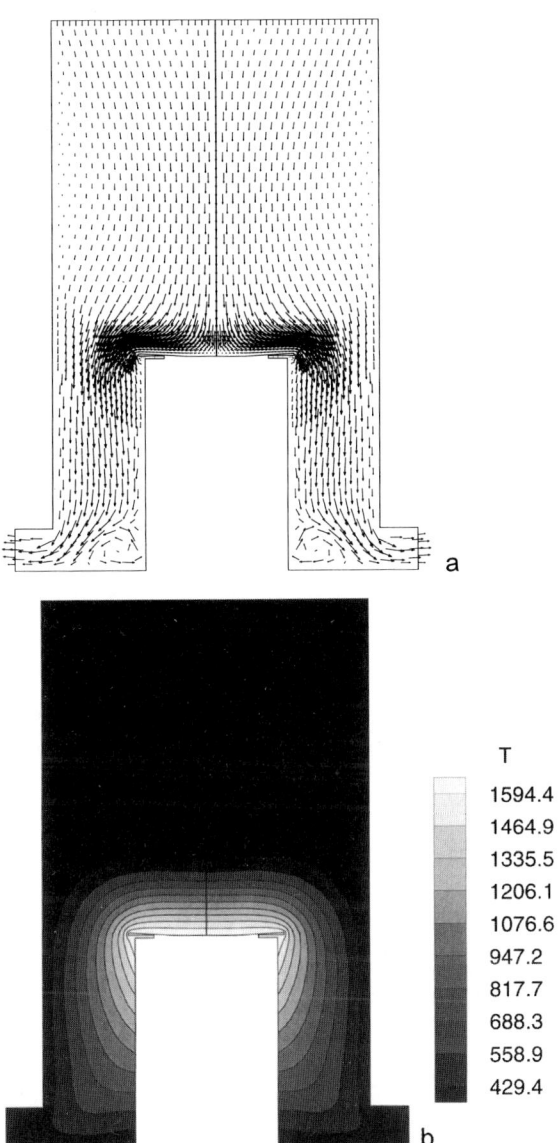

Fig. 5. Calculated gas flow velocity and temperature field in our reactor chamber. Parameters: rotation speed 750 rpm, pressure 100 Torr ($\approx 133 \times 10^2$ Pa), total flow 30 slm hydrogen, maximum temperature 1600 °C. No recirculation is observable. a) Velocity field, the length of the velocity vectors is proportional to the velocity. b) Temperature field, the temperatures are given in °C

T

1594.4
1464.9
1335.5
1206.1
1076.6
947.2
817.7
688.3
558.9
429.4

cesses on the growing surface but by the species transport. It was shown in [11] that the growth process is limited by diffusive transport if the off-orientation of the wafer is high enough (3.5° to 8° in this study).

4.2 Results of the process simulation

Calculated flow and temperature distributions in the reactor are shown in Fig. 4 for the following conditions: Total flow rate of hydrogen $F = 30$ slm, rotation speed $r = 750$ rpm and pressure $P = 200$ Torr ($\approx 266 \times 10^2$ Pa).

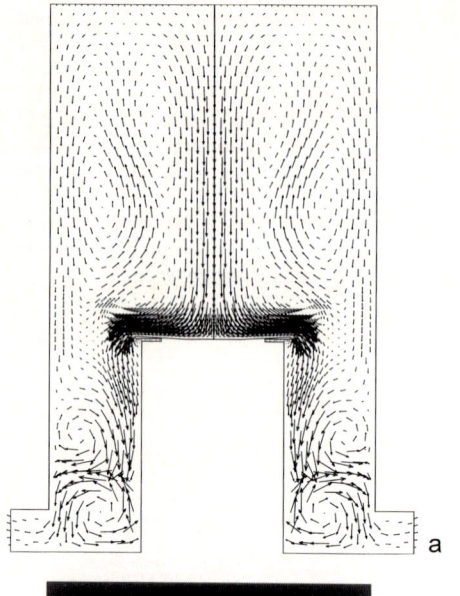

a

Fig. 6. Calculated gas flow velocity and temperature field in our reactor chamber. Parameters: rotation speed 750 rpm, pressure 300 Torr ($\approx 400 \times 10^2$ Pa), total flow 30 slm hydrogen, maximum temperature 1600 °C. Heavy recirculation and disturbation of the temperature field is observable. a) Velocity field, the length of the velocity vectors is proportional to the velocity. b) Temperature field, the temperatures are given in °C

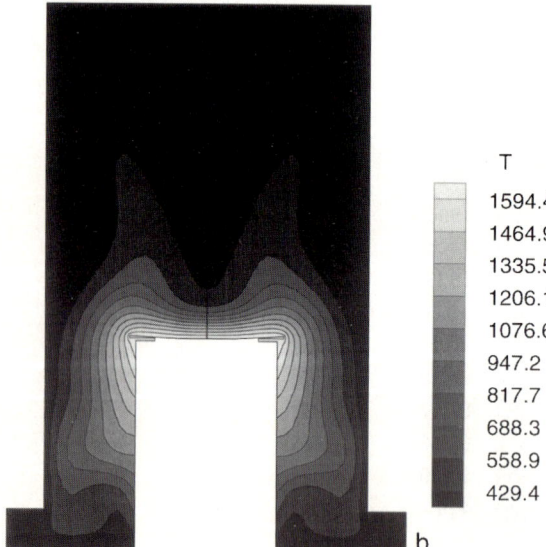

b

T

1594.4
1464.9
1335.5
1206.1
1076.6
947.2
817.7
688.3
558.9
429.4

These pictures reveal that, in general, the flow above the susceptor is almost stable apart from some weak recirculation above the susceptor and close to the reactor wall. Temperature gradients are almost parallel to the susceptor and are highest in its vicinity. The recirculation disturbs the planarity of the temperature isolines on the periphery of the susceptor. There also exists a recirculation area near the reactor outlet, but it has no influence on the deposition process.

Decrease of the total pressure improves the flow stability. For comparison the flow pattern for the case $P = 100$ Torr ($\approx 133 \times 10^2$ Pa) is shown in Fig. 5. In this regime the flow exhibits no recirculation at all and the temperature isolines are more smooth. In the opposite case, when the pressure is increased, a strongly unstable flow occurs. This is

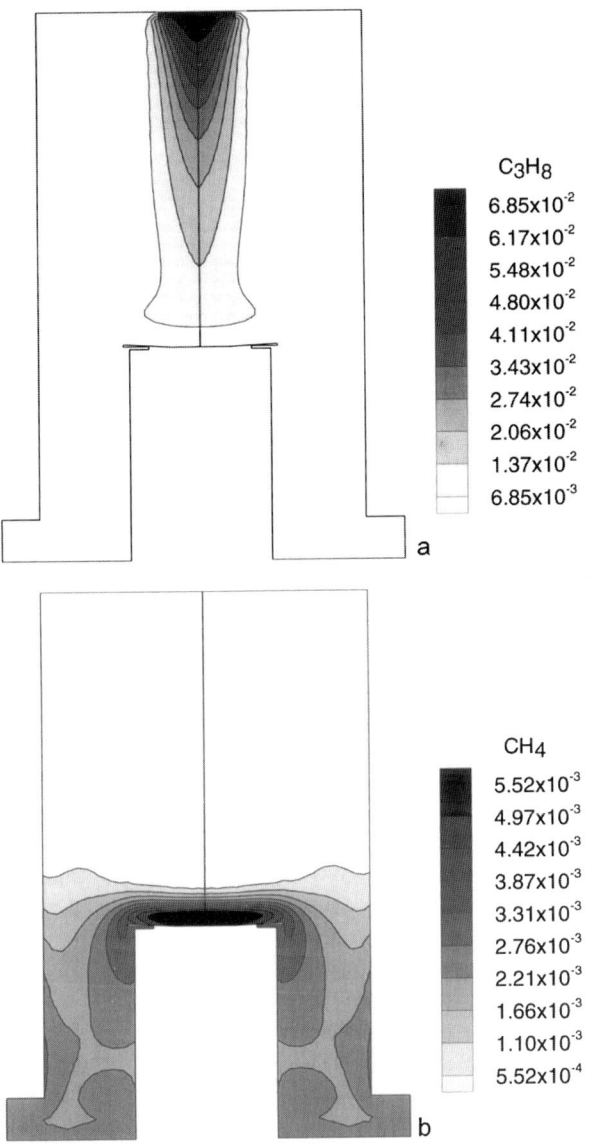

Fig. 7. Isolines of the concentration field of a) propane and b) its main decomposition product methane. Parameters: rotation speed 750 rpm, pressure 200 Torr ($\approx 266 \times 10^2$ Pa), total flow 30 slm hydrogen + process gases, maximum temperature 1600 °C

C_3H_8

6.85x10^{-2}
6.17x10^{-2}
5.48x10^{-2}
4.80x10^{-2}
4.11x10^{-2}
3.43x10^{-2}
2.74x10^{-2}
2.06x10^{-2}
1.37x10^{-2}
6.85x10^{-3}

CH_4

5.52x10^{-3}
4.97x10^{-3}
4.42x10^{-3}
3.87x10^{-3}
3.31x10^{-3}
2.76x10^{-3}
2.21x10^{-3}
1.66x10^{-3}
1.10x10^{-3}
5.52x10^{-4}

displayed in Fig. 6. The existence of several recirculation patterns above the susceptor can be observed. By further increase of pressure the flow gradually becomes three-dimensional and time-dependent, which can be observed through the reactor viewport by the time-dependent behavior of the irradiant layer (see Section 3). The rotation of the susceptor stabilizes the flow. Decrease of the rotation speed under the considered conditions results in loss of the flow stability.

The flow computations are useful to predict the boundary of instability of the axially symmetric flow. Verification of the calculations is performed using the observed boundaries for the appearance of instabilities of the irradiant layer by now experimentally varying flow rate, pressure and rotation speed. By this comparison we found that the

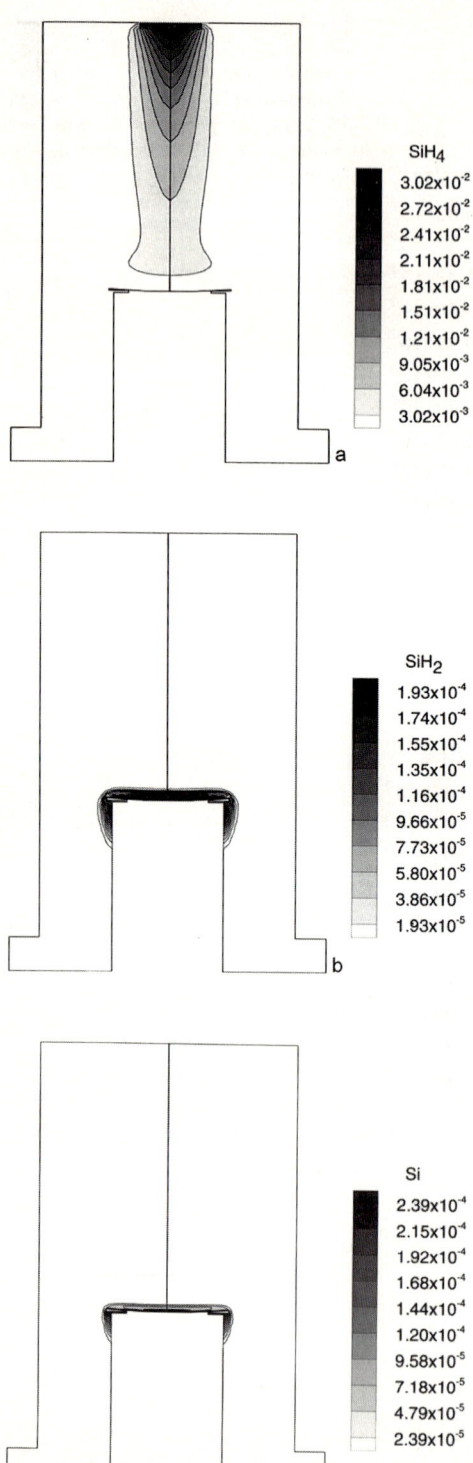

SiH$_4$

3.02x10^{-2}
2.72x10^{-2}
2.41x10^{-2}
2.11x10^{-2}
1.81x10^{-2}
1.51x10^{-2}
1.21x10^{-2}
9.05x10^{-3}
6.04x10^{-3}
3.02x10^{-3}

a

SiH$_2$

1.93x10^{-4}
1.74x10^{-4}
1.55x10^{-4}
1.35x10^{-4}
1.16x10^{-4}
9.66x10^{-5}
7.73x10^{-5}
5.80x10^{-5}
3.86x10^{-5}
1.93x10^{-5}

b

Si

2.39x10^{-4}
2.15x10^{-4}
1.92x10^{-4}
1.68x10^{-4}
1.44x10^{-4}
1.20x10^{-4}
9.58x10^{-5}
7.18x10^{-5}
4.79x10^{-5}
2.39x10^{-5}

c

Fig. 8. Isolines of the concentration field of a) silane and its sequence of decomposition products, b) SiH$_2$ and c) elemental Si. Parameters: rotation speed 750 rpm, pressure 200 Torr ($\approx 266 \times 10^2$ Pa), total flow 30 slm hydrogen + process gases, maximum temperature 1600 °C

two-dimensional flow simulations predict the experimentally observed boundaries of the flow instability with an accuracy of about 10%.

The flow and temperature distributions in the reactor determine the pathways of the chemical processes which occur during the deposition. The input species are silane and propane which decompose at the growth temperature. Decomposition of the species starts when the gas enters the hot area of the reactor. Kinetics of the gas phase decomposition is influenced by the input flow velocity and the pressure.

In Fig. 7 isolines of the mass fractions of propane C$_3$H$_8$ and the key product of its decomposition — methane CH$_4$ are shown. Propane is almost totally decomposed before it arrives at the surface of the SiC wafer and the growth is provided mainly by methane and some other sub-products of the propane decomposition — the so-called unsaturated hydrocarbons or radicals like CH$_3$, C$_2$H$_2$ and C$_2$H$_5$.

In Fig. 8 decomposition of silane and formation of SiH$_2$ and Si vapor species are illustrated. Obviously silane is totally decomposed in the gas phase. The transport of Si to the growing SiC layer occurs visathe sub-products SiH$_2$ and Si. Atomic Si is the dominant vapor species in the vicinity of the wafer surface.

The calculated radial growth rate for standard growth conditions yields a value of 6.2 µm/h in the center of a 30 mm wafer and is slightly decreased towards the edge (5.85 µm/h). The calculated total variation of growth rate across the wafer surface is about 6%, a value which corresponds quite well to the experimentally observed thickness inhomogeneity of about 5% (see Fig. 10 in Section 5.1). The experimental values of the growth rate itself are in the range 4.2 to 4.35 µm/h versus about 6 µm/h achieved by the calculations for comparable conditions. The reason for this discrepancy between theory and experiment concerning the growth rate is, in our opinion, related to a formation of Si clusters in the vapor phase and will be discussed further in the following section of this paper.

4.3 Explanation and discussion of mechanisms for Si cluster formation

In the following section an explanation on the nature of the irradiant layer over the susceptor based on Si cluster formation in the gas phase is proposed.

Historically, formation of clusters during CVD of tungsten from SiH_4 and WF_6 was reported in [12]. The rate of formation of these clusters was measured and it was shown that the process is influenced significantly by the total pressure and SiH_4/WF_6 ratio. However, this case does not directly apply to the conditions of CVD of SiC. We explain the existence of the irradiant layer during CVD of SiC in the vertical reactor by a scattering of the heat radiation coming from the hot susceptor at clusters and droplets of Si formed in the gas phase. We think that the reason for a formation of these clusters is the existence of Si vapor formed by the chemical decomposition of SiH_4 and SiH_2 in which the equilibrium vapor pressure of elemental Si is exceeded. To prove this, the spatial distribution of the ratio between the partial pressure of the Si vapor generated in the process and the saturated pressure of Si over liquid Si is shown in Fig. 9a to c. It is obvious that the supersaturation of Si vapor is quite significant. Such supersaturation should result in a fast formation of Si clusters or precipitates in the gas phase, and the velocity of this process should be proportional to the frequency of collisions between Si atoms [13]. First numerical estimations give a size of the clusters in the range of several ten Å [13] depending on the process conditions. Practically, the supersaturation should be replaced very soon by Si droplets or clusters formed via a condensation of the excess Si. These clusters have lower diffusivity and move mainly in accordance with the streamlines of the flow.

After a droplet is formed, two possible mechanisms can occur in dependence on the streamline contour in the flow:

A) If the droplet moves into the area with higher temperature, the saturated vapor pressure of Si over liquid Si increases and the droplet starts to re-evaporate. This results in a decreasing size of the droplet and, finally, it can disappear totally. With this mechanism it can be explained that these clusters do not have any influence on the morphology of the growing layer.

B) In the opposite case when the droplet moves into a colder area, it collects more and more Si atoms and grows in size. Its diffusivity decreases and it moves with the flow away from the wafer to the exhaust of the reactor. In this case the material collected in the droplet will be ineffective for the growth process. With this mechanism the fact can be explained that theoretically predicted growth rates (without taking this cluster formation into account) are always higher than the experimental values. If the droplet comes into the area where temperature is much lower than the melting point of Si, a solid Si particle can be formed. Such particles then can be found inside the exhaust filter of the reactor as well as at the water-cooled bottom of the reactor chamber.

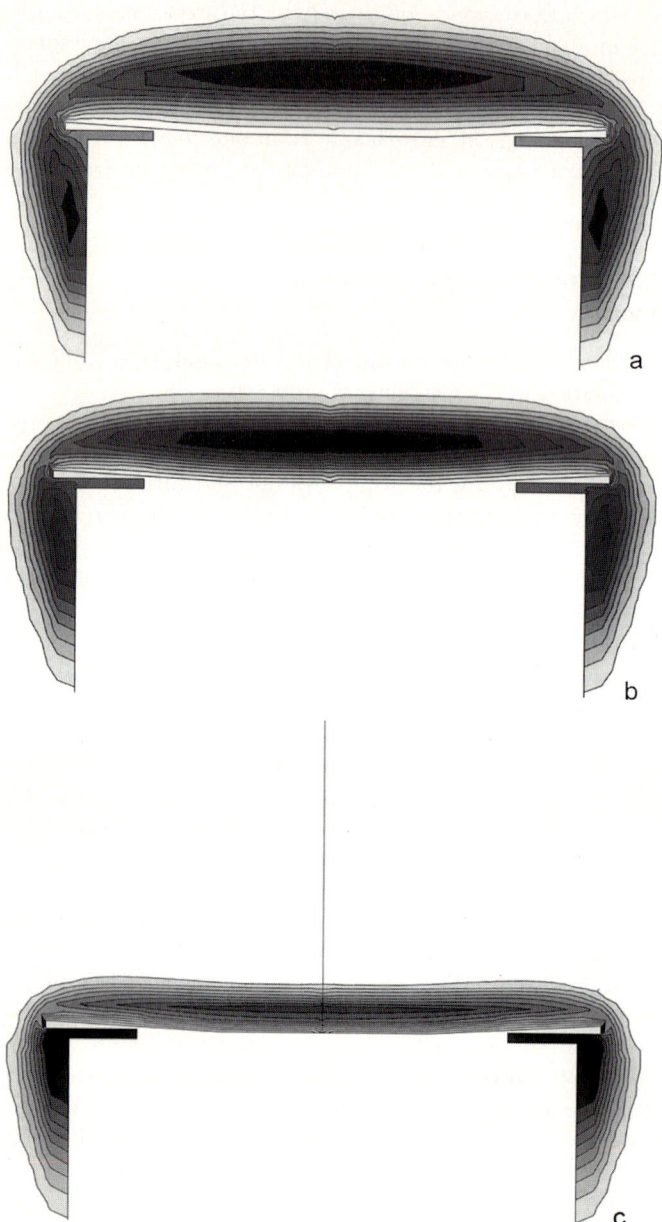

Fig. 9. Silicon supersaturation field: Ratio between Si concentration generated by the decomposition of silane (compare Fig. 8) and equilibrium vapor pressure of Si at the local temperature. The gray scale is linearly varied between the maximal supersaturation value (black) and zero (white). Parameters: a) Rotation speed 750 rpm, pressure 50 Torr ($\approx 67 \times 10^2$ Pa), total flow 30 slm hydrogen + process gases, maximum temperature 1600 °C, maximum supersaturation ratio: 3.51; b) rotation speed 750 rpm, pressure 100 Torr ($\approx 133 \times 10^2$ Pa), total flow 30 slm hydrogen + process gases, maximum temperature 1600 °C, maximum supersaturation ratio: 4.45; c) rotation speed 750 rpm, pressure 200 Torr ($\approx 267 \times 10^2$ Pa), total flow 30 slm hydrogen + process gases, maximum temperature 1600 °C, maximum supersaturation ratio: 4.06

Fig. 9a to c also exhibits the effect of total pressure on the location and value of the supersaturation. A clear correlation between the calculated maximum supersaturation in Fig. 9b, c and the appearance of the irradiant layer taken from the experiment (Fig. 2 and 3) is observable.

Comparable results on silane decomposition and Si cluster formation are described in [14]. The growth conditions used in this work are significantly different from ours (much lower growth temperatures of around 625 °C and pressures less than 1 Torr (133 Pa)) but nevertheless some significant similarities are reported: When the input partial pressure of silane was higher than a critical value, a strongly increased light scattering from the crystalline Si particles was observed. The typical sizes of the particles were described as 50 to 100 Å. The mechanism for this particle formation proposed in [14] was based on the assumption that linear Si–H polymers are formed in the vapor phase. From our point of view, this is unlikely because in this case the polymers usually will not form single-crystalline particles. Further, the obtained correlation between the localization of the irradiant layer and the maximum supersaturation of Si vapor in the vertical reactor cannot be explained by the polymerization mechanism, because this process is a slow chemical process which is determined by the concentration of SiH_2 molecules. We think that the mechanism of the homogeneous nucleation is substantially different from that proposed in [14] and is driven by the local supersaturation in Si vapor.

5. Description and Discussion of Experimental Results

5.1 Growth behavior

Fig. 10 shows SIMS depth profiles at different radial positions on a 1″ wafer with four alternately N-doped and undoped epilayers. A total thickness of 1.92 μm was determined by the weight increase of the sample, whereas the SIMS profile revealed a thickness of

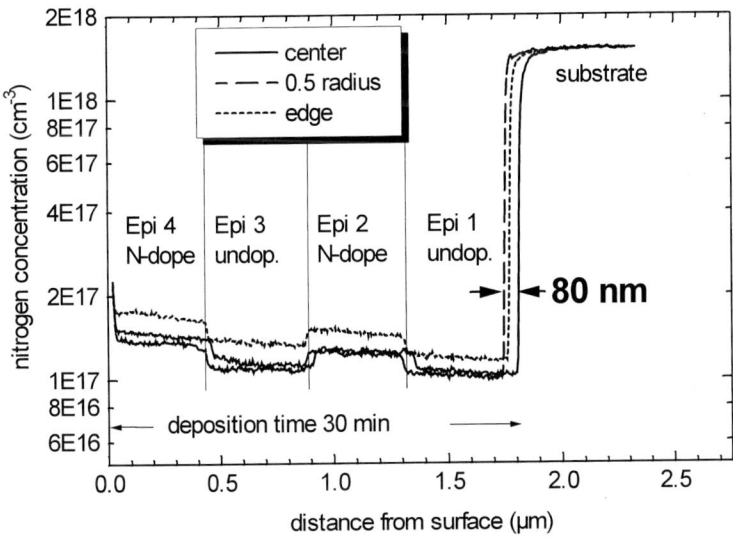

Fig. 10. Nitrogen depth profiles revealed with SIMS on different positions on a SiC wafer with a sequence of doped and undoped epilayers

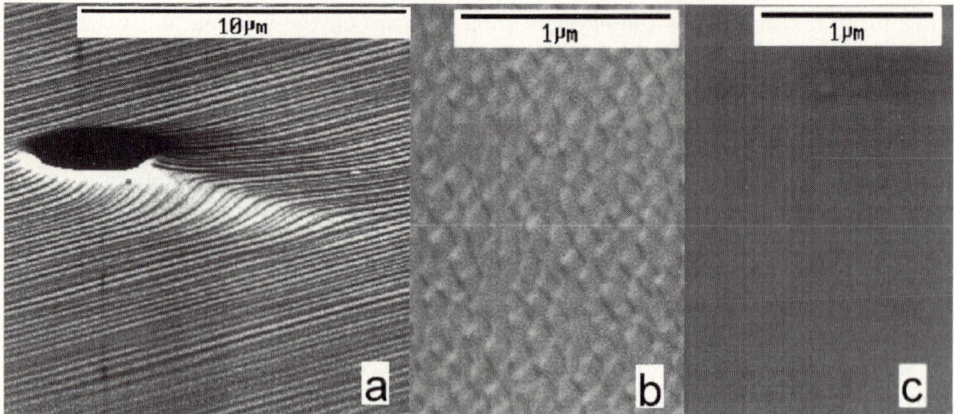

Fig. 11. SEM pictures of epitaxial grown SiC surfaces. a) Macrosteps (height 20 to 30 nm) in the vicinity of a micropipe, which causes a depression. b) Small steps (2 to 5 nm) typical for C/Si ≥ 1 and 3.5° off-orientation. c) No steps are visible on 8° off-oriented surfaces

1.72 to 1.80 µm. This indicates that on one hand the weight method really is a reasonable tool for a fast determination of layer thickness and that, on the other hand, the thickness inhomogeneity over a 1″ wafer is less than 5%. The comparatively big differences in the absolute N concentrations of the different profiles is due to the varying nitrogen background of the employed SIMS equipment, which is in the order of 10^{17} cm^{-3}.

Growth rate is the important issue in order to meet the productivity goals of an industrial VPE process, because it directly influences the process and accumulated production costs of potential devices. Therefore one of the aims of our work is to maximize the growth rates.

At a temperature of 1500 °C we have achieved growth rates up to 6 µm/h on 3.5° off-oriented 4H surfaces without any 3C inclusions while maintaining smooth surface morphology. Under the optical microscope the surface of such a layer looks featureless apart from very few of the well-known "amphitheater"-like disturbances [15, 16] originating at substrate defects and from Si droplets formed before the growth process starts, although we use a thermal pretreatment in pure hydrogen ambient. This is particulary due to the reduced pressure (below 300 Torr (≈ 40 kPa)) which allows a fast evaporation of excess Si from the surface of the wafer and therefore droplet formation is suppressed (compare [16]).

Using electron microscopy (SEM), we have recognized two different types of steps on these surfaces (Fig. 11): large steps with a typical height of 20 to 30 nm, which can effectively be eliminated by increasing the C/Si ratio in the source gas, and much smaller steps (2 to 5 nm), which seem to be nearly independent of variations of growth conditions. No steps are visible with SEM on 8° off-oriented surfaces even at growth rates of 7 to 8 µm/h but "amphitheater" formation is still observable in small quantities.

A precondition for the achievement of the surface morphology described above is the surface preparation before the growth process. If the surface damage resulting from the mechanical polishing procedure is not totally removed, a severe influence on the surface quality of the epitaxial layer is observable. Even if the surface looks totally smooth

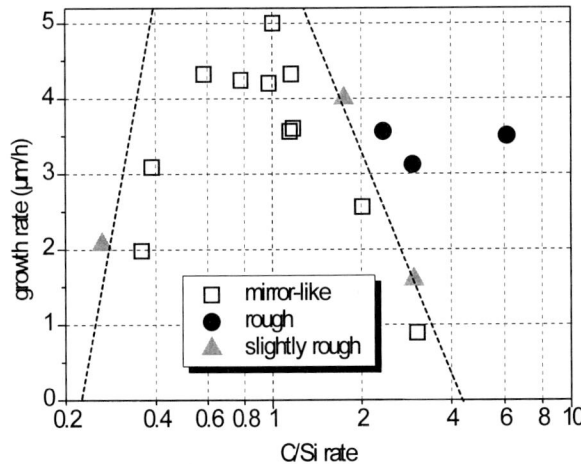

Fig. 12. Dependence of surface quality from C/Si ratio and growth rate

before the growth, all the hidden scratches reappear after the growth and are visible on the wafer with the naked eye. Microscopically, these scratches are decorated by a closed row of shallow depressios and — especially on wafers with 3.5° off-orientation — 3C inclusions.

Varying both growth rate and C/Si ratio we found a diagram of stable surface morphology as displayed in Fig. 12 for a reactor pressure of 50 Torr ($\approx$6.6 kPa). Increasing the process pressure leads to a narrowing of the region with good surface quality on the silicon excess side of the diagram, i.e. silicon droplet formation is enhanced by the increasing pressure.

At growth rates exceeding 6 μm/h (with 3.5° off wafer) we frequently observe triangular defects on the epitaxial surfaces, which can be identified by photoluminescence at 4 K and by the oxidation behavior to be of the 3C polytype. No differences could be found in the electrical behavior (breakdown field, see Section 5.5; conductivity) of the layers grown at different growth rates and C/Si ratios as long as such inclusions are not affecting the active area of the test structures. PL spectra are similar in the near-band region and do not show any additional lines pointing to defects generated by the high growth rates.

With 8° off-oriented wafers even growth rates of 7 to 8 μm/h do not cause problems with surface morphology and heterogeneous nucleation of 3C. This is in accordance to the predictions made by Kimoto and Matsunami [17], concerning the decreasing probability for two-dimensional nucleation on surfaces with increasing off orientation. This nucleation usually leads to 3C seeds on the surface.

5.2 Background impurities and nitrogen incorporation

In contrast to many other groups, we do not use SiC-coated parts to place our wafer on. There are two reasons why this is possible allowing reasonable layer purity:

A) The carrier and reagent gases are not passing over the hot uncoated parts before reaching the wafer surface.

B) The high gas flow velocity generated by the high speed rotation of the wafer holder counteracts the impurity diffusion from the surrounding parts.

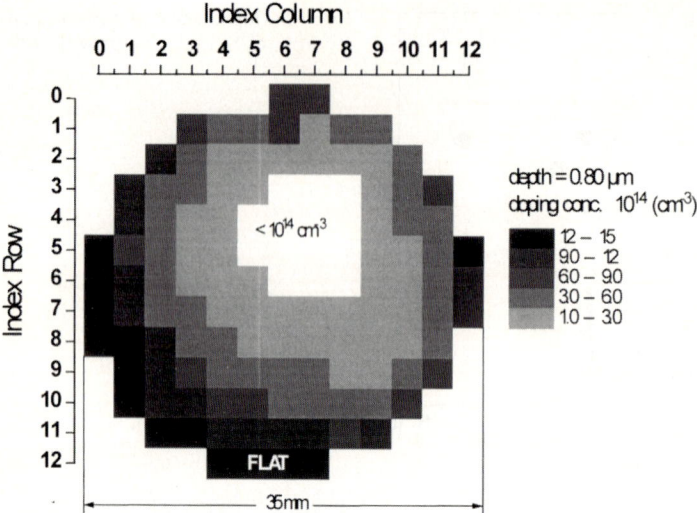

Fig. 13. Mapping of the background $N_A - N_D$ (in cm^{-3}) at a C/Si ratio of unity (determined by CV). The dominant impurity is boron

Nevertheless, the main impurity in the system, dominating the electrically active background, is boron (determined by Hall and admittance measurements). A typical distribution of the background acceptor level on a 35 mm diameter wafer with an epitaxial layer grown with C/Si = 1 shown in Fig. 13. In the central region of the wafer the background doping is below 10^{14} cm^{-3}. The bowl-like distribution indicates that especially at the wafer edge the diffusion of the boron impurity (released by the dissolution of carbon in the process atmosphere) against the main flow direction still is not negligible. A further reduction of these background impurities should be possible by silicon excess in the source gases, but we do not make use of this in our standard processing, in order to avoid increased macrostep formation as described in the previous section.

A comparison of the background doping distribution between wafers of 30 and 35 mm diameter shows that the minimum level in the center is about a factor of 5 lower for the bigger wafer. Nevertheless, the edge value is comparable, leading to a similar bowl-like distribution, where the bigger wafer shows a lower bottom of the bowl. This is another strong indication that the main source of impurities is located in the setup materials (susceptor, substrate holder) directly surrounding the wafer.

Similar to the boron background the nitrogen dopant incorporation is strongly influenced by the C/Si ratio as shown in Fig. 14. Between C/Si = 2 and 0.5 (all growth parameters constant, only propane flow varied; see figure caption) the average net donor concentration is increased by one order of magnitude. Interestingly, the lateral doping distribution along the linescan across the wafer has nearly the same shape for all four samples: There is a minimum in the center and the total variation between minimum and maximum value is about 20% for this series of growth runs.

This indicates that this typical lateral profile is not due to a shift of the C/Si ratio in the gas phase passing over the wafer (as would be expected for most of the horizontal reactors). In the above experiments we measured a temperature linescan across the wafer, which has essentially the same shape, exhibiting a total ΔT of about 15 K for these

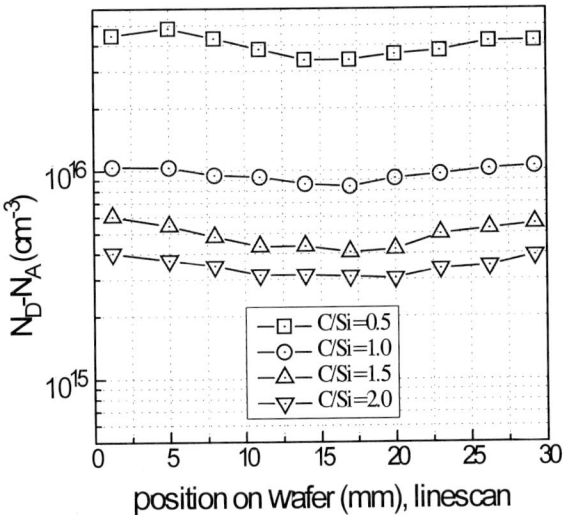

Fig. 14. Dependence of the net donor concentration on the carbon to silicon ratio in the source gas (N_2 flow 2.5 sccm, chamber pressure 300 Torr ($\approx 400 \times 10^2$ Pa), silane flow constant, propane varied)

runs (minimum temperature in the center). A readjustment of our rf coil reduces this value to 7 K and leads to an increased doping homogeneity ($\Delta(N_D - N_A) \approx 12\%$) as displayed in Fig. 15.

The dependence of the nitrogen incorporation on the nitrogen flow was found to be linear from 0.1 to 100 sccm (= available flow range for our nitrogen dilution configuration, see Fig. 16). With the help of the site-competition effect, we can gain another order of magnitude, leading to a total controllable n-doping range from about (6 to 8) $\times 10^{14}$ to (1 to 2) $\times 10^{18}$ cm^{-3} with a single dopant source. The lower limit is caused by the background impurity concentration (comp. Fig. 13).

In the vertical direction the doping concentration is constant within margins of $\pm 2\%$ even in the 10^{15} cm^{-3} range, which is about the error of the C–V measurement. This is

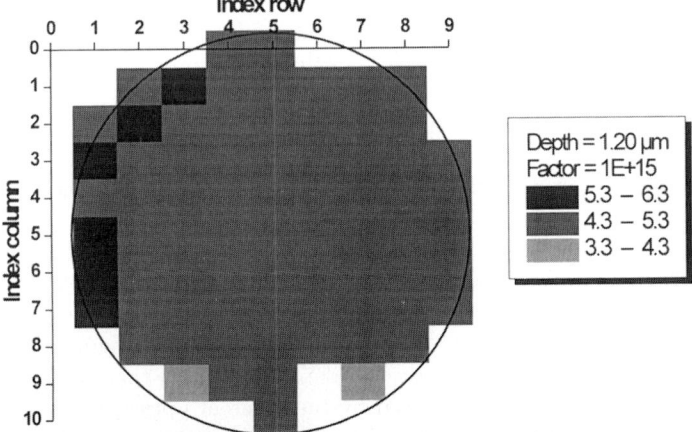

Fig. 15. Two-dimensional map of the lateral distribution of the net donor concentration (in cm^{-3}) of a 4H epitaxial layer (determined by CV)

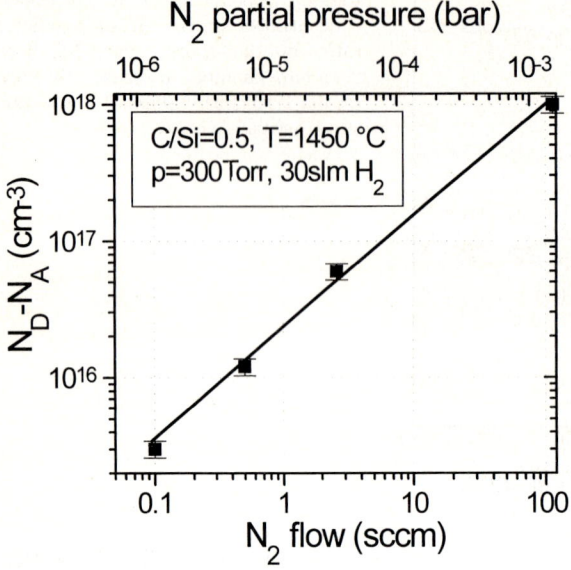

Fig. 16. Dependence between nitrogen flow into the reactor and net donor concentration (the error bars are due to variations on the wafer, compare Fig. 14, 15)

made possible by an accurate and time stable automatic control of all relevant process parameters.

The reproducibility of the doping value from one run to the other is excellent, as long as the wafer thickness and doping are not varied too much: For a series of ten consecutive runs the average doping value in the layers can be kept constant in margins of less than ±10%.

5.3 Growth rate and nitrogen incorporation in different polytypes

In the literature we have not found any data about the dependence of epitaxial growth velocity and dopant incorporation on the polytype of the substrate. Larkin [4] has reported that site competition is applicable for the Si face of both 4H and 6H, but does not give absolute values for the doping efficiency. Therefore we want to present here some results on this topic achieved very recently in our laboratory. The following types of substrates were used in this experiments: 1. standard 4H, 2. standard 6H, 3. mixed polytype substrates (consisting of at least two of the following polytypes: 4H, 6H, 15R, see Fig. 17). All these substrates were 3.5° off-oriented. Only polytype inclusions in the type 3 substrates having also this orientation were used for this analysis.

Seven substrates of type 1 (3) and 2 (4) were processed alternately under the same conditions. The average growth rate in these experiments was (3.9 ± 0.05) μm/h for 4H and (4.3 ± 0.1) μm/h for 6H. The difference in the dopant incorporation efficiency was less than 10% at a level of about 10^{16} cm^{-3} (i.e. comparable to the scatter from one run to the other). In interpreting these results, one has to take into account that temperature measurement and distribution may be influenced by the different optical properties of these polytypes. Therefore the results were checked by growth runs with type 3 substrates, where at least at the boundary between two polytypes both should face exactly the same growth conditions. Fig. 18 shows a linescan of the surface topography of the

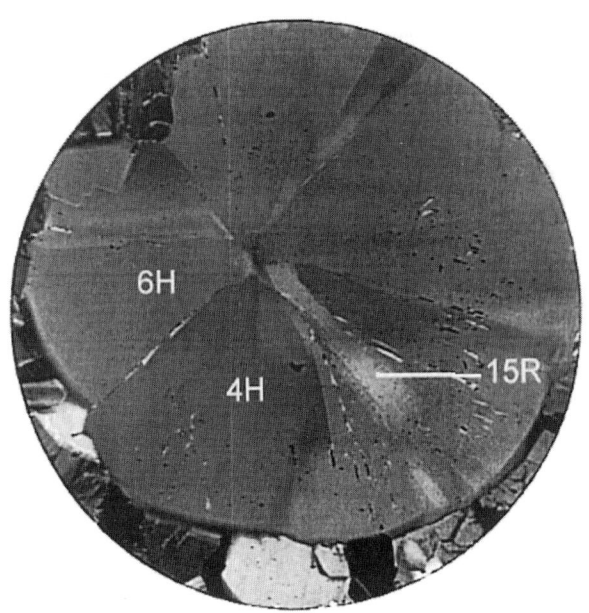

Fig. 17. Example of a wafer with inclusions of different polytypes used for epitaxial growth experiments (see text, here 6H and 15R in a 4H wafer). This wafer was contributed to this study by the Institute of Material Science at the University Erlangen

epitaxial layer across this boundary. It is most striking in this picture that there is a local deformation of the surface at the right boundary. On the other hand, there is only a shallow step on the left edge of the 15R region and — looking at the "bulk" areas — there is only a very small difference between the surface levels in the 15R and 6H regions (less than 100 nm at a total layer thickness of 10 μm, i.e. <1%). A similar behavior was found at 4H/15R and 6H/4H boundaries.

Concerning the doping concentration on type 3 wafers, differences not bigger than 5% are observable between 6H and 4H at a concentration of about 10^{16} cm^3. This is about the accuracy of our measuring method (resulting from uncertainties in the absolute area of the Schottky contact used for the C–V measurement). On the other hand, a small but

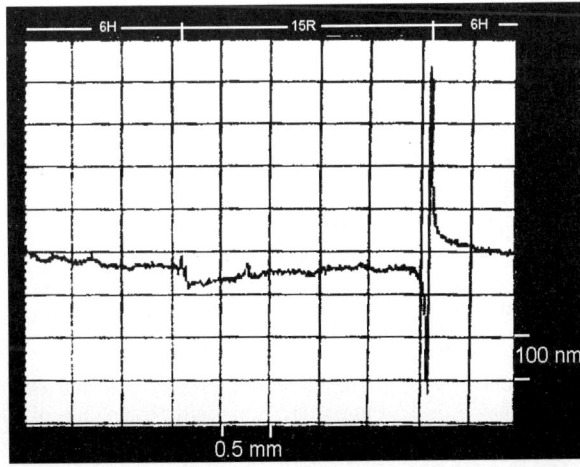

Fig. 18. Surface morphology across a 15R inclusion in a 6H wafer having an epitaxial layer with a thickness of 10 μm

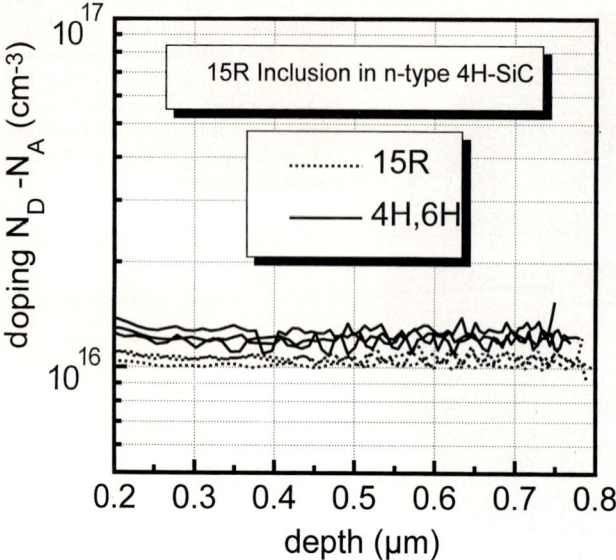

Fig. 19. Depth profile of the doping concentration in the epitaxial layer on a 15R inclusion in a 4H wafer and directly beneath that inclusion

obvious difference is observable between 4H and 15R as shown in Fig. 19. The average value is 1.06×10^{16} cm^{-3} in 15R and 1.25×10^{16} cm^{-3} in 4H in this specific experiment.

From these experiments we can conclude because of the identical growth conditions that the incorporation efficiency of nitrogen in the rhomboedral 15R polytype is about 20% less than in the hexagonal polytypes 4H and 6H. This has to be confirmed for a wider range of doping concentrations.

5.4 Surface constitution after the epitaxial growth

We have investigated how the surface structure and composition are influenced by the gas composition during the cooling procedure following the epitaxial growth of SiC. A detailed description of these experiments can be found in [6]. The most important results

Table 2

Dependence of the surface properties of epitaxial layers from the gas ambient ($p = 400 \times 10^2$ Pa) during the cooling process from growth temperature (1500 °C) to room temperature

preparation technique	wafer as delivered (Cree)	cooling after epitaxial growth done in		
		H$_2$ atmosphere	H$_2$/silane atmosphere	Ar atmosphere
surface composition	nearly stochio-metric	nearly stochio-metric	Si-rich, nearly amorphous	well ordered graphite top layer 0.8 nm (sp^2)
natural oxide formation	natural oxide 2 nm	natural oxide 1 nm after few hours	natural oxide 1 nm after few hours	nearly no oxide even after some days of exposure to air

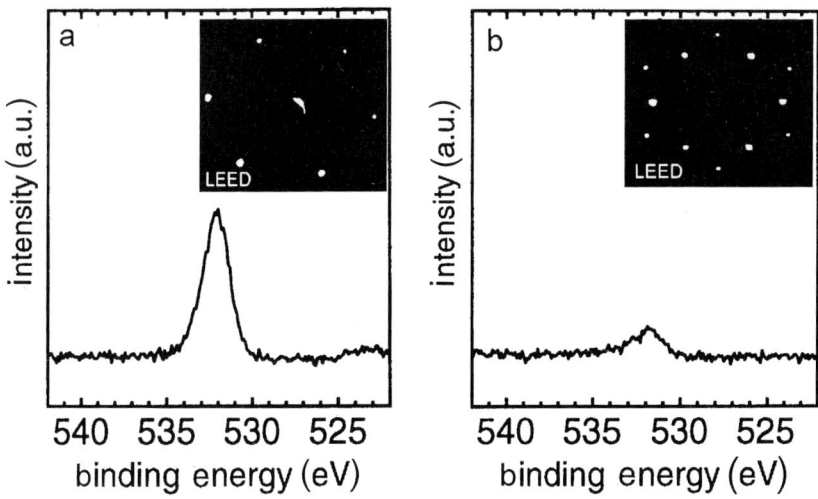

Fig. 20. O1s XP spectra and LEED patterns (inserts) of epitaxial surfaces cooled in different gas atmospheres (see text and Table 2). a) Hydrogen, b) argon (a.u. means arb. units)

are summarized in Table 2. A hydrogen atmosphere produces a very stochiometric surface, which exhibits the typical bright and sharp hexagonal LEED pattern of an undisturbed α-SiC surface (see Fig. 20a), whereas a small silane partial pressure (5 to 10% of the flow during growth) leads to a nearly amorphous surface with very weak diffraction patterns. In both cases the surface spontaneously forms a natural oxide at room temperature (about 1 to 2 nm within a few hours). If the cool-down process is carried out in a vacuum or an inert gas (e.g. Ar), a completely different surface shows up (Fig. 20b). The LEED pattern now consists of two simultaneously visible hexagons rotated by 30° against each other. One hexagon is similar to the diffraction pattern of the typical α-SiC surface (periodicity in real space 0.3 nm), while the other hexagon, having a larger unit cell in the reciprocal space, exhibits a real space lattice constant of 0.25 nm, which is very similar to the lattice constant of graphite (0.246 nm). The photoelectron spectrum of this surface shows an additional line at a binding energy of 284.3 eV, which is characteristic of sp^2-bonded graphite species. As a whole, we can conclude from the surface analysis [6] of this sample that the silicon evaporation at the high temperatures in the beginning of the cooling process leads to a carbon-enriched surface, which reconstructs itself in a well ordered manner. The thickness of the graphite layer was found to be about 0.8 nm.

The technical importance of this behavior becomes obvious after a closer look to Table 2 and Fig. 20: In contradiction to all other samples, the carbon-rich surface shows nearly no spontaneous oxidation, even after an exposure to air for several days. This in situ surface passivation can be of high importance especially for the formation of ohmic and rectifying contacts on this surface, where an oxide interfacial layer usually leads to a degradation of the contact properties. Furthermore, it has to be expected that silicon or carbon excess at the surface controls the formation of metal carbides or silicides for metals which can form both.

Table 3

Reverse characteristics of different rectifying structures (edge terminated [18])

	pn-diode	Schottky diode
structure	top p$^+$-layer implanted	barrier height ≈ 1.1 eV
net donor concentration	5×10^{15} cm^{-3}	8×10^{15} cm^{-3}
thickness of the n$^-$-layer	13 μm	10 μm
breakdown voltage	1800 V	>1200 V
electrical field	2.1 MV/cm at breakdown	1.9 MV/cm at 1200 V
leakage current at 1100 V	$<10^{-7}$ A/cm^{-2}	$\approx 10^{-4}$ A/cm^{-2}

5.5 Electrical behavior of the epitaxial layers

The final test for the epitaxial layer is the performance of a device with this layer as an active region. From the analysis of the differential resistance of Schottky diodes in forward direction we can estimate the electron mobility ($\parallel$ c-axis, i.e. direction of current flow) in our epitaxial layers to be in the range of 700 to 800 cm^2/Vs ($N_D - N_A = (5$ to $10) \times 10^{15}$ cm^{-3}), even at growth rates of 5 to 6 μm/h. The properties in the reverse direction are given in Table 3 and Fig. 21. Both Schottky and pn-diodes have an edge termination [18] to avoid an enhancement of the electrical field at the edge of the devices and are measured in Fluorinert^{TM} oil. Breakdown fields of more than 2 MV/cm can reproducibly be achieved in pn-diodes [19], which is close to the theoretical limit of SiC.

The yield of this diodes (active area 0.75 mm^2) exhibiting this high breakdown field primarily depends on the wafer quality, i.e. micropipe density.

6. Conclusions

The use of a vertical reactor with high gas velocities and rotation speed has both advantages and disadvantages compared to the well known horizontal reactor type. It is the most striking advantage that good homogeneity in growth rate and dopant incorpora-

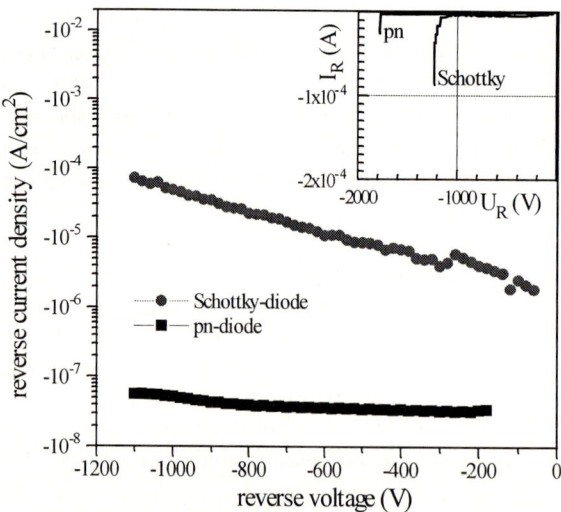

Fig. 21. Typical blocking characteristic of low-doped epi-layers (see text and Table 3). Reverse current measurement done with a Keithley SMU 237 (voltage limit 1100 V) in Fluorinert^{TM} oil. The insert shows the breakdown behavior (taken from a Sony curve tracer 370 A, limit 2000 V)

tion can easily be achieved, whereas in horizontal reactors either a separate rotation of the wafer has to be established (for example with gas foil rotation [20], which may cause additional technical difficulties) or a significant change in the gas composition along the substrate has to be taken into account, which may especially influence the homogeneity of the doping concentration. On the other hand, stable flow at reasonable flow rates can only be achieved at low pressure (in our case below 350 Torr (≈ 46 kPa)). Even if this reduced pressure increases the usable range of C/Si ratios towards Si excess conditions by allowing fast evaporation of excess Si from the substrate surface, it also results in a higher gas diffusion constant (almost indirectly proportional to the reciprocal pressure). Therefore diffusion of impurities emitted from the surrounding parts of the wafer towards the growing surface is a critical problem and the optimization of the design of susceptor and sample holder is very important and time consuming.

Nevertheless, within less than two years we have developed a stable and reliable process which satisfies many demands for an industrial SiC-VPE process. These are, especially high growth rates (>5 μm/h) while maintaining quality, accurate and reproducible process control, excellent homogeneity of all relevant properties and reasonable electrical characteristics and yields for upcoming, commercially interesting SiC devices like Schottky diodes in the 1200 V range.

7. Outlook

To make SiC-VPE really cost effective and therefore production suited, it is especially necessary, to scale up the process to a multiple wafer system and/or to use larger wafers (e.g. 4″). Due to the complicated heat/mass transfer and flow dynamics problems, we believe that it will be extremely helpful to solve this task with the aid of a numerical process simulation. This will enable a fast optimization of the reactor design and the process parameters. We are confirmed in this approach by the good agreement between experimental observations and numerical results for our existing CVD reactor.

Acknowledgements One of the authors would like to thank A. Vorobev, Yu. Egorov and A. Galyukov for their assistance in software development and computations, and S. Yu. Karpov for the analysis of the Si nucleation mechanism. Further thanks go to the members of the SiC research team at Siemens for their help on device realization and measurement techniques. This work was in part supported by the German Ministry of Education and Research under grant number 03M2746.

References

[1] H. MATSUNAMI, T. UEDA, and H. NISHINO, Mater. Res. Soc. Symp. Proc. **162**, 397 (1990).
[2] A. ITOH, H. AKITA, T. KIMOTO, and H. MATSUNAMI, Appl. Phys. Lett. **65**, 1400 (1994).
[3] K. ROTTNER and R. HELBIG, J. Crystal Growth **144**, 258 (1994).
[4] D. J. LARKIN, Mater. Res. Soc. Symp. Proc. **410**, 337 (1996).
[5] D. J. LARKIN, P. G. NEUDECK, J. A. POWELL, and L. G. MATUS, Appl. Phys. Lett. **65**, 1659 (1994).
[6] H. BEHNER and R. RUPP, Appl. Surf. Sci. **99**, 27 (1996).
[7] R. RUPP, P. LANIG, J. VÖLKL, and D. STEPHANI, Mater. Res. Soc. Symp. Proc. **423**, 253 (1996).
[8] YU. N. MAKAROV, Abstr. MRS Spring Meeting, San Francisco, 1996 (p. 333).
[9] S. YU. KARPOV, YU. N. MAKAROV, and M. S. RAMM, Inst. Phys. Conf. Ser. No. 142, 177 (1996).

[10] M. D. ALLENDORF and R. J. KEE, J. Electrochem. Soc. **138**, 841 (1991).
[11] T. KIMOTO, H. NISHINO, W. S. YOO, and H. J. MATSUNAMI, J. Appl. Phys. **73**, 726 (1993).
[12] E. J. MCINERNEY, T. W. MOUNTSIER, B. L. CHIN, and E. K. BROADBENT, Mater. Res. Soc. Symp. Proc. ULSI-VII, 69 (1992).
[13] S. YU. KARPOV, private communication.
[14] C. H. J. VAN DEN BREKEL and J. J. M. BOLLEN, J. Crystal Growth **54**, 310 (1981).
[15] J. A. POWELL, D. J. LARKIN, and P. B. ABEL, J. Electronic Mater. **24**, 295 (1995).
[16] A. A. BURK and L. B. ROWLAND, J. Crystal Growth **167**, 586 (1996).
[17] T. KIMOTO and H. MATSUNAMI, J. Appl. Phys. **75**, 850 (1994).
[18] H. MITLEHNER, U. WEINERT, W. BARTSCH, K. DOHNKE, and D. STEPHANI, Proc. Internat. Conf. EDPE, Dubrovnik (Croatia), Oct. 1996 (p. 64).
[19] K. HÖLZLEIN, H. MITLEHNER, R. RUPP, R. STEIN, D. PETERS, J. VÖLKL, and D. STEPHANI, Inst. Phys, Conf. Ser. No. 142, 561 (1996).
[20] A. A. BURK and L. B. ROWLAND, phys. stat. sol. (b) **202**, 263 (1997).

phys. stat. sol. (b) **202**, 305 (1997)

Subject classification: 68.55.Ln; S6

SiC Dopant Incorporation Control Using Site-Competition CVD

D. J. Larkin

NASA Lewis Research Center, 21000 Brookpark Road, Cleveland, OH 44135, USA

(Received January 31, 1997)

The use of site-competition epitaxy, which is based on intentional variation of the Si/C ratio during epitaxy, has now been reproduced in numerous national and international laboratories. Presented in this paper is a summary of the site-competition technique as a comparison of controlled doping on C-face 6H-SiC(000$\bar{1}$) versus Si-face 6H-SiC(0001) substrates for phosphorus (P), aluminum (Al), boron (B), and nitrogen (N). Also reported herein is the detection of hydrogen in boron-doped CVD SiC epilayers and hydrogen-passivation of the boron-acceptors. Results from low temperature photoluminescence (LTPL) spectroscopy indicate that the hydrogen content increased as the C–V measured net hole concentration increased. Secondary ion mass spectrometry (SIMS) analysis revealed that the boron and the hydrogen incorporation both increased as the Si/C ratio was sequentially decreased within the CVD reactor during epilayer growth. Boron-doped epilayers that were annealed at 1700 °C in argon no longer exhibited hydrogen-related LTPL lines, and subsequent SIMS analysis confirmed the outdiffusion of hydrogen from the boron-doped SiC epilayers. The C–V measured net hole concentration for the B-doped epitaxial layers increased more than threefold as a result of the 1700 °C anneal, which is consistent with hydrogen passivation of the boron-acceptors. For N-doped epitaxy, N incorporation into C-sites is favored on the Si-face whereas N incorporation into the Si-site is apparently the preferred lattice site on the C-face. Both P and N exhibit preferred incorporation on the C-face while Al and B incorporation is more efficient on the Si-face.

1. Introduction

In order to ensure reproducible and reliable SiC semiconductor device characteristics, controlled dopant incorporation must be accomplished. Unlike the doping technology routinely used in the silicon semiconductor industry, it is well known that SiC cannot be efficiently doped by diffusion at typical SiC epitaxial growth temperatures. Doping of SiC epitaxial layers of a device structure is accomplished in situ during the crystal growth of each epitaxial layer by flowing either a specific p-type or n-type dopant source into the CVD reactor. The most common p-type sources are trimethylaluminum (TMA) and diborane (B_2H_6) for obtaining Al-doped and B-doped SiC, respectively. Nitrogen (N_2) is the most common n-type dopant whereas phosphorus-doping (using PH_3) for producing n-type SiC epitaxial layers is less common. However, control over net dopant incorporation for CVD SiC epilayers by simply increasing or decreasing dopant-source flow (i.e. reactor concentration of dopant species) during epitaxial layer growth is limited both in reproducibility and in attainable doping range. This doping range was typically limited between $N_D \cong 2 \times 10^{16}$ to 5×10^{18} cm^{-3} for n-type and from $N_A \approx 2 \times 10^{16}$ cm^{-3} to 1×10^{18} cm^{-3} for p-type 6H-SiC epilayers [1].

Both the reproducibility and doping range has been greatly improved with the discovery of the site-competition epitaxial technique [2], which is based on varying the

Si-source to C-source ratio (Si/C ratio) within the crystal growth reactor to control the dopant incorporation during SiC epitaxial layer (epilayer) growth. Control of dopant incorporation is accomplished by varying the Si/C ratio within the growth reactor to effectively exclude or enhance a particular dopant atom from either a Si lattice site (Si-site) or C lattice site (C-site) of the growing SiC epilayer. As previously reported [3], the proposed mechanism for site-competition epitaxy on 6H-SiC(0001) Si-face substrates is based on the experimental evidence that Al dopant incorporation is inversely releated to the Si/C ratio, whereas the N dopant incorporation is directly related to the Si/C ratio within the reactor during epilayer growth. Site-competition epitaxy has been reported for control of nitrogen, phosphorus, aluminum, and boron dopant incorporation for the CVD of SiC(0001) Si-face and C-face epilayers [2 to 5]. The use of this technique has also been reproduced on SiC(0001) Si-face substrates in other laboratories using either propane [6, 7] or methane [8] in atmospheric and low pressure CVD systems, as well as reproduced [9] on SiC(1$\bar{2}$10) a-face substrates. Use of this technique has enabled an expansion of the reproducible doping ranges to include decreased concentrations ($N_D < 1 \times 10^{14}$ cm^{-3}) for the fabrication of multi-kilovolt SiC power devices, whereas the availability of greater doping concentrations ($N_D > 5 \times 10^{19}$ cm^{-3}) has resulted in devices with increased performance because of lower parasitic resistances. For example, use of site-competition epitaxy has led to improved device performance which includes the first multi-kilovolt rectifiers [10], ohmic as deposited contacts [11], and high temperature JFETs [12].

2. Experimental

The 6H-SiC epilayers were grown on commercially available n-type 6H- and 4H-(0001)SiC Si-face and C-face 6H-(000$\bar{1}$)SiC boule-derived wafers [13] in an atmospheric pressure CVD system [14, 15], with a typical growth rate of 3 to 4 μm/h. The SiC substrates were precleaned using a standard degreasing solution, followed by immersion in boiling sulfuric acid for 10 min, with a final deionized-water rinse and then dried with filtered nitrogen. The cleaned substrates were placed onto a SiC-coated graphite susceptor and then loaded into a water-cooled fused-silica reactor using a fused-silica-carrier. The samples were heated via the RF-coupled susceptor which was temperature controlled at 1500 °C using an optical pyrometer. Silane (gas cylinder containing 3% in H$_2$) and propane (3% in H$_2$) were used with a 3 sLpm flow of H$_2$ carrier/co-reactant for SiC epilayer growth. A flow of ultrapure hydrogen chloride gas (90 sccm) in a 3 sLpm flow of hydrogen was used during a 1350 °C in situ etch just prior to epilayer growth. All gases were mass flow controlled, including the ultra-pure hydrogen carrier-gas which was purified by using a heated-palladium diffusion cell. The epilayers were doped n-type by the addition of phosphine (200 ppm PH$_3$ in H$_2$) or nitrogen (0.1% N$_2$ in H$_2$) and p-type by the addition of diborane (100 ppm B$_2$H$_6$ in H$_2$) or trimethylaluminum (bubbler configuration) into the reactor during epilayer growth. Secondary ion mass spectrometry (SIMS) was performed [16] using a CAMECA IMS-4f double-focussing, magnetic sector ion microanalyzer. Cesium bombardment was used for determination of hydrogen, boron, phosphorus, and nitrogen atomic concentration profiles by using the detector in a negative secondary ion detection mode to monitor H-, P- and the diatomic species B(+C)-, N(+C)-, respectively. The aluminum and higher accuracy boron elemental concentrations were determined by using oxygen bombardment and a positive secondary ion detection mode. The sputter-time data was converted into depth data by measurement of crater depths, by using a stylus profilometer. Capacitance–voltage (C–V) measurements for B-doped epitaxial layers were obtained at 100 kHz on a mercury-probe instrument with a mercury-Schottky (contact area $= 1.64 \times 10^{-3}$ cm^2), using the mechanical sample-hold-down paddle as the electrical ground on the back of the sample. The low temperature photoluminescence (LTPL) was performed at 2K by immersing the SiC samples in pumped liquid helium, and excited using the 3250 Å radiation from a He–Cd laser. The LTPL spectra were recorded on a 0.75 meter Spex monochromator (10 Å/mm), using a thermoelectrically cooled ($-30\,^{\circ}$C) GaAs/CsO photomultiplier and a photon counting system.

3. Site-Competition Epitaxial CVD

As already stated, the Si/C ratio within the growth reactor has a strong influence on intentional and unintentional dopant incorporation during the epitaxial CVD growth of 6H, 3C (on 6H), 15R, and 4H-SiC. Specifically, the active n-type (nitrogen) carrier concentration was found to be directly proportional to the Si/C ratio, whereas, the active p-type (aluminum) concentration was found to be inversely proportional to the Si/C ratio for epitaxial growth on the SiC(0001)Si-face basal plane. For example, as the Si/C ratio within the growth reactor was decreased by increasing the propane flow, the active nitrogen concentration in the subsequently grown SiC epitaxial layers decreased. When the Si/C ratio was decreased from Si/C = 0.44 to 0.1, the unintentionally doped epitaxial layers changed from n-type to p-type. As a result, both p-type and n-type epilayers have been produced with room temperature carrier concentrations of $<1 \times 10^{14}$ cm^{-3}, as measured by both mercury-probed C–V and LTPL [17]. Previously, the unintentionally doped epilayers produced in our laboratory were exclusively n-type with the lowest net carrier concentrations typically limited to about $n = 2 \times 10^{16}$ cm^{-3}.

Dopants in SiC are believed to occupy specific lattice sites, specifically nitrogen occupies the carbon site (C-site) while aluminum (Al) occupies the silicon site (Si-site) of the SiC lattice [18, 19]. Based upon this information, and on our experimental results of decreasing the Si/C ratio, the relative increase in C concentration is proposed to "outcompete" the N for the C-sites of the growing SiC lattice. The analogous situation exists for an increased Si/C ratio, in which the relative increase in Si concentration "outcompetes" the Al for the Si-sites of the growing SiC lattice. This model has been previously named "site-competition epitaxy" and was initially used to rationalize the experimental results of active dopant dependence on the Si/C ratio. For example, a N$_2$ flow resulting in a reactor concentration of 100 ppm of N was introduced into the growth reactor where the majority of the N was excluded from the growing epilayer by only decreasing the Si/C ratio from about Si/C = 0.44 to 0.1. Subsequent experiments using a Si/C = 0.1 with a 100 ppm of N resulted in consistently producing an intentionally N-doped n-type SiC epilayer with a net carrier concentration of $n = 1 \times 10^{15}$ cm^{-3}. In contrast, growth using the more typical Si/C = 0.44 with 100 ppm of N yields an n-type epilayers of $n = 2 \times 10^{17}$ cm^{-3}. For the epilayers grown using a Si/C = 0.1 ($n = 1 \times 10^{15}$ cm^{-3}), the increased amount of C is believed to have outcompeted the N for the C-sites of the growing SiC lattice.

To determine if the electrically measured results corresponded to actual atomic N incorporation, a series of N doping profiles were grown and then characterized using

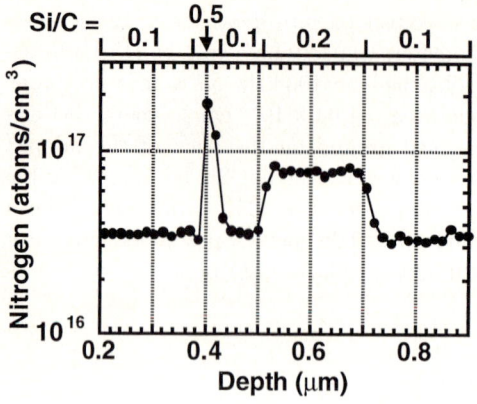

Fig. 1. A direct comparison between a SIMS profile analysis (top) and a mercury-probe C–V electrical result for a sample grown by varying only the carbon-source concentration (400 to 1800 at. ppm) while maintaining a constant silicon-source concentration (200 at. ppm) and a constant nitrogen concentration (180 at. ppm) within the CVD reactor during SiC epilayer growth

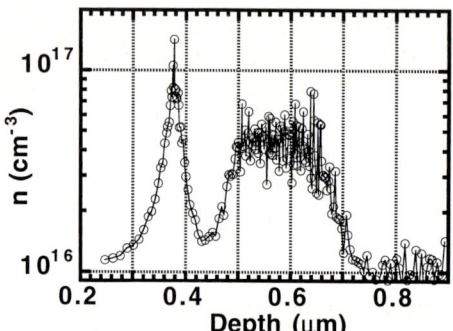

mercury-probe C–V and SIMS. During the growth of n-type doped epilayers, the silane (200 ppm) and molecular nitrogen (90 ppm) concentrations were held constant while only the propane concentration was varied between 130 and 660 ppm. The N dopant profiles were then characterized using SIMS and mercury-probe C–V. The results from one such experiment are shown in Fig. 1. The SIMS profile (Fig. 1 top) exhibits variations of atomic N concentration within the grown epilayer from varying only the Si/C ratio (between 0.5 and 0.1) during epilayer growth. The net carrier concentration profile obtained from mercury-probe C–V for the same sample (bottom of Fig. 1) correlates well with the SIMS profile. These results indicate that site-competition epitaxy can be used to control N-dopant incorporation into electrically active crystal sites of the growing SiC epilayer. To date, use of the site-competition epitaxial technique has been reported for control of phosphorus (P), Al, boron (B), and N dopant incorporation for the CVD of SiC(0001) Si-face and C-face epilayers of the various polytypes including 6H, 4H, 15R and 3C grown on 6H (3C/6H) [2, 3].

The remainder of the discussion of the site-competition effect is organized into sections for each specific dopant atom. This organization facilitates discussions of peculiarities associated with each dopant atom, such as the co-incorporation and hole passivation effect of hydrogen for B, the apparent reversal in site preference for N on the Si-face (C-site) versus the C-face (Si-site), and the significant difference in dopant incorporation efficiency for each dopant atom when simultaneously incorporated into Si-face versus adjacent C-face SiC epilayers. Also for the following sections, 6H-SiC is referenced for clarity and serves only as an example because the other common polytypes (i.e. 4H, 15R, and 3C/6H) exhibit similar dopant incorporation behavior to the site-competition effect.

4. Phosphorus Doping

Phosphorus doping was studied to provide more data on the site-competition effect. This is because phosphorus presently appears to have limited interest as a potential

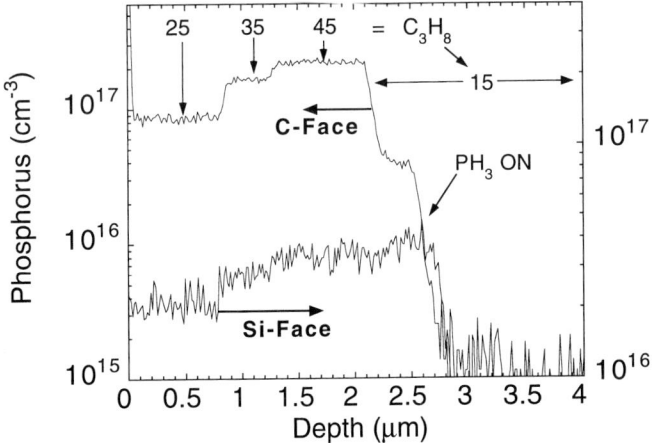

Fig. 2. SIMS depth profile of a phosphorus-doped epilayer simultaneously grown on a 6H-SiC(0001) Si-face versus a C-face substrate. Initially, the propane flow is constant then the phosphine flow is established. Then both the silane (20 sccm) and phosphine (100 sccm) flows are maintained constant while the propane flow is varied (stepwise)

replacement for nitrogen, especially considering that the phosphorus donor is reported to have only a slightly lower ionization energy (80 and 110 meV) [20] compared to that of the nitrogen donor (82 and 140 meV) [21] in 6H-SiC. For n-type doping with P, the site-competition effect is most consistent with the P atom mainly occupying the Si-sites. The P dopant incorporation was determined to be inversely related to the Si/C ratio as displayed in Fig. 2. As the propane flow was decreased, while maintaining a constant silane (20 sccm) and phosphine (100 sccm) flow, the SIMS determined P incorporation also decreased within the Si-face 6H-SiC epilayer. These changes in P incorporation with variation in the Si/C ratio are consistent with the P atom occupying the Si-sites of the SiC crystal lattice. As the propane flow is increased, the amount of vacant Si-sites increases which could allow enhanced P incorporation into the SiC epilayer.

Site-competition epitaxy had a pronounced effect for control of P dopant incorporation on both C-face and Si-face epilayers. As the propane flow was decreased, the P incorporation also decreased (see Fig. 2). As the silane flow was increased and then decreased, the P incorporation decreased then increased, respectively (not shown). It is believed that P is competing with Si for available Si-sites on the growing C-face SiC epilayer. This effect can be rationalized by considering that the lattice site occupied by each dopant atom is partly determined by the size of the dopant atom as compared to the size of the Si or C atom. Specifically, the non-polar covalent radii for the elements of interest are: Si (1.17 Å); C (0.77 Å); Al (1.26 Å); B (0.82 Å); P (1.10 Å); and N (0.74 Å); [22]. Using atomic size as a first approximation (i.e. by neglecting chemical bonding arguments), P should substitute for Si and not for C in the SiC lattice, which is consistent with our experimental results.

5. Aluminum and Boron Doping

It was previously reported [3] that the concentration of Al dopant atoms incorporated into a growing SiC epilayer can be decreased by increasing the silane concentration so that Si outcompetes Al for the Si-sites. For comparison, two different experiments that were conducted used identical flows of TMA introduced into the reactor but employed different Si/C ratios during epilayer growth. The resulting p-type epilayer grown with a Si/C ratio of 0.44 was measured by $C\text{–}V$ mercury probe to be 5×10^{16} cm^{-3}, while the epilayer grown using a Si/C ratio of 0.11 yielded a degenerately doped p-type epilayer with an estimated net carrier concentration of 1×10^{19} cm^{-3}. These experimental results support the hypothesis that as the Si/C ratio is decreased from 0.44 to 0.1, the relative amount of Si competing with the Al for the Si-sites of the SiC lattice also decreases, which results in an increased Al incorporation.

The site-competition effect remains active even at such highly degenerate Al dopant levels (i.e. although this effect is significantly diminished compared to the orders of magnitude change observed at lower dopant concentrations). As clearly demonstrated in Fig. 3, the concentration of Al incorporated into a growing SiC epilayer can be increased by decreasing only the Si/C ratio within the reactor during epilayer growth. This is accomplished by either increasing the C-source concentration (i.e. by increasing the propane flow into the reactor) or by decreasing the Si-source concentration in the CVD reactor. In Fig. 3, the SIMS determined Al incorporation increased as the propane flow was increased and also when the silane flow was decreased within the CVD reactor, which clearly demonstrates competition between Si and Al for the Si-sites on the growing Si-face 6H-SiC epilayer.

In general, p-type doping of epilayers grown on the C-face 6H-SiC($000\bar{1}$) samples was similar to that determined for the Si-face substrates. Both Al and B dopant incorporation on the C-face exhibited a similar dependence on the Si/C ratio, as previously described for the Si-face substrates. However, both Al and B incorporation

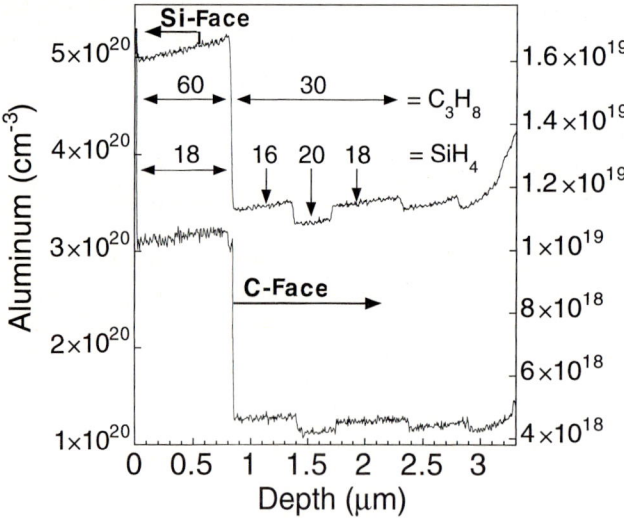

Fig. 3. SIMS depth profile of an aluminum-doped epilayer simultaneously grown on a 6H-SiC(0001) Si-face versus a C-face substrate. As the silane flow is varied during a constant propane flow, the Al is outcompeted by Si for available Si-sites. As the propane flow is increased to 60 sccm, Al incorporation increases because of the increase in available Si-sites

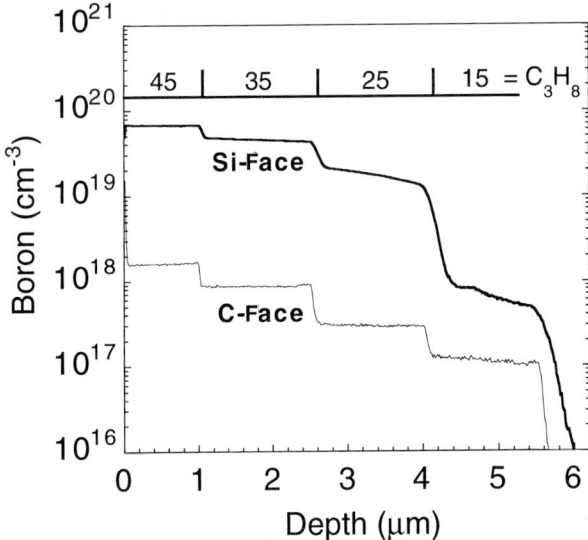

Fig. 4. SIMS depth profile of a boron-doped Si-face vs. C-face 6H-SiC(0001) epilayer from the same growth run. The silane (20 sccm) and diborane (50 sccm) flows were maintained constant as the propane flow was varied (sccm) during the epilayer growth

were markedly less efficient on the C-face substrates when compared to Si-face substrates that were doped during the same growth run. For Al doped C-face epilayers, the SIMS determined Al dopant incorporation was approximately 50× less than that of the corresponding Si-face epilayer. Similarly, B-doped C-face epilayers also exhibited a 50× lower atomic B concentration compared to simultaneously B-doped Si-face epilayers (see Fig. 4).

The similar behavior of B to Al is somewhat unexpected because B has been reported to occupy both the Si-site and the C-site [19, 23 to 26], which would preclude effective use of site-competition epitaxy for control of B doping in SiC. However, previous experimental results support that B preferentially occupies the Si-site of the SiC lattice [4, 26]. This is because B incorporation into the SiC epilayer decreases as the silane concentration increases (i.e. increasing the Si/C ratio), which is consistent with the increased amount of Si outcompeting the B for available Si-sites during growth of the SiC epilayer.

Conversely, boron incorporation can be increased by decreasing the Si/C ratio by decreasing the silane concentration or, alternatively, increasing only the propane concentration. Varying only the propane concentration for effective control of the boron doping is illustrated in Fig. 4, in which the Si/C ratio was decreased stepwise during epilayer growth by successive increases in propane flow while maintaining a constant silane (200 at. ppm) and diborane (3.3 at. ppm) concentration. Here, the boron incorporation in the epilayer increases as the propane concentration is increased because of the relative decrease in the Si/C ratio within the growth reactor. This relative decrease in silicon concentration enables the boron atoms to outcompete the silicon atoms for more of the available Si-sites on the surface of the growing SiC epilayer, resulting in increased boron incorporation.

To determine the reproducibility of the doping control, numerous B-doped SiC epilayers were grown during separate 2 h growth experiments, each using a constant Si/C ratio ranging from Si/C = 0.1 to 0.5. For selected epilayers, the SIMS determined elemental boron concentration ([B]) was compared to the net carrier concentration measured using mercury-probe C–V. A typical epilayer grown using a Si/C = 0.51 had a C–V measured net carrier concentration of $p = 5 \times 10^{15}$ cm^{-3} as compared to a net carrier concentration of $p = 3.5 \times 10^{17}$ cm^{-3} for an epilayer grown using a Si/C = 0.11 in a separate experiment but with an identical reactor concentration of diborane (1.6 ppm). SIMS analysis revealed an elemental B concentration of [B] = 6.5×10^{16} cm^{-3} for the lower doped ($p = 5 \times 10^{15}$ cm^{-3}) epilayer and [B] = 1×10^{18} cm^{-3} for the more highly boron-doped ($p = 3.5 \times 10^{17}$ cm^{-3}) SiC epilayer. The large increase in B incorporation resulting solely from a change in the Si/C ratio illustrates 1. the strong dependency of the B incorporation on the Si/C ratio used during epilayer growth and 2. the preferential B occupancy of the Si-site versus the C-site.[1]

6. Hydrogen Incorporation with Boron

6.1 Low temperature photoluminescence

Hydrogen incorporation was also revealed in the B-doped Si-face and C-face epilayers via low temperature photoluminescence and by SIMS analysis [4]. Three B-doped SiC epilayers, grown using identical Si/C ratios but with different diborane concentrations, were subsequently examined using LTPL. Identical Si/C ratios were used in order to eliminate potential effects of different propane concentrations on hydrogen incorporation, thereby isolating the hydrogen incorporation effect to only the change in diborane concentration. The resulting spectra, shown in Fig. 5a, reveal that a significant amount of hydrogen was contained in each of the B-doped epilayers. The most prominent line (at 4193 Å and labelled H$_3$), observed both in hydrogen-ion implanted SiC [27, 28] and in CVD grown SiC epilayers [29], has been previously assigned to a hydrogen center. The H$_3$ line intensity increased, relative to the two N bound exciton lines (S$_0$ and R$_0$), as the C–V net carrier concentration increased for the sample series $p = 2 \times 10^{15}$, 7×10^{15}, and 2×10^{16} cm^{-3}. This correlation of increased H$_3$ intensity with increasing p-type character suggests that the hydrogen incorporation is directly proportional to the amount of B incorporated into the SiC epilayer.

The intensity of all hydrogen related spectra decreases after the samples are subjected to a 1700 °C anneal in argon for 0.5 h (see Fig. 5b). In comparing the results shown in Fig. 5b with those of a, it becomes evident that the H$_3$ line intensity has severely decreased relative to S$_0$, a nitrogen bound exciton line. This indicates that either most of the hydrogen has diffused out of the SiC epilayer or the optical activity of the hydrogen has been altered. Also note that lines related to the boron center in 6H-SiC, as recently published for B implanted 6H-SiC [30], were not observed before or after the high temperature anneal, which is less understood and still under investigation.

---

[1] Note: More quantitative electrical comparison of these samples is not reliable because of the differing unintentional elemental nitrogen concentrations contained in these epilayers resulting from the relative decrease in propane concentration when changing the Si/C ratio from Si/C = 0.51 to 0.11 ($N_D = 4.5 \times 10^{16}$ versus $N_D = 5 \times 10^{15}$ cm^{-3}) combined with the uncertainty in SIMS determinations of these relatively low nitrogen concentrations.

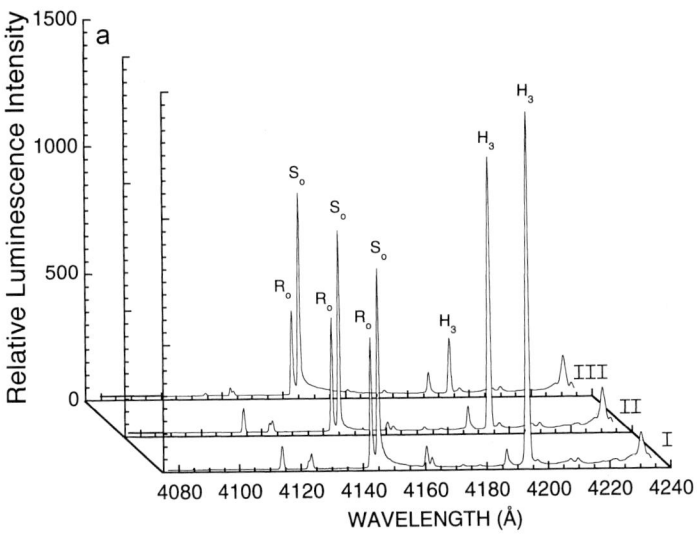

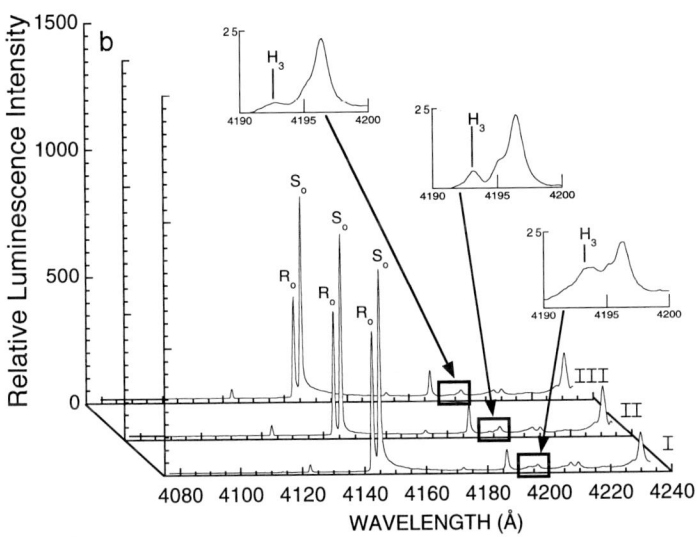

Fig. 5. a) LTPL spectra of three 6H-SiC epilayers before anneal, with decreasing hydrogen-related line intensities (labelled H_3). Corresponding to the decrease in H_3 line intensities, the samples had decreasing measured net carrier concentrations of I: $p = 2 \times 10^{16}$ cm^{-3}, II: 7×10^{15} cm^{-3}, and III: 2×10^{15} cm^{-3}. The LTPL spectra were normalized relative to the intensity of S_0, one of the nitrogen bound exciton LTPL lines. b) LTPL spectroscopy of the samples after a 1700 °C anneal in argon for 0.5 h revealed a relative decrease in the intensity of the hydrogen related line (labelled H_3). The insets of the LTPL spectra from 4190 to 4200 Å illustrate that a small amount of hydrogen remained in the epilayers following the high temperature anneal

6.2 SIMS analysis for hydrogen

Epilayers containing stepped increases in B concentration [B], from only varying the Si/C ratio, were prepared as described in Fig. 6 and subsequently analyzed for hydrogen using

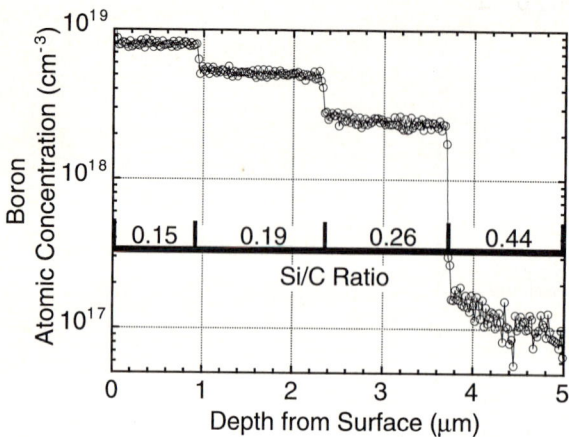

Fig. 6. The SIMS determined atomic boron concentration increased as the Si/C ratio was decreased by changing only the propane flow (stepwise in 10 sccm increments from 15 sccm to 45 sccm) while maintaining a constant silane (20 sccm) and diborane (50 sccm) flow during the CVD 6H-SiC epilayer growth

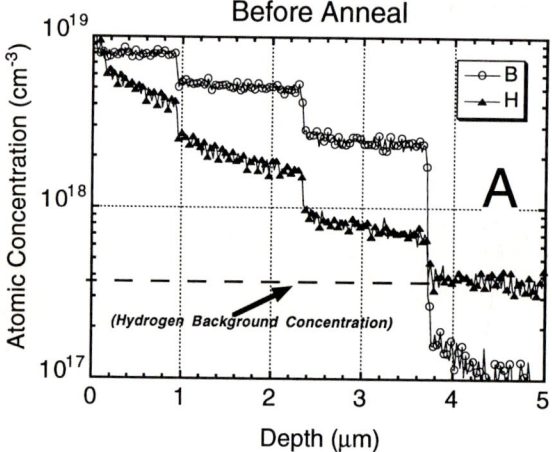

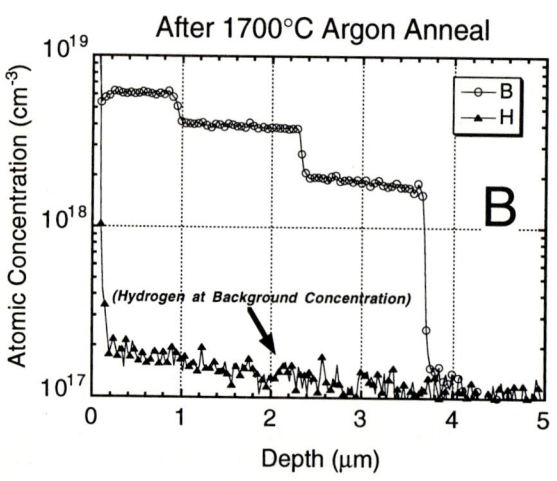

Fig. 7. A) SIMS analysis of a boron-doped, Si-face SiC epilayer revealed that prior to anneal the hydrogen concentration increased as the boron concentration increased within the CVD SiC epilayer. B) SIMS analysis after an anneal in argon at 1700 °C (0.5 h) revealed that the hydrogen concentration is at or below the H-detection limit of the SIMS instrument, indicating the outdiffusion of hydrogen during the high temperature anneal. Also, no significant solid state diffusion of boron was detected, as evidenced by the continued sharpness of the boron concentration profile

SIMS. This was done to 1. confirm that hydrogen and B incorporation are related and 2. determine whether the hydrogen was removed or if its optical activity was simply altered as a result of the 1700 °C anneal. The results of SIMS analysis of the [B]-stepped epilayer, prior to the high temperature anneal, are displayed in Fig. 7A. The increase in hydrogen concentration is observed to correspond with the increase in B concentration

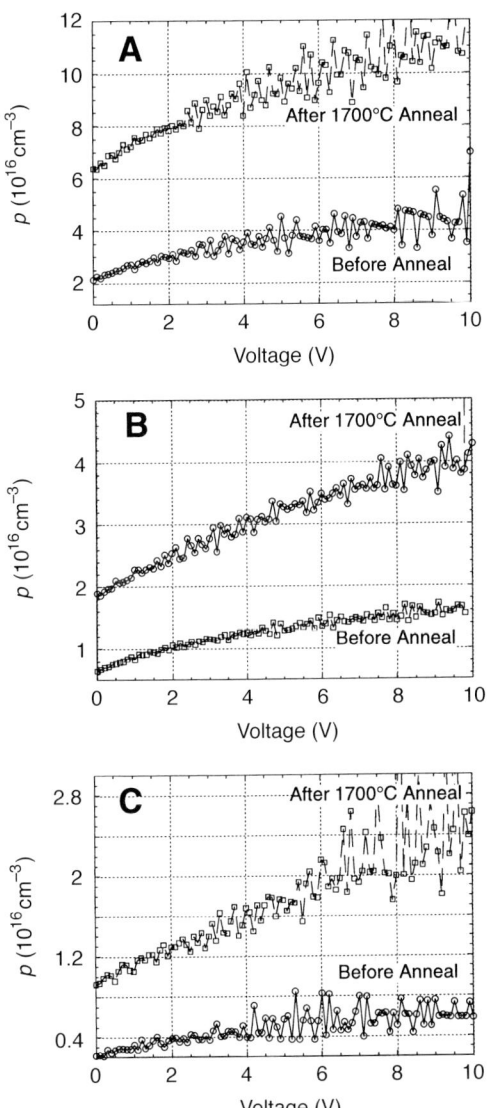

within the epilayer. This indicates that the hydrogen incorporation is directly related to the B incorporation in the 6H-SiC epilayers. It is not know whether the hydrogen atom is substitutional or defect related [31] as was previously reported for hydrogen in other semiconductor materials (the hydrogen concentration profile displayed in Fig. 7A is artificially less than the B concentration and each step in the hydrogen concentration profile also contains a downward slope, both of which are artifacts of the SIMS measurement.[2])

After a 1700 °C anneal for 0.5 h in argon, the sample was again SIMS depth profiled for determination of B and hydrogen concentration. The results are displayed in Fig. 7B which indicates that the hydrogen has diffused out of the SiC epilayer as a result of the 1700 °C anneal. The amount of hydrogen remaining within the B-doped epi-

Fig. 8. The C–V measured net carrier concentrations before (lower plot in each graph) versus after (upper plot) a 1700 °C anneal indicates more than a threefold increase in the net carrier concentration for all three doping levels. Identical Si/C ratios (0.44) were used during the epilayer growth in order to minimize potential effects of different propane concentrations on hydrogen incorporation. The diborane flows were A) 35, B) 26, and C) 24 sccm

---

[2]) Note: This can be explained by considering that lower mass elements are more affected by the geometric effects of deep craters (i.e. greater than 6 μm sputter-depth) and that the resulting field perturbations "scatter" the lower mass ions more readily than the relatively heavier ions (e.g. in comparing B(+C)- to H-signals as a function of sputter depth).

layer is below the hydrogen background concentration ($<2 \times 10^{17}\,\mathrm{cm}^{-3}$) in the SIMS instrument. Also, it is interesting to note that B does not seem to undergo appreciable solid state diffusion as a result of the 1700 °C anneal, which is evidenced by the continued sharpness of the [B]-profile displayed in Fig. 7B.

6.3 Hydrogen passivation of acceptors

A significant concentration of hole-passivating hydrogen is also incorporated during the growth of the B-doped epilayer. To determine the extent of hydrogen passivation of dopants, the sample series discussed in Fig. 5 was examined using mercury-probe C–V before and after a high temperature anneal in argon. The C–V results after a 1700 °C anneal, versus before the anneal, indicate a three- to fourfold increase in the net carrier concentration for all three doping levels (see Fig. 8). This increase indicates that, prior to the anneal, hydrogen was passivating the acceptor atoms of the B-doped SiC epilayers. The amount of hydrogen-passivated acceptors was estimated for each sample by a comparison of the measured net carrier concentration (at $V = 0$) before and after the 1700 °C anneal. The increase in carrier concentration, due to the reduction in hydrogen-

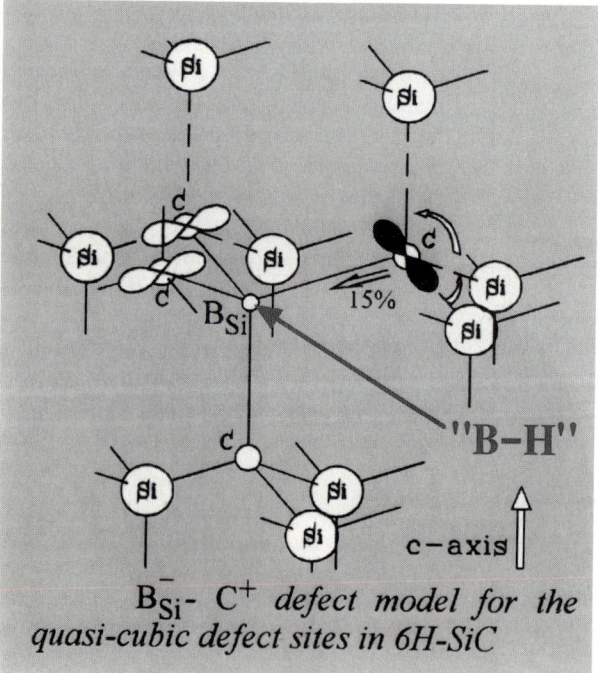

Fig. 9. After Reinke et al. [26] in which they propose that B substitutes into a Si-site and that the hole resides on an adjacent C atom p-orbital as a result of an electron transfer from the adjacent C to the B atom. The "B–H" species is now proposed as the source of the hole-passivating H which is simultaneously incorporated into the growing B doped CVD epilayer. In this way, the H could be considered to passivate the B, thereby preventing the formation of a hole by interfering in the electron transfer process from the adjacent C to the B atom. This transfer could then be enabled by the removal of H from the crystal via the high temperature anneal, consistent with post-anneal C–V measurements

passivation with post-anneal, was approximately $4.5\times, 3\times$, and $3.25\times$ for the three samples with post-anneal net carrier concentrations of $p = 9 \times 10^{15}, 1.9 \times 10^{16}$, and 6.5×10^{16} cm^{-3}, respectively [4].

By using atomic size as a first approximation to explain the preference for B incorporation into the Si-site (i.e. by neglecting chemical bonding arguments), the nonpolar covalent radii for the elements of interest are: Si (1.17 Å); C (0.77 Å); and B (0.82 Å); [22]. The more closely matched atomic size of B to C suggests that B should substitute mainly for C in the C-sites, which is in direct conflict with experimental evidence [4, 32]. However, a considerable amount of hole-passivating hydrogen is simultaneously incorporated into the growing B doped CVD epilayer and therefore a "B–H" species could be considered as the substitutional species. In this case, a "B–H" species would have a size (≈ 1.10 Å) more closely matched with that of Si (1.17 Å) and therefore should occupy a Si-site, in agreement with the experimental results. Because this is only a first approximation, it does not imply that B substitutes exclusively into the Si-sites but rather that Si-site substitution is favored because of the larger effective size of the proposed "B–H" species.

This rationale is also consistent with the results reported by Reinke et al. [26] in which they propose that B substitutes into a Si-site and that the hole resides on a adjacent C atom as a result of an electron transfer from an adjacent C to the B atom. The H atom could be considered to passivate the B, thereby preventing the formation of a hole by interfering in the electron transfer from the adjacent C to the B atom (see Fig. 9). This transfer could then be enabled by the removal of hydrogen from the crystal via the high temperature anneal which is consistent with the C–V measured three-fold increase in net hole concentration as previously discussed above.

7. Nitrogen Doping

In contrast to the similar doping results on the C-face and Si-face for the other three dopants, N doped epilayers on the 6H-SiC(000$\bar{1}$) C-face revealed a more complex response of N incorporation with a change in the Si/C ratio. As shown in Fig. 10, the SIMS determined atomic N concentration contained in the C-face epilayer initially decreases as the unintentional background nitrogen in the CVD system slowly decreases during the start of epilayer growth. The N incorporation then decreases as the propane flow is decreased (from 50 to 15 sccm), which is directly opposite to the site-competition effect for the Si-face epilayers. It is important to note that the (C-face) N incorporation decreases despite the simultaneous introduction of 100 sccm of nitrogen into the CVD reactor, which demonstrates the dominant effect of the Si/C ratio on dopant incorporation. As the propane flow is subsequently increased stepwise, the C-face N incorporation initially increases from 3×10^{18} to 6×10^{19} cm^{-3} and then decreases stepwise to 3×10^{19} cm^{-3}, coinciding with the stepwise increases in propane flow.

The N doping on the Si-face is consistent with the proposed site-competition mechanism, in which the relatively small N atoms compete with C atoms for available C-sites. However, the results from using site-competition epitaxy on the C-face are not as easily understood. SIMS analysis of N-doped C-face epilayers indicate that N-incorporation initially increases with increasing propane flow. This increase in N-incorporation for the C-face reaches a maximum and subsequent increases in propane flow result in a decrease in N-incorporation, which is similar to site-competition for Si-face epilayers.

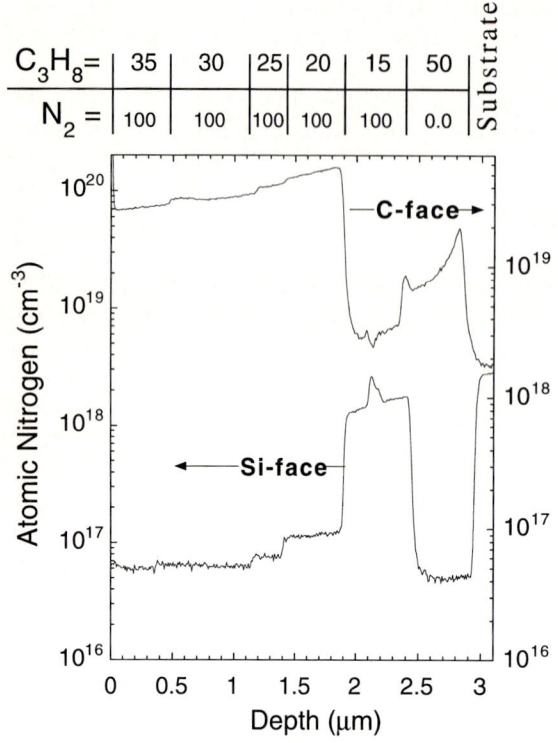

$C_3H_8=$	35	30	25	20	15	50	Substrate
$N_2 =$	100	100	100	100	100	0.0	

Fig. 10. SIMS depth profile of a C-face compared to a Si-face 6H-SiC(0001) N-doped epilayer grown during the same growth run. Note that the (C-face) N incorporation decreases, as the propane flow decreases (from 50 to 15), despite the simultaneous introduction of 100 sccm of nitrogen into the CVD reactor, which demonstrates the dominant effect of the Si/C ratio on dopant incorporation. Subsequent propane flows were varied, from 15 to 35 (sccm), to change the Si/C ratio and effectively control N incorporation with constant nitrogen (100 sccm) and silane (20 sccm) flows

This can be understood by considering the relatively small size of the N atom (0.74 Å) which would allow possible substitution into both the C-site and the Si-site. For example, in Fig. 10 the Si-face substrates have very low unintentional (i.e. at $N_2 = 0$ sccm) N-dopant incorporation when grown in excess propane ($C_3H_8 = 50$ sccm) at the start of the epilayer growth. In contrast, the C-face 6H-SiC epilayer initially demonstrated a large amount of unintentional N incorporation (also at $C_3H_8 = 50$ sccm). The propane flow was then decreased ($C_3H_8 = 15$ sccm) resulting in a decreased N incorporation for the C-face despite the simultaneous introduction of nitrogen (100 sccm). This is consistent with N mainly occupying the Si-sites on the C-face. As the propane flow was then increased, as shown in Fig. 10 ($C_3H_8 = 20$ sccm), the N-dopant incorporation increases for the C-face which is consistent with N substituting into Si-sites. However, further stepwise increases in propane flow (in excess of 20 sccm) appears to reverse this effect for the C-face, resulting in a stepwise decrease in N-dopant incorporation (for $C_3H_8 = 25$ to 35 sccm) which is consistent with the excess C competing with N for the C-sites.

These results of using site-competition on the C-face are consistent with the N-dopant atoms preferentially substituting into the Si-sites for propane flows less than 20 sccm. As the propane flow is then increased further from $C_3H_8 = 20$ to 35 sccm, the decrease in SIMS-determined N incorporation is more consistent with N substitution into the C-sites, which is similar to the Si-face results. Work is continuing to further elucidate the site-competition mechanism on both the Si-face as compared to the C-face of SiC.

8. Conclusion

Site-competition epitaxy has been successfully used for p-type and n-type dopant control on both the Si-face and C-face of 6H-SiC(0001) off-axis substrates. The doping results for the C-face provides supporting evidence that the site-competition effect is highly dependent on the SiC polarity but not significantly affected by the polytype. The results for both the 6H- and 4H-SiC Si-face samples were similar whereas the 6H-SiC C-face samples yielded significantly different results, most dramatically for N-doped epilayers. It also becomes obvious, in comparing the results from simultaneous growth of N-doped epilayers on the C-face and Si-face, that the site-competition mechanism is dominated by surface chemistry of the SiC substrate and relatively less on gas phase interactions.

For p-type doping, the Al and B dopant incorporation increased as the Si/C ratio was decreased by increasing the propane flow on both polar faces. Similarly, the P dopant incorporation also increased on both the Si-face and C-face as the Si/C ratio was intentionally decreased from either a decrease in the silane flow or increase in the flow of propane during epilayer growth. This is consistent with the P, Al, and B competing with Si for the available Si-sites.

The apparent site preference for each dopant can be partially rationalized by considering their atomic sizes relative to the sizes of Si (Si-site) and C (C-site). Here, the P and Al atoms are clearly too large for incorporation into the C-site without significantly distorting the lattice spacing, whereas B needs further explanation. Because a significant amount of H is incorporated with B we must consider the relative size of B to possibly be of a "B–H" species. For this situation, the effective size of the "B–H" species is also too large and will not easily fit into the C-sites without significantly altering the SiC lattice dimensions. In contrast, the relatively small size of the N dopant atom theoretically allows incorporation into either of the lattice sites. Therefore, it is now proposed that N doping on the Si-face could result in N incorporation into both the Si-site and the C-site, with N potentially having a preferred incorporation into the C-site. In contrast, for N doping on the C-face, N appears to mainly incorporate into Si-sites with relatively less incorporation into C-sites. Therefore, the originally proposed site-competition mechanism needs modification to possibly account for N dopant incorporation into both Si-sites and C-sites, in which the degree of incorporation for each site (i.e. Si-site versus C-site) could be dependent upon the growth surface polarity (i.e. Si-farce or C-face).

Acknowledgements The author would like to recognize the contributions of colleagues at NASA Lewis Research Center and the University of Pittsburgh. The author would also like to thank the reviewer for the excellent comments. This work was supported using internal NASA funding.

References

[1] Cree Research, Inc., Data Sheet 4/91.
[2] D. J. LARKIN, P. G. NEUDECK, J. A. POWELL, and L. G. MATUS, Inst. Phys. Conf. Ser. No. 137, 51 (1994).
[3] D. J. LARKIN, P. G. NEUDECK, J. A. POWELL, and L. G. MATUS, Appl. Phys. Lett. **65**, 1659 (1994).
[4] D. J. LARKIN, S. G. SRIDHARA, R. P. DEVATY, and W. J. CHOYKE, J. Electronic Mater. **24**, 289 (1995).

[5] D. J. LARKIN, Inst. Phys. Conf. Ser. No. 142, 23 (1995).
[6] S. KARMANN, L. D. CIOCCIO, B. BLANCHARD, T. OUISSE, D. MUYARD, and C. JAUSSAUD, Mater. Sci. Engng. B **29**, 134 (1995).
[7] R. RUPP, P. LANIG, R. SCHORNER, K.-O. DOHNKE, J. VOLKL, and D. STEPHANI, Inst. Phys. Conf. Ser. No. 142, 185 (1995).
[8] M. A. TISCHLER, private communication.
[9] A. A. BURK, JR., et al., Inst. Phys. Conf. Ser. No. 137, 29 (1994).
[10] P. G. NEUDECK, D. J. LARKIN, J. A. POWELL, L. G. MATUS, and C. S. SALUPO, Appl. Phys. Lett. **64**, 1386 (1994).
[11] J. B. PETIT, P. G. NEUDECK, C. S. SALUPO, D. J. LARKIN, and J. A. POWELL, Inst. Phys. Conf. Ser. No. 137, 679 (1994).
[12] P. G. NEUDECK, J. B. PETIT, and C. S. SALUPO, 2nd Internat. High Temperature Electronic Conference, Vol. 1, Eds. D. B. KING and F. V. THOMES, Sandia National Laboratories, Charlotte (NC), 1994 (p. X).
[13] Cree Research, Inc., Durham, NC, 27713.
[14] J. A. POWELL, L. G. MATUS, and M. A. KUCZMARSKI, J. Electrochem. Soc. **134**, 1558 (1987).
[15] J. A. POWELL, D. J. LARKIN, L. G. MATUS, W. J. CHOYKE, J. L. BRADSHAW, L. HENDERSON, M. YOGANATHAN, J. YANG, and P. PIROUZ, Appl. Phys. Lett. **56**, 1442 (1990).
[16] Charles Evans & Associates, Redwood City, CA, 94063.
[17] M. YOGANATHAN, W. J. CHOYKE, R. P. DEVATY, and P. G. NEUDECK, J. Appl. Phys. **80**, 1763 (1996).
[18] W. J. CHOYKE, in: The Physics and Chemistry of Carbides, Nitrides, and Borides, Vol. 185, Kluwer Academic Publ., Dordrecht 1990 (p. 863).
[19] R. F. DAVIS and J. T. GLASS, in: Advances in Solid-State Chemistry, Vol. 2, Ed. C. R. A. CATLOW, JAI Press, Ltd., London 1991 (p. 1).
[20] T. TROFFER, C. PEPPERMULLER, G. PENSL, K. ROTTNER, and A. SCHONER, J. Appl. Phys. **80**, 3739 (1996).
[21] C. RAYNAUD, F. DUCROQUET, G. GUILLOT, L. M. PORTER, and R. F. DAVIS, J. Appl. Phys. **76**, 1956 (1994).
[22] K. F. PURCELL and J. C. KOTZ, Inorganic Chemistry, W. B. Saunders Co., Philadelphia (PA) 1977.
[23] H. H. WOODBURY and G. W. LUDWIG, Phys. Rev. **124**, 1083 (1961).
[24] A. I. VEINGER, G. A. VODAKOV, Y. I. KOZLOV, G. A. LOMAKINA, E. I. MOKHOV, V. G. ODING, and V. I. SOKOLOV, Soviet Phys. − J. Tech. Phys. Lett. **6**, 566 (1980).
[25] A. G. ZUBATOV, I. M. ZARITSKII, S. N. LUKIN, E. N. MOKHOV, and V. G. STEPANOV, Soviet. Phys. − Solid State **27**, 197 (1985).
[26] J. REINKE, S. GREULICH-WEBER, J.-M. SPAETH, E. N. KALABUKHOVA, S. N. LUKIN, and E. N. MOKHOV, Inst. Phys. Conf. Ser. No. 137, 211 (1994).
[27] L. PATRICK, and W. J. CHOYKE, Phys. Rev. B **9**, 1997 (1974).
[28] W. J. CHOYKE, P. J. DEAN, and L. PATRICK, Phys. Rev. B **10**, 2554 (1974).
[29] L. L. CLEMEN, W. J. CHOYKE, A. A. BURK, JR., D. J. LARKIN, and J. A. POWELL, Inst. Phys. Conf. Ser. No. 137, 227 (1993).
[30] C. PEPPERMULLER, R. HELBIG, K. ROTTNER, and A. SCHONER, Appl. Phys. Lett. **70**, 1014 (1997).
[31] S. A. STOCKMAN and G. E. STILLMAN, Mater. Sci. Forum **148/149**, 501 (1994).
[32] J. REINKE, R. MULLER, M. FEEGE, S. GREULICH-WEBER, and J.-M. SPAETH, Mater. Sci. Forum **143/147**, 63 (1994).

phys. stat. sol. (b) **202**, 321 (1997)

Subject classification: 68.55.Jk; 73.50.Gr; 78.55.Hx; S6

Growth of SiC by "Hot-Wall" CVD and HTCVD

O. Kordina (a, b, c), C. Hallin (a), A. Henry (a, b), J. P. Bergman (a, b),
I. Ivanov (a), A. Ellison (a), N. T. Son (a), and E. Janzén (a, b)

(a) Department of Physics and Measurement Technology, Linköping University,
S-58183 Linköping, Sweden

(b) ABB Corporate Research, S-72178 Västerås, Sweden

(c) Okmetic Ltd., Sinimäentie 12, P.O.Box 44, 02631 Espoo, Finland

(Received January 31, 1997)

A reactor concept for the growth of high-quality epitaxial SiC films has been investigated. The reactor concept is based on a hot-wall type susceptor which, due to the unique design, is very power efficient. Four different susceptors are discussed in terms of quality and uniformity of the grown material. The films are grown using the silane–propane–hydrogen system on off-axis (0001) 6H- and 4H-SiC substrates. Layers with doping levels in the low 10^{14} cm^{-3} showing strong free exciton emission in the photoluminescence spectra may readily be grown reproducibly in this system. The quality of the grown layers is also confirmed by the room temperature minority carrier lifetimes in the microsecond range and the optically detected cyclotron resonance data which give mobilities in excess of 100000 cm^2/Vs at 6 K. Finally, a brief description will be given of the HTCVD technique which shows promising results in terms of high quality material grown at high growth rates.

1. Introduction

SiC has in comparison with Si superior properties regarding high-power, high-frequency and high-temperature electronics. The material has extremely high thermal conductivity, can withstand high electric fields before breakdown and also high current densities. The wide bandgap results in a low leakage current even at high temperatures. The above-mentioned properties make SiC promising as a power device material. The electrotechnical industry, with applications at high voltages could thus in the future advantageously replace Si power transistors, thyristors and rectifiers with SiC devices.

There are several fundamental material problems that need to be solved before SiC devices can be of any substantial commercial value. The most severe problem is the existence of the micropipes. The micropipe density is today 50 to 200 cm^{-2} in high-quality substrate material. This makes the yield for devices with areas larger than a few mm^2 extremely small especially in power applications where the device areas are often as large as several cm^2. To make large area devices at all possible and improve the yield of small area devices the micropipe density must be reduced considerably.

High-voltage bipolar devices require thick epitaxial layers with a low doping concentration and a long carrier lifetime. A 5 kV rectifier is, for instance, projected to require a 40 μm thick active layer with a doping concentration in the low 10^{15} cm^{-3} range. In the bipolar case the rectifier will additionally require a high-injection lifetime of about 1 μs. In order to obtain a reasonable yield from the material it is essential that the morphology is good.

The material for such devices is, at present, most conveniently grown by chemical vapour deposition (CVD). In this paper we will very simply describe the construction of the hot-wall susceptors and the growth behaviour in these. The object of the epitaxial work has been to produce material for power devices rather than an academic study of the epitaxy itself therefore much focus will be on the power device material and its properties.

For power devices the 4H polytype is preferred but we will also present some results on 6H epitaxial layers. The 3C polytype is of less technological interest but during some growth conditions it will occur and deteriorate the wanted 4H or 6H layer.

2. Experimental

Photoluminescence (PL) measurements were made at 2 K using an Ar-ion laser emitting in the 334 nm line as excitation source. The PL was dispersed by a 0.85 m double grating monochromator fitted with two 1200 grooves/mm gratings blazed at 5000 Å and detected by a UV sensitive photomultiplier tube.

Time resolved measurements were measured between 300 and 500 K using pulsed excitation from a dye laser synchronously pumped by a mode-locked Ar-ion laser. The light pulses, with a duration of less than 10 ps and a wavelength of approximately 680 nm were frequency doubled by a $LiIO_3$ crystal to a wavelength of approximately 340 nm (3.65 eV). The photoluminescence (PL) decays were dispersed by a 0.25 m monochromator equipped with two 1200 grooves/mm gratings blazed at 0.5 μm and detected by a photomultiplier tube. Using a time correlated photon counting system, the luminescence decays could be measured with a total time resolution better than 200 ps [1].

Optically Detected Cyclotron Resonance (ODCR) measurements were performed on a modified Bruker ER-200D X-band (9.23 GHz) ESR spectrometer using a TE_{011} cavity fitted with a continuous flow helium cryostat. The sample temperature used for these measurements was 6 K. The multi UV-lines (351.1 to 363.8 nm) of an Ar-ion laser were used as excitation source and the PL emissions were detected by a LN_2-cooled Ge detector.

3. The Hot-Wall CVD Concept

The idea to choose a hot-wall susceptor was originally to obtain a very good heating efficiency together with a high cracking efficiency of the hydrocarbons since the gas will be much more efficiently heated in a hot-wall susceptor. At the intended growth temperatures between 1500 to 1650 °C, much of the thermal losses are in the form of radiation. These losses will be greatly reduced by the thermal insulation which is wrapped around the susceptors.

Three different hot-wall susceptors will be described in detail in this paper whereas a fourth will only be mentioned, each with its particular behaviour. The feature common for all four susceptors is a rectangular-shaped hole which runs along the entire length of the susceptors which we choose to call the reaction chamber inside which the substrates are placed. The susceptors are, as mentioned, wrapped by graphite felt alternatively by graphite foam for thermal insulation and the whole package is pushed inside an air cooled quartz tube. The gas stream is thus being forced through the reaction chamber of the susceptors.

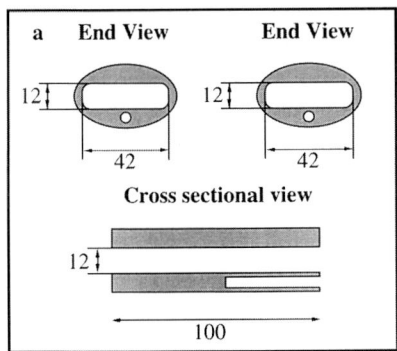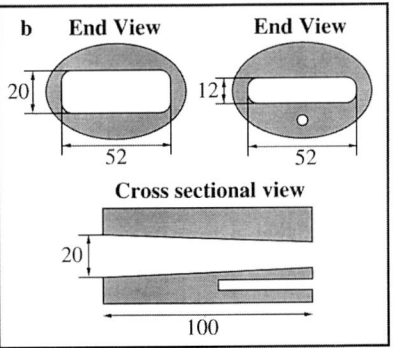

Fig. 1. The 'hot-wall' susceptors a) S:12 and b) T shown in end views and cross-sectional view. The tapered shape in susceptor T is made to compensate for the depletion. Dimensions in mm

The three first susceptors described in this paper are called susceptor S:6, S:12 and T, respectively: Susceptors S:6 and S:12 (S as in Straight) are both straight susceptors, i.e. the height of the reaction chamber does not change along the length of the susceptor. For these two susceptors, the height of the reaction chamber is 6 and 12 mm, respectively. Susceptor S:12 is shown in Fig. 1a. Susceptor T (T as in Tapered) in Fig. 1b is a tapered susceptor where the height of the reaction chamber decreases from 20 to 12 mm over a total length of 100 mm. All susceptors give a good temperature homogeneity, a very fast but well controlled laminar flow and an efficient heating.

The susceptors are not taken out from the quartz tube until their lifetime is served. Loading and unloading is simplified by a SiC plate, on top of which the substrates are placed. The SiC plate with the substrate on it is then simply placed on the 'floor' of the reaction chamber. This plate is typically 1 mm thick and extends over the whole length of the reaction chamber but not the entire width. In susceptor T the plate is, for instance, 32 mm wide whereas the widest part of the reaction chamber is 52 mm wide.

The SiC plate serves another purpose which is to protect the growing film from impurities and additional hydrocarbons emanating from various parts of the susceptor. In cold-wall reactors it is particularly common that the SiC coating from underneath the substrates is etched away and deposited on the backside of the substrate due to the small temperature difference between the substrate and the susceptor. In a hot-wall susceptor this temperature gradient is much smaller hence the deposition on the backside is also smaller, however, the effect is still there and after a few 10 h runs the SiC coating will be completely removed and the bare graphite is revealed. When this happens unwanted impurities from the graphite may incorporate into the growing films and, additionally, there will be a limited control of the C/Si ratio since a major part of the hydrocarbons will come from the revealed graphite through reactions with the hydrogen carrier gas. Good control of the C/Si ratio is, of course, essential in order to obtain a uniform doping and a good morphology of the thick layers [2]. Thus, the SiC plate serves to protect the growing film against the exposure of bare graphite.

Surprisingly, the material transfer from the plate to the substrate appears to be smaller than from the SiC susceptor coating to the substrate had the plate not been there. Possibly this can be attributed to the fact that the plate has much less thermal stress whereas a coating on graphite inevitably has significant amounts of stress which may make it more apt to react with the hydrogen. The backside growth on the plate is also less pronounced as compared to the previously observed backside growth to a substrate without the plate. This fact is somewhat puzzling, however, the most realistic explanation we can find is that the exchange of hydrogen is lower underneath the much larger plate than underneath a substrate.

3.1 Susceptors S:6 and S:12

The straight susceptor S:12 was a further development of a susceptor S:6 with only a 6 mm high reaction chamber (see Fig. 1a). Susceptor S:6 was not practically useful due to the severe depletion which could readily result in layer thickness variations of more than an order of magnitude over a distance of 30 mm along the flow direction [3]. The big drawback with hot-wall reactors is the depletion which is a result of the growth on the hot walls and homogeneous nucleation. The most common way to solve this is to make the susceptors slightly tapered which will accelerate the gas mixture as it passes over the susceptor surface and thereby push down the boundary layer or Blasius profile, see e.g. [4] making the diffusion process more efficient. With a depletion as large as that previously mentioned a suitable angle will be difficult to choose. A simple way to reduce the depletion is to ensure that the gas mixture is less efficiently heated and thereby suppress the homogeneous nucleation and the growth on the walls. This can be done by, for instance, adding a gas with a low thermal conductivity such as Ar into the hydrogen carrier gas. Alternatively the height of the reaction chamber could be increased which also was the preferred solution.

The straight susceptor S:12, having a height of the reaction chamber of 12 mm, showed significantly less depletion when compared to susceptor S:6 due to the gas mixture being less efficiently heated. This is caused by the fact that, at corresponding velocities in the two susceptors, the area of the hot walls compared to the volume these enclose is smaller in susceptor S:12. Even at equal flow rates, i.e. the velocity in S:6 is (at least) twice that of S:12, the influence of the walls is significantly larger in S:6 than in S:12. Once the 1 mm thick SiC plate was placed inside the reaction chamber of susceptor S:12, the height naturally decreased to only 11 mm, which rendered in a thickness variation of approximately a factor of two over a length of 30 mm depending on the growth conditions. The material grown in this susceptor was, however, very good but the thickness variation was unfortunately still too large.

3.2 Susceptor T

In order to improve the thickness uniformity a susceptor with a taper was constructed (Fig. 1b). The uniformity in this susceptor was indeed very good, with a thickness variation of $\pm 3\%$ over a 30 mm diameter wafer. The thickness closest to the wafer edges is, however, significantly higher as may be seen in Fig. 2. The cause for this epi-crown is not completely revealed, however, the substrates are not placed in a recess. This will not only result in a flow disturbance but also in significant radiative losses from the wafer edges which will, thus, become slightly cooler. Had the substrates been recessed, the flow would be less disturbed and the thermal losses would be smaller and, hence, the layer thickness uniformity better.

Both the straight and the tapered susceptors are manufactured from a single piece of graphite. This will provide for efficient coupling to the rf field, however, due to the

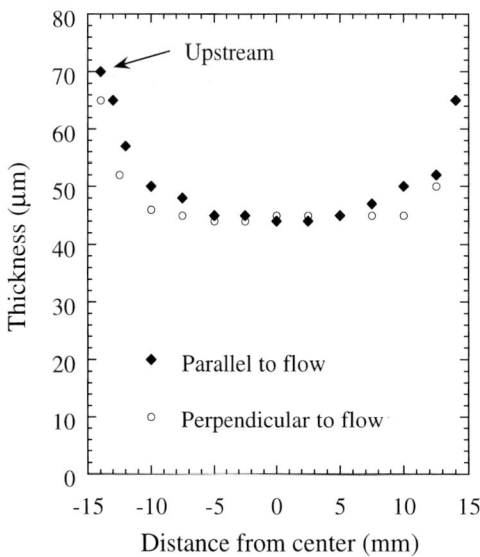

Fig. 2. Thickness variations of a nominally undoped 45 μm thick 4H film grown on a 30 mm diameter wafer. As is seen, the thickness increases dramatically towards the edge of the wafer, however, in the central part of the wafer the thickness is uniform

geometry of the susceptors, hot spots will be formed along the sides of the reaction chamber where the graphite susceptors are thinnest and also closest to the coil. Etching and sublimation of the SiC coating at these hot spots will inevitably be higher and especially when the SiC plate is used the etching takes place predominantly along the sides of the plate. The coating will be removed at these parts, however, as long as the substrate is not too close to the edge of the plate this is fairly well protected from impurities and hydrocarbons leeching out from the susceptor at these places. After a prolonged use, 300 h or so, the etching may have gone so far as to produce a hole throughout the side of the susceptor which easily can be noticed by an increase in the power consumption and a change in the radiation pattern seen through the felt along the sides of the susceptor.

The lifetime of the previously described susceptors S and T was chiefly limited by the hydrogen etching at hot spots. The idea of making the susceptors in a single piece was to make them efficient to heat, however, due to the problems with the hot spots it seemed much more tempting to break the current loops and make the susceptor in several parts at the price of a slightly higher power consumption. Such susceptors have been made although at this time, not fully evaluated. The lifetime of this susceptor, which will further be referred to as susceptor A (A as in assembly), is significantly longer (at present >500 h) and only limited by the parasitic depositions which may become very thick after a prolonged period of use.

4. Growth

The SiC growth was normally made using the silane–propane–hydrogen system at atmospheric pressure. Some runs were made using methane as C precursor. Typically, a hydrogen carrier flow of 10 l/min was used in susceptor S. The temperature range was between 1500 to 1600 °C and C/Si ratios between 1.8 to 5 were used. Growth rates were between 1.5 to 3 μm/h.

The material in susceptor S:12 shows very good electrical and optical properties [5] especially after the SiC loading plate had been introduced. When the SiC plate was not used, the normal doping concentration of a film grown using propane as the carbon precursor was n-type approximately 1×10^{15} cm^{-3} more or less independent of the C/Si ratio. When methane was used the doping concentration could easily be decreased down to the low 10^{14} cm^{-3} range and could even, when the C/Si ratio was high, flip to p-type. Once the SiC plate was introduced, the normal outcome of a layer grown with propane was n-type in the low 10^{14} cm^{-3} range, i.e. one order of magnitude lower than before using the same conditions. The morphology was also improved as expected, however, the depletion increased somewhat. It is difficult to understand what difference there may be between the SiC coating of the susceptor and a free standing SiC plate which could cause such a difference. Possibly the answer lies in the change of geometry since the height of the reaction chamber has decreased from 12 to 11 mm after the introduction of the plate.

Growing in susceptor T proved to be much different. Thick epitaxial layers (between 30 to 100 µm) which readily could be grown in susceptor S:12 were no longer possible using the exact same conditions. Once the thickness started to increase above 25 µm or so, the surface became very step bunched and in the worst case some inclusions of 3C could evolve. The step bunching could be controlled to some extent by changing the temperature, C/Si ratio and growth rate, however, it was not until we realized that the carrier gas flow needed to be increased that we could grow thick layers again. The carrier gas flow was thus increased from 10 to 14 to 16 l/min which gave good morphology again. The increase in flow also gave some new effects which had not been observed previously. The incorporation of Al in the layers was now significantly more pronounced and more sensitive to changes in C/Si ratio. The morphologies around micropipes of the layers grown at 1550 °C in susceptor T showed strong similarities to those of the layers grown at 1500 °C in susceptor S. It is obvious that the heating of the gas is not as efficient in susceptor T as in susceptor S, in fact, it appears that the gas flow actually cools the substrate somewhat in susceptor T.

Susceptor A is in addition to being made in several parts significantly larger than susceptor T which makes a proper comparison difficult, however, the preliminary results show a reasonably similar behaviour between these two susceptors.

5. Results

One of the most difficult tasks to overcome was to grow thick layers (more than 40 µm) of good morphology and without any 3C inclusions. Normally, as the layers become thicker, defects (mainly substrate induced) may cause step retardation and as a consequence step bunching. This could prove hazardous for the active layer since each of the high steps will have a large on-axis terrace at the down-step side on top of which 3C may nucleate. The step bunching is, however, only significant when the C/Si ratios are large. Low C/Si ratios always render very smooth layers with only small signs of step bunching.

A significantly better morphology of the finished epitaxial layer can be obtained if the substrate surface is etched in hydrogen prior to growth at a temperature between 1500 and 1600 °C for 10 to 30 min [6]. The hydrogen etch will remove most of the polish induced damage and produce a very smooth surface. When off-axis substrates are used, as is normally the case, a small addition of propane and/or HCl may also be suitable as suggested by Burk and Rowland [7].

One type of defects which are often seen and which are detrimental for the devices are the triangular-shaped defects most commonly seen on 4H-SiC epitaxial layers. The triangular defect is caused by a slip along the [0001] crystal plane presumably initiated by a disturbance in the substrate e.g. a micropipe. The continuous step flow does not cross the boundaries caused by the slip which results in a triangular-shaped on-axis plateau on the down-step side (relative to the step-flow direction) of the micropipe. After a while a negative pyramid may be formed, however, there can still be growth inside this pyramid but due to the on-axis surface the probability of growing 3C-SiC inside this area is high. Details on this may be found in [8, 9].

The doping concentration is at present normally n-type in the low 10^{14} cm^{-3} depending on the growth conditions used. Varying the C/Si ratio influences mainly the morphology and the impurity incorporation. The impurity incorporation, however, also depends much on the other conditions chosen for the growth such as the surface temperature of the growing film.

Under normal conditions, the 6H-SiC films show no signs of any compensating Al acceptors, 4H-SiC can show signs of minor amounts of compensating Al acceptors. Fig. 3 shows the photoluminescence (PL) spectrum of a typical low-doped n-type 4H-SiC layer grown using susceptor T. The presence of strong free exciton (FE) lines underlines the high purity of the layer.

Noteworthy is also the very high mobility derived from the ODCR measurements both from 6H- and 4H-SiC as seen in Fig. 4 for 4H-SiC with the magnetic field **B** parallel to the c-axis. This result together with a more complete study of the effective masses are discussed in [10]. By assuming $\mu = e\tau/m^*$, the mobility in the basal plane is in this particular example for 4H-SiC estimated as $\mu_\perp \approx 1.8 \times 10^5$ cm^2/Vs. The dominating scattering processes at these temperatures are assumed to be ionized impurity scattering and neutral impurity scattering. The measurements presented by Son et al. [11] on n-type 6H-SiC samples clearly demonstrate that ODCR cannot be observed should a too high concentration of compensating acceptors be present in the layers, or should the nitrogen concentration be too high. Merely to obtain an ODCR spectrum requires, thus,

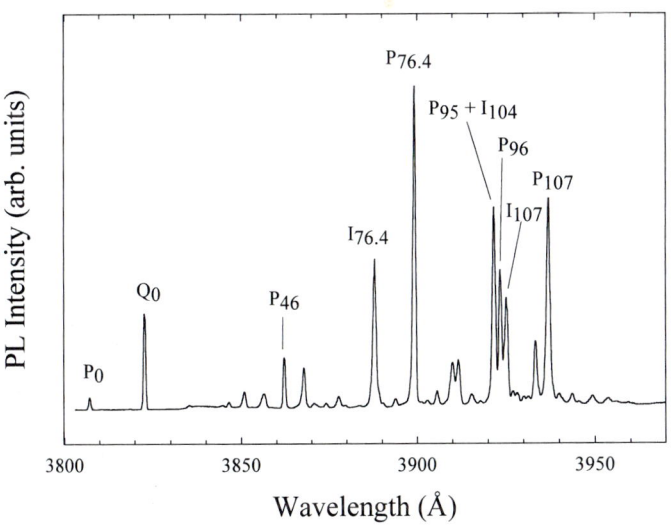

Fig. 3. Low temperature ($T = 4.2$ K) PL spectrum of the near bandgap region from a 4H-SiC epitaxial film. The N_{BE} related lines are marked P$_{h\nu}$ and Q$_{h\nu}$ where $h\nu$ denotes the phonon replica energy in meV involved in the transition. FE related lines are marked I$_{h\nu}$

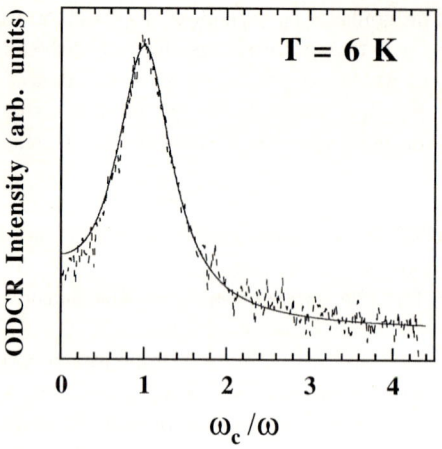

Fig. 4. Optically detected cyclotron resonance measurements performed at 6 K with the magnetic field parallel to the c-axis, using the multi UV lines of an Ar-ion laser as excitation source. The dashed line represents the experimental data and the solid line is the fitted curve. The mobility $\mu_\perp$ may thus be estimated to $\mu_\perp \approx 1.8 \times 10^5 \text{ cm}^2/\text{Vs}$

very high purity material. It must, however, be pointed out that compensating acceptors may partly be neutralized by free holes, produced by the UV laser excitation. The degree of the influence by this is still not clear and requires further studies.

The map in Fig. 5 shows the room temperature Minority Carrier Lifetime (MCL) variation over an epilayer grown on a 4H-SiC wafer in susceptor T. It is interesting to note that the MCL is in excess of one μs over an area larger than half the wafer. The lifetime peaks at 2.1 μs close to the center of the wafer but drops rapidly towards the sides of the wafer. Along the flow direction, the lifetime is more uniform, however, towards the leading edge (upstream) the lifetime is slightly lower. The fact that the MCL drops rapidly towards the sides of the wafer may partly be attributed to the fact that the substrates are of lower quality close to the edges. To this comes also the fact that the substrates are less protected towards the sides perpendicular to the flow direction due to the size of the protecting SiC plate which only extends 1 mm on each side of the 30 mm

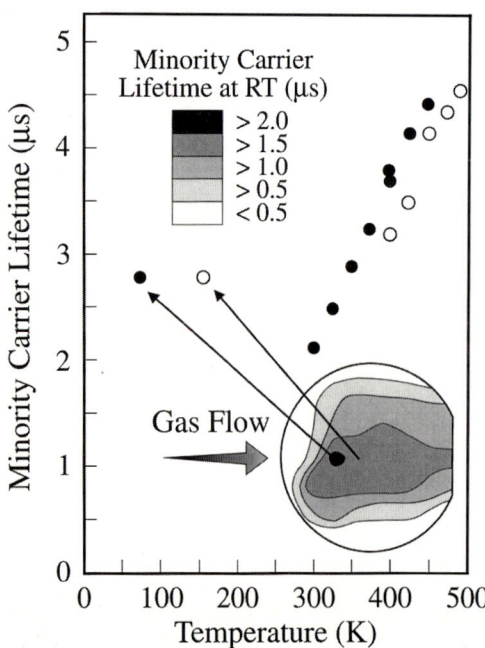

Fig. 5. Map of the minority carrier lifetime of a 30 mm diameter 4H-SiC epitaxial film. The lifetime reaches its highest value close to the center of the wafer. The temperature dependence of the minority carrier lifetime is also shown for two points of the wafer. The minority carrier lifetime is seen to increase with increasing temperature and is at a tentative operating temperature of a device close to 5 μs

diameter wafer. As already mentioned, after a prolonged use of the susceptor, significant etching of the SiC coating of the susceptor can be detected along the sides of the SiC plate. This etching will eventually reveal the bare graphite and an unwanted addition of hydrocarbons and impurities will add to the gas flow. These additions are seen as areas of pronounced step bunching of the grown layer close to the sides perpendicular to the gas flow direction. The maximum size wafer for uniform growth in this susceptor is, thus, only approximately 20 mm. This is, as mentioned, clearly seen on the morphology which is significantly rougher (more step bunched) along the sides of the wafer where the lifetimes also are shorter. It appears, thus, that in order to have a long MCL, at least the morphology must be good which makes it likely to presume that the lifetime limiting defect is related to some type of crystalline imperfection [12].

The temperature dependence of the MCL is also shown in the same figure (Fig. 5) for two spots on the same 4H-SiC wafer. The MCL increases rapidly with increasing temperature similar to 6H-SiC. The MCL at 500 K is, as seen in the figure, approximately 5 μs.

We have observed evidence that the MCL is reduced in some samples containing larger amounts of compensating Al acceptors whereas others, which are grown using the SiC plate, appear not to be limited in lifetime by compensating Al acceptors although these may be seen in the PL spectra. Possibly the concentration of compensating Al acceptors is lower than in previous growth runs before the SiC plate was introduced. It may, however, also be argued that some other defect, for instance boron, was introduced together with the Al in the previous runs. Fig. 6 shows, for instance, the MCL of a set of n-type films grown without the SiC plate. It is clearly seen that the films which contain larger amounts of compensating Al acceptors have significantly reduced lifetimes although these have a lower net doping concentration. The uncompensated films also show a weak trend with respect to the nitrogen doping concentration [13].

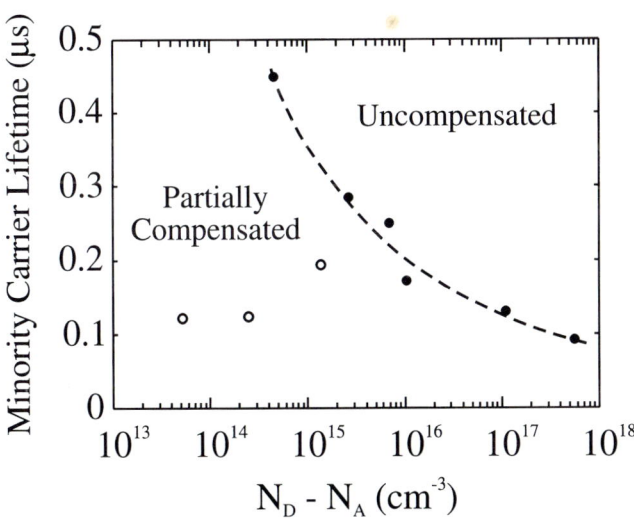

Fig. 6. Minority carrier lifetimes plotted versus the residual doping concentration of n-type samples. The lifetime appears weakly dependent on the nitrogen concentration of the uncompensated layers (filled circles). Compensating acceptors in n-type samples clearly decrease the minority carrier lifetime (unfilled circles). The dashed line should only be regarded as a guide for the eye

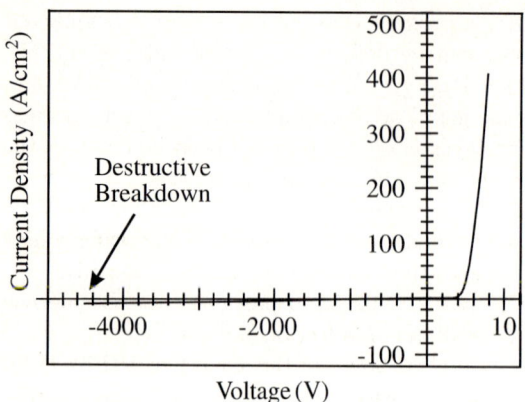

Fig. 7. *I–V* characteristics obtained from a processed diode in SF$_6$ atmosphere. The reverse blocking voltage was as high as 4.5 kV before catastrophic failure occurred at the periphery of the component. The current leakage was below the detection limit of 1 μA of the instrument for reverse voltages up to 1.1 kV. Close to 4.5 kV the current leakage was about 1.8 mA

The model proposed for the recombination is that an electron is captured by a deep defect which subsequently recombines with a hole at an acceptor. This transfer process is shown to be much more efficient than the direct transition of an electron from the deep defect to a hole in the valence band [13, 14]. Once the compensating acceptors are reduced significantly in concentration, the direct transition may become important.

A general trend seen in the measured samples is that the MCL increases with increasing layer thicknesses. This tells us that the lifetime limiting defect is a defect at or close to the substrate–epilayer interface or an extending defect propagating from the substrate into the epilayer where it is slowly healed as the growth continues. It is not likely that the lifetime limiting defect is an impurity which is gradually 'baked out' as the growth proceeds. If this were the case, a strong correlation would be seen with respect to the pump time and susceptor lifetime.

A noteworthy detail which we have seen is that the MCL in 4H-SiC is consistently higher than in 6H-SiC for samples grown in the same run. There appears to be some difference in the dominating recombination mechanisms between 6H-SiC and 4H-SiC [14].

The final proof of the material quality comes once a properly working device is made. A growth run was made on an n$^+$ substrate comprising a 45 μm thick low-doped n-type active layer followed by a 2 μm thick p$^+$ layer to form the pn junction. A set of 160 μm diameter mesa diodes were subsequently processed from this material[1]) [5].

The *I–V* curve of one of the diodes is shown in Fig. 7. The diode could block a voltage as high as 4.5 kV before a catastrophic failure occurred. In the forward direction, a typical voltage drop of 6 V at 100 A/cm^2 was registered, which is a little too high with respect to the material properties. This comparatively high voltage drop is largely caused by rather poor quality contacts which were not properly annealed and therefore not fully ohmic.

6. HTCVD

As mentioned above, high-voltage devices require thick epitaxial layers. Since the growth rate of the CVD technique is a few μm/h, the layers for a 5 kV rectifier will take 10 to 25 h to grow. If the voltage ratings increase, it is essential that the growth rates

---

[1]) The diodes were processed at the Industrial Microelectronics Center in Stockholm, Sweden.

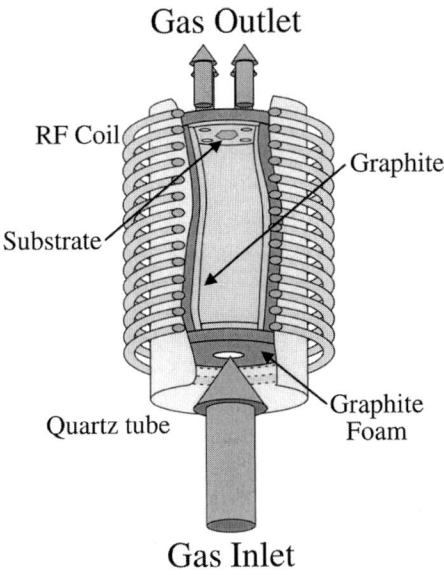

Gas Outlet

RF Coil

Graphite

Substrate

Quartz tube

Graphite Foam

Gas Inlet

Fig. 8. Schematic drawing of the HTCVD crucible. The gases flow through holes at the bottom and top parts of the crucible. The seed crystal is attached to the top part by molten glucose

are increased to make it commercially interesting. A possibility which at present appears to be promising in this respect is the High-Temperature CVD (HTCVD) technique where growth rates of several hundred µm/h already have been demonstrated [15].

The system which is schematically depicted in Fig. 8 consists of a cylindrical graphite crucible with a lid and a bottom. The gases enter through a hole at the bottom and are heated in the crucible to exit through four small holes around the perimeter of the lid. The substrate is glued to the lid by molten glucose. The 'bottom up' configuration of the flow is shown to give a very high efficiency in terms of gas consumption due to the drag caused by the buoyancy. The growth takes place in a temperature regime between 1800 to 2300 °C where the sublimation and etching of the SiC are significant. The idea is that the high temperature will stimulate all chemical reactions in the gas phase and on the surface as well as will increase the surface mobility which will allow high growth rates. Moreover, the sublimation and etching will be more pronounced on areas where the crystalline quality is worse which thus would provide a better crystal.

The precursor gases are silane and (when necessary) propane. In most cases, additions of propane prove to be detrimental to the growth since significant amounts of carbon emanating from the walls are already available. The carrier gas must be chosen with extreme care (see also [15]). Pure hydrogen is a bad choice in this system due to the very high etch rates at the temperatures in use.

The carrier gas flow must also not be too high. In fact, it is not possible to increase the carrier gas flow too much especially if the crucible length is small since the reactive gases will not be properly heated and also, more importantly, a jet of cold gas may hit the sample surface and cool the sample. The former may result in an inefficient usage of the precursor gases which may be acceptable, however, the latter may give rise to a polycrystalline layer or in the best case, a small area of interesting morphological features as may be seen in Fig. 9. At times, when the growth has been conducted under Si-rich conditions, droplets of Si may be seen on the surface especially on places where the impinging gas jet has lowered the surface temperature. Generally, the cooling jet effect can be avoided by a proper choice of carrier gas flow or an increase of the temperature.

The actual growth process is in some ways similar to that of CVD, however, due to the high concentrations of precursor gases added, there must inevitably occur some homogeneous nucleation. In CVD, homogeneous nucleation will set a limit to the growth rate, however, in HTCVD, the case may be different. For the sake of argument, let us

Fig. 9. The morphology of a 6H-SiC layer grown at 2200 °C at a rate of 200 μm/h. The morphology on the left-hand side is influenced by the impinging gas jet

assume that nuclei are formed in the gas phase. The nuclei thus formed will dissociate into smaller species determined by thermodynamics under the conditions chosen for the growth. In this respect the HTCVD process is more similar to the seeded sublimation growth with the difference that the source is formed and sublimed in the gas phase and transported to the cooler substrate by the carrier gas.

At present a full evaluation of the parameter space has not been possible to make, however, some general observations have been made which are of interest. The growth rate is chiefly determined by the temperature. Typically, a growth rate of 200 μm/h is obtained around 2200 °C. At 2300 °C the growth rate is increased to approximately

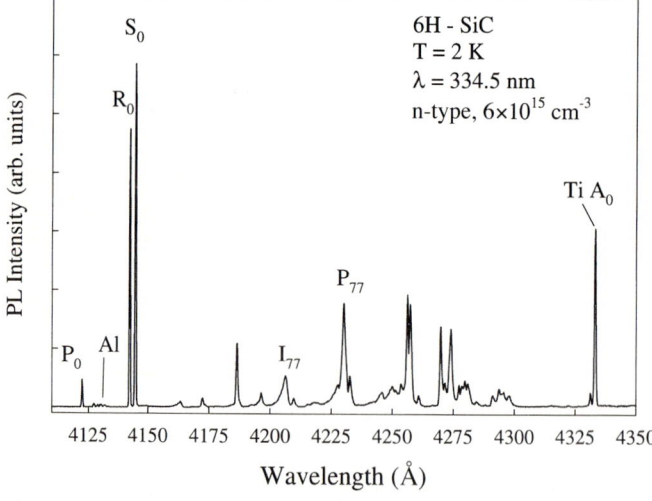

Fig. 10. PL spectrum of a low-doped 6H-SiC layer grown at 2200 °C at a rate of 200 μm/h. The linewidths are only limited by the spectral resolution of the measurement

Fig. 11. The cleaved edge of a wafer with a 100 µm thick layer grown on top as observed through an optical microscope using transmitted light. For comparison, the whole thickness of the wafer is shown still attached to some remnants of the graphite lid

500 µm/h. The silane flow has importance too, however, it appears that after a certain limit, a larger input of silane does not affect the growth rate. This may not mean that the growth rate cannot be increased further at the chosen temperature since the growth rate may be limited by carbon instead. This has been observed several times when the growth rate increased by an addition of propane. The material quality is, however, generally better when the growth is conducted under Si-rich conditions or conditions close to what we believe are stoichiometric conditions in the gas phase.

Very thick layers up to 2 mm in total thickness have been grown. Both 4H- and 6H-SiC layers are generally mirror like by eye, however, step bunching effects are normally observed especially if the growth is conducted under a too high C/Si ratio.

Typically, the grown layers are n-type in the 10^{16} to 10^{17} cm^{-3} range. In some cases where the cell has been conditioned properly and the growth parameters chosen with some care the doping level can be made lower. Even unintentionally very low-doped p-type samples have been grown. Fig. 10 shows for instance, a PL spectrum of a 6H-SiC layer grown at 2200 °C at a growth rate of 200 µm/h. The doping level of this particular sample is n-type 6×10^{15} cm^{-3} as determined by a comparison of the free exciton luminescence to the nitrogen bound exciton luminescence [16]. It may be pointed out that the linewidths of the nitrogen bound exciton lines are only limited by the spectral resolution of the measurement. What is more encouraging to see is that there is no evidence of any boron related luminescence and that the aluminium bound exciton lines are of such low intensity considering that the sample was grown in an uncoated graphite crucible. Titanium appears, however, to be omnipresent in the grown samples and the titanium lines are also clearly seen in the spectrum. The titanium is most likely coming from the graphite.

The layer thicknesses are measured by studying a cleaved edge through an optical microscope using transmitted illumination. The layers are very transparent and may easily be discernible from the substrate as can be seen in Fig. 11 which shows a 100 µm thick 6H layer grown on a highly doped n-type substrate.

7. Summary

In this paper a hot-wall CVD growth concept has been described for the growth of 6H- and 4H-SiC epitaxial layers suitable as material for power devices. Various susceptors have been described: two straight, one tapered, and an assembled tapered susceptor. The material grown in the straight susceptors is hampered by depletion problems particularly in susceptor S:6. In the tapered susceptor the thickness uniformity of the grown layers is adequate for further processing into power devices.

The material grown in susceptors S:12, T, and A is of good quality with an uncompensated n-type doping concentration in the low 10^{14} cm^{-3}. Photoluminescence measurements of nominally undoped layers show intense FE emission. Mobilities amounting to more than 100000 cm^2/Vs have been estimated from ODCR measurements at 6 K. Room temperature minority carrier lifetimes in excess of 1 µs from n-type 4H-SiC layers have also been measured. The lifetime appears correlated to the thickness of the films and to some extent also to the morphology. A higher lifetime may also be observed from 4H-SiC layers as compared to 6H-SiC layers.

Finally, a growth concept for higher growth rates called HTCVD has briefly been described. With this type of system, layers or even crystals of high purity may be grown at growth rates which are comparable to those of the seeded sublimation growth. Although the material quality still has to improve somewhat to reach the high standards of CVD grown material, it shows promising results enough to stimulate further research.

Acknowledgements Financial support for this work was provided by the Swedish Board for Industrial and Technical Development (NUTEK)/Asea Brown Boveri (ABB) Power Device Program, the Swedish Council for Engineering Sciences (TFR), NUTEK, the Swedish Natural Science Research Council (NFR), the NUTEK/NFR Material Consortium on Thin Film Growth.

References

[1] J. P. Bergman, phys. stat. sol. (a) **162**, No. 1 (1997).
[2] C. Hallin, I. G. Ivanov, A. Henry, T. Egilsson, O. Kordina, and E. Janzén, submitted to J. Crystal Growth.
[3] A. Henry, O. Kordina, C. Hallin, R. C. Glass, and E. Janzén, Inst. Phys. Conf. Ser. No. 137, 305 (1994).
[4] D. J. Tritton, Physical Fluid Dynamics, Oxford University Press, Oxford 1988.
[5] O. Kordina, J. P. Bergman, A. Henry, E. Janzén, S. Savage, J. André, L. P. Ramberg, U. Lindefelt, W. Hermansson, and K. Bergman, Appl. Phys. Lett. **67**, 1561 (1995).
[6] F. Owman, C. Hallin, Per Mårtensson, and E. Janzén, J. Crystal Growth, accepted.
[7] A. A. Burk, Jr. and L. B. Rowland, Appl. Phys. Lett. **68**, 382 (1996).
[8] C. Hallin, A. O. Konstantinov, O. Kordina, and E. Janzén, Inst. Phys. Conf. Ser. No. 142, 85 (1996).
[9] E. N. Mokhov, I. L. Shulpina, A. S. Tregubova, and Y. A. Vodakov, Cryst. Res. and Technology **16**, 879 (1981).
[10] N. T. Son, W. M. Chen, O. Kordina, A. O. Konstantinov, B. Monemar, E. Janzén, D. M. Hofmann, D. Volm, M. Drechsler, and B. K. Meyer, Appl. Phys. Lett. **66**, 1074 (1995).
[11] N. T. Son, O. Kordina, A. O. Konstantinov, W. M. Chen, E. Sörman, B. Monemar, and E. Janzén, Appl. Phys. Lett. **65**, 3209 (1994).
[12] O. Kordina, J. P. Bergman, C. Hallin, and E. Janzén, Appl. Phys. Lett. **69**, 679 (1996).
[13] O. Kordina, J. P. Bergman, A. Henry, and E. Janzén, Appl. Phys. Lett. **66**, 189 (1995).
[14] N. T. Son, E. Sörman, W. M. Chen, O. Kordina, and E. Janzén, Appl. Phys. Lett. **65**, 2687 (1994).
[15] O. Kordina, C. Hallin, A. Ellison, A. S. Bakin, I. G. Ivanov, A. Henry, R. Yakimova, M. Tuominen, A. Vehanen, and E. Janzén, Appl. Phys. Lett. **69**, 1456 (1996).
[16] I. G. Ivanov, C. Hallin, A. Henry, O. Kordina, and E. Janzén, J. Appl. Phys. **80**, 3504 (1996).

phys. stat. sol. (b) **202**, 335 (1997)

Subject classification: 68.55.Jk; 68.35.Ct; 73.61.Le; S6

3C-SiC Single-Crystal Films Grown on 6-Inch Si Substrates

H. NAGASAWA [1]) and K. YAGI

Hoya Corporation, R&D Center, 3-3-1 Musashino, Akishima, Tokyo 196, Japan

(Received January 31, 1997)

The heteroepitaxial growth of 3C-SiC on Si(001) substrates has been studied in a hot-wall-type low-pressure reactor. The Si substrates were carbonized by C_2H_2 prior to the SiC growth process to suppress the undesirable effects of lattice mismatching between Si and 3C-SiC. A single-crystal carbonized layer (3C-SiC) was obtained from 500 °C to higher than 1000 °C in an C_2H_2 environment. Following the carbonization process, SiH_2Cl_2 and C_2H_2 were alternately supplied into the reaction tube to grow an epitaxial 3C-SiC film. The growth rate of 3C-SiC depended on the amount of Si incorporated into the surface of the substrates by H_2 reduction of $SiCl_2$ as a Si precursor. The "H_2 intermittent flow" method employed during the SiC growth process efficiently suppressed the reduction of $SiCl_2$ and induced a constant growth rate of the SiC. The crystallinity of the grown 3C-SiC films on Si substrates was evaluated using transmission electron microscopy, selected-area electron diffraction, and X-ray diffraction methods. The grown 3C-SiC films included anti-phase boundaries and twins. The concentration of these plane defects decreased due to coalescence with each other during SiC growth and resulted in an improvement in crystallinity and electrical properties.

1. Introduction

Silicon carbide (SiC) is a wide band-gap semiconductor with several outstanding properties, including stability at high temperatures, a high ability to withstand irradiation, and mechanical toughness. SiC is not only available for power devices, but can also be reliably used with microsensors and microactuators employed at high temperatures and/or in severe environments due to its excellent mechanical properties and chemical stabilty. Of the major polytypes, the cubic SiC (3C-SiC) has potentially the best electrical properties. 3C-SiC is marked by high electron mobility (1000 cm^2 V^{-1} s^{-1}) and high electron saturation velocity (2.7×10^7 cm s^{-1}) [1, 2]. The above electrical properties are a great advantage in the development of high-speed electrical devices. However, the lack of mono-crystalline 3C-SiC has limited the development of electronic devices and sensors composed of this material. Although 3C-SiC ingots have been made by sublimation methods, they are very small as substrates.

As 3C-SiC is the lowest-temperature phase among the SiC polytypes, only 3C-SiC can be grown on Si substrates. The chemical vapor deposition (CVD) method is the most common means of growing 3C-SiC films on Si substrates [3 to 9]. The growth of 3C-SiC on Si substrates using CVD has great economic advantages over sublimation methods due to its large-area formation and high productivity. In addition, stacked SiC on Si substrate is suitable for the fabrication of hetero-bipolar transistors or sensors.

---

[1]) Telephone: +81-425-46-2756; Fax: +81-425-46-2709; e-mail: nags@rdc.hoya.co.jp

In recent years, excellent results have been reported for SiC epitaxial growth by the atomic layer epitaxy (ALE) method using the molecular beam epitaxy (MBE) system [10 to 13]. As the ALE method employs a self-limiting process of molecular adsorption on the surface of a substrate, high controllability and high accuracy can be realized through the ALE mechanism. Furthermore, it is generally recognized that the ALE has great potential for large-area formation with high uniformity. However, there are no evidences about the large-area formation of 3C-SiC films using the ALE method.

The purpose of our study is to implement the large-area formation of 3C-SiC films on Si substrates using an analogy of the ALE method. A hot-wall-type low-pressure reactor was employed to realize the large-area formation of 3C-SiC with high productivity. In addition, an alternating supply of dichlorosilane (SiH_2Cl_2) and acetylene (C_2H_2) as source gases was provided to realize an effective surface reaction and maintain a high thickness uniformity.

This paper concerns three topics: first, the carbonization of a Si surface will be discussed. Carbonizing the Si surface prior to the SiC growth process is a profitable method of obtaining high-quality epitaxial 3C-SiC films on Si substrates with good reproducibility [6, 14 to 16]. A carbonization mechanism and the optimum carbonization condition will be discussed. Second, SiC growth processes using an alternating supply of SiH_2Cl_2 and C_2H_2 will be described in detail. The self-limiting adsorption process of precursors on the surface of a substrate was realized through the alternating supply of the source gases. The 3C-SiC growth mechanism will be clarified through the use of the results of quadrupole mass spectroscopy. The advantages of growth mechanisms with self-limiting adsorption of precursors will be clearly demonstrated by the remarkably high uniformity of SiC thickness over a 6-inch diameter. Finally, we will report on the crystallinity and electrical properties of grown 3C-SiC determined through the use of the X-ray diffraction (XRD) method, selected-area electron diffraction (SAED), transmission electron microscopy (TEM) and Hall measurements.

2. Experiments

Fig. 1 shows a schematic diagram of a hot-wall-type low-pressure reactor. A reaction tube made with fused silica was uniformly heated through the use of a surrounding heater. The reaction tube was connected to three gas inlets and an evacuation system

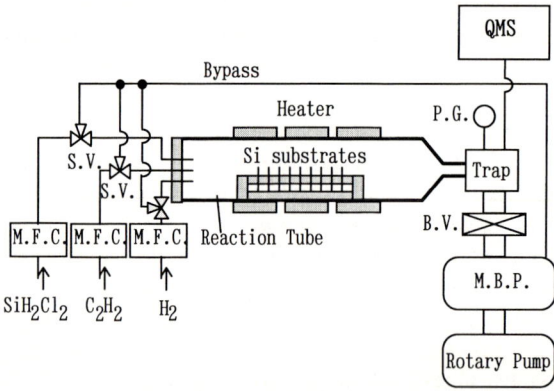

Fig. 1. Hot-wall-type low-pressure reactor for vapor-phase epitaxy of 3C-SiC. M.F.C. mass flow controller, S.V. switching valve, P.G. Pirani gauge, B.V. butterfly valve, M.B.P. mechanical booster pump, QMS quadrupole mass spectrometer

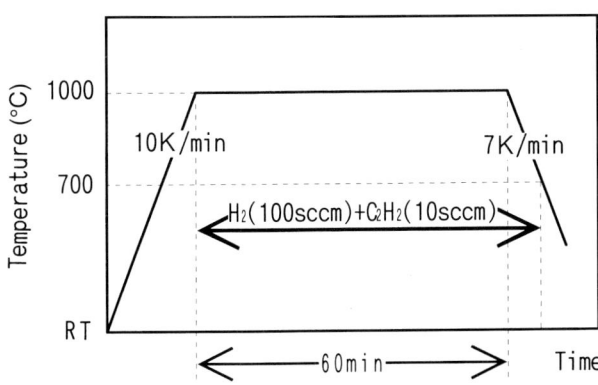

Fig. 2. Quick carbonization program. P_{total} ($= 66$ m Torr ≈ 8.78 Pa) means the total pressure in the reaction tube, while $P_{C_2H_2}$ ($= 6$ m Torr ≈ 0.798 Pa) means the C_2H_2 partial pressure in the reaction tube

through water-cooled flanges. The base pressure of the reaction tube evacuated by a mechanical booster pump (M.B.P.) was $\approx 1.33 \times 10^{-1}$ Pa.

C_2H_2 (99.9999%) and SiH_2Cl_2 (99.999%) were employed as the C source and Si source, respectively, Hydrogen (H_2: 99.999 999 9%) was used as a reducing agent. Each gas flow rate was precisely controlled through the use of a mass-flow controller (M.F.C.). The flow directions of the source gases were changed so that they were fed into the reaction tube or bypass line through the use of a computer-controlled switching valve (S.V.) installed between the M.F.C. and the gas inlet. To control the pressure in the reaction tube, a butterfly valve (B.V.) was installed between the reaction tube and the M.B.P. A Pirani gauge (P.G.) monitored the total pressure (P_{total}) in the reaction tube. In order to analyze products in the gas phase a quadrupole mass spectrometer (QMS) was attached to the end of the reaction tube.

Single-crystal (001) Si wafers with 6-inch diameters were used as substrates. These wafers were treated with HF 5% solution before being loaded into the reaction tube.

3. Carbonization of Si Surfaces

The crystallinity of the carbonized layer on Si surfaces strongly influences that of the epitaxial SiC layer grown on it, so it is very important to determine the carbonization

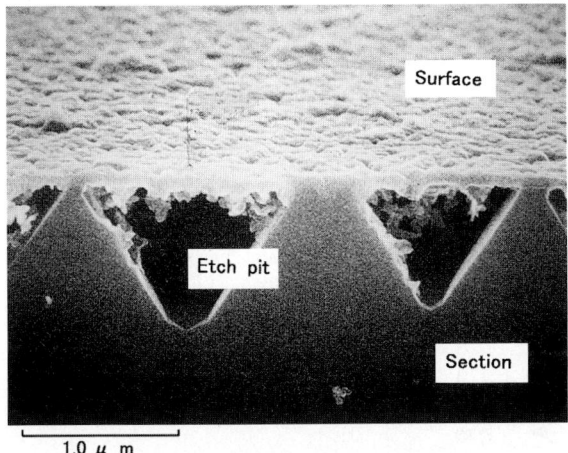

Fig. 3. Cross-sectional SEM image of the carbonized Si surface prepared by the carbonization program shown in Fig. 2

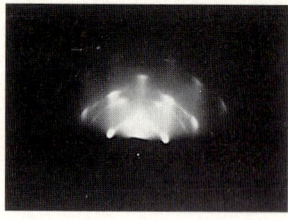

Fig. 4. RHEED image of the carbonized Si surface prepared by the carbonization program shown in Fig. 2

mechanism of the Si surface and alter the crystallinity of the carbonized layer. Although many studies have been conducted on the carbonization mechanism of Si surfaces [17 to 23], little is known about the carbonization behavior with an C_2H_2 partial-pressure ($P_{C_2H_2}$) range higher than $\approx 1.33 \times 10^{-1}$ Pa, which is suitable for our reactor. Therefore, we studied the carbonization mechanism through the use of our hot-wall-type reactor with a $P_{C_2H_2}$ range higher than $\approx 1.33 \times 10^{-1}$ Pa.

It is generally recognized that Si surfaces are converted into a thin SiC layer through heating in hydrocarbon environments [23]. However, quick carbonization at a higher temperature, as shown in Fig. 2, resulted in the formation of an undesirable carbonized layer. As shown in the carbonization program in Fig. 2, the reaction tube was first heated to 1000 °C with no gas flow (in the vicinity of 1.33×10^{-1} Pa in total pressure). When the temperature of the reaction tube reached at 1000 °C, H_2 and C_2H_2 were introduced into the reaction tube. The reaction tube was kept at 1000 °C for 60 min with H_2 and C_2H_2 flowing into it. The reaction tube was then cooled to 700 °C with continuous flow of the gases, and was finally cooled to room temperature with no gas flow. In this program, the H_2 and C_2H_2 flow rates were fixed at 100 and 10 sccm, respectively. The obtained $P_{C_2H_2}$ was ≈ 0.798 Pa, and P_{total} was ≈ 8.78 Pa.

As shown in the scanning electron microscope (SEM) image in Fig. 3, a SiC layer approximately 1000 Å in thickness on the Si substrate and large etch pits under the SiC layer were abruptly formed by the quick carbonization program shown in Fig. 2. The reflection high-energy electron diffraction (RHEED) pattern of the quickly carbonized layer showed a ring-like shape that suggested polycrystallinity, as shown in Fig. 4. Therefore, the quickly carbonized layer is not a suitable substrate for single-crystal 3C-SiC epitaxy.

A gradual carbonization program during heating enables a high-quality carbonized layer on a Si substrate to be obtained. Fig. 5 shows a typical carbonization program for obtaining a high-quality carbonized layer. H_2 and C_2H_2 were first introduced at 500 °C. The reaction tube was then heated to 1000 °C with continuous flows of C_2H_2 and H_2.

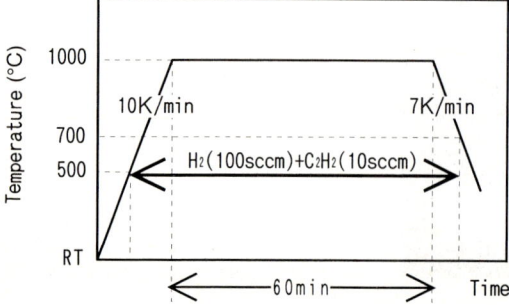

Fig. 5. A typical carbonization program to obtain a single-crystal carbonized layer on the Si substrate. P_{total} (= 66 m Torr ≈ 8.78 Pa) means the total pressure in the reaction tube, while $P_{C_2H_2}$ (= 6 m Torr ≈ 0.798 Pa) means the C_2H_2 partial pressure in the reaction tube

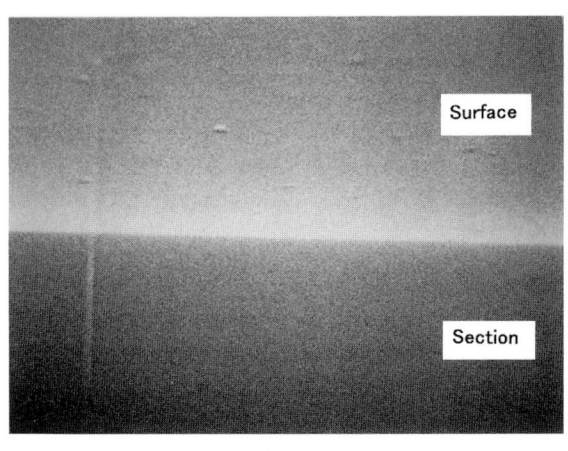

Fig. 6. SEM image of the carbonized Si surface prepared by the carbonization program shown in Fig. 5

1.0 μ m

The heating rate of the reaction tube was 10 K/min. The reaction tube was kept at 1000 °C for 60 min with H_2 and C_2H_2 flows. It was then cooled to 700 °C with a continuous flow of the gases, and finally to room temperature with no gas flow. Refer to Fig. 5 for more detail.

As the SEM image in Fig. 6 shows, no etch pits were observed in the surface treated with the gradual carbonization program shown in Fig. 5. Although the thickness of the carbonized layer measured using an ellipsometer is very thin at 84 Å, the RHEED pattern of the carbonized layer was clearly streaked, suggesting the (2×3) superstructure of the single-crystalline 3C-SiC surface shown in Fig. 7.

The carbonization temperature (T_C) influenced the thickness of the carbonized layer. Fig. 8 shows a carbonization program at various carbonization temperatures. The carbonization conditions in Fig. 8 are the same as those in Fig. 5, except for the carbonization temperature. Fig. 9 shows the change of the thickness of the carbonized layer and of the refraction index as a function of T_C in Fig. 8. The formation of the carbonized layer was observed at T_C higher than 800 °C, but the carbonized layer was not formed at T_C below 750 °C as suggested by the low refraction indexes (1.96).

Cheng et al. [24] and Bozack et al. [25] reported that the clean Si surface was carbonized at approximately 500 °C in an C_2H_2 environment. It seems reasonable that the large difference in temperature for carbonized-layer formation between our results and Cheng's results is caused by a difference in the state of the Si surface during the carbonization program. It has been reported that the C_2H_2 molecules are chemically adsorbed on the (2×1) superstructure of the Si(001) surface forming the di-σ bond between C_2H_2

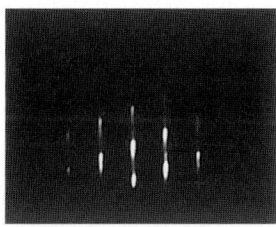

Fig. 7. RHEED image of carbonized Si surface prepared by the carbonization program shown in Fig. 5

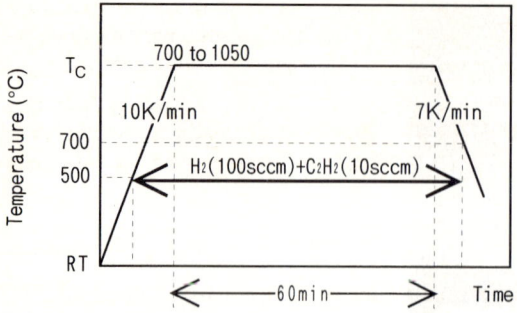

Fig. 8. Carbonization program at various carbonization temperatures (T_C). P_{total} (= 66 m Torr $\approx$ 8.78 Pa) means the total pressure in the reaction tube, while $P_{C_2H_2}$ (= 6 m Torr $\approx$ 0.798 Pa) means the C_2H_2 partial pressure in the reaction tube

and Si-dimer. As the adsorbed C_2H_2 on the Si-dimer is easily decomposed even at 500 °C, the formation temperature of the carbonized layer strongly depend on the concentration of Si-dimers on the Si surface. Cheng et al. carried out their carbonization program using an ultra-high vacuum system with a background pressure of $\approx 10^{-8}$ Pa in order to realize a clean Si(001) surface which maintains a (2 × 1) superstructure with a Si-dimer concentration. On the other hand, we carried out our carbonization program using a fused-silica reaction tube with a background pressure of 1.33×10^{-1} Pa. The residual molecules in our reaction tube detected by the QMS were confirmed as H_2O and O_2, so we can assume that the clean Si surface with a (2 × 1) superstructure was hindered by the thin SiO_2 layer during the heating process in our carbonization program. The formed SiO_2 layer seemed to prevent the Si surface from chemically adsorbing C_2H_2 and resulted in a higher carbonization temperature.

It is likely that the concentration of Si-dimers increased with a T_C above 800 °C due to elimination of a SiO_2 layer. In general, O_2 and H_2O react with the Si surface to form a SiO_2 layer as the temperature increases, but the formation of SiO_2 is suppressed and the formation of SiO is enhanced below a critical pressure of O_2 and H_2O [26]. The formed SiO is easily vaporized from the Si surface as the temperature increases. There-

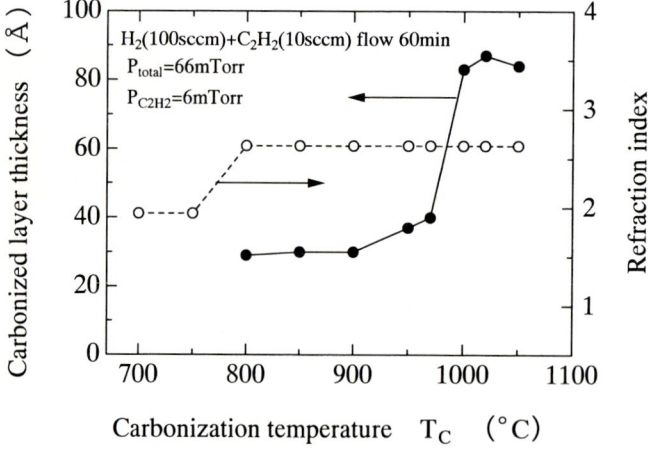

Fig. 9. Change of the thickness of the carbonized layer and refraction index as a function of carbonization temperature (T_c); see Fig. 8 for detailed carbonization condition

fore, in our carbonization program, the SiO_2 layer may be removed from the Si surface, resulting in the adsorption of C_2H_2 on Si-dimer at T_C higher than 800 °C.

A change in the thickness of the carbonized layer as a function of T_C shown in Fig. 9 falls into three temperature stages: in the T_C range from 800 to 970 °C, the thickness of the carbonized layer gradually increased along with T_C, while a remarkable increase in the thickness of the carbonized layer was observed in the T_C range from 970 to 1000 °C. Finally, however, the thickness of the carbonized layer became independent of T_C at temperatures higher than 1000 °C.

In order to examine the possibility that changes in the thickness of the carbonized layer may elevate T_C, we must consider both the dependence of the amount of supplied C_2H_2 and the diffused Si atoms on T_C, as well as the dependence of reaction between C_2H_2 and Si atoms on T_C. However, it does not have to be taken account for the individual decomposition of C_2H_2 due to its own strong triple bond (σ bond: 230 kcal/mol) [27]. Furthermore, it is also unnecessary to consider the probability of the reaction between C_2H_2 and Si atoms at T_C higher than 800 °C, as C_2H_2 molecules easily can be decomposed by interation with Si even at 500 °C. It is therefore clear that the mutual migration on C_2H_2 and Si atoms via the carbonized layer was the predominant influence on the thickness of the carbonized layer at T_C higher than 800 °C.

The gradual increase in the thickness of the carbonized layer in the T_C range between 800 and 970 °C may be explained by the penetration of C into the Si substrate. As reported by Weiner et al. [28] and Cheng et al. [24], the carbon atoms of C_2H_2 penetrate into the Si substrate because they are the most energetically stable, rather than being simply adsorbed into the Si surface. Furthermore, the most disturbed surfaces that enhance the carbon diffusion are expected to be the most probable structures at high temperatures.

The remarkable increase in the thickness of the carbonized layer at T_C higher than 970 °C can no longer be considered to account for the C penetration. It appears that the vertical migration of Si atoms through porous defects in the carbonized layer promoted the formation of the carbonized layer in a T_C range righer than 970 °C [15].

The constant thickness of the carbonized layer at T_C higher than 1000 °C can be explained by the sealing-off of the porous defects [23]. At T_C higher than 1000 °C, upon reaching the top of the porous defects, Si atoms might react with C_2H_2 resulting in the preferential formation of a SiC layer only at the top of porous defects. Accordingly, the formed SiC layer shut Si atoms in porous defects and suppressed the growth of the carbonized layer [15, 23].

The crystallinity of the carbonized layer was strongly affected by T_C. Halo images that suggested the existence of an amorphous layer were observed from RHEED patterns on the carbonized surface in a T_C range from 800 to 900 °C. At T_C higher than 900 °C, the RHEED patterns changes in semicircular images, indicating the existence of a polycrystal layer. Finally, the RHEED patterns became streaked, suggesting the existence of a single-crystal layer at T_C higher than 1000 °C.

Fig. 10 shows the schematic progress of the formation of a carbonized layer on the Si surface during the heating process in an C_2H_2 environment. At T_C below 800 °C, residual gases such as H_2O or O_2 seem to react with the Si surface to form a thin SiO_2 layer preventing the chemical adsorption of C_2H_2, as shown in Fig. 10a. Therefore, the formation of the carbonized layer is suppressed at T_C below 800 °C. In the T_C range between 800 and 900 °C, the SiO_2 layer is likely to begin to vaporize, resulting in an increase in

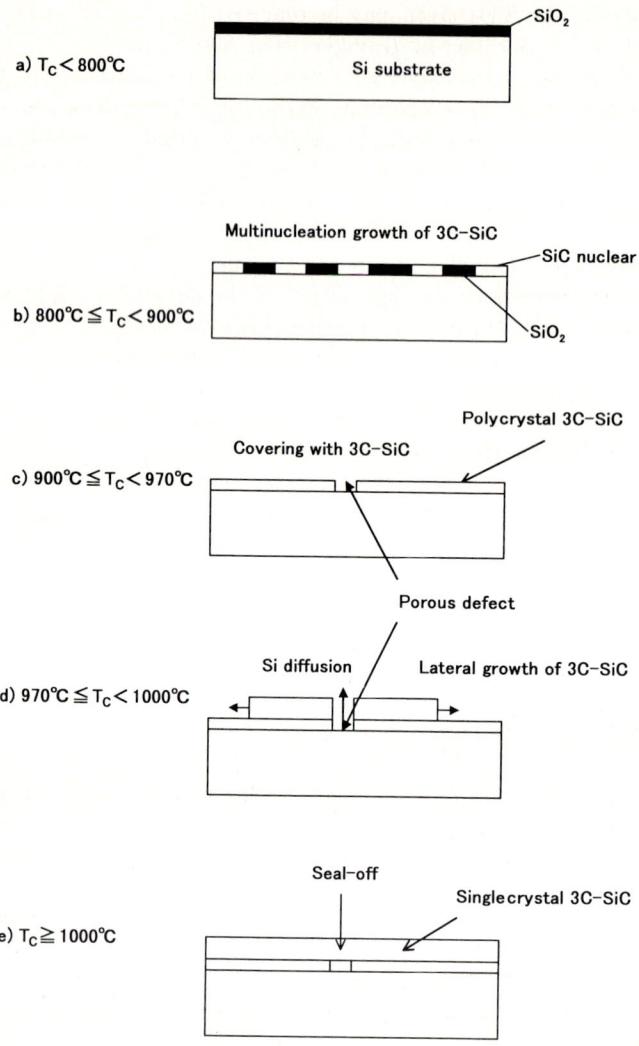

a) $T_C < 800°C$

Multinucleation growth of 3C–SiC

SiC nuclear

b) $800°C \leqq T_C < 900°C$

SiO$_2$

Polycrystal 3C–SiC

Covering with 3C–SiC

c) $900°C \leqq T_C < 970°C$

Porous defect

Si diffusion   Lateral growth of 3C–SiC

d) $970°C \leqq T_C < 1000°C$

Seal–off

Singlecrystal 3C–SiC

e) $T_C \geqq 1000°C$

Fig. 10. Schematic progress of the formation of a carbonized layer on the Si surface during the heating process in an C_2H_2 environment

the Si-dimer concentration. Multi-nucleation growth of thin 3C-SiC then seems to occur, as shown in Fig. 10b. The reason for the amorphous RHEED pattern in this temperature range is thought to be the high strain on the thin carbonized layer [28]. At 900 °C in T_C the entire Si surface is covered with the carbonized layer due to the high probability of the appearance of the Si-dimer. The thickness of the carbonized layer gradually increases along with T_C in the range from 900 to 970 °C due to the penetration of C. The C penetration seems to relieve the strain in the carbonized layer and form the polycrystalline carbonized layer. However, many porous defects are likely to be still being included in the carbonized layer in this T_C range, as shown in Fig. 10c. During the heating process from 970 to 1000 °C, Si atoms vertically migrate from the substrates toward the surface through the porous defects in the carbonized layer, bringing about a lateral growth of 3C-SiC on the surface of the carbonized layer. Thus, remarkable growth of the carbonized layer and an improvement in its crystallinity occur in this T_C range, as shown in Fig. 10d. Finally, at T_C higher than 1000 °C, the growth of the carbonized layer is suppressed due to the sealing-off of porous defects in the carbonized layer, as shown in Fig. 10e.

It can be concluded, from what has been discussed above, that an ideal carbonized layer for 3C-SiC epitaxial growth can be obtained on the Si(001) surface through the use of a heating process from 500 to higher than 1000 °C in an C_2H_2 environment.

4. Growth Mechanism of 3C-SiC Resulting from an Alternating Supply of Gases

SiC growth processes were carried out following the carbonization process of the Si surface. The first issue to be discussed is why it is necessary to employ an alternating gas

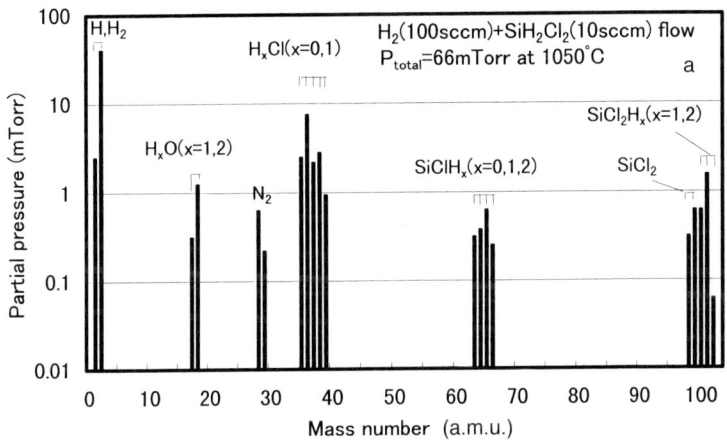

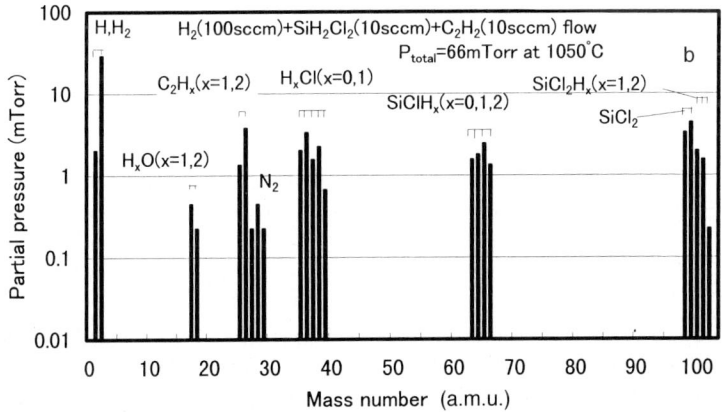

Fig. 11. Partial pressures (1 m Torr ≈ 0.133 Pa) in the reaction tube during the supply of SiH_2Cl_2 and C_2H_2 at 1050 °C: a) H_2 (100 sccm) + SiH_2Cl_2 (10 sccm) supply; b) H_2 (100 sccm) + SiH_2Cl_2 (10 sccm) + C_2H_2 (10 sccm) supply (a.m.u. means atomic mass units)

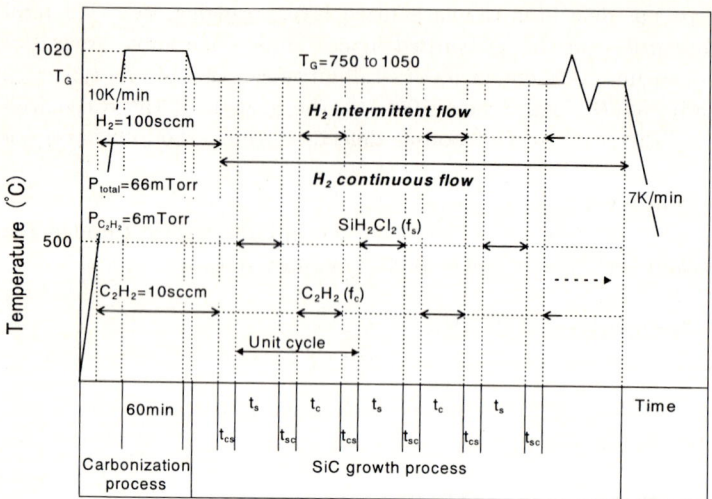

Fig. 12. A typical SiC growth program using the alternating supply of SiH$_2$Cl$_2$ and C$_2$H$_2$. T_G growth temperature, t_s duration of SiH$_2$Cl$_2$ supply, t_c duration of C$_2$H$_2$ supply, f_s flow rate of SiH$_2$Cl$_2$, f_c flow rate of C$_2$H$_2$, t_{cs} interval after stopping C$_2$H$_2$ supply, t_{sc} interval after stopping SiH$_2$Cl$_2$ supply

supply instead of a simultaneous gas supply. The first important point is that an alternating supply of source gases is a mean to implement the atomic layer epitaxy (ALE). Even more important is that an alternating supply of source gases is necessary to promote SiC growth when SiH$_2$Cl$_2$ and C$_2$H$_2$ are used. In other words, the SiC growth rate is considerably suppressed by the simultaneous supply of SiH$_2$Cl$_2$ and C$_2$H$_2$ due to the high sticking probability of C$_2$H$_2$ on the surface of substrate.

It is recognized that the precursors concerned with SiC growth are SiCl$_2$ and C$_2$H$_2$ in our gas system [9]. The SiCl$_2$ molecules are generated by the thermal decomposition of SiH$_2$Cl$_2$. SiC growth is promoted by the reduction of adsorbed SiCl$_2$ on the surface of substrates and the subsequent reaction between generated Si and C$_2$H$_2$. However, C$_2$H$_2$ molecules supplied simultaneously with SiH$_2$Cl$_2$ predominantly adsorb on the surface of substrates and suppress SiCl$_2$ adsorption.

Fig. 11 shows the partial pressures in the reaction tube during the supply of SiH$_2$Cl$_2$ and C$_2$H$_2$. Refer to Fig. 11 for details. The SiCl$_2$ partial pressure with simultaneous supply of SiH$_2$Cl$_2$ and C$_2$H$_2$ increased to eight times that of the supply of SiH$_2$Cl$_2$ alone. It is likely that the drastic increase in SiCl$_2$ partial pressure with the simultaneous supply of SiH$_2$Cl$_2$ and C$_2$H$_2$ was caused by the obstruction of SiCl$_2$ adsorption by the adsorbed C$_2$H$_2$ molecules on the surface of substrates. Practically, it has been confirmed that the simultaneous supply of SiH$_2$Cl$_2$ and C$_2$H$_2$ could not bring about SiC film formation, but rather SiC whisker formation [29].

It is clear that the suppression of C$_2$H$_2$ adsorption on the surface of substrates during SiH$_2$Cl$_2$ supply may bring about the stable adsorption of the SiCl$_2$ molecules. Accordingly, the supply of SiH$_2$Cl$_2$ alone must be maintained until a stable SiCl$_2$ molecular layer is formed on the surface of substrates prior to the supply of C$_2$H$_2$. The supplied C$_2$H$_2$ molecules following the supply of SiH$_2$Cl$_2$ will effectively react with Si atoms on the surface resulting in SiC film growth.

Fig. 12 shows a typical SiC growth program using the alternating supply of source gases. This program consists of two processes: carbonization and growth. First, the carbonization process was carried out to convert the Si surface into a thin epitaxial SiC layer, in which H_2 and C_2H_2 were supplied into the reaction tube at 500 °C, as described in the previous section. The reaction tube was then heated to 1020 °C while maintaining the continuous flows of C_2H_2 and H_2. The flow rates of H_2 and C_2H_2 were fixed at 100 and 10 sccm, respectively.

After the carbonization process was carried out for 60 min at 1020 °C, the temperature of the reaction tube was changed to the desired growth temperature (T_G), while keeping the H_2 and C_2H_2 flows constant. The growth temperatures were varied from 1000 to 1050 °C. The SiC growth processes begun with the introduction of SiH_2Cl_2. Following the desired duration of SiH_2Cl_2 supply (t_s), C_2H_2 supply was conducted. A delay following SiH_2Cl_2 supply (t_{sc}) and a delay following C_2H_2 supply (t_{cs}) were employed to prevent the mixing of gas in the reaction tube. The t_s were varied from 3 to 30 s, while the duration of C_2H_2 supply (t_c) was varied from 1 to 30 s. The flow rate of SiH_2Cl_2 (f_s) and the flow rate of C_2H_2 (f_c) were independently varied from 1 to 30 sccm.

The SiC growth mechanism is strongly influenced by the decomposition process of Si species. In order to clarify the decomposition process of the adsorbed Si species on the surface of substrates, two methods of H_2 supply were employed (see Fig. 12 for details): one was the continuous flow of 100 sccm H_2 during the SiC growth process, and the other was the synchronized supply of 100 sccm of H_2 with C_2H_2 supply. We will use the term "H_2 continuous flow" to refer to the former and "H_2 intermittent flow" for the latter.

The gas-supply cycle, defined as the period from the start of SiH_2Cl_2 supply to the beginning of the next SiH_2Cl_2 supply, was repeated from 500 to 2500 times. Following the SiC growth process, the reaction tube was cooled to room temperature with no gas flow.

The thickness of the grown SiC films was measured using an ellipsometer with a He–Ne laser. The thickness of the measured SiC film divided by the number of gas-supply cycles gives the growth rate (r_g) of the SiC.

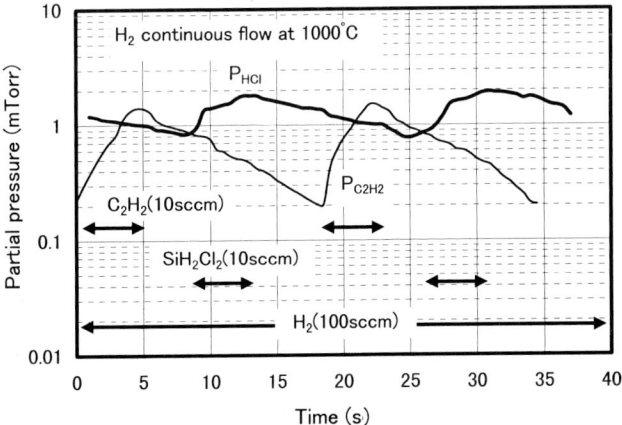

Fig. 13. Changes in C_2H_2 partial pressure ($P_{C_2H_2}$) and HCl partial pressure (P_{HCl}) during the SiC growth process with the "H_2 continuous flow" method at 1000 °C. Refer to Table 1 for details

Table 1

3C-SiC growth condition by the method of alternating supply of source gases using "H$_2$ continuous flow"

T_G:	growth temperature	1000 °C
f_s:	flow rate of SiH$_2$Cl$_2$	10 sccm
f_c:	flow rate of C$_2$H$_2$	10 sccm
t_s:	duration of SiH$_2$Cl$_2$ supply	5 s
t_c:	duration of C$_2$H$_2$ supply	5 s
t_{cs}:	delay following C$_2$H$_2$ supply	3 s
t_{sc}:	delay following SiH$_2$Cl$_2$ supply	5 s

The decomposition processes of source gases in the reaction tube were observed through the use of the QMS. Fig. 13 shows a change in partial pressures in the reaction tube during the SiC growth process with the "H$_2$ continuous flow" method at 1000 °C. For details on gas supply, see Table 1. The partial pressures of HCl (P_{HCl}) and C$_2$H$_2$ ($P_{C_2H_2}$) exhibited a synchronized change with the supply of gas.

Most of the SiCl$_2$ molecules generated by the thermal decomposition of SiH$_2$Cl$_2$ were adsorbed on the surface of the substrates [30]. The top surface covered with SiCl$_2$ molecules was terminated by Cl [31, 32]. However, the Cl atoms on the top surface were immediately removed by H$_2$ as a reducing agent with the "H$_2$ continuous flow" method. There is considerable evidence indicating that the reduction of SiCl$_2$ was caused during the supply of SiH$_2$Cl$_2$ with the "H$_2$ continuous flow" method. As shown in Fig. 13, there was a P_{HCl} peak during SiH$_2$Cl$_2$ supply. It is clear that the immediate reduction of the adsorbed SiCl$_2$ was induced by H$_2$ supply accompanied by the SiH$_2$Cl$_2$ supply. Accordingly, the self-limiting adsorption of SiCl$_2$ was impossible by the SiC growth process with the "H$_2$ continuous flow" method.

It is necessary for the realization of the self-limited adsorption of SiCl$_2$ by stable Cl termination on the surface of the substrate to stop the supply of H$_2$ during SiH$_2$Cl$_2$ supply. In this respect, "H$_2$ intermittent flow" is an effective method of inducing the

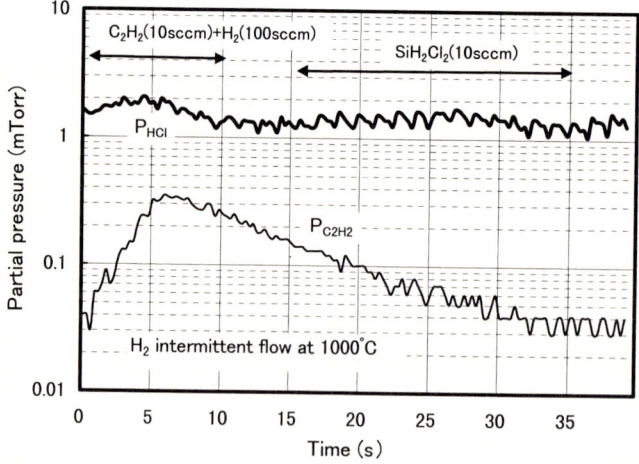

Fig. 14. Changes in C$_2$H$_2$ partial pressure ($P_{C_2H_2}$) and HCl partial pressure (P_{HCl}) during the SiC growth process with the "H$_2$ intermittent flow" method at 1000 °C. Refer to Table 2 for details

Table 2

3C-SiC growth condition by the method of alternating supply of source gases using "H$_2$ intermittent flow"

T_G:	growth temperature	1000 °C
f_s:	flow rate of SiH$_2$Cl$_2$	10 sccm
f_c:	flow rate of C$_2$H$_2$	10 sccm
t_s:	duration of SiH$_2$Cl$_2$ supply	20 s
t_c:	duration of C$_2$H$_2$ supply	10 s
t_{cs}:	delay following C$_2$H$_2$ supply	5 s
t_{sc}:	delay following SiH$_2$Cl$_2$ supply	5 s

self-limiting adsorption of SiCl$_2$ [31]. Fig. 14 shows the change in partial pressures in the reaction tube during the SiC growth process with the "H$_2$ intermittent flow" method. Refer to Table 2 for details on the SiC growth process with the "H$_2$ intermittent flow" method. The change in P_{HCl} with the "H$_2$ intermittent flow" method is quite different from that of the "H$_2$ continuous flow" method. The P_{HCl} peak occurred during C$_2$H$_2$ and H$_2$ supply. The displacement of the P_{HCl} peak suggested the suppression of SiCl$_2$ reduction during SiH$_2$Cl$_2$ supply.

Fig. 15 shows the changes in the SiC growth rate (r_g) as a function of the duration of SiH$_2$Cl$_2$ supply (t_s). For the details on the SiC growth condition, see Table 1 for the "H$_2$ continuous flow" method and Table 2 for the "H$_2$ intermittent flow" method, except for T_G and t_s.

In the case of the "H$_2$ continuous flow" method, r_g was increased by t_s [30 to 32]. Furthermore, at each T_G, SiC growth shows different rates as a function of t_s. The above characteristics of the SiC growth rate can be explained by the reduction process

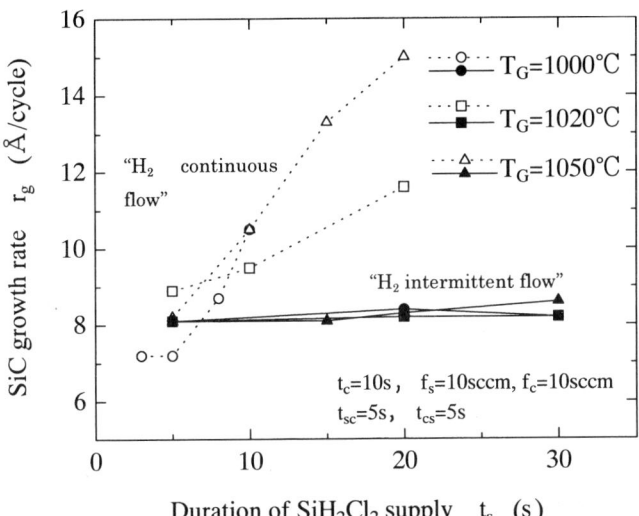

Fig. 15. Changes in the SiC growth rate (r_g) as a function of the duration of SiH$_2$Cl$_2$ supply (t_s) at various growth temperatures (T_G). For the details on the SiC growth condition, refer to Table 1 for the "H$_2$ continuous flow" method and to Table 2 for the "H$_2$ intermittent flow" method except for T_G and t_s

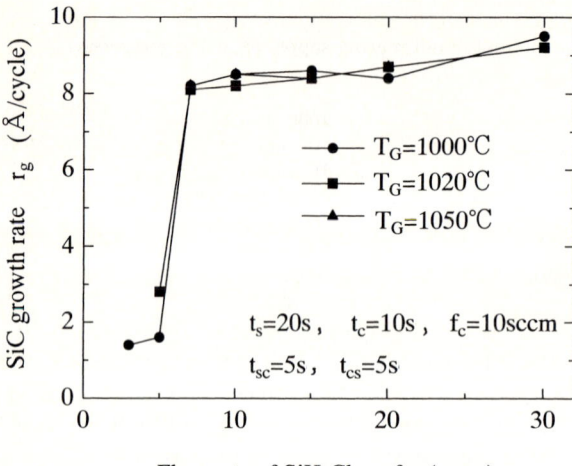

Fig. 16. Changes in the SiC growth rate (r_g) as a function of SiH_2Cl_2 flow rate (f_s) with the "H_2 intermittent flow" method at various growth temperatures (T_G). Refer to Table 2 for detailed growth condition except for T_G and f_s

of $SiCl_2$ molecules by the H_2 on the surface of substrates. In the case of the "H_2 continuous flow", the adsorbed $SiCl_2$ molecules were immediately reduced by H_2 during SiH_2Cl_2 supply, resulting in the incorporation of Si atoms on the surface of substrates. The generated Si atoms will react with C_2H_2 upon its introduction into the reaction tube. Since the number of reduced $SiCl_2$ molecules — that is, the number of Si atoms incorporated into the surface of substrates — increased with t_s, the SiC growth rate also increased with t_s [33]. There are two likely reasons for the discrepancy between the SiC

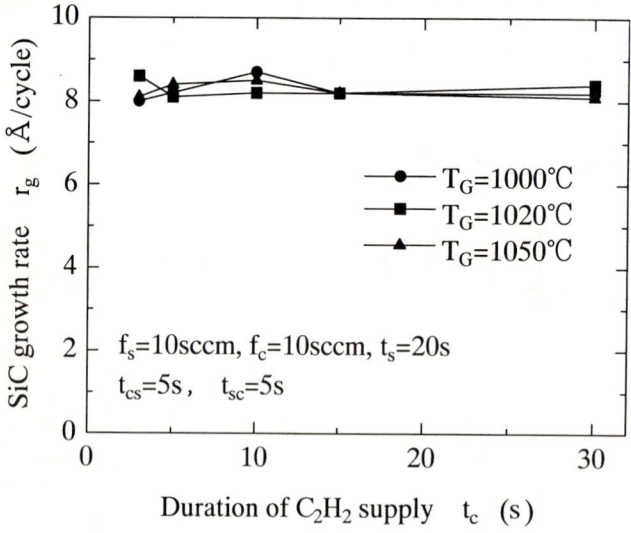

Fig. 17. Changes in the SiC growth rate (r_g) as a function of duration of C_2H_2 supply (t_c) with the "H_2 intermittent flow" method at various growth temperatures (T_G). Refer to Table 2 for detailed growth condition except for T_G and t_c

growth rates at different T_G: one is that the decomposition rate of SiH_2Cl_2 — the generation rate of $SiCl_2$ — was affected by T_G, and the other is that the reduction rate of adsorbed $SiCl_2$ molecules was also a function of T_G.

On the other hand, during 5 to 30 s in t_s, r_g was steady at 8 Å/cycle at any T_G with the "H_2 intermittent flow" method, as can be clearly seen in Fig. 15. It can be said that the constant SiC growth rate suggests the realization of the self-limiting process of $SiCl_2$ adsorption. The reduction of the adsorbed $SiCl_2$ was suppressed by the lack of H_2 during SiH_2Cl_2 supply. The stable $SiCl_2$ layer on the surface of the substrates will not adsorb excess $SiCl_2$ molecules at any temperature.

The r_g was independent of the flow rate of SiH_2Cl_2 (f_s) between 7 and 30 sccm with the "H_2 intermittent flow" method, as shown in Fig. 16. See Table 2 for details, except for T_G and f_s. The constant growth rate of SiC also suggests that the number of adsorbed $SiCl_2$ was independent of the $SiCl_2$ partial pressure in the reaction tube. The reason for the rapid increases in r_g with f_s below 7 sccm is that the surface termination by Cl was not completed due to the lack of adsorbed $SiCl_2$ molecules on the surface of the substrate.

The C_2H_2 molecules never thermally decomposed at any temperature in our experiment [27], but C_2H_2 molecules can easily be decomposed by interaction with Si dangling bonds on the surface of substrates to bring about SiC formation. From this viewpoint, it can be said that the SiC growth rate may be independent of the amount of C_2H_2 supplied during the SiC growth process. Actually, r_g was independent for duration of C_2H_2 supply (t_c) from 3 to 30 s with the "H_2 intermittent flow" method, as shown in Fig. 17 (see Table 2 for details, except for T_G and t_c). While the dependence of r_g on flow rate of C_2H_2 (f_c) was quite different from the above prediction, r_g decreased with f_c higher than 15 sccm with the "H_2 intermittent flow" method, as shown in Fig. 18 (see Table 2 for details, except for T_G and f_c). The decreases in r_g at f_c higher than 15 sccm can be

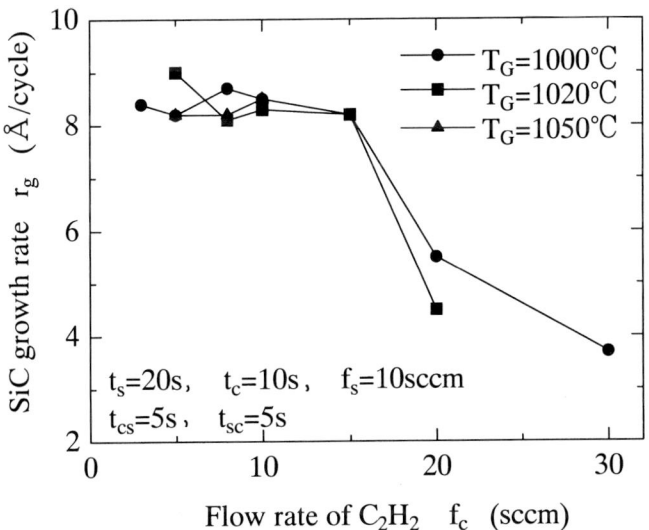

Fig. 18. Changes in the SiC growth rate (r_g) as a function of C_2H_2 flow rate (f_c) with the "H_2 intermittent flow" method at various growth temperatures (T_G). Refer to Table 2 for detailed growth condition except for T_G and f_c

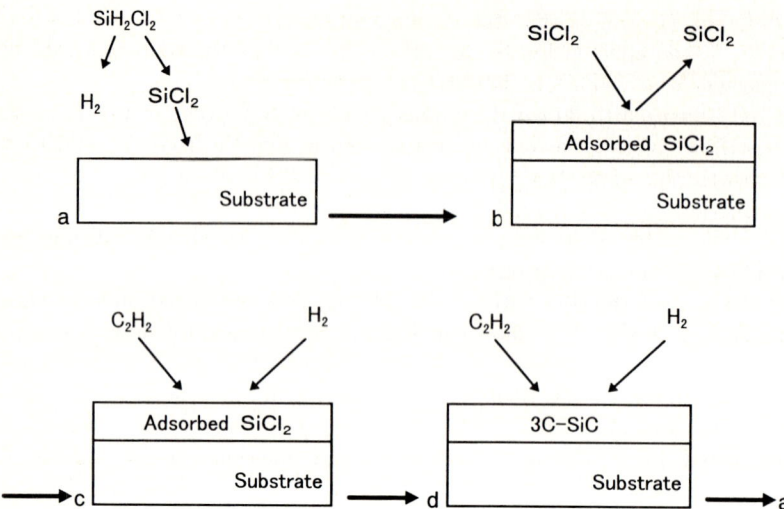

Fig. 19. Mechanism of SiC growth with the alternating supply of gases with the "H_2 intermittent flow" method

explained by the obstruction of $SiCl_2$ adsorption on the surface of the substrate by the residual C_2H_2 layer on it during SiH_2Cl_2 supply. It seems reasonable to suppose that C_2H_2 could not bring about self-limiting adsorption, but C_2H_2 decomposition might be caused only by the interaction with Si dangling bonds on the surface. Therefore, SiC growth rates were predominantly determined by the amount of adsorbed $SiCl_2$ molecules on the surface of the substrate. If a sufficient interval (t_{cs}) was employed after stopping the C_2H_2 supply, r_g might be independent of f_c due to the disappearance of the residual C_2H_2 layer on the surface of substrates during SiH_2Cl_2 supply.

We will conclude this discussion of the mechanism of SiC growth with the method of alternating supply of gases with "H_2 intermittent flow" shown in Fig. 19. At first, the introduced SiH_2Cl_2 molecules are thermally decomposed, causing the generation of H_2 and $SiCl_2$ (Fig. 19a). The generated $SiCl_2$ molecules adsorb on the surface of the substrate as Si precursor. The constant amount of $SiCl_2$ molecules on the surface is maintained until the beginning of the supply of H_2 and C_2H_2 due to lack of H_2 (Fig. 19b). When H_2 and C_2H_2 are introduced into the reaction tube, the $SiCl_2$ layer is immediately reduced by H_2, and Si atoms are incorporated into the surface (Fig. 19c). Simultaneously, the introduced C_2H_2 molecules react with the new Si surface resulting in the SiC formation (Fig. 19d). Thus, the SiC growth rates are determined by the amount of adsorbed $SiCl_2$ molecules on the surface, and become independent of the growth condition. Accordingly, excellent controllability of the SiC growth rate can be realized by the method of alternating supply of source gases with "H_2 intermittent flow".

The occurrence of the self-limiting adsorption of $SiCl_2$ also induced an excellent uniformity of SiC thickness over a large area [31]. Fig. 20 shows the appearance and thickness distribution of the 3C-SiC film grown on a Si wafer six inches in diameter by the "H_2 intermittent flow" method (refer to Table 2 for details). The standard deviation of SiC thickness was less than 0.5% of the average SiC thickness. This high uniformity of SiC thickness suggests the implementation of self-limited $SiCl_2$ adsorption over a 6-inch diameter by the alternating supply of source gases with the "H_2 intermittent flow" method.

Fig. 20. Appearance and thickness distribution of the 3C-SiC film grown on a Si wafer of 6-inch diameter by the "H_2 intermittent flow" method. The standard deviation of SiC thickness (σ) corresponds to 0.44% of thickness average. Refer to Table 2 for detailed growth condition

5. Crystallinity of 3C-SiC Films

The crystallinity of 3C-SiC films grown on Si substrates by the method of alternating supply of gases with "H_2 intermittent flow" was evaluated using SAED, XRD, and TEM. The TEM image and SAED pattern were observed using incident electrons along the [$\bar{1}$10] direction with an acceleration voltage of 200 kV. A pole figure of the XRD was observed to correspond with the crystal orientation by keeping the detector position fixed at the diffraction angle of the 3C-SiC(111) plane and rotating the sample about an axis normal to the substrate (azimuth angle φ) and about an axis parallel to the substrate and parallel to the plane of dispersion (polar angle χ) [34]. The surface morphologies of 3C-SiC films were observed through the use of atomic force microscope (AFM).

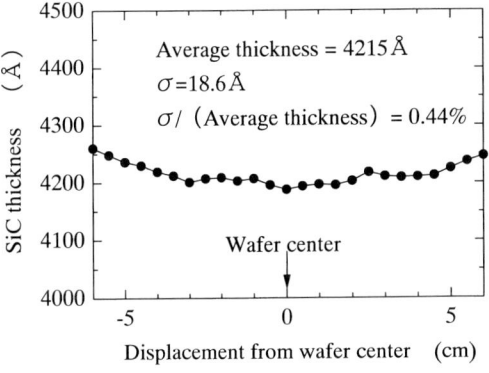

Average thickness = 4215 Å
$\sigma = 18.6$ Å
$\sigma /$ (Average thickness) = 0.44%

Wafer center

SiC thickness (Å)

Displacement from wafer center (cm)

The SAED pattern of the grown SiC film exhibited spots that corresponded only to the reciprocal lattice of 3C-SiC, as shown in Fig. 21. This seems to indicate that single-crystal 3C-SiC films were grown on Si(001) substrates with the alternating supply of source gases.

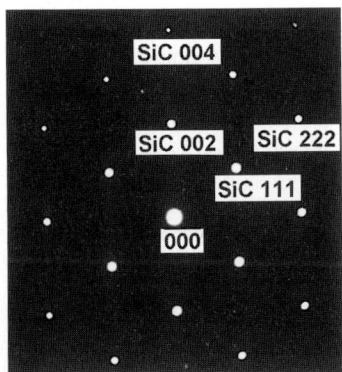

SiC 004

SiC 002   SiC 222

SiC 111

000

Fig. 21. SAED pattern of a 3C-SiC film observed along the [$\bar{1}$10] direction of electron incidence

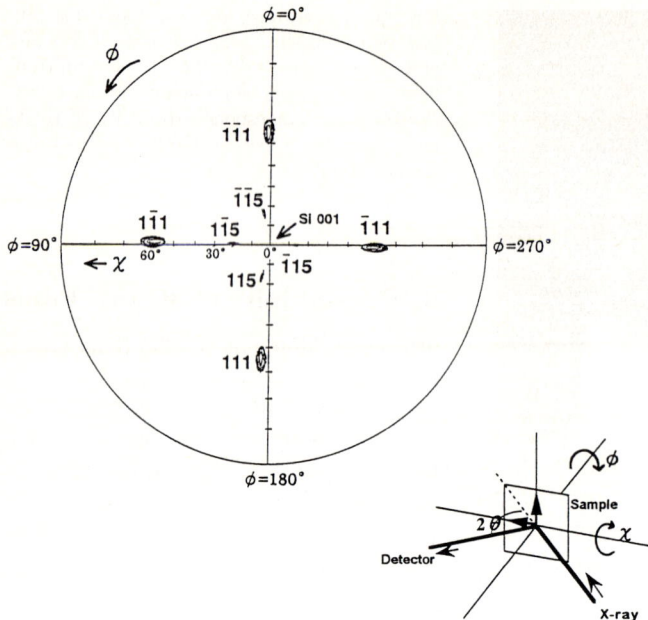

Fig. 22. {111} pole figure of the 3C-SiC film. The indexes represent the reciprocal lattice points

Fig. 22 shows an XRD pole figure of the 3C-SiC film on Si(001) substrate. Four strong peaks crossing the 3C-SiC(001) pole with at approximately 55° in χ were observed in Fig. 22. These four peaks corresponded to the diffraction from the 3C-SiC{111} face. Four other weak peaks crossing the 3C-SiC(001) pole at approximately 16° in χ were also observed. These four weak peaks corresponded to the 3C-SiC{115} face and suggested the existence of a symmetrical twin band around the ⟨111⟩ axis, as shown in Fig. 23.

Fig. 24 shows cross-sectional TEM images of the 3C-SiC film on Si substrate. Many boundaries along the {111} plane can be clearly observed in Fig. 24a. These boundaries can be classified as anti-phase boundaries (APBs) and twin boundaries (TBs). Fig. 24b shows that the width of twin bands is a few atomic layers. All of these boundaries are derived from an interface between the 3C-SiC film and the Si substrate.

A cross-sectional TEM image of a carbonized layer on the Si substrate in Fig. 25 shows many steps with a height of few atoms at the interface between the 3C-SiC film and the Si substrate. If an odd-atom

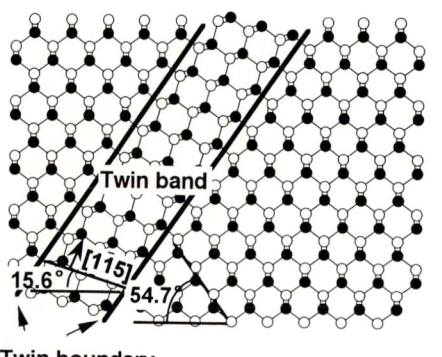

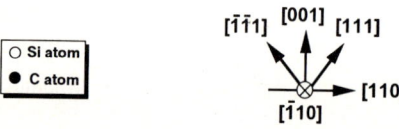

Fig. 23. Schematic diagram of the twin structure

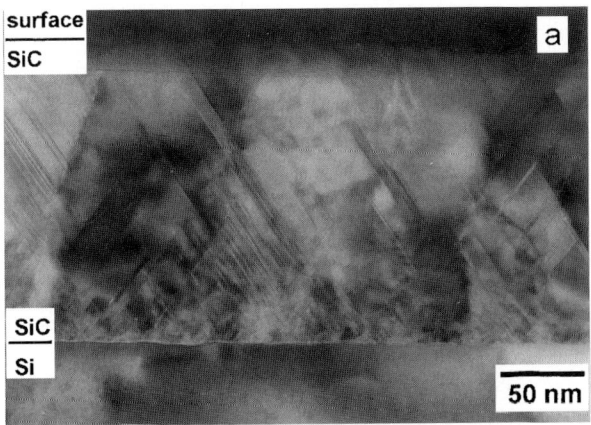

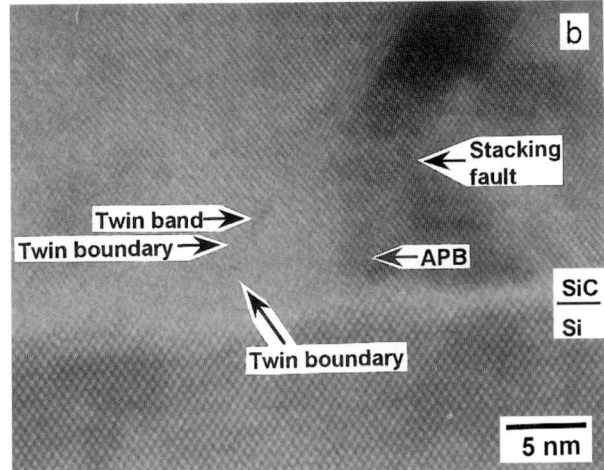

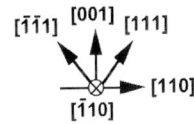

Fig. 24. Cross-sectional TEM images of 3C-SiC films observed along the [Ī10] direction of electron incidence. a) Low magnification image, b) high magnification image around the SiC/Si interface

height step exists in the interface, APB will be generated at the step, as shown in Fig. 26. The generated APB will propagate in the ⟨111⟩ direction in a straight line due to the discrepancy of priority for Si or C disposition on the top surface of the APB [35].

Fig. 27 shows an AFM image of a 3C-SiC surface 1 μm in thickness. The ditch lines observed on the surface correspond to those of the APB. Fig. 28 shows the change in APB density and maximum domain size as a function of SiC film thickness, which was estimated through the use of AFM images. The APB density decreased along with the SiC film thickness.

There are two factors contributing to the elimination of APB: the existence of stacking faults and the annihilation of boundaries. If the APB and the stacking fault cross

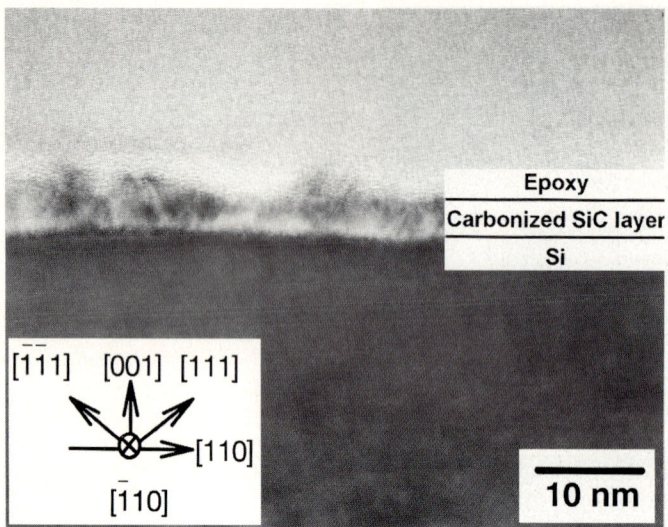

Fig. 25. Cross-sectional TEM image of a carbonized layer on Si substrate along the [$\bar{1}$10] direction of electron incidence

perpendicularly, the stacking fault reverses the stacking order of Si and C. Therefore, the APB will disappear at the junction with the stacking faults as shown in Fig. 24b. On the other hand, if the APBs cross each other, they will annihilate at the junction (X), as indicated in Fig. 26.

6. Electrical Properties of 3C-SiC Film

The SiC films grown using the alternating supply of gases with the "H$_2$ intermittent flow" method were cut in 10×10 mm^2 square chips to fabricate specimens for Hall measurements. These specimens were treated in 3% HF solution to remove the native oxide

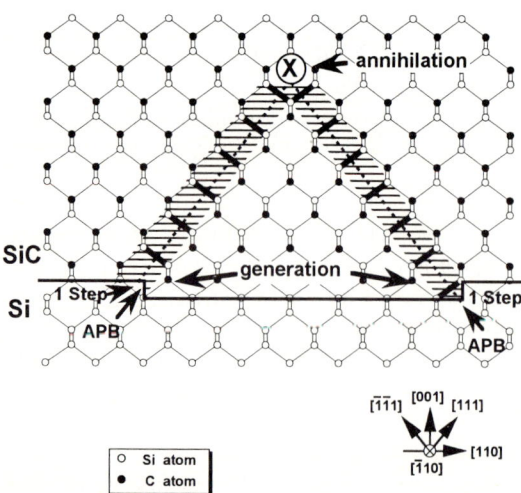

Fig. 26. Schematic structure of APBs derived from an interface between 3C-SiC film and Si substrate

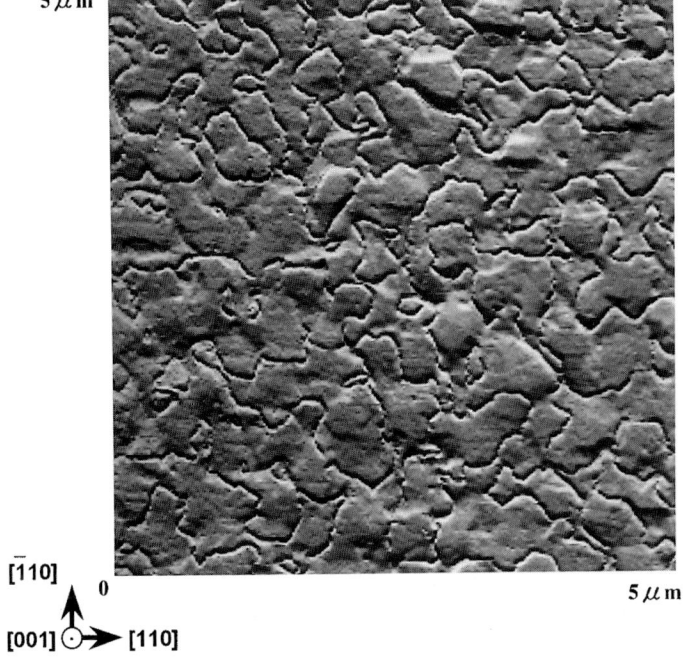

5 μm

[1̄10]

[001] ⊙ → [110]

Fig. 27. AFM image of a 3C-SiC surface

layer of the surface just before the formation of electrodes. Ni 700 Å in thickness was evaporated at the four corners of each SiC chip to form electrodes. Following the evaporation of Ni electrodes, the SiC chips were annealed at 700 °C for 5 min to create ohmic contacts. The resistivity of 3C-SiC film was measured by the Van der Pauw method, and the Hall measurements were carried out in a magnetic field of 5000 G. The changes in electrical properties that accompanied changes in SiC thickness were measured at room temperature (25.7 °C).

Although no intentional doping was employed during the SiC growth process, the 3C-SiC films grown by the alternating supply of SiH_2Cl and C_2H_2 exhibited n-type con-

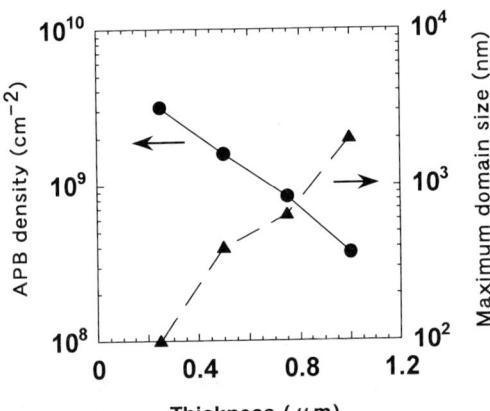

Fig. 28. Change in APB density and maximum domain size as a function of SiC film thickness

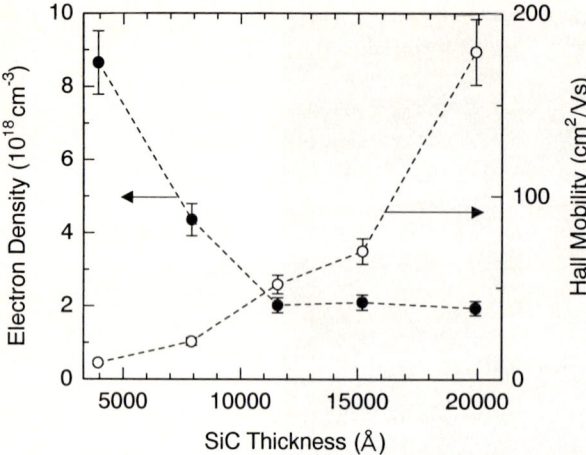

Fig. 29. Thickness dependence of the electron density and Hall mobility of the grown 3C-SiC film measured at 25.7 °C

duction. The electron density changed from 8.7×10^{18} to 2.0×10^{18} cm^{-3} with the change in SiC thickness at room temperature. Fig. 29 shows the thickness dependence of the electron density and Hall mobility of the grown 3C-SiC film. The thin SiC films have extremely high electron density. The electron density exhibits a decrease with SiC thickness below 11500 Å, while it exhibits no change in the range from 11500 to 19900 Å. This result suggests that a high donor density exists within a SiC layer approximately 10000 Å in thickness near the SiC/Si interface.

It is important to clarify the origin of the donor in 3C-SiC films. It is possible that some donor impurities, such as nitrogen, were mixed with source gases during the SiC growth process due to residual gases in the reaction tube. However, it may be incorrect to assume that the main origin of the donor is the mixed impurities during the SiC growth process. If the main origin is the mixed impurities during the SiC growth process, the electron density must be independent of SiC thickness. We believe that the main origin of high electron density is crystal defects such as APBs or twins.

The electron mobility in SiC film is also affected by its thickness. The Hall mobility increases along with SiC thickness, as shown in Fig. 29. A drastic increase in the Hall mobility is observed at a film thickness above 15000 Å.

The behavior of the electron density and mobility in relation to SiC thickness suggests that a part of the epitaxial layer near the Si substrate has a low mobility and high donor density resulting from the high density of crystal defects [36]. As mentioned previously, it is clear that there is a high density of APBs and twins exist in the vicinity of the SiC/Si interface. It could be argued that the decrease in APBs and twins accompanying an increase in SiC film thickness results in an increase in the electron mobility and a decrease in electron density in SiC films.

7. Conclusion

We have attempted to grow single-crystal 3C-SiC films on large-area Si substrates using SiH$_2$Cl$_2$ and C$_2$H$_2$ as source gases.

Prior to the SiC growth process, the carbonization of the Si surface by C$_2$H$_2$ was optimized to suppress the undesirable effect of lattice mismatching in the interface between SiC and the Si substrate. An ideal carbonized layer for 3C-SiC epitaxial growth could be obtained on the Si(001) surface by a heating process from 500 °C to higher than 1000 °C in an C_2H_2 environment.

3C-SiC growth following the carbonization process was realized by an alternating supply of source gases. The SiC growth rates with the alternating supply of source gases were predominantly determined by the amount of $SiCl_2$ incorporated into the surface of the substrate as a Si species. The suppression of $SiCl_2$ reduction and the subsequent occurrence of self-limiting $SiCl_2$ adsorption were induced with the "H_2 intermittent flow" method due to a lack of H_2 during SiH_2Cl_2 supply. Remarkably, high uniformity over a 6-inch diameter and excellent controllability of the SiC growth rate were realized by the self-limiting adsorption of $SiCl_2$, which maintained Cl termination on the top surface of the substrate.

The 3C-SiC films grown on Si(001) substrates included anti-phase boundaries and twin boundaries derived from the steps on the 3C-SiC/Si interface. The anti-phase boundary density gradually decreased along with the SiC growth due to coalescence of the crossing anti-phase boundaries or arrangement of the stacking order by the stacking faults.

The decrease in anti-phase boundaries and twins accompanying an increase in SiC film thickness resulted in an increase in electron mobility and a decrease in electron density.

References

[1] W. E. NELSON, F. A. HALDEN, and A. ROSENGREEN, J. Appl. Phys. **37**, 333 (1966).
[2] D. K. FERRY, Phys. Rev. B **12**, 2361 (1975).
[3] A. SUZUKI, Y. FUJII, H. SAITO, Y. TAJIMA, K. FURUKAWA, and S. NAKAJIMA, J. Cryst. Growth **115**, 623 (1991).
[4] H. MATSUNAHI, Amorphous and Crystalline Silicon Carbide, Vol. 3, Ed. C. Y. YANG, M. M. RAHMAN, and G. L. HARRIS, Springer-Verlag, Berlin 1992.
[5] H. KIMOTO, H. NISHINO, A. YAMASHITA, W. S. YOO, and H. MATSUNAMI, Amorphous and Crystalline Silicon Carbide, Vol. 31, Ed. C. Y. YANG, M. M. RAHMAN, and G. L. HARRIS, Springer-Verlag, Berlin 1992.
[6] S. NISHINO, J. A. POWELL, and H. A. WILL, Appl. Phys. Lett. **42**, 460 (1983).
[7] M. YAMANAKA, H. DAIMON, E. SAKUMA, S. MISAWA, and S. YOSHIDA, J. Appl. Phys. **61**, 599 (1987).
[8] M. IWAI, M. HIRAI, M. KUSAKA, Y. YOKOTA, and H. MATSUNAMI, Jpn. J. Appl. Phys. **28**, 293 (1989).
[9] H. NAGASAWA and Y. YAMAGUCHI, Thin Solid Films **225**, 230 (1993).
[10] T. FUYUKI, M. NAKAYAMA, T. YOSHINOBU, H. SHIOMI, and H. MATSUNAMI, J. Cryst. Growth **95**, 461 (1989).
[11] T. YOSHINOBU, M. NAKAYAMA, H. SHIOMI, T. FUYUKI, and H. MATSUNAMI, J. Cryst. Growth **99**, 520 (1990).
[12] S. MOTOYAMA and S. KANEDA, Appl. Phys. Lett. **54**, 242 (1989).
[13] S. MOTOYAMA, N. MORIKAWA, M. NASU, and S. KANEDA, J. Appl. Phys. **68**, 101 (1990).
[14] H. MATSUNAMI, S. NISHINO, and H. ONO, IEEE Trans. Electron Devices **28**, 1235 (1981).
[15] H. NAGASAWA and Y. YAMAGUCHI, J. Cryst. Growth **115**, 612 (1991).
[16] T. SUGII, T. ITO, Y. FURUMURA, M. DOKI, F. MIENO, and M. MAEDA, IEEE Trans. Electron Device Lett. **9**, 87 (1988).
[17] W. G. SPITZER, D. A. KLEINMAN, and C. J. FROSCH, Phys. Rev. **113**, 133 (1959).
[18] H. NAKASHIMA, T. SUGANO, and H. YANAI, Jpn. J. Appl. Phys. **5**, 874 (1966).
[19] J. GRAUL and E. WAGNER, Appl. Phys. Lett. **21**, 67 (1972).
[20] I. H. KHAN and R. N. SUMMERGRAD, Appl. Phys. Lett. **11**, 12 (1967).
[21] K. E. HAQ and A. J. LEARN, J. Appl. Phys. **40**, 431 (1969).
[22] A. J. LEARN and I. H. KHAN, Thin Solid Films **5**, 145 (1970).

[23] C. J. MOGAB and H. J. LEAMY, J. Appl. Phys. **45**, 1075 (1974).
[24] C. CHENG, P. A. TAYLOR, R. M. WALLACE, H. GUTLEBEN, L. CLEMEN, M. L. COLAIANNI, P. J. CHEN, W. H. WEINBERG, W. J. CHOYKE, and J. T. YATES, JR., Thin Solid Films **225**, 196 (1993).
[25] M. J. BOZACK, W. J. CHOYKE, L. MUEHLOFF, and J. T. YATES, JR., Surf. Sci. **176**, 547 (1986).
[26] G. GHIDINI and F. W. SMITH, J. Electrochem. Soc. **131**, 2924 (1984).
[27] J. A. KERR, in: Handbook Chemistry and Physics, 67th ed., Ed. R. C. WEAST, CRC Press, Florida 1986 (F167).
[28] B. WEINER, C. S. CARMER, and M. FRENKLACH, Phys. Rev. **43**, 1678 (1991).
[29] C. Y. YANG, M. M. RAHMAN, and G. L. HARRIS, Springer Proc. Phys. **71**, 40 (1992).
[30] H. NAGASAWA, Y. YAMAGUCHI, T. IZUMI, and K. TONOSAKI, Appl. Surf. Sci. **70/71**, 542 (1993).
[31] H. NAGASAWA and Y. YAMAGUCHI, Inst. Phys. Conf. Ser. No. 137, 71 (1993).
[32] H. NAGASAWA, Y. YAMAGUCHI, and T. IZUMI, Proc. Fac. Engng. Tokai University **33**, 61 (1993).
[33] H. NAGASAWA and Y. YAMAGUCHI, Appl. Surf. Sci. **82/83**, 405 (1994).
[34] L. G. SCHULZ, J. Appl. Phys. **20**, 1033 (1949).
[35] M. KITABATAKE and J. E. GREENE, Jpn. J. Appl. Phys. **35**, 5261 (1996).
[36] H. NAGASAWA and Y. YAMAGUCHI, Inst. Phys. Conf. Ser. No. 142, 141 (1995).

phys. stat. sol. (b) **202**, 359 (1997)

Subject classification: 68.55.Jk; 68.35.Bs; 68.35.Ct; S6

Heterointerface Control and Epitaxial Growth of 3C-SiC on Si by Gas Source Molecular Beam Epitaxy

T. Fuyuki, T. Hatayama, and H. Matsunami

Department of Electronic Science and Engineering, Kyoto University, Yoshidahonmachi, Sakyo, Kyoto 606-01, Japan

(Received January 31, 1997)

Heterointerface modification and epitaxial growth of 3C-SiC on Si by gas source molecular beam epitaxy (MBE) are surveyed. A Si surface was carbonized by the use of C_2H_2, thermal cracking of C_3H_8, and dimethylgermane $(CH_3)_2GeH_2$ (DMGe) to chemically convert the surface region into single crystalline 3C-SiC prior to crystal growth. It was found that a Si surface can be carbonized reproducibly by the use of hydrocarbon radicals at a temperature as low as 750 °C. The initial stage of carbonization is discussed based on the time-resolved reflection high-energy electron diffraction analysis. Low-temperature heterointerface modification by DMGe is described. As an advanced epitaxial growth, atomic-level control in SiC crystal growth by gas source MBE is given. Crystallinity and surface morphology of low-temperature 3C-SiC homoepitaxy on a carbonized layer is presented.

1. Introduction

Silicon carbide (SiC) is a wide bandgap semiconductor material with numerous crystalline forms (polytypes) and attractive electrical properties. Among various polytypes, 3C-SiC and 6H-, 4H-SiC are most popular, having cubic (zincblende) and hexagonal structures, respectively.

3C-SiC has a relatively narrower bandgap of 2.23 eV at room temperature, and is considered to be a candidate material for high-temperature and high-power electronic devices because of its electron mobility of 1000 cm^2/Vs. On the other hand, 6H- and 4H-SiC have been studied for blue-light-emitting diodes (LEDs) and high-power electronic devices due to wide bandgaps of 3.02 and 3.26 eV, respectively. Since 6H- and 4H-SiC wafers above 1 inch in diameter come to the market in early 1990s, device applications of 6H- and 4H-SiC have been developed.

Even though the superior electrical properties of 3C-SiC have been well known for a long time, the progress in applications has been delayed because 3C-SiC crystals grown by sublimation methods are too small (a few mm^2) for device fabrication. To obtain large-size single-crystalline 3C-SiC, there have been numerous studies on heteroepitaxy of 3C-SiC. A large-area Si wafer is quite useful as a heteroepitaxial substrate for 3C-SiC growth. Successful heteroepitaxy of 3C-SiC has been achieved by chemical vapor deposition (CVD) at 1350 °C [1]. Single-crystalline 3C-SiC could be grown reproducibly on a carbonized layer in which a Si surface was chemically converted to very thin 3C-SiC. With the introduction of a carbonization process prior to thick SiC growth in CVD, large-area single-crystalline 3C-SiC films could be obtained [2].

By the fundamental investigations in heteroepitaxy of 3C-SiC on Si [3 to 7], electronic devices based on 3C-SiC heteroepilayers were reported, e.g., an inversion-type metal–oxide–semiconductor field effect transistor (MOSFET) [8], a junction-gate FET [9], and a heterojunction bipolar transistor (HBT) [10]. Possibility of 3C-SiC heteroepilayers as an electronic material became clear. In addition, investigation on the structure and composition of the 3C-SiC heteroepilayer surface was carried out by the electron energy loss spectroscopy (EELS) [11 to 15], Auger electron spectroscopy (AES) [12, 16], and low-energy electron diffraction (LEED) [11 to 16] analysis.

However, in the 3C-SiC/Si heteroepitaxial system, the difference in the thermal expansion coefficients (8%) and the lattice mismatch (20%) seem to cause serious problems of crystal defects such as dislocations and stacking faults. The crystal defects in the heterointerface are thought to be reduced by low-temperature carbonization. Gas source molecular beam epitaxy (MBE) is a powerful technique for this purpose. Gas source MBE is also expected to analyze surface structures during carbonization and epitaxial growth of 3C-SiC and to produce high-quality epilayers.

In this paper, recent progress in heterointerface modifications and epitaxial growth of 3C-SiC on Si by gas source MBE is presented. First, growth of SiC by MBE methods is reviewed. Then, low-temperature carbonization of Si surfaces by several hydrocarbon sources is described. In particular, quantitative analysis of hydrocarbon radicals in thermally cracked C_3H_8 and the initial stage of carbonization on a clean Si surface are discussed based on the results of threshold ionization quadrupole mass spectrometer (QMS) analysis and time-resolved reflection high-energy electron diffraction (RHEED) analysis, respectively. As a proposal in a low-temperature heterointerface modification, carbonization of a clean Si surface by dimethylgermane $(CH_3)_2GeH_2$ (DMGe) is described. Finally, utilizing the reconstruction of surface structures on a 3C-SiC surface, the epitaxial growth by an alternate supply of source gases is given as an advanced epitaxial growth.

2. Review of SiC Growth by Molecular Beam Epitaxy

Growth of 3C-SiC on Si in ultrahigh vacuum was tried when MBE became popular in the 1980s. Low-temperature growth of 3C-SiC heteroepilayers were reported by several combinations of sources for Si and C, e.g., solid Si source/C^+ ion beam [17], reactive ion beam (SiH_4/CH_4) [18], $SiHCl_3/C_2H_4$ [19], solid Si source/C_2H_2 [20, 21], and Si_2H_6/C_2H_4 [22] at 820, 770, 750 to 1000, 900, and 975 °C, respectively, where single-crystalline 3C-SiC heteroepilayers were grown for optimum Si/C flux ratios. To improve the crystal quality of the 3C-SiC heteroepilayers, growth of a Si buffer layer on a Si substrate [18, 23] and carbonization by the exposure to C sources [19 to 21] were carried out prior to the 3C-SiC growth. Although low-temperature growth of 3C-SiC by MBE methods could be presented, reproducible growth and smooth epitaxial surfaces were very difficult to obtain.

In the late 1980s and early 1990s, fundamental studies of SiC growth by MBE methods were initiated. Atomic level epitaxy of 3C-SiC was demonstrated by gas source MBE utilizing the reconstruction of the surface superstructure by an alternate supply of Si_2H_6 and C_2H_2 [24, 25] and the possibility of atomic layer epitaxy was proposed. The surface structures of 3C-SiC under Si_2H_6 supply were analyzed in detail by dynamic RHEED observation [26]. The details of the atomic level epitaxy are discussed in Section 5.

Crystal growth of 6H-SiC to get lattice-matched heteroepitaxy in gas source MBE was investigated. On 6H-SiC(000$\bar{1}$) substrates, 3C-SiC($\bar{1}\bar{1}\bar{1}$) with double-positioning twin structures was grown at 1000 °C [27]. However, on a 6H-SiC(0$\bar{1}$14) substrate, a single-crystalline 3C-SiC(001) layer without twin structures and anti-phase domain free growth was reported [27]. This epitaxial growth was obtained at even lower temperatures down to 850 °C. Recently, a polytype formation of 3C-SiC or 6H-SiC on a 6H-SiC(0001) off-axis substrate by the control of the Si_2H_6/C_2H_4 flow ratio was reported [28].

In the heteroepitaxy of 3C-SiC on Si, a gradual temperature rise during carbonization together with the existence of a surface oxide layer was effective to moderate the reaction between Si and C_2H_2 [29, 30] and a single-crystalline 3C-SiC thin layer could be obtained reproducibly at 970 °C [30]. By the use of chemically active hydrocarbon radicals from thermal cracking of C_3H_8, a Si surface could be carbonized reproducibly at temperatures as low as 750°C [31, 32]. A single-crystalline 3C-SiC layer showing a (3×2) structure was obtained at 1000 °C [33, 34].

Mechanism of the 3C-SiC heteroepitaxy was studied by a combination of molecular dynamics simulations and solid source MBE experiments [35]. A (3×2) structure was most stable on the Si-terminated 3C-SiC surface, which has a low dangling bond density of 0.67 per unit cell [36 to 38].

Though the crystallinity of 3C-SiC heteroepilayers became better with the accumulation of many efforts, 3C-SiC heteroepilayers of satisfactorily high quality have still not been obtained: they include a number of crystal defects such as stacking faults and dislocations.

3. Apparatus for Gas Source Molecular Beam Epitaxy

Growth of SiC was carried out in an ultrahigh vacuum MBE system which consisted of a growth chamber and a loading chamber. The growth chamber is equipped with a sputter ion pump to achieve an ultrahigh vacuum prior to growth, and a turbo molecular pump for evacuation during source gas supply. The base pressure of the growth chamber was about 1×10^{-7} Pa, and the working pressure was kept below 1×10^{-3} Pa during the growth.

A Si(001) substrate was chemically cleaned with an oxidation solution ($NH_4OH : H_2O_2 : H_2O = 1 : 1 : 6$). Then, the substrate was attached to a tantalum holder with tantalum nails, and it was immediately loaded into a loading chamber. The loading chamber was pumped down to about 10^{-5} Pa, before the substrate was transferred to the growth chamber. The substrate was heated from the backside by a resistive heating element. The substrate temperature was controlled by referring to the output of a thermocouple located behind the substrate. The absolute value of the substrate temperature was measured by an optical pyrometer. The manipulator is surrounded by a liquid nitrogen shroud, by which most of the gas molecules are adsorbed giving a high vacuum below 10^{-3} Pa during growth.

Pure hydrocarbon gases of C_2H_2 and C_3H_8 were mainly employed as carbon sources in both the carbonization process and crystal growth. Pure Si_2H_6 was employed as a silicon source in the growth. The hydrocarbon gases and Si_2H_6 were supplied to the substrate through individual nozzles each located 7 cm distant from the substrate. The flow rate of the source gases was precisely controlled by a mass flow controller.

The growth chamber is equipped with a 25 kV RHEED system for in-situ observation of the surface. The growth chamber is also equipped with a QMS for detection of gas molecules and decomposed species.

During heteroepitaxy of 3C-SiC on a Si clean surface, the surface structure was analyzed by the RHEED system. Diffraction patterns on a RHEED screen were monitored and recorded with a charge-coupled device (CCD) camera and a video recorder. The recorded pattern was processed by a microcomputer, and the change of the diffracted intensity related to each structure was acquired with a sampling rate of 10 Hz.

4. Carbonization of Si Surface

In order to realize reproducible growth of single-crystalline 3C-SiC on Si by gas source MBE, carbonization of the Si surface was carried out by several hydrocarbon sources. Since the chemical activity is different in each source, an adequate carbonization process for each source is proposed. First, a carbonization process for a Si(001) surface using C_2H_2 is described. The carbonization parameters were optimized to obtain single-crystalline 3C-SiC layers. A cracking technique is employed for carbonization of Si surfaces using C_3H_8 as a carbon source, by which the carbonization temperature can be reduced. Finally, low-temperature heterointerface modification by DMGe is presented.

4.1 Carbonization by C_2H_2

In the case of a clean Si surface, control of carbonization by substrate temperature was not sufficient to reproducibly obtain single-crystalline 3C-SiC layers [29]. The control of reactivity between Si and C_2H_2 seems to be a key to obtain a single-crystalline 3C-SiC layer with a smooth surface. In order to form a carbonized layer, the reactivity was moderated by the existence of an oxide layer (< 10 Å) on the substrate surface [29, 30].

Fig. 1 shows the temperature program for the carbonization with a modified process and in-situ RHEED patterns observed at different stages in the carbonization process [30]. The flow rate of C_2H_2 was 1.6 sccm. When C_2H_2 supply was started at 400 °C without oxide removal, the RHEED pattern was unchanged, showing a Si(1×1) structure. The substrate temperature was elevated at a rate of 7 K/min until a halo pattern was observed by RHEED at about 870 °C. The halo pattern indicates the existence of an amorphous-like layer on the surface, most probably a Si–C–O layer [29, 30]. Above this temperature, the rate was reduced to 2 K/min until the substrate was heated up to 970 °C. A diffused spotty pattern gradually appears and becomes clear after 60 min of carbonization at 970 °C, which indicates single-crystalline 3C-SiC. With the slow increase of substrate temperature, the reaction between Si and C_2H_2 takes place simultaneously with evaporation of oxide and crystallization of 3C-SiC. When the increase rate of the substrate temperature above 870 °C was kept at 7 K/min, the crystallinity of the 3C-SiC layer became poor. Since the oxide in this case will evaporate before the formation of the 3C-SiC layer, the bare Si surface will be exposed to C_2H_2 at high temperatures.

Fig. 2 shows RHEED patterns and scanning electron microscope (SEM) photographs of the layers carbonized for 60 min with different values of C_2H_2 flow rate after heating up to 970 °C according to the time program in Fig. 1. As shown in Fig. 2, for a small amount (0.8 sccm) of C_2H_2 supply, extra spots are observed in the diffraction pattern,

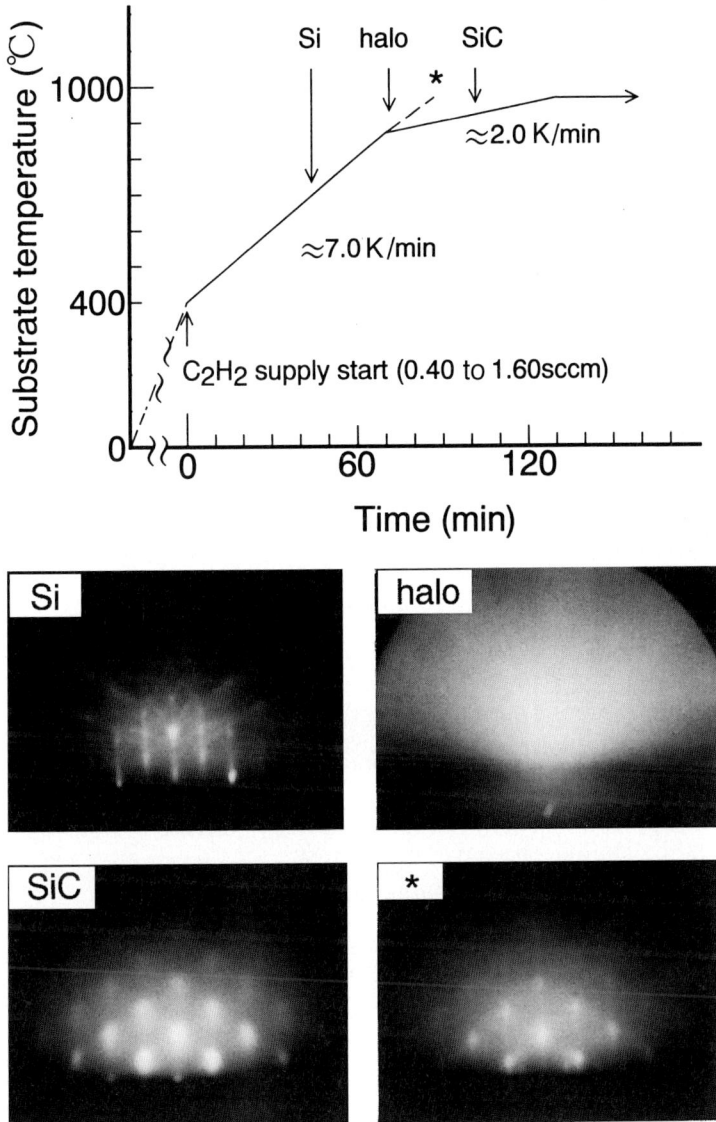

Fig. 1. Carbonization by C_2H_2. Temperature program for the process and RHEED patterns during carbonization

which indicates the existence of 3C-SiC twin structures. As shown in the SEM photograph (Fig. 2a), there exist circular hillocks on the surface and pits in the substrate. A single-crystalline 3C-SiC layer with a very smooth surface was obtained when the flow rate of C_2H_2 was increased up to 1.6 sccm (Fig. 2b). When the C_2H_2 flow rate is small, the 3C-SiC layer is not thick enough to prevent outdiffusion of Si atoms from the substrate. With the increase of C_2H_2 flow rate, the defects are sealed off at an early stage and the Si surface is covered with a single-crystalline 3C-SiC layer that prevents outdiffusion of Si atoms, and thus a degradation of the crystallinity does not occur.

(a) 0.8 sccm

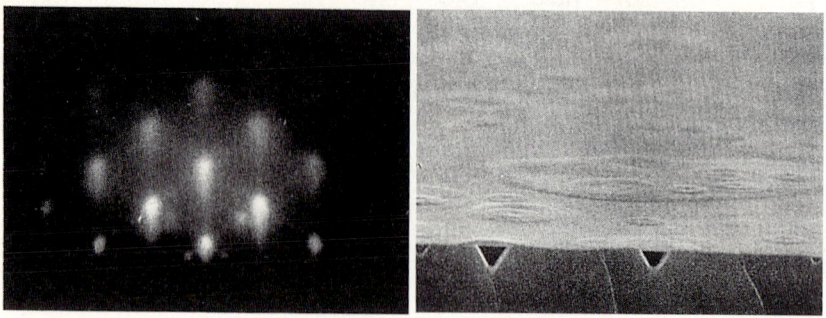

(b) 1.6 sccm

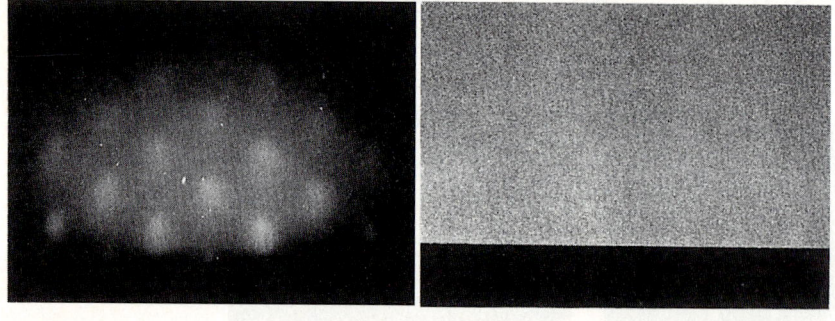

1μm

Fig. 2. RHEED patterns and SEM photographs of 3C-SiC layers obtained by carbonization of Si(001) at 970 °C for 60 min with C_2H_2 flow rates of a) 0.8 sccm and b) 1.6 sccm

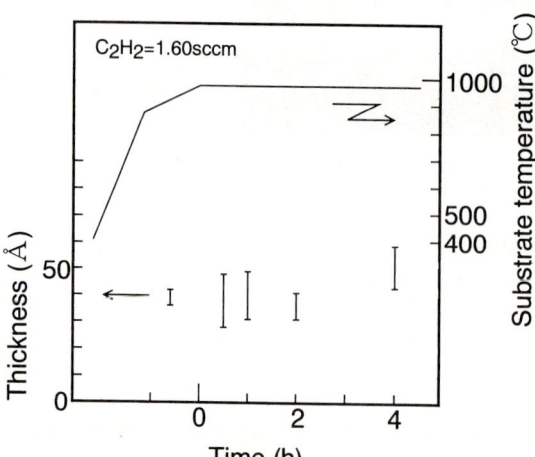

Fig. 3. Thickness of 3C-SiC layers obtained by carbonization of Si(001) at 970 °C with a C_2H_2 flow rate of 1.6 sccm versus time

Fig. 3 shows the thickness of a carbonized layer versus carbonization time. The thickness was saturated at about 50 Å at an early stage of carbonization, and did not increase for prolonged carbonization times as long as 4 h [30]. This means that the surface is covered with a single-crystalline 3C-SiC layer at an early stage of carbonization, preventing any further reactions between Si and C_2H_2.

4.2 Carbonization by cracked saturated hydrocarbon

Carbonization of a Si surface by thermal cracking of C_3H_8 in a gas source MBE system is obtained reproducibly at a temperature as low as 750 °C [31]. C_3H_8 was decomposed by a cracker cell comprised of a tantalum heater in a quartz guiding tube. The typical C_3H_8 flow rate and cracker temperature were 0.20 sccm and 1200 °C, respectively. The supply of cracked C_3H_8 was started at temperatures of 400 °C, and elevated to 750 °C at a rate of 7 to 10 K/min. At 750 °C, a 3C-SiC diffused spotty pattern gradually appears with a diminishing Si pattern. The thickness of the carbonized layer was controlled easily by cracker temperature and exposure time [32].

Fig. 4 shows a RHEED pattern and a SEM photograph of a layer carbonized for 10 min at 750 °C. The thickness of the carbonized layer is about 30 Å, and shows a smooth uniform surface. Mirror-like surface morphology was obtained in the initial stage of exposure time at 750 °C or using lower cracker temperatures [32, 34]. The carbonized layers could be also obtained by cracked C_2H_6 with the same condition for C_3H_8. However, a SiC RHEED pattern did not appear, when a Si substrate was exposed to either uncracked C_3H_8 or C_2H_6 molecular beams. The success of carbonization at such low temperatures is considered to be due to reactions of chemically-active hydrocarbon radicals, generated by the cracking of C_3H_8 and C_2H_6, with Si atoms on the surface. Hydrocarbon radicals are indispensable for the carbonization at 750 °C.

Generation of hydrocarbon radicals was investigated by QMS analysis. The signal intensity for $C_3H_8^+$ decreased with the increase of cracker temperature, showing pyrolysis of C_3H_8. Radical species generated by thermal cracking of C_3H_8 were characterized using a threshold ionization technique avoiding the influence of fragmentation caused by the QMS ionizer [34, 39]. To measure the number of CH_3 radicals, the C_3H_8 nozzle was aligned along the direction of the QMS ionizer. The distance between the cracker cell

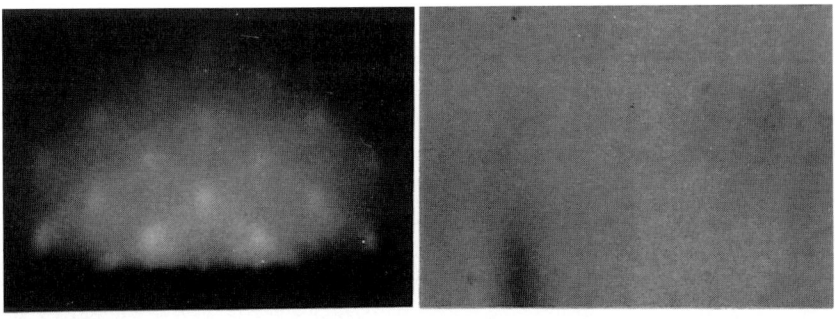

1 μm

Fig. 4. RHEED pattern and SEM photograph of the 3C-SiC layer obtained by carbonization of Si(001) at 750 °C for 10 min with cracked C_3H_8

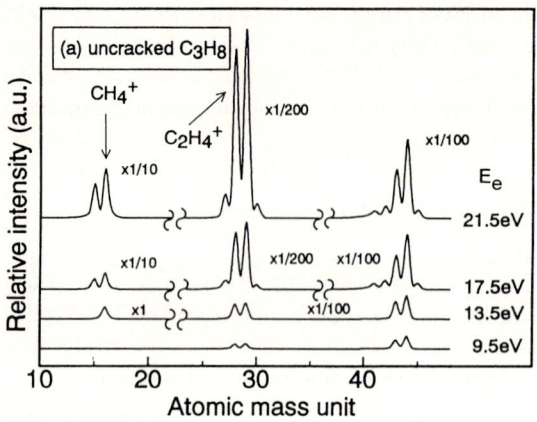

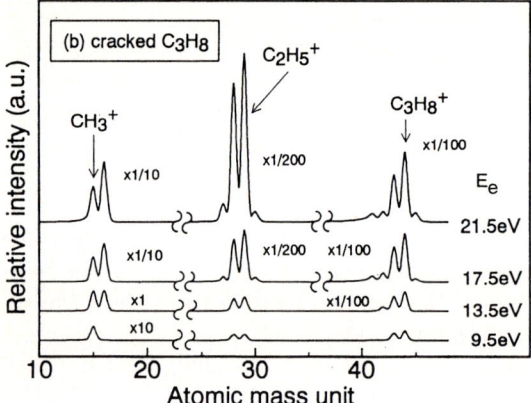

Fig. 5. QMS spectra from a) uncracked and b) cracked C_3H_8 for various ionization energies

and the QMS ionizer is the same as that between the cracker cell and the substrate during growth.

Fig. 5a, b shows the mass spectra of uncracked and cracked C_3H_8 for various electron energies E_e in the QMS ionizer [39], respectively. The flow rate of C_3H_8 was kept at 0.1 sccm, and the cracker temperature was settled at 1050 °C. The intensities of the mass signals of uncracked and cracked C_3H_8 are normalized by each $C_3H_8^+$ intensity for $E_e = 9.5$ eV. In Fig. 5a for $E_e = 21.5$ eV, several mass signals corresponding to CH_3^+, CH_4^+, $C_2H_4^+$, $C_2H_5^+$, and $C_3H_7^+$ come from fragmentation of C_3H_8 in the QMS ionizer. With the decrease of ionization energy, the intensities of mass signals are lowered due to the decrease of dissociative ionization cross sections.

It should be noted that the CH_3^+ signals are observed even for $E_e = 9.5$ eV in Fig. 5b. Since the threshold ionization energy of CH_3 radicals is reported as 9.8 eV [40, 41], CH_3 radicals can be formed by the cracking of C_3H_8 in the cracker cell, taking account of the electron energy spread of 0.5 eV. The dominant reaction and decomposition energy are reported as [42]

$$C_3H_8 \rightarrow CH_3 + C_2H_5 : 85.3\,\text{kcal/mol}. \tag{1}$$

This decomposition energy is lower than that of C_3H_8 pyrolysis to $H + C_3H_7$ radicals (94.9 kcal/mol) [42]. Although C_3H_8 was pyrolyzed in the cracker cell and the CH_3^+

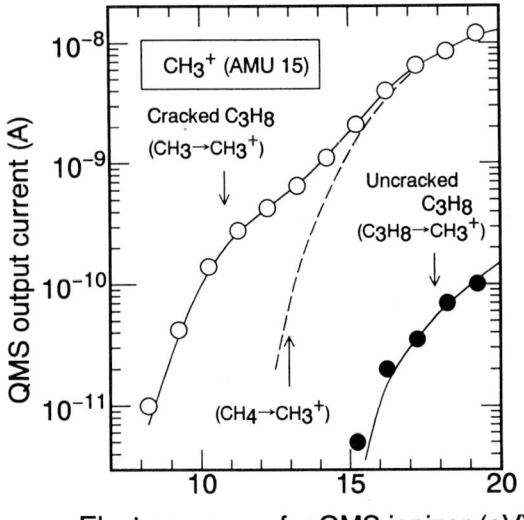

Fig. 6. QMS output current for CH_3^+ (atomic mass units: AMU 15) as a function of electron energy of the QMS ionizer. Open and closed circles show cracked and uncracked C_3H_8, respectively. The dashed line shows the dissociative ionization process for $CH_4 \rightarrow CH_3^+$

signals increased, the $C_2H_5^+$ signals in Fig. 5b did not remarkably increase. C_2H_5 radicals given in Eq. (1) might be further decomposed into C_2H_m ($m = 1, 2, 3, 4$) and/or CH_n ($n = 1, 2, 3, 4$). Since the CH_3 radicals are chemically active, a carbonized layer is formed at temperature as low as 750 °C.

Fig. 6 shows the QMS output current I_q for CH_3^+ signals as a function of electron energy E_e in the QMS ionizer. Open and closed circles correspond to cracked and uncracked C_3H_8, respectively. The output current for cracked C_3H_8 is higher than that for uncracked C_3H_8, probably due to generation of CH_3 radicals by cracking of C_3H_8. The output current for cracked C_3H_8 can be observed for E_e above 9.5 eV, and it increases with the electron energy in the QMS ionizer. The increase of output current for E_e near 14 eV is considered to be caused by the decomposition of other species in the QMS ionizer. Since the dissociative ionization energy for $CH_4 \rightarrow CH_3^+$ is 14.3 eV [40], the decomposed species from cracked C_3H_8 are thought to be CH_4 molecules. The output current for cracked C_3H_8 below $E_e = 14$ eV demonstrates the ionization of neutral CH_3 radicals ($CH_3 \rightarrow CH_3^+$) in the QMS ionizer.

The number of CH_3 radicals is estimated by the QMS output current at an ionization energy lower than 14 eV, taking account of the ionization cross section ($CH_3 \rightarrow CH_3^+$) and the calibrated sensitivity of the QMS [39].

For a C_3H_8 flow rate of 0.1 sccm and a cracker temperature of 1200 °C, the number of CH_3 radicals is about 3.3×10^9 cm^{-3}. This value corresponds to 1.6% of total C_3H_8 molecules (which is 2.1×10^{11} cm^{-3} taking account of the QMS output current for C_3H_8 and the ionization cross section of $C_3H_8 \rightarrow C_3H_8^+$ [43]). The number of CH_3 radicals can be controlled by the flow rate of C_3H_8 and/or the cracker temperature, and reaches up to 3.0×10^{11} cm^{-3} for a C_3H_8 flow rate of 0.8 sccm and a cracker temperature of 1200 °C.

4.3 Initial stage of carbonization

The initial stage in heteroepitaxy of 3C-SiC on a clean Si(001) surface was revealed by time-resolved RHEED analysis [44]. Starting from a clean Si(001) surface showing a (2×1) structure, the surface structure changed according to: Si$(2 \times 1) \rightarrow$ a mixed structure of Si(2×1) and Si c$(4 \times 4) \rightarrow$ 3C-SiC with continuous supply of cracked C_3H_8.

A Si(001) on-axis substrate was chemically cleaned with a solution ($NH_4OH : H_2O_2 : H_2O = 1 : 1 : 6$), by which the surface was eventually covered with a thin oxide layer. The oxide layer was removed at 840 °C in ultrahigh vacuum. The generation of a Si(2 × 1) structure, which indicates the formation of Si dimers on the surface, was confirmed. After deposition of a 600 Å thick Si buffer layer using Si_2H_6, the substrate temperature was settled at a given value to form a 3C-SiC layer by exposure to cracked C_3H_8.

Fig. 7 shows the intensity change from each structure and RHEED-pattern photographs during the carbonization at 750 °C. The intensity change of Si(0 1/2) and Si(1 1/4) in Fig. 7 indicates the change of Si(2 × 1) and Si c(4 × 4) structures, respectively. Marks (a) to (c) correspond to RHEED-pattern photographs of a) Si(2 × 1) from the ⟨110⟩ azimuth, b) Si c(4 × 4) from the ⟨100⟩ azimuth, and c) 3C-SiC from the ⟨110⟩ azimuth, respectively. The 3C-SiC diffraction spots do not appear immediately after the start of cracked-C_3H_8 supply (marked with "⇓"), thus, the Si surface becomes 3C-SiC after an incubation time. In every experiment, the intensity of the Si-related surface reconstructions is changed in the incubation time. The change of the RHEED intensities is be-

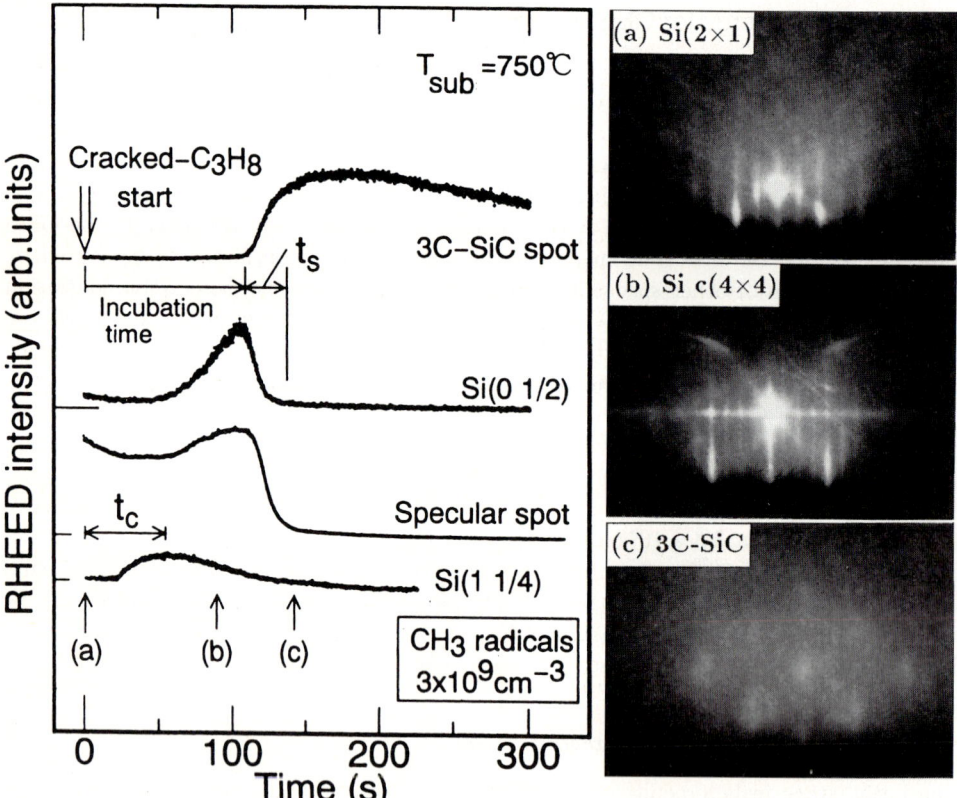

Fig. 7. Change of RHEED intensities and RHEED-pattern photographs during carbonization at 750 °C. The profiles of 3C-SiC, Si(0 1/2) and Si(1 1/4) indicate 3C-SiC spot, Si(2 × 1) and Si c(4 × 4) structures, respectively. The start of cracked-C_3H_8 supply is marked with "⇓". Marks (a) to (c) correspond to RHEED-pattern photographs of a) Si(2 × 1) from the ⟨110⟩ azimuth, b) Si c(4 × 4) from the ⟨100⟩ azimuth, and c) 3C-SiC from the ⟨110⟩ azimuth, respectively

lieved to be related to the amount of adsorbed hydrocarbon species. Chemical reactions between the adsorbed species and Si atoms lead to the surface reconstruction.

The diffraction intensity of the 3C-SiC structure increases after the start of the decrease in the diffraction intensity of the $Si(2 \times 1)$ structure. At the disappearance of the $Si(2 \times 1)$ structure, Si dimers on the surface are almost broken, and dangling bonds of top Si atoms may be terminated by CH_3 radicals and/or other hydrocarbon species. By successive supply of cracked C_3H_8, the diffraction intensity of 3C-SiC decreases monotonically, showing the increase of surface roughness on a growing surface in comparison with the initially-grown 3C-SiC layer.

In order to characterize the growth rate in the initial stage, we assume that the inverse of time t_s in Fig. 7 between the appearance of 3C-SiC spots and the disappearance of $Si(2 \times 1)$ structure corresponds to the average growth rate of 3C-SiC. Fig. 8 shows the Arrhenius plots of the average growth rate of the initially grown 3C-SiC layers. Closed circles and squares correspond to carbonizations with CH_3 radicals of 3×10^{10} and 3×10^9 cm^{-3}, respectively. Two distinct temperature regions exist. In the low-temperature region, the average growth rate depends on the substrate temperature, which indicates surface-reaction-limited growth. The activation energy calculated from the slope in this region is about 46.9 kcal/mol. This energy is close to a reported value of 47.7 kcal/mol for hydrogen desorption from the surface [45]. The adsorbed CH_3 radicals are decomposed into lower CH_n species ($n = 0, 1, 2$) above 360 °C [46] which leads to the formation of Si–H species [47]. In addition, the Si surface is partly covered with hydrogen atoms in the cracked-C_3H_8 molecular beam.

In the high-temperature region, the average speed is not very dependent on the substrate temperature. The average speed decreases gently at higher carbonization temperatures for each supply of CH_3 radicals, probably due to the increase in the desorption rate of adsorbed CH_3 radicals and in the etching rate of carbon atoms by hydrogen atoms on the surface [48].

The RHEED intensity of the Si $c(4 \times 4)$ structure reached a maximum value before the appearance of the 3C-SiC diffraction spots. Here, we define a transition time t_c in

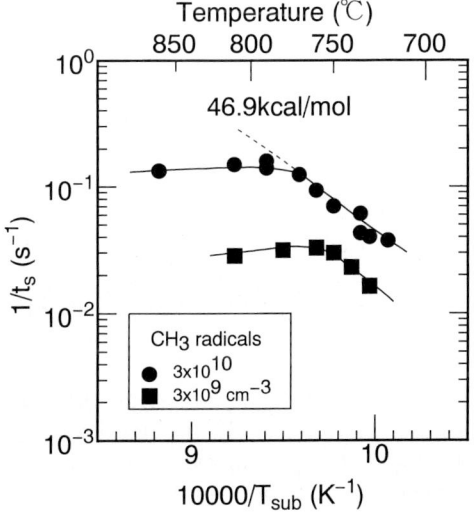

Fig. 8. Arrhenius plot of the average speed of initially grown 3C-SiC layers with the cracked C_3H_8 supply

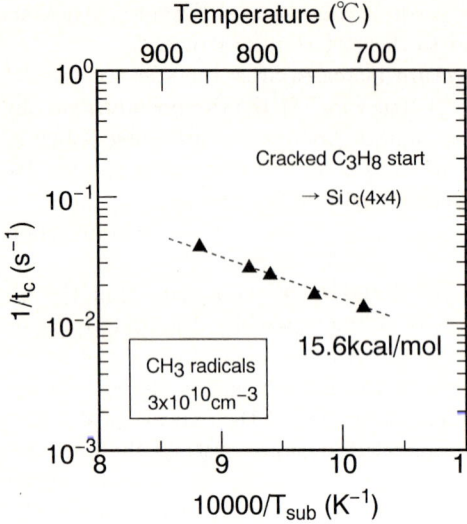

Temperature (℃)

Fig. 9. Arrhenius plot of the reciprocal transition time between the start of cracked C_3H_8 supply and the maximum RHEED intensity of Si c(4×4)

Fig. 7 between the start of cracked-C_3H_8 supply and the maximum RHEED intensity of the Si c(4×4) structure. Fig. 9 shows the reciprocal transition time (t_c^{-1}) as a function of the reciprocal substrate temperature (T_{sub}^{-1}). The activation energy calculated from the slope is about 15.6 kcal/mol. This value is smaller than that of the initially-grown 3C-SiC layer. The mixed structure of Si(2×1) and Si c(4×4) can also be observed before the appearance of 3C-SiC when using C_2H_2 as a source gas, giving an activation energy of 11.1 kcal/mol [39]. The activation energy of about 15.6 kcal/mol may be related with the formation of a Si c(4×4) structure with the adsorption of hydrocarbon species.

4.4 Heterointerface modification by $(CH_3)_2GeH_2$

To improve the crystallinity of the 3C-SiC layers, our group investigated a heterointerface modification by a large-size atom bigger than silicon (atomic radius 1.32 Å) and carbon (0.91 Å) atoms. A low-temperature interface modification by $(CH_3)_2GeH_2$ (DMGe), involving CH_3 radicals and germanium (atomic radius 1.37 Å) atoms, was carried out [49, 50].

Pure DMGe (99.9999%, vapor pressure $\approx 1.6 \times 10^5$ Pa at 25 °C) was used as a source gas, which was supplied continuously to the substrate. Quantitative analysis of a DMGe molecular beam was achieved with a threshold ionization QMS technique [34, 39]. The number of DMGe molecules was varied in the range of 3×10^9 to 3×10^{10} cm^{-3}.

A clean Si surface showing a (2×1) structure can be carbonized reproducibly even at a temperature as low as 650 °C [49, 50]. The carbonized layer had good crystallinity, as indicated by RHEED patterns showing single-crystalline 3C-SiC without any 3C-SiC twin spots and Ge-related diffraction patterns. Adsorbed DMGe molecules on a Si surface are thermally decomposed even at 350 °C, and then some of the CH_3 radicals and Ge atoms are desorbed [51, 52]. The success of carbonization at such a low temperature in this experiment is due to the fact that most of Ge atoms will be desorbed and active CH_3 radicals will react with Si atoms on the surface to form a carbonized layer.

The change of surface reconstructions takes the same sequence as in the case of cracked C_3H_8. Starting from a clean Si surface showing a (2×1) structure, the surface structure changed according to: Si(2×1) $\rightarrow$ a mixed structure of Si(2×1) + Si c(4×4) $\rightarrow$ 3C-SiC. Thus, the change of RHEED pattern and their

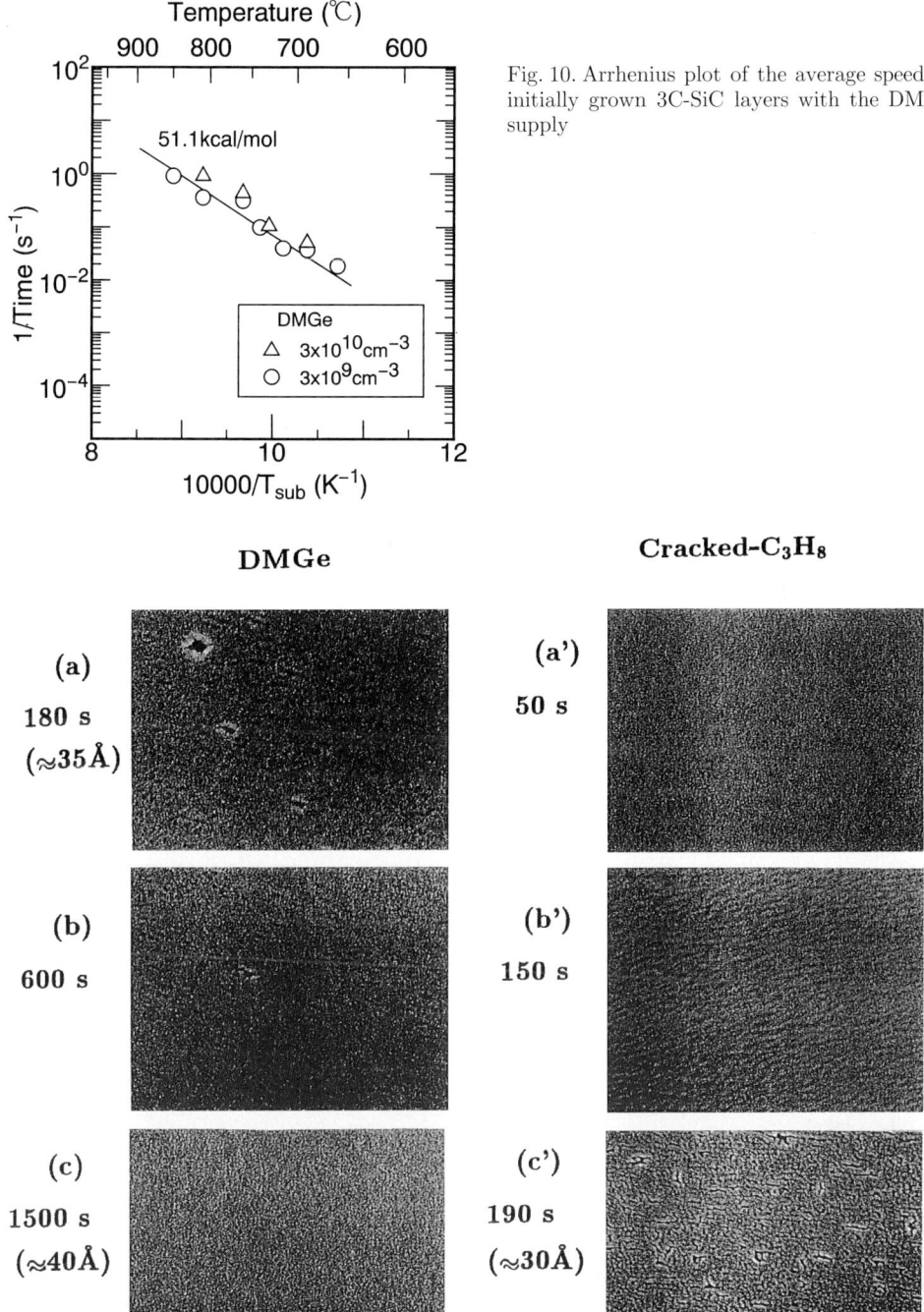

Fig. 10. Arrhenius plot of the average speed of initially grown 3C-SiC layers with the DMGe supply

Fig. 11. High-resolution SEM photographs showing the surface morphology for different carbonization times of a) to c) DMGe and a') to c') cracked-C_3H_8

intensities is due to chemical reactions between CH_3 radicals from DMGe molecules and Si atoms on the surface.

To discuss the growth mechanism in the initial stage, the average growth rate of 3C-SiC is defined in a similar way as in the case of cracked C_3H_8. The average growth rate strongly depends on the substrate temperature [50, 53]. Fig. 10 shows the Arrhenius plots of the average speed of initially grown 3C-SiC layers. The average speed shows no significant change under the DMGe supply in the range of 3×10^9 to 3×10^{10} cm^{-3}. The activation energy calculated from the slope is about 51.1 kcal/mol, close to 46.9 kcal/mol obtained in the case of cracked C_3H_8. These values seem to be the formation energy of 3C-SiC growth in the initial stage.

Fig. 11 shows high resolution SEM photographs of 3C-SiC layers after different carbonization times for a) to c) DMGe and a') to c') cracked-C_3H_8 [50]. The numbers of both DMGe molecules and CH_3 radicals in cracked C_3H_8 were 3×10^9 cm^{-3}. Carbonization was carried out at 720 °C for DMGe and at 750 °C for cracked C_3H_8. In each case, SiC nuclei and some pits are observed in the initial stage. The coalescence of 3C-SiC nuclei takes place on the surface with the continued growth. However, after growing a carbonized layer of about 40 Å thickness, a smooth surface without pits can be obtained only in the case of the DMGe supply (Fig. 11c). The surface morphology of the carbonized layer formed by DMGe was better than that of cracked C_3H_8, because the DMGe molecules may cover the growing surface more uniformly and closely. The thickness of the carbonized layer tends to saturate at about 40 Å.

Fig. 12 shows Rutherford backscattering spectroscopy (RBS) spectra for a carbonized layer formed at 720 °C [53]. From RBS analysis, it turned out that Ge atoms exist in the carbonized layer. The Ge concentration in the carbonized layer is calculated from the backscattering yield ratio of Ge (1.4×10^2 counts) to Si (7.0×10^3 counts), taking account of the sensitivity of RBS to atomic weights [54]. For the data shown in Fig. 12, the Ge concentration is about 0.3%. Ge atoms in the carbonized layer are distributed in interstitial positions in the SiC lattice, because the backscattering yield of the Ge spectrum in the aligned geometry was unchanged compared with that in the random geometry.

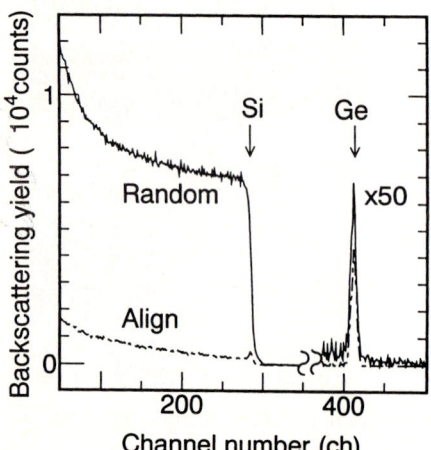

Fig. 12. RBS spectra for a carbonized layer formed at 720 °C for 3 min. The number of DMGe molecules is 3×10^9 cm^{-3}

5. Advanced Epitaxial Growth of 3C-SiC Using Gas Source MBE

In this section, epitaxy of 3C-SiC on both CVD grown 3C-SiC layers and carbonized layers is described.

5.1 Surface superstructures

Several types of surface superstructure have been reported for the 3C-SiC(001) surface [16, 25, 26, 36, 37]. From RHEED [24 to 26], LEED [16, 37], AES [16], and medium-energy ion scattering spectroscopy (MEIS) [16], 3C-SiC(3×2) and c(2×2) structures were identified to be Si- and C-terminated surfaces, respectively. The 3C-SiC(2×1) and c(2×2) structures are terminated by only one monolayer (ML) of Si and C atoms, respectively. The transition of a 3C-SiC(001) surface superstructure under Si_2H_6 supply was sequentially observed by in-situ RHEED.

The substrate temperature was kept at 1000 °C in high vacuum, and a C_2H_2 beam was supplied to the substrate in order to obtain the C-terminated surface of the 3C-SiC c(2×2) structure. The C_2H_2 beam was stopped and a Si_2H_6 beam was supplied after the recovery of a high vacuum of 10^{-5} Pa by several minutes of evacuation. The flow rate of Si_2H_6 was 0.07 sccm.

Fig. 13 shows typical profiles of RHEED intensities of the 3C-SiC fundamental and superstructure-related diffraction streaks as a function of the Si_2H_6 exposure time. The C-terminated c(2×2) structure changes according to: c(2×2) $\rightarrow$ (2×1) $\rightarrow$ (5×2) $\rightarrow$ (3×2) with continuous Si_2H_6 supply [26]. It was found that the amount of Si atoms on the 3C-SiC surface increases in the order of (2×1) < (5×2) < (3×2).

By a careful monitoring of 3C-SiC superstructures, the intensity of the (3×2) structure begins to increase with the appearance of the (2×1) structure. It is possible to consider the coexistence of (2×1) and (3×2) structures on the surface during the Si_2H_6 supply. The appearance of the structure of mixed (2×1) and (3×2) implies that the (3×2) structure is the more stable structure compared with Si-adsorbed superstructures.

Recently, atomic-resolution images of 3C-SiC surfaces were reported by a scanning tunneling microscope (STM) observation [36] and molecular dynamics simulations [37]. Though the (3×2) structure was indicated with the

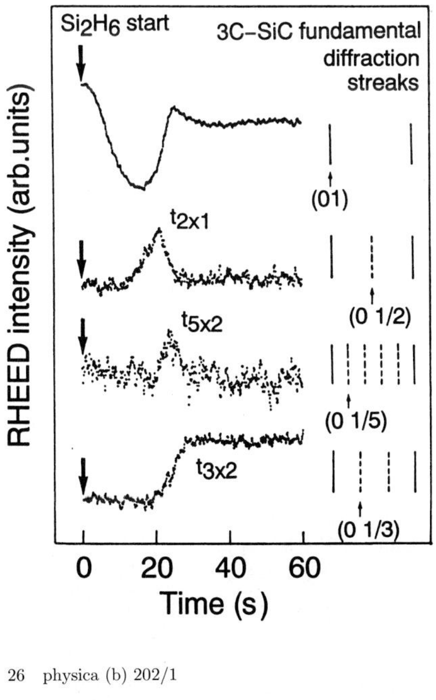

Fig. 13. RHEED intensity of fundamental and superstructure-related diffraction streaks under the supply of Si_2H_6 molecular beam

LEED analysis, coexistence of (3×2) and (5×2) structures was observed by the STM analysis. The (3×2) structure consists of only one missing Si dimer row in every three rows above one complete single layer of Si atoms [36, 37]. When the 3C-SiC surface is completely covered with the (3×2) structure, an adlayer of Si atoms corresponds to 1.67 ML.

5.2 Atomic-level control in 3C-SiC growth

Epitaxial 3C-SiC on CVD grown 3C-SiC layers was successfully deposited using gas source MBE at 1000 °C, utilizing the change in surface reconstructions when the source gases of Si_2H_6 and C_2H_2 were supplied alternately [24, 25].

When Si_2H_6 is supplied, thermal decomposition of Si_2H_6 molecules generates Si atoms which adsorb and construct surface superstructures. Superstructures are reconstructed corresponding to the number of adsorbed Si atoms [16, 24, 25, 36, 37]. When C_2H_2 is supplied, the adsorbed Si atoms react with C_2H_2 to form SiC, and the structure of the surface changes in the reverse order. In this method, the amount of crystal growth in a cycle is determined by the number of Si atoms adsorbed on the surface, and the number of Si atoms on the surface is reflected in the structure of the surface.

Fig. 14 shows the growth rates (grown layer thickness per cycle) as a function of Si_2H_6 supply (product of Si_2H_6 flux intensity and its duration) [24, 25]. As for the temperature dependence, higher substrate temperatures resulted in the reduction of the growth rate, which implies reduced sticking coefficients or enhanced desorption of Si atoms at higher temperatures.

If all the Si atoms forming the (3×2) structure react with C_2H_2 and crystallize in 3C-SiC, the resulting growth rate should be 1.7 ML/cycle, taking into account the above-mentioned model. However, the experimental value of the growth rate exceeded this expected value, and increased with the Si_2H_6 supply. The additional thickness above 1.7 ML/cycle is attributed to physically adsorbed Si atoms not contributing to the superstructure. Precise control of Si_2H_6 supply to form the (3×2) structure will give ideal monolayer growth.

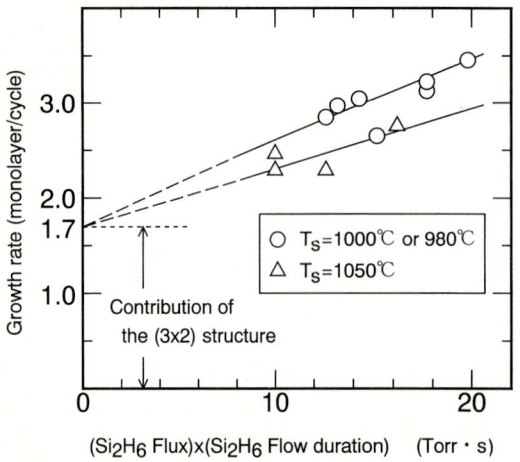

Fig. 14. Dependence of growth rate for the amount of Si_2H_6 supply

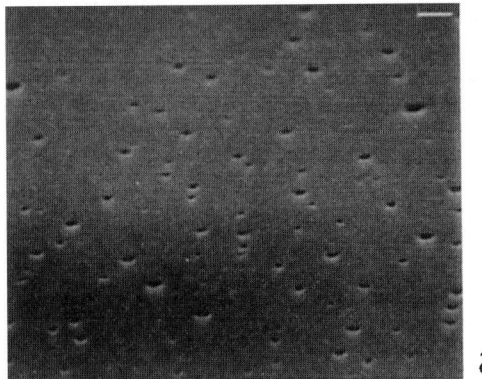

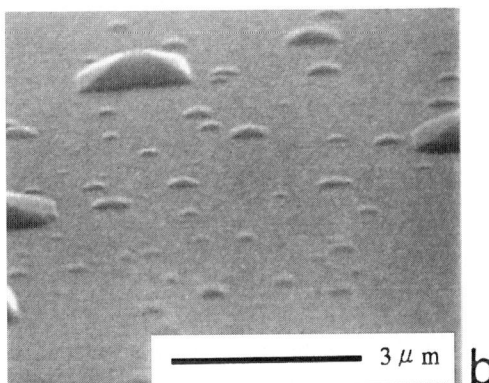

a

b

$3\,\mu\text{m}$

Fig. 15. Surface morphology of 3C-SiC epitaxial layers of about 250 Å thickness for Si_2H_6 flow rates of a) 0.06 sccm and b) 0.08 sccm

Recently, by the use of alternate supply of SiH_2Cl_2 and C_2H_2, atomic-level epitaxy of 3C-SiC on Si was carried out by hot-wall low-pressure CVD at 1020 °C [55]. A single-crystalline 3C-SiC heteroepilayer with high thickness uniformity over 6 inches in diameter and a mirror surface was realized [56].

5.3 Low-temperature growth of 3C-SiC

Single-crystalline 3C-SiC on a carbonized layer with mirror-like flat morphology can be grown at 1000 °C. Following the success in single-crystal growth of 3C-SiC by alternate supply of C_2H_2 and Si_2H_6 [24, 25, 30], Si_2H_6 was supplied intermittently for 3 s within a period of 2 min repeating 30 to 90 cycles. Here, C_3H_8 was supplied continuously to increase the number of CH_3 radicals, since the number becomes scarce due to cracking of C_3H_8 molecules [33, 34].

With an optimum flow condition (0.06 sccm Si_2H_6, 0.1 sccm C_3H_8, cracker temperature 1190 °C), the epitaxial layer had good crystallinity and a smooth surface, because the RHEED pattern showed a 3C-SiC (3 × 2) structure [33, 34].

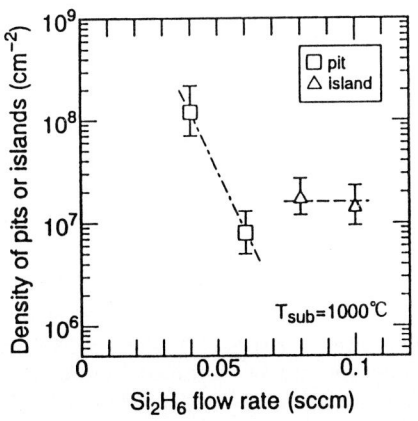

Fig. 16. Si_2H_6 flow rate dependence of the density of pits and islands

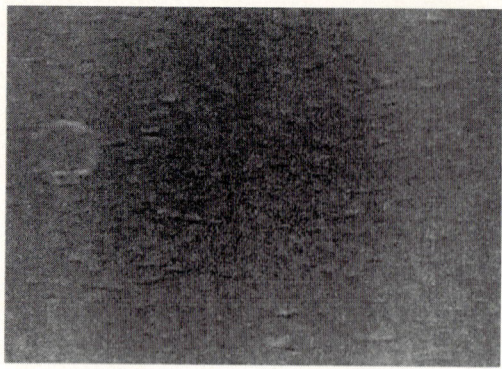

━━━━━ 600nm

Fig. 15 shows SEM photographs for epitaxial layers of about 250 Å thickness grown at 1000 °C under different Si_2H_6 flow rates [34]. The features of surface morphology can be classified into two cases. Rectangular pits are observed for low Si_2H_6 flow rates (Fig. 15a). On the other hand, islands (pyramidal shapes) appeared on the 3C-SiC epitaxial layers for a Si_2H_6 flow rate of 0.08 sccm or more (Fig. 15b). From RHEED and AES analysis, the island has $Si\{113\}$ facets [34].

Fig. 16 shows the density of pits and islands on the 3C-SiC epitaxial layers [34]. The density of pits decreased with the increase of Si_2H_6 flow rate. In the range above a Si_2H_6 flow rate of 0.08 sccm, the pits could not be observed, and the epitaxial layer contained Si islands of about $10^7 \, cm^{-2}$. It can be considered that surplus Si atoms were generated by thermal decomposition of Si_2H_6 on the substrate, forming islands with $\{113\}_{Si}$ facets. The above results suggest that the increase in the density of pits and islands would be caused by deficient and surplus Si_2H_6 supply, respectively. By a precise supply of Si_2H_6, a very smooth surface without pits and islands will be expected.

Recently, a single-domain 3C-SiC heteroepilayer has been grown at a temperature as low as 910 °C by the precise control of the Si_2H_6/DMGe flux ratio [57]. The high-resolution SEM photograph in Fig. 17 shows the surface morphology of the grown layer of about 560 Å thickness. The precise control of the Si_2H_6/DMGe flux ratio shows much potential for low-temperature 3C-SiC heteroepitaxy with good crystallinity.

6. Summary

Recent progress in gas source MBE of 3C-SiC on Si was surveyed. In the case of C_2H_2, the reaction between Si and C_2H_2 can be controlled successfully by a modified carbonization process, obtaining single-crystalline 3C-SiC thin layers reproducibly. Low-temperature heterointerface modification was achieved by chemically active hydrocarbon sources of cracked C_3H_8 and DMGe at 750 and 650 °C, respectively.

In the homoepitaxial growth on 3C-SiC layers, the precise control of Si_2H_6 supply was needed for atomic-layer epitaxy of 3C-SiC. In addition, the Si_2H_6 flow rate affected the surface morphology and crystallinity of the 3C-SiC epitaxial layer. The good crystallinity of 3C-SiC layer with single-domain (3×2) structure can be obtained by an optimum Si_2H_6/DMGe flow rate at a temperature as low as 910 °C. Heteroepitaxial growth of 3C-SiC on Si has reached a certain level and further research should be oriented to growth of layers with low defect density.

Acknowledgements   The authors would like to thank the members of this research group, Dr. H. Shiomi, Mr. M. Nakayama, Dr. T. Yoshinobu, Mr. Y. Tarui and Mr. N. Tanaka, for their efforts in the experimental and analytical works. This work was partly supported by Grant-in-Aid for Scintific Research from the Ministry of Education, Science and Culture of Japan, and facilities of Kyoto University Venture Business Laboratory.

References

[1] H. MATSUNAMI, S. NISHINO, and H. ONO, IEEE Trans. Electron Devices **28**, 1235 (1981).
[2] S. NISHINO, J. A. POWELL, and H. A. WILL, Appl. Phys. Lett. **42**, 460 (1983).
[3] K. SHIBAHARA, S. NISHINO, and H. MATSUNAMI, Jpn. J. Appl. Phys. **23**, L862 (1984).
[4] K. SHIBAHARA, S. NISHINO, and H. MATSUNAMI, J. Cryst. Growth **78**, 538 (1986).
[5] K. SHIBAHARA, S. NISHINO, and H. MATSUNAMI, Appl. Phys. Lett. **50**, 1888 (1987).
[6] S. NISHINO, H. SUHARA, H. ONO, and H. MATSUNAMI, J. Appl. Phys. **61**, 4889 (1987).
[7] K. SHIBAHARA, T. TAKEUCHI, S. NISHINO, and H. MATSUNAMI, Jpn. J. Appl. Phys. **28**, 1341 (1989).
[8] K. SHIBAHARA, T. SAITO, S. NISHINO, and H. MATSUNAMI, IEEE Electron Devices Lett. **7**, 692 (1986).
[9] K. FURUKAWA, A. HATANO, A. UEMOTO, Y. FUJII, K. NAKANISHI, M. SHIGETA, A. SUZUKI, and S. NAKAJIMA, IEEE Electron Devices Lett. **8**, 48 (1987).
[10] T. SUGII, T. ITO, Y. FURUMURA, M. DOKI, F. MIENO, and M. MAEDA, J. Electrochem. Soc. **134**, 2545 (1987).
[11] M. DAYAN, Surf. Sci. **149**, L33 (1985).
[12] M. DAYAN, J. Vacuum Sci. Technol. A **4**, 38 (1986).
[13] R. KAPLAN and T. M. PARRIL, Surf. Sci. **165**, L45 (1986).
[14] R. KAPLAN, J. Vacuum Sci. Technol. A **6**, 829 (1988).
[15] R. KAPLAN, Surf. Sci. **215**, 111 (1989).
[16] S. HARA, W. F. J. SLIJKERMAN, J. F. VAN DER VEEN, I. OHDOMARI, S. MISAWA, E. SAKUMA, and S. YOSHIDA, Surf. Sci. Lett. **231**, L196 (1990).
[17] T. MIYAZAWA, S. YOSHIDA, S. MISAWA, S. GONDA, and I. OHDOMARI, Appl. Phys. Lett. **45**, 380 (1984).
[18] H. YAMADA, J. Appl. Phys. **65**, 2084 (1989).
[19] S. MOTOYAMA, N. MORIKAWA, M. NASU, and S. KANEDA, J. Appl. Phys. **68**, 101 (1990).
[20] T. SUGII, T. AOYAMA, and T. ITO, J. Electrochem. Sci. **137**, 989 (1990).
[21] K. KIM, S. CHOI, and K. L. WANG, J. Vacuum Sci. Technol. B **12**, 930 (1992).
[22] L. B. ROWLAND, S. TANAKA, R. S. KERN, and R. F. DAVIS, Proc. 4th Intern. Conf. Amorphous and Crystalline Silicon Carbide IV, Santa Clara 1992 (p. 84).
[23] G. L. ZHOU, Z. MA, M. E. LIN, T. C. SHEN, L. H. ALLEN, and H. MORKOÇ, J. Cryst. Growth **134**, 167 (1993).
[24] T. FUYUKI, M. NAKAYAMA, T. YOSHINOBU, H. SHIOMI, and H. MATSUNAMI, J. Cryst. Growth **95**, 461 (1989).
[25] T. YOSHINOBU, N. NAKAYAMA, H. SHIOMI, T. FUYUKI, and H. MATSUNAMI, J. Cryst. Growth **99**, 520 (1990).
[26] T. YOSHINOBU, I. IZUMIKAWA, H. MITSUI, T. FUYUKI, and H. MATSUNAMI, Appl. Phys. Lett. **59**, 2844 (1991).
[27] T. YOSHINOBU, H. MITSUI, I. IZUMIKAWA, T. FUYUKI, and H. MATSUNAMI, Appl. Phys. Lett. **60**, 824 (1992).
[28] S. TANAKA, R. S. KERN, and R. F. DAVIS, Appl. Phys. Lett. **65**, 2851 (1994).
[29] T. YOSHINOBU, T. FUYUKI, and H. MATSUNAMI, Jpn. J. Appl. Phys. **30**, L1086 (1991).
[30] T. YOSHINOBU, H. MITSUI, Y. TARUI, T. FUYUKI, and H. MATSUNAMI, J. Appl. Phys. **72**, 2006 (1992).
[31] T. YOSHINOBU, H. MITSUI, Y. TARUI, T. FUYUKI, and H. MATSUNAMI, Jpn. J. Appl. Phys. **31**, L1580 (1992).

[32] T. Hatayama, Y. Tarui, T. Yoshinobu, T. Fuyuki, and H. Matsunami, J. Cryst. Growth **136**, 333 (1994).
[33] T. Fuyuki, Y. Tarui, T. Hatayama, and H. Matsunami, Mater. Res. Soc. Symp. Proc. **318**, 201 (1994).
[34] T. Hatayama, Y. Tarui, T. Fuyuki, and H. Matsunami, J. Cryst. Growth **150**, 934 (1995).
[35] M. Kitabatake, J. Appl. Phys. **74**, 4438 (1993).
[36] S. Hara, S. Misawa, S. Yoshida, and Y. Aoyagi, Phys. Rev. B **50**, 4548 (1994).
[37] M. Kitabatake and J. E. Greene, Jpn. J. Appl. Phys. **35**, 5261 (1996).
[38] M. Kitabatake and J. E. Greene, Appl. Phys. Lett. **69**, 2048 (1996).
[39] T. Hatayama, T. Fuyuki, and H. Matsunami, Jpn. J. Appl. Phys. **34**, L1117 (1995).
[40] H. Toyoda, H. Kojima, and H. Sugai, Appl. Phys. Lett. **54**, 1507 (1989).
[41] H. Toyoda, H. Kojima, and H. Sugai, Jpn. J. Appl. Phys. **30**, 2912 (1991).
[42] D. J. Hautman, R. J. Santoro, F. L. Dryer, and I. Glassman, Internat. J. Chem. Kinet. **13**, 149 (1981).
[43] V. Grill, G. Walder, D. Margreiter, T. Rauth, H. U. Poll, P, Scheier, and T. D. Märk, Z. Phys. D **25**, 217 (1993).
[44] T. Hatayama, T. Fuyuki, and H. Matsunami, Jpn. J. Appl. Phys. **35**, 5255 (1996).
[45] T. Yoshinobu, T. Fuyuki, and H. Matsunami, Jpn. J. Appl. Phys. **31**, L1213 (1992).
[46] C. C. Cheng, P. A. Taylor, R. M. Wallace, H. Gutleben, L. Clemen, M. L. Colaiann, P. L. Chen, W. H. Weinberg, W. J. Choyke, and J. T. Yates, Jr., Thin Solid Films **225**, 196 (1993).
[47] M. L. Colaianni, P. J. Chen, H. Gutleben, and J. T. Yates, Jr., Chem. Phys. Lett. **191**, 561 (1992).
[48] C. C. Cheng, S. R. Lucas, H. Gutleben, W. J. Choyke, and J. T. Yates, Jr., Surf. Sci. Lett. **273**, L441 (1992).
[49] T. Hatayama, N. Tanaka, T. Fuyuki, and H. Matsunami, Proc. 38th Electronic Materials Conf., Santa Barbara (California) 1996 (p. 20).
[50] T. Hatayama, N. Tanaka, T. Fuyuki, and H. Matsunami, to be published in J. Electronic Mater. (1997).
[51] H. Ishi, Y. Takahashi, and K. Fujinaga, JSAP-MRS Internat. Conf. Electronic Materials, Tokyo 1989 (p. 137).
[52] H. Namba, T. Yamaguchi, and H. Kuroda, Appl. Surf. Sci. **79/80**, 449 (1994).
[53] T. Hatayama, N. Tanaka, T. Fuyuki, and H. Matsunami, Appl. Phys. Lett. **70**, 1411 (1997).
[54] J. M. Walls, Methods of Surface Analysis, Cambridge University Press, 1989.
[55] H. Nagasawa and Y. Yamaguchi, Thin Solid Films **225**, 230 (1993).
[56] H. Nagasawa, K. Yagi, and T. Kawahara, Proc. 38th Electronic Materials Conf., Late News Paper, Santa Barbara (California) 1996 (J9).
[57] T. Hatayama, T. Fuyuki, and H. Matsunami, in preparation.

phys. stat. sol. (b) **202**, 379 (1997)

Subject classification: 68.55.Jk; 68.55.Ln; S6

Homoepitaxial SiC Growth by Molecular Beam Epitaxy

R. S. Kern[1]), K. Järrendahl, S. Tanaka[2]), and R. F. Davis

Department of Materials Science and Engineering, North Carolina State University, NC-27695-7907, USA

(Received January 31, 1997)

The homoepitaxial growth of SiC thin films by solid- and gas-source molecular beam epitaxy is reviewed and discussed. Our recent results regarding the homoepitaxial growth of single crystal 3C-SiC(111) and 6H-SiC(0001) thin films are also presented. The 3C-SiC(111) films were grown on both vicinal and on-axis 6H-SiC(0001) substrates at temperatures between 1000 and 1500 °C using SiH_4 and C_2H_4. They contained double positioning boundaries and stacking faults and the surface morphology and growth rate depended strongly on temperature. Films of 6H-SiC(0001) with low defect densities were deposited at high growth rates on vicinal 6H-SiC(0001) substrates by adding H_2 to the reactant mixture at temperatures between 1350 and 1500 °C. At temperatures below 1350 °C, only the cubic phase was formed. A kinetic analysis of the SiC deposition process is also presented. The SiC films were resistive with an n-type character and a lower N concentration than the p-type CVD-grown epilayers of the substrate. Undoped 6H-SiC films with the lowest atomic nitrogen and electron concentration had a mobility of 434 cm^2 V^{-1} s^{-1}, the highest room temperature value ever reported for this polytype. Both the 6H-SiC(0001) and the 3C-SiC(111) epilayers were controllably doped using a NH_3/H_2 mixture (for lighly n-doped films), pure N_2 (for heavily n-doped SiC epilayers) and Al evaporated from a standard effusion cell (for p-type doping).

1. Introduction

Thin films of SiC have been grown by molecular beam epitaxy (MBE) on several substrate materials including sapphire (Al_2O_3), aluminum nitride (AlN), titanium carbide (TiC), silicon (Si) and silicon carbide (SiC). Even if most of the research previously has been done on Si, the use of SiC substrates is now gaining more attention since homoepitaxial growth results in higher quality films. The dominate method for epitaxial SiC growth is chemical vapor deposition (CVD). However, MBE has also attracted interest, since it is a growth method that can offer a cleaner ambient and lower deposition temperatures.

There is a limited number of studies of homoepitaxial SiC growth by solid source MBE. Kaneda et al. [1] made p–n junction diodes between 3C- and 6H-SiC. This constitutes one of the first reports of homoepitaxial device structures being fabricated by this technique. The 6H-SiC(0001) substrate was n-type ($n = 5.6 \times 10^{16}$ cm^{-3}). Epitaxial 3C-SiC(111) was grown at 1150 °C using electron beam heated solid sources of C and mixed Si/B. The Si/B mix was formed by dissolving B (a p-type dopant in SiC) in Si by

---

[1]) Present address: Hewlett-Packard Optoelectronics Division, San Jose, CA 95131, USA.
[2]) Present address: The Institute of Physical and Chemical Research (RIKEN), Saitama 351-01, Japan.

electron beam heating. The resulting films were found to be p-type ($p = 1 \times 10^{18}$ cm^{-3}). The breakdown field was 6.7×10^5 V cm^{-1} which is comparable with values determined from p–n junctions formed by other growth methods. Building on their previous work on Si(111), Fissel et al. [2 to 4] used their "Si-stabilized" growth method to grow 3C-SiC on vicinal 6H-SiC(0001) substrates at modest growth rates (1 nm min^{-1}) below 1000 °C. The experiment was conducted with an excessive Si flux and monitored by reflection high energy electron diffraction (RHEED). They reported the growth of mixtures of 3C- and 2H-SiC at temperatures of 850 °C; however, the HRTEM images presented indicated that a 3C-SiC layer with a high density of ⟨111⟩ stacking faults would be a more appropriate description. At higher temperatures, layer-by-layer growth was achieved, as monitored by RHEED intensity oscillations. Although the growth rates obtained in this research were promising, the quality of the presented material displayed many stacking faults, double positioning boundaries (DPBs), rough surfaces and non-uniform surface coverage. Recently, the same group has grown 6H-SiC starting on ($\sqrt{3} \times \sqrt{3}$)R30° reconstructed 6H-SiC(0001) substrates (3 to 4° off-axis) [5, 6]. The growth process was controlled on the atomic level by monitoring surface superstructures occurring during the alternate supply of silicon and carbon.

Gas-source molecular beam epitaxy (GSMBE) has also been used for homoepitaxial SiC growth. Single crystalline 3C-SiC was grown by Fuyuki et al. [7, 8] and Yoshinobu et al. [9] using an alternating supply of disilane (Si$_2$H$_6$) and ethylene (C$_2$H$_4$) and a substrate temperature of 1000 °C. Films of 3C-SiC grown on Si(100) by CVD were used as substrates. Monolayer controlled SiC growth (2 to 6 monolayer/growth cycle) was demonstrated. The same group also grew epilayers of 3C-SiC on 6H-SiC(000$\bar{1}$) and 6H-SiC(0$\bar{1}$1$\bar{4}$) using alternating beams of Si$_2$H$_6$ and C$_2$H$_2$ [10]. Growth on the carbon-faced (000$\bar{1}$) substrates, which were angle-lapped to a 5° misorientation toward [11$\bar{2}$0], at 1000 and 1150 °C resulted in 3C-SiC($\bar{1}\bar{1}\bar{1}$) with a high density of DPBs. The films resulting from growth at temperatures as low as 850 °C on 6H-Si(0$\bar{1}$1$\bar{4}$), which is 54.7° off (000$\bar{1}$), were 3C-SiC(100) with no DPBs. This encouraging result must be tempered, however, since no commercial sources of such an unusual orientation are available nor is current production expected to expand to include this orientation.

In our group considerable SiC deposition studies by GSMBE on vicinal 6H-SiC(0001) substrates (cut $3.5 \pm 0.5°$ toward [11$\bar{2}$0]) using Si$_2$H$_6$ and C$_2$H$_4$ have been performed [11 to 13]. Initially, these studies focused on the influence of total flow rate, flow ratios and growth temperature on film quality and formation. The ratio of C$_2$H$_4$ to Si$_2$H$_6$ was varied between 1:1 and 10:1 and the growth temperature between 1000 and 1250 °C. The resulting films were epitaxial and monocrystalline 3C-SiC with rough surfaces and defects including DPBs and stacking faults. Lowering the total flow rate produced smoother films at the expense of slower growth rate. Deposition at 1050 °C produced sufficient nuclei to form continuous layers and resulted in the smoothest films. Under all conditions, step flow and step bunching mechanisms were observed to occur initially. When high C$_2$H$_4$:Si$_2$H$_6$ ratios were used, deposition rates and the nucleated island density increased. Further studies [14] of the growth mode of SiC films on vicinal 6H-SiC substrates was conducted to determine the role of Si-to-C flux ratio on growth rate and polytype formation. Major conclusions included the preference for the growth of 3C-SiC at high C-rich ratios while growth under excess Si resulted in a surface reconstruction that was believed to enhance the diffusion lengths of the adatoms promoting step flow. While the latter result presented a method to form 6H-SiC homoepitaxially, the growth rates (5 to 10 nm h^{-1}) obtained by this method were to low for practical study of the structural and electronic character of the films.

In Sections 2 and 3 below, our more recent research regarding SiC homoepitaxial films deposited in the same GSMBE system is presented [15 to 17].

2. Experimental Details

The GSMBE system used in this study was designed for moderate temperature growth and doping of monocrystalline SiC [11, 15]. The primary advantages of using gas sources are that they can be replenished without breaking vacuum, and their flux can be controlled and changed very quickly and accurately. After several investigations using disilane (Si$_2$H$_6$) as the source of Si [11 to 14], silane (SiH$_4$) was ultimately chosen for its higher purity (99.999%). A primary obstacle to low temperature growth of SiC is the difficulty in providing a suitable source of reactive carbon. For gaseous sources, species such as ethylene (C$_2$H$_4$) and acetylene (C$_2$H$_2$) are favored over species such as methane (CH$_4$) and propane (C$_3$H$_8$) due to the positive free energies of formation of the former two gases at the temperatures used in this research. Acetylene is the hydrocarbon with the highest free energy at the temperatures used in this research, but it is unfortunately difficult to obtain in purities greater than 99.7%. Thus, ethylene (99.99% pure) was chosen for use as the C source. The gases were delivered through a 0.75″ orifice in the source flange via a servo driven leak valve maintained by means of a pressure/flow controller. For doping purposes, sources were added to provide the n- and p-type dopants of N and Al, respectively. A NH$_3$/H$_2$ (99.999% pure) mixture containing 300 ppm NH$_3$ delivered through a mass flow controller and a 30 cm^3 effusion cell (filled with 99.999% pure aluminum pellets) were used. In addition, 99.9999% pure hydrogen (H$_2$), 99.9995% pure nitrogen (N$_2$) and 99.9995% pure argon (Ar) could be delivered to the system via mass flow controllers.

Background pressures of less than $\approx 6.65 \times 10^{-7}$ Pa were routinely achieved; however, pressure below $\approx 10^{-7}$ Pa were achieved when the three cryoshrouds located inside the reactor vessel and on the source flange were filled with liquid nitrogen. The system also contained a 100 amu mass spectrometer used mainly as a residual gas analyzer and helium leak detector. RHEED analysis was performed using a 10 kV HEED gun which was directed at about a 1° angle to the substrate surface. The samples were mounted with clips and heated resistively. The clips, substrate carrier and heater were all made of SiC coated graphite. All temperature measurements were made using an optical pyrometer.

The dominate substrates used in the current research were pieces of single crystal, 1.0 and 1.375″ diameter α(6H)-SiC(0001) wafers obtained from Cree Research. The first wafer orientation was oriented 3 to 4° off (0001) towards [11$\bar{2}$0]. These wafers are hereafter denoted as off-axis or vicinal. After polishing, epitaxial layers of lightly doped ($\approx 5 \times 10^{16}$ carriers cm^{-3}) single phase 6H-SiC, approximately 0.8 to 1.0 μm in thickness, either n- or p-type, were grown by CVD at Cree on these off-axis wafers. The second type of wafer was oriented nominally on-axis with no or very few steps on the surface. These wafers, referred to as on-axis, were used without epitaxial layers. The substrate cleaning procedure involved a 5 min dip in 10% HF prior to loading the sample into the chamber via a load-lock. This was followed by the formation and desorption of an excess Si layer on the surface using SiH$_4$ [15]. The substrate surface and its reconstructions were monitored by RHEED. At the completion of each SiC deposition experiment, considerable care was taken to ensure that the SiC surface was not damaged or destroyed. The method for surface stabilization was obtained using mixtures of Ar and SiH_4 during the cooling process [15].

A scanning electron microscope (SEM) equipped with a field emission electron source was employed to observe the surface morphology. Both Auger electron spectroscopy (AES) and secondary ion mass spectrometry (SIMS) were used extensively to determine the chemical composition and impurity concentration in the grown films. Depth profiled AES and SIMS characterizations were vital in determining the uniformity of composition and doping profiles.

In order to determine information regarding crystallinity, polytype and orientation relationships at interfaces and to observe defects present in the grown films, transmission electron microscopy (TEM) and reflection high energy electron diffraction (RHEED) were used. High resolution electron microscopy (HREM) analysis was performed operating at 200 kV, which provided a minimum resolution of 1.8 Å.

3. Results and Discussion

3.1 Growth of 3C-SiC on vicinal and on-axis 6H-SiC(0001)

Growth of SiC by GSMBE on vicinal and on-axis 6H-SiC(0001) was recently performed using the experimental conditions described in Section 2. When only the precursor gas species (SiH_4 and C_2H_4) were introduced into the deposition system environment, the resulting epilayers were always 3C-SiC(111) at deposition temperatures from 1000 to 1500 °C and reactant flows from 0.15 to 2.5 sccm (the SiH_4:C_2H_4 ratio varied from 5:1 to 1:5). The previous results obtained using Si_2H_6 and C_2H_4 revealed the optimal reactant input at a substrate temperature of 1050 °C to be at the Si_2H_6:C_2H_4 ratio of 1:2 (i.e., for each Si atom in the vapor phase, there were two C atoms) with a total reactant flow of 1.5 sccm [11]. To maintain similar pressure and chemical conditions, the typical SiH_4:C_2H_4 flow ratio was 1:1 (using 0.75 sccm of each reactant gas) during this study; however, the SiH_4 and C_2H_4 flow rates were varied independently from 0.5 to 1.5 sccm to study the effect on the SiC growth rate. All films produced using these conditions appeared smooth and specular to the naked eye. Fig. 1 shows a representative RHEED pattern ([1$\bar{1}$0] zone axis) for a typical 3C-SiC(111) film. This pattern is very indicative of a cubic film containing DPBs as described by Rowland et al. [12]. Because the presence of DPBs in a film produces extra streaks and spots in the electron diffraction pattern [12], RHEED was used extensively for polytype monitoring and determination of

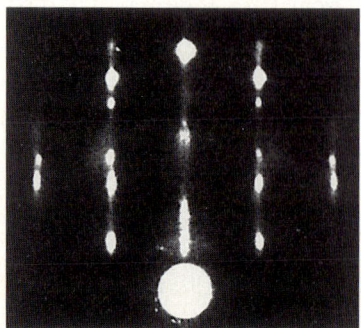

Fig. 1. RHEED pattern of the [1$\bar{1}$0]-azimuth of a 3C-SiC(111) layer

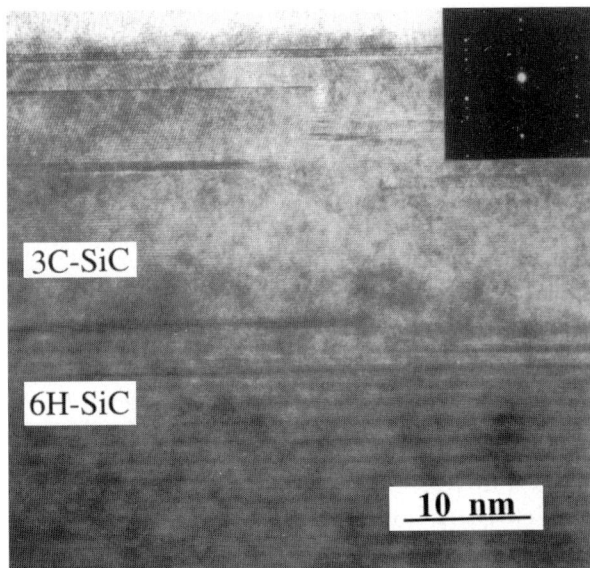

Fig. 2. HREM image of a typical 3C-SiC(111) film deposited on a vicinal 6H-SiC(0001) substrate. The inset shows a selected area electron diffraction pattern ($[1\bar{1}0]$ zone axis) indicative of 3C-SiC

SiC epilayers. In all cases where the 3C-SiC polytype was grown, these boundaries were very distinct in the $[1\bar{1}0]$-azimuth RHEED pattern making this pattern the most commonly utilized image for surface and film characterization. Fig. 2 shows a HREM image of a typical 3C-SiC(111) film and its corresponding selected area electron diffraction pattern ($[1\bar{1}0]$ zone axis), deposited on a vicinal 6H-SiC(0001) surface. The planar defects (predominately stacking faults and twins) seen in the image are characteristic of these films.

The effect of the growth temperature on 3C-films was investigated using 0.75 sccm SiH_4 and 0.75 sccm C_2H_4 between 1000 and 1500 °C on vicinal 6H-SiC(0001) substrates. Fig. 3 shows the RHEED patterns of films grown at 1000, 1300 and 1500 °C. Each of these patterns was taken from films that were grown for 5 to 10 h. The extra spots, dim and diffuse regions and the diamond-hatched shapes seen primarily in the films grown at lower temperatures are indicative of rough surfaces, twin and DPB defects and poor crystallinity. As the temperature was increased, the RHEED images showed fewer of these irregularities indicating an improvement in the surfaces and overall crystal quality. The improvements in surface smoothness and uniformity as the deposition temperature

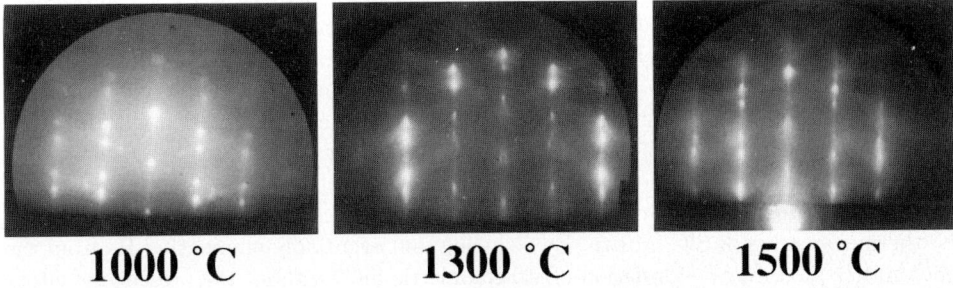

Fig. 3. RHEED patterns of 3C-SiC(111) grown at various temperatures

increased were also readily observed by SEM. The progression from very rough, mosaic-like structures for films grown at 1000 °C to the smooth, almost featureless surface for films grown at 1500 °C demonstrated the improvements in surface quality as the temperature was increased. The explanation for these observed improvements was the increase in surface mobility which resulted from increasing the deposition temperature. The increase in surface mobility apparently caused the more rapid coalescence of the individual nuclei that formed on the substrate. Thus, the film surface displayed an improved smoothness and uniformity as the substrate temperature increased. The kinetic effect of temperature on growth rate will be presented in Section 3.3.

3.2 Growth of 6H-SiC on vicinal 6H-SiC(0001)

Although expected for growth on the on-axis substrates, the occurrence of 3C-SiC on vicinal substrates was somewhat unexpected and was at variance with results obtained from related CVD research. The accepted model for epitaxial CVD growth of SiC on 6H-SiC{0001} was independently proposed by Kuroda et al. [18] and Kong et al. [19]. In this model, the density and orientation of the surface steps determine the resultant SiC polytype. When few steps are present, as in the case of nominally on-axis SiC, the growth conditions determine the resultant polytype, and 3C-SiC is usually, but not always, formed. However, if there exists a high density of $[11\bar{2}0]$ surface steps and if the mobility of adsorbed species is sufficiently large, then these species should easily find steps and/or kinks and jogs on these steps which can act as incorporation sites for lateral growth. At each step, the incorporation site is uniquely defined by the bonding in the neighborhood of the step. In contrast, incorporation that occurs on the terraces away from the step sites is not uniquely defined and can take either of two possible configurations. Sufficient incorporation at one or more of the step sites should lead to lateral growth of the steps. Thus, these steps serve as a template for SiC growth, and the stacking sequence of the 6H (or 4H) polytype is preserved by means of a special type of two-dimensional mode referred to as lateral (step flow) growth and is believed to be the cause of the formation of 6H-SiC on 6H-SiC{0001} cut 3 to 4° towards $[11\bar{2}0]$ [18, 19]. This step flow phenomenon has been reported previously in SiC growth by CVD for a variety of reactant gas sources including those used in this research [19, 20].

Since the GSMBE growth formed 3C-SiC films on vicinal 6H-SiC(0001) substrates in contradiction to the results from CVD-grown films using the same gas sources and similar temperatures, a study was performed in order to determine the mode of SiC film deposition. The growth of 3C-SiC on vicinal 6H-SiC(0001) substrates despite the high density of surface step sites was attributed to the differences between the processing conditions present in typical CVD growth reactors and those present in the MBE system used during this research. The primary differences between CVD and MBE are, 1. the amount of H_2 present during growth, 2. the amount of gas phase collisions and 3. the growth temperature. Since the second factor could not be modified without compromising the beam nature of the MBE sources, the other two factors were adjusted in an attempt to alter the character of the SiC films.

Thermodynamic analysis by Kingon et al. [21] determined the primary effect of H_2 on the thermodynamics of SiC growth. By reducing and eventually eliminating H_2 from the reactant gas mixture, the formation of stoichiometric SiC (without the presence of either excess C or Si as a second phase) was deemed a virtual impossibility unless precise gas

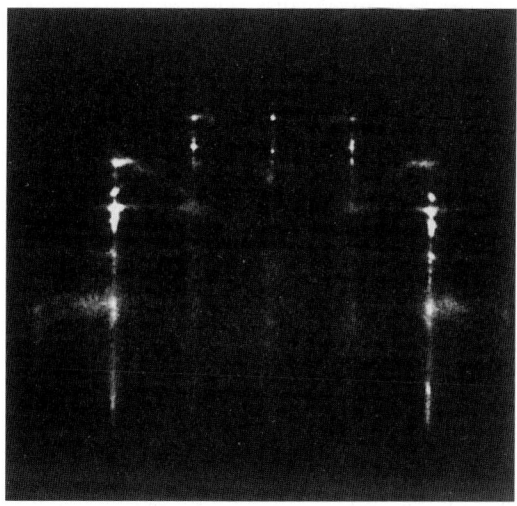

Fig. 4. RHEED pattern of the [11$\bar{2}$0]-azimuth of a 6H-SiC(0001) layer

mixture compositions and deposition temperatures were employed. Although this study could not be translated directly to MBE growth, it did elucidate the importance of H_2 as a carrier gas in SiC growth. The results at variable chemical compositions and pressures indicated that the parameter window for growing SiC was very small at the pressure and temperatures used in MBE or CVD. However, the addition of H_2 was expected to broaden this formation regime. Chaudhry et al. [22, 23] found that using Ar/H_2 mixtures rather than pure H_2 reduced the growth rate of 3C-SiC on Si by CVD at 1100 to 1350 °C. Interestingly, the lowest deposition temperature for monocrystalline SiC was reduced by the use of Ar/H_2 mixtures. However, the general structural character of the films and their electrical measurements were compromised by additions of Ar. Kruaval and Parsons [24, 25] observed considerable homogeneous reaction in the gas phase when Ar and He were substituted for the H_2 carrier at 1100 to 1630 °C. As a result, polymeric films were deposited rather than crystalline ones. Both gas phase chemistry and surface process (reactant decomposition and adatom surface mobility) were postulated to be the agents responsible for these observations. Similar results were obtained by Kordina [26] for 6H-SiC growth on 6H-SiC(0001) by CVD. The substitution of Ar for H_2 as the carrier gas resulted in a reduced growth rate. In fact, when H_2 was completely removed from the growth reactor, only 3C-SiC and 3C/6H mixtures were deposited below 1800 °C. Based on the work of these researchers, H_2 was added to this MBE system as a reactant source gas. In all cases described, 5 sccm H_2 was used because this was the lowest input flow rate that allowed for reproducible results.

For substrate temperatures below 1350 °C, the addition H_2 to the reactant gas input (at various flow rates of 0.5 to 1.5 sccm for SiH_4 and C_2H_4) resulted in smoother film surfaces; however, only 3C-SiC films were produced despite the use of vicinal 6H-SiC(0001) substrates. Moderate growth rate increases were also observed for films grown under these conditions. The discussion of these factors and the kinetics of the SiC deposition process are presented in Section 3.3.

At temperatures above 1350 °C, 6H-SiC(0001) films were deposited on vicinal 6H-SiC substrates. RHEED was again the primary method used to determine the crystallinity, surface quality and polytype of the 6H-SiC layers. Fig. 4 shows the RHEED pattern of a

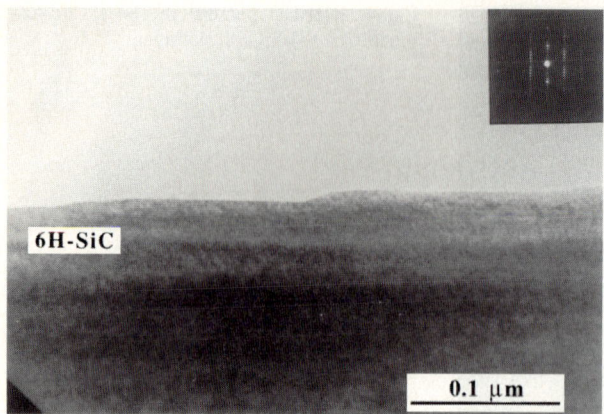

Fig. 5. HREM image of a typical 6H-SiC(0001) film deposited on a vicinal 6H-SiC(0001) substrate. Note the considerable step bunching present in the sample and the absence of 3C-SiC nuclei indicative of the stabilization of the step flow growth mechanism. The inset shows a selected area electron diffraction pattern ($[11\bar{2}0]$ zone axis) indicative of 6H-SiC

homoepitaxial 6H-SiC(0001) film grown using 0.75 sccm SiH$_4$, 0.75 C$_2$H$_4$ and 5 sccm H$_2$ at 1350 °C. This image indicated that the film was 6H-SiC similar to the image of a freshly cleaned substrate before growth. This assertion was confirmed by the cross-sectional HREM image and selected area electron diffraction pattern of the same film shown in Fig. 5. Note the extreme degree of step bunching–as much as six unit cells (i.e., $\approx$90 Å) high–that was present in this film as well as the absence of any evidence of 3C-SiC nuclei. This extreme degree of step bunching was also observed via SEM where the step flow appeared to be pinned by the presence of micropipes in the SiC substrate [15].

Since the deposition of SiC on vicinal 6H-SiC(0001) substrates at temperatures $\geq$1350 °C with 0.75 sccm SiH$_4$, 0.75 sccm C$_2$H$_4$ and 5 sccm H$_2$ produced films of the 6H polytype, attempts were made to study the range of gas flows under which 6H-SiC formed. Changes in the C$_2$H$_4$ flow rate were the most studied and found to have the least impact on the growth rate and surface morphology. Under similar conditions of temperature as well as SiH$_4$ and H$_2$ flow rates, it was determined that the C$_2$H$_4$ input could be varied from 0.375 to 0.75 sccm without affecting the growth rate or surface morphology. However, when the C$_2$H$_4$ input was decreased to less than 0.375 sccm, the result was a decrease in the growth rate. These results appear to indicate that at these temperatures and flow rates, the growth process was independent of C$_2$H$_4$ until the supply ratio of C$_2$H$_4$:SiH$_4$ decreased below 2:1. At this point, the reactant input (namely C$_2$H$_4$) flow rate controlled the deposition process. At increased C$_2$H$_4$ flow rates, a mixture of cubic and hexagonal polytypes of SiC resulted. This process occurred for all conditions where the C$_2$H$_4$ flow was increased above 1.0 sccm.

3.3 Kinetics of SiC growth

MBE is generally recognized as a thin film process that is kinetically-controlled rather than thermodynamically-controlled. Under certain assumptions the deposition rate R_g (Å h^{-1}) of such a process can be expressed as [15, 16]

$$R_g = k_{dep} f_{SiH_4}^x f_{C_2H_4}^y, \qquad (1)$$

where k_{dep} is the rate constant for the deposition reaction (in Å h^{-1} sccm$^{-(x+y)}$), f_{SiH_4} and $f_{C_2H_4}$ are the flow rates of the reactants (in sccm), and x and y are the exponential

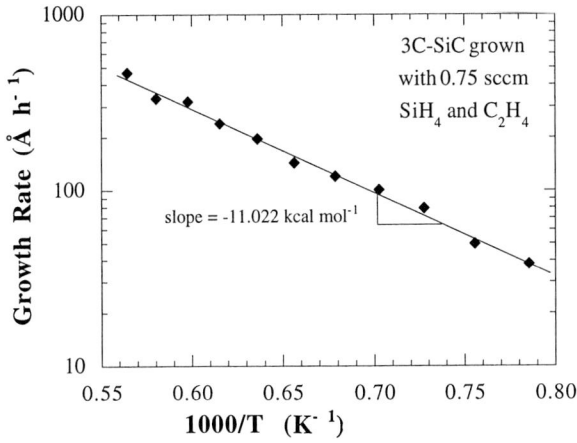

Fig. 6. Arrhenius plot of the growth rate for 3C-SiC(111) films grown between 1000 and 1500 °C with 0.75 sccm SiH_4 and 0.75 sccm C_2H_4

dependencies of the deposition process on the individual reactant flow rates. Similar to analysis of the kinetics in other open tube systems [27], the kinetics of SiC deposition via the decomposition of SiH_4 and C_2H_4 were studied [15, 16] with respect to the four most significant variables in the deposition process, 1. substrate temperature, 2. SiH_4 flow rate at a constant C_2H_4 flow rate, 3. C_2H_4 flow rate at a constant SiH_4 flow rate and 4. addition of 5 sccm H_2 at constant SiH_4 and C_2H_4 flow rates. Below, the nature of the rate determining step by adjusting the various input characteristics will be investigated. Although the exact nature of the rate determining step is beyond the scope of the present study, some general comments and observations will be presented.

3.3.1 Substrate temperature

In addition to improving the surface character of the films, elevated growth temperatures were also observed to produce increased SiC epilayer growth rates. Fig. 6 shows an Arrhenius plot of $\ln R_g$ versus T^{-1} (in K^{-1}) for the growth of 3C-SiC(111) on vicinal 6H-SiC(0001) in a non-hydrogen ambient obtained at various temperatures for reactant input flows of 0.75 sccm for SiH_4 and C_2H_4.

Arrhenius plots typically have different shapes which provide an indication of the reaction limiting constraint. All vapor phase reaction processes can be considered as either mass input (equilibrium), mass transport or surface reaction controlled. Reactions that are mass input, or equilibrium controlled can be generally characterized using the guidelines that the partial pressures of reactants and products in the vicinity of the substrate crystal are at their equilibrium value. Thus, an Arrhenius curve that has a very low slope indicating a nearly constant rate of reaction with respect to temperature results. The slope of such a curve will be related to the equilibrium constant for the reaction. As a result, the predominate means of increasing the formation of reaction products is by increasing the input of reactants. For vapor phase epitaxy, the attainment of mass input limited growth is very difficult because the linear gas stream velocity and reactant input into the reactor are generally sufficiently high to prevent the reaction from attaining equilibrium. However, it must be noted that Karmann et al. [28] have apparently achieved mass input controlled SiC deposition on 6H-SiC(0001) cut 1.6 to 1.8° off-axis toward [1$\bar{1}$00] using SiH_4 and C_3H_8 at 1405 to 1600 °C.

Generally, two temperature regimes are recognized in the CVD growth of materials [29]. For growth at high temperature, gas phase diffusion through boundary layers that are assumed to be present in CVD controls the reaction rate and is defined as the rate determining step. At lower temperatures, some type of surface reaction is the rate determining step in the deposition. The former process generally exhibits a weak dependence on temperature; however, the latter is normally associated with a strong dependence on temperature. From the arguments presented by Kong et al. [30] and Jones and Shaw [31], the slopes of Arrhenius curves that describe mass transport controlled reactions should have values of 3.5 to 4.5 kcal mol^{-1}($= 14.65$ to 18.84 kJ mol^{-1}). Surface reaction controlled processes have a larger apparent activation energy which is related to the rate constant of the rate determining process. In this case, the rate determining step is generally one of the many processes that occur during the adsorption of reactant gas species, decomposition reaction(s) on the surface and desorption of product species.

Analysis of the linear curve in Fig. 6 indicated that it follows the general Arrhenius equation

$$R_g = R_0 \exp\left(-\frac{\Delta H_a}{RT}\right),$$
$$(2)$$

where R_g is the growth rate (in Å hr^{-1}) perpendicular to the substrate surface, R_0 is a pre-exponential factor, ΔH_a is the effective activation energy (in kcal mol^{-1}), R is the ideal gas constant (1.987 cal mol^{-1} K$^{-1} = 8.319$ J mol^{-1} K^{-1}) and T is the absolute temperature (in K). From the slope of the curve, the apparent activation energy, ΔH_a, was determined to be 21.9 kcal mol^{-1} ($= 91.69$ kJ mol^{-1}) rate expression describing the data in this curve was given by

$$\ln R_g = 12.3 - \frac{11.022}{T}.$$
$$(3)$$

From the shape (linear) of the fitted curve and the size of the activation barrier, the reaction appears to be surface reaction limited. The size of this activation barrier was determined to be independent of reactant input partial pressures. Additions of up to 5 sccm Ar into the gas flow also produced no observable changes in growth rate or morphology.

3.3.2 SiH_4 flow rate

As noted above, the effect of substrate temperature on the growth rate of SiC was studied at a specific gas composition, namely, 0.75 sccm SiH$_4$ and 0.75 sccm C$_2$H$_4$. The observed growth rates were very low. Based on studies by Buss et al. [32] on SiH$_4$ and Bozso et al. [33] on C$_2$H$_4$, the rate limiting factor was assumed to be the decomposition of C$_2$H$_4$ into reactant species that were suitable for SiC growth at the temperatures studied. As such, the flow rate of each reactant was varied to study the kinetic effect on growth rate and to verify this assertion. In order to study the effect of SiH$_4$ on growth rate, the flow rate of that reactant was varied between 0.5 and 1.25 sccm while the C$_2$H$_4$ flow rate was maintained at 0.75 sccm. For SiH$_4$ flow rates of 0.5, 0.75 and 1.0 sccm, no apparent change in growth rate was observed. When the flow rates were ≥ 1.25 sccm, considerable increases in the deposition rate were observed. However, in an indication of the reactive nature of SiH$_4$, these SiC films were not stoichiometric (being Si-rich in composition), and data from these samples were not useful for analysis. A

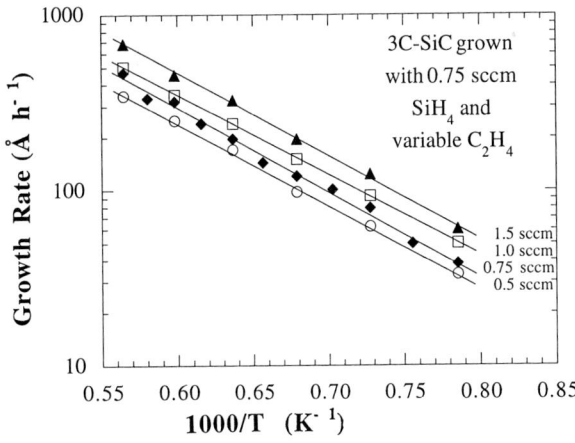

possible explanation for this occurrence was that the thermodynamcis favored Si deposition in addition to SiC deposition at the temperatures studied here.

Since the growth rate was constant for a given deposition temperature between 1000 and 1500 °C for all conditions where $0.67 \leq f_{\mathrm{SiH_4}}/f_{\mathrm{C_2H_4}} \leq 1.33$, the SiC deposition process was deemed to be independent of the SiH$_4$ flow rate within this regime of temperature and SiH$_4$ flow rate. This indicates that the deposition reaction was of zeroth order with respect to the silane flow rate. As a result, the value of the rate constant for the SiC growth rate at 0.75 sccm C$_2$H$_4$ was identical to that in Eq. (1) and the growth rate was simplified to

$$R_{\mathrm{g}} = k_{\mathrm{dep}} f_{\mathrm{C_2H_4}}^y \, . \tag{4}$$

These results indicate that the deposition reaction as presented in Eq. (4) is most likely governed by the decomposition of C$_2$H$_4$ into suitable species to form SiC in the presence of the reaction products from SiH$_4$.

3.3.3 C$_2$H$_4$ flow rate

The flow rate of C$_2$H$_4$ was varied between 0.5 and 1.5 sccm with the SiH$_4$ flow rate maintained at 0.75 sccm. Fig. 7 shows a number of $\ln R_{\mathrm{g}}$ versus T^{-1} curves, similar to the plot presented in Fig. 6, with each line representing a different input flow rate of C$_2$H$_4$. Within experimental error, the slopes of these curves were identical, indicating that the reaction mechanism was unchanged as a function of the C$_2$H$_4$ flow rate.

Fig. 8 is a series of plots of $\ln R_{\mathrm{g}}$ versus $\ln f_{\mathrm{C_2H_4}}$ in the range of 0.5 to 1.5 sccm at 1100, 1200 and 1300 °C. In each case, the data appeared to be linear on a logarithmic plot. Each curve had a slope of ≈ 0.63 which indicated that the deposition reaction was a two-thirds order power reaction which was limited by the decomposition of the C$_2$H$_4$. The fractional order of the reaction may be indicative of a series of elementary step reaction(s) (i.e., the adsorption and decomposition of the C$_2$H$_4$ into more reactive species). The value of k_{dep} was determined at each temperature using Eq. (4), and the temperature dependence was determined from the Arrhenius equation

$$\ln k_{\mathrm{dep}} = 12.48 - \frac{11.000}{T} \, . \tag{5}$$

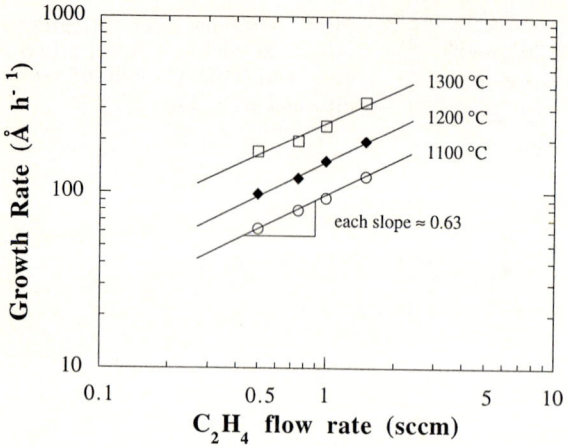

Fig. 8. Plot of $\ln R_g$ vs. $\ln f_{C_2H_4}$ for 3C-SiC(111) films grown at 1100, 1200 and 1300 °C with 0.75 sccm SiH$_4$ and 0.5 to 1.5 sccm C$_2$H$_4$. Note that the slope of each curve is ≈0.63 indicating that the reaction is a 2/3 order power reaction with respect to $f_{C_2H_4}$

From Eqs. (4) and (5), the deposition rate as a function of temperature and C$_2$H$_4$ input flow rate was determined to be

$$\ln R_g = 12.48 - \frac{11.000}{T} + 0.63 \ln f_{C_2H_4} \,. \tag{6}$$

Again, the apparent activation barrier for this process was ≈22 kcal mol^{-1} (≈92.11 kJ mol^{-1}). Since the curves presented in Fig. 8 were linear over the range studied and had a dependence on the C$_2$H$_4$ flow rate, the deposition process appeared to be controlled by surface reaction, and the rate of species formation, reaction and desorption from C$_2$H$_4$ appeared to be the rate limiting step.

From existing studies of hydrocarbon reactions with silicon surfaces, several general comments can be made concerning the reaction mechanism of ethylene with semiconductor surfaces. Firstly, C$_2$H$_4$ is non-dissociatively chemisorbed onto Si(100) and (111) surfaces to about 330 °C in a di-σ bonding configuration [34 to 41]. Because the Si surface acts as a good π donor [42], and ethylene can easily accept electrons into its empty π* orbital, the result is a rehybridization of the C in the adsorbed molecule to the sp^3 state that saturates the Si dangling bonds. Secondly, further heating of the surface on which ethylene is adsorbed results in the desorption of both H$_2$ and ≈40% of the adsorbed C$_2$H$_4$ at about 390 °C leaving behind CH$_x$ species which form SiC on the surface when heated above 660°C [35, 37]. Thirdly, the exposed bonding sites on these adsorbed molecules can then act as thermally reversible adsorption sites for atomic H between 800 and 1100 °C [34]. However, this adsorption and decomposition of C$_2$H$_4$ at elevated temperatures is a relatively slow process requiring an incubation time before Si converts to SiC [34]. For growth temperatures of 800 to 1200 °C, previous UHV studies of the Si–C$_2$H$_4$ interaction show that SiC forms by a two-dimensional layer-by-layer growth process with a C$_2$H$_4$ reaction probability on the order of 2×10^{-3} [33, 43]. Improvements in the reaction kinetics for Si conversion and SiC deposition were observed by several researchers [34, 44 to 48] when CH$_3$ radicals were supplied by cracking C$_2$H$_4$ or C$_2$H$_8$ in atomic H or by a W filament. These results appear to confirm the assumption that the rate determining step is related to forming activated CH$_x$ radicals.

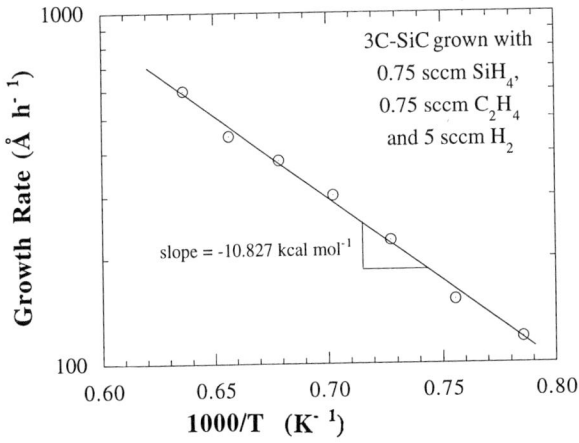

Fig. 9. Arrhenius plot of the growth rate for 3C-SiC(111) films grown between 1000 and 1300 °C with 0.75 sccm SiH$_4$, 0.75 sccm C$_2$H$_4$ and 5 sccm H$_2$. Note that the apparent activation energy is nearly the same as in Figs. 6 and 7 indicating the same rate determining step in each case

3.3.4 Addition of H$_2$

Fig. 9 shows the growth rate dependence on temperature for 3C-SiC films grown with 0.75 sccm SiH$_4$, 0.75 sccm C$_2$H$_4$ and 5 sccm H$_2$ at 1000 to 1300 °C. Similar to the previous example, the ΔH_a was calculated from the graph using Eq. (2). The value of the activation barrier was 21.6 kcal mol^{-1} (=90.43 kJ mol^{-1}), indicating that the presence of hydrogen did not change the magnitude of the activation barrier over this temperature range. However, since the slopes of the curves shown in Figs. 7 and 9 were nearly identical, the reaction mechanism was apparently unchanged and was independent of the input flow of C$_2$H$_4$.

Fig. 10 shows an extrapolation of the curves in the plot of $\ln R_g$ versus $\ln f_{C_2H_4}$ (Fig. 8) and a plot of constant lines representing the growth rate at each of the individual temperatures (1100, 1200 and 1300 °C) from Fig. 9. In each case, the intersection between the isothermal curves occurred at ≈ 4 sccm. Subsequent increases in the $f_{C_2H_4}$ did not result in an increased growth rate. These observations are attributed to the presence of H$_2$ which apparently provided the impetus for at least one of the following processes, 1. adsorption of C$_2$H$_4$ onto the surface of the growing film, 2. sweeping of the unreacted source gas species or unwanted product species from the growth surface and/or 3. formation of a suitable reactant specie in tandem with the decomposition or reaction of C$_2$H$_4$ in the presence of H$_2$.

Hydrogen may affect the growth process by facilitating the removal of various adsorbed product species which occupy potential bonding sites on the growing crystal surface. In effect, H$_2$ may adsorb onto the SiC surface, react with the various species that have "poisoned" the various incorporation sites on the surface and desorb as volatile SiH$_x$ or CH$_x$ species leaving behind unterminated dangling bonds on the surface which act as incorporation sites for SiC growth. This desorption process may be thermodynamically favorable at $T > 1300$ °C. While this process should occur from the evolution of H$_2$ from the decomposition of the various source gases, the quantity of H$_2$ produced via this process may be insufficient to adequately purge the "poisons" from the surface. Likewise, additions of H$_2$ significantly increase the total flow rate and the linear gas stream velocity. The impact of these increases may provide the driving force for not only reactant delivery, but also product removal. However, neither of these mechanisms is expected be

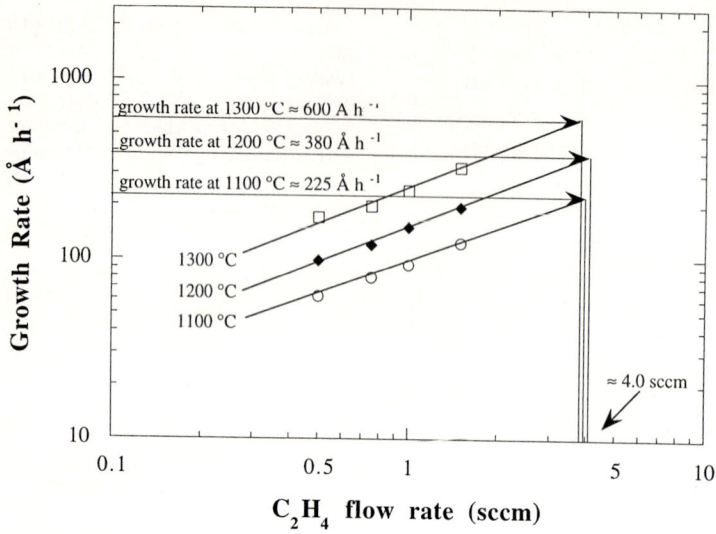

Fig. 10. Plot of $\ln R_g$ vs. $\ln f_{C_2H_4}$ for 3C-SiC(111) films grown at 1100, 1200 and 1300 °C with 0.75 sccm SiH_4 and 0.5 to 1.5 sccm C_2H_4. Horizontal lines represent the growth rates achieved at 0.75 sccm SiH_4, 0.75 C_2H_4 and 5 sccm H_2 at the same temperatures. From the graph, the addition of H_2 produces a reaction coordinate equivalent to that with ≈ 4.0 sccm C_2H_4

the rate controlling process in the deposition because the delivery of C_2H_4 also was shown above (see Fig. 7) to produce analogous results in terms of growth rate in the absence of extra H_2. Thus, the rate controlling step was assumed to be related to the production of a suitable active carbon species from C_2H_4 decomposition. The work of several researchers [34, 44 to 48] indicating the improved C incorporation when CH_3 radicals were supplied to the substrate appears to also favor this assumption.

For the growth of 6H-SiC films on vicinal 6H-SiC(0001) substrates at temperatures between 1350 and 1500 °C, a considerable increase in growth rate, and an apparent decrease in the magnitude of the activation barrier were obtained with the addition of H_2 to the gas stream. The temperature dependence of the growth rate of 6H-SiC films grown using 0.75 sccm for both SiH_4 and C_2H_4 and 5 sccm H_2 appear to obey an Arrhenius relationship, as shown in Fig. 11. The value of ΔH_a was calculated from the graph using Eq. (2). Although a smaller than desirable temperature range was studied because of the stability limitations of the materials used to heat the substrate, the activation barrier was determined to be 12.6 kcal mol^{-1} ($= 52.75$ kJ mol^{-1}), which is in excellent agreement with values calculated from CVD research (13.0 kcal mol^{-1} = 54.43 kJ mol^{-1}) using the same reactant sources, substrate orientation and crystallographic face [20]. Although the growth rate was still strongly dependent on the deposition temperature, the decrease in the value of this activation barrier was an excellent indicator of the effect of the H_2 gas on the gas chemistry, growth kinetics and gas flow dynamics. Even though the rate limiting step and the controlling factor could not be ascertained, limited studies showed that varying the C_2H_4 input from 0.375 to 1.0 sccm did not result in a change in the growth rate, thus indicating that the rate controlling factor at these temperatures in the presence of H_2 had most likely changed. It is important to note that growth on both vicinal and on-axis substrates proceeded at approximately the same rate

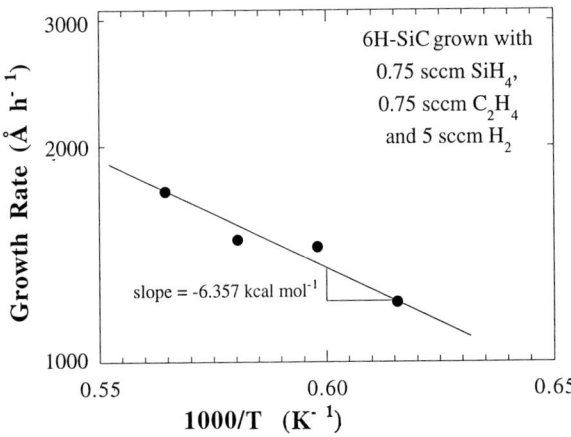

Fig. 11. Arrhenius plot of the growth rate for 6H-SiC(0001) films grown between 1350 and 1500 °C with 0.75 sccm SiH$_4$, 0.75 sccm C$_2$H$_4$ and 5 sccm H$_2$. Note that the apparent activation energy is lower than previously

(within 5%) under similar growth conditions, however, 3C-SiC was always produced on the on-axis substrates, regardless of growth conditions.

Sublimation growth of bulk single crystal SiC material represents the vapor phase formation process that is the closest to equilibrium. When performed in a moderate vacuum, boules of 6H-SiC can be reproducible grown at temperatures $\geq$1800 °C [49]. By introducing 10% H$_2$ to his processing environment, Rutz [50] observed an increase in the growth rate and a decrease in the temperature necessary to grow 6H-SiC to 1700 °C. Kimoto and Hatsunami [51] have reported that growth of 6H-SiC thin films can be achieved by "step-controlled" VPE at temperatures as low as 1200 °C under conditions where copious amounts of H$_2$ (ratios of H$_2$:reactants >2000:1) were present.

Data from the present study indicated a formation temperature which was intermediate to the two extremes reported by Tairov [49] and Kimoto and Matsunami [51]. In the foregoing paragraphs, the effect of H$_2$ was discussed on the basis of the production of a reactive species from the C$_2$H$_4$ source gas. When considering the various shifts in the stabilization regimes of the 6H polytype and the growth mechanism responsible for its formation, this work indicated that H$_2$ has an enhanced role in SiC deposition at higher temperatures (>1300 °C).

Although the exact nature of this effect cannot be determined, several plausible explanations and the consequence of this effect can be discussed. Firstly, at these conditions of temperature and reactant partial pressure, the thermodynamics may favor the decomposition of SiH$_4$ and/or C$_2$H$_4$ into intermediate species that are more reactive or form adducts on the surface (such as vinylsilane, H$_2$C=CHSiH$_3$ [52]) which results in both lower activation energy barriers and higher surface mobilities for adatom species. The result is that the surface species react and form species that are sufficiently mobile to reach the step incorporation sites, producing for 6H-SiC deposition via the step flow model.

Secondly, previous studies [53 to 56] of Ge deposition on Si(001) and (111) have suggested that surface hydrogen can act as a surfactant, thus changing the surface energy and the growth mode of the deposited epilayer. Although the behavior of H$_2$ on both Si and diamond surfaces has been well studied, very little study of the chemisorption of H$_2$ onto the surface of SiC onto the surface of SiC has been performed. At temperatures below 800 °C, the interaction between H$_2$ and SiC is negligible [57, 58], however, reactions between H_2 and SiC have been reported via observations of C removal and etching of high surface energy features [59 to 61] as well as H incorporation and passivation of dopants [62, 63]. When considering Si surfaces, the surface reactivity toward hydrocarbons can either be reduced by H capping of active sites [64] or enhanced by formation of hydrocarbon radicals [65]. In order to adequately describe both the increased reactivity observed with H_2 additions and the stabilization of the step flow mechanism (by disallowing incorporation except at step sites), both mechanisms may actually be present in the depositions of this research.

Finally, Powell et al. [66 to 68] have recently shown that the formation of 3C-SiC and the DPB defects in these layers may be more related to the presence of damage and dislocations than the density of step sites. These defects, which are related to polishing, oxidation, cutting and HF etching, produce irregularities on the substrate surface which serve as nucleation sites for the deposition of defective 3C-SiC layers. By removing these features with surface etching, 6H-SiC films could be grown on substrates with tilt angles as small as 0.1° off-axis. These researchers, in an attempt to confirm their hypothesis, intentionally introduced dislocations onto the surfaces of SiC by pressing the tip of a diamond scribe onto the substrate surface prior to growth. Although the mechanism remains unexplained, these dislocations apparently acted as 3C-SiC nucleation sites since 3C-SiC grew in locations where the diamond tip was applied. However, when in situ etching was performed as a cleaning procedure, the 3C-SiC nucleation sites were eliminated and only 6H-SiC resulted. Since all of the samples used during the course of the present study were cut into relatively small squares using a diamond wafering blade, the likelihood of inducing dislocations onto the surface of this material was high. Additionally, although the silane cleaning procedure that was used on each of the samples prior to growth was demonstrated to clean the SiC surface, dislocations and surface irregularities would most likely be unaffected by this cleaning procedure. However, at the higher growth temperatures, H_2 has been observed to slowly etch areas of high surface energy, thereby removing the damaged near-surface layers from the substrate [60]. Thus, the introduction of H_2 into the growth environment may facilitate the removal of these 3C-SiC nucleation sites and may provide a more defect-free growth plane, allowing only step flow growth to occur. As an additional consequence of this etching process, H_2 may etch some of the stray 3C-SiC nuclei that form during MBE growth in the high C/Si environments [14] in a manner analogous to that used to explain the role of H in diamond growth or produced a surface that is always Si-rich by preferential C removal [59, 61].

From the data provided, the rate determining step and the controlling mechanism for the deposition process could not be determined. The apparent change in the activation barrier indicates that the rate determining step has changed from the previously described cases, however, the nature of the new controlling process is unknown. Since changes in the C_2H_4 flow rate had no effect on the growth rate, and the deposition process was always performed in an atmosphere where C/Si $\geq$ 1, the new rate determining step may involve 1. the production or delivery of Si species to the crystal surface, 2. the adsorption of adatoms onto the surface or 3. the competition between finding surface incorporation sites and the desorption of unincorporated adatoms. However, the change in the rate controlling step coincides with the change in growth mode from island nucleation to step flow indicating a possible relation between the two observations.

Since the deposition of different polytypes of SiC is identical chemically (as observed by similar growth rates for 6H-SiC on vicinal substrates and 3C-SiC on on-axis wafers), the shift in the slope of the curve may be indicative of a process that is mass transport limited. Typically, reactions have a transition point between the surface reaction controlled and mass transport (or diffusion) controlled regimes. Generally, the slopes of the reaction rate curves on either side of this transition point are related to each other by a factor of two [69]. The relationship between the activation energy of the surface reaction controlled regime ΔH_a^{scr}, and the activation energy of the mass transport controlled regime ΔH_a^{mtc}, is given by

$$\Delta H_a^{\text{scr}} = 2\,\Delta H_a^{\text{mtc}}\,. \tag{7}$$

Since the calculated activation barriers between the regions where 6H- and 3C-SiC are formed are related by a factor of nearly 2 (12.6 versus 21.9 kcal mol^{-1} = 52.75 versus 91.69 kJ mol^{-1}), mass transport controlled growth may be present under conditions of high temperature in a hydrogen ambient. However, this rate controlling mechanism is generally attributed to the diffusion of reactants through a static boundary layer at the substrate surface. Since the employment of MBE necessitates the continuous flux of precursor species onto a heated substrate, thus negating the concept of a boundary layer, the interpretation of mass transport control must be adjusted for the present study and for MBE in general. While diffusion of reactants through the mixing zone where reactants and products as well as hydrogen are present may suffice, the mass transport process may be best explained by considering the nature of the source beams that are produced and their transport route to the substrate when the system is operating under only moderate ($\approx 10^{-2}$ Pa) vacuum conditions due to the hydrogen ambient. If this reaction can be considered mass transport controlled, this new definition of the transport/diffusion mechanism may explain the deviation from the predictions of Jones and Shaw [31] for the slope of Arrhenius curves describing mass transport controlled reactions. It is important to note that Kimoto et al. [70 to 72] has calculated activation barriers of 2.8 to 3.0 kcal mol^{-1} (11.72 to 12.56 kJ mol^{-1}) for VPE deposition of SiC in the mass transport limited regime ($T \geq 1200\,^{\circ}\text{C}$) using SiH$_4$, C$_2H_8$ and H$_2$.

3.4 Chemical and electrical characterization

3.4.1 Undoped 3C- and 6H-SiC films

Chemical and electrical analysis was performed on selected films of both 6H-SiC(0001) and 3C-SiC(111) thin films grown on 6H-SiC(0001) substrates. Unintentionally doped films of both 6H-SiH and 3C-SiC were grown using a range of Si-to-C flux ratios [17]. These undoped films were studied to determine the background levels of contaminants and carriers. Since even the most pure SiC grown from the vapor phase contains some quantity of N, it was not surprising that the major contaminant in these films was N. This impurity was particularly significant, because it is the most shallow donor impurity in SiC. Consequently, it was the agent responsible for the n-type character of unintentionally doped SiC films.

Although the growth rate was not observed to change when the C$_2$H$_4$ flow rate was modulated between 0.375 and 0.75 sccm while the SiH$_4$ flow was maintained at 0.75 sccm, considerable differences in the background atomic nitrogen and electron concentrations in the SiC films was measured. Similar to the "site-competition epitaxy"

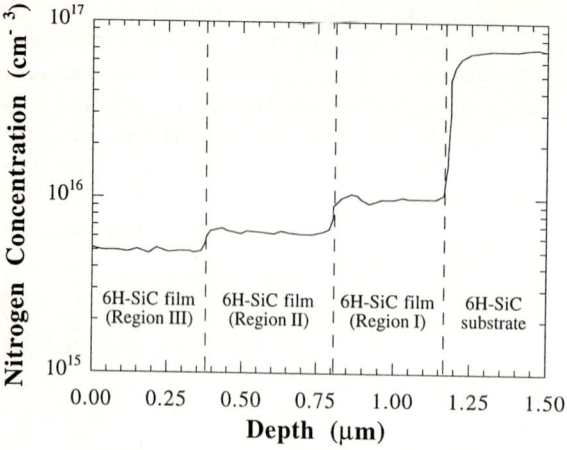

Fig.12. SIMS profile of 6H-SiC(0001) films grown at 1400 °C with 0.75 sccm SiH$_4$, 5 sccm H$_2$ and a variable C$_2$H$_4$ flow, 0.375 sccm in Region I, 0.5 sccm in Region II and 0.75 in Region III. Note the changes in the nitrogen incorporation that resulted from changing the ethylene input due to site-competition epitaxy between C and N

arguments of Larkin et al. [73 to 76], the N contamination was significantly decreased by increasing the amount of C$_2$H$_4$ delivered in the gas phase. Site-competition epitaxy is a technique based on appropriately adjusting the Si-to-C ratio within the growth reactor to effectively control the amount of dopant incorporated substitutionally into the SiC crystal lattice. The model of site-competition epitaxy is based on the principle of competition between N and C for the C sites and between Al and Si for the Si sites in the growing SiC epilayer. Thus, the concentration of N impurities incorporated into a growing epilayer is decreased by increasing the C source concentration, as the C successfully competes with N for the available C sites on the basis of the overwhelming partial pressure difference between the two components. Using this logic, the quantity of N in the SiC layer was expected to decrease as the amount of C source gas was increased.

Analysis by SIMS depth profiling and electrical measurements (C–V and Hall effect) were made to confirm the site-competition model and its relevance to MBE. Three different C$_2$H$_4$ flow rates (0.75, 0.5 and 0.375 sccm) were used. Fig. 12 shows a depth profiled 6H-SiC(0001) film grown at 1400 °C using 0.75 sccm SiH$_4$ and a variable C$_2$H$_4$ flow. The C$_2$H$_4$ flow was varied using the following schedule, 0.375 sccm for 3 h (Region I), 0.5 sccm for 3 h (Region II) and 0.75 sccm for 3 h (Region III). The total thickness of the film was ≈1.25 μm. The change in N content with C source supply is very apparent from the abrupt changes that occur in the N depth profile. A flow of 0.75 sccm C$_2$H$_4$ resulted in the incorporation of N at the detection limit for N (5×10^{15} cm^{-3}) in the SIMS system.

Table 1
Electrical properties of undoped 3C-SiC films

	C–V measurement	Hall measurement	
C$_2$H$_4$ flow (sccm)	$N_D - N_A$ (cm^{-3})	n (cm^{-3})	μ_n (cm^2 V^{-1} s^{-1})
0.375	9.1×10^{15}	5.3×10^{15}	608
0.5	3.7×10^{15}	9.8×10^{14}	681
0.75	9.2×10^{14}	4.6×10^{14}	772

Table 2

Electrical properties of undoped 6C-SiC films

	C–V measurement	Hall measurement	
C_2H_4 flow (sccm)	$N_D - N_A$ (cm^{-3})	n (cm^{-3})	μ_n (cm^2 V^{-1} s^{-1})
0.375	4.3×10^{15}	2.2×10^{15}	371
0.5	1.3×10^{15}	5.4×10^{14}	398
0.75	7.1×10^{14}	3.6×10^{14}	434

Several films of both 6H-SiC (grown on vicinal 6H-SiC) and 3C-SiC (grown on the on-axis 6H-SiC) were also grown at 1400 °C using the three different ethylene flow rates noted above. The results of C–V and Hall effect characterization are listed in Tables 1 and 2. The increase in the impurity and electron concentrations with decreasing C_2H_4 input confirmed the site-competition model for N incorporation by MBE. To our best knowledge, the mobility of 434 cm^2 V^{-1} s^{-1}, presented in Table 2 is the highest room temperature value ever reported for 6H-SiC.

Fig. 13 shows a SIMS depth profile of the concentrations of N, O and H present in the 6H-SiC film which exhibited the lowest electron concentration (grown with 0.75 sccm C_2H_4). This plot is indicative of the high-quality, high-purity SiC layers that can be grown using the MBE technique, since the O, H and N levels are at or below the background measurement level. Note that the CVD-grown p-type substrate epilayer has more residual N in it than the undoped MBE layer grown on it.

3.4.2 Doped C- and 6H-SiC films

After achieving high-quality films and the ability to control polytype by either growth conditions or substrate orientation, intentional doping of the MBE SiC epilayers was performed. Both n- and p-type doping were achieved using N and Al, respectively [17]. For ease and consistency, films of both polytypes with the same doping level were grown simultaneously. In this case, polytype control was established by controlling the orientation of the substrate crystal. Films of 6H-SiC were grown on vicinal substrates and

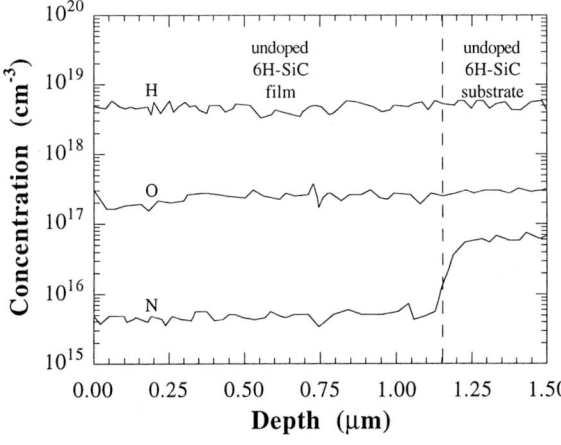

Fig. 13. SIMS profile of the 6H-SiC(0001) film with the highest mobility in Table 2 which was grown at 1400 °C with 0.75 sccm SiH$_4$, 0.75 sccm C$_2$H$_4$ and 5 sccm H$_2$. All contaminants (O, H and N) are at the background level of the instrument

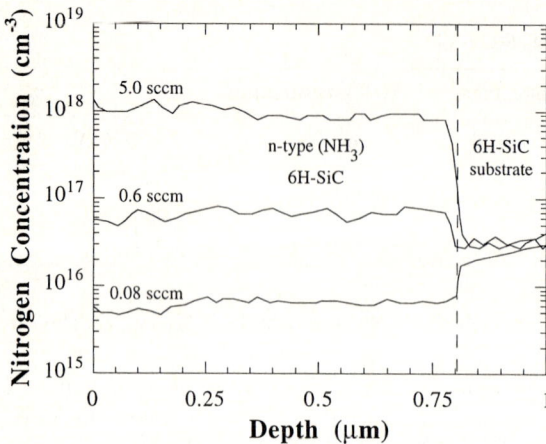

Fig. 14. SIMS profile for a series of n-type 6H-SiC(0001) films grown at 1400 °C with 0.75 sccm SiH_4, 0.375 sccm C_2H_4 and 5 sccm H_2. Films were doped with NH_3 using a mixture of NH_3/H_2

3C-SiC was grown on on-axis substrates. Below, the results of donor doping (with NH_3 and N_2) and acceptor doping (with Al) of both 6H-SiC(0001) and 3C-SiC(111) epilayers will be presented.

Despite the fact that unintentionally doped SiC films were n-type as grown, n-type doping was studied in order to establish a controllable electron population within the grown films. In order to perform this study, N was introduced controllably into the SiC lattice during growth of the SiC epilayers [17]. Since the quantity of N incorporated into the SiC lattice from residual gas molecules in the growth reactor was not a concern when n-type doping was performed, the reactant mixture with the lowest total input of source gases (0.75 sccm SiH_4, 0.375 sccm C_2H_4 and 5 sccm H_2) was chosen.

The thin films of 6H- and 3C-SiC were doped n-type by substituting an NH_3/H_2 mixture for some or all of the pure H_2 gas used in the growth of high-quality 6H- and 3C-SiC epilayers. Similar to the case for the undoped films, these films were grown on p-type substrates. Fig. 14 shows a SIMS plot of the N concentration versus depth for a series of n-type doped 6H-SiC films. The numbers on the plot indicate the gas mixture inplut flow supplied to the reactor. In cases where the NH_3/H_2 flow was <5 sccm, pure H_2 was added such that the sum of the flows of NH_3/H_2 and pure H_2 equaled 5 sccm. From the SIMS profile, it is evident that the epilayers could be doped controllably with N to concentration levels below that of the p-type CVD epilayer indicating the precision available using MBE. The results of $C–V$ and Hall effect (measured using the heavily-doped mesa contact described previously) characterization are listed in Tables 3 and 4. The increase in the atomic concentration of nitrogen and electron concentration with

Table 3

Electrical properties of n-doped (NH_3) 3C-SiC films

	$C–V$ measurement	Hall measurement	
NH_3/H_2 flow (sccm)	$N_D - N_A$ (cm^{-3})	n (cm^{-3})	μ_n (cm^2 V^{-1} s^{-1})
0.08	1.3×10^{16}	7.6×10^{15}	592
0.6	1.1×10^{17}	6.8×10^{16}	482
5.0	9.7×10^{18}	7.2×10^{17}	361

Table 4
Electrical properties of n-doped (NH_3) 6C-SiC films

	C–V measurement	Hall measurement	
NH_3/H_2 flow (sccm)	$N_D - N_A$ (cm^{-3})	n (cm^{-3})	μ_n (cm^2 V^{-1} s^{-1})
0.08	9.4×10^{15}	5.5×10^{15}	366
0.6	8.2×10^{16}	5.6×10^{16}	302
5.0	7.9×10^{17}	4.8×10^{17}	215

increasing NH_3 input confirms that the gas mixture was the dominant source of N incorporated in the films.

Although NH_3 was shown to be an attractive dopant for SiC epilayers, the accessibility of inputs ports on the system limited use to only one dilution which limited the range over which doping levels could be controlled. Thus, molecular nitrogen was added to the gas mixture (in addition to the SiH_4, C_2H_4 and H_2) to achieve higher doping levels than were achieved by using the NH_3/H_2 mixture. Similar to the case for the undoped films, these films were grown on p-type substrates to prevent parallel conduction. Thin films of 6H- and 3C-SiC were doped n-type by adding 0.5 and 5.0 sccm N_2 to the typcial reactant gas input (0.75 sccm SiH_4, 0.375 C_2H_4 and 5 sccm H_2) used in the growth of high-quality SiC epilayers. Fig. 15 shows a SIMS plot of the N concentration versus depth or the two different N_2 flows used to heavily dope 6H-SiC films n-type. The numbers in the plot indicate the quantity of N_2 applied into the reactor. It should be noted that 0.5 sccm was the lowest flow of gas that could be repeated controllably. The results of C–V and Hall effect characterization are listed in Tables 5 and 6. The increase in the electron concentration with increasing N_2 input confirms that the N_2 gas was the dominant source of N incorporated in the films.

Thin films of 6H- and 3C-SiC were doped p-type by evaporating Al from a standard MBE effusion cell during the growth of SiC epilayers [17]. All growth experiments were performed at 1450 °C using 0.75 sccm SiH_4, 0.75 sccm C_2H_4 and 5 sccm H_2. The higher C_2H_4 flow rate was used to take advantage of the site-competition process which resulted in a decrease in the concentration of background N, a compensating impurity in

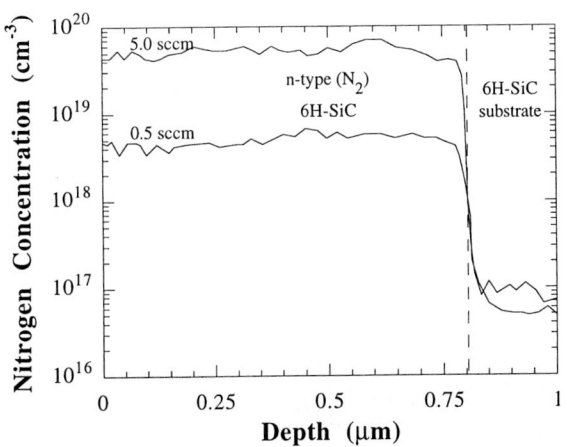

Fig. 15. SIMS profile for a series of n-type 6H-SiC(0001) films grown at 1400 °C with 0.75 sccm SiH_4, 0.375 sccm C_2H_4 and 5 sccm H_2. Films were doped by adding N_2 to the gas flow

Table 5

Electrical properties of n-doped (N_2) 3C-SiC films

N$_2$ flow (sccm)	C–V measurement	Hall measurement	
	$N_D - N_A$ (cm^{-3})	n (cm^{-3})	μ_n (cm^2 V^{-1} s^{-1})
0.5	3.0×10^{18}	1.9×10^{18}	308
5.0	4.0×10^{19}	3.6×10^{19}	121

Table 6

Electrical properties of n-doped (N_2) 6H-SiC films

N$_2$ flow (sccm)	C–V measurement	Hall measurement	
	$N_D - N_A$ (cm^{-3})	n (cm^{-3})	μ_n (cm^2 V^{-1} s^{-1})
0.5	2.2×10^{18}	1.5×10^{18}	157
5.0	3.1×10^{19}	2.3×10^{19}	88

p-type SiC. The higher temperature was used in an attempt to aid in dopant activation. Similar to the case for the undoped and n-type films, the intentionally doped films were grown on substrates of the opposite conductivity type. In this instance, the substrates used were n-type. Fig. 16 shows a SIMS profile of the Al concentration versus depth for a series of p-type doped 6H-SiC films. The numbers on the plot indicate the temperature of the effusion cell during the in situ doping of 6H- and 3C-SiC films. The results of C–V and Hall effect characterization measurements are listed in Tables 7 and 8. The

Table 7

Electrical properties of p-doped (Al) 3C-SiC films

Al temperature (°C)	C–V measurement	Hall measurement	
	$N_A - N_D$ (cm^{-3})	p (cm^{-3})	μ_p (cm^2 V^{-1} s^{-1})
770	3.1×10^{15}	2.1×10^{15}	31
840	2.6×10^{16}	1.7×10^{16}	25
920	4.3×10^{17}	3.0×10^{17}	19
1010	6.9×10^{18}	5.4×10^{18}	7

Table 8

Electrical properties of p-doped (Al) 6H-SiC films

Al temperature (°C)	C–V measurement	Hall measurement	
	$N_A - N_D$ (cm^{-3})	p (cm^{-3})	μ_p (cm^2 V^{-1} s^{-1})
770	4.3×10^{15}	2.6×10^{15}	86
840	3.4×10^{16}	2.3×10^{16}	69
920	4.9×10^{17}	3.5×10^{17}	56
1010	7.5×10^{18}	6.3×10^{18}	47

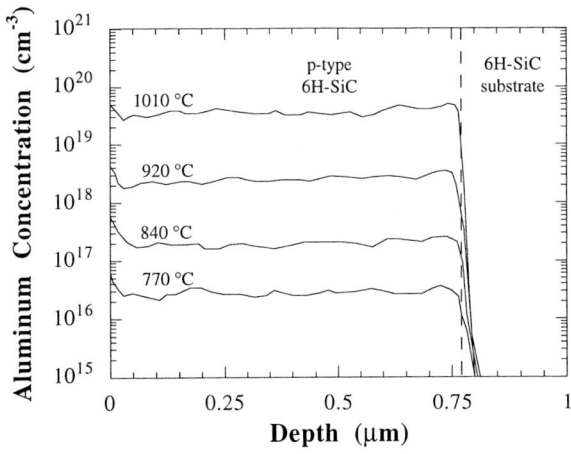

Fig. 16. SIMS profile for a series of p-type 6H-SiC(0001) films grown at 1450 °C with 0.75 sccm SiH$_4$, 0.75 sccm C$_2$H$_4$ and 5 sccm H$_2$. Films were doped with Al from a standard MBE effusion cell

increase in the hole concentration with increasing Al operating temperature indicates that the Al effusion cell temperature is the major factor affecting the Al incorporation and resultant hole population in the films.

4. Summary

Monocrystalline thin films of 3C-SiC(111) containing DPBs and stacking faults were grown on both vicinal and nominally on-axis 6H-SiC(0001) substrates at temperatures between 1000 and 1500 °C using SiH$_4$ and C$_2$H$_4$. While the surface morphology and growth rate were found to depend strongly on temperature, only the cubic polytype was formed. The various phases in the evolution of SiC film growth were also studied and presented.

Monocrystalline thin films of homoepitaxial 6H-SiC(0001) with low defect densities were deposited at high growth rates on vicinal 6H-SiC(0001) substrates by adding H$_2$ to the reactant mixture at temperatures between 1350 and 1500 °C. At temperatures below 1350 °C, only the cubic phase was formed, however, a small increase in the growth rate was achieved compared to films grown without the H$_2$ addition. Step bunching was also observed in the MBE grown epilayers, especially in the vicinity of micropipes in the SIC substrate. However, the addition of H$_2$ and the high deposition temperature stabilized the step flow growth process and only the 6H polytype was formed for growth at $\geq$1350 °C.

Kinetic analysis of the 3C-SiC(111) deposition process revealed that the growth rate was dependent on temperature and C$_2$H$_4$ flow rate. The rate limiting step in the process was demonstrated to be the decomposition of the C$_2$H$_4$ into suitable species to react with the cracking products of the SiH$_4$. Additions of H$_2$ resulted in an increasing growth rate at all temperatures. At lower temperatures ($<$1350 °C), only 3C-SiC was deposited; however, the addition of H$_2$ was shown to have an effect on the C$_2$H$_4$ decomposition into reactive radical species. At higher temperatures, the growth mechanism that controlled the growth rate was observed to change as indicated by a decrease in the activation barrier. High resolution TEM and SEM were employed to determine the location of the film/substrate interface. This allowed the accurate determination of the growth rate of the homoepitaxial SiC films.

Both 6H-SiC(0001) and 3C-SiC(111) epilayers were controllably doped in situ using NH_3, N_2 and Al. Undoped films were resistive with an n-type character; however, they had a lower N concentration than the p-type CVD-grown epilayers of the substrate. Undoped 6H-SiC films with the lowest atomic nitrogen and electron concentration had a mobility of $434 \, cm^2 \, V^{-1} \, s^{-1}$, the highest room temperature value ever reported for this polytype. Films were doped n-type using a NH_3/H_2 mixture (for lightly doped films) and using pure N_2 (for heavily doped SiC epilayers). Layers were also doped p-type using Al evaporated from a standard MBE effusion cell. Films of one conductivity type were always grown on substrates of the opposite conductivity type to prevent parallel conduction phenomenon and substrate effects. Substrate orientation was used to control the polytype of the doped films.

Acknowledgements The authors wish to acknowledge the Office of Naval Research for the sponsorships of this research under Grant. No. N00014-92-J-1500 and Cree Research, Inc. for providing the 6H-SiC(0001) substrates. K. Järrendahl acknowledges The Swedish Foundation for International Cooperation in Research and Higher Education (STINT) for financial support. We also would like to express our appreciation to R. G. Wilson (Hughes Research Laboratories) and Z. Radzimski (NCSU) for SIMS analysis, and A. Reisman (NCSU), H. H. Lamb (NCSU), C. Wolden (NCSU) and J. C. Angus (CWRU) for helpful discussions.

References

[1] S. Kaneda, Y. Sakamoto, T. Mihara, and T. Tanaka, J. crystal Growth **81**, 536 (1987).
[2] A. Fissel, B. Schröter, and W. Richter, Appl. Phys. Lett. **66**, 3182 (1995).
[3] A. Fissel, U. Kaiser, E. Ducke, B. Schröter, and W. Richter, J. Crystal Growth **154**, 72 (1995).
[4] A. Fissel, U. Kaiser, K. Pfennighaus, E. Ducke, B. Schröter, and W. Rcihter, in: Silicon Carbide and Related Materials 1995, Proc. of the sixth International Conf. Kyoto, Japan, Sept. 18–21, 1995. Eds. S. Nakashima, H. Matsunami, S. Yoshida, and H. Marima (p. 21).
[5] A. Fissel, U. Kaiser, K. Pfennighaus, B. Schröter, and W. Richter, Appl. Phys. Lett. **68**, 1204 (1996).
[6] A. Fissel, U. Kaiser, K. Pfennighaus, E. Ducke, B. Schröter, and W. Richter, Inst. Phys. Conf. Ser. No. 142, 121 (1996).
[7] T. Fuyuki, M. Nakayama, T. Yoshinobu, H. Shiomi, and H. Matsunami, J. Crystal Growth **95**, 461 (1989).
[8] T. Fuyuki, T. Yoshinobu, and H. Matsunami, Thin Solid Films **225**, 225 (1993).
[9] T. Yoshinobu, M. Nakayama, H. Shiomi, T. Fuyuki, and H. Matsunami, J. Crystal Growth **99**, 520 (1990).
[10] T. Yoshinobu, H. Mitsui, I. Izumikawa, T. Fuyuki, and H. Matsunami, Appl. Phys. Lett. **60**, 824 (1992).
[11] L. B. Rowland, PhD Thesis, North Carolina State University, Raleigh (NC), 1992.
[12] L. B. Rowland, R. S. Kern, S. Tanaka, and R. F. Davis, J. Mater. Res. **8**, 2753 (1993).
[13] S. Tanaka, PhD Thesis, North Carolina State University, Raleigh (NC), 1995.
[14] S. Tanaka, R. S. Kern, and R. F. Davis, Appl. Phys. Lett. **65**, 2851 (1994).
[15] R. S. Kern, PhD Thesis, North Carolina State University, Raleigh (NC), 1996.
[16] R. S. Kern, S. Tanaka, L. B. Rowland, and R. F. Davis, submitted to J. Crystal Growth.
[17] R. S. Kern and R. F. Davis, to appear in Appl. Phys. Lett.
[18] N. Kuroda, K. Shibahara, W. Yoo, S. Nishino, and H. Matsunami, in: Extended Abstracts 19th Conf. Solid State Devices and Materials, 1987 (p. 227).
[19] H. S. Kong, J. T. Glass, and R. F. Davis, J. Appl. Phys. **64**, 2672 (1988).
[20] Y. C. Wang and R. F. Davis, J. Electronic Mater. **20**, 869 (1991).

[21] A. I. KINGON, L. J. LUTZ, P. LIAW, and R. F. DAVIS, J. Amer. Ceram. Soc. **66**, 558 (1983).
[22] M. I. CHAUDHRY and R. L. WRIGHT, J. Mater. Res. **5**, 1595 (1990).
[23] M. I. CHAUDHRY, R. J. MCCLUSKEY, and R. L. WRIGHT, J. Crystal. Growth **113**, 120 (1991).
[24] G. B. KRUAVAL and J. D. PARSONS, J. Electrochem. Soc. **141**, 765 (1994).
[25] J. D. PARSONS and G. B. KRUAVAL, J. Electrochem. Soc. **141**, 771 (1994).
[26] O. KORDINA, Linköping University, private communication.
[27] A. REISMAN and M. BERKENBLIT, J. Electrochem. Soc. **113**, 146 (1966).
[28] S. KARMANN, C. HABERSTROH, F. ENGELBRECHT, W. SUTTROP, A. SCHÖNER, M. SCHADT, R. HELBIG, G. PENSL, R. A. STEIN, and S. LEIBENZEDER, Physica **185B**, 75 (1993).
[29] C. H. J. VAN DER BERKEL, Philips Res. Rep. **32**, 118 (1977).
[30] H. S. KONG, J. T. GLASS, and R. F. DAVIS, J. Mater. Res. **4**, 204 (1989).
[31] M. E. JONES and D. W. SHAW, in: Treatise on Solid State Chemistry, Ed. N. B. HANNEY, Plenum Press, New York 1975 (p. 283).
[32] R. J. BUSS, P. HO, W. G. BREILAND, and M. E. COLTRIN, J. Appl. Phys. **63**, 2808 (1987).
[33] F. BOZSO, J. T. YATES, JR., W. J. CHOYKE, and L. MUEHLHOFF, J. Appl. Phys. **57**, 2771 (1985).
[34] V. M. BERMUDEZ and R. KAPLAN, Phys. Rev. B **44**, 11149 (1991).
[35] H. FROITZHEIM, U. KÖHLER, and H. LAMMERING, J. Phys. C **19**, 2767 (1986).
[36] T. YOSHINOBU, H. TSUDA, M. ONCHI, and M. NISHIJIMA, Solid State Commun. **60**, 801 (1986).
[37] T. YOSHINOBU, H. TSUDA, M. ONCHI, and M. NISHIJIMA, J. Chem. Phys. **87**, 7332 (1987).
[38] C. C. CHENG, R. M. WALLACE, P. A. TAYLOR, W. J. CHOYKE, and J. T. YATES, JR., J. Appl. Phys. **67**, 3693 (1990).
[39] P. BADZIAG, Phys. Rev. B **44**, 11143 (1991).
[40] J. M. POWERS, A. WANDER, P. J. ROUS, M. A. VAN HOVE, and G. A. SOMORJAI, Phys. Rev. B **44**, 11159 (1991).
[41] B. I. CRAIG and P. V. SMITH, Surf. Sci. **276**, 174 (1992).
[42] M. N. PIANCASTELLI, M. K. KELLY, D. G. KILDAY, G. MARGARITONDO, D. J. FRANKEL, and G. J. LAPEYRE, Phys. Rev. B **35**, 1461 (1987).
[43] C. D. STINESPRING and J. C. WORMHOUDT, J. Appl. Phys. **65**, 1377 (1989).
[44] S. MOTOYAMA and S. KANEDA, Appl. Phys. Lett. **54**, 242 (1989).
[45] S. MOTOYAMA, N. MORIKAWA, M. NASU, and S. KANEDA, J. Appl. Phys. **68**, 101 (1990).
[46] T. HATAYAMA, Y. TARUI, T. FUYUKI, and H. MATSUNAMI, J. Crystal Growth **150**, 934 (1995).
[47] S. MOTOYAMA, N. MORIKAWA, and S. KANEDA, J. Crystal Growth **100**, 615 (1990).
[48] T. HATAYAMA, T. FUYUKI, and H. MATSUNAMI, Jpn. J. Appl. Phys. **34**, L1117 (1995).
[49] YU. M. TAIROV, Mater. Sci. Engng. B **29**, 83 (1995).
[50] R. F. RUTZ, U.S. Patent 3,577,285, May 4, 1971.
[51] T. KIMOTO and H. MATSUNAMI, J. Appl. Phys. **76**, 7322 (1994).
[52] S. TANAKA and H. KOMIYAMA, J. Amer. Ceram. Soc. **73**, 3046 (1990).
[53] M. COPEL and R. M. TROMP, Appl. Phys. Lett. **58**, 2648 (1991).
[54] N. OHTANI, S. M. MOKLER, M. H. XIE, J. ZHANG, and B. A. JOYCE, Surf. Sci. **284**, 305 (1993).
[55] K. SUMITOMO, T. KOBYAHI, F. SHOJI, K. OURA, and I. KATAYAMA, Phys. Rev. Lett. **66**, 1193 (1991).
[56] A. SAKAI and T. TATSUMI, Appl. Phys. Lett. **64**, 52 (1994).
[57] M. D. ALLENDORF and D. A. OUTKA, Surf. Sci. **258**, 177 (1991).
[58] Y. KIM and D. R. OLANDER, Surf. Sci. **313**, 399 (1994).
[59] A. A. BURK, JR. and L. B. ROWLAND, J. Crystal Growth, to be published.
[60] C. HALLIN, A. S. BAKIN, F. OWMAN, P. MÅRTENSSON, O. KORDINA, and E. JANZÉN, see [4] (p. 613).
[61] A. A. BURK, JR., L. B. ROWLAND, A. K. AGARWAL, S. SRIRAM, R. C. GLASS, and C. D. BRANDT, see [4] (p. 201).
[62] F. GENDRON, L. M. PORTER, C. PORTE, and E. BRINGUIER, Appl. Phys. Lett. **67**, 1253 (1995).
[63] B. CLERJAUD, F. GENDRON, C. PORTE, and W. WILKENING, Solid State Commun. **93**, 463 (1995).

[64] M. J. BOZACK, W. J. CHOYKE, L. MUEHLHOFF, and J. T. YATES, JR., J. Appl. Phys. **60**, 3750 (1986).
[65] M. J. BOZACK, P. A. TAYLOR, W. J. CHOYKE, and J. T. YATES, JR., Surf. Sci. **179**, 132 (1987).
[66] J. A. POWELL, D. J. LARKIN, L. G. MATUS, W. J. CHOYKE, J. L. BRADSHAW, L. HENDERSON, M. YOGANATHAN, J. YANG, and P. PIROUZ, Appl. Phys. Lett. **56**, 1353 (1990).
[67] J. A. POWELL, J. B. PETIT, J. H. EDGAR, I. G. JENKINS, L. G. MATUS, J. W. YANG, P. PIROUZ, W. J. CHOYKE, L. CLEMEN, and M. YOGANATHAN, Appl. Phys. Lett. **59**, 333 (1991).
[68] J. A. POWELL, D. J. LARKIN, J. B. PETIT, and J. H. EDGAR, in: Amorphous and Crystalline Silicon Carbide IV, Ed. C. Y. YANG, M. M. RAHMAN, and G. L. HARRIS, Springer-Verlag, Berlin 1992 (p. 23).
[69] G. F. FROMENT and K. B. BISCHOFF, in: Chemical Reactor Analysis and Design, John Wiley & Sons, New York 1990.
[70] T. KIMOTO, PhD Thesis, Kyoto University, Kyoto (Japan), 1995.
[71] T. KIMOTO, H. NISHINO, A. YAMASHITA, W. S. YOO, and H. MATSUNAMI, see [68] (p. 31).
[72] T. KIMOTO, A. YAMASHITA, A. ITOH, and H. MATSUNAMI, Jpn. J. Appl. Phys. **32**, 1045 (1993).
[73] D. J. LARKIN, S. G. SRIDHARA, R. P. DEVATY, and W. J. CHOYKE, J. Electronic Mater. **24**, 289 (1995).
[74] D. J. LARKIN, P. G. NEUDECK, J. A. POWELL, and L. G. MATUS, Appl. Phys. Lett. **65**, 1659 (1994).
[75] D. J. LARKIN, P. G. NEUDECK, J. A. POWELL, and L. G. MATUS, in: Silicon Carbide and Related Materials, Ed. M. G. SPENCER, R. P. DEVATY, J. A. EDMOND, M. A. KHAN, R. KAPLAN, and M. RAHMAN, Institute of Physics, Bristol 1994 (p. 51).
[76] D. J. LARKIN, see [4] (p. 23).

phys. stat. sol. (b) **202**, 405 (1997)

Subject classification: 68.55.Jk; 61.14.Hg; 68.35.Ct; S6

Simulations and Experiments of 3C-SiC/Si Heteroepitaxial Growth

M. KITABATAKE

Central Research Laboratories, Matsushita Electric Industrial Co. Ltd., Hikaridai 3-4, Seika-cho, Kyoto 619-02, Japan

(Received January 31, 1997)

Mechanistic reaction paths for the heteroepitaxial growth of 3C-SiC on carbonized Si(001) were investigated using a combination of molecular dynamics (MD) simulations and molecular beam epitaxy (MBE) experiments. Possible mechanisms of 3C-SiC heteroepitaxial growth on the Si(001) surface by carbonization was derived by MD [carbonization] simulation as the shrinkage of the [110] row of the Si lattice atoms with C adatoms. The stable Si-terminated 3C-SiC(001) surface was found by MD [surface] simulations to exhibit $h \times 2$ (where $h = \ldots, 7, 5, 3$ with increasing Si adatom coverage) reconstructions with Si adatoms on 3C-SiC(001) 2×1. The most stable surface structure is 3C-SiC(001)-Si 3×2 with a dangling bond density of 0.67 per 3C-SiC(001) 1×1 unit cell. Good quality layers were obtained by surface-structure-controlled epitaxy in which in-situ reflection high-energy electron diffraction was used as a feedback signal to adjust J_C/J_{Si} during growth to maintain a 3×2 surface reconstruction. A model involving asymmetric shrinkage and asymmetric growth kinetics parallel and perpendicular to step edges on a miscut substrate is presented to suppress the growth of antiphase boundaries (APB). 3C-SiC(001) 3×2 surface-structure-controlled epitaxial MBE growth on the miscut Si(001)-4°[110] substrate results in single-phase 3C-SiC with low density of APB.

1. Introduction

SiC is the only known solid-phase Si–C binary compound wide-bandgap semiconductor which exhibits a high breakdown field, a high electron saturated drift velocity, and good thermal conductivity [1, 2]. SiC_4 and CSi_4 tetrahedra are fundamental building blocks for all SiC polytypes, the most common of which are 3C (or β) zincblende-structure SiC, 4H SiC, 6H SiC, and 15R SiC [3]. Interest in the growth, microstructure, and optoelectronic properties of SiC thin films has increased markedly over the past several years driven by potential applications including high-temperature, high-power, high-frequency and radiation-hard transistors [4 to 7]. The electron mobility of high-purity undoped 3C-SiC has been postulated from theoretical calculations to be greater than that of other polytypes over the temperature range 300 to 1000 K due to a smaller contribution from phonon scattering [8].

Much of published literatures on epitaxial growth of 3C-SiC is concentrated on the use of Si(001) as a substrate [1, 9]. The lattice constant mismatch $\Delta a/a_{Si}$ between 3C-SiC ($a_{SiC} = 4.3589$ Å) and Si ($a_{Si} = 5.430$ Å) is enormous, 19.7% in tension. Most of the crystal growth experiments have been carried out using chemical vapor deposition (CVD) with CH_4 or C_3H_8 and SiH_4 as precursor gases [10], solid-source molecular-beam epitaxy (MBE) [11], or gas-source MBE using C_2H_2 or C_3H_8 and SiH_4 [12]. In almost all cases, the Si surface is first carbonized, following the initial work of Nishino et al. [10] using one of an array of empirically-developed recipes involving the manipulation of incident C-containing flux during heating of a Si wafer at various rates to the film growth temperature [9]. The carbonized interfacial region is a highly defective SiC layer with a typical thickness of several tens to several hundreds Å. Epitaxial SiC films grown on such interfacial layers contain high concentrations of microtwins, stacking faults, dislocations, and antiphase domains [13, 14]. The growth of very thick films, several tens of μm, has been shown to substantially decrease the number density of extended defects [15]. The primary residual defects in such thick layers are antiphase boundaries (APBs). Along the 3C-SiC[001] direction, the crystal is composed of alternating Si and C layers in which $\langle 110 \rangle$ $\overset{\text{Si}}{\underset{\text{Si}}{\diagup\diagdown}}$ and $\overset{\text{C}}{\underset{\text{C}}{\diagup\diagdown}}$ bond directions rotate by 90°. The intersection of coplanar $\overset{\text{C C}}{\underset{\text{C C}}{\diagup\diagdown}}$ and $\overset{\text{Si Si}}{\underset{\text{Si Si}}{\diagup\diagdown}}$ bonding chains along the $\langle 110 \rangle$ direction leads to the formation of APBs along {111} planes. Thus, the growth of 3C-SiC(001) on higher-symmetry diamond-structure substrates such as Si(001), whose surface consists of A and B terraces separated by single-atom-height steps [16], will lead to the formation of domains separated by APBs consisting of either Si–Si or C–C bonds. APBs can form on a given terrace, since the substrate provides no sublattice template, as well as across terrace boundaries as SiC islands coalesce. Uneven carbonization depths can also lead to the formation of APBs. It is also reported that the number density of APBs can be reduced through the use of Si(001) substrates miscut toward [110] [17, 18].

Reported epitaxial 3C-SiC(001) surface reconstructions include 1×1, c 2×2, 2×1, 3×2, 5×2, and 7×2 [1, 19 to 21]. The Si terminated 3C-SiC(001) surface is known to reconstruct and form 2×1 domains [22] in a manner similar to Si(001). The atomic structure of the stable 3C-SiC(001) 3×2 reconstruction was proposed by Dayan [23] as each set of two ad-atom dimer pairs along [110] was separated by a single missing ad-atom dimer. Hara et al. [21] reported observing, using scanning tunneling microscopy, 3×2, 5×2, and 7×2 surface composed of unified-additional-dimer rows on 3C-SiC(001) films grown on Si(001)-0.5°[110].

In this paper, we summarize the results of a series of investigations combining MD simulations and MBE experiments to probe reaction paths and mechanisms of SiC formation during carbonization [11] and 3C-SiC epitaxial growth on carbonized 3C-SiC/Si [24]. The MD [carbonization] [11] simulation results show that the reaction of incident C atoms with Si surface atoms causes Si–Si bond breaking between the (001) layers which in turn allows Si atomic row spacings to shrink along $\langle 110 \rangle$ directions in order to accommodate the smaller SiC bond lengths. The MD [surface] [24, 25] simulations show the stable surface reconstructions of 3C-SiC(001). The 2×1 Si-terminated 3C-SiC(001) surface consists of surface Si (s-Si) dimers. The adidition of Si adatoms (ad-Si) to the 3C-SiC(001) 2×1 surface results in the formation of a series of missing-dimer-row type reconstructions of $h \times 2$ where $h = \ldots, 7, 5, 3$ with increasing ad-Si coverage. The most energetically stable 3C-SiC(001) surface was found to be the 3×2 which has a dangling bond density of 0.67 per 3C-SiC(001) 1×1 unit cell.

A model involving the asymmetric $\langle 110 \rangle$ shrinkage during the carbonization and the asymmetric 3C-SiC$\langle 110 \rangle$ growth kinetics parallel and perpendicular to step edges on miscut Si(001) is presented. Film growth experiments carried out as a function of incident C/Si flux ratio show that good quality epitaxial SiC layers are obtained through growth under conditions of surface-structure-controlled epitaxy which maintain the 3×2 surface reconstruction during deposition. Surface-structure-controlled epitaxial growth of SiC following the carbonization of miscut Si(001) results in the formation of single-phase 3C-SiC with low APB density. Electron spin resonance (ESR) measurements show that the use of surface-structure-controlled growth on the miscut Si(001) substrate decreases APB-related dangling bond densities by more than two orders of magnitude [24].

2. MD Simulations

MD and quasidynamics (QD) simulations [26 to 29] employing the Tersoff many-body empirical potential [30], were used to calculate the structural change of the Si(001) surface with C adatoms (MD [carbonization] [11]) and stable surface structures for Si-terminated 3C-SiC(001) as a function of ad-Si atom coverage (MD [surface] [24, 25]). The Tersoff potential has been shown to provide good descriptions of energies and bond lengths in Si, SiC, diamond, graphite, and other C polytypes [30, 31]. In the QD simulations, the potential energy, velocity, and force associated with each atom in the computational cell, as well as the total potential energy and the total kinetic energy, are calculated for each time step and atomic configurations are allowed to relax. Atom positions and velocities are computed in a fully dynamic mode until the total kinetic energy of the ensemble reaches a maximum at which point the velocity of each atom is set to zero and the system is allowed to evolve further. This procedure is repeated until stable atom positions with approximately zero net force are obtained. Further details regarding the calculational procedure may be found in [11] and [26 to 29].

2.1 MD [carbonization] simulations [11]

The starting atomic configurations for the MD [carbonization] calculations included 800 Si-lattice atoms, eight layers of (001) planes with dimensions 10×10 atoms and 36 carbon atoms on the Si(001) surface. Two layers of Si(001) planes from the bottom were fixed and all other layers were free. Periodic boundary conditions were applied to the four {110} boundary planes. The Si(001) surface was not 2×1 reconstructed (dimerized) but 1×1 relaxed.

Fig. 1a shows the initial configurations of C adatoms and Si lattice atoms for the MD [carbonization] calculations. Each Si(001) 1×1 surface atom exhibited two dangling bonds which could capture adatoms. All dangling bonds of Si(001) surface atoms were aligned in the [110] direction indicated as a [110] row in the top view of Si(001) surface (Fig. 1a). 36C adatoms with dimensions 6×6 were placed on the Si(001) surface. Each C adatom was placed 0.89 Å ((001) layer spacing of SiC single crystal) above the middle point between the Si surface atoms aligned along the [110] direction ([110] row). The side view of one of the [110] $\overset{C}{\underset{Si \quad Si}{\diagup\diagdown}}$ row is also shown in Fig. 1a.

Fig. 1b shows the atomic configurations of the relaxed Si(001) surface with the C adatoms after 15000 steps of the MD [carbonization] calculation. In the carbonized region, the [110] rows of the (001) surface Si atoms shrink with C atoms. The side view of the third [110] $\overset{C}{\underset{Si \quad Si}{\diagup\diagdown}}$ row from the top in the top view is also shown in Fig. 1b. Some of the Si surface layer atoms (for example; atoms 1 and 2) are unconnected from the second layer Si atoms (3, 4 and 5, 6, respectively). The breaking of the Si–Si bonds between first and second layers enables the shrinkage of the [110] row of Si surface atoms with C atoms. Si atom 1 is displaced along the arrow in Fig. 1b by C atom α and makes new bonds with Si atoms 5 and 6 while atom 1 initially has bonds with Si atoms 3 and 4.

Two-layer two-dimensional configurations $\overset{C}{\underset{Si \quad Si}{\diagup\diagdown}}$ of C-SiC were formed under the orientation relationship of 3C-SiC[001] ‖ Si[001] and 3C-SiC[110] ‖ Si[110]. This relationship

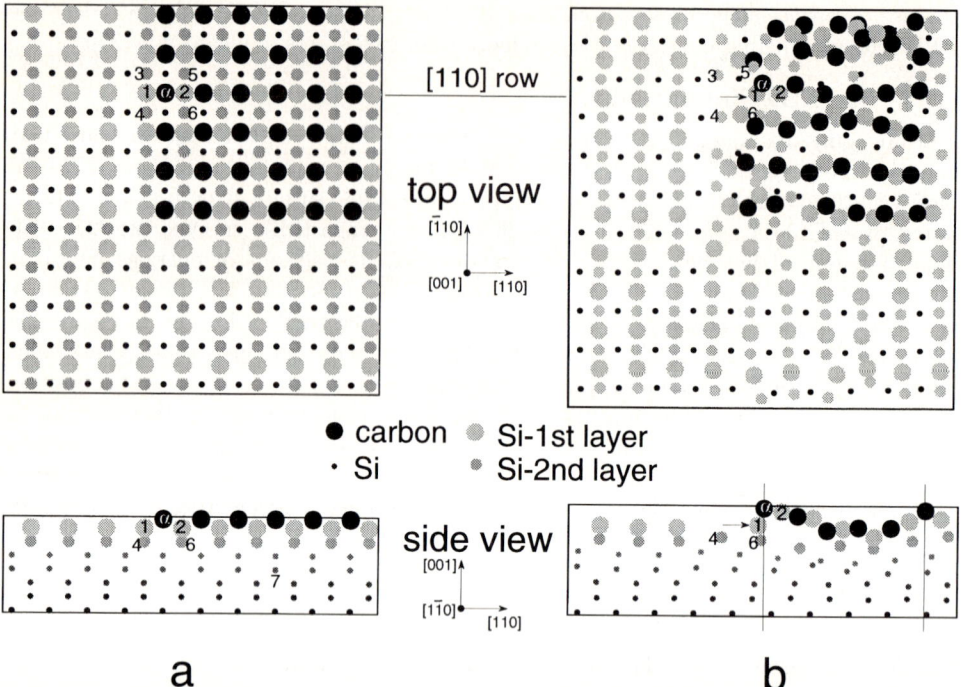

Fig. 1. Top and side view of the MD [carbonization] simulation results. a) Initial configurations of the C ad-atoms and the Si lattice atoms of the 1×1 Si(001) surface for the QD calculations. b) Atomic configurations of the relaxed Si(001) surface with the adsorbed C atoms after 15 000 steps of the QD calculation

between epitaxial 3C-SiC and the Si substrate agree with reported experimental results [1, 9, 10]. Five atom spacings of carbons in the SiC configurations coincide with four-atom spacings of the Si lattice as illustrated by vertical lines in Fig. 1b. This relationship agrees with the bond length ratio of the crystalline lattice; $Si/SiC = 2.35/1.89 \approx 5/4$. Possible heteroepitaxial growth mechanisms of 3C-SiC on the Si(001) surface are elucidated by the MD [carbonization] simulation as the breaking of the Si–Si bonds between the (001) layer atoms in the Si lattice and the shrinkage of the [110] row of the Si lattice atoms with the C adatoms.

2.2 MD surface simulations [24, 25]

The computational cell for MD [surface] included more than 800 3C-SiC lattice atoms, eight (001) layers of Si and C planes (four layers each) with dimensions 10×10 atoms together with, when appropriate, ad-Si atoms. The two lowest 3C-SiC(001) planes were fixed and periodic boundary conditions were applied to the four {110} boundary planes.

The Si-terminated 3C-SiC(001) surface is reconstructed to form 2×1 domains [22] in a manner similar to Si(001) with s-Si dimerization. The stable s-Si dimer length is calculated to be 2.49 Å. Energy gain upon s-Si dimerization on 3C-SiC(001) is 1.4 eV/dimer.

Ad-Si atoms were added to the Si-terminated 3C-SiC(001) 2×1 surface and came to rest at bridge positions between s-Si dimer rows. Further addition of ad-Si atoms re-

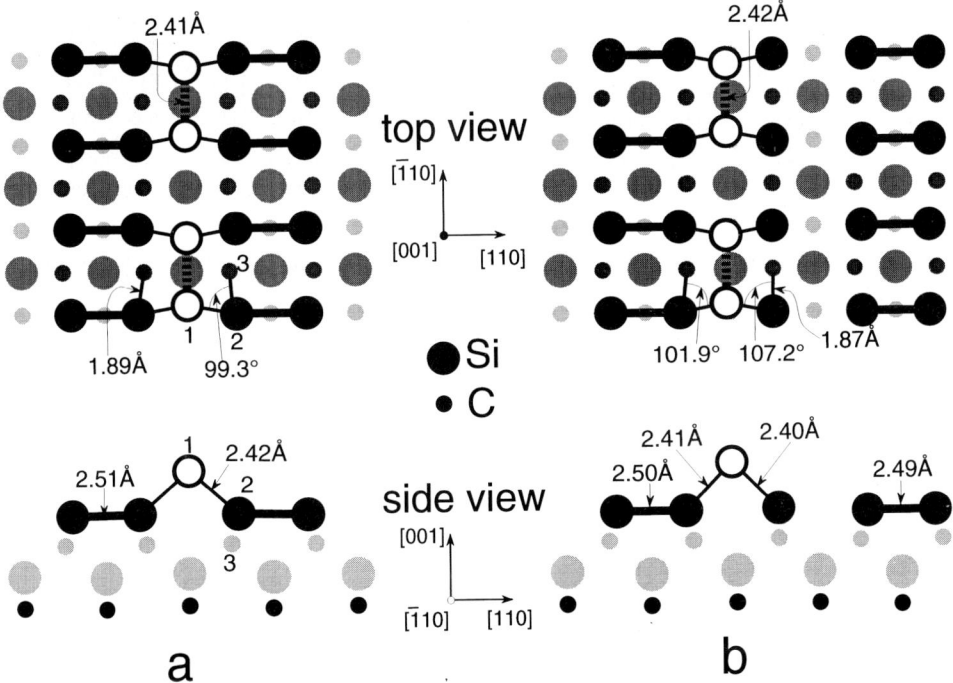

Fig. 2. Calculated atom positions after depositing ad-Si atoms, shown in white, on the Si-terminated 3C-SiC(001) surface. a) Symmetric ad-Si dimers and b) stable asymmetric ad-Si dimers. Black atoms are in the plane of the drawing while lighter grey atoms occupy successively deeper planes

sulted in the formation of ad-Si dimers rotated 90° with respect to the underlying 2×1 s-Si dimers as shown in Fig. 2a. The ad-Si dimer length and dimerization energy gain were calculated to be 2.41 Å and 1.6 eV/dimer, respectively. The energy gain upon ad-Si dimerization is larger than that upon s-Si dimerization.

The s-Si dimers attached to the ad-Si dimer are stretched and have a dimer bond length of 2.51 Å while the angle (see Fig. 2a), 99.3°, between, for example, the bond formed by ad-Si atom 1 with s-Si atom 2 and that formed by atom 2 and the C atom 3 in the next layer is highly strained compared to the relaxed bulk bond angle of 109.5°. The overall system consisting of adjacent ad-Si dimers on top of s-Si dimers on Si-terminated SiC(001) was allowed to relax and the lowest energy structure obtained is shown in Fig. 2b. Each of the ad-Si dimers become asymmetric as one of the attached s-Si dimers breaks and reforms out of phase. This partially relaxes the strain in the remaining s-Si dimers bonded to ad-Si atoms such that the s-Si dimer length decreases slightly to 2.50 Å with Si–C backbound angles of 101.9 and 107.2°.

Subsequent simulations revealed that, as ad-Si atoms are added, the total energy of the asymmetric structure in Fig. 2b can be further decreased through the formation of the two-ad-dimer unit shown in Fig. 3. In fact, the lowest-energy ad-Si surface structure formed was obtained by filling the Si-terminated 3C-SiC(001) surface with the close packed array of the two-ad-dimer units resulting in the 3×2 reconstruction (also shown in Fig. 3) in which each set of two ad-Si dimer pairs along [110] is separated by a single

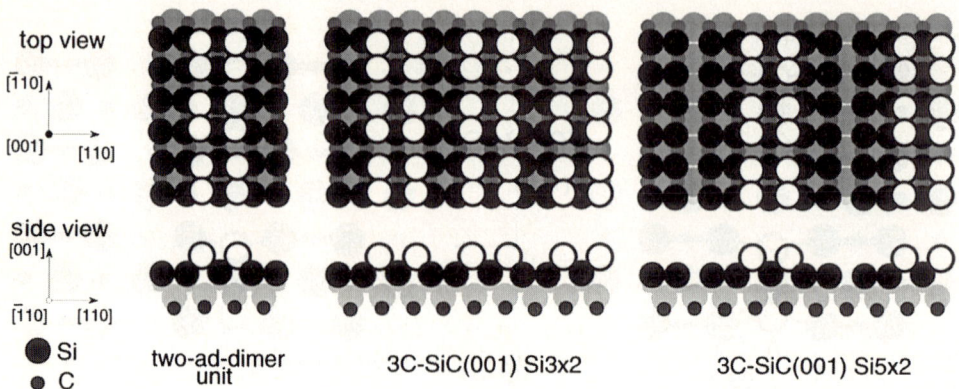

Fig. 3. Calculated relaxed atom positions for the two-ad-dimer unit on Si-terminated 3C-SiC(001) 2×1, the Si-terminated 3C-SiC(001) 3×2 surface, and the Si-terminated 3C-SiC(001) 5×2 surface. The atom grey scale code is as in Fig. 2

missing ad-Si dimer consistent with the model proposed by Dayan [23]. The ad-Si surface coverage for the 3×2 structure is 0.67. The two-ad-dimer unit (in Fig. 3) is also the basic building block for a family of higher-order less-dense $h \times 2$ surface superstructures where $h = 5, 7, \ldots$ The relaxed 5×2 surface is also shown in Fig. 3. This structure was constructed by filling the surface with the two-ad-dimer units, as in the 3×2 case, except that the underlying s-Si dimers do not overlap. Thus, there is a translation of one [110] lattice spacing between each two-ad-dimer unit giving rise to three missing ad-Si dimers between each set of ad-dimer pairs. The ad-Si coverage in this structure is 0.4. Similarly, separating neighboring two-ad-dimer units by a row of s-Si dimers (corresponding to a translation distance of three [110] lattice spacings) yields the 7×2 ad layer structure with five missing ad-Si dimers between each set of ad-Si dimer pairs. The adlayer coverage in this 7×2 structure is 0.29. These adatom structures of the 3C-SiC(001) $h \times 2$ surface reconstructions agree with those reported by Hara et al. [21]. The general expression for ad-Si coverage $\theta_{\text{ad-Si}}$ on $h \times 2$ reconstructed Si-terminated SiC(001) surfaces is

$$\theta_{\text{ad-Si}} = \frac{2}{n+2} = \frac{2}{h}, \tag{1}$$

where n is the number of missing ad-Si dimers between each ad-Si dimer pair. Two of the possible $(n+2) = h$ sites are filled by ad-Si dimer pairs in each $h \times 2$ superstructure as shown in Fig. 3. The structural models obtained from our MD simulations, Fig. 3, yield dangling bond densities θ_{db} for $h \times 2$ reconstructed surfaces of

$$\theta_{\text{db}} = \frac{2 + (n-1)}{n+2} = \frac{n+1}{n+2} = \frac{h-1}{h}. \tag{2}$$

That is, there are two ad-Si and $(n-1)$ s-Si dangling bonds for each set of $(n+2)$ possible sites. Thus, $\theta_{\text{db}} = 0.67$, 0.80, and 0.86 for 3×2, 5×2, and 7×2 surfaces in agreement with the reported order of stability [21] and the finding that 3×2 is the lowest energy ad-Si reconstruction [23].

2.3 Model for the heteroepitaxial 3C-SiC/Si growth

Based upon the MD [carbonization] simulations [11] (Section 2.1), the SiC/Si interface is expected to be locally abrupt as adsorbed C atoms break Si–Si backbonds in the upper layers, thereby allowing Si atomic row spacings to shrink along $\langle 110 \rangle$ directions in order to accommodate the smaller SiC bond lengths. The shrinkage of the [110] row of the Si lattice atoms resulted in the formation of the 3C-SiC[110] $\overset{C}{\underset{Si \quad Si}{\wedge}}$ row. The single-phase 3C-SiC growth with low APB density is expected when all of the carbonized 3C-SiC crystallites exhibit aligned 3C-SiC[110] $\overset{C}{\underset{Si \quad Si}{\wedge}}$ direction. Large asymmetry in terrace shape on the miscut Si(001) may cause the preferred shrinkage in the narrow terrace direction. The miscut Si(001) substrates have a possibility to promote the single-phase 3C-SiC/Si heteroepitaxial growth.

From the MD [surface] simulation results [24, 25] (Section 2.2), the substantial energy gain upon formation of ad-Si dimers and the stability of the 3×2 reconstructed 3C-SiC(001) surface suggest that growth under beam flux conditions which favor this surface reconstruction (surface-structure-controlled epitaxy) may well lead to higher quality films. This implies, for a given growth temperature, a minimum incident Si/C flux ratioa, below which mixed 3×2, 5×2, $7 \times 2 \ldots$ and/or 2×1 surface superstructures are formed.

3. Experimental Procedure

All films were grown in a three chamber MBE system. Substrates were introduced through a separately-pumped load-lock chamber, which was then evacuated to $\lesssim 1.0 \times 10^{-4}$ Pa, before being transported into the growth chamber which has a base pressure in the 10^{-8} Pa range. A separately-pumped analytical chamber (base pressure $\lesssim 6.0 \times 10^{-8}$ Pa) contains facilities for Auger electron spectroscopy (AES) and low-energy electron diffraction (LEED).

The growth chamber was equipped with an electron-beam evaporator, an effusion cell, a quadrupole mass spectrometer, and a reflection high-energy electron diffraction (RHEED) system. Substrate temperatures T_s were determined using both optical pyrometry and Pt–Rh thermocouples. During film growth, C was supplied primarily by electron-beam evaporation of 99.99% pure graphite chunks which had been thoroughly outgassed while Si was obtained by thermal evaporation of 99.99999999% pure Si chunks. Typical C and Si beam during steady-state 3C-SiC(001) growth at $T_s = 1050\,^\circ$C where $J_C \approx J_{Si} \approx 4 \times 10^{13}$ cm^{-2} s^{-1} corresponding to a deposition rate of 300 Å h^{-1}. The substrates used in these experiments were 14×14 mm^2 Si(001) or Si(001)-4°[110] plates cleaved from 0.5 mm thick n-type (resistivity $\approx 10\,\Omega$ cm) wafers.

Substrates were introduced as-received into the vacuum system, degassed at $500\,^\circ$C for 10 min, and rapidly heated to $>900\,^\circ$C for 30 s. RHEED patterns from thermally cleaned wafers were 2×1 with sharp Kikuchi lines and no residual C or O was detected by AES. Following cleaning, the substrates were allowed to cool to less than $200\,^\circ$C and then heated again to proceed the carbonization and SiC growth processes. The carbonization process was carried out with the Si(001) substrates heated at a linear ramp rate of 100 to 250 K min^{-1} with the evaporated C flux ($J_C \approx 4 \times 10^{13}$ cm^{-2} s^{-1}) which was added when T_s reached $400\,^\circ$C. When T_s reached $1050\,^\circ$C, the Si beam flux ($J_{Si} \approx 4 \times 10^{13}$ cm^{-2} s^{-1}) was initiated and 3C-SiC(001) growth was carried out at constant $T_s = 1050\,^\circ$C.

Microstructures and surface topographies of as-deposited layers were analyzed using a combination of X-ray diffraction (XRD), scanning electron microscopy (SEM), transmission electron microscopy (TEM), and cross-sectional TEM (XTEM). Room-temperature electron spin resonance (ESR) measurements were carried out on as-deposited 3C-SiC films.

4. SiC Growth on Nominally Singular Si(001)

Initial 3C-SiC MBE film growth experiments were carried out on nominally singular Si(001) substrates. During the carbonization process, the intensity and sharpness of the initial Si(001) 2×1 substrate RHEED pattern continuously decreases. Fundamental diffraction rods broaden as half-order rod intensities rapidly diminish to eventually disappear while diffuse scattering increases. At temperatures above approximately 900 °C during the carbonization process, weak diffuse 3C-SiC RHEED spots begin to appear.

During the 3C-SiC(001) growth, a streaky 3×2 pattern is obtained after about 50 Å of film growth with $J_C/J_{Si} \approx 1$. The fundamental diffraction rods are typically sharp and intense while the half-order rods, although sharp, are much weaker in the intensity and the one-third order rods are broadened but intense. The half-order rods are weaker because it is attributed to the ad-Si dimer which is possibly broken at high temperature. Formation of the mixed 3×2, 5×2, 7×2, ... surface superstructures on the 3C-SiC(001), as discussed in Sections 2.2 and 2.3, results in the broadened one-third order rods.

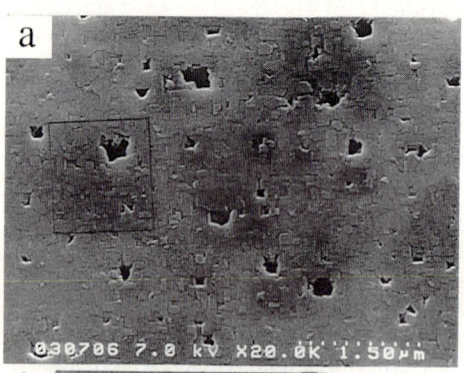

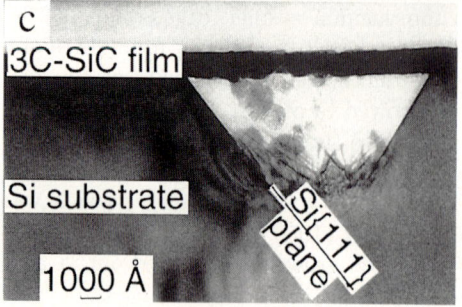

4.1 Pit formation below the SiC/Si interface

Once a 3C-SiC(001) 3×2 reconstruction is established during deposition, the surface remains in that structure for some period even if the incident deposition flux ratio J_C/J_{Si} is Si-deficient. This is due to the strong driving force for Si atoms from the substrate to diffuse to the growth surface and react with the excess C ad-atom

Fig. 4. a) Plan-view low-resolution SEM, b) higher-resolution SEM, and c) XTEM micrographs of a pit formed at the 3C-SiC/Si interface during carbonization and 3C-SiC film growth on Si(001) under C-rich conditions

population. However, continued growth in this mode leaves pits [32] which originate
below the 3C-SiC/Si interface. Low and higher resolution SEM and XTEM micrographs
illustrating such pits in a 1500 Å thick 3C-SiC film are shown in Figs. 4a, b, and c,
respectively. The pits, which nucleate between 3C-SiC crystallites during the early stage
of carbonization in order to supply Si to the 3C-SiC growth surface, expand from the
3C-SiC/Si interface into the Si substrate during subsequent 3C-SiC growth under Si-
deficient conditions. The XTEM image in Fig. 4c shows that the pits are bounded by
close-packed Si{111} planes in order to minimize the dangling bond density. With continued film growth, the pits are eventually laterally overgrown, but often with SiC crystallites having orientations different than that of the bulk film.

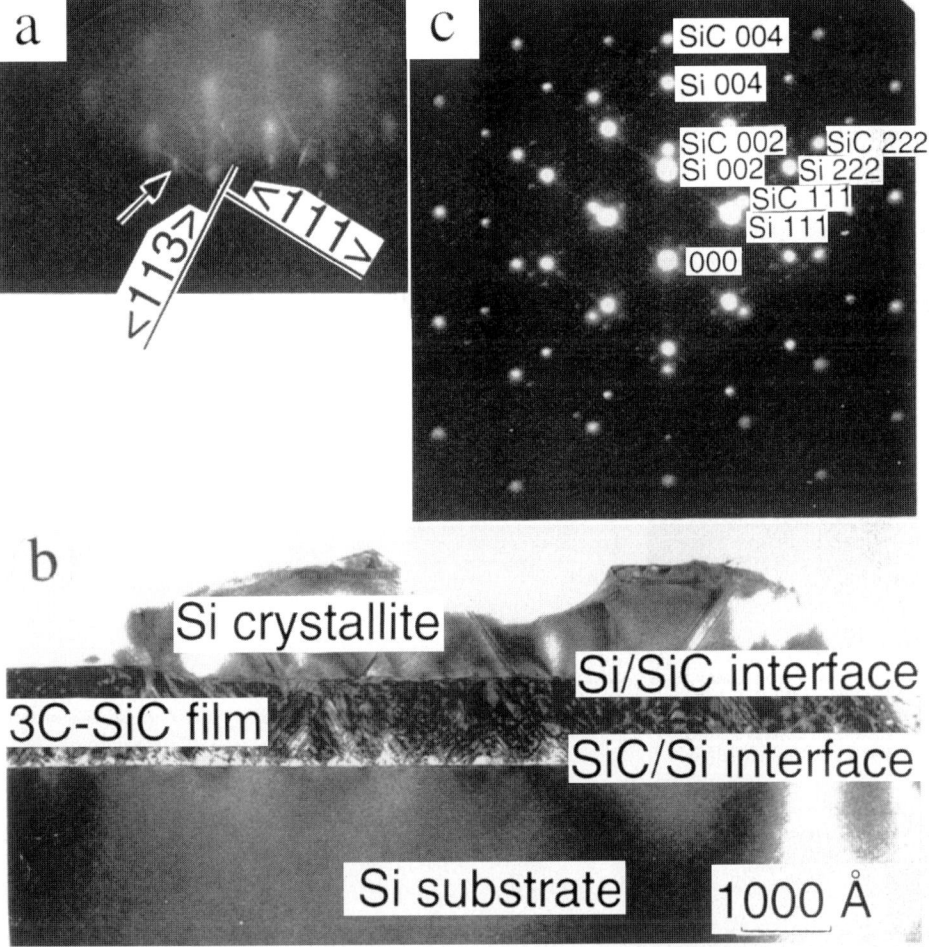

Fig. 5. a) RHEED pattern, b) XTEM micrograph, and c) SAED pattern from an ≈1000 Å thick
3C-SiC film grown on Si(001) under Si-rich conditions. The RHEED pattern contains Si reflections
with ⟨111⟩ and ⟨113⟩ streaks superimposed on a 3C-SiC(001) structure. The XTEM and SAED
pattern show the growth of an epitaxial Si crystallite on top of the 3C-SiC layer

4.2 Si crystallite formation on the 3C-SiC(001) surface

In contrast, film growth under conditions in which the J_C/J_{Si} flux ratio is significantly less than the value required to substrain the optimum 3×2 RHEED pattern leads to the formation of second-phase Si crystallites through precipitation of excess Si at the growth surface. The first indication of this occurring is the appearance of three-dimensional Si diffraction spots (indicated by an arrow in Fig. 5a) superimposed on the SiC(001) 3×2 RHEED pattern. Further growth in this mode gives rise to $\langle 111 \rangle$ and $\langle 113 \rangle$ streaks in the RHEED pattern as shown in Fig. 5a. The Si crystallites are bounded by $\{111\}$ and $\{113\}$ facet planes as would be expected for competitive growth in which the (001)-oriented crystallites serve as a sink for the local excess Si population which is rapidly depleted [33]. Fig. 5b is an XTEM micrograph showing a Si crystallite on an ≈ 1000 Å thick 3C-SiC film. A corresponding selected-area diffraction pattern from a region including both the 3C-SiC film and the Si crystallite is presented in Fig. 5c. Indexing the pattern shows that the crystallite grew heteroepitaxially on 3C-SiC(001) with an orientation relationship given by Si[001] ∥ 3C-SiC[001] and Si[110]$_{Si}$ ∥ 3C-SiC[110].

4.3 Surface-structure-controlled 3C-SiC(001) 3 × 2 growth

The results of film microstructure investigations under $J_C/J_{Si} > 1$ in Section 4.1 and under $J_C/J_{Si} < 1$ in Section 4.2 suggest the problems of pit formation below the 3C-SiC/Si interface and Si crystallite formation on the 3C-SiC(001) surface, respectively. Based upon the results of the MD [surface] simulations presented in Section 2.2 showing that the 3×2 reconstruction is the lowest energy 3C-SiC(001) surface, we have carried out a series of experiments (surface-structure-controlled 3C-SiC(001) 3×2 growth) in which in-situ RHEED was used as a feedback signal to control J_C/J_{Si} in order to maintain a sharp 3×2 pattern during 3C-SiC(001) growth on Si(001). During deposition, the Si flux was maintained constant while the evaporated C flux was continuously adjusted.

Fig. 6a, b, and c are typical RHEED, LEED and plan-view SEM results obtained from the 3×2 surface-structure-controlled 3C-SiC(001) film with a thickness of 1500 Å. The RHEED and LEED patterns are both 90°-rotated $3 \times 2 + 2 \times 3$ structures with no

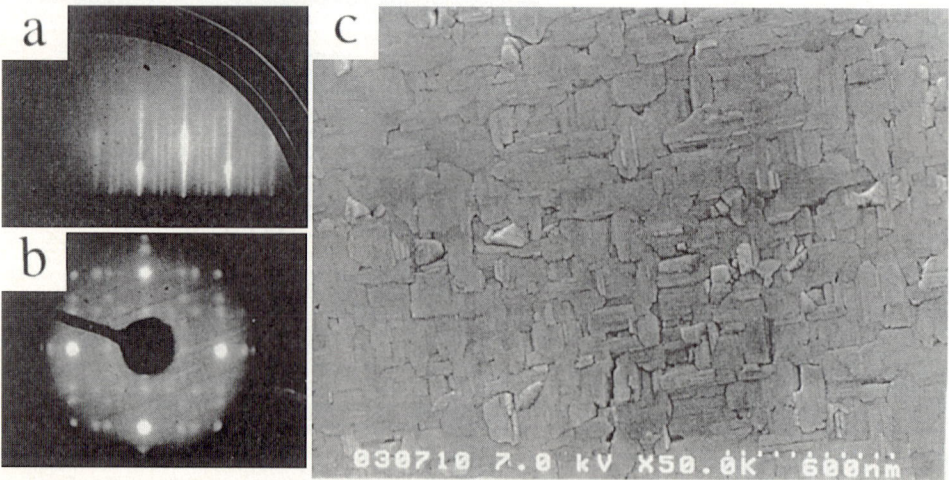

Fig. 6. Tyical a) RHEED, b) LEED and c) plan-view SEM results obtained from a 3C-SiC(001) film grown on normally singular Si(001) by 3×2 surface-structure-controlled epitaxy

extra reflections due to faceting or crystallite formation. All RHEED diffraction rods, including fundamental, 1/2-order, and 1/3-order ones, are sharper and considerably more intense than those observed during carbonization. The intensities of the $3 \times 2 + 2 \times 3$ LEED spots are symmetrically distributed. The SEM micrograph in Fig. 6c also shows no evidence of interfacial pitting and Si crystallite formation.

The antiphase domains in 3×2 stabilized films, observed in Fig. 6c are asymmetric, in contrast to the relatively square domains (see, for example, Fig. 4b) commonly observed under excess-C growth conditions leading to interfacial pit formation. The asymmetric domains in the 3×2-stabilized films have an average length to width ratio along $\langle 110 \rangle$ directions of ≈ 3 and an average dimension along the long axis of ≈ 3000 Å. The domain anisotropy indicates the existence of preferential $\langle 110 \rangle$ growth directions on 3×2 terraces during surface-structure-controlled epitaxy. In zincblende-cubic 3C-SiC, (001)Si and C planes alternate such that the $\overset{\text{Si}}{\underset{\text{C} \quad \text{C}}{\wedge}}$ bonding direction is along $[1\bar{1}0]$ then the $\overset{\text{C}}{\underset{\text{Si} \quad \text{Si}}{\wedge}}$ bonding direction in the next (001) plane down is rotated 90° and along [110]. The marked asymmetry in antiphase domain shape suggests that growth in this Si-rich mode proceeds along different mechanistic pathways than for C-rich 3C-SiC growth. That is, the ad-Si atoms in the 3×2 surface Si adlayer — which is continuously resupplied from the incident Si vapor beam — serve as the Si supply flux to bond with incident C atoms causing growth to occur more rapidly in the $\overset{\text{Si}}{\underset{\text{C} \quad \text{C}}{\wedge}}$ $[1\bar{1}0]$ direction than the $\overset{\text{C}}{\underset{\text{Si} \quad \text{Si}}{\wedge}}$ [110] direction. At $T_s = 1050\,°\text{C}$, the ad-Si atoms are very mobile and immediately cover adsorbed incident C atoms such that the surface remains Si terminated. This results in preferential growth in the $\overset{\text{Si}}{\underset{\text{C} \quad \text{C}}{\wedge}}$ bonding direction leading to the pronounced domain anisotropy observed. In contrast, growth under Si deficient conditions results in excess surface C which enhances the growth rate along the $\overset{\text{C}}{\underset{\text{Si} \quad \text{Si}}{\wedge}}$ bonding direction, reduces the $\langle 110 \rangle$ growth anisotropy, and gives rise to the more symmetric domains — corresponding to essentially equal probability of growth along $\overset{\text{Si}}{\underset{\text{C} \quad \text{C}}{\wedge}}$ and $\overset{\text{C}}{\underset{\text{Si} \quad \text{Si}}{\wedge}}$ bonding directions — shown in the SEM micrograph in Fig. 4b.

5. Single-Phase 3C-SiC(001) 3×2 Growth on Si(001)-4°[110]

In this section, single-phase 3C-SiC/Si heteroepitaxial growth is investigated using the MD simulation results and the 3×2 surface-structure-controlled growth technique. As discussed in Section 2.3, miscut Si(001) 2×1-4°[110] substrates was introduced in order to form aligned carbonized 3C-SiC crystallites and grow single-phase 3C-SiC with low density of APBs. The average terrace length, bounded by double-height steps [34, 35], of this substrate surface was ≈ 40 Å (10 atoms). Due to the large asymmetry in terrace shape, it is energetically easier for the shrinkage to occur in the narrow terrace direction perpendicular to step edges than along the extended direction parallel to step edges in the carbonization process. Thus, the initially formed 3C-SiC is C-terminated and preferentially forms with the $\overset{\text{C}}{\underset{\text{Si} \quad \text{Si}}{\wedge}}$ bonding direction along [110] as shown schematically in Fig. 7. The carbonization step described above, in turn, biases the direction of 3C-SiC island growth during subsequent 3×2 surface-structure-controlled epitaxy which proceeds in a Si-rich environment along $\overset{\text{Si}}{\underset{\text{C} \quad \text{C}}{\wedge}}$ bonding directions, parallel to $[1\bar{1}0]$ step edges

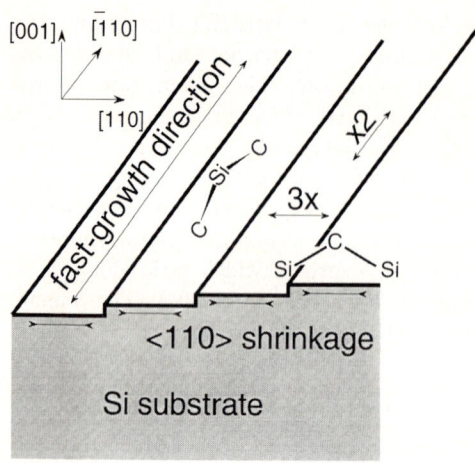

Fig. 7. Schematic diagram illustrating bonding, shrinkage (in carbonization process), and fast-growth directions (during 3×2 surface-structure-controlled epitaxy) of 3C-SiC on miscut Si(001)

(Fig. 7). This anisotropy favoring Si-terminated growth parallel to step edges greatly suppresses APB formation as demonstrated by the typical steady-state RHEED patterns shown in Fig. 8a obtained along [1$\bar{1}$0] and [110] azimuths. The [1$\bar{1}$0] pattern contains only fundamental and one-third-order diffraction rods while the [110] pattern consists of only fundamental and one-half-order streaks. Similarly, the LEED pattern in Fig. 8b shows a single-domain 3×2 surface structure with no evidence of the antiphase 2×3 reflections obtained from films grown on nominally-singular substrates (see Fig. 6b). In the (004) XRD pole figure from the same 1000 Å thick 3C-SiC layer, the film peak intensity occurs at the same position as that of the substrate, tilted $\omega = 4°$ off the 004 pole toward [110]. Thus, the orientation relationship between the film and miscut-Si substrate is identical to that on the singular surface, 3C-SiC[001] ∥ Si[001] with 3C-SiC[110] ∥ Si[110]. The full width at half maximum intensity of the (004) ω-rocking curve along [110] is 1.64°.

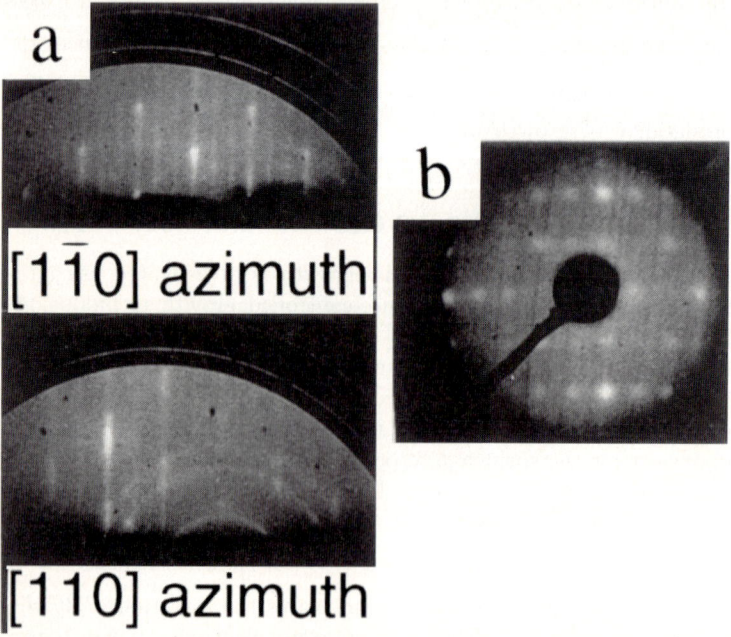

Fig. 8. a) RHEED patterns along the [1$\bar{1}$0] and [110] azimuths and b) LEED pattern obtained from a 3C-SiC(001) film grown on miscut Si(001)-4°[110] by 3×2 surface-structure-controlled epitaxy

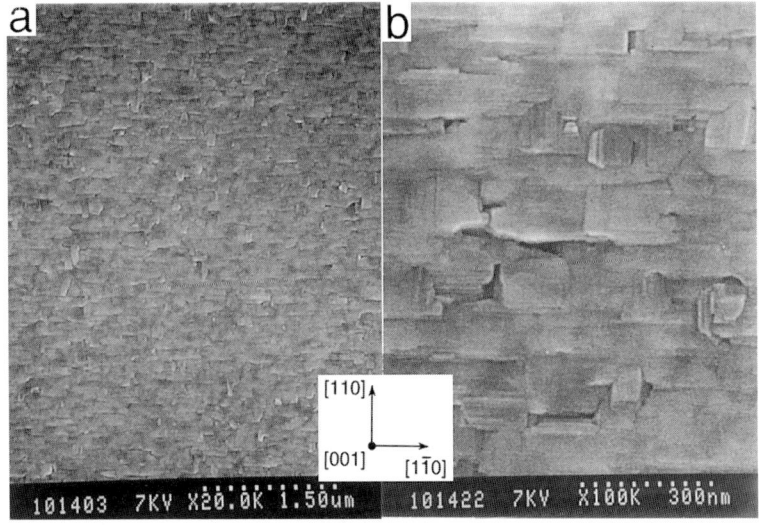

Fig. 9. a) Low-resolution SEM and b) higher-resolution SEM micrographs from a 1000 Å 3C-SiC layer grown by 3×2 surface-structure-controlled epitaxy on miscut Si(001)-4°[110]

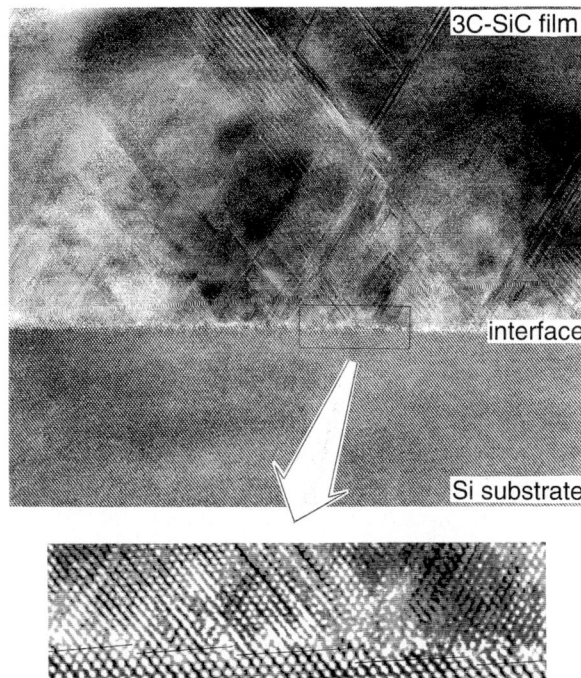

Fig. 10. Typical $[1\bar{1}0]$ XTEM image of the interfacial region of a 3C-SiC(001) film grown by 3×2 surface-structure-controlled epitaxy on Si(001)-4°[110]

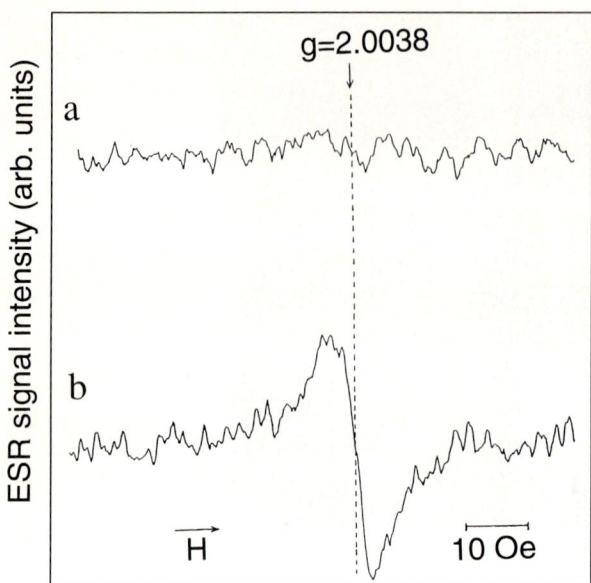

Fig. 9a and b are low and higher-resolution SEM micrographs from a relatively thin 1000 Å 3C-SiC layer grown by 3×2 surface-structure-controlled epitaxy on miscut Si(001)-4°[110]. The micrographs show evidence of island coalescence along the fast growth [$\bar{1}$10] direction (see particularly Fig. 9b) while average [110] terrace widths have increased from $\approx$40 Å for the underlying Si(001)-4°[110] substrate to more than 1000 Å. The [$\bar{1}$10] step edges remain relatively straight over very long distances.

The high-resolution [1$\bar{1}$0] XTEM in Fig. 10 reveals that the interface along the [110] narrow terrace direction is smooth and flat. The film/substrate interface appears abrupt with no evidence of pitting. A series of low resolution micrographs, each imaging lateral distances >3 μm, also showed no evidence of interfacial pitting. The misfit dislocation density along the 3C-SiC/Si interface was found to be $\approx 6.5 \times 10^6$ cm^{-1}, corresponding to five [110] SiC lattice spacings for each four [110] Si spacings. However, [110] XTEM micrographs show that along the [$\bar{1}$10] fast growth direction, the interface is rougher (not shown) due to surface strain associated with Si–Si bond shrinkage during carbonization.

The ESR signal intensity from 150 nm thick surface-structure-controlled 3C-SiC(001) films grown on miscut Si(001)-4°[110] and nominally-singular substrates are plotted in Figs. 11a and b, respectively. The strong ESR feature at $g = 2.0038$ is indicative of the existence of Si-dangling bonds with C atom neighbors [36]. While such defects can arise from a variety of sources including the 3C-SiC/Si interface, we believe that the primary source is defects in 3C-SiC APBs. The intensity of the $g = 2.0038$ feature in Fig. 11b corresponds to an unpaired spin density of $\approx 1.6 \times 10^{18}$ cm^{-3} assuming a uniform distribution in the 3C-SiC film. This result agrees well with reported data from 3C-SiC films grown on singular Si(001) by low-pressure CVD [37]. For the 3×2 surface-structure-controlled MBE film grown on miscut Si, however, the ESR signal at $g = 2.0038$ is within the measurement noise indicating that the unpaired spin density has been reduced to less than 1×10^{17} cm^{-3} consistent with SEM and XTEM results revealing a dramatic decrease in the number density of antiphase domains.

6. Conclusions

MD [carbonization] simulations elucidate possible heteroepitaxial growth mechanisms of 3C-SiC on a Si(001) surface by carbonization as the breaking of Si–Si bonds between the (001) layer atoms in the Si lattice and the shrinkage of the [110] row of the Si lattice atoms with C adatoms.

MD [surface] simulations were used to show that the adidition of Si adatoms on a Si-terminated SiC(001) 2×1 surface results in the formation of a series of reconstructions of the type $h \times 2$ where $h = \ldots, 7, 5, 3$ with increasing ad-Si coverage. The basic building block for the $h \times 2$ ad-Si reconstruction is a two-ad-dimer unit. The lowest-energy 3×2 ad-Si surface structure is obtained by filling the Si-terminated 3C-SiC(001) surface with the close-packed array of these units. The 5×2 and 7×2 superstructures are formed by separating each two-ad-dimer unit by one and three [110] lattice spacings, respectively. The present results demonstrate that the most energetically stable 3C-SiC(001) surface being the 3×2 which has a dangling bond density of 0.67 per 3C-SiC(001) 1×1 unit cell. The higher-order 5×2 and 7×2 ad-Si reconstructions have dangling bond densities of 0.80 and 0.86.

Based upon the MD [surface] simulation results, 3C-SiC films were grown at 1050 °C by 3×2 surface-structure-controlled epitaxial MBE growth using the intensity of one-third-order RHEED diffraction rods as the feedback element to control the J_C/J_{Si} flux ratio. Growth with 3×2 RHEED feedback virtually eliminated interfacial pit and Si crystallite formation and provided films of high structural quality with relatively abrupt 3C-SiC/Si interfaces and flat film surfaces.

The MD [carbonization] results suggest that the antiphase domain volume was greatly decreased through the use of [110] miscut substrates. During the carbonization step, 3C-SiC is formed under excess-C conditions through Si–Si bond breaking and shrinking along $\langle 110 \rangle$. Due to the large asymmetry in terrace shape, it is energetically easier for the shrinkage to occur in the narrow terrace direction perpendicular to step edges than along the extended direction parallel to steps edges. Thus, 3C-SiC forms preferentially with the $\overset{\text{C}}{\underset{\text{Si}\quad\text{Si}}{\diagup\diagdown}}$ bonding direction perpendicular to step edges. This process, in turn, biases the direction of 3C-SiC island growth during subsequent 3×2 surface-structure-controlled epitaxy, which proceeds in a Si-rich environment, along $\overset{\text{Si}}{\underset{\text{C}\quad\text{C}}{\diagup\diagdown}}$ bonding directions parallel to step edges on miscut substrates. The anisotropy favoring Si-terminated growth parallel to steps greatly suppresses APB formation. ESR results show that the density of Si dangling bond defects are reduced to less than the detection limit, $\approx 1 \times 10^{17}$ cm^{-3}.

Acknowledgements The author appreciates the support of the National Center for Supercomputing Applications at the University of Illinois, USA. Part of this work was supported by NEDO (New Energy and Industrial Technology Development Organization) Japan. MBE in Ion Engineering Center (Osaka, Japan) was used for the 3C-SiC growth experiments.

References

[1] R. F. DAVIS, G. KELNER, M. SHUR, J. W. PALMOUR, and J. A. EDMOND, Proc. IEEE **79**, 677 (1991).
[2] R. F. DAVIS, Thin Solid Films **181**, 1 (1989).
[3] Y. M. TAIROV and Y. A. VODAKOV, Topics Appl. Phys. **17**, 31 (1977).
[4] E. O. JOHNSON, RCA Rev. **26**, 163 (1965).
[5] R. W. KEYES, Proc. IEEE **60**, 225 (1972).
[6] B. J. BALIGA, IEEE Electron Device Lett. **10**, 455 (1989).
[7] M. BHATNAGAR and B. J. BALIGA, IEEE Trans. Electron Devices **40**, 645 (1993).
[8] P. DAS and D. K. FERRY, Solid State Electronics **19**, 851 (1976).
[9] K. NISHINO, T. KIMOTO, and H. MATSUNAMI, Mem. Fac. Engng. Kyoto Univ. **54**, 299 (1992).
[10] S. NISHINO, J. A. POWELL, and H. A. WILL, Appl. Phys. Lett. **42**, 460 (1983).
[11] M. KITABATAKE, M. DEGUCHI, and T. HIRAO, J. Appl. Phys. **74**, 4438 (1993).
[12] T. YOSHINOBU, H. MITSUI, Y. TARUI, T. FUYUKI, and H. MATSUNAMI, J. Appl. Phys. **72**, 2006 (1992).
[13] P. PIROUZ, C. M. CHOREY, and J. A. POWELL, Appl. Phys. Lett. **50**, 221 (1987).
[14] C. H. CARTER, JR., R. F. DAVIS, and S. R. NUTT, J. Mater. Res. **1**, 811 (1987).
[15] K. SHIBAHARA, S. NISHINO, and H. MATSUNAMI, J. Cryst. Growth **78**, 538 (1986).
[16] P. DESJARDINS and J. E. GREENE, J. Appl. Phys. **79**, 1423 (1996).
[17] K. SHIBAHARA, S. NISHINO, and H. MATSUNAMI, Appl. Phys. Lett. **50**, 1888 (1987).
[18] H. S. KONG, Y. C. WANG, J. T. GLASS, and R. F. DAVIS, J. Mater. Res. **3**, 521 (1988).
[19] C. CHANG, N. ZHENG, I. S. T. TSONG, Y. WANG, and R. F. DAVIS, J. Amer. Ceram. Soc. **73**, 3264 (1990).
[20] T. YOSHINOBU, I. IZUMIKAWA, H. MITSUI, Y. TARUI, T. FUYUKI, and H. MATSUNAMI, Appl. Phys. Lett. **59**, 2844 (1991).
[21] S. HARA, S. MISAWA, S. YOSHIDA, and T. AOYAGI, Phys. Rev. B **50**, 4548 (1994).
[22] M. DAYAN, J. Vacuum Sci. Technol. A **3**, 361 (1985).
[23] M. DAYAN, J. Vacuum Sci. Technol. A **4**, 38 (1986).
[24] M. KITABATAKE and J. E. GREENE, Appl. Phys. Lett. **69**, 2048 (1996).
[25] M. KITABATAKE and J. E. GREENE, Jpn. J. Appl. Phys. **35**, 5261 (1996).
[26] M. KITABATAKE, P. FONS, and J. E. GREENE, J. Vacuum Sci. Technol. A **8**, 3726 (1990).
[27] M. KITABATAKE, P. FONS, and J. E. GREENE, J. Vacuum Sci. Technol. A **9**, 91 (1991).
[28] M. KITABATAKE and J. E. GREENE, J. Appl. Phys. **73**, 3183 (1993).
[29] M. KITABATAKE and J. E. GREENE, Thin Solid Films **272**, 271 (1996).
[30] J. TERSOFF, Phys. Rev. B **38**, 9902 (1988).
[31] J. TERSOFF, Phys. Rev. B **39**, 5566 (1989).
[32] C. J. MOGAB and H. J. LEAMY, J. Appl. Phys. **45**, 1075 (1974).
[33] G. XUE, H. Z. XIAO, M.-A. HASAN, J. E. GREENE, and H. K. BIRNBAUM, J. Appl. Phys. **74**, 2512 (1993).
[34] O. L. ALERHAND, A. N. BERKER, J. D. JOHANNOPOULOS, D. VANDERBILT, R. J. HAMERS, and J. E. DEMUTH, Phys. Rev. Lett. **64**, 2406 (1990).
[35] E. PEHKLE and J. TERSOFF, Phys. Rev. Lett. **67**, 1290 (1991).
[36] G. W. WAGNER, B. NA, and M. A. VANNICE, J. Phys. Chem. **93**, 5061 (1983).
[37] H. NAGASAWA, Y. YAMAGUCHI, T. IZUMI, and K. TONOSAKI, Appl. Surf. Sci. **70/71**, 542 (1993).

phys. stat. sol. (b) **202**, 421 (1997)

Subject classification: 68.35.Bs; 73.20.At; S6

Atomic and Electronic Structure of SiC Surfaces from ab-initio Calculations

J. POLLMANN, P. KRÜGER, and M. SABISCH

*Institut für Theoretische Physik II – Festkörperphysik, Universität Münster,
D-48149 Münster, Germany*

(Received January 31, 1997)

We briefly review recent results of first-principles electronic structure calculations of polar and non-polar surfaces of cubic and hexagonal SiC polytypes. The structural properties of these systems as resulting from ab-initio total energy and grand canonical potential calculations are presented. A general picture of the surface relaxation and reconstruction behaviour derives from these results. Electronic properties of prototype surfaces are presented, as well. The theoretical results are discussed in comparison with available atomic and electronic structure data from experiment. The systems addressed, in particular, comprise (110), (001), $(10\bar{1}0)$ and (0001) surfaces of β- and α-SiC, respectively.

1. Introduction

SiC is a wide-band-gap compound semiconductor with very promising potential for high-temperature, high-power and high-frequency applications in microelectronic and electro-optical devices [1 to 6]. It occurs in an extremely large number of polytypes, the cubic 3C and the hexagonal 6H of which seem to be the most important for technological applications. In view of the extraordinary application potential, a thorough knowledge of the physical properties of SiC polytypes and their surfaces is a matter of both fundamental interest and technological importance. Bulk properties of SiC polytypes are discussed in several other contributions to this volume. We concentrate on structural and electronic properties of SiC surfaces, therefore.

Surfaces of SiC polytypes are currently investigated very intensively both in experiment and theory. The topic is very exciting, indeed, since many basic questions related to the atomic and electronic structure of these surfaces are far from being conclusively resolved, to date. The results of advanced experimental and theoretical techniques, like tensor low-energy electron diffraction (TLEED) analyses, photoelectron spectroscopy (PES) techniques or scanning tunneling microscopy (STM) and ab-initio total energy or molecular dynamics studies are controversial, in cases. The intriguing and stimulating disagreement between experiment and theory calls for new ideas, models and explanations making the field truely interesting from a basic scientific point of view. For some of the technologically most important SiC surfaces there is not even agreement on the gross features of their reconstructions. New conflicting and contradictory results keep coming up in the literature almost on a monthly rate so that conclusive interpretations are still lacking in many cases. In this short review, therefore, we refrain from trying to give "final" answers which would have to be premature, anyway, but rather try to summarize the current state of the theoretical research on SiC surfaces to the best of our knowledge.

A number of SiC surfaces has been investigated within the last two decades by empirical and semi-empirical methods [7 to 15]. Only very recently, prototype SiC surfaces have been addressed by "state-of-the-art" first-principles local density approximation (LDA) [16 to 31], GW quasiparticle [26], and Car-Parrinello molecular dynamics [32] calculations. We will mostly concentrate on first-principles results in our discussions. By now, relaxed nonpolar β-SiC(110) [16, 17, 25, 29, 31] and 2H-SiC(10$\bar{1}$0) [31] surfaces, as well as β-SiC(111) [18, 19], β-SiC(001) [20 to 23, 26, 29, 31, 32] and 6H-SiC(0001) [24, 27 to 30] surfaces have been investigated employing the LDA together with norm-conserving smooth pseudopotentials and plane wave [16 to 24] or Gaussian orbital [25 to 31] basis sets. The different first-principles approaches yield results in very good general agreement but these results considerably disagree with experimental data, in some cases.

Experimentally, SiC surfaces have been investigated by LEED [33 to 42], Auger electron spectroscopy (AES) [34 to 38, 43 to 45], STM [37, 39, 46 to 53], atomic force microscopy (AFM) [54], electron-energy loss spectroscopy (EELS) [34, 37, 43], X-ray photoelectron spectroscopy (XPS) [36, 43, 55], near-edge X-ray absorption fine structure (NEXAFS) [56], core-level spectroscopy (CLS) [55, 57, 58], photoelectron spectroscopy (PES) [57], angle-resolved photoelectron spectroscopy (ARPES) [58 to 60] and $\mathbf{k}_\parallel$-resolved inverse photoelectron spectroscopy (KRIPES) [61]. The current status of experimental research is reviewed in the contributions of Bermudez [62] and Starke [63].

We first briefly summarize the theoretical framework of current days ab-initio atomic and electronic structure calculations. Next, we address some key features of bulk SiC polytypes to the extent they are important for our discussion of the surfaces. Finally, we discuss nonpolar and polar surfaces of cubic 3C and hexagonal 2H and 6H polytypes of SiC in some detail. A short summary concludes the paper.

2. Theoretical Framework

Most current days "state-of-the-art" calculations of structural and electronic properties of SiC surfaces are carried out in the framework of density-functional theory within the local density approximation (LDA) [64] employing nonlocal, norm-conserving pseudopotentials in separable form, as suggested by Kleinman and Bylander [65]. Smooth pseudopotentials for Si and C atoms, constructed following the prescriptions given by Troullier and Martins [66], by Vanderbilt [67] or by Hamann et al. [68] are being used. Plane wave basis sets or linear combinations of Gaussian orbitals with s, p, d and s* symmetry are employed to expand the wavefunctions. When plane wave basis sets are employed, a very large number of basis states is necessary for good convergence (cf. [17, 20, 23, 24, 32]) because of the very localized nature of the C pseudopotential and the C orbitals, in particular. When Gaussian basis sets are used, a relatively small number of Gaussians per atom yields already sufficient accuracy (cf. [25, 26, 30, 69]). For the exchange and correlation potential most often the functional of Ceperley and Alder [70], as parametrized by Perdew and Zunger [71], is used. Most authors employ the supercell method for treating SiC surfaces saturating the broken sp^3 bonds at the bottom layer atoms in each supercell by hydrogen. The total energy is usually calculated self-consistently within the momentum space formalism of Ihm et al. [72]. Fixing a sufficiently large number of SiC layers in the center of the supercell in the bulk configuration, the surface structure is optimized by moving the atoms on sufficiently many surface layers until all forces on these atoms vanish, typically within 10^{-3} Ry/a.u. Iterative elimination of the forces is, e.g., achieved by employing the scheme of Broyden [73]. When a Gaussian basis set is employed in the calculations, Pulay forces have to be taken into account in addition to the Hellmann-Feynman forces [74]. It has turned out that employing a sufficiently large number of $k_\parallel$ points is crucial for convergent surface-structure calculations. Using Γ-point sampling only is often not sufficient.

3. Bulk Properties of SiC

Cubic β-SiC, or 3C-SiC, has one Si and one C atom per bulk unit cell. Hexagonal α-SiC occurs in 2H, 4H and 6H modifications depending on the stacking sequence of SiC bi-layers along the crystal c-axis. They have nSi and nC atoms (with $n = 2$, 4 or 6) per bulk unit cell, respectively. In these cubic and hexagonal polytypes, the nearest-neighbor configuration of Si and C atoms is tetrahedral and the SiC bond length is close to 1.89 Å. Though being a group-IV semiconductor, SiC is fairly ionic. The ionicity of cubic SiC amounts to $g = 0.475$ on the Garcia-Cohen scale (see [75, 25]) placing SiC between, e.g., GaAs ($g = 0.302$) and ZnS ($g = 0.676$) in this respect. The ionicity of SiC gives rise to an ionic gap within the valence bands of the bulk band structure of all polytypes, very much like in heteropolar III–V or ionic II–VI compound semiconductors. The pronounced ionicity of the SiC bond stems from the different strengths of the C and Si potentials, giving rise to the very different covalent radii of C ($r_C = 0.77$ Å) and Si ($r_{Si} = 1.17$ Å). In addition, the electronegativity of C ($e_C = 2.5$) is considerably larger than that of Si ($e_{Si} = 1.7$). The stronger C potential, as compared to that of Si, leads, e.g., in β-SiC to a charge transfer $\delta\varrho_{Si \to C} = 0.14e$ from Si to C so that the electronic charge density distribution along the SiC bond is strongly asymmetric (see, e.g. [25]). Therefore, Si atoms act as cations while C atoms act as anions in SiC. In addition, the SiC bulk bond length of $d = 1.89$ Å is about 3% smaller than the sum of the covalent radii of Si and C amounting to 1.94 Å. The charge transfer obviously reduces the effective covalent radius of the Si cations more strongly than it increases the effective covalent radius of the C anions. In consequence of the ionicity of SiC, there are nonpolar and polar SiC surfaces. Si and C layers alternate, e.g., along the [001] direction of β-SiC and along the [0001] direction of α-SiC. Thus, there are two distinctly different SiC surfaces in each case which are usually referred to as Si- or C-terminated surfaces.

Structural and electronic properties of SiC surfaces are often discussed in conjunction with those of the related Si or diamond surfaces. In this context it is important to have in mind, however, that the bulk lattice constant $a_0 = 4.36$ Å of SiC is some 20% smaller than that of bulk Si ($a_0 = 5.43$ Å) and some 22% larger than that of bulk diamond ($a_0 = 3.57$ Å). In consequence, at Si- or C-terminated surfaces one encounters Si(C) orbitals on a two-dimensional lattice with a lattice constant that is much smaller (larger) than that of related Si (diamond) surfaces, respectively. This is of considerable relevance for the particular reconstruction of some SiC surfaces, as compared to those of related Si and diamond surfaces, respectively. Another important point to be noted in this context is the large difference in angular forces occurring at Si and C atoms when the tetrahedral bonds to their nearest neighbors become bent upon surface relaxation or reconstruction. They are much larger at C than at Si atoms so that structural changes of the bulk tetrahedral configuration around C atoms involve a much larger contribution to the reconstruction energy than for Si atoms. This fact strongly discerns Si- from C-terminated surfaces of cubic or hexagonal SiC and is one of the reasons for their distinctively different reconstruction behaviour.

29*

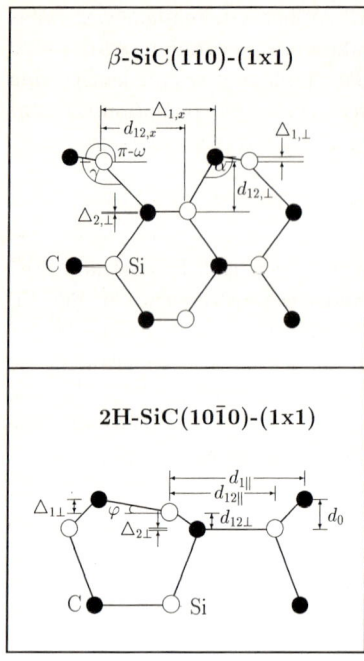

β-SiC(110)-(1x1)

2H-SiC(10$\bar{1}$0)-(1x1)

Fig. 1. Side view of the relaxed β-SiC(110)-(1 × 1) (top panel) and the relaxed 2H-SiC(10$\bar{1}$0)-(1 × 1) surface (bottom panel)

4. Nonpolar β-SiC(110) and 2H-SiC(10$\bar{1}$0) Surfaces

Both the cubic and the hexagonal modifications of SiC exhibit nonpolar surfaces like, e.g., the β-SiC(110) or the 2H-SiC(10$\bar{1}$0) surface. These nonpolar surfaces, in general, are largely similar to the related GaAs(110) surface or the (10$\bar{1}$0) surfaces of II–VI compounds, respectively (cf. [25, 31]). Nonpolar SiC surfaces have not been investigated experimentally, to date. First-principles calculations on the nonpolar β-SiC(110) surface have been reported in [16, 17, 25, 29, 31]. These calculations yield excellent mutual agreement concerning the atomic structure. The nonpolar 2H-SiC(10$\bar{1}$0) surface has been studied as well [31]. In Fig. 1 we show side views of the optimized structures of both

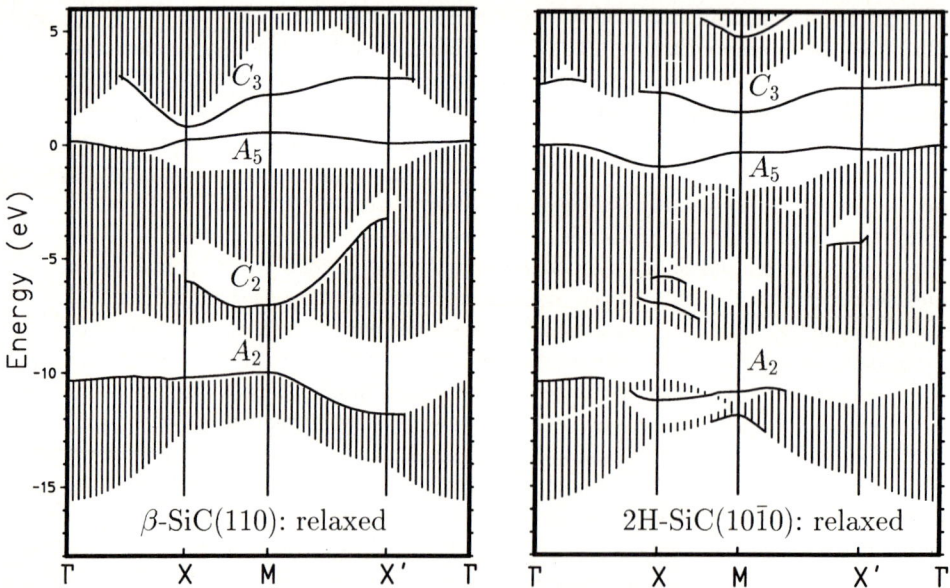

Fig. 2. Surface band structure of the relaxed β-SiC(110)-(1 × 1) (left panel) and the relaxed 2H-SiC(10$\bar{1}$0)-(1 × 1) surface (right panel). The projected bulk band structure is shown by the vertically shaded areas in each case

surfaces which are characterized by a bond length contracting rotation-relaxation. The surface-bond contractions amount to 6% for β-SiC(110) and 9% for 2H-SiC(10$\bar{1}$0) with respect to the SiC bulk bond length. A full account of the structure parameters is given in [25, 31]. The relaxation-induced energy gains are 0.64 and 0.71 eV for the two surfaces, respectively [31]. They are very similar since the nearest-neighbor configuration of Si and C atoms is the same in both polytypes and only the second-nearest-neighbor configurations discern these structures. The relaxation-induced bond rotation at the surfaces is characterized by the tilt angle φ and the relaxation angle ω, respectively. The tilt angle φ is the angle between the SiC surface layer bonds and the surface plane. The relaxation angle ω results from a projection of the SiC bonds onto the drawing plane in Fig. 1. Since the SiC surface-layer bonds lie in the drawing plane for 2H-SiC(10$\bar{1}$0), ω and φ are identical for the hexagonal surface while they are different for the cubic surface. For β-SiC(110) we find 8.2° and 16.9° for φ and ω, respectively, while they are equal and amount to only 3.8° at the 2H-SiC(10$\bar{1}$0) surface (see [31]). Thus, ω is much smaller at β-SiC(110) than at GaAs(110) where it amounts to about 30°. This is mainly due to the larger ionicity of SiC and to the more pronounced asymmetry of the charge

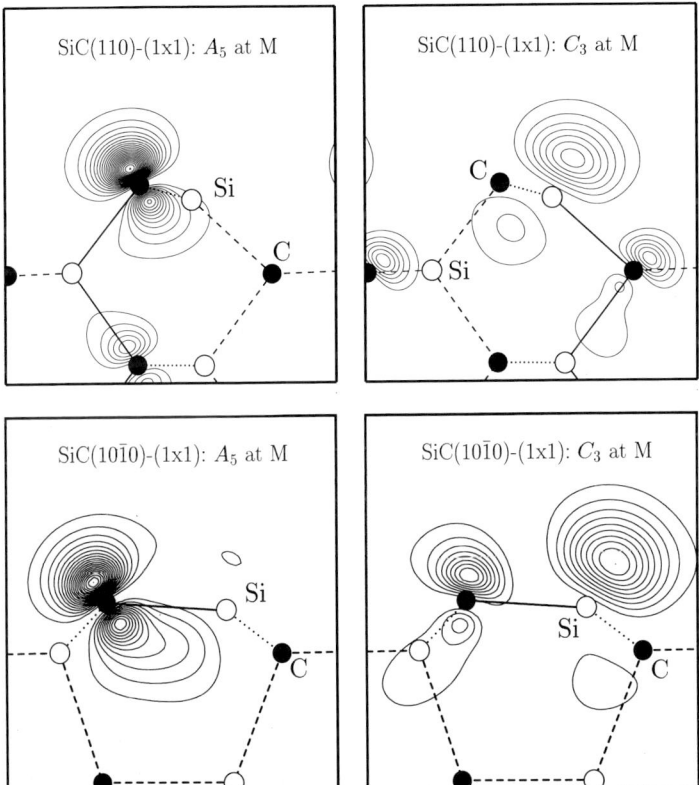

Fig. 3. Charge densities of the C- and Si-derived dangling-bond states A$_5$ and C$_3$ at the M-point of the relaxed β-SiC(110)-(1 × 1) (top panels) and the relaxed 2H-SiC(10$\bar{1}$0)-(1 × 1) surface (bottom panels). Bonds within (parallel to) the drawing plane are shown by full (dashed) lines. Bonds forming an angle with the drawing plane are shown by dotted lines

density along SiC bonds, as compared to GaAs. In addition, the C anion in SiC is a first row element whose covalent radius is much smaller than that of the cation Si. In GaAs both ions have similar covalent radii.

The surface electronic structure of the two nonpolar surfaces is shown in Fig. 2 together with the projected bulk band structure (PBS). The calculated gap is some 50% smaller than the experimental gap as is usual in LDA results. This is obtained for all LDA surface band structures shown in this paper. In this and all following figures of surface band structures the PBS is shown by vertically shaded regions. Both surfaces exhibit pronounced anion-derived (A_5) and cation-derived (C_3) surface-state bands within the gap-energy region. Their energetic separation is larger for the hexagonal than for the cubic surface since the bulk gap of 2H-SiC is considerably larger than that of 3C-SiC. Additional anion- and cation-derived surface-state bands occur within the projected bulk valence bands. As is obvious from Fig. 2, both nonpolar surfaces are semiconducting. Fig. 3 reveals the origin and nature of the A_5 and C_3 dangling-bond states at the two surfaces. Clearly, A_5 is a C-derived dangling-bond state while C_3 is a Si-derived dangling-bond state. The respective states at the cubic and at the hexagonal surface are amazingly similar which is again due to the identical nearest-neighbor configuration of the two polytypes. For more details concerning nonpolar SiC surfaces, the interested reader is referred to [16, 17, 25, 29, 31].

5. Polar (001) Surfaces of β-SiC

The current status of experimental research on structural and electronic properties of β-SiC(001) surfaces is beautifully and exhaustively discussed in the contribution of Bermudez [62]. Within the past two decades several semi-empirical structure studies of Si(001) surfaces have been carried out [7 to 11, 13, 14]. More recently, a number of first-principles calculations have been reported [20 to 23, 26, 29, 31, 32], as well. We first address the Si-terminated SiC(001)-(2 × 1) surface and then discuss (2 × 1), (1 × 2) and c(2 × 2) reconstruction models for the C-terminated β-SiC(001) surface.

5.1 Si-terminated β-SiC(001)-(2 × 1)

Experimental data show (2 × 1), c(4 × 2), (3 × 2), and (5 × 2) reconstructions of the Si-terminated β-SiC(001) surface [33, 34, 40, 42, 50 to 53, 62]. We restrict ourselves to a discussion of the (2 × 1) surface. A side view of the optimized geometry resulting from first-principles calculations [26] is shown in the top panel of Fig. 4 in direct comparison with that of the related Si(001)-(2 × 1) surface [76]. As is most obvious from the figure, the reconstruction of the Si-terminated SiC(001)-(2 × 1) surface is largely different from that of Si(001)-(2 × 1). This is related to the charge transfer from Si to C in SiC and to the fact that the lattice constant of SiC is about 20% smaller than that of Si. Dimer formation at the SiC surface involves much larger back-bond repulsions, therefore, as compared to the case of the Si surface. In addition, when Si surface dimers are to be formed at the Si-terminated SiC(001) surface, angular forces on the second layer C atoms are involved. These are much larger at the C atoms of the SiC surface than at the Si atoms of the Si surface so that strong dimer formation is prevented at the SiC surface. In the optimized structure, the Si surface layer atoms have moved slightly towards each other with respect to the ideal surface, their distance amounting to 2.73 Å. *No Si*

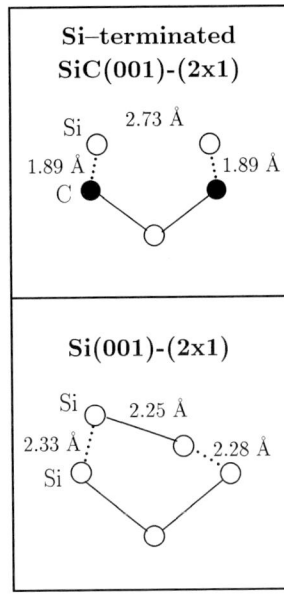

Si–terminated SiC(001)-(2x1)

Si 2.73 Å

1.89 Å 1.89 Å

C

Si(001)-(2x1)

Si 2.25 Å

2.33 Å 2.28 Å

Si

Fig. 4. Side views of the surface structure of Si-terminated β-SiC(001)-(2 × 1) (top panel) and of Si(001)-(2 × 1) (bottom panel)

surface dimers are formed in marked contrast to the case of the Si(001)-(2 × 1) surface. The reconstruction-induced energy gain is only 0.01 eV per unit cell. This type of a very weak reconstruction agreeingly results from all convergent first-principles calculations [20 to 22, 26, 29, 31, 32]. The structural parameters for the reconstruction are given in [26].

A section of the surface electronic structure of the Si-terminated β-SiC(001)-(2 × 1) surface is shown in Fig. 5. There are four salient surface state bands in the gap-energy region. They originate from the dangling- and bridge-bond bands originally present at the ideal SiC(001) surface (cf. [26]). The nature and origin of these four bands become most apparent from the charge densities of the respective states shown in Fig. 5, as well. The occupied states π and π* mainly result from symmetric and antisymmetric combinations of the former dangling-bond orbitals at the ideal surface while the empty states σ and σ* result from symmetric and antisymmetric combinations of the former bridge-bond orbitals. Our optimized structure leads to a semiconducting surface already within LDA. A reconstruction of the Si-terminated SiC(001)-(2 × 1) surface, with dimers fully buckled as at the Si(001)-(2 × 1) surface, is energetically less favorable by 0.67 eV and has to be dis-

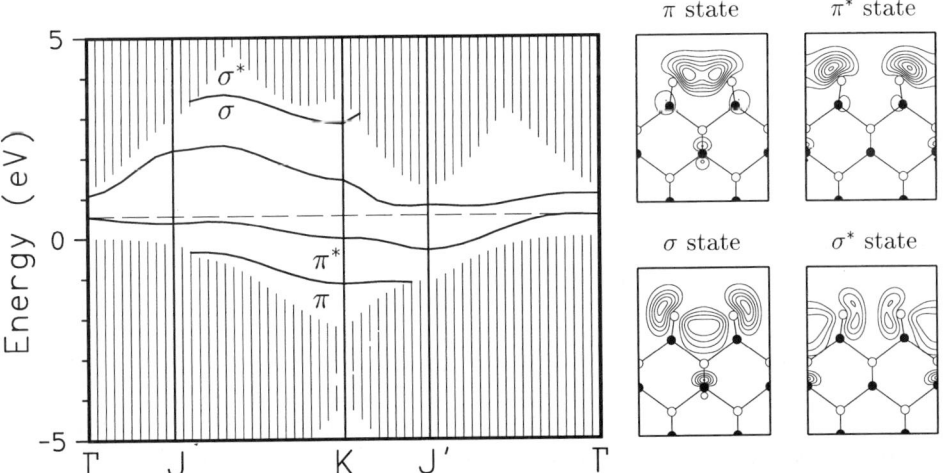

Fig. 5. Section of the surface band structure of β-SiC(001)-(2 × 1) (left panel) and charge densities of salient surface states at the K point of the (2 × 1) surface Brillouin zone shown in the x–z plane containing Si surface atoms. All atoms are connected by solid lines also if the indicated bonds do not lie in the drawing plane

carded as the optimal structure, therefore. Furthermore, it gives rise to a strongly metallic surface (cf. [26]) in contrast to experiment (cf. [62]).

Powers et al. [40] conclude from their TLEED data that buckled dimers with a bond length of 2.31 Å are formed at the surface. This interpretation was supported by semi-empirical results [9, 13, 15] and by the results of first-principles calculations [23] which, however, were obtained with Γ-point sampling, only. The results of more recent first-principles calculations [20 to 22, 26, 29, 31, 32] contradict this interpretation.

To the best of our knowledge, ARPES or KRIPES data are not available on this surface, to date. They would certainly be most useful to shed more light on the question of the actual surface reconstruction and to resolve the above-mentioned discrepancy between theory and experiment. Interestingly enough, very recent STM measurements on the $c(4 \times 2)$ surface [53] and accompanying calculations have confirmed that the reconstruction of this SiC surface is significantly different from the reconstruction of the respective Si(001)-$c(4 \times 2)$ surface. In their STM calculations, the authors have employed the bond length between Si atoms, as calculated for the (2×1) surface [26], and have obtained excellent agreement between their measured and calculated STM pictures. This lends further support to the notion that Si-terminated SiC surfaces are strongly different from related Si surfaces and do not show strong Si surface dimers with a bond length as small as 2.31 Å.

PES data clearly indicate the existence of two occupied surface-state bands which are referred to as "V_1" and "V_2" features (see [57] and the discussion in [62]). V_1 occurs above the bulk valence band maximum and V_2 is observed roughly 1 eV below V_1. Our calculated π^* and π bands (see Fig. 5) appear to be closely related to these measured features. Both the absolute energy positions and, in particular, the energetic separation of about 1 eV between the calculated bands are well in accord with the data.

First- and second-derivative EELS data on the SiC(001)-(2×1) surface have been presented, e.g., in [34, 62]. In the second-derivative spectra, three features are observed at relatively low transition energies of 1.8, 3.1 and 4.7 eV. The latter feature seems to correspond to a peak observed in the first-derivative spectrum at 5.3 eV (see [62]). The resolution in these spectra is of the order of 0.5 eV, at best [77]. Our theoretical results are compatible with these data. When comparing theory and experiment one has to have in mind, however, that LDA badly underestimates gap energies. For β-SiC our calculated LDA gap is 1.28 eV while the measured gap is 2.41 eV. According to our related GW results (cf. [26] and Fig. 10), it is to be expected that the occupied surface-state bands π and π^* (see Fig. 5) are not affected by quasiparticle corrections while the empty surface-state bands σ and σ^* are shifted up in energy by some 1.1 eV (difference between measured and calculated gap energy) due to quasiparticle corrections. Certainly, so far we did neither calculate joint densities of states (JDOS) nor even EELS spectra for this surface. However, from the surface band structure in Fig. 5 one can easily extract ranges of possible transition energies. Taking into account the above-mentioned upward shift of the empty bands by quasiparticle corrections we can read off from Fig. 5 the following energy ranges for the four possible transitions: (1) 1.6 to 2.8 eV for $\pi^* \to \sigma$ transitions, (2) 3.0 to 3.8 eV for $\pi \to \sigma$ transitions, (3) 4.0 to 4.2 eV for $\pi^* \to \sigma^*$ transitions, and (4) 4.9 to 5.1 eV for $\pi \to \sigma^*$ transitions. The first and second of these ranges appear to have a higher spectral weight on their low-energy side, respectively. Thus, the first could be related to the measured peak at 1.8 eV, the second to the measured peak at 3.1 eV and the fourth to the measured peaks at 4.7 or 5.3 eV in the second- or first-derivative spectra, respectively.

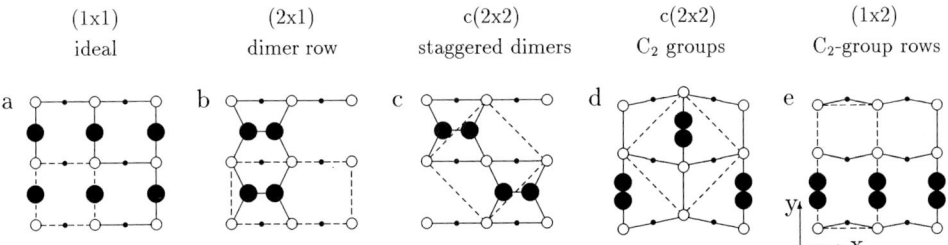

Fig. 6. Top views, of the a) (1×1) ideal, b) (2×1) dimer-row, c) $c(2 \times 2)$ staggered dimer, d) $c(2 \times 2)$ staggered C_2-group, and e) (1×2) C_2-group row configurations of the C-terminated β-SiC(001) surface. The unit cell is indicated by dashed lines in each case

5.2 C-terminated β-SiC(001)

A number of structural models for the C-terminated β-SiC(001) surface have been suggested in the literature [40, 44, 50, 56, 62]. They comprise (2×1) or (1×2) row and $c(2 \times 2)$ staggered configurations of dimers or C_2 groups. The C_2-group reconstructions are often referred to as "bridging-dimer" reconstructions (cf. [62]). We will not adopt that nomenclature here. The atomic structure of these models has been optimized by first-principles calculations [20 to 23, 26, 29, 31, 32]. Our results are shown in Fig. 6. A full account of structural parameters from the literature is compiled in [26].

Fig. 7 shows a side view of the optimized C-terminated SiC(001)-(2×1) surface (cf. Fig. 6b) in direct comparison with the structure of the related C(001)-(2×1) surface [76]. There is amazing similarity between these two reconstructions. In both cases strong surface dimers are formed. Their bond lengths of 1.36 and 1.37 Å, respectively, are very close to the C=C double-bond length in molecules like, e.g., C_2H_4. All arguments given in Section 5.1 against dimer formation at the Si-terminated surface work in favor of dimer formation at the C-terminated SiC(001)-(2×1) surface. At the latter surface, charge is transferred from the second layer Si atoms to the surface layer C atoms which reside at a surface lattice whose lattice constant is some 22% larger than that of C(001). So there is plenty of space for a full dimerization without invoking strong back-bond repulsions. In addition, surface dimer formation now involves angular forces on the second-layer Si atoms which are much smaller than those at the second-layer C atoms of the Si-terminated surface. In consequence, C surface dimers are easily formed and the reconstruction-induced energy gain turns out to be as large as 4.88 eV. This very strong surface reconstruction results, as well, from semi-empirical MINDO calculations [9].

The $c(2 \times 2)$ staggered dimer structure of Fig. 6c is also characterized by C=C surface dimers with a double-bond

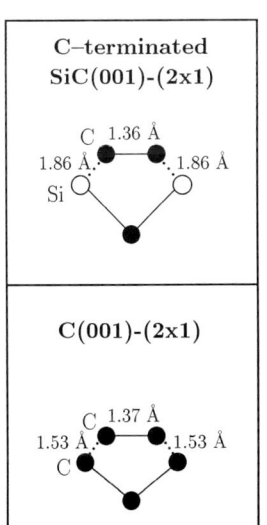

Fig. 7. Side views of the surface structure of C-terminated β-SiC(001)-(2×1) (top panel) and of C(001)-(2×1) (bottom panel)

length of 1.36 Å. The structure of this type of reconstruction results from all calculations [9, 20 to 23, 26, 29, 31, 32] in very close mutual agreement (cf. [26]). The energy gains resulting from the different calculations differ to a certain extent. We have found a value of 4.73 eV and Käckell et al. [20, 21] report a value of 4.36 eV, while Yan et al. [23] have obtained 3 eV. Craig and Smith [9] even find the c(2 × 2) surface to be 0.79 eV higher in total energy than the (2 × 1) dimer-row structure from their semi-empirical calculations.

The c(2 × 2) staggered C_2-group reconstruction shown in Fig. 6d, which is referred to as bridging-dimer configuration in [62], yields a reconstruction-induced energy gain of 4.76 eV. This value is only 0.12 eV smaller than that for our (2 × 1) dimer-row structure. In the staggered C_2-group structure triple bonds between surface C atoms with a bond length of only 1.22 Å are formed. The structure parameters resulting from different calculations for this configuration [20 to 23, 26, 29] are in excellent agreement with one another (cf. [26]).

The (1 × 2) C_2-group-row structure of Fig. 6e also exhibits surface triple bonds with a bond length of 1.22 Å and Si dimers on the second layer. This structure yields an energy gain of 4.58 eV relative to the ideal surface and thus turns out in our calculations to be the least favorable of the four models.

Powers et al. [40] have favored the c(2 × 2) C_2-group reconstruction on the basis of their TLEED data. Long et al. [56] have confirmed this conclusion by a polarization analysis of NEXAFS data. From the first-principles results for the C-terminated surfaces it appeares that theory slightly favors the (2 × 1) dimer-row structure. The energy gains for the different reconstructions are fairly close, however, their mutual difference being much smaller than the absolute gain with respect to the ideal surface. Thus a coexistence of domains of different reconstructions, as observed in experiment [40], depending on the particular sample preparation method used, is compatible with the theoretical results.

The electronic structure of these four models has been discussed in great detail in [20 to 22, 26, 29, 31]. Here we only address the surface band structure of the two most probable reconstructions, namely the (2 × 1) dimer-row and the c(2 × 2) staggered C_2-group reconstructions. The band structures are shown in Fig. 8. Both show a number of bands in the gap-energy region and back-bond bands within the PBS. A pronounced band S occurs below the PBS in both cases originating from s orbitals on the C surface layer atoms. The P_1' and P_2' bands (see left panel of Fig. 8) and the P_1 band (see right panel of Fig. 8) originate from C–Si backbonds having predominantly p wavefunction character. The π and π^* bands in the gap-energy region of the (2 × 1) surface (see left panel of Fig. 8) are very similar to the related bands at the C(001)-(2 × 1) surface [76]. They are separated in energy by roughly 1 eV and originate from symmetric (π) and antisymmetric (π^*) linear combinations of the dangling-bond orbitals at the dimer atoms, as can clearly be seen in Fig. 9 (two upper left panels) which shows charge densities for some of the gap states. The P_5' band is mostly occupied and it originates from p states at the surface-layer C atoms (see upper right panel of Fig. 9). The (2 × 1) surface is marginally metallic in our LDA results. Interestingly enough, the c(2 × 2) C_2-group reconstruction gives rise to a semiconducting surface already within LDA. The π_1^* and π_1 states are antibonding and bonding states of the triply-bonded C_2 groups (see Fig. 9, two lower left panels). Their charge densities show amazing similarities with those of the π^* and π states at the (2 × 1) surface (see Fig. 9) in spite of the fact that the lattice

C-terminated β-SiC(001)

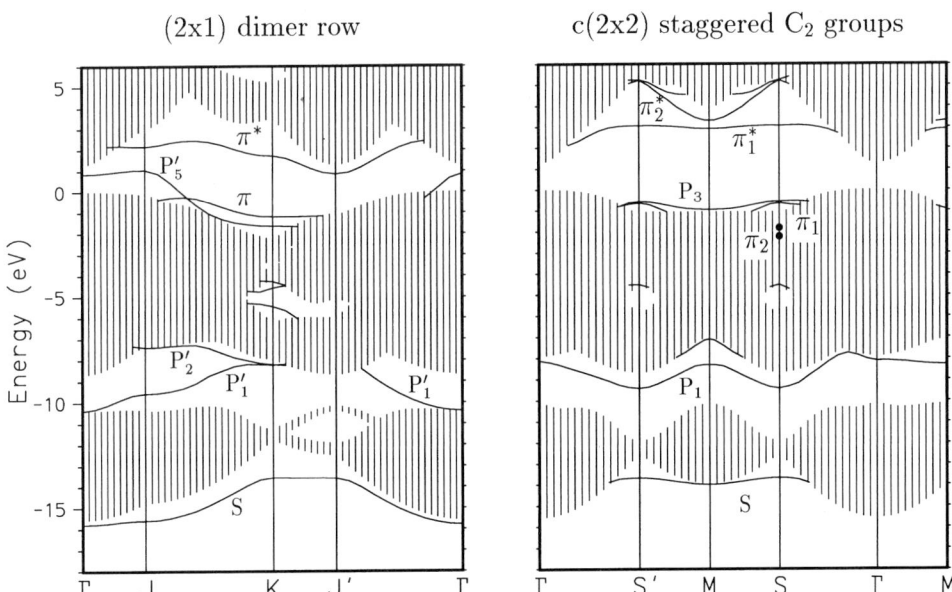

Fig. 8. Surface band structure of the (2×1) dimer row reconstructed (left panel) and the $c(2 \times 2)$ staggered C_2-group reconstructed (right panel) C-terminated β-SiC(001) surfaces. The projected bulk band structure is shown by the vertically shaded areas in each case

topology is quite different for the two surfaces. The former exhibits five-membered rings while the latter has seven-membered rings. The similarity of the respective states is a consequence of the fact that they are highly localized surface states whose properties are basically determined by the surface-layer atoms. The P_3 state originates from p_y and p_z orbitals at the C_2 groups and the second-layer Si atoms (see lower right panel of Fig. 9). The π_2 state lies in the surface plane and is a bonding state related to the C_2 groups at the surface.

While the $c(2 \times 2)$ surface results already as semiconducting within LDA, the (2×1) surface is marginally metallic due to the occurrence of the P_5' band. This metallicity, however, is an artefact of the LDA calculations (cf. [26]). When quasiparticle corrections are included within the GW approximation, the gap opens up (see Fig. 10) and this surface becomes semiconducting, as well. Including respective quasiparticle corrections in the calculation of the surface band structure of the $c(2 \times 2)$ staggered C_2-group structure would increase the surface gap by about 1 eV, in addition. Thus, both reconstructions give rise to semiconducting surfaces with largely similar surface states (cf. Fig. 9).

The fact that the gap of the $c(2 \times 2)$ surface is clear from surface states is in good accord with photoemission measurements by Bermudez and Long [44] and Semond et al. [52] who did not observe surface states in the gap-energy region. In addition, there is no clear indication of surface states in the band gap in EELS data (cf. [34, 62]). H-sensitive structure in EELS data has been observed [45] at about 4 eV and between 8 to 11 eV which is clearly due to surface excitations but an assignment of these features has not yet been given. From the right panel of Fig. 8, i.e., for the experimentally favored recon-

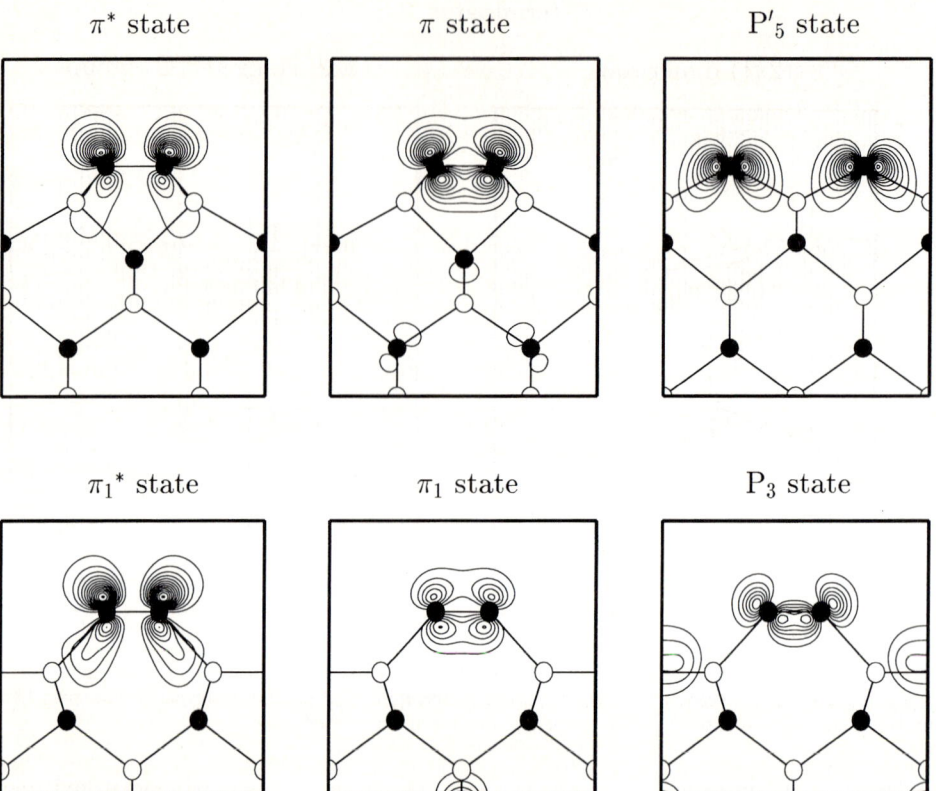

Fig. 9. Charge density contours of salient surface states at the C-terminated (2×1) dimer-row re-constructed (top panels) and the $c(2 \times 2)$ staggered-C_2-group reconstructed (bottom panels) C-ter-minated β-SiC(001) surfaces. The states π^* and π at the K point of the (2×1) SBZ are shown in the x–z plane and P'_5 at the Γ-point is shown in the y–z plane. The states π_1^*, π_1 and P_3 at the S point of the $c(2 \times 2)$ SBZ are drawn in the y–z plane which is perpendicular to the surface and contains the C_2 groups

struction, we read off possible transition energies of about 5 eV for $P_3 \rightarrow \pi_1^*$ transitions (including the gap correction of about 1 eV discussed in Section 5.1) and 12 eV for $P_1 \rightarrow \pi_1^*$ transitions. Both are not very close to the measured peak positions. The respec-tive transition energies for the (2×1) dimer-row reconstruction can directly be inferred from the left panel of Fig. 8 in conjunction with our quasiparticle surface band structure in Fig. 10 showing the correct experimental gap. Possible $\pi \rightarrow \pi^*$ transitions range from 3.5 to 4.4 eV. The maximum in the JDOS is to be expected near 4.2 eV. In addition, $P'_2 \rightarrow \pi^*$ transitions range from 10 to 11 eV. These values are fairly close to the peak positions in the EELS data. However, since we have merely estimated these transition energies in a very rough way, more or less good agreement with the EELS data should not be considered as a proof or disproof of one structural model as compared to the other.

To fully resolve the structure of the C-terminated SiC(001) surfaces, it would be ex-tremely useful to have a host of ARPES and KRIPES data available for detailed com-parisons with the calculated electronic structure.

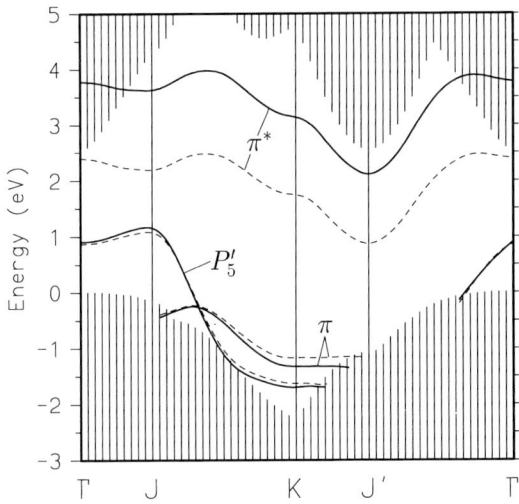

Fig. 10. Section of the surface band structure of the dimer-row reconstruction of the C-terminated β-SiC(001) surface (see Fig. 6b) as resulting from LDA (dashed lines) and GWA (full lines) calculations

6. Polar (0001) Surfaces of 6H-SiC

The current status of experimental research on structural and electronic properties of hexagonal SiC surfaces is discussed in detail in the contribution of Starke [63]. Most reconstruction models for polar 6H-SiC(0001) surfaces involve Si or C adatoms or adsorbed Si or C trimers and are called adsorption-induced reconstructions, therefore. The respective Si- or C-terminated *substrate* surfaces are characterized by Si(C) top layer atoms with one dangling bond and three back bonds connecting them with their three nearest-neighbour C(Si) atoms on the second substrate layer. LEED, AES and EELS results are almost identical for corresponding reconstructions of the β-SiC(111) and the 6H-SiC(0001) surface [34] since the stacking sequence of Si–C bilayers in the cubic [111] direction of β-SiC and in the hexagonal [0001] direction of 6H-SiC are identical down to the eighth layer. Based on the indistinguishable LEED results for these two surfaces one can conclude that they are characterized by the same reconstruction geometry. Among the structures reported are (1×1), $(\sqrt{3} \times \sqrt{3})R30°$, (3×3), $(6\sqrt{3} \times 6\sqrt{3})R30°$ and (9×9) configurations depending sensitively on temperature and on sample preparation. Different preparation methods appear to yield different reconstructions and, in general, the experimental structure data seem not to be entirely conclusive, yet (for details see [6, 34, 36 to 39, 46, 63]).

First-principles investigations of polar hexagonal 6H-SiC(0001) surfaces have been reported recently [24, 27 to 30]. Northrup and Neugebauer [24] have studied Si-terminated β-SiC(111)-$(\sqrt{3} \times \sqrt{3})R30°$ surfaces which are largely equivalent to the respective hexagonal surfaces, as mentioned above. Badziag [12, 13] has studied some of the (0001) surfaces by semi-empirical calculations.

6.1 Relaxed 6H-SiC(0001)-(1 × 1) surfaces

At the unreconstructed Si-terminated (1 × 1) surface there is usually a disordered layer of impurities like O which can be removed by annealing in UHV. The (1 × 1) structure of the C-terminated surface results from impurities at the surface as well [6]. Well-ordered unreconstructed 6H-SiC(0001) surfaces seem to have been investigated experimentally in some more detail only very recently [63, 78].

Atomic relaxations can occur only along the surface normal (z-direction) due to the hexagonal symmetry of the lattice. The most significant effect resulting from first-principles calculations is a pronounced inward relaxation of the top layer atoms for both surfaces. The respective decrease in the distance of the first two surface layers, as compared to its value at the ideal surfaces, amounts to -0.15 Å for the Si- and -0.25 Å for the C-terminated surface in our calculations [27, 29, 30]. This is in good general accord with the structural results of [78]. The calculated atomic relaxation on lower-lying layers is very small. We obtain a relaxation-induced energy gain of 0.09 eV at the Si-terminated surface and a more than three times larger gain of 0.30 eV at the C-terminated surface. This energy difference originates from the larger relaxation of the C-face, as compared to that of the Si-face, and is again related to the difference in bond-bending angular forces at second-layer C or Si atoms, respectively. Structure parameters for the optimally relaxed Si- and C-terminated substrate surfaces are given in [27, 29, 30].

Fig. 11 shows the surface band structure for the relaxed Si- and C-terminated 6H-SiC(0001)-(1 × 1) surfaces. For both surfaces there results one dangling-bond band in the gap-energy region. These bands D_{Si} and D_C originate from dangling bonds localized at the Si or C top-layer atoms, respectively. Each of these dangling-bond bands is half filled since there is only one top layer atom per (1 × 1) unit cell in both cases. The resulting band structures are thus metallic. Comparing the energetic positions of D_{Si} and D_C it becomes obvious that D_C occurs roughly 1.5 eV lower in energy than D_{Si} which is due to the stronger C potential, as compared to that of Si. The dispersion of D_{Si} is more pronounced than that of D_C, because D_{Si} is laterally more extended as can be seen in

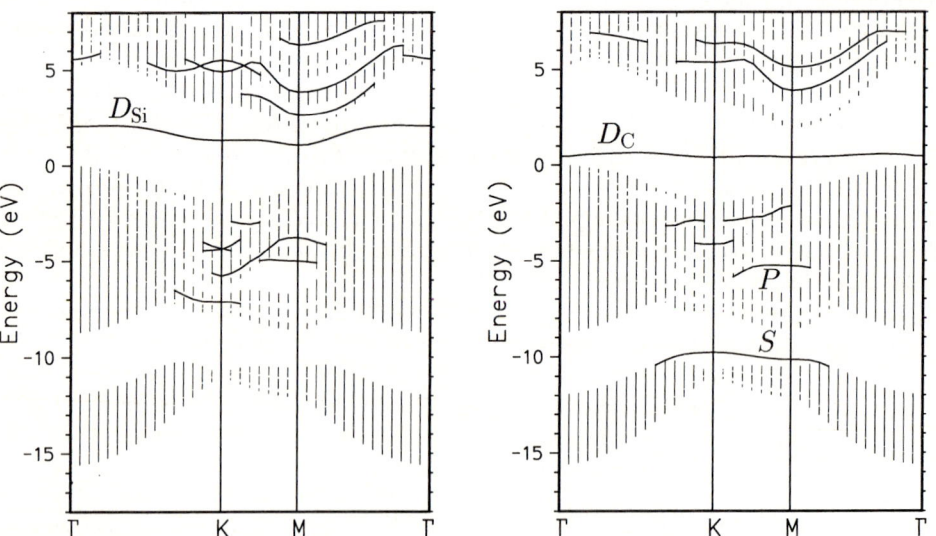

Fig. 11. Surface band structures of the relaxed Si-terminated (left panel) and C-terminated (right panel) 6H-SiC(0001)-(1 × 1) surfaces. The projected bulk band structure is shown by vertically shaded areas

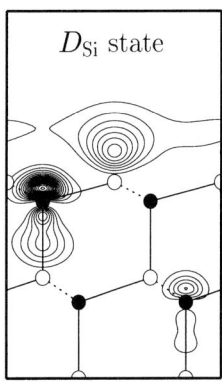

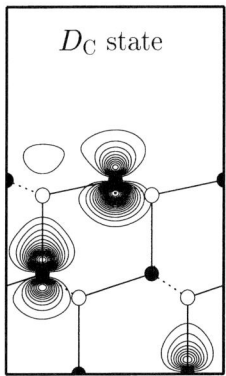

Fig. 12. Charge density contours of the dangling-bond states at the Γ-point of the relaxed Si-terminated (left panel) and C-terminated (right panel) 6H-SiC(0001)-(1×1) surfaces. The charge densites are presented for a side view of the relaxed structures

Fig. 12 showing charge density contours of these dangling-bond states. Obviously, the dangling bonds are predominantly localized at the top layer atoms and are oriented perpendicularly to the surface.

6.2 Si-terminated 6H-SiC(0001)-$\sqrt{3} \times \sqrt{3}$ surfaces

Owman and Mårtensson [39] have investigated Si-terminated 6H-SiC(0001)-$(\sqrt{3} \times \sqrt{3})$ R30° surfaces by STM. The authors observed images consistent with a structural model composed of 1/3 layer of Si or C adatoms in threefold-symmetric sites above the outermost Si–C bilayer, similar to the reconstructions observed for 1/3 monolayer of, e.g., Al, Ga, In or Pb on the Si(111) surface [79 to 82]. Similar structural models had been suggested by Kaplan [34] before. From STM data alone, it is neither possible to identify which one of the elements (Si or C) constitutes the adatoms nor to determine in which of the two symmetry-allowed sites (T_4 or H_3) the adatoms are located. A mixture of adsorbed Si and C adatoms was excluded in [39]. But more complex structures like, e.g., trimers could not be excluded.

The $\sqrt{3} \times \sqrt{3}$ unit cell is three times as large as that of the relaxed surfaces and contains three atoms per layer unit cell in the substrate. Motivated by the STM results of Owman and Mårtensson [39] five adsorption models of the Si-terminated 6H-SiC(0001)-$(\sqrt{3} \times \sqrt{3})$R30° surface have been studied by first-principles calculations [24, 27 to 30]. They are shown in Fig. 13 by top and side views. The structure parameters characterizing the different configurations are introduced in Fig. 13 as well. Si adatoms in T_4 (Fig. 13a) and H_3 (Fig. 13b) and C adatoms in T_4 positions (Fig. 13c) have been considered. In addition, Si and C trimers in T_4 positions (Figs. 13d, e) have been optimized. Adatoms in $T_4(H_3)$ positions reside above second (fourth) substrate layer atoms (see, e.g., Figs. 13a, b). For shortness sake, we refer to the different configurations as Si(T_4), Si(H_3), C(T_4), Si$_3$(T_4) and C$_3$(T_4), respectively. The optimal structures as resulting from our total energy minimization are drawn to scale in Figs. 13a to e. The structure parameters are compiled in Table 1.

In all five adsorption configurations the three substrate surface dangling bonds per $\sqrt{3} \times \sqrt{3}$ unit cell become fully saturated by the adatoms. But, now the adatoms have dangling bonds. Their number, however, is smaller than the number of the original dangling bonds at the clean substrate surface. In the configurations of Figs. 13a to 13c the number of dangling bonds is reduced by adatom adsorption to one third. In the case of

top view   side view

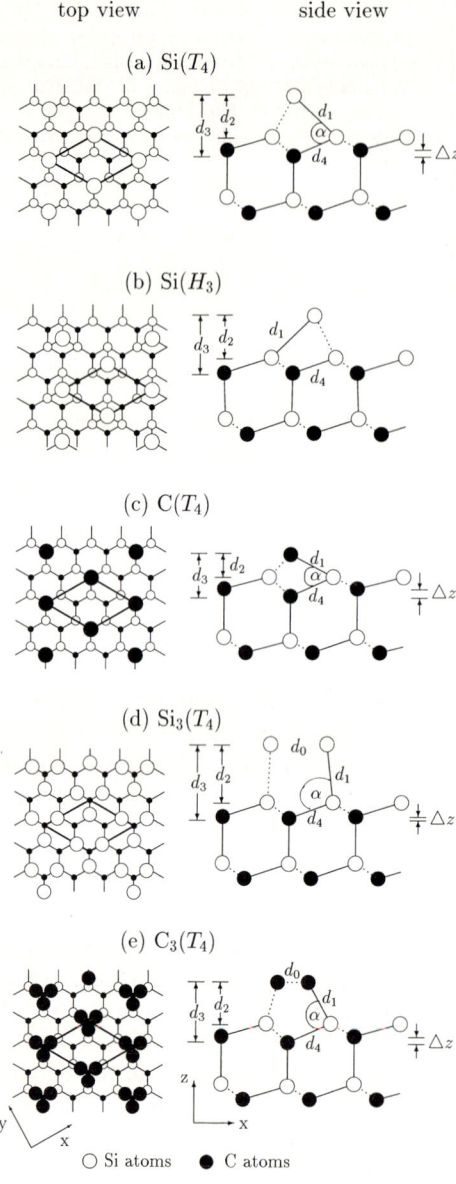

(a) Si(T_4)

(b) Si(H_3)

(c) C(T_4)

(d) Si$_3$(T_4)

(e) C$_3$(T_4)

○ Si atoms   ● C atoms

Fig. 13. Top and side views of the optimized ($\sqrt{3} \times \sqrt{3}$)R 30° reconstructions of Si-terminated 6H-SiC(0001) surfaces. The $\sqrt{3} \times \sqrt{3}$ unit cells are indicated by heavy lines in the left panels. The structure parameters are introduced in the side views and their actual values are compiled in Table 1

the Si and C trimers (see Figs. 13d, e) the dangling bond reduction sensitively depends on to which extent the dangling bonds at the trimer atoms become involved in chemical bonding. The newly established surface bonds d_1 in all five configurations have a bond length which turns out to be slightly larger than respective bulk bonds (2.33 Å for Si and 1.89 Å for SiC) and close to the sum of the covalent radii of the involved atoms ($r_C = 0.77$ Å and $r_{Si} = 1.17$ Å). The new surface bonds are significantly strained, however, since the respective bond angle α (see Fig. 13 and Table 1) is far from the tetrahedral angle of 109.5°. In all five configurations, the bond length d_4 between the first and second substrate layer atoms result as bulk-like. There is an important difference between the T_4 and H_3 configurations to be noted at this point. In the T_4 adatom configurations, the system can reduce its strain and increase the bond angle α by pushing down the second-layer C atom residing vertically below the adatom and lifting the other two C atoms in the $\sqrt{3} \times \sqrt{3}$ unit cell on the second substrate layer. In consequence, a buckling of the second substrate layer, characterized by Δz (see Fig. 13 and Table 1) occurs. In the H_3 configuration, on the contrary, all three C atoms per unit cell on the second substrate layer are equivalent so that no buckling of the second layer can occur. Thus, strain relief can be accomplished much more efficiently in the T_4 than in the H_3 configuration.

In the optimized Si(T_4) structure (see Fig. 13a) each Si adatom forms a bond of length $d_1 = 2.41$ Å (see Table 1) with each of the three Si substrate surface layer atoms. By pushing down the second-layer C atom residing below the adatom and lifting the other two C atoms on the second substrate layer the bond angle α results as 70°. So a

Table 1

Optimized structure parameters (as defined in Fig. 13) for the Si and C adatom and Si and C trimer reconstructions of the Si-terminated 6H-SiC(0001)-($\sqrt{3} \times \sqrt{3}$)R 30° surfaces discussed in this work. The values in parentheses are the results of [24]

| | adsorption models for the Si-terminated $\sqrt{3} \times \sqrt{3}$ surface | | | | |
	Si(T$_4$)	Si(H$_3$)	C(T$_4$)	Si trimer (T$_4$)	C trimer (T$_4$)
d_0 (Å)	—	—	—	2.59	1.41
d_1 (Å)	2.41(2.42)	2.44	1.93(1.98)	2.39	1.92
d_2 (Å)	1.71(1.75)	1.73	0.92(1.22)	2.37	1.69
d_3 (Å)	2.50	2.36	1.70	3.03	2.46
d_4 (Å)	1.88	1.88	1.87	1.87	1.89
Δz (Å)	0.25(0.22)	—	0.31(0.52)	0.07	0.25
α	70°	—	53°	105°	86°

buckling of $\Delta z = 0.25$ Å occurs. The results of Northrup and Neugebauer [24] for this configuration are in very close agreement with our results (see Table 1).

In the optimized Si(H$_3$) structure the bond lengths d_1, d_2 and d_4 are very similar to those at the optimized Si(T$_4$) configuration. The discerning feature of the two is the buckling Δz of the second substrate layer which occurs only in the latter configuration. The Si(T$_3$) configuration turns out to be energetically less favorable by $\Delta E = 0.60$ eV per $\sqrt{3} \times \sqrt{3}$ unit cell than the Si(T$_4$) configuration in our results in very close agreement with the respective energy difference of $\Delta E = 0.54$ eV reported in [24].

The bond length $d_1 = 1.93$ Å between C adatoms and Si substrate surface layer atoms in the optimized C(T$_4$) configuration turns out to be smaller than the respective bond length $d_1 = 2.41$ Å in the Si(T$_4$) configuration because of the smaller covalent radius of carbon. The distance of the C adatoms to the second-layer C atoms turns out to be only $d_3 = 1.70$ Å. In consequence, the buckling of the second substrate layer $\Delta z = 0.31$ Å is larger than that at the Si(T$_4$) surface. Northrup and Neugebauer [24] have obtained a similar bond length d_1 and even find a buckling of $\Delta z = 0.52$ Å in this case (cf. Table 1). The surface bonds in the C(T$_4$) configuration are very strongly strained since d_3 is very small and α turns out to be only 53°. Because of their smaller covalent radius the C adatoms in C(T$_4$) reside nearer to the Si substrate surface layer atoms than the Si adatoms in Si(T$_4$).

For Si trimers adsorbed in T$_4$ positions at the substrate surface, we observe that the individual Si atoms of the trimers move into new positions vertically above the Si substrate surface atoms (see Fig. 13d). In the optimized configuration their distance d_0 of 2.59 Å is considerably larger than twice the covalent radius of Si or the Si bulk bond length of 2.33 Å, respectively. This behaviour is very similar, in general, to that observed at the Si-terminated β-SiC(001)-(2 × 1) surface having the same physical origin (cf. Section 5.1). The surface bonding configuration is now nearly tetrahedral resulting in an angle of $\alpha = 105°$ and a concomitantly small buckling $\Delta z = 0.07$ Å of the second substrate layer. The bond length d_1 between the Si adatoms and the Si substrate surface layer atoms of 2.39 Å is close to the bulk bond length of Si and to the sum of the covalent radii of two Si atoms.

For C trimers adsorbed in T$_4$ positions the resulting bond length $d_1 = 1.92$ Å is again close to the sum of the covalent radii of C and Si. The three C trimer atoms saturate the substrate surface dangling bonds and form C bonds with a bond length of $d_0 = 1.41$ Å. This value is between the length of C=C double bonds (1.36 Å) in molecules and C–C single bonds (1.52 Å) in bulk diamond. The bond angle $\alpha = 86°$ is closer to the tetrahedral angle than for the $Si(T_4)$ and $C(H_3)$ adatom geometries. Because of the smaller strain in the bonds, the buckling of the C atoms in the second substrate layer ($\Delta z = 0.25$ Å) is smaller than for the $C(T_4)$ structure ($\Delta z = 0.31$ Å).

To decide which one of these configurations is the most favorable surface structure, one cannot directly compare the total energies of the different optimized adsorption models in a meaningful way since the number (one or three) and the species (Si or C) of the adatoms are different. Nevertheless, one can compare the different structures by referring to the grand canonical potential Ω at $T = 0$ K, as suggested by Qian et al. [83] and by Northrup and Froyen [84]. Application of this scheme to SiC surfaces has been discussed in detail in [24, 27, 29, 30]. In Fig. 14 we show respective changes in Ω relative to its value for the ideal surface as a function of the chemical potential μ_{Si} of the Si atoms for the different configurations studied. The chemical potential μ_{Si} is restricted [83, 84] to the range $\mu_{Si(bulk)} - \Delta H_f \leq \mu_{Si} \leq \mu_{Si(bulk)}$. ΔH_f is the formation enthalpy of 6H-SiC. Obviously, C adatoms and C trimers are even less favorable than the ideal Si-terminated surface. This result derives from the fact that C adatoms are too small to efficiently saturate the dangling bonds on the Si substrate layer atoms, as has been pointed out by Northrup and Neugebauer [24]. In particular, the bonds between C adatoms and Si substrate surface layer atoms are far from tetrahedral and are thus strongly strained. The relaxed surface is lower in energy than the ideal surface, as discussed already in Section 6.1 but the $Si_3(T_4)$, $Si(H_3)$ and $Si(T_4)$ configurations are more favorable. The $Si(T_4)$ configuration is the lowest energy structure for all μ_{Si} values both in our as well as in the results of [24]. From this analysis we conclude that the Si-terminated 6H-SiC(0001) surface should show a $\sqrt{3} \times \sqrt{3}$ reconstruction if ample Si and C atoms are offered in the gas phase. This conclusion is fully compatible with experimental data [6, 34, 36, 37, 39, 46].

The surface electronic structure for all discussed adsorption models of the Si-terminated 6H-SiC(0001)-($\sqrt{3} \times \sqrt{3}$)R30° surface has been calculated and is discussed in detail in [24, 28 to 30]. Here we restrict ourselves, for shortness sake, to the surface band structure and salient charge densities of the energeti-

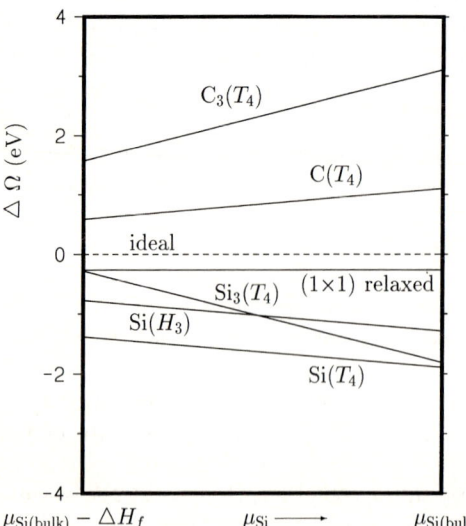

Fig. 14. Comparison of grand canonical potentials (relative to that of the ideal surface) for six different structural models (relaxation and five different reconstructions) of the Si-terminated 6H-SiC(0001)-($\sqrt{3} \times \sqrt{3}$)R 30° surface as a function of the Si chemical potential μ_{Si} for the allowed range

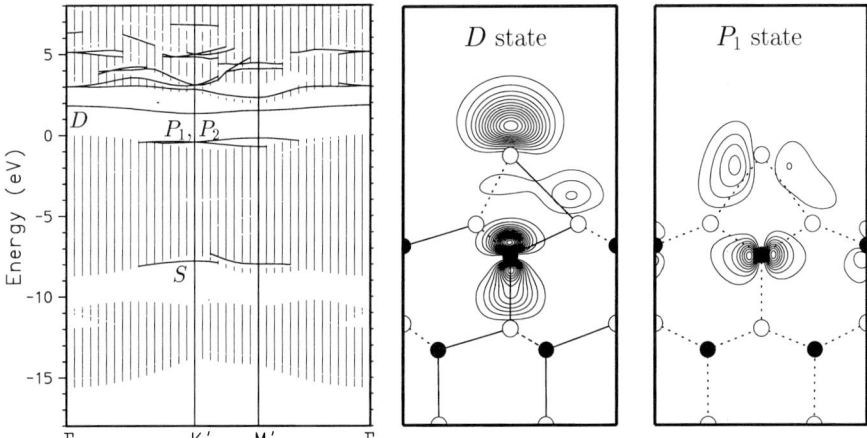

Fig. 15. Surface band structure of the Si-terminated 6H-SiC(0001)-($\sqrt{3} \times \sqrt{3}$)R 30° surface for Si adatoms in T_4 position (see Fig. 13a) and charge density contours of salient surface states at the M′ point of the SBZ. The charge density of the P_1 state is presented in the y–z plane. The charge density of the D state is presented in the x–z plane containing the middle Si–C zig-zag chain for Si adatoms in T_4 position (see Fig. 13a). The projected bulk band structure for the $\sqrt{3} \times \sqrt{3}$ SBZ is shown by vertically shaded areas

cally most favorable Si(T_4) configuration, which are shown in Fig. 15. Within the gap-energy region there are three bands of localized states. The bands P_1 and P_2 with p_x and p_y symmetry originate from the interaction of the Si(T_4) adatoms with the Si substrate surface layer atoms. Their charge densities are largely similar in nature so that we show only the P_1 state in the right panel of Fig. 15. The band D having mostly p_z wave-function character originates from the dangling-bond states localized at the Si(T_4) adatoms (see middle panel of Fig. 15). These bands show a very small dispersion because the Si adatoms interact only very weakly since they have a large distance of 5.32 Å. The D band is half filled since there is only one adatom per $\sqrt{3} \times \sqrt{3}$ unit cell so that the surface band structure is metallic (see also [24]). Contrary to the theoretical results, recent ARPES [58] and KRIPES [61] investigations have observed one band of occupied and one band of empty dangling-bond states in the gap-energy region. The authors of both references conclude on the basis of their data that the surface is semiconducting. This is an obvious contradiction between theory and experiment which calls for further investigations. One possible reason for the discrepancy could be the fact that the actual structure of this surface is more complex than was anticipated in the calculations, so far. If there were adatoms with three or five valence electrons at the surface instead of Si adatoms this would lead to semiconducting surfaces. In the first case one empty and in the second case one fully occupied dangling-bond band would result within the gap but certainly not both of them as observed in experiment [58, 61]. Another possible explanation could be related to many-body correlation effects of the Hubbard-type in the D band of Fig. 15 giving rise to a splitting of this band. This would result in a fully occupied and an empty dangling-bond band within the gap-energy region, as observed in experiment [58, 61]. Certainly, more work is needed to resolve this issue.

6.3 C-terminated 6H-SiC(0001)-$\sqrt{3} \times \sqrt{3}$ surfaces

Several $(\sqrt{3} \times \sqrt{3})R30°$ reconstruction models of the C-terminated substrate surface have been investigated, to date. Si and C adatoms and adsorbed Si trimers have been considered [27 to 30]. The optimal structures as obtained from total energy minimization are shown in Figs. 16a to c by top and side views. The relevant structure parameters are defined in the figure and their actual values are compiled in Table 2.

In the Si(T_4) and C(T_4) configurations (see Fig. 16a, b), the adatoms are bound to three substrate surface layer atoms. The bond lengths d_1 (see Fig. 16 and Table 2) turn out to be somewhat larger than the respective bulk bond lengths of SiC (1.89 Å) and of diamond (1.52 Å). In both configurations the resulting distances d_2 between adatoms and Si atoms in the second substrate layer are relatively small because of the very small covalent radius of C. In consequence, a large buckling Δz of the second substrate layer results. The Si atoms on this layer show a much larger buckling than the C atoms on the corresponding second layer of the Si-terminated surface (see Table 1). This difference is again related to the differences in angular forces at C or Si atoms involved in the reconstruction-induced bond bending.

Si trimers originally adsorbed in T_4 positions move upon energy minimization to ideal on top positions above the C atoms at the substrate layer and a geometrically ideal (1×1) structure results (see Fig. 16c). One should note, however, that three Si adatoms are only onefold coordinated to the substrate in this structure having the unsaturated dangling bonds, each.

In Fig. 17 we show the change in the grand canonical potential Ω relative to its value at the ideal surface for the investigated structures. As in the case of the Si-terminated substrate surface (cf. Fig. 14), C adatoms turn out to be less favorable than the ideal and the relaxed surface. The Si(T_4) configuration is considerably more stable for all values of μ_{Si}. The Si monolayer resulting from adsorbed Si trimers after energy minimization amazingly enough is energetically most favorable. In this case there are three equivalent adatoms per $\sqrt{3} \times \sqrt{3}$ unit cell while in the Si(T_4) configuration there is only one adatom per unit cell. Comparing these two structures, one has to have in mind that in the first config-

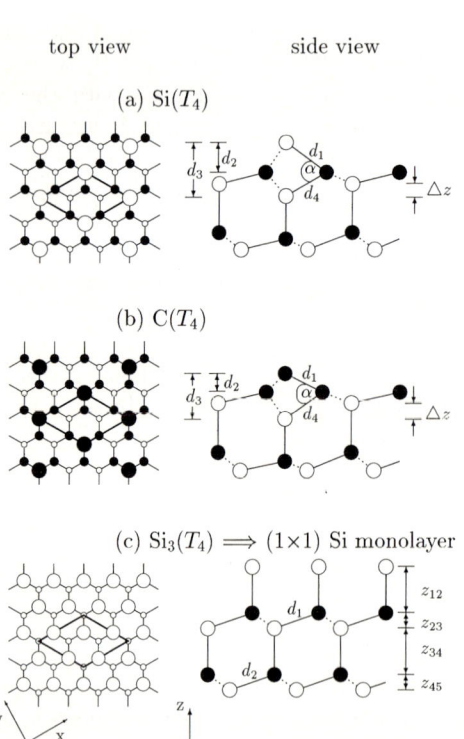

top view   side view

(a) Si(T_4)

(b) C(T_4)

(c) Si$_3$(T_4) $\Longrightarrow$ (1×1) Si monolayer

○ Si atoms   ● C atoms

Fig. 16. Top and side views of the optimized $(\sqrt{3} \times \sqrt{3})$R 30° reconstructions of the C-terminated 6H-SiC(0001) surfaces. The structure parameters are introduced in the side views (see also caption of Fig. 13). Their actual values are compiled in Table 2

Table 2

Optimized structure parameters (as defined in Fig. 16) for the Si and C adatom and Si trimer reconstructions of the C-terminated 6H-SiC(0001)-($\sqrt{3} \times \sqrt{3}$) R30° surfaces discussed in this work. Adsorption of Si trimers leads to an ideal (1×1) structure in our calculations, in which the Si surface atoms are one-fold coordinated with the C substrate surface layer atoms and have three unsaturated dangling bonds each

| | adsorption models for the C-terminated $\sqrt{3} \times \sqrt{3}$ surface | | |
	Si(T_4)	C(T_4)	Si trimer (T_4)
d_0 (Å)	–	–	3.07
d_1 (Å)	2.04	1.66	1.88
d_2 (Å)	1.24	0.73	1.88
d_3 (Å)	2.21	1.86	2.51
d_4 (Å)	1.89	1.87	1.88
Δz (Å)	0.50	0.68	0.00
α	68°	63°	109.5°

uration three Si adatoms are adsorbed in ideal onefold-coordinated sites, having three dangling bonds each, while in the second configuration one Si adatom is adsorbed in a T_4 site of the $\sqrt{3} \times \sqrt{3}$ unit cell and two free Si atoms are in the gas phase. Although the Si(T_4) adatom is stronger bound to the substrate than each Si adatom in the ideal (1×1) configuration, the monolayer system has the lower grand canonical potential. Our results clearly show that the investigated $\sqrt{3} \times \sqrt{3}$ structures of the C-terminated surface are no minimum configurations. Instead, an adsorbed Si monolayer turns out to be energetically much more favorable. Nevertheless, it seems obvious that such a monolayer cannot be the stable structure of this surface since each adlayer Si atom has three unsaturated dangling bonds in this (1×1) configuration. A more complicated reconstruction is instead to be expected. This conclusion is in accord with experiment which has observed (3×3) reconstructions [36 to 38] of this surface. Only Li and Tsong [46] have observed a $\sqrt{3} \times \sqrt{3}$ reconstruction in their STM investigations. They have annealed a (3×3) sample under a Si flux at 850 °C which changed into a $\sqrt{3} \times \sqrt{3}$ reconstruction after further annealing at 950 °C. Another possibility could be a $\sqrt{3} \times \sqrt{3}$ B_5 model, as dis-

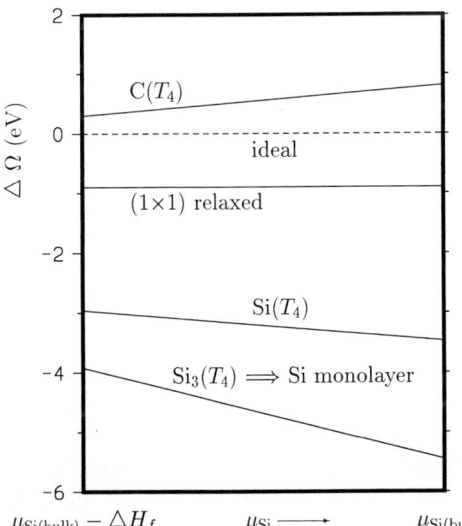

Fig. 17. Comparison of grand canonical potentials (relative to that of the ideal surface) for four different structural models (relaxation and three different reconstructions) of the C-terminated 6H-SiC(0001)-($\sqrt{3} \times \sqrt{3}$) R 30° surface as a function of the chemical potential μ_{Si} for the allowed range (for details, see text)

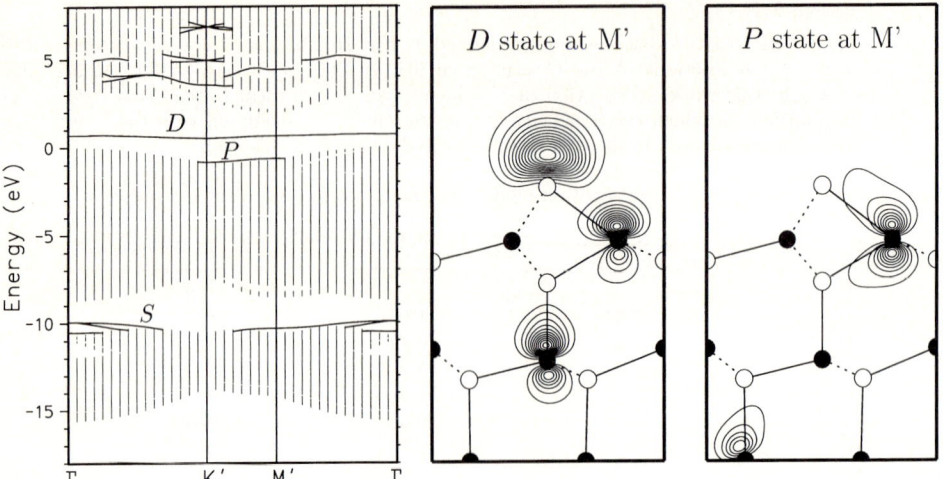

Fig. 18. Surface band structure of the C-terminated 6H-SiC(0001)-($\sqrt{3} \times \sqrt{3}$)R 30° surface for Si adatoms in T_4 position (see Fig. 16a) and charge density contours of salient surface states at the M' point of the SBZ. The charge densities are presented in the $x-z$ plane containing the middle Si–C zig-zag chain (see Fig. 16)

cussed in [12] and [46], in which Si adatoms are adsorbed in T_4 positions above the C-terminated surface and all second-layer substrate Si atoms are replaced by C atoms. Also impurity-stabilized (1×1) structures, as observed by Bermudez [6], should be considered in this context.

The surface band structure of the optimized Si(T_4) adatom configuration is shown in Fig. 18 together with charge density contours of the gap surface states. Surface band structures for the other models of the C-terminated 6H-SiC(0001) surface are given in [28 to 30]. We address this surface band structure for the sake of comparison with Fig. 15. In the gap energy region there occurs a fully occupied band P and a half-filled dangling-bond band D. The D band in Fig. 18 is roughly 1 eV lower in energy than that in Fig. 15 which originates from the stronger C potentials on the substrate surface layer of the C-terminated substrate as compared to the Si potentials on the substrate surface layer of the Si-terminated substrate. The related charge density of the dangling-bond state D in Fig. 18 is localized at the Si(T_4) adatoms and shows a strong coupling to the C atoms on the first and third substrate layers. This is significantly different from the character of the D state in Fig. 15.

7. Summary

Structural and electronic properties of cubic and hexagonal SiC surfaces as resulting from first-principles calculations have briefly been reviewed. Relaxed nonpolar β-SiC(110) and 2H-SiC(10$\bar{1}$0) show bond length contracting rotation relaxations very much like the surfaces of other ionic wide-band-gap semiconductors. Cubic Si- or C-terminated polar β-SiC(001) surfaces show significantly different reconstructions. The C-terminated surfaces exhibit double-bonded C=C surface dimers very much like the C(001)-(2 × 1) surface while the Si-terminated surface shows a reconstruction that differs considerably from that of Si(001)-(2 × 1). The polar hexagonal 6H-SiC(0001)-(1 × 1) surfaces show a strong inward relaxation of the top layer towards the substrate, the relaxation being larger for the C- than for the Si-terminated surface. For the reconstructed polar Si-terminated 6H-SiC(0001)-$\sqrt{3} \times \sqrt{3}$ surface, adsorption of Si adatoms in T_4 positions turns out to be most favorable. The polar C-terminated 6H-SiC(0001)-$\sqrt{3} \times \sqrt{3}$ surface appears to be no stable energy minimum configuration.

The surface electronic structure for a number of these surfaces has been presented and discussed in comparison with PES and EELS data. Nevertheless, many of the theoretical results are predictions, at the time being, since ARPES and KRIPES data on SiC surfaces are still fairly scarce, to date. Some of the theoretical results are in good accordance with structure and electronic structure data but significant deviations between theory and experiment are found in some cases, most noticeably for the structure of the Si-terminated β-SiC(001)-(2 × 1) surface and for the electronic structure of the 6H-SiC(0001)-$\sqrt{3} \times \sqrt{3}$ surfaces. More work seems necessary to resolve these issues.

Acknowledgements We would like to acknowledge A. Mazur and M. Rohlfing for their contributions to the results discussed in this paper. Furthermore we acknowledge V. M. Bermudez for making his contribution to this volume available to us prior to publication and for many useful comments on the subject.

References

[1] R. F. DAVIS, Z. SITAR, B. E. WILLIAMS, H. S. KONG, H. J. KIM, J. W. PALMOUR, J. A. EDMOND, J. RYU, J. T. GLASS, and C. H. CARTER, JR., Mater. Sci. Engng. **1**, 77 (1988).

[2] W. J. CHOYKE, in: The Physics and Chemistry of Carbides, Nitrides and Borides, Vol. 183, NATO Advanced Study Institute, Ser. E, Ed. R. FREER, Kluwer, Dordrecht 1990 (p. 563).

[3] G. PENSL and R. HELBIG, Festkörperprobleme/Adv. Solid State Phys. **30**, 133 (1990).

[4] P. A. IVANOV and V. E. CHELNOKOV, Semicond. Sci. Technol. **7**, 863 (1992).

[5] G. L. HARRIS, in: Properties of Silicon Carbide, Ed. G. L. HARRIS, EMIS Datareviews Ser. No. 13, INSPEC, London 1995.

[6] V. M. BERMUDEZ, Appl. Surf. Sci. **84**, 45 (1995).

[7] D. H. LEE and J. D. JANNOPOULOS, J. Vac. Sci. Technol. **21**, 351 (1982).

[8] S. P. MEHANDRU and A. B. ANDERSON, Phys. Rev. B **42**, 9040 (1990).

[9] B. I. CRAIG and P. V. SMITH, Surf. Sci. **233**, 255 (1990).

[10] B. I. CRAIG and P. V. SMITH, Surf. Sci. Lett. **256**, L609 (1991).

[11] W. LU, W. YANG, and K. ZHANG, J. Phys.: Condensed Matter **3**, 9079 (1991).

[12] P. BADZIAG, Surf. Sci. **236**, 48 (1990).

[13] P. BADZIAG, Surf. Sci. **269/270**, 1152 (1992).

[14] P. BADZIAG, Surf. Sci. **337**, 1 (1995).

[15] P. BADZIAG, Phys. Rev. B **54**, 11143 (1996).

[16] B. WENZIEN, P. KÄCKELL, and F. BECHSTEDT, in: IOP Conf. Proc. No. 137, Institute of Physics and Physical Society, London 1994 (p. 223).

[17] B. WENZIEN, P. KÄCKELL, and F. BECHSTEDT, Surf. Sci. **307/309**, 989 (1994).

[18] B. WENZIEN, P. KÄCKELL, and F. BECHSTEDT, Proc. ICPS-22, Vancouver, 1994, World Scientific Publ. Co., Singapore 1994 (p. 389).

[19] B. WENZIEN, P. KÄCKELL, and F. BECHSTEDT, Surf. Sci. **331/333**, 1105 (1995).

[20] P. KÄCKELL, J. FURTHMÜLLER, and F. BECHSTEDT, Appl. Surf. Sci. **104/105**, 45 (1996); Surf. Sci. **352/354**, 55 (1996).

[21] P. KÄCKELL, J. FURTHMÜLLER, F. BECHSTEDT, G. KRESSE, and J. HAFNER, Phys. Rev. B **54**, 10304 (1996).

[22] P. KÄCKELL, PhD Thesis, University Jena, 1996.

[23] H. YAN, A. P. SMITH, and H. JÓNSSON, Surf. Sci. **330**, 265 (1995).

[24] J. E. NORTHRUP and J. NEUGEBAUER, Phys. Rev. B **52**, R17001 (1995).

[25] M. Sabisch, P. Krüger, and J. Pollmann, Phys. Rev. B **51**, 13367 (1995).

[26] M. Sabisch, P. Krüger, A. Mazur, M. Rohlfing, and J. Pollmann, Phys. Rev. B **53**, 13121 (1996).

[27] M. Sabisch, P. Krüger, A. Mazur, and J. Pollmann, Surface Review and Letters (1996), in press.

[28] M. Sabisch, P. Krüger, A. Mazur, and J. Pollmann, in: Proc. ICPS-23, Ed. M. Scheffler and R. Zimmermann, World Scientific Publ. Co., Singapore 1996 (p. 819).

[29] M. Sabisch, PhD Thesis, University Münster, 1996.

[30] M. Sabisch, P. Krüger, and J. Pollmann, Phys. Rev. B **55**, in print (1997).

[31] J. Pollmann, P. Krüger, M. Rohlfing, M. Sabisch, and D. Vogel, Appl. Surf. Sci. **104/105**, 1 (1996).

[32] A. Catellani, G. Galli, and F. Gygi, Phys. Rev. Lett. **77**, 5090 (1996).

[33] M. Dayan, J. Vac. Sci. Technol. A **4**, 38 (1986).

[34] R. Kaplan, Surf. Sci. **215**, 111 (1989).

[35] A. J. van Bommel, J. E. Crombeen, and A. van Tooren, Surf. Sci. **48**, 463 (1975).

[36] S. Nakanishi, H. Tokutaka, K. Nishimori, S. Kishida and N. Ishihara, Appl. Surf. Sci. **41/42**, 44 (1989).

[37] U. Starke, Ch. Bram, P.-R. Steiner, W. Hartner, L. Hammer, K. Heinz, and K. Müller, Appl. Surf. Sci. **89**, 175 (1995).

[38] J. Schardt, Ch. Bram, S. Müller, U. Starke, K. Heinz, and K. Müller, Surf. Sci. **337**, 232 (1995).

[39] F. Owman and P. Mårtensson, Surf. Sci. **330**, L639 (1995).

[40] J. M. Powers, A. Wander, P. J. Rous, M. A. van Hove, and G. A. Somorjai, Phys. Rev. B **44**, 11159 (1991).

[41] J. M. Powers, A. Wander, M. A. van Hove, and G. A. Somorjai, Surf. Sci. Lett. **260**, L7 (1992).

[42] T. M. Parill and Y. W. Chung, Surf. Sci. **243**, 96 (1991).

[43] L. Muehlhoff, W. J. Choyke, M. J. Bozack, and J. T. Yates, Jr., J. Appl. Phys. **60**. 2842 (1986).

[44] V. M. Bermudez and J. P. Long, Appl. Phys. Lett. **66**, 475 (1995).

[45] V. M. Bermudez and R. Kaplan, Phys. Rev. B **44**, 11149 (1991).

[46] L. Li and I. S. T. Tsong, Surf. Sci. **351**, 141 (1996).

[47] C. S. Chang, I. S. T. Tsong, Y. C. Wang, and R. F. Davis, Surf. Sci. **256**, 354 (1991).

[48] M. A. Kulakov, P. Heuell, V. F. Tsvetkov, and B. Bullemer, Surf. Sci. **315**, 248 (1994).

[49] M. A. Kulakov, G. Henn, and B. Bullemer, Surf. Sci. **346**, 49 (1996).

[50] S. Hara, W. F. J. Slijkerman, J. F. van der Veen, I. Ohdomari, S. Misawa, E. Sakuma, and S. Yoshida, Surf. Sci. Lett. **231**, L196 (1990).

[51] S. Hara, S. Misawa, S. Yoshida, and Y. Aoyagi, Phys. Rev. B **50**, 4548 (1994).

[52] F. Semond, P. Soukiassian, A. Mayne, G. Dujardin, L. Douillard, and C. Jaussaud, Phys. Rev. Lett. **77**, 2013 (1996).

[53] P. Soukiassian, F. Semond, L. Douillard, A. Mayne, G. Dujardin, L. Pizzagalli, and C. Joachim, Phys. Rev. Lett. 1997, in print.

[54] S. Tyc, J. Physique **4**, 617 (1994).

[55] L. I. Johansson, F. Owman, and P. Mårtensson, Phys. Rev. B **53**, 13793 (1996).

[56] J. P. Long, V. M. Bermudez, and D. E. Ramaker, Phys. Rev. Lett. **76**, 991 (1996).

[57] M. L. Shek, K. E. Miyano, Q.-Y. Dong, T. A. Callcott, and D. L. Ederer, J. Vac. Sci. Technol. A **12**, 1079 (1994).

[58] L. I. Johansson, F. Owman, and P. Mårtensson, Surf. Sci. Lett. **360**, L483 (1996).

[59] L. I. Johansson, F. Owman, and P. Mårtensson, Surf. Sci. Lett. **360**, L478 (1996).

[60] L. I. Johansson, F. Owman, P. Mårtensson, C. Persson, and U. Lindefelt, Phys. Rev. B 353, 13803 (1996).

[61] J.-M. Themlin, I. Forbeaux, V. Langlais, H. Belkhir, and J.-M. Debever, submitted to Phys. Rev. B.

[62] V. M. Bermudez, phys. stat. sol. (b) **202**, 447 (1997).

[63] U. Starke, phys. stat. sol. (b) **202**, 475 (1997).

[64] P. Hohenberg and W. Kohn, Phys. Rev. **136**, B864 (1964). W. Kohn and L. J. Sham, Phys. Rev. **140**, A1133 (1965).

[65] L. KLEINMAN and D. M. BYLANDER, Phys. Rev. Lett. **48**, 1425 (1982).

[66] N. TROULLIER and J. L. MARTINS, Phys. Rev. B **43**, 1993 (1990).

[67] D. VANDERBILT, Phys. Rev. B **41**, 7892 (1990).

[68] D. R. HAMANN, M. SCHLÜTER. and C. CHIANG, Phys. Rev. Lett. **43**, 1494 (1979).

[69] M. SABISCH, Diploma Thesis, University Münster, 1993.

[70] D. M. CEPERLEY and B. I. ALDER, Phys. Rev. Lett. **45**, 566 (1980).

[71] J. P. PERDEW and A. ZUNGER, Phys. Rev. B **23**, 5048 (1981).

[72] J. IHM, A. ZUNGER, and M. L. COHEN, J. Phys. C **12**, 4409 (1979).

[73] C. G. BROYDEN, Math. Comp. **19**, 577 (1965).
See also D. D. JOHNSON, Phys. Rev. B **38**, 12807 (1988).

[74] M. SCHEFFLER, J. P. VIGNERON, and G. BACHELET, Phys. Rev. B **31**, 6541 (1985).
P. KRÜGER and J. POLLMANN, Physica **172B**, 155 (1991).

[75] A. GARCIA and M. L. COHEN, Phys. Rev. B **47**, 4215 (1993); **47**, 4221 (1993).

[76] P. KRÜGER and J. POLLMANN, Phys. Rev. Lett. **74**, 1155 (1995).

[77] V. M. BERMUDEZ, private communication.

[78] M. HOLLERING, A. ZIEGLER, R. GRAUPNER, B. MATTERN, L. LEY, A. P. J. STAMPFL, J. D. RILEY, R. C. G. LECKEY, J. BERNHARDT, J. SCHARDT, U. STARKE, and K. HEINZ, to be published.

[79] R. J. HAMERS, Phys. Rev. B **40**, 1657 (1989).

[80] J. NOGAMI, S.-I. PARK, and C. F. QUATE, Phys. Rev. B **36**, 6221 (1987).

[81] J. NOGAMI, S.-I. PARK, and C. F. QUATE, Surf. Sci. **203**, L631 (1988).

[82] E. GANZ, I.-S. HWANG, F. XIONG, S. K. THEISS, and J. GOLOVCHENKO, Surf. Sci. **257**, 259 (1991).

[83] G.-X. QIAN, R. M. MARTIN, and D. J. CHADI, Phys. Rev. B **38**, 7649 (1988).

[84] J. E. NORTHRUP and S. FROYEN, Phys. Rev. Lett. **71**, 2276 (1993).

phys. stat. sol. (b) **202**, 447 (1997)

Subject classification: 68.35.Bs; 68.35.Dv; 68.55.Jk; S6

Structure and Properties of Cubic Silicon Carbide (100) Surfaces: A Review

V. M. Bermudez[1])

Electronics Science and Technology Division, Naval Research Laboratory, Washington, DC 20375-5347, USA

(Received January 31, 1997)

A review is presented of recent experimental and theoretical work on the structure and properties of the clean, ordered and well-characterized (100) surfaces of cubic SiC.

1. Introduction

The many uses for SiC in electronics have been well documented [1 to 3]. Many of these applications, as well as the growth of the material itself, are critically dependent on the structural and electronic properties of surfaces and interfaces. From a basic perspective, SiC is unique in that it is a IV–IV compound semiconductor. As such it forms a "bridge" between the elemental column-IV materials Si and Ge and the III–V's GaN, GaAs, etc. These considerations have led to many experimental and theoretical studies of SiC surfaces, a significant fraction of which have appeared in the past few years.

This paper reviews recent experimental results for clean, ordered and well-characterized (100) surfaces. Cubic (111) surfaces are excluded since they are more properly discussed [4] together with those of hexagonal SiC(0001) and (000$\bar{1}$), to which they are essentially identical. There have, to date, been no experimental data reported for SiC(110) surfaces. Work up to about mid-1993 has already been briefly reviewed [5]. A discussion of metal/SiC interfaces was also given [5], but more recent work will be omitted. Mention is also made here of theoretical results, which are discussed extensively elsewhere [6].

2. Cubic SiC(100)

2.1 General remarks

To our knowledge, all surface studies of cubic SiC (also termed "β-SiC" or "3C-SiC") have been done on films grown on single-crystal Si by chemical vapor deposition (CVD). This has involved largely the (100) surface, with less work being done on the (111) and none on (110), probably due to a lack of readily-available samples. The CVD growth is not epitaxial, since the β-SiC lattice constant is smaller than that of Si by ≈20%, but the substrate orientation determines that of the film.

In the USA, virtually all surface data have been obtained on samples grown by Powell et al. [7] at the NASA-Lewis Research Center. Typically, these films are ≈3 μm

---

[1]) e-mail: bermudez@estd.nrl.navy.mil.us

thick and n-type, with carrier densities in the 6×10^{16} to $8 \times 10^{17}/cm^3$ range. Nitrogen is the (unintentional) n-type dopant [7]. It has proven more difficult to obtain good p-type material, possibly because of the need to eliminate traces of N_2 from the CVD process.

There is an extensive literature on macroscopic bulk defects (twins, stacking faults, grain boundaries, etc.) in CVD β-SiC films, as revealed by microscopy. A few references are given in [8], but there has been little effort in correlating the quality of reconstructed surfaces with these defects. Surface texture is also important. For example, one very smooth sample exhibited an unusual c(4 × 4) Si-terminated structure [9], and rough surfaces are more difficult to clean [10]. If sufficiently large bulk single crystals of β-SiC become available in the future [11], surfaces of these presumably more nearly perfect samples could be studied.

Films grown on Si oriented to within 1/4° of (100) show two-domain reconstructions in low-energy electron diffraction (LEED). The two orthogonal domains result from mono-atomic steps at the Si surface. In many experiments (e.g., polarized X-ray absorption) single-domain structures are needed. These can be grown on substrates "misoriented" by $\geq 1/2°$ away from [100] toward [01$\bar{1}$]. The favored domain is the one in which Si dangling bonds at the surface lie parallel to the step edges (i.e., in the [011] direction). This domain can be stabilized by Si–Si dimer formation; whereas, the other requires dimers to cross steps. There is also evidence [12 to 14] that β-SiC films with a lower density of macroscopic bulk defects are grown on such "off-axis" substrates.

Clean, well-ordered and stoichiometric SiC(100) surfaces generally cannot be prepared by ion bombardment and annealing. In one case [15] a (100)-(1 × 1) LEED pattern was seen after 1 keV ion bombardment at 650 °C. In another [16], the c(2 × 2) pattern of a fully C-terminated surface was found after 0.5 keV ion bombardment at 600 °C followed by flashing to 900 °C. In neither case were well-ordered surfaces formed reliably. In addition to the problem of different Si and C sputtering rates, Si rapidly desorbs at temperatures well below that needed to repair the lattice damage [17], leading to a high concentration of graphitic C as an impurity [18]. The term "graphitic C", as opposed to "carbidic C", is applied to C not bonded to Si. Some authors have used "heat cleaning", annealing at high temperature in ultra-high vacuum (UHV). Unless the starting surface consists of excess Si remaining from growth, this must necessarily lead to a Si-deficient surface since impurity O desorbs as SiO. Even in preparing a C-terminated surface by heat cleaning, it is difficult to avoid some amount of graphitic C.

Fortunately, lower-temperature in-situ chemical cleaning techniques have been developed by Kaplan [9, 10]. In one method, Ga metal is deposited from a Knudsen cell then thermally desorbed. Under these conditions, Ga forms a volatile oxide but not a stable carbide. The exact cleaning mechanism is unknown, but it is believed that graphitic C is displaced from the SiC surface and "floats" to the Ga metal surface. It is then carried away during desorption, along with oxides formed by reaction of Ga with chemisorbed O. The cleaning process is probably aided by the high heat of formation [19] of Ga_2O_3 ($\Delta H_F = -1084$ kJ/mol) versus SiO_2 and SiO ($\Delta H_F = -912$ and -96.3 kJ/mol, respectively).

A widely-used approach [9], termed "Si flux-annealing", is to heat the sample in the vapor from a nearby resistively-heated Si wafer. Impurity O probably desorbs as SiO,

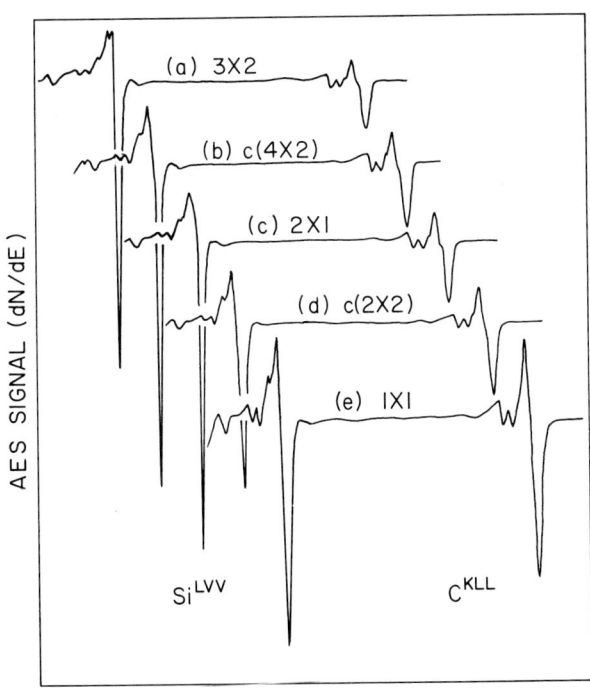

Fig. 1. Auger data [9] for the commonly-observed SiC(100) reconstructions. The spectra have been displaced vertically and horizontally for clarity, and relative intensities of different spectra are not quantitative

with the lost Si being replaced from the incident flux, while graphitic C reacts to form SiC. Typically, no O is detectable on the final surface by either Auger electron (AES) or X-ray photoemission (XPS) spectroscopies. Depending on post-cleaning thermal and chemical treatment, various reconstructed surfaces can be reproducibly formed. A dis-

Table 1
(Si LVV)/(C KLL) PPH ratio versus reconstruction[a])

surface	Kaplan [9]	Dayan [20]	Hara [21][c])	Parrill and Chung [22][d])
(3×2)	5 ± 0.3		6.5 to 7	≈ 3.5 to 4.7
c(4×2), (2×1)[b])	4 ± 0.5	1.8 to 3.5	≈ 2 to 6.5	≈ 1.0 to 3.3
c(2×2)	3 ± 0.5	1.38 to 1.75	0.8	0.8 to 1.0
(1×1)	≈ 1.4			< 0.8

[a]) The results are intended only to indicate trends seen in each study, as well as the variability for nominally identical surfaces. The ratio for a given surface depends on primary beam energy, modulation amplitude, electron detector, etc., and results from different laboratories cannot be compared without additional information.
[b]) The (2×1) and c(4×2) are considered chemically equivalent for the purpose of this table.
[c]) These results were obtained for a series of Si_2H_6 doses, starting with a c(2×2) surface. The axis of the electron energy analyzer made a large angle ($\approx 66°$) with the surface normal. Due to the increased surface sensitivity in this configuration, the spread in PPH ratios between C- and Si-terminated surfaces is larger than in other studies.
[d]) These values indicate the ranges over which the respective fractional-order spots could be detected in LEED. The ranges over which the patterns were "good" were narrower.

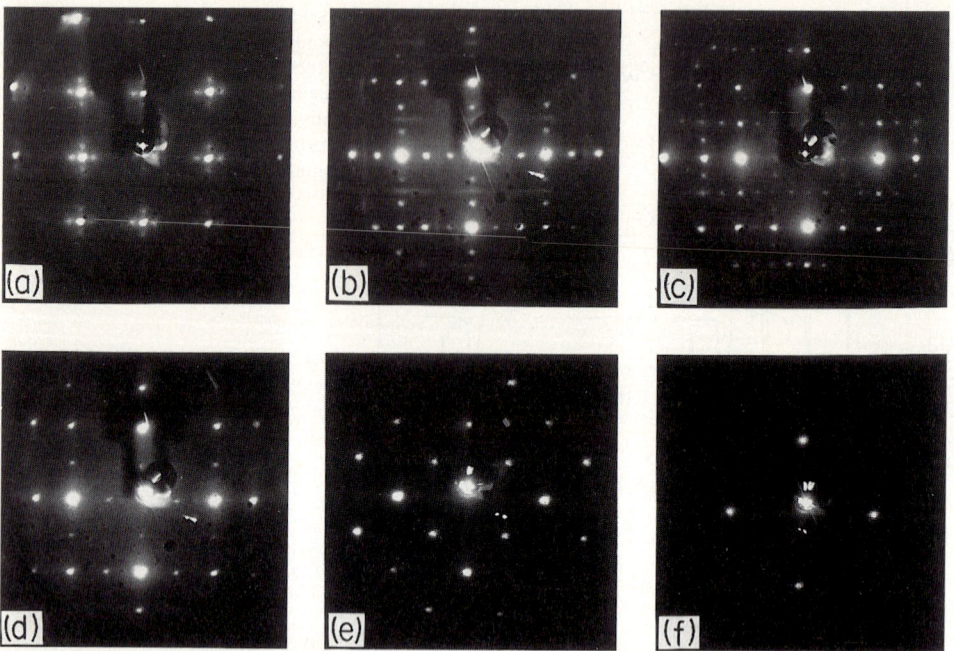

Fig. 2. Two-domain LEED patterns [9], corresponding to Fig. 1, at a primary beam energy of 90 eV. a) An ordered partial monolayer of oxygen, b) (3×2), c) $c(4 \times 2)$, d) (2×1), e) $c(2 \times 2)$, f) (1×1)

tinct advantage is that surfaces graphitized or otherwise contaminated can be easily restored.

Bellina et al. [16] have used a technique involving metallic Cr. Here, SiC(100) is ion bombarded to remove O; then Cr is deposited in situ. Annealing forms Cr carbide which desorbs above 750 °C, leaving a well-ordered Si-terminated (2×1) surface.

Most of the work reviewed below was performed on surfaces prepared by Si flux-annealing or by heat-cleaning. These structures range from the Si-rich (3×2) to a Si-depleted, graphite-like (1×1). Fig. 1 shows AES surveys for this full range [9], and Table 1 lists representative (Si LVV)/(C KLL) peak-to-peak height (PPH) ratios from various sources [9, 20 to 22]. Fig. 2 and 3 show corresponding LEED and electron energy loss spectroscopy (ELS) data, all of which are discussed below. A general characteristic of these structures is their thermal stability; the various LEED patterns are observable [9] up to at least 600 °C. Presumably, this is due to a high β-SiC Debye temperature ($\Theta_D = 1157$ °C [23], versus 363 °C for Si [24]) and to a strong interaction between surface atoms.

Temperature measurement during annealing is important, especially as it affects comparison of results from different laboratories. In most work the resistively-heated Si substrate is viewed with an optical or infrared pyrometer since SiC is transparent over the relevant wavelength range. Since the thin CVD film is in good contact with the substrate, the Si and SiC temperatures are assumed to be identical. Methods for attaching thermocouples to non-metallic materials have been described recently [25].

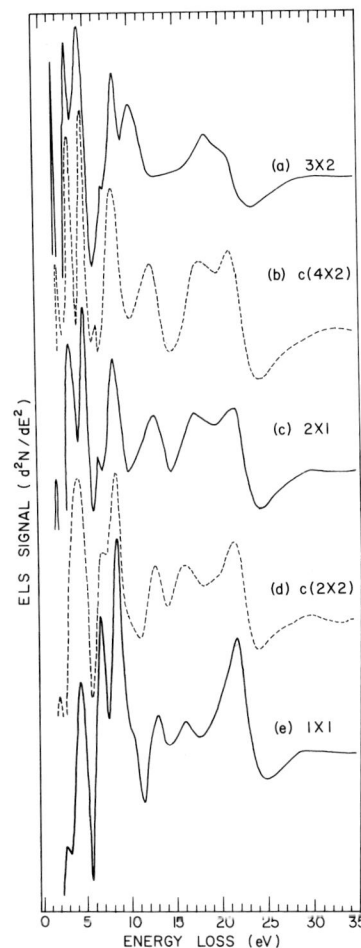

ELS SIGNAL (d²N/dE²)

ENERGY LOSS (eV)

Fig. 3. Second-derivative ELS data [9], at a primary beam energy of 100 eV, corresponding to Fig. 1. Data are displaced vertically for clarity, and relative intensities of different spectra are not quantitative. Transition energies correspond to positive extrema

2.2 The (100)-(2 × 1) surface

Like the (100) surface of Si, that of SiC exhibits a (2×1) LEED pattern. For Si, the interpretation in terms of Si–Si dimers is well established [26]. The dimers are generally agreed to be asymmetric, or "buckled", i.e., inclined with respect to the surface plane.

The SiC(100)-(2×1) was first reported, for heat-cleaned surfaces, by Dayan [20]. Bellina et al. [16] and Kaplan and Parrill [10] also formed this surface by in-situ chemical cleaning. Kaplan [9] prepared surfaces by Si flux-annealing, followed by further annealing in UHV, and studied the (2×1) for both single- and two-domain samples. This structure can also be formed [27 to 29] by dosing the C-terminated $c(2 \times 2)$ surface with Si_2H_6 (disilane) at high temperature.

Dayan [20] also formed a (2×1) by thermal desorption of O in UHV from initially Si- or C-terminated surfaces which had been given O_2 exposures, at elevated temperature, of from 3×10^4 to 10^{10} L (1 Langmuir [L] = 1×10^{-6} Torr s). Such treatment is expected to remove Si, in the form of SiO, and may produce a Si-deficient surface. One should note the existence of a C-terminated (1×2) phase (discussed below) which is indistinguishable from the Si-terminated (2×1) in a two-domain LEED pattern.

Hara et al. [30] formed the (2×1) (and other structures) by etching Si from an initially Si-rich (3×2) surface via heating in 1×10^{-7} Torr of O_2 at 1050 °C. The resulting surface exhibited a few percent of a monolayer (ML) of residual O in AES. With continued etching, a second "(2×1)" phase appeared, with a lower (Si LVV)/(C KLL) PPH ratio than the first. This is probably the (1×2) C-terminated structure mentioned above which, again, is indistinguishable from a (2×1) in a two-domain LEED pattern.

Hara et al. [21] studied the full range of (100) reconstructions using medium-energy ion scattering (MEIS) and found that the (2×1) corresponds to termination in a full ML of Si. This has been supported by quantitative growth studies [27 to 29] measuring the amount of Si required to convert a C-terminated surface to a series of Si-terminated structures.

Powers et al. [31] have done a LEED structural analysis for this surface and obtained the buckled-dimer model shown in Fig. 4. Soukiassian et al. [32] have reported scanning tunneling microscopy (STM) results indicating that the (2×1) occurs only in regions

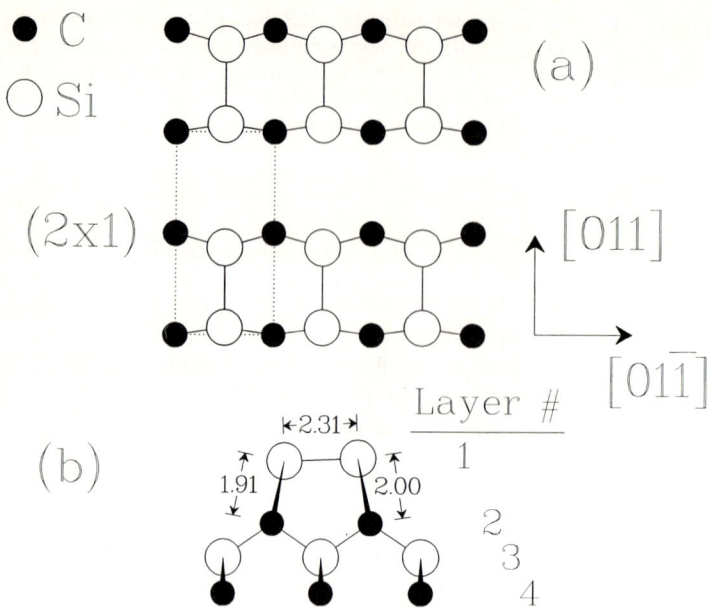

Fig. 4. a) A model for the SiC(100)-(2 × 1) surface (single-domain) viewed along the surface-normal [100] direction. Only the surface Si (Layer #1) and first-underlayer C (Layer #2) are shown. The dotted lines show the unit cell. Atom sizes and distances are approximately to scale, but atomic covalent radii have been reduced by a factor of two for clarity. The axis convention is such that the (111) plane is Si-terminated. b) The buckled Si–Si dimer, viewed along the [01$\bar{1}$] direction, found by Powers et al. [31] from the analysis of LEED data. Some bond distances are shown; others are given in the original reference. The bulk Si–Si distance is 3.08 Å

with a high density of defects, such as missing dimers. The c(4 × 2) (see below), comprised of symmetric dimers, is found to be the natural structure of a well-formed surface.

The idea that the (2 × 1) is "defective" is consistent with other observations. The LEED pattern appears over a wider range of surface composition than is the case for other structures [33] (see also Table 1). The c(4 × 2), thought to be characteristic of a very well-ordered terminating Si ML [9, 32, 34], can be formed from the (2 × 1) by controlled deposition of Si [34]. In XPS, the (2 × 1) shows a larger inhomogeneous broadening of the Si 2p core level, and a larger band bending, than do other reconstructions [8]. The larger band bending implies a higher density of surface defects. Other observations will be noted below in connection with the C 1s XPS.

Several groups [8, 22, 33 to 36] have reported Si 2p XPS data for the (2 × 1). Surface-induced binding energy (BE) shifts were first shown clearly by Shek et al. [36][2]), and representative data [8] are given in Fig. 5. The bulk and surface features are identified by the dependence of relative intensity on $h\nu$ (i.e., on photoelectron escape depth). All components show substantial inhomogeneous broadening. Some is due to phonon coupling, as shown by the temperature dependence of the widths, and some to strain or other imperfection [8]. The large Gaussian widths obscure any fine-structure related to

---

[2]) This paper reports data for a (3 × 2) surface; however, subsequent work [34] indicated that it was essentially a c(4 × 2).

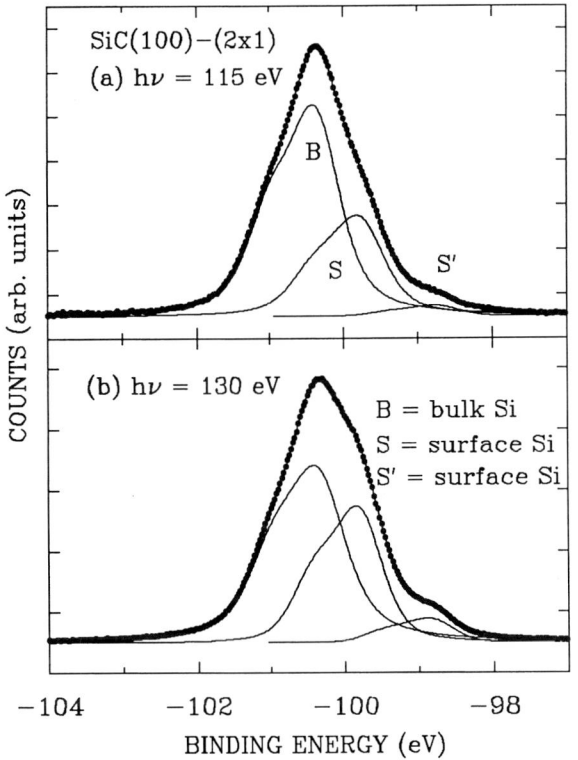

Fig. 5. SiC(100)-(2 × 1) Si 2p XPS for a) bulk- and b) surface-sensitive conditions [8]. The points are data, and the lines are least-squares fits of a sum of Gaussian-broadened Lorentzian (Voigt) functions and a polynomial background (which has been subtracted from the data and the fit for display). B labels the bulk spin–orbit doublet and S and S′ the surface-shifted components. The relative intensity of a) and b) is not quantitative. The resolution is ≈0.35 eV, and binding energy is referenced to the Fermi level

reconstruction, as seen for elemental Si(100) [37, 38]. Fig. 5 also shows an S′ feature at very low BE, first observed by Shek et al. [34, 36], which will be discussed below in connection with the c(4 × 2).

Many groups have used XPS of the C 1s level (BE ≈ 284 eV), excited by MgKα ($h\nu = 1253.6$ eV) or AlKα ($h\nu = 1486.6$ eV) radiation, to study near-surface compositional changes in SiC. However, these spectra typically have neither the surface sensitivity nor the resolution to provide detailed information concerning reconstruction. The only such data for the (2 × 1) have been reported by Douillard et al. [33] at $h\nu = 340$ eV and by Shek [34] at $h\nu = 332$ eV. Asymmetry is seen to higher BE, suggesting a second component in addition to the bulk C 1s. Since a "perfect" (2 × 1) surface, terminated in a full ML of Si, is not expected to show a significant surface-induced C 1s BE shift, this suggests Si vacancies in the "real" (2 × 1). This extra feature is absent for the c(4 × 2) [34].

Surface-sensitive ultraviolet photoemission spectroscopy (UPS) and XPS data for the (2 × 1) valence band (VB) [8, 22, 35][3]) are in general agreement. Fig. 6 shows typical data, for good resolution, which are in excellent agreement with the theoretical atom- and momentum-resolved density-of-states (DOS) results of Li and Lin-Chung [39] for

---

3) In [35], the (3 × 2) surface is referred to as "(3 × 1)" because the 1/2-order LEED spots were streaked and relatively weak, in comparison to the 1/3-order spots, as is usually the case for this surface (e.g., Fig. 2). Also, the c(2 × 2) model used was the most widely accepted one at the time.

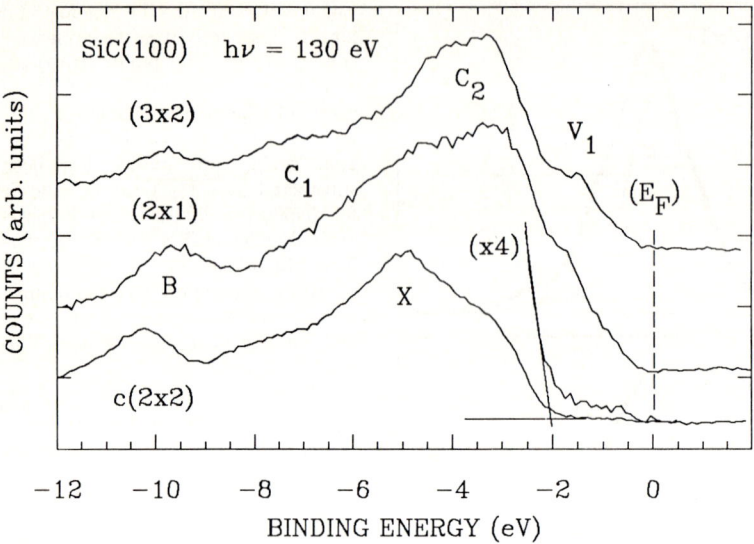

Fig. 6. Surface-sensitive XPS data showing the upper part of the valence band [8]. V_1 marks a surface state [36] and E_F the Fermi level. The $c(2 \times 2)$ VBM is shown at fourfold magnification with an extrapolation of the edge to zero intensity. The labels B, C_1, etc. conform to those in most published data. A deep valence state at about -14 eV (labelled A in [22, 35]) is omitted. The resolution is ≈ 0.35 eV. Upward band bendings of 0.25 ± 0.10, 0.85 ± 0.15 and 0.40 ± 0.15 eV, respectively, are seen [8] for the $c(2 \times 2)$, (2×1) and (3×2)

bulk β-SiC. An upward band bending [8] of $\delta \varphi_B = 0.85 \pm 0.15$ eV is found for the (2×1), significantly larger than for other surfaces. The variation in $\delta \varphi_B$ with surface preparation is also seen [22] in the bulk Si 2p and C 1s BEs.

The V_1 feature was first seen by Shek et al. [36]. This is thought to be an intrinsic surface state, based on its intensity relative to the VB, its location above the bulk valence band maximum (VBM) and its sensitivity to oxygen adsorption. The surface is seen to be semiconducting, rather than metallic, since the V_1 intensity goes to zero below the Fermi energy. This is an important observation, which can be compared with surface band-structure results. Another surface state (V_2) [36], at ≈ 1 eV below V_1 for $h\nu = 65$ eV, is not clearly evident at $h\nu = 130$ eV but may cause the apparent splitting of the C_2 peak in Fig. 6.

For the (2×1), several groups have given surface-sensitive ELS data for valence and plasmon excitation [9, 10, 22, 35, 40] and for excitation of the Si 2p [20, 40] and C 1s [40] levels. Data have been shown in either first- or second-derivative form, and examples of the latter are given in Fig. 3. Peak positions are easily determined in second-derivative data, but such spectra are subject to artifacts [41]. Although not well-resolved in all data, some spectra [9, 10, 35] show a peak (S_1, not labelled in Fig. 3) at about 1.8 eV versus the bulk β-SiC bandgap of $E_g = 2.39$ eV. This feature is contamination-sensitive, suggesting that the initial state may be the V_1 surface state detected in UPS. A second peak (S_2, not labelled in Fig. 3), at ≈ 5.3 eV, is sensitive to adsorbates [35, 40] and is assigned to another surface state transition. No effort has been made to assign other, bulk-related, features to structure in the VB and conduction band (CB).

Surface-sensitive Si 2p ELS data have been reported [20, 40] for the (2×1). Losses at >101 eV may reasonably be assigned to transitions into the CB. Another feature, at ≈98.5 eV, was tentatively ascribed [40] to patches of excess Si and may be related in some way to the Si 2p "S'" peak (Fig. 5). Dayan [20] has related bulk features in the Si 2p ELS of the (2×1), and other SiC(100) surfaces, to structure in the CB DOS. The surface-sensitive (2×1) C 1s ELS [40] shows no obviously surface-related features and resembles that of bulk diamond, as expected for a fully Si-terminated SiC surface.

The Si LVV and C KLL AES lineshapes have been shown for the (2×1) [9, 16, 20, 22, 40], and typical results are given in Fig. 7 and 8. Although both spectra show characteristic features for the (2×1), and for other surfaces, these have not been analyzed quantitatively in terms of structure in the surface VB DOS.

The (2×1) has been studied theoretically using semi-empirical [42 to 52] or ab initio [53 to 57] methods, and the results are controversial. A critical evaluation of the different approaches is beyond the scope of this review, and the various results will simply be summarized. With one exception [45], semi-empirical studies find strong Si–Si dimers (bond energy ≈1 eV) with a length of 2.16 to 2.46 Å, in accord with LEED results [31]. However, only one [43] finds asymmetric dimers. On the other hand, all ab initio studies except one find weak, symmetric dimers with a length of 2.58 to 2.75 Å and a bond energy of ≤10 meV. One group [53] finds strong dimers in a $p(2 \times 2)$ unit cell, rather than a (2×1) or $c(4 \times 2)$. The small dimer energy found in most ab initio work appears in conflict with the observation [9] of (2×1) and $c(4 \times 2)$ LEED patterns up to 600 °C. Integrating the Debye heat capacity (C_V) [24] up to 873 K, with $\Theta_D = 1430$ K [23], gives ~116 meV/atom for the lattice thermal energy.

Craig and Smith [43, 58] calculated the effect of H-atom chemisorption on the (2×1) to form the monohydride phase. This involves a single Si–H bond at each surface site with

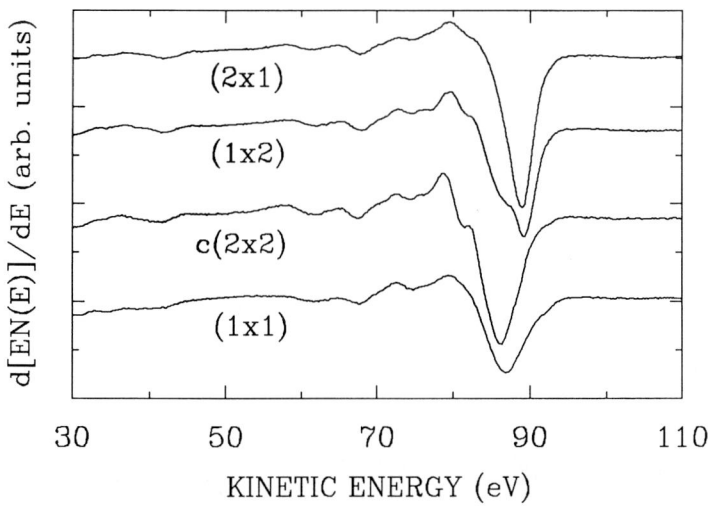

Fig. 7. Evolution of the Si LVV Auger spectrum from Si-terminated (2×1) through C-terminated $c(2 \times 2)$ to graphitic (1×1) [40]. The (1×2) is an intermediate phase (see text). Spectra have been displaced vertically for clarity, and relative intensities are not quantitative. The resolution is ≈1 eV and the primary beam energy 3 keV

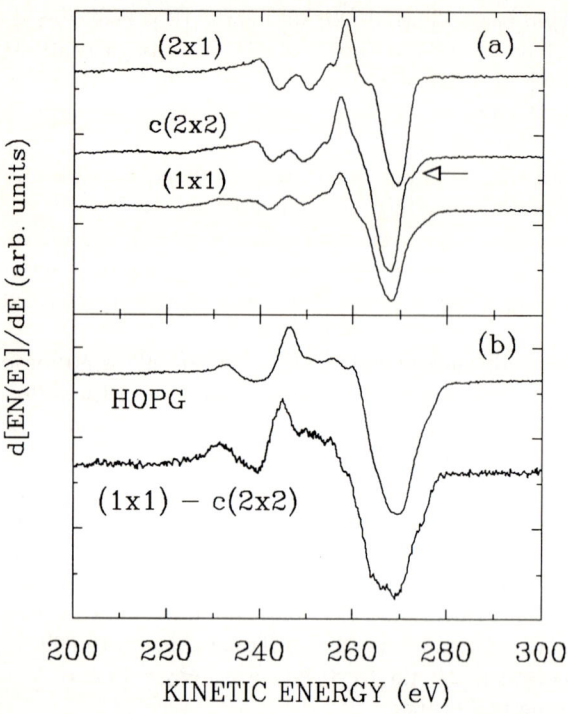

Fig. 8. a) Similar to Fig. 7 but showing C KLL Auger data [40]. The arrow marks a c(2 × 2) feature discussed in the text. It is also seen for the intermediate (1 × 2) surface (not shown). b) Data for highly-oriented pyrolytic graphite (HOPG) and the result of the subtracting the c(2 × 2) spectrum from that of the (1 × 1). Spectra have been displaced vertically for clarity, and relative intensities are not quantitative. The resolution is ≈1.8 eV and the primary beam energy 3 keV

the Si–Si dimer remaining intact. They find a weakened, symmetric dimer (length 2.476 Å versus 2.329 Å when free of H). A local DOS is also given [58] showing the distribution of H-derived valence states. Although no experimental results have been reported for this (2 × 1)H monohydride phase, a comparison of theory with structural (LEED and STM) or UPS data could be useful in evaluating various models for the clean (2 × 1).

In addition to the dimer parameters, theory and experiment can be compared with regard to surface states. Of those works reporting band structures, both semi-empirical [44, 45, 48] and ab initio [54] results agree as to the existence of a Si dangling-bond state near the VBM (i.e., V_1, Fig. 6). The various results differ throughout the surface Brillouin zone, but angle-resolved UPS data are needed for a critical test.

The observation that the (2 × 1) surface is semiconducting can be compared with band-structure results. For elemental Si(100)-(2 × 1), symmetric dimers result in a metallic surface-state band formed from the half-filled dangling orbitals. Buckling leads to charge asymmetry [26], resulting in a surface-state bandgap. For SiC(100), on the other hand, Wenchang et al. [45] find the ideal Si-terminated (1 × 1) to be metallic but the (2 × 1) to be semiconducting even for symmetric dimers; whereas, Hu et al. [48] find a metallic (2 × 1) surface with symmetric dimers. Sabisch et al. [54], using ab initio theory, find their weakly- and symmetrically-dimerized (2 × 1) to be semiconducting; whereas, strong, buckled dimers give a metallic surface.

2.3 The (100)-c(4 × 2) and p(2 × 2) surfaces

Until recently, the c(4 × 2) surface was assumed to be derived from the (2 × 1), in analogy with Si(100). Such a c(4 × 2) can be modelled [9] as an "antiferromagnetic" arrange-

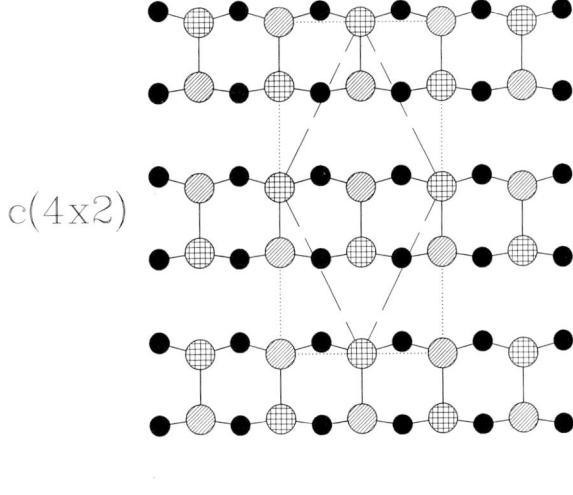

c(4x2)

Fig. 9. Similar to Fig. 4 but showing a "buckled-dimer" model proposed [9] for the c(4 × 2). Different cross-hatchings of the dimer atoms indicate up and down displacement along the surface normal. Note the "antiferromagnetic" ordering along the [011] and [01Ī] directions (cf. Fig. 4). Dashed and dotted lines show the primitive and non-primitive unit cells, respectively. The lower diagram shows the corresponding LEED pattern with the integral- and fractional-order spots and the primitive vectors indicated

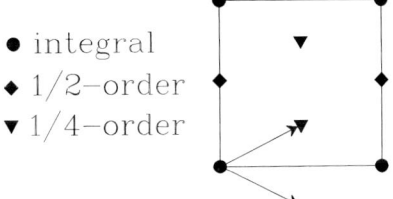

- integral
- 1/2–order
- 1/4–order

ment of buckled dimers (Fig. 9) in which each dimer is surrounded by oppositely-tilted nearest neighbors. Experimentally, this structure is not easy to form, being sensitive to contamination by adsorption from the UHV background [21, 34]. Many samples never show a c(4 × 2), perhaps due to strain or surface roughness [9]. For those that do, it can be reproducibly obtained and is seen in LEED up to ≈600 °C [9]. Kaplan [9] compared ELS and AES data for c(4 × 2) and (2 × 1) and found few, if any, differences, suggesting that the two are chemically equivalent.

Soukiassian et al. [32, 59] have recently given STM data supporting a different model. Here, dimers are symmetric but displaced vertically, either up or down along the surface normal in the "antiferromagnetic" pattern shown in Fig. 9. The height difference in this "alternatively up- and down-dimer" model is ≈0.1 Å. The c(4 × 2) is found in defect-free areas of the surface; whereas the (2 × 1) is observed in Si-deficient regions. Kitamura et al. [60] have independently proposed a similar c(4 × 2) model, based also on STM. A p(2 × 2) structure was sometimes observed, in agreement with the results of Shek [34], and assigned to a "layered antiferromagnetic" arrangement in which all (symmetric) dimers in a given [011] row are either "up" or "down", with up/down alternation in the [01Ī] direction (cf. Fig. 4).

An extensive set of XPS and LEED data for the "c(4 × 2)*" has been given by Shek [34, 36]. This surface shows a c(4 × 2) pattern with weak additional spots suggestive of a p(2 × 2). Si 2p near-edge X-ray absorption fine-structure (NEXAFS) results (also termed "partial photoyield" or "constant final state" spectroscopy) are also given. Here, one photoexcites a core electron into an empty final state and measures, versus $h\nu$, the yield of secondaries due to inelastic scattering of Auger electrons produced by core-hole decay. Bulk or surface sensitivity is obtained by appropriate choice of the detected electron kinetic energy (KE).

The $c(4 \times 2)^*$ is obtained by adsorption of Si on a (2×1), consistent with the latter being Si-deficient [32, 33]. The Si 2p XPS data are similar to those in Fig. 5. The S and S' intensities are correlated with that of the $c(4 \times 2)^*$ LEED pattern, and the S'/S relative intensity remains fixed during a series of Si doses. This indicates that S and S' are intrinsic to the $c(4 \times 2)^*$ and that S' may not be due to excess Si, as suggested earlier [8]. However, the low S' intensity and BE, relative to S, remain unexplained. The Si 2p NEXAFS shows an empty Si-derived surface state near the conduction band minimum (CBM).

Some theoretical results have been given [52, 57] for the buckled-dimer $c(4 \times 2)$. Although they differ as to the length and energy of the Si–Si dimer, both agree that this $c(4 \times 2)$ is less stable than the (2×1). Some work has also been reported [52, 53, 56] for $p(2 \times 2)$ surfaces, but the results differ as to the structure and stability of this phase.

2.4 The (100)-(N × 2) surfaces: (3 × 2), (5 × 2) and (7 × 2)

The (100)-$(N \times 2)$ structures, of which the most widely studied is the (3×2), are formed by chemisorption of a partial ML of "excess" Si on the terminating Si layer of the (2×1) or $c(4 \times 2)$. The (3×2) was first reported by Dyan [20] for some heat-cleaned samples which presumably had a layer of excess Si remaining from CVD growth. The (3×2) is reliably produced on any SiC(100) surface cleaned by Si flux-annealing [9]. The (5×2) can be formed [21] by carefully annealing the (3×2) in UHV or by dosing, at high temperature [27, 29], the (3×2) with C_2H_2 (acetylene) or the C-terminated $c(2 \times 2)$ with Si_2H_6. It has also been prepared [34] by Si deposition on the $c(4 \times 2)$ at 1000 °C. Etching of Si from the (3×2) by heating in O_2 [30] gives a (5×2) but with a small amount of residual O. The (7×2) has been formed by desorption of Si from the (3×2) by annealing in UHV [59] or by exposure to hydrogen at high temperature [61].

The (Si LVV)/(C KLL) PPH ratios [21, 61] show that Si coverage decreases in the order $(3 \times 2) > (5 \times 2) > (7 \times 2) > (2 \times 1)$, consistent with MEIS results [21] and with quantitative studies [27 to 29] of the interconversion of the different surface structures. The (3×2) and (5×2) are reported [27] to involve ≈ 0.36 and ≈ 0.16 MLs, respectively, of Si chemisorbed on the (2×1) terminating layer. Further adsorption of Si in stable, well-defined sites did not occur beyond the (3×2) phase [21, 27, 29].

Various STM studies [59, 61 to 65] have been performed for the $(N \times 2)$ series. Hara et al. [61, 63, 65] proposed the (3×2) structure, shown in Fig. 10a, set forth first by Dayan [20] and subsequently by Kaplan [9]. This can be termed the "double dimer-row model". Here, Si atoms chemisorb as bridges between atoms in the (2×1) terminating layer, forming rows of dimers with every third adatom row missing. The Si adlayer coverage is $\Theta_{Si}^{ad} = 2/3$ ML. The (5×2) and (7×2) are seen as having three and five consecutive missing-adatom rows, respectively, with dimerization between terminating-layer Si atoms along the missing-adatom rows. Conversion of the (3×2) to a (3×1) by chemisorption of H was observed [61] and could be reversed by desorption. A similar effect was seen by Dayan [20] for O_2 chemisorption on the (3×2).

Semond et al. [64] analyzed STM data for high-quality single-domain surfaces in terms of a model with two missing dimers per unit cell, giving $\Theta_{Si}^{ad} = 1/3$ ML. This structure,

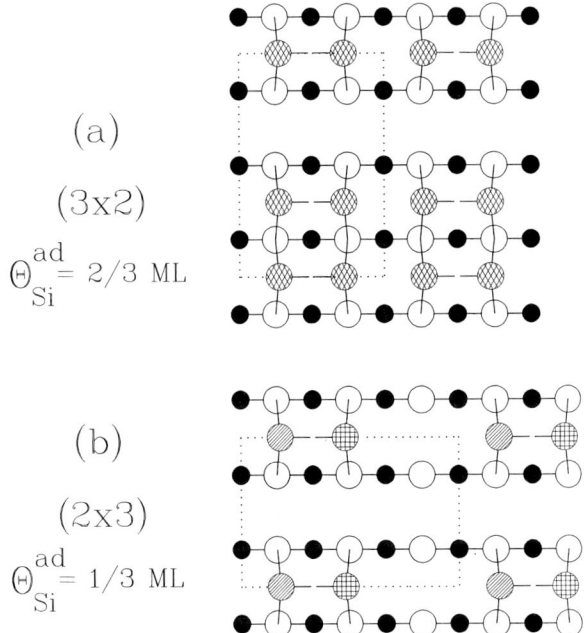

(a)

(3x2)

$\Theta_{Si}^{ad} = 2/3$ ML

(b)

(2x3)

$\Theta_{Si}^{ad} = 1/3$ ML

Fig. 10. Models proposed by a) Dayan [20] and Hara et al. [61] and by b) Yan et al. [47] and Semond et al. [64] for the SiC(100)-(3 × 2). The latter is properly termed (2 × 3) when referenced to the single-domain (2 × 1), Fig. 4. Cross-hatched atoms represent the Si adlayer and dotted lines the unit cells. Adlayer Si coverages (Θ_{Si}^{ad}) for either model are indicated. For simplicity, adlayer dimers in a), the "double dimer-row model", are shown as symmetric; although various forms of buckling have been deduced from LEED [9] and STM [63]. In b), the "alternate dimer-row model", adlayer dimers are shown with "ferromagnetic" buckling [64]. In b) Si–Si dimerization is possible between first-underlayer atoms along the [011] missing-dimer row, but is not shown

properly termed "(2 × 3)" when referenced to the bare single-domain (2 × 1), is shown in Fig. 10b. It is sometimes referred to as the "alternate dimer-row model" and was first suggested in the theoretical work of Yan et al. [47, 53]. Surfaces were prepared [64] by a modified flux-anneal method with Si deposition near room temperature followed by annealing in UHV. Chemisorption was found not to be self-limiting, in this case, since epitaxial growth of Si islands was seen. The adlayer dimers were found to be buckled, all in the same direction and at the same angle. Various defects — missing, paired or "incorrectly-tilted" dimers — were identified. In Fig. 10b, Si–Si dimerization is possible along the [001] missing-adatom row, but has not been shown. Soukiassian et al. [59] have also used STM to study structures formed during the annealing-induced transition from (3 × 2) to c(4 × 2).

There are conflicts between the two ($N \times 2$) models and other data. In the Dayan-Hara model [61, 63, 65] an ($N \times 2$) requires a Θ_{Si}^{ad} of $2/N$ MLs, in conflict with the experimental [21, 27] (3 × 2) and (5 × 2) values of $\approx 1/3$ and $\approx 1/5$ MLs, respectively. The Yan-Semond model [64] gives the correct Θ_{Si}^{ad} for the (2 × 3) and, implicitly, for the (2 × 5) and (2 × 7). However, the "3×" direction of the (2 × 3) is perpendicular to the "2×" direction of the (2 × 1) underlayer; whereas, LEED [9] for a single-domain surface shows these to be parallel, in agreement with the Dayan-Hara model. Furthermore, breaking the adlayer Si–Si dimers in the Yan-Semond model, by chemisorption of H or O, would not give the (3 × 1) found by Hara et al. [61] and by Dayan [20]. Instead a different (2 × 3) structure — e.g., (2 × 3)H — would be seen.

In principle there could be two surfaces, the "low-coverage" (2 × 3) Yan-Semond and "high-coverage" (3 × 2) Dayan-Hara structures. These are indistinguishable in LEED for two-domain substrates. However, the (Si LVV)/(C KLL) PPH ratio versus Si_2H_6 dose [28], for an initially C-terminated surface, implies that the Si adlayer corresponds to a single well-defined coverage; whereas, a two-fold variation would be required to accommodate both models.

An obvious way to reconcile the Dayan-Hara model with the data is to replace the double-dimer row in Fig. 10a with a single-dimer row to form a (3×2) with two consecutive $[01\bar{1}]$ vacancy rows and $\Theta_{Si}^{ad} = 1/3$ ML. This structure, termed the "added (or additional) dimer-row model", was proposed in an earlier study by Hara et al. [21]. However, it has not been reported in STM, and theory (see below) finds it to be energetically unfavourable compared to the structure in Fig. 10b. In any case, resolution of the apparent uncertainty in the value for Θ_{Si}^{ad} is essential.

Another aspect concerns streaking of the 1/2-order spots in the "3×" direction, seen in Fig. 2b and more clearly in single-domain data [9]. Within the Dayan-Hara model, Fig. 10a, Kaplan [9] related this to dimer buckling. All dimers in a $[01\bar{1}]$ adlayer row are suggested to be buckled in the same direction, with the adjacent row buckled either in the same or in the opposing direction. Adjacent double-dimer rows are decoupled, due to the intervening vacancy row, and the resulting disorder then causes streaking.

Hara et al. [61, 63, 65] have proposed a "×1" shift in the "×2" direction as the cause of the streaking. Here, adjacent double dimer-rows, Fig. 10a, undergo a relative displacement by one bulk unit-cell distance ("×1") in the $[01\bar{1}]$ ("×2") direction. This was seen in STM to occur mainly near the edges of (3×2) terraces, and a similar "×1" shift was also observed by Semond et al. [64]. Recent STM results of Hara et al. [63, 65], imaging both filled and empty states, indicate both the "×1" shift and disordered inter-dimer buckling, i.e., a relative displacement of dimers along the surface normal.

Hara et al. [61] find that streaking disappears for a fully-formed (3×2) surface which has, by definition, no terraces. Hara et al. [28] also find that streaking is absent for the final (3×2) surface resulting from repeatedly cycling between the $c(2 \times 2)$ and (3×2) using alternating doses of C_2H_2 and Si_2H_6. Thus, streaking is not intrinsic to the (3×2) but represents disorder which can be avoided by careful surface preparation. Presumably it is the "×1" shift which produces the streaking. The separate effect, on the LEED pattern, of disordered inter-dimer buckling — which occurs even on a defect-free (3×2) [65] — has not yet been determined.

Surface-induced Si 2p BE shifts for the (3×2) have been reported in XPS [8, 66, 67]. As for the (2×1), Fig. 5, the Gaussian widths are too large to permit any conclusions regarding surface fine-structure. Surface-sensitive C 1s data [66, 67] indicate a single component, as expected for a Si-terminated structure.

Surface-sensitive UPS data have been given [8, 22, 35, 66, 67] for the (3×2) (see Fig. 6). A decrease is found [22, 35] in the intensity of the deep-valence state near -14 eV (not shown in Fig. 6) upon going from (2×1) to (3×2), consistent with the bulk DOS [39] showing this feature to be largely derived from C 2s orbitals. The V_1 surface state — and possibly also the V_2 [36], noted above in connection with the (2×1) — is evident. The (3×2) surface, like the (2×1), is semiconducting.

Surface-sensitive ELS data for valence and plasmon excitation have been reported [9, 22, 35] (see Fig. 3). As for the (2×1), an S_1 surface-state transition is seen at ≈ 1.8 eV. An S_2 surface-state transition, as evidenced by its sensitivity to O_2 adsorption [35], is seen at ≈ 3.0 eV. Considering the (2×1) and (3×2) data together, it appears likely that the common S_1 transition is due to excitation of the common V_1 surface state (Fig. 6) which originates in Si dangling orbitals. The elemental Si(100)-(2×1) surface also shows an S_1 transition at 1.8 eV [68]. The S_2 transition — at 3.0 and 5.3 eV, respectively, for (3×2) and (2×1) − is not much higher in energy that the bulk bandgap $(E_g = 2.39 \text{ eV})$. Hence, S_2 might be due to an excitation originating in the Si−Si dimer bond. For elemental Si(100)-(2×1), on the other hand, an "S_2" peak appears in ELS at ≈8.6 eV [68]. Si 2p ELS data have been given for the (3×2) [20], but these show no clear indication of surface-related features.

The Si LVV and C KLL AES lineshapes have been shown in detail [9, 22]. For the (3×2), the C KLL is essentially identical to that of the (2×1), Fig. 8, as expected for a Si-terminated surface. However, the Si LVV shows structure related to Si−Si bonding, as expected for a surface with chemisorbed Si.

Semi-empirical [47, 48, 50, 52] and ab initio [53] studies of the (3×2) have been reported. Yan et al. [47, 53] compared the alternate-dimer, Fig. 10b, with the added-dimer model described above. Both structures correspond to $\Theta_{Si}^{ad} = 1/3 \text{ ML}$. The former is found to be more stable, in agreement with other results [52]. Hu et al. [48] note that the added-dimer unit cell involves four types of inequivalent Si atoms and obtained an atom-resolved VB DOS for each. They point out that the relatively large Gaussian width for the (3×2) surface-shifted Si 2p [8, 22] may result from this multiplicity of sites, which is common to all the (3×2) models. Molecular dynamics calculations [50] indicate that the Dayan-Hara model, Fig. 10a, is energetically favored over either of the $\Theta_{Si}^{ad} = 1/3 \text{ ML}$ models and that the adlayer dimers are buckled, with Si−Si dimerization along the missing-adatom row.

2.5 The (100)-c(2 × 2) surface

The (100)-c(2×2) surface can be formed in various ways. The (2×1) can be annealed in UHV to desorb Si [9, 20, 22] or exposed at $\geq 800 \,°\text{C}$ to C_2H_4 (ethylene) [40]. It is also formed by reacting the (2×1) or (3×2) at ≈1000 °C with C_2H_2 [27 to 29]. However, for unknown reasons, C_2H_4 is inert [40] toward the (3×2) under conditions leading to reaction with the (2×1). Surfaces formed by Si desorption are contaminated by graphitic C (see below). Thermal desorption is not self-limiting in this case; hence, it is difficult to remove precisely one ML of Si by this means. The c(2×2) has also been formed [16] by 0.5 keV ion bombardment at 600 °C followed by flashing to 900 °C.

The c(2×2) was first reported by Dayan [20] and subsequently by Bellina et al. [16] and by Kaplan [9]. The MEIS results of Hara et al. [21] show it to be terminated in a full ML of carbidic C. This is supported by quantitative studies [27 to 29] of the inter-conversion of the different (100) surface structures. Hasegawa et al. [69] also suggested termination in a complete ML of C by showing that Ti deposited on the c(2×2) bonds only to C and not to Si. The MEIS results rule out early models based on termination in a half [9] or a full [20] ML of Si.

Two very different models, both consistent with a full-ML C coverage, were suggested for the c(2×2). These are termed (somewhat cryptically) the "staggered-dimer" [20, 40] and the "bridging-dimer" [70] and are shown in simplified form in Figs. 11a and 12, respectively (see also Fig. 13). In the former, C atoms are sp^3-hybridized, each with σ-bonds to two Si atoms and a σ-bond to another C. There is one dangling orbital per C, no dangling Si orbital and the direction of the C−C σ-bond is perpendicular to that of the Si−Si dimers on the initial (2×1) surface. In the bridging-dimer, the C atoms are sp^2-hybridized, forming a π-bond to another C and a σ-bond to one Si atom. The C=C bridge is parallel to the Si−Si dimer direction on the (2×1), and there is one dangling orbital on each surface Si and C atom.

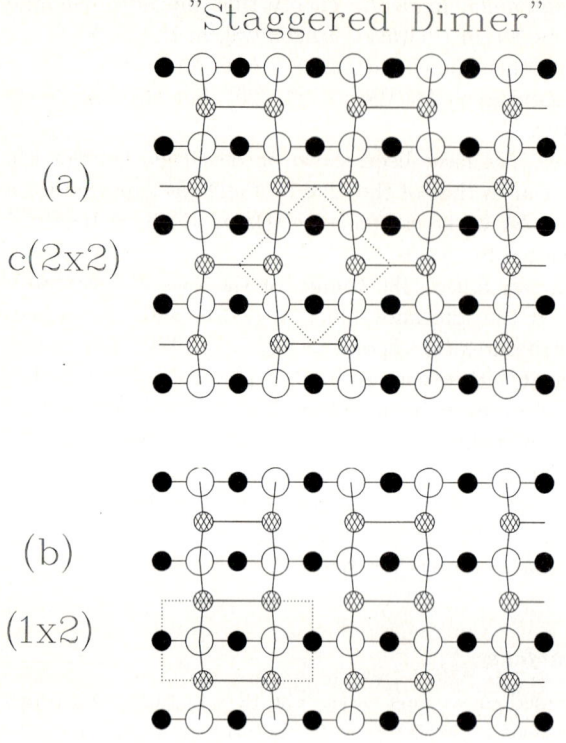

"Staggered Dimer"

(a)

c(2x2)

(b)

(1x2)

Fig. 11. Models for a) the c(2 × 2) and b) the (1 × 2) proposed by Bermudez and Kaplan [40]. The cross-hatched atoms represent the adlayer formed from σ-bonded C pairs. No buckling is shown, since none has been documented. The dotted lines show the primitive unit cells. See also Fig. 13 "σ"

The bridging-dimer, Fig. 12, is the correct c(2 × 2) model (see below). Applying the label "dimer" to this structure is misleading; although, it will be done here to maintain consistency with the literature. The C pair is not "dimerized", like the Si pairs on (2 × 1) and (3 × 2). Rather, a formal chemical bond exists, as in a free molecule. This is shown by the fact that H chemisorption [40, 70] does not destroy the c(2 × 2) LEED pattern; although, relaxation is indicated [40] by a change in the dependence of 1/2-order beam intensity on primary energy. It is the molecular nature of this unusual reconstruction that leads to the unique properties of the c(2 × 2).

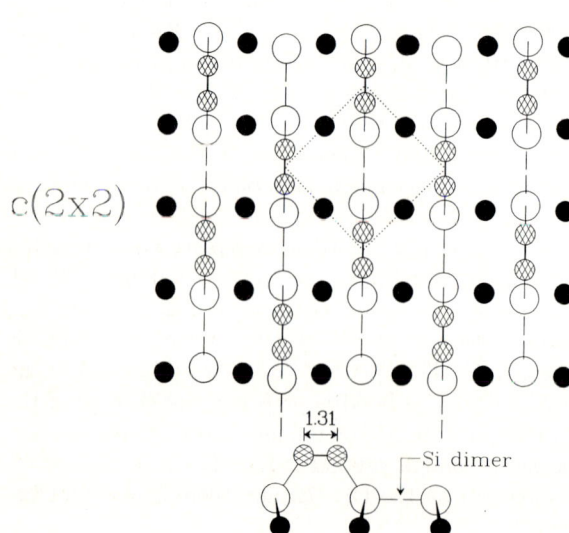

"Bridging Dimer"

c(2x2)

1.31

Si dimer

Fig. 12. The model for the thermally-generated c(2 × 2) determined by Powers et al. [70]. (See also Fig. 13 "π".) The cross-hatched atoms represent the adlayer formed from π-bonded C pairs and the dashed lines the Si–Si dimers. The lower diagram shows the structure viewed along the [01$\bar{1}$] direction with the bridge length indicated. Other dimensions are given in [70]. The model for the surface formed by C_2H_4 reaction differs slightly in that Si–Si dimers are not seen and the C=C bridge is shorter (1.25 Å)

Early STM data [62] suggested C dimers but could not distinguish between the two models. Powers et al. [70] analyzed LEED data for surfaces prepared from a (2×1) either by reaction with C_2H_4 or by thermal desorption of Si. The results are similar, other than that the former is better ordered. This is expected, given the almost inevitable formation of a graphite impurity for a thermally-generated surface. It is also significant that both preparations yield essentially the same reconstruction. The analysis indicates better agreement (i.e., a smaller Pendry R-factor) for the bridging-dimer model, and the results are shown schematically in Fig. 12. In addition to the C=C bridge, thermally-generated surfaces show evidence for Si–Si dimer pairing. This was absent on surfaces formed by C_2H_4 reaction, perhaps due to the lower annealing temperature (850 versus 1030 °C) which may have left H adsorbed at Si dangling bonds. The length of the

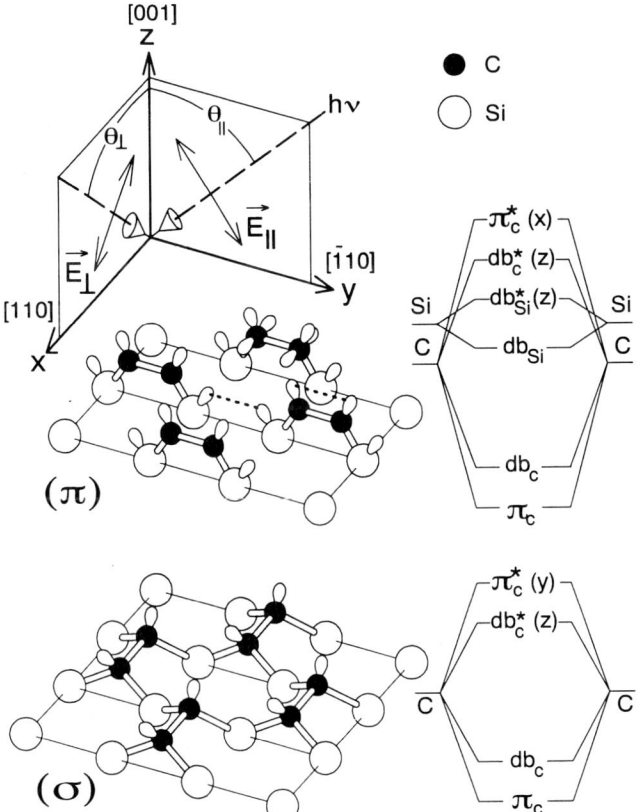

Fig. 13. Schematic diagram defining angles and polarizations for the $c(2 \times 2)$ C 1s NEXAFS data ([71]). The diagrams labelled "σ" and "π" correspond to Figs. 11a and 12, respectively. In "π", the dashed lines show the interactions between sp^2-hybridized C dangling orbitals, leading to a partial "acetylene-like" bond (see text), and between Si dangling orbitals. For clarity, only one of the C pairs in the "π" diagram shows the C p_x orbitals which overlap to form the ethylenic C=C π-bond. The energy-level diagrams summarize qualitatively the results of semi-empirical cluster calculations for the two models. These show the energetic ordering of antibonding excited states (labelled "*") derived from Si and C dangling bond (db) and π orbitals. For each, the dominant electric-dipole polarization is shown in parentheses (e.g., (x))

π-bonded bridge is intermediate between typical C=C and C≡C distances (1.33 and 1.21 Å, respectively).

Long et al. [71] reported C 1s and Si 2p NEXAFS data for surfaces formed by C_2H_4 reaction at 1100 °C. The absence of chemisorbed H on such surfaces can be verified [40] using H^+ electron stimulated desorption. Due to the use of single-domain samples, the incident polarization vector could be related to the Si–Si dimer direction on the starting (2×1) surface and, therefore, to the C–C or C=C directions required by the competing models (Fig. 13). The data were analyzed using a semi-empirical cluster calculation giving the orbital composition and energetic ordering of the excited states.

The C 1s NEXAFS data are shown in Fig. 14. The angular variations of the intensities indicate dipole-active processes. The dependence on E_k and also on H adsorption (not shown) reveals the structure near 284 eV to be surface-derived, unlike the bulk feature at ≈ 286.7 eV. In accord with the bridging-dimer model, an x-polarized absorption, identified with the π^*_C excited state, lies above a z-polarized feature assigned to the db^*_C. No surface feature is seen in y-polarization ($\mathbf{E}_\parallel$, $\Theta_i = 0$). The results conflict with the staggered dimer model, which predicts one z- and one y-polarized transition but nothing in x-polarization.

The Si 2p NEXAFS data, Fig. 15, are similar to those given by Shek [34] for the $c(4 \times 2)$ and show a z-polarized transition to an H-sensitive Si surface state near the

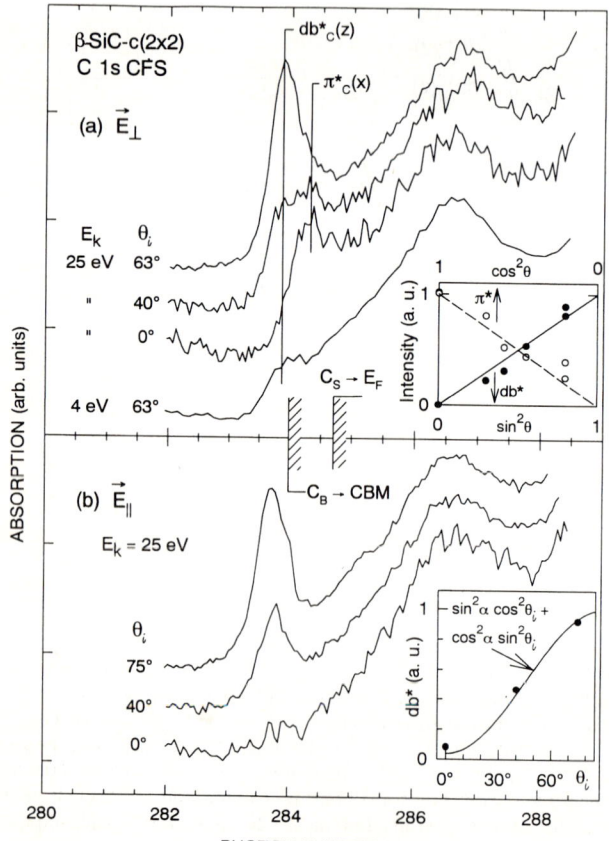

Fig. 14. C 1s NEXAFS (or CFS) data [71] for $c(2 \times 2)$ versus polarization ($\mathbf{E}_\parallel$ or $\mathbf{E}_\perp$) and angle of incidence (Θ_i) (see Fig. 13). The spectra are normalized for equal amplitudes in the bulk absorption at ≈ 287 eV and are offset vertically for clarity. The detected electron kinetic energy (E_k) is given for each trace. Thresholds are marked for absorption from the bulk (B) C 1s to the CBM and from the surface (S) C 1s to the Fermi level. The insets show the angular dependence of intensities (points) compared to dipole theory (lines). For $\mathbf{E}_\perp$, a) shows the Θ_i-dependence of transitions involving the db^*_C and π^*_C orbitals. The weak signal for $E_k = 4$ eV versus 25 eV shows the surface character of these transitions. For $\mathbf{E}_\parallel$, b) shows the db^*_C transition as the only surface feature

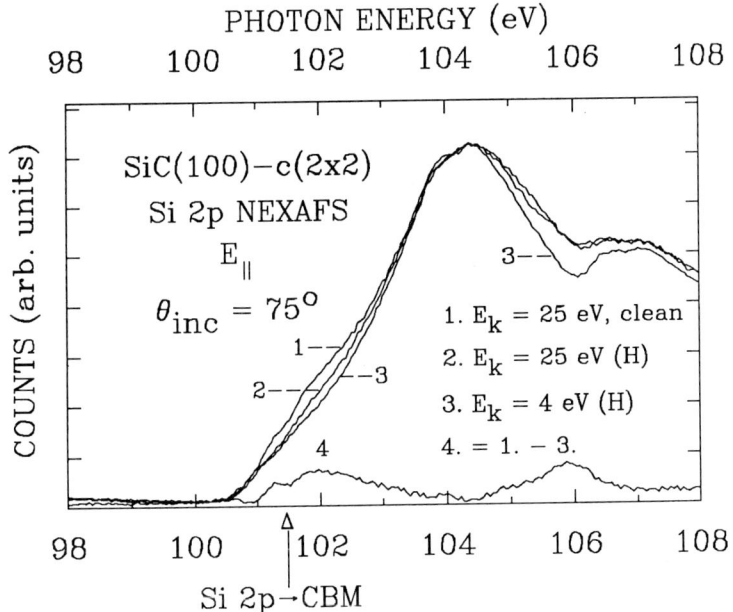

Fig. 15. Similar to Fig. 14 but showing Si 2p NEXAFS [71]. Curves 1, 2 and 3 are surface-sensitive clean, surface-sensitive H-saturated and bulk-sensitive H-saturated, respectively. Curve 4 is a difference spectrum (#1 minus #3) showing surface-state features on the clean surface

CBM. This is derived from the antibonding combination of dangling orbitals (db$^*_{Si}$, Fig. 13 (π)). It was not seen previously in Si 2p ELS [40], probably due to inferior resolution, and its presence further confirms the bridging-dimer model and excludes the staggered dimer. A second Si-related surface state, as shown by its dependence on E_k, is seen at 4.5 eV above the CBM. It appears relatively insensitive to H chemisorption and is, therefore, assigned to the C–Si surface backbond.

Surface-sensitive, high-resolution Si 2p XPS data [8, 72] show a single spin–orbit doublet, as expected for C-termination. Similar data [73] for the C 1s, Fig. 16, show a surface component shifted to higher BE. This is identified by its response to H adsorption and by the dependence of S/B relative intensity on $h\nu$ (i.e., photoelectron escape depth). The S feature is not eliminated by H adsorption, and, in fact, the shift and linewidth increase. Hence, it is inferred that the surface shift arises from π-bonding of surface C atoms (versus σ-bonding for bulk C) and that the π-bonding persists after H adsorption. The higher "S" BE is consistent with theory [53] indicating a smaller valence charge density on the π-bonded surface atoms, which is further decreased by adsorbed H. These data were obtained for a surface prepared by reacting a (2 × 1) with C_2H_4 at 1100 °C. Data for a thermally-generated c(2 × 2) [72], at $h\nu = 340$ eV, show only a single broad peak (full-width at half-maximum ≈ 2.5 eV). A graphite impurity on this surface is suggested by the streaking and relatively high diffuse background seen in LEED.

Surface-sensitive XPS data have been given [8, 22, 35, 67, 72] for the c(2 × 2) VB, and results are shown in Fig. 6 for a surface formed by reacting a (2 × 1) with C_2H_4 at ≈ 1000 °C. An increase is seen [22, 35] in the intensity of the deep valence state near -14 eV (not shown in Fig. 6) upon forming the c(2 × 2), consistent with the bulk DOS

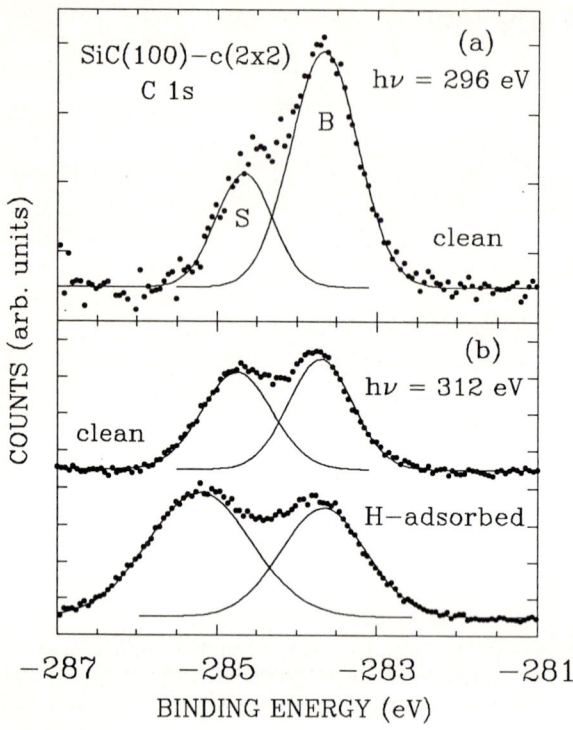

COUNTS (arb. units)

SiC(100)–c(2x2)
C 1s

(a)
hν = 296 eV

B

S

clean

(b)
hν = 312 eV

clean

H–adsorbed

−287   −285   −283   −281

BINDING ENERGY (eV)

Fig. 16. C 1s XPS for the c(2 × 2) under a) bulk- and b) surface-sensitive conditions [73]. The points are data, and the lines are least-squares fits of sums of Gaussians and a polynomial background (which has been subtracted from the data and fits for display). The high noise level in a) is due to the intense slow secondary electron background for low photoelectron energies. Surface-sensitive spectra are shown before and after H chemisorption. B and S indicate the bulk and surface components, respectively. Relative intensities of different spectra are not quantitative, and linewidths may be limited by the experimental resolution. Binding energy is referenced to the Fermi level

[39] showing this feature to be largely derived from C 2s orbitals. The upward band bending is smallest for the c(2 × 2) ($\delta\varphi_B = 0.25 \pm 0.10$ eV [8]) as can be seen in the relative positions of the "B" peak (Fig. 6).

The most striking effect on the VB of the change from Si- to C-termination is the attenuation of the C_2 peak near the VBM and the appearance of the feature labelled "X". The origin of this effect has not yet been established, but it will be discussed below in connection with the Si LVV AES spectrum. Unlike Si-terminated surfaces, the c(2 × 2) shows no indication of occupied surface states above the bulk VBM, consistent with the absence of unpaired Si dangling orbitals in the bridging-dimer model. The weak tail extending into the gap may be due in part to bulk defects since it is not completely eliminated by H chemisorption. Some contribution from surface defect sites with undimerized Si dangling orbitals is also possible. The c(2 × 2) VB XPS is quite complex [73], being sensitive both to H chemisorption and to the photoemission angles. There are indications of surface states below the VBM, but these have not been studied in detail.

Surface-sensitive ELS data in the valence- and plasmon-excitation range (Fig. 3) have been reported for the c(2 × 2) formed by C_2H_4 reaction [40] or by Si desorption [9, 10, 22], and the results are in essential agreement. H-sensitive structure [40] at ≈4 and ≈8 to 11 eV is clearly due to surface excitation, but an assignment of these features has not been given yet. A very weak feature at 1.6 eV [9] may be a second-derivative artifact [41]. There is no clear indication in ELS of surface states in the bulk bandgap. Surface-sensitive Si 2p ELS data have been given for surfaces formed by C_2H_4 reaction [40] or by Si desorption [20]. The results are in essential agreement but show no evidence for the surface state seen in Si 2p NEXAFS, Fig. 15. However, H adsorption was not included in these ELS experiments.

Fig. 17 shows C 1s ELS data [40], and the results are not well understood. A surface feature appears at 281.5 eV, but it is ≈2.3 eV lower in energy than the db_C^* peak in C 1s NEXAFS, Fig. 14. Forming the (1×1) yields π^* and σ^* losses characteristic of graphite and verifies the ELS energy calibration. Adsorption of H removes this feature, suggesting that only the C 1s $\rightarrow$ db_C^* and not the C 1s $\rightarrow$ π_C^* transition (Fig. 13) is seen in ELS. Based on C 1s XPS (Fig. 16) and on theoretical results (see below), H chemisorption is believed to saturate C dangling orbitals but not to remove the C=C π-bond. However, the continued presence of C=C π-bonds after H adsorption has not yet been shown directly. Further study is needed to resolve this apparent conflict between C 1s ELS and NEXAFS.

Si LVV Auger data [40] are shown in Fig. 7 for a $c(2 \times 2)$ formed by reacting C_2H_4 with a (2×1), and similar results were reported [22] for a thermally-generated $c(2 \times 2)$. A shift is seen in the main $L_{2,3}VV$ peak (negative extremum) from ≈89 to ≈86 eV on going from (2×1) to $c(2 \times 2)$. At higher gain, a small shift to lower energy is also seen in the Si $L_1L_{2,3}V$ near 40 eV. Chemisorption of H [40] that causes subtle changes in lineshape but little or no further energy shift. For the elemental-Si $L_{2,3}VV$ [74], the main peak is derived from p-like states comprising the uppermost VB peak. It has been suggested [22] that the ≈3 eV shift in the SiC $L_{2,3}VV$ is due to removal of Si dangling-bond states at the VBM. It has also been suggested [40] the shift reflects the ≈1.5 eV shift in the uppermost peak in the SiC VB (i.e., X versus C_2 in Fig. 6).

The basis for this latter suggestion is that the energy of an Auger transition, $KE(WYZ)$, involving levels W, Y and Z is given by

$$KE(WYZ) = E(W) - E(Y) - E(Z) - U_{\text{eff}}(YZ), \qquad (1)$$

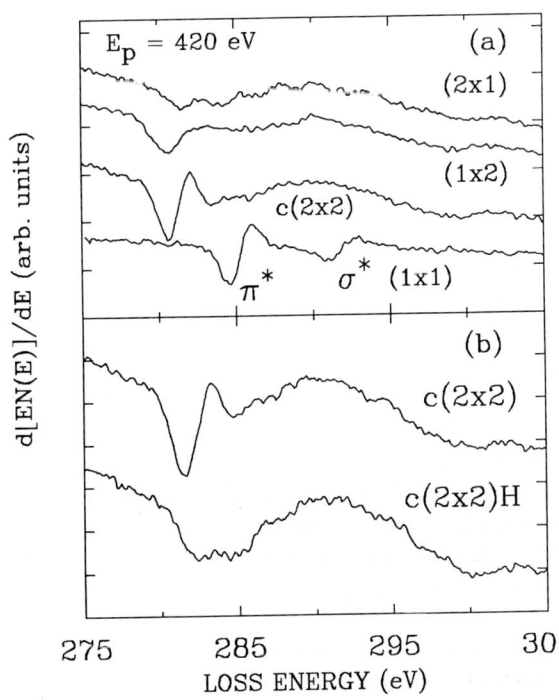

Fig. 17. Similar to Figs. 7 and 8 but showing surface-sensitive C 1s ELS. The (1×1) spectrum is essentially identical to that of bulk graphite, and π^* and σ^* mark transitions to states in the graphite CB. b) Effect of adsorbing H on the $c(2 \times 2)$ to form the $c(2 \times 2)H$ surface. The resolution is 0.8 eV in a) and 2 eV in b). Peak energies correspond to inflection points in the first-derivative spectra

where $E(Q)$ is the one-electron BE of level Q and $U_{\text{eff}}(YZ)$ the effective hole–hole interaction and polarization energy in the final state. For constant $E(W)$ and $U_{\text{eff}}(YZ)$, a shift of ΔE in the uppermost VB peak produces a shift of $2\,\Delta E$ in the Si LVV energy. This interpretation identifies the X-feature in Fig. 6 as receiving a significant contribution from Si p-like orbitals. Resonant photoemission [71] also reveals a C contribution to this state. Of course, this rough estimate of $\Delta KE(\text{LVV})$ overlooks differences in the Si 2p BEs and the fact that $U_{\text{eff}}(YZ)$ can depend strongly [74] on orbital configuration.

C KLL Auger data [40] are shown in Fig. 8 for a c(2 × 2) formed by reacting C_2H_4 with a (2 × 1). As for the Si LVV, the (2 × 1) → c(2 × 2) transformation shifts the main peak a few eV to lower energy. A shoulder (marked by the arrow) appears at the high-energy edge of the spectrum. This feature is removed by H adsorption and is assigned [40] to an Auger transition in which one of the electrons is initially in a C dangling orbital. With reference to Eq. (1), two factors account for the high KE. First, the BE of a dangling orbital should be the lowest of any C-derived valence state. Second, due to limited overlap between dangling and valence orbitals, the $U_{\text{eff}}(YZ)$ term should be relatively small. Similar data [22] for a thermally-generated c(2 × 2), however, show neither the energy shift nor the high-energy shoulder.

Many semi-empirical [46 to 49, 51, 52, 75, 76] and ab initio [53, 54, 56, 57, 77] studies of the c(2 × 2) have been done. Most of the former support the staggered dimer. Others [46] – and all ab initio studies – favor the bridging dimer with a bond length approaching that of a C≡C triple bond. The energy difference between the two structures, and between the c(2 × 2) and (1 × 2) (see below), is typically much less than between the c(2 × 2) and unreconstructed surfaces. Studies [46, 53, 77] of H adsorption find that the c(2 × 2) persists with a change in relaxation, as seen in LEED [40], and a bridge length increased to that of a C=C π-bond. However, theory indicates H adsorption only at C, not Si, dangling orbitals.

A simple description of the c(2 × 2) can be given in summary of the theoretical and experimental [70, 71] results. The C orbital configuration is sp^2 with some sp^1 character, making the Si–C=C–Si bridge more nearly linear than for pure sp^2. This permits interaction between carbon dangling orbitals, resulting in a bridge-bond of order greater than two, and forces Si atoms closer together so that Si–Si dimers can form. This removes occupied intrinsic surface states from the gap [54, 77] (see Fig. 6) and produces a stable surface that chemisorbs O_2 much more slowly than do Si-terminated surfaces (see below). Much of the c(2 × 2) electron spectroscopy can be understood in terms of excitation of this "molecular" Si–C=C–Si bridge.

2.6 The (100)-(1 × 2) surface

During conversion of the (2 × 1) to the c(2 × 2), by thermal desorption [20] or by reaction with C_2H_4 [40], another structure is seen prior to the c(2 × 2). For a single-domain starting surface [40], this appears in LEED as a (1 × 2), i.e., as a 90°-rotated (2 × 1). The half-order spots are weak and difficult to see, either in situ or in photographs of the LEED screen [40]. To date, there have been no reports of well-ordered surfaces of this nature. It is unknown whether this reflects an intrinsic instability or simply an inadequacy in the surface preparation technique. Given the probable structure of this surface (see below) reacting C_2H_6 (ethane) with the Si-terminated (2 × 1) might favor (1 × 2) formation. However, CH_4 (methane) has been found [40] to react very slowly with the (2 × 1), and the same might be true for other saturated hydrocarbons.

Hara et al. [30] have etched Si from a (3×2) surface, by exposure to 1×10^{-7} Torr O_2 at 1050 °C, and report a "partially Si-terminated (2×1)" phase. This is believed to be the same as the present (1×2) structure, but the two are indistinguishable in two-domain LEED patterns. The surface prepared in this way exhibited a few percent of an ML of residual O, and no indication was given of the quality of the LEED pattern.

The (1×2) has not been the subject of much experimental study. The Si LVV (Fig. 7) and C 1s ELS (Fig. 17) data for samples showing a faint (1×2) pattern appear intermediate between those for the (2×1) and $c(2 \times 2)$. It is unknown whether these spectra represent characteristic features of a distinct (1×2) phase or simply an incomplete $c(2 \times 2)$.

The (1×2) is treated in semi-empirical [42 to 47, 49, 51, 52, 75, 78] and ab initio [54, 57] studies, where it is usually referred to as "(2×1) C-terminated" since the starting point is a hypothetical ideal (1×1) C-terminated surface. We will continue to use "(2×1)" in reference to the surface terminated in an ML of Si and "(1×2)" for this new C-terminated phase. One should be aware of this notational difference to avoid confusion.

Given the lack of data, only a brief summary of theoretical results is given. The calculations agree that the dimers are symmetric and arranged as in Fig. 11b. Generally, only small energy differences are found between the (1×2) and $c(2 \times 2)$ surfaces. Other than that, the results fall into two groups. In one [42, 43, 46, 47, 51, 54, 57, 75, 78], the dimer is intermediate in length between C=C π- (1.33 Å) and C–C σ-bonds (1.54 Å). Due to the partial sp^2 hybridization the dimer relaxes inwardly along the surface normal, relative to the ideally-terminated position. Tying up C dangling orbitals by H adsorption [43, 46, 75] reverses this. The dimer length approaches that of a C–C σ-bond, and it relaxes outwardly relative to the H-free position. In the other group [44, 45, 49], the H-free dimer length is found to be comparable to or greater than that of a C–C σ-bond, and one study [52] finds a mix of short (1.37 Å) and long (1.82 Å) dimers in a (4×2) unit cell. There are also results [46, 52, 54] suggesting formation of an actual (2×1), not (1×2), C-terminated surface by placing [01$\bar{1}$] rows of π-bonded C=C bridges between the [01$\bar{1}$] Si–Si dimer rows on the Si-terminated (2×1), Fig. 4.

2.7 The (100)-(1 × 1) surfaces

A true (1×1) surface — consisting of a clean, complete ML of Si or C — may not exist on SiC(100) under normal conditions. Most theoretical studies agree that such a surface is unstable with respect to reconstruction, even if relaxation is included. Some ab initio results [54, 56, 57] predict only very weak Si dimer bonding on the (2×1), implying the theoretical possibility of a true Si-terminated (1×1). In any case, there has — with one exception [15] — been no reported experimental observation of a true SiC(100)-(1×1). On one very smooth sample cleaned by 1 keV Ar- or Kr-ion bombardment at 650 °C, Kaplan [15] observed an apparently well-ordered C-terminated (1×1) with little evidence of graphite impurity. Auger and ELS data were reported in detail for this surface.

Several pseudo-(1×1) surfaces have been observed. A (1×1) pattern with well-defined satellite spots, shown in Fig. 2a, was seen by Kaplan [9] for SiC(100) incompletely cleaned by Si flux-annealing. Similar satellites were found for Si substrates processed in the same manner. Dayan [20] also observed these features for incompletely heat-cleaned samples. The origin of the satellites is unknown; although, an incommensurate (1×1) layer [79] of chemisorbed O appears likely [20] (see below).

Another (1×1) pattern (Fig. 2f) occurs [9, 22, 28, 40] when the $c(2 \times 2)$ is subjected to further high-temperature annealing or C adsorption, leading to an increased diffuse background in LEED and a decreased (Si LVV)/(C KLL) PPH ratio. This is due to a poorly-ordered layer of graphitic C, with the (1×1) pattern being that of bulk SiC. One indication of graphitic C is the change in the C KLL lineshape (Fig. 8). Others are seen [40] in valence-excitation ELS (Fig. 3), which shows the characteristic graphite π-plasmon at ≈ 7 eV, and in C is ELS (Fig. 17) which shows graphitic π^* and σ^* losses. The C 1s XPS also shows [22] a high-BE satellite due to graphitic C.

A (1×1) surface was reported by Hara et al. [28] during dosing of a $c(2 \times 2)$ at 1050 °C with Si_2H_6. This appeared just before the onset of the (2×1) but with the same (Si LVV)/(C KLL) PPH ratio as the $c(2 \times 2)$. This was interpreted as the result of cleaving carbon–carbon bonds, with H then saturating the dangling C orbitals. However, it should be noted that an H-saturated $c(2 \times 2)$ surface [40, 70] retains a $c(2 \times 2)$ LEED pattern, indicating that exposure to H atoms alone does not break the carbon–carbon bond. This structure might instead indicate disordering of the $c(2 \times 2)$.

Hara et al. [30] etched Si from a (3×2) surface by heating to 1050 °C in 1×10^{-7} Torr O_2. The surface passes through a "C-terminated (2×1) phase", believed to be the (1×2) structure discussed above, and ends in a (1×1) without ever showing a $c(2 \times 2)$. However, the (Si LVV)/(C KLL) PPH ratio for this (1×1) is the same as typically seen for the $c(2 \times 2)$. Even though there was a few percent of an ML of residual O, no satellites were reported in LEED. This supports the suggestion [20] of an incommensurate Si oxide layer as the source of the satellites seen for O-contaminated, Si-terminated surfaces.

The results of Hara et al. [28, 30] suggest the existence of a C-terminated pseudo-(1×1) phase derived from a disordering of the $c(2 \times 2)$ with no change in stoichiometry. This would be distinct from the graphite-contaminated pseudo-(1×1) obtained by high-temperature annealing.

Van Elsbergen et al. [80] have recently reported that the (3×2), (2×1) and thermally-generated $c(2 \times 2)$ are all converted into pseudo-(1×1) structures by H-atom chemisorption and that the process is reversed by thermal desorption of H. This effect was not observed in previous work on the (3×2) and $c(2 \times 2)$ (see above) and is worthy of further study.

2.8 Chemisorption studies

In this section we briefly review experimental results for chemisorption on well-characterized, ordered SiC(100) surfaces. There have been only a few such studies, other than the work involving H, C_2H_4, etc. discussed above.

Chemisorption of N on the (2×1) surface has been observed [81] via exposure to N_2 in the presence of a hot W filament. For a single-domain surface a (1×3) structure is found at an N coverage of $\approx 2/3$ ML. With reference to Fig. 4, this has been ascribed to rows of Si–N–Si bridges in the [011] direction with every third row missing. At low coverage, either $c(4 \times 4)$ or $c(2 \times 6)$ LEED patterns were observed, depending on the sample. The reason for the sample dependence, and the nature of the low-coverage adsorption, were not determined. This study was complicated by the need for very large N_2 exposures ($> 1 \times 10^6$ L), as a result of the low efficiency for filament-assisted N_2 dissociation. Adsorbed O contamination, due to traces of O_2 or H_2O either in the N_2 or displaced from the chamber walls, had to be removed from the chemisorbed N layer by post-exposure Si flux annealing.

The initial chemisorption of O_2 at room temperature has been compared [35] for different SiC(100) surfaces for exposures of $\leq 10^3$ L. The rate is higher for Si(100)-(2×1) than for any SiC(100) surface, due to the relative ease of O insertion into Si–Si versus Si–C backbonds. On Si-terminated SiC(100) surfaces, Si–O–Si bridges form easily between surface atoms, but at an initially slower rate for (3×2) than for (2×1). This was suggested to reflect stronger Si–Si dimer pairing on the (3×2). The thermally-generated $c(2 \times 2)$ is nearly inert under the same conditions.

Theoretical works specifically comparing the (2×1) and (3×2) [47, 50, 53] agree that Si–Si dimer pairing is stronger on the (3×2). All these studies are among those which also predict a strong dimer pairing on the (2×1) surface, and similar conclusions are reached for all three of the possible (3×2) models discussed above. One study [50] also finds that a Si–C backbond on the SiC(100)-(2×1) is stronger than a backbond on the Si(100) surface, consistent with the above O_2 chemisorption results.

Janzen et al. [80, 82] performed similar experiments over a wide range of O_2 exposure ($\leq 10^{13}$ L), with care to avoid electronic excitation of the O_2 by the ion pump or by hot ionization gauge filaments. They observed an essentially equal rate of initial chemisorption, up to $\approx 10^2$ L, for (3×2) and (2×1). Above $\approx 10^2$ L, the (3×2) showed a second chemisorption phase, up to the onset of oxidation at $\approx 10^5$ L. No SiO_2 formation and no change in the Si/C atom ratio was seen in XPS. Pre-adsorption of H on the (2×1) produced a completely passivated surface.

Semond et al. [66] studied the effect on the (3×2) of a 10^4 L O_2 exposure at room temperature and above. They found evidence for direct formation of SiO_2, rather than the mix of suboxides formed under these conditions on elemental Si surfaces. The C 1s XPS indicates Si–O–C bonding as in an oxycarbide. The (2×1) did not show a corresponding formation of SiO_2.

Acknowledgements During my involvement with SiC surfaces I have benefitted greatly from collaborations with Ray Kaplan, Jim Long, Tom Parrill and Dave Ramaker. Our work, and that of others, on cubic SiC could not have been done without the excellent samples generously provided by Tony Powell, Larry Matus and colleagues at the NASA-Lewis Research Center. I am grateful to Shiro Hara, Johannes Pollmann, Patrick Soukiassian and Volker van Elsbergen for helpful discussions, critical reading of the manuscript and/or preprints of unpublished work. The work at NRL has been supported by the Office of Naval Research.

References

[1] H. MORKOÇ, S. STRITE, G. B. GAO, M. E. LIN, B. SVERDLOV, and M. BURNS, J. Appl. Phys. **76**, 1363 (1994).

[2] P. A. IVANOV and V. E. CHELNOKOV, Fiz. Tekh. Poluprov. **29**, 1921 (1995) (Soviet Phys. – Semiconductors **29**, 1003 (1995)).

[3] C. E. WEITZEL, J. W. PALMOUR, C. H. CARTER, JR., K. MOORE, K. J. NORDQUIST, S. ALLEN, C. THERO, and M. BHATNAGAR, IEEE Trans. Electron Devices **43**, 1732 (1996).

[4] U. STARKE, phys. stat. sol. (b) **202**, 475 (1997).

[5] R. KAPLAN and V. M. BERMUDEZ, in: Properties of Silicon Carbide – EMIS Data Reviews Series No. 13, Ed. G. L. HARRIS, INSPEC, London 1995 (p. 101).

[6] J. POLLMANN, P. KRÜGER, and M. SABISCH, phys. stat. sol. (b) **202**, 421 (1997).

[7] J. A. POWELL, L. G. MATUS, and M. A. KUCZMARSKI, J. Electrochem. Soc. **134**, 1558 (1987).

[8] V. M. BERMUDEZ and J. P. LONG, Appl. Phys. Lett. **66**, 475 (1995).

[9] R. Kaplan, Surf. Sci. **215**, 111 (1989); J. Vacuum Sci. Technol. A **6**, 829 (1989).

[10] R. Kaplan and T. M. Parrill, Surf. Sci. **165**, L45 (1986).

[11] K. Furukawa, Y. Tajima, H. Saito, Y. Fujii, A. Suzuki, and S. Nakajima, Jpn. J. Appl. Phys. **32**, L645 (1993).

[12] K. Shibahara, S. Nishino, and H. Matsunami, Appl. Phys. Lett. **50**, 1888 (1987).

[13] J. A. Powell, L. G. Matus, M. A. Kuczmarski, C. M. Chorey, T. T. Cheng, and P. Pirouz, Appl. Phys. Lett. **51**, 823 (1987).

[14] H. S. Kong, Y. C. Wang, J. T. Glass, and R. F. Davis, J. Mater. Res. **3**, 521 (1988).

[15] R. Kaplan, J. Appl. Phys. **56**, 1636 (1984).

[16] J. J. Bellina, Jr., J. Ferrante, and M. V. Zeller, J. Vacuum Sci. Technol. A **4**, 1692 (1986).

[17] J. J. Bellina, Jr. and M. V. Zeller, Appl. Surf. Sci. **25**, 380 (1986).

[18] D. R. Wheeler and S. V. Pepper, Surf. Interface Anal. **10**, 153 (1987).

[19] O. Kubaschewski and C. B. Alcock, Metallurgical Thermochemistry, 5th ed., Pergamon Press, Oxford 1979.

[20] M. Dayan, J. Vacuum Sci. Technol. A **3**, 361 (1985); **4**, 38 (1986).

[21] S. Hara, W. F. J. Slijkerman, J. F. van der Veen, I. Ohdomari, S. Misawa, E. Sakuma, and S. Yoshida, Surf. Sci. **231**, L196 (1990).

[22] T. M. Parrill and Y. W. Chung, Surf. Sci. **243**, 96 (1991).

[23] G. L. Harris, see [5] (Chap. 1).

[24] M. W. Zemansky, Heat and Thermodynamics, Chap. 11, 5th ed., McGraw-Hill Publ. Co., New York 1968.

[25] H. Nishino, W. Yang, Z. Dohnálek, V. A. Ukraintsev, W. J. Choyke, and J. T. Yates, Jr., J. Vacuum Sci. Technol. A **15**, 182 (1997).

[26] W. Mönch, Semiconductor Surfaces and Interfaces, Springer-Verlag, Berlin 1993.

[27] T. Yoshinobu, I. Izumikawa, H. Mitsui, T. Fuyuki, and H. Matsunami, Appl. Phys. Lett. **59**, 2844 (1991).

[28] S. Hara, Y. Aoyagi, M. Kawai, S. Misawa, E. Sakuma, and S. Yoshida, Surf. Sci. **273**, 437 (1992).
S. Hara, T. Meguro, Y. Aoyagi, M. Kawai, S. Misawa, E. Sakuma, and S. Yoshida, Thin Solid Films **225**, 240 (1993).

[29] T. Fuyuki, T. Yoshinobu, and H. Matsunami, Thin Solid Films **225**, 225 (1993).

[30] S. Hara, Y. Aoyagi, M. Kawai, S. Misawa, E. Sakuma, and S. Yoshida, Surf. Sci. **278**, L141 (1992).

[31] J. M. Powers, A. Wander, M. A. Van Hove, and G. A. Somorjai, Surf. Sci. **260**, L7 (1992).

[32] P. Soukiassian, F. Semond, L. Douillard, A. Mayne, G. Dujardin, L. Pizzagalli, and C. Joachim, Phys. Rev. Lett. **78**, 907 (1997).

[33] L. Douillard, F. Semond, P. Soukiassian, D. Dunham, F. Amy, S. Rivillon, and Z. Hurych, Surf. Rev. Lett., in press.

[34] M. L. Shek, Surf. Sci. **349**, 317 (1996).

[35] V. M. Bermudez J. Appl. Phys. **66**, 6084 (1989).

[36] M. L. Shek, K. E. Miyano, Q.-Y. Dong, T. A. Callcott, and D. L. Ederer, J. Vacuum Sci. Technol. A **12**, 1079 (1994).

[37] E. Landemark, C. J. Karlsson, Y.-C. Chao, and R. I. G. Uhrberg, Phys. Rev. Lett. **69**, 1588 (1992).

[38] G. K. Wertheim, D. M. Riffe, J. E. Rowe, and P. H. Citrin, Phys. Rev. Lett. **67**, 120 (1991).

[39] Y. Li and P. J. Lin-Chung, Phys. Rev. B **36**, 1130 (1987).

[40] V. M. Bermudez and R. Kaplan, Phys. Rev. B **44**, 11149 (1991).

[41] V. E. Henrich, Appl. Surf. Sci. **6**, 87 (1980).

[42] J. N. Carter, Solid State Commun. **72**, 671 (1989).

[43] B. I. Craig and P. V. Smith, Surf. Sci. **233**, 255 (1990); in: The Structure of Surfaces III, Eds. S. Y. Tong, M. A. Van Hove, K. Takayanagi, and X. D. Xie, Springer-Verlag, Berlin 1991 (p. 545).

[44] S. P. Mehandru and A. B. Anderson, Phys. Rev. B **42**, 9040 (1990).

[45] L. Wenchang, Y. Weidong, and Z. Kaiming, J. Phys.: Condensed Matter **3**, 9079 (1991).

[46] P. BADZIAG, Phys. Rev. B **44**, 11143 (1991); Surf. Sci. **269/270**, 1152 (1992); Diamond Relat. Mater. **1**, 285 (1992).
[47] H. YAN, X. HU, and H. JÓNSSON, Surf. Sci. **316**, 181 (1994).
[48] X. HU, H. YAN, M. KOHYAMA, and F. S. OHUCHI, J. Phys.: Condensed Matter **7**, 1069 (1995).
[49] T. HALICIOGLU, Phys. Rev. B **51**, 7217 (1995).
[50] M. KITABATAKE and J. E. GREENE, Appl. Phys. Lett. **69**, 2048 (1996); Jpn. J. Appl. Phys. **35**, 5261 (1996).
[51] T. HALICIOGLU, Thin Solid Films **286**, 184 (1996).
[52] R. GUTIERREZ and TH. FRAUENHEIM, Mater. Res. Soc. Symp. Proc. **423**, 427 (1996).
[53] H. YAN, A. P. SMITH, and H. JÓNSSON, Surf. Sci. **330**, 265 (1995).
[54] M. SABISCH, P. KRÜGER, A. MAZUR, M. ROHLFING, and J. POLLMANN, Phys. Rev. B **53**, 13121 (1996).
[55] J. POLLMANN, P. KRÜGER, M. ROHLFING, M. SABISCH, and D. VOGEL, Appl. Surf. Sci. **104/105**, 1 (1996).
[56] P. KÄCKELL, J. FURTHMÜLLER, and F. BECHSTEDT, Surf. Sci. **352/354**, 55 (1996); Appl. Surf. Sci. **104/105**, 45 (1996).
[57] A. CATELLANI, G. GALLI, and F. GYGI, Phys. Rev. Lett. **77**, 5090 (1996).
[58] B. I. CRAIG and P. V. SMITH, Physica **170B**, 518 (1991).
[59] P. SOUKIASSIAN, F. SEMOND, A. MAYNE, and G. DUJARDIN, submitted for publication.
[60] J. KITAMURA, S. HARA, H. OKUSHI, S. YOSHIDA, S. MISAWA, and K. KAJIMURA, in: Proc. 15th Symp. Materials Science and Engineering, Hosei Univ., 1997, in press (contact S. HARA at: shara@etl.go.jp).
[61] S. HARA, S. MISAWA, S. YOSHIDA, and Y. AOYAGI, Phys. Rev. B **50**, 4548 (1994); Inst. Phys. Conf. Ser. No. 137, 177 (1994).
[62] C.-S. CHANG, N.-J. ZHENG, I. S. T. TSONG, Y.-C. WANG, and R. F. DAVIS, J. Amer. Ceram. Soc. **73**, 3264 (1990); J. Vacuum Sci. Technol. B **9**, 681 (1991).
[63] S. HARA, J. KITAMURA, H. OKUSHI, S. MISAWA, S. YOSHIDA, and Y. TOKUMARU, Surf. Sci. **357/358**, 436 (1996).
[64] F. SEMOND, P. SOUKIASSIAN, A. MAYNE, G. DUJARDIN, L. DOUILLARD, and C. JAUSSAUD, Phys. Rev. Lett. **77**, 2013 (1996).
[65] S. HARA, J. KITAMURA, H. OKUSHI, S. MISAWA, S. YOSHIDA, and Y. TOKUMARU, submitted for publication.
[66] F. SEMOND, L. DOUILLARD, P. SOUKIASSIAN, D. DUNHAM, F. AMY, and S. RIVILLON, Appl. Phys. Lett. **68**, 2144 (1996).
[67] F. SEMOND, P. SOUKIASSIAN, P. S. MANGAT, Z. HURYCH, L. DI CIOCCIO, and C. JAUSSAUD, Appl. Surf. Sci. **104/105**, 79 (1996).
[68] Y. SUDA, D. LUBBEN, T. MOTOOKA, and J. E. GREENE, J. Vacuum Sci. Technol. B **7**, 1171 (1989).
[69] S. HASEGAWA, S. NAKAMURA, N. KAWAMOTO, H. KISHIBE, and Y. MIZOKAWA, Surf. Sci. **206**, L851 (1988).
[70] J. M. POWERS, A. WANDER, P. J. ROUS, M. A. VAN HOVE, and G. A. SOMORJAI, Phys. Rev. B **44**, 11159 (1991).
[71] J. P. LONG, V. M. BERMUDEZ, and D. E. RAMAKER, Phys. Rev. Lett. **76**, 991 (1996).
[72] F. SEMOND, P. SOUKIASSIAN, P. S. MANGAT, and L. DI CIOCCIO, J. Vacuum Sci. Technol. B **13**, 1591 (1995).
[73] J. P. LONG and V. M. BERMUDEZ, unpublished.
[74] D. E. RAMAKER, Crit. Rev. Solid State Mater. Sci. **17**, 211 (1991).
[75] B. I. CRAIG and P. V. SMITH, Surf. Sci. **256**, L609 (1991).
[76] B. I. CRAIG and P. V. SMITH, Surf. Sci. **276**, 174 (1992); Erratum, Surf. Sci. **285**, 295 (1993).
[77] P. KÄCKELL, J. FURTHMÜLLER, F. BECHSTEDT, G. KRESSE, and J. HAFNER, Phys. Rev. B **54**, 10304 (1996).
[78] Q. A. BHATTI, G. J. MORAN, and C. C. MATTHAI, Mater. Res. Soc. Symp. Proc. **423**, 439 (1996).
[79] L. J. CLARKE, Surface Crystallography, Chap. 1, Wiley, Chichester 1985.
[80] V. VAN ELSBERGEN, O. JANZEN, and W. MÖNCH, Mater. Sci. Engng. B, in press.
[81] V. M. BERMUDEZ, Surf. Sci. **276**, 59 (1992).
[82] O. JANZEN, V. VAN ELSBERGEN, and W. MÖNCH, in: Proc. 23rd Internat. Conf. Physics of Semiconductors, Berlin 1996, World Scientific, Singapore 1996 (p. 915).

phys. stat. sol. (b) **202**, 475 (1997)

Subject classification: 68.35.Bs; 61.14.Hg; 61.16.Ch; 68.55.Jk; S6

Atomic Structure of Hexagonal SiC Surfaces

U. STARKE

Lehrstuhl für Festkörperphysik, Universität Erlangen-Nürnberg,

Staudtstr. 7, D-91080 Erlangen, Germany

(Received January 31, 1997)

A review of the atomic structures found on hexagonal SiC surfaces is presented. Ex situ preparation methods such as oxidation and subsequent etching with hydrofluoric acid or hydrogen etching generate nearly bulk truncated surfaces whose dangling bonds are saturated by atomic or molecular adspecies. These surfaces exhibit the two-dimensional periodicity of a SiC bulk bilayer. The topmost surface layer arrangement and thus the surface morphology seem to be dependent on polytype and orientation of the sample, and on the particular preparation treatment. In several cases of nominally Si rich orientated samples such as 6H-SiC(0001) step bunching can be found. Then a linear surface layer stacking is preferred. On the nominally carbon rich orientation, for example 6H-SiC(000$\bar{1}$), mostly single steps are present. Heating the samples in vacuum in combination with Si evaporation leads to different reconstructed phases with a stoichiometry depending on annealing temperature and Si flux. The atomic structure of these phases is still under debate. On the Si rich orientation a (3×3) phase is generated upon large Si exposure. Its surface presumably consists of a Si bilayer and additional Si adatoms. Lower silicon flux during annealing or heating the (3×3) phase leads to the development of a $(\sqrt{3} \times \sqrt{3})$-R30° phase. This phase can also be obtained starting on an ex situ prepared sample by heating alone. Adatom, adcluster and vacancy models have been proposed in the literature for the $(\sqrt{3} \times \sqrt{3})$-R30° phase. On the opposite sample orientation the development of a (1×1) phase was reported. Further annealing causes silicon depletion for both surface polarities. On the Si side a phase of apparent $(6\sqrt{3} \times 6\sqrt{3})$-R30° periodicity develops while on the carbon side a carbon rich (3×3) phase was found. Eventually, heating leads to the development of graphitic overlayers.

1. Introduction

Silicon carbide is a semiconductor material that in several aspects allows unique technological applications. Not only that it is an extremely hard and inert material, it also has electric properties which are the basis for electronic device performances that cannot be achieved using other semiconductor materials. The high breakdown field and thermal conductivity [1, 2] allow high power and high frequency applications in regimes that are out of reach for silicon or GaAs based devices. Due to the large electronic band gap, varying between 2.4 and 3.3 eV for different polytypes, SiC devices can be operated at temperatures above 1000 K. Unfortunately, it is still difficult to grow SiC material of electronic and crystalline quality sufficient for a large scale industrial application. Several reasons are responsible for this problem among which one is the lack of understanding and control of the surface properties of SiC. The atomic structure and morphology of the surface influences the incorporation of defects into the grown material which in fact represents one of the most severe problems in SiC growth at present. Several types of defects are commonly observed in SiC crystals. At the interface between the seed crystal and the newly grown material different crystallites may be formed and from their boundaries or interfacial dislocations so called micropipes emerge. Grains of different polytypes can develop although this has been successfully suppressed under certain growth conditions. At least, the polytype cannot be chosen at will, and the atomic origin of the development of one or the other polytype is not understood. Most growth experiments at present concentrate on the natural growth direction of SiC which is the direction perpendicular to the hexagonal bilayers, i.e. the [0001] direction in hexagonal polytypes and the [111] direction in the cubic modification. This paper presents a review of investigations that have been performed to understand the properties of hexagonal surfaces of both possible orientations and several different polytypes of silicon carbide. After this introduction the crystal structure of SiC is presented in Section 2 both with respect to the bulk unit cell of different polytypes and to possible atomic configurations at the crystal surface. In Section 3 a brief description follows of the relevant surface science techniques used in the reviewed work. In Section 4 preparation methods and their influence on surface order and stoichiometry are discussed including chemical methods, heating and evaporation of silicon. The atomic surface structure of SiC is presented in Section 5. This section is divided into a part dealing with ex situ prepared unreconstructed surfaces of 6H-, 4H- and 3C-polytypes and a part describing the reconstruction phases obtained by Si evaporation and thermal treatment of the samples. A short conclusion finishes this review.

2. Crystal Structure

Silicon carbide contains carbon and silicon in $1:1$ stoichiometry. Each atom is covalently bonded to four atoms of the other chemical species in a tetrahedral coordination. These tetrahedral Si–C bonds are arranged in a hexagonal bilayer with carbon and silicon in alternating position with single bonds pointing upwards and downwards perpendicular to the bilayer plane as displayed in Fig. 1a. Along this perpendicular direction the bilayers are stacked on top of each other. Corresponding to the two possibilities to continue the tetrahedral arrangement of Si–C bonds, differing by a $60°$ (or $180°$) rotation, two possibilities exist for the mutual orientation of adjacent bilayers. The different stacking and orientation sequences of the bilayers correspond to different crystal modifications, so-called polytypes [3]. In the two extreme cases all bilayers may be oriented in the same direction or bilayers may be mutually rotated. In the first case (shown in Fig. 1b) we have a zinkblende structure which corresponds to the cubic SiC modification (historically called β-SiC). The second case corresponds to a wurtzite structure which is displayed in Fig. 1c. Here the crystal displays hexagonal symmetry. The three-dimensional bulk unit cells of the cubic zinkblende and the hexagonal wurtzite structure have a completely different spatial orientation. Nevertheless, SiC allotropes are commonly referred to in a terminology based on the bilayer stacking sequence. The cubic modification can be described by a hexagonal unit cell with a periodicity along the c-axis of three bilayers. It is called 3C-SiC whereby 3 indicates the number of bilayers per unit cell and C emphasizes the cubic symmetry. The stacking sequence of the bilayers is ABC(ABC...). In the wurtzite structure the bilayer arrangement is repeated every two bilayers with a stacking sequence AB(AB...). Accordingly, this structure is called 2H where H denotes hexagonal symmetry. It should be noted that the 2H polytype has not been observed to be stable in nature. One of the most intriguing features of SiC is that a large number of polytypes seems to be stable. More than 170 polytypes have been identia

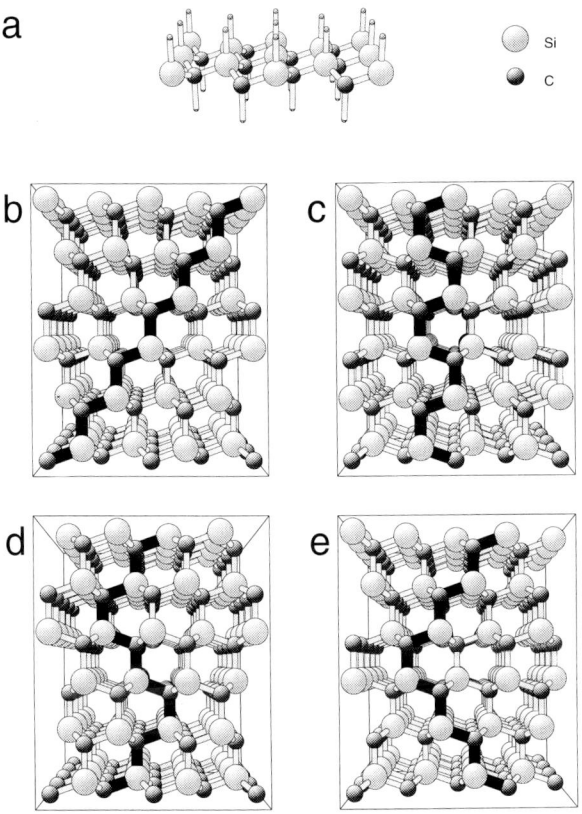

Si
C

b

c

d

e

Fig. 1. Crystal structure of silicon carbide, side view along $(11\bar{2}0)$ direction (hexagonal terminology): a) hexagonal bilayer; b) zinkblende structure, 3C-SiC; c) wurtzite structure, 2H-SiC; d) 4H-SiC; e) 6H-SiC. The stacking direction is depicted by the enhanced Si–C bond train parallel to the $(11\bar{2}0)$ direction. The $(11\bar{2}0)$-plane corresponds to the $(\bar{1}10)$-plane in cubic terminology

fied experimentally with periodicities along the c-axis as large as 1000 Å [4]. They all differ by the arrangement of cubic and hexagonal stacking sequences. Apart from cubic material most growth experiments and electrical applications have been reported for the 4H- and 6H-polytypes. These two modifications are based upon hexagonal unit cells of a height (along the c-axis) of 4 and 6 bilayers, respectively. In the 4H-structure two bilayers of identical orientation are followed by two bilayers in the opposite orientation, i.e. cubic and hexagonal stacking alternate with each other. The stacking sequence then is ABCB(AB...) (cf. Fig. 1d). In the 6H-structure, stacks of 3 linearly stacked bilayers are arranged in a periodically changing orientation resulting in a stacking sequence AB-CACB(AB...) as shown in Fig. 1e. When the alternating bilayer stacks are of mutually different height the crystal has rhombohedral symmetry. Nevertheless, it still can be described by a hexagonal unit cell. The smallest rhombohedral polytype which is called 15R has a stacking sequence of ABCACBCABACABCB(AB...) which is repeated after 15 bilayers (not shown in Fig. 1). These many different allotropes of SiC seem to be more or less energetically equivalent as shown by different theoretical calculations [5 to 11] which certainly is the reason for their natural appearance. It also seems feasible to manufacture stacking sequences by design that may then have tailored electrical or physical properties [12].

On surfaces of hexagonal orientation the translational symmetry along the c-axis is broken. Accordingly, the difference of the vertical periodicity of polytypes is projected to

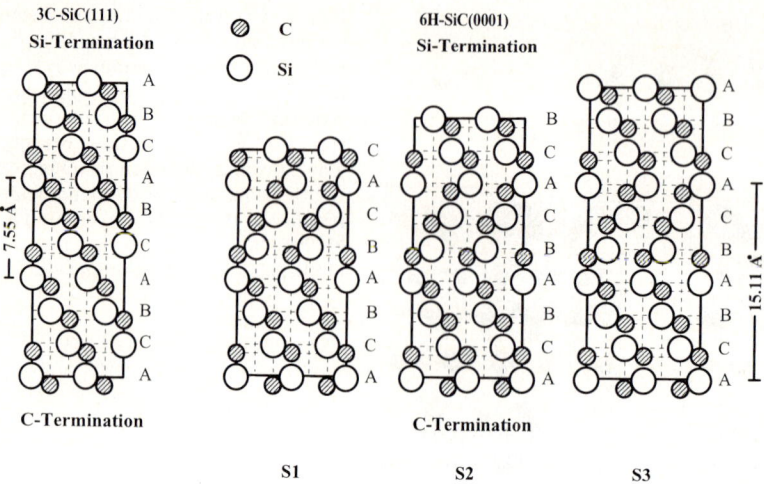

Fig. 2. Cross sectional view of the linear stacking present in the cubic (3C) polytype (ϑ-SiC) and three different stacking sequences possible at the surface of the 6H polytype. The different surface terminations are labeled according to the depth of the orientation change (S1 = CACBABC, S2 = BCACBAB and S3 = ABCACBA)

different surface stacking sequences. The relevant number of layers to be considered in this respect is related to the depth sensitivity of the probe used. For typical surface science methods this probing depth is rather low. Low energy electron based techniques such as Auger electron spectroscopy (AES), X-ray photoelectron spectroscopy (XPS) and low energy electron diffraction (LEED) are sensitive to a few, maybe four or five atomic layers due to the strong damping of the electrons. In scanning tunneling micros-copy (STM) only the topmost atomic layer is probed. So, instead of 170 or more poly-types, in a surface investigation only a few polytypes, namely 2H, 4H, 6H and 3C exhi-bit structural differences. The surfaces of more complicated modifications are equivalent to one of these polytypes. Yet, a single polytype may exhibit different surface termina-tions, depending on the position of the topmost bilayer in the stacking sequence of the bulk unit cell. For a 6H-polytype six different bulk truncations are distinguished by the depth of the bilayer orientation change (stacking fault) and by the orientation of the topmost bilayer. In Fig. 2 three of these different surface layer arrangements are plotted and denoted S1, S2 and S3 according to the depth of the stacking fault. The other three arrangements which represent a truncation plane within the other half of the bulk unit cell, can be constructed from those shown in Fig. 2 by a 60° rotation. For comparison the stacking arrangement of a 3C-SiC surface is also displayed in Fig. 2. Here, obviously only one orientation is possible. Furthermore, we have to distinguish between the two possible polarities of the crystal, where in case of a bulk truncation with intact bilayers the (0001) surface (hexagonal terminology) would be silicon terminated while the (000$\bar{1}$) surface would be carbon terminated. Of course, on a real sample an extended surface area may be composed of a mixture of domains with different terminations. In addition, surfaces are not necessarily bulk truncated. It is the purpose of this article to show that the topmost layers can be relaxed, and the dangling bonds tend to be saturated by an atomic or molecular adlayer or by a surface reconstruction pattern.

3. Techniques

SiC surfaces have been investigated using a variety of surface science techniques [13] ranging from electron spectroscopies to crystallographic methods such as LEED and STM. The techniques used in the work reviewed here are discussed in the present section with emphasis put on aspects of particular importance to the investigation of SiC. Some textbook references arc provided for more detailed descriptions of the physical principles and technical implementations of each method.

3.1 Electron spectroscopy

AES and XPS are commonly used to determine the chemical composition of SiC surfaces. In XPS the kinetic energy of core electrons excited by the X-ray photon is detected. This energy is characteristic of the atomic element excited by the photon [14]. In AES the recombination of a core hole is observed by means of the Auger electron whose kinetic energy is determined by the electronic configuration of the excited atom and, again, is element specific [14, 15]. For the primary excitation typically electrons, yet in some cases X-ray photons (XAES) are used. In both techniques, AES and XPS the chemical environment of the excited atom can be determined by analyzing energy shifts in the spectrum caused by chemically induced differences in the exact binding energies of the electronic states involved which typically include valence states in the case of AES [15, 16]. The energy of valence states can also be analyzed directly by XPS or, more commonly, by means of ultra-violet photoelectron spectroscopy (UPS) [17]. In the latter case, by angular resolved detection of the photoelectrons the electronic band structure of the surface can be investigated [18].

In a few investigations high resolution electron energy loss spectroscopy (HREELS) was used to detect the chemical nature of adspecies present on the SiC surfaces. In this technique the excitation of interatomic vibrations is monitored by the spectroscopy of inelastically scattered electrons. Vibrational modes of molecules as well as vibrations between adatoms and the substrate can be analyzed. In addition phonon modes of substrate and adlayers can be investigated [19, 20]. Electron energy loss spectroscopy (ELS) in lower resolution can be used to probe the frequencies of surface and bulk phonons [13].

3.2 Low energy electron diffraction (LEED)

The degree of order is an important parameter to characterize and identify different phases on a solid surface. In many cases the two-dimensional translational symmetry of the surface is directly correlated to a certain stoichiometry and can be determined by means of the spot pattern observed in low energy electron diffraction (LEED). The quality of the LEED pattern can be used to optimize the preparation procedure for the respective surface phase. The LEED pattern contains information about the atomic arrangement on the surface. Yet, the two-dimensional periodicity of the LEED pattern only reflects the real space translational symmetry of the surface while information about the detailed atomic geometry within the surface unit cell itself is concealed in the spot intensities and their dependence on the electron wavelength. On the one hand, this is due to the fact that the LEED electron has a limited penetration depth and a certain number of layers give rise to residual diffraction conditions along the surface normal.

Corresponding to the ill-defined crystal periodicity in the third, i.e. surface normal direction the spots of the two-dimensional LEED pattern exhibit smooth intensity maxima and minima in analogy to the sharp diffraction spots observed in three dimensions in X-ray diffraction. On the other hand, due to the large elastic scattering cross section of the electrons, multiple scattering events are the rule rather than the exception in LEED. This leads to additional variations of the spot intensities with varying electron wavelengths which are directly correlated to interatomic distances in the surface region. In a detailed analysis of the LEED spot intensities taking these and more subtle effects into account the complete atomic geometry of the surface can be determined by comparing experimental and calculated data [21 to 24].

Experimentally, $I(E)$ spectra are acquired by integrating the diffraction spot intensities as a function of the electron energy using a computer controlled video digitizing data acquisition system [23, 25 to 27]. In some cases, when the LEED pattern is sensitive to electron irradiation the full LEED pattern has to be stored in a single run on video tape in order to minimize the electron beam exposure. Subsequently, the intensities of the different spots can be digitized in several runs from the tape-recorded LEED patterns. For the geometry determination LEED intensities are generated using full dynamical calculations based on standard computer programs [21, 28] for several trial surface models. After a coarse model search the tensor LEED approximation [29] can be used for a refinement of the fit in order to save computer time. Deviation from the bulk stoichiometry and occupation density of adatom sites can be tested using the average t-matrix approximation (ATA) where the scattering amplitudes of two elements, or for diluted adatom coverages the element and a vacancy, are averaged according to the atomic concentration [30 to 32]. The correct surface geometry is found by optimizing the agreement between experimental and theoretical intensity spectra. For a quantitative measure the Pendry R-factor R_p [33] is used. The error margin for the different parameters fitted can be estimated by using the variance of the R-factor [33] which is derived

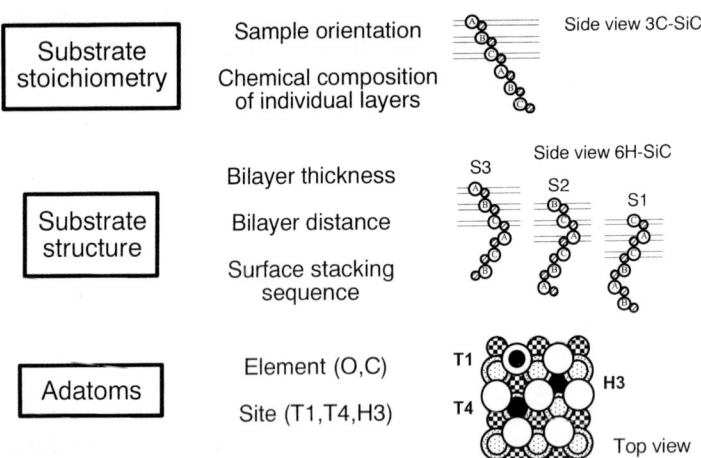

Fig. 3. Geometry variation in a SiC surface structure analysis by LEED. By varying the stoichiometry the sample orientation as well as the chemical composition of the individual layers can be determined. For each surface layer stacking sequence shown in Fig. 2 a structural model is examined. Within each model the thickness and distance of the bilayers as well as the possible presence, element and site of adatoms or admolecules are tested

statistically and depends on the quality of the fit and on the size of the data base used. For surfaces involving complicated crystal structures such as SiC it is necessary to employ an automated search algorithm [34, 35] in order to be able to scan through the large parameter space which would otherwise be practically impossible to handle. In Fig. 3 the different structural aspects included in the SiC calculations are indicated. In our analyses [36 to 40] we tried all possible surface layer stackings and mixtures as outlined in Section 2 and verified the orientation of the crystal and the layer stoichiometry. We typically varied the first two or three atomic layer distances and tested the chemical nature and different high symmetry sites of adlayer species, denoted T1, T4 and H3 as shown in the bottom of the figure.

3.3 STM

The scanning tunneling microscope (STM) provides a real space image of the surface. A small metallic tip is brought into close distance to the surface, a distance sufficiently small such that a quantum mechanical tunneling current between sample and tip can be measured [41, 42]. By scanning the tip across the surface while simultaneously adjusting the tip position to keep the tunneling current constant by means of a feedback loop, the morphology of the surface can be monitored. It is even possible to observe a variation of the tunneling current with atomic resolution. However, it should be noted that it is not the atomic positions that are imaged by the STM but rather the electronic states at the tip position several Ångströms above the surface. This may cause the appearance of the STM images to be dependent on the tip–sample bias voltage as then electronic states at different energies are involved in the tunneling process.

4. Surface Preparation

As early as in the late fifties work on the hexagonal surfaces of SiC with different surface sensitive methods has been reported and since then it has become clear that the preparation of clean and ordered SiC surfaces requires methods that are more complicated than for many other materials. Dillon et al. [43] using LEED and work function measurements investigated samples after different ways of chemical preparation. After degreasing in acetone their samples were either cleaned in fuming nitric acid or in a mixture of nitric acid, hydrofluoric acid (HF) and acetic acid. However, differences of the surface properties related to the chemical method applied were not observed. In vacuum the samples were ion bombarded and annealed, which resulted in surfaces that the authors concluded to be carbon rich, i.e. deviating from the bulk stoichiometry. In LEED an ordered surface with a two-dimensional periodicity corresponding to the SiC bulk unit cell was observed, although the diffraction spots were rather weak. Only after the SiC crystals were heated in oxygen (4×10^{-7} Torr, 1000 °C, 2 h) a stronger LEED pattern appeared. Later it was shown by other groups that ion bombardment can provide an atomically clean surface with a stoichiometry similar to that in bulk SiC, i.e. a 1 : 1 ratio of carbon and silicon [44, 45] (this situation is commonly denoted as a "stoichiometric" surface). However, after ion bombardment the surface is disordered [46, 47] and has to be annealed which causes a depletion in silicon [48, 49] due to the larger vapor pressure of Si in comparison to carbon. So, it turned out to be impossible to obtain stoichiometric, well ordered surfaces by a sputter/anneal preparation method.

Table 1
Preparation of hexagonal SiC surfaces

method	result	references
sputtering	stoichiometric, disordered surface	[44 to 47]
annealing	silicon depletion	[46, 48, 49]
thermal oxidation	thick oxide layer	e.g. [54]
hydrogen etching	SiC removal, (1×1) phase	[56, 58]
sublimation etching	SiC removal	[56]
etching in molten hydroxides/peroxides	SiC removal, rough surface	[50 to 52]
HF-treatment	thick oxide removal, (1×1) phase	[55]
heating in Ga flux	oxide removal	[53]
heating in Si flux	oxide removal, different surface phases (see Table 2)	[60, 71, 82]

4.1 Chemical methods

It was shown by different groups that well ordered surfaces can be prepared by certain methods of pretreatment before introducing the samples into vacuum. Sample slices obtained from boules of bulk grown material are covered with a thick disordered layer from the cutting and polishing procedure. As SiC is quite inert, it is difficult to remove the contaminated or disordered layer at least by chemical methods. Only by etching in molten hydroxides (e.g. NaOH, KOH) or peroxides (e.g. Na_2O_2) material can be removed. However, the surfaces tend to be rough after such an etching procedure [50 to 52]. A complementary method of removing SiC from the surface is prolonged heating in oxygen at atmospheric pressures and temperatures around 1000 °C. The surface is then covered with an oxide layer ("sacrificial" oxide) that in turn can be removed by HF [53, 54]. This method was shown to yield well ordered, stoichiometric surfaces with a two-dimensional periodicity corresponding to the bilayer structure [36, 55]. SiC material can also be removed by etching in hydrogen at elevated temperatures [56] or a sublimation etching procedure carried out in a growth reactor with the pressure and temperature parameters reversed as compared to the growth conditions [57]. If epitaxial films are grown on bulk grown substrates, e.g. by chemical vapor deposition (CVD), by carefully handling the samples after the growth process, more or less well ordered, bulk truncated surfaces can be obtained. This can in particular be forced when the sample is etched under hydrogen exposure at elevated temperatures before removing it from the growth reactor [58]. Yet, in all cases it seems inevitable that the surface is contaminated with spurious oxides due to exposure to the ambient atmosphere. The different preparation methods described in the literature and their effect on the properties of hexagonal SiC surfaces are compiled in Table 1.

4.2 Thermal treatment and silicon evaporation

The residual surface oxide can be removed by annealing the samples in ultra-high vacuum (UHV). However, heating leads to a carbon enrichment as mentioned above. So, Kaplan and Parrill successfully applied a Ga or As beam as reducing agent [53] at temperatures low enough (≈ 850 °C) to prevent Si depletion. Later, it turned out that thin oxide layers can also, and even easier, be removed by evaporating silicon on the sample surface at temperatures of around 1000 °C in order to compensate for the Si loss during annealing [59, 60]. However, the composition and structure of the surface after this preparation is strongly dependent on the balance between the Si gain from the evaporation and the Si loss caused by heating the sample. Several different ordered phases can be obtained, and are well distinguished by their surface stoichiometry. These phases obviously develop independent of the sample polytype. On the (111) or (0001) orientation the stable, Si rich phase has a (3×3) periodicity. A $(\sqrt{3} \times \sqrt{3})$-R30° phase develops upon lower silicon flux and/or higher annealing temperatures or by annealing a (3×3) surface. This phase has also been prepared by annealing a native, chemically pretreated surface [61]. Additional annealing produces a phase with is commonly referred to as $(6\sqrt{3} \times 6\sqrt{3})$-R30° because it shows a LEED pattern of that reciprocal space periodicity. The phase has been associated to a graphitic overlayer by several authors. However, it has been shown by STM that the phase actually consists of several coexisting phases of different surface order [62, 63] while a graphitic phase develops only upon even further annealing. In the latter phase a graphite overlayer produces a (6×6) surface periodicity as reported in different publications based on LEED and STM [64 to 68]. Details of the nature of the $(\sqrt{3} \times \sqrt{3})$-R30°, the $(6\sqrt{3} \times 6\sqrt{3})$-R30° and the graphitic phases are presented in a separate article in this review volume by Martensson et al. [69]. On the opposite sample surface, i.e. the $(000\bar{1})$ and $(\bar{1}\bar{1}\bar{1})$ orientation of hexagonal and cubic SiC, respectively, contrasting observations were published for surface phases developing upon heat and Si flux treatment. Kaplan and Parrill [53] reported the development of a (1×1) phase after preparation under Ga flux at elevated temperatures. Upon further annealing they observed a $(\sqrt{3} \times \sqrt{3})$ phase. Yet, most other authors found a carbon rich (3×3) phase to be stable on this side of hexagonal SiC

Table 2
Surface phase on hexagonal SiC surfaces

reference	method	Si orientation (0001), (111)	C orientation (0001), $(\bar{1}\bar{1}\bar{1})$
van Bommel et al. [48]	heating	$(\sqrt{3} \times \sqrt{3})$-R30°, $(6\sqrt{3} \times 6\sqrt{3})$-R30°, graphite	(2×2), (3×3)
Nakanishi et al. [70]	heating	$(\sqrt{3} \times \sqrt{3})$-R30°	(3×3)
Kaplan and Parrill [53]	annealing in Ga flux	(3×3)	(1×1)
	further annealing	(3×3)-R30°, (1×1)	$(\sqrt{3} \times \sqrt{3})$-R30°
Kaplan [60]	annealing in Si flux	(3×3)	
	further annealing	$(\sqrt{3} \times \sqrt{3})$-R30°, (1×1)	
Bermudez [71]	annealing in Si flux	(3×3)	(1×1)
	further annealing	$(\sqrt{3} \times \sqrt{3})$-R30°, (1×1)	(3×3)
Li and Tsong [68]	annealing in Si flux	(3×3)	(3×3)
	further annealing	$(\sqrt{3} \times \sqrt{3})$-R30°	$(\sqrt{3} \times \sqrt{3})$-R30°
Owman and Martensson [62, 63]	heating	$(\sqrt{3} \times \sqrt{3})$-R30°, mixed phases producing apparent $(6\sqrt{3} \times 6\sqrt{3})$-R30° LEED pattern	
STM work [62 to 68]	heating	graphitic overlayer with (6×6) honeycomb structure	

[48, 68, 70, 71]. The phase could be prepared by annealing with and without additional Si evaporation. Only Li and Tsong [68] observed the development of a $(\sqrt{3} \times \sqrt{3})$-R30° phase upon heating the (3×3) one also on 6H-SiC(000$\bar{1}$). Table 2 contains a summary of the surface phases observed on hexagonal SiC after heat treatment and/or Si evaporation.

5. Surface Structure

5.1 Unreconstructed surfaces

As described in the previous section, samples prepared by oxidation and HF treatment exhibit an unreconstructed surface with a two-dimensional periodicity corresponding to the SiC bilayer structure. This bulk-like structure is referred to as the (1×1) phase in the remainder of this paper. The samples show a sharp (1×1) LEED pattern immediately after introduction into UHV. No further treatment and in particular no annealing or outgassing is necessary. We have shown this in our group for both hexagonal orientations of 6H-SiC, i.e. the (0001) [36 to 38, 52, 55] and the (000$\bar{1}$) surfaces [37, 38, 72] and for a 3C-SiC(111) sample [37, 38, 73] (see the LEED patterns in Fig. 4a, b and c). Using 4H-SiC(0001) samples we could recently confirm that the (1×1) phase can also be obtained on hydrogen etched samples and on as grown epitaxial layers [58]. In Fig. 4d and e the respective LEED patterns of two 4H-SiC(0001) samples are shown. In spite of the good long range order attempts to image this native (1×1) phase by STM with atomic resolution were not successful [55, 57]. Tunneling required a large bias of $\approx$3V and in addition the current was rather unstable, so that only the morphology could be investigated in a few cases [52, 55]. The large bias was attributed to a chemical passivation of the surface, the unstable tunneling conditions were presumably caused by atmospheric contaminations on the surface. Indeed, oxygen is always present on HF etched surfaces as shown by several authors using AES and XPS [36, 37, 44, 53, 55, 58, 60, 70, 72 to 75]. The atomic structure of this native (1×1) phase was studied by our group for different polytypes and orientations of SiC. We determined the layer stacking sequence at the surface and relaxations of atomic distances, i.e. layer spacings as discussed in Sections 2 and 3. For the topmost atomic layers deviations from bulk-like stoichiometry were considered. In addition, the presence of adatoms and admolecules was tested in order to account for a possible saturation of dangling bonds. The samples used in these studies were obtained from different sources[1]) and included bulk grown material (modified Lely technique) and epitaxial layers grown by CVD [76, 77]. For both orientations of 6H-SiC we verified that the results of our studies were independent of the particular growth technique.

The 3C-SiC(111) sample was the only one of our samples grown heteroepitaxially on Si(111) [77]. Due to the large lattice misfit such a sample is very fragile. Therefore an extended chemical treatment or oxidation process was avoided. Solely, the native oxide was removed using buffered HF (pH = 5). Nevertheless, the sample was well ordered exhibiting a sharp (1×1) LEED pattern after introduction into vacuum. The pattern had an unperturbed threefold rotational symmetry (see Fig. 4c) which immediately indi-

---

[1]) Most samples were provided by Industrial Microelectronics Center (IMC), Kista, Sweden and Prof. Helbig, Lehrstuhl für Angewandte Physik, Universität Erlangen-Nürnberg, Germany. A few samples were bought from Cree Research, Inc., Durham, U.S.A.

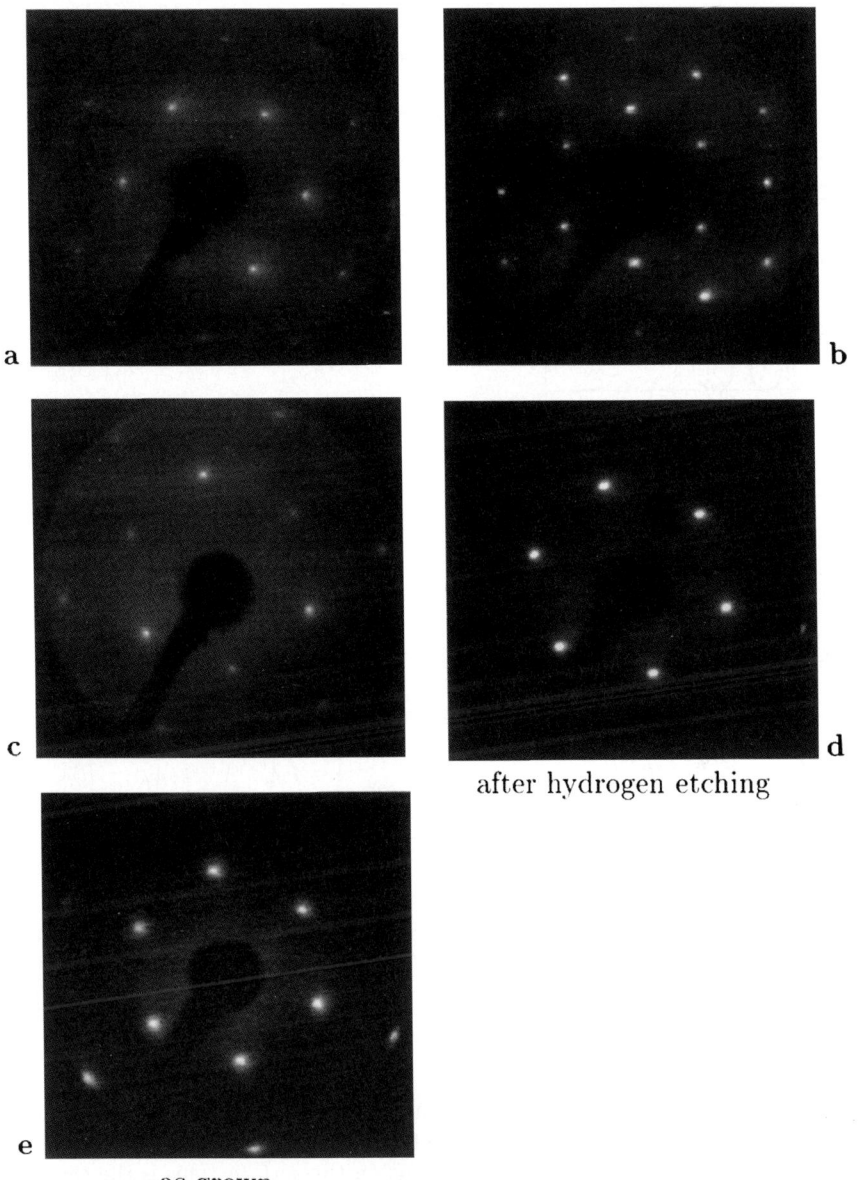

after hydrogen etching

as grown

Fig. 4. LEED pattern for the hexagonal SiC surfaces, a) 6H-SiC(0001), 120 eV; b) 6H-SiC(000$\bar{1}$), 242 eV; c) 3C-SiC(111), 120 eV and d) 4H-SiC(0001), 98 eV prepared by d) hydrogen or e) investigated as grown

cated untwinned growth of the epitaxial layer. This can be understood by recalling that a single surface domain of any polytype has a threefold rotational (2D) symmetry, and generates a diffraction pattern with a similar threefold rotational symmetry. So, a domain rotated by 60° which corresponds to the twin crystal in the case of cubic SiC would yield a diffraction pattern rotated by 60° and symmetry independent spots such

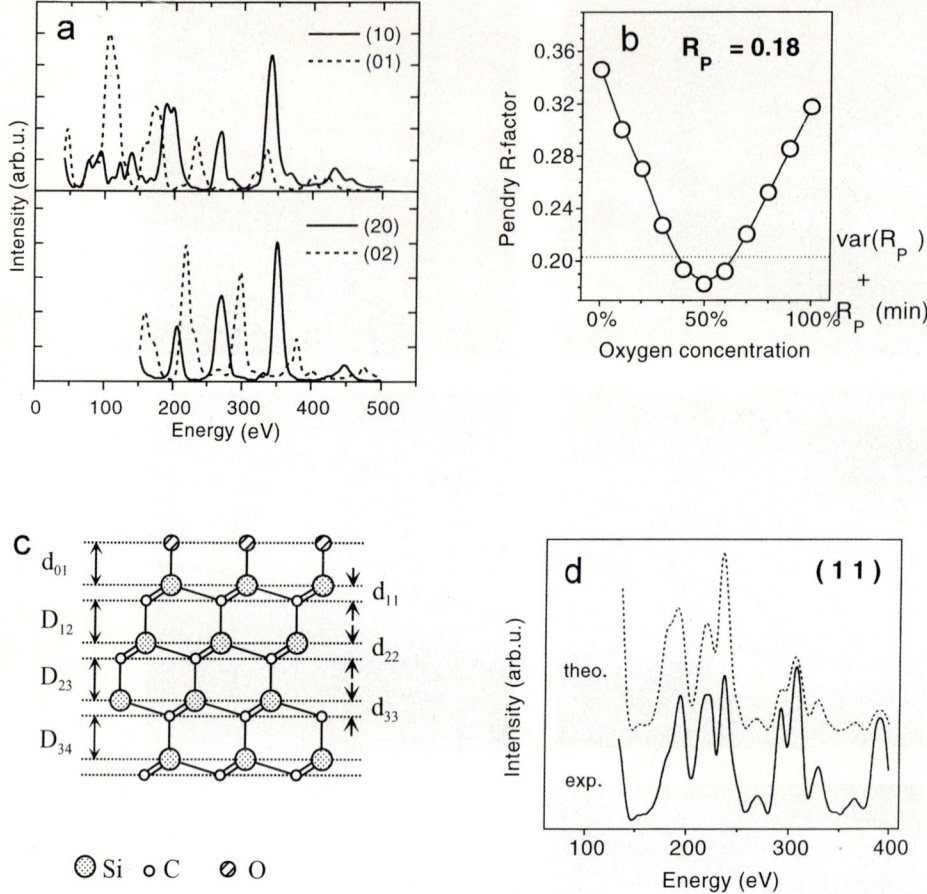

Fig. 5. a) Comparison of experimental LEED $I(E)$ spectra for first and second order diffraction measured from a 3C-SiC(111) surface. The absence of symmetry degeneracy within a diffraction order confirms untwinned film growth. b) R-factor plot depicting the best-fit oxygen coverage of 50%. c) Structural model of the 3C-SiC(111) surface (side view, projection parallel to the $(\bar{1}10)$-plane). d) Comparison of experimental and calculated (11) beam

as (10) and (01), would be superimposed one upon another. When for a certain electron energy one of the two inequivalent spots is completely extinct due to multiple scattering it is clear that only one domain orientation is present on the surface. This is the case for several electron energies. The $I(E)$ spectra of (10) and (01) spots and (20) and (02) spots, respectively, are compared in Fig. 5a. The LEED structure analysis [37, 38, 73] revealed that the surface of this domain indeed is composed of linearly stacked bilayers, i.e. has cubic crystal structure. Oxygen adatoms were found in T1 coordination bound to the topmost Si atoms. However, only 50% of the available T1 sites are occupied as depicted in Fig. 5b by a plot of the best-fit R_p-factor for different oxygen coverages. Fig. 5c displays a vertical cut through the surface in a projection parallel to the $(\bar{1}10)$ plane and indicates the terminology used for the geometry parameters considered in the analysis. In the best fit the Si–O bond length was found to be $d_{01} = 1.61$ Å. Bilayer

Table 3

Best-fit values of domain mixing percentages, layer spacing, adatom coverage and R-factor for the different polytype samples analyzed. Individual values for each domain and an average evaluated according to the mixing ratio is given

sample domain	3C-SiC(111) O adatoms (T1)	6H-SiC(0001) 3 clean samples (type 1) average values for			6H-SiC(0001) (#2), O adatoms			6H-SiC(000$\bar{1}$) H adatoms (T1)			
		#1	#1b	#1c	S3	S2	S1	S3	S2	S1	av
mix (%)	—				80	5	15	30	45	25	
d_{11} (Å)	0.63	0.65	0.61	0.64	0.55	0.66	0.72	0.57	0.63	0.60	0.60
D_{12} (Å)	1.86	1.83	1.89	1.89	1.94	1.80	1.77	1.94	1.94	1.84	1.91
d_{22} (Å)	0.67	—	—	—	—	—	—	—	—	—	—
D_{23} (Å)	1.86	1.95	1.89	1.93	1.88	1.95	1.85	1.89	1.89	1.84	1.88
d_{33} (Å)	0.64	—	—	—	—	—	—	—	—	—	—
D_{34} (Å)	1.93	1.89	1.90	1.89	—	1.89	1.95	—	—	—	—
d_{01} (Å)	1.61	—	—	—	1.66	—	—	1.03	1.03	1.00	1.02
ad (%)	50	—	—	—	100	—	—	100	100	100	100
R_{p}	0.18	0.23	0.25	0.28			0.13	0.184			

thickness and spacing between bilayers turned out to be more or less bulk-like (see Table 3 for a complete list of the best-fit results). The largest deviation of 0.05 Å cannot be considered significant according to the error analysis. It should be noted that total energy calculations for hexagonal SiC surfaces [78 to 80] showed that on a clean (1×1) surface the topmost bilayer would be strongly compressed (≈ 0.20 Å for the Si orientation) while in our analysis this topmost bilayer is exactly unrelaxed. Yet, this corresponds to an average value for the oxygen saturated and unoccupied sites. Saturation of the dangling bond with oxygen partly removes the contraction as we will see for 6H-SiC in the next paragraph. So, if 50% of the sites would have an unsaturated bond, the average bilayer would still be contracted. Thus, it seems that these sites cannot be uncovered. They may be covered by hydrogen which unfortunately would not be detectable in the LEED study due to its weak scattering cross section. Fig. 5d demonstrates the good reproduction of the experimental data by the model calculations which is also expressed by the R_{p}-factor of 0.18.

6H-SiC is a most common material, and samples in (0001) orientation are readily available in various kinds. In our group we have investigated a variety of samples including bulk grown material and epitaxial layers, and tested different recipes for the preparation treatment [36 to 38, 47, 50, 52, 55, 81, 82]. The treatment was varied from a simple dip in buffered HF at pH = 7.8 to a full application of the RCA procedure [83]. When samples were used several times a thermal oxidation step was always included. For new samples, however, this step was sometimes omitted. For one particular sample an etching step in molten NaOH was included before the chemical cleaning (see STM results below). A well ordered (1×1) surface could be obtained in all cases as shown in Fig. 4a. Yet, the LEED pattern of this native surface turned out to be extremely sensitive to electron irradiation, so that the intensities had to be stored on video tape as described in Section 3.2. In the course of the experiments $I(E)$ spectra of different samples were compared, but only small differences were found. The overall slope of the spectra was

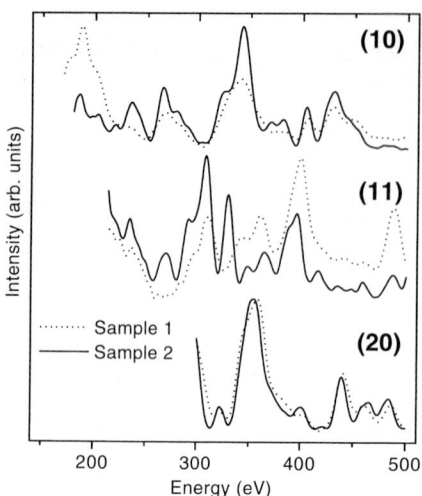

Fig. 6. Experimental $I(E)$ spectra of two 6H-SiC(0001) samples

dependent on the electron dose already applied to the sample, because irradiation caused an intensity drop for energies below 200 eV mainly. However, the structural elements of the spectra such as peak positions and minima were rather similar indicating similar atomic surface geometries. Slight differences could be detected in the sharpness of the spectral features and for a particular peak of the (11) spot at about 330 eV. As example of these differences Fig. 6 shows two extreme intensity data sets that are still looking fairly similar. Yet, while sample 1 produces spectra of a very smooth course, the structural elements in the intensities obtained from sample 2 are more pronounced including shoulders that are smeared out in sample 1. In addition, the peak mentioned above at 330 eV is only present in the (11) spectrum of sample 2. Data of other samples were similar to or in between the two examples shown, whereby the type of sample 1 was found more often. However, these differences could not be clearly attributed to different growth methods or certain preparation methods.

For a LEED analysis the situation is more complicated on 6H-SiC(0001) than on 3C-SiC(111) because one has to consider several domains of different surface layer stacking as described in Section 2. The best-fit geometries obtained for the two samples (1 and 2) were indeed different with respect to the surface stacking sequence. While sample 1 contained all three possible domains S1, S2 and S3 in equal amounts, for sample 2 mainly S3 domains were found [36]. The diffraction patterns exhibited sixfold rotational symmetry for all 6H samples. Accordingly, each domain type exists in both possible orientations, rotated by 60° with respect to each other. For three different samples of type 1 subjected to a LEED analysis a domain mixture was found with the domain occupation ratio varying between 15% and 45% in the best-fit structures [39, 82]. It should be noted that a 1 : 1 : 1 mixture still yielded a rather good fit being within the R-factor variance. The presence of adatoms, although feasible, could not be analyzed due to the excessive number of geometry parameters then to be included. The best-fit results for layer spacings were fluctuating around the bulk distances by up to 0.10 Å. Still, the average layer spacings of the best-fit structures were within 0.05 Å of the bulk values in all cases of sample 1 type data, which indicates that the fluctuations are of statistical origin. Contrary to sample 1 for sample 2 the S3 domain was found to be predominant with an

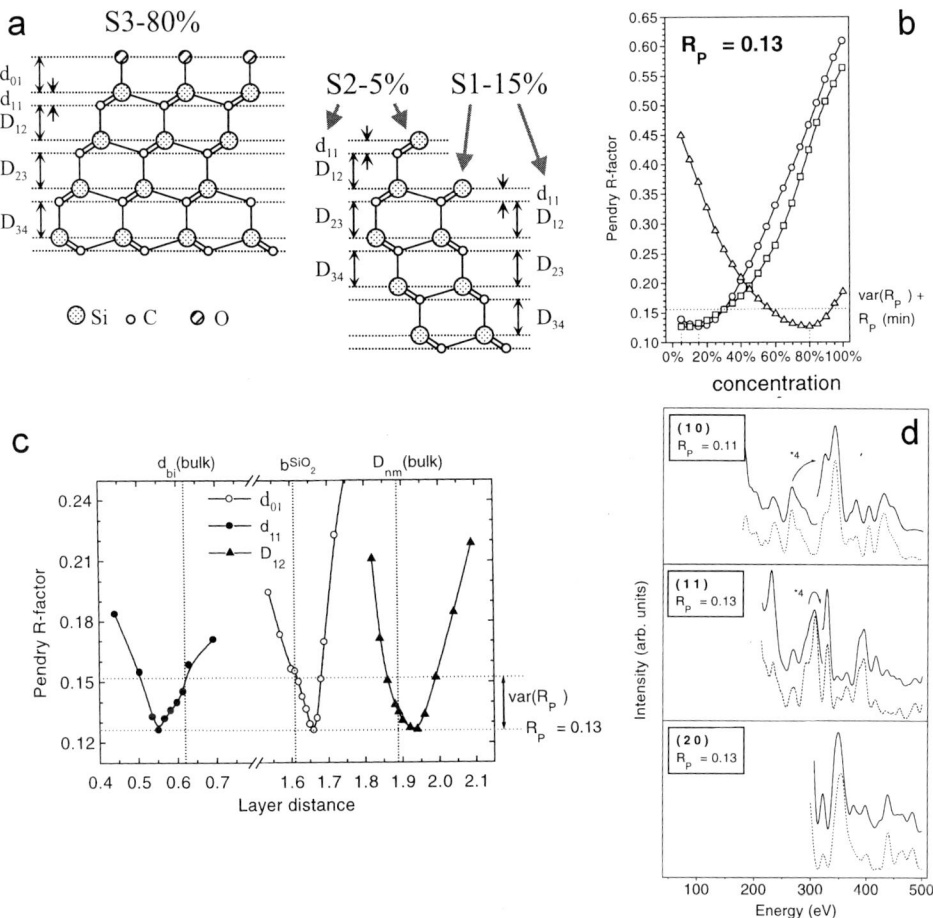

Fig. 7. Structure of the 6H-SiC(0001) (sample 2): a) Structural model (side view, projection parallel to the (11$\bar{2}$0) plane) depicting the geometry parameters included in the calculation for the three different stacking domains. b) R-factor plot representing the best-fit domain ratios. (○ S1-6H, □ S2-6H, △ S3-6H). c) R-factor plot for best-fit layer spacings of the S3 domain covering 80% of the surface (layer distance in Å). d) Comparison of experimental (solid lines) and calculated (dotted lines) $I(E)$ spectra

occupied surface area of 80% [37]. For the other two domains 15% (S1) and 5% (S2) occupation were found. On these S1 and S2 domains the best-fit interlayer spacing values are probably not very accurate due to the small domain percentages and large error bars, and therefore are not discussed further on. On the S3 domain interlayer spacings and the presence of adatoms could be determined. Oxygen and carbon were tested in T1, H3 and T4 coordination. The R-factor clearly preferred oxygen on T1 sites with a Si–O bond length of 1.66 Å. The substrate interlayer spacings seemed to be affected by the different bond coordinations of the topmost Si atoms. The topmost bilayer was found to be compressed by 13% to 0.55 Å while the distance between first and second bilayers was slightly expanded to 1.94 Å. Fig. 7 and Table 3 summarize the best-fit geometry and domain ratios. The parameter terminology used in Table 3 is indicated in

Fig. 7a for each stacking domain in a $(11\bar{2}0)$ cross sectional projection. The best-fit domain ratios are displayed in an R-factor plot in Fig. 7b where each data point represents the best-fit structure obtained when the surface area (percentage) of a single domain is preset to the corresponding value on the x-axis. Fig. 7c shows R-factor plots for the different interlayer spacings on the S3 domain while a comparison of experimental and calculated spectra is shown in Fig. 7d.

The presence of oxygen on the 6H-SiC(0001) surfaces, also confirmed by AES, was further illuminated by HREELS experiments [55, 81]. The vibrational spectra (see Fig. 8) of SiC samples are dominated by the large optical surface phonon ("Fuchs-Kliever phonon") and its multiple excitations as also shown by Nienhaus et al. [84]. On native samples investigated before any electron damage by LEED additional vibrational modes can be seen. At a loss energy of 443 meV a peak identified as O–H stretching mode indicates that the oxygen found in the LEED analysis represents hydroxyl groups linearly bound to the topmost Si atoms of the first bilayer. A detailed intensity analysis of the peak around 360 meV reveals that besides the triple loss of the Fuchs-Kliewer phonon (FK$_3$) it contains a C–H mode indicating the presence of hydrocarbon species on the surface [81]. The absence of carbon as adatom in the LEED results suggests that these hydrocarbons are disordered on the surface. A possible presence of hydrogen could only be suspected because the respective Si–H stretching vibration (257 meV) is concealed by the base of the FK$_2$ peak.

The dominance of one domain type (S3) on the surface requires a bunching of steps so that the other two domains are suppressed. In STM experiments on native 6H-SiC(0001) surfaces this type of morphology could be confirmed. Although the image quality was not good for reasons discussed in the beginning of this section line scans

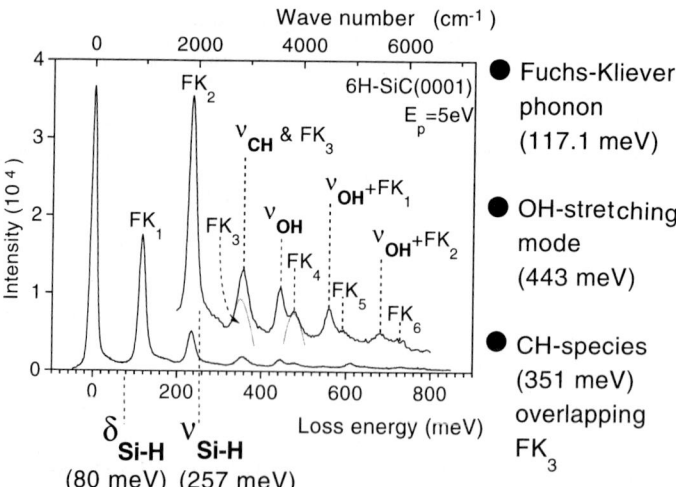

Fig. 8. Vibrational spectrum of a chemically prepared 6H-SiC(0001) sample by HREELS. Single and multiple excitation losses of the optical surface phonon ("Fuchs-Kliever" phonon) are indicated FK$_N$ ($\nu_1 = 117.1$ meV). The FK$_3$ loss overlaps with the C–H stretching vibration at 351 meV. An additional loss at $\nu = 443$ meV is interpreted as the O–H stretching mode of hydroxyl species. The potential frequencies of stretching and bending modes of the Si–H bond are indicated

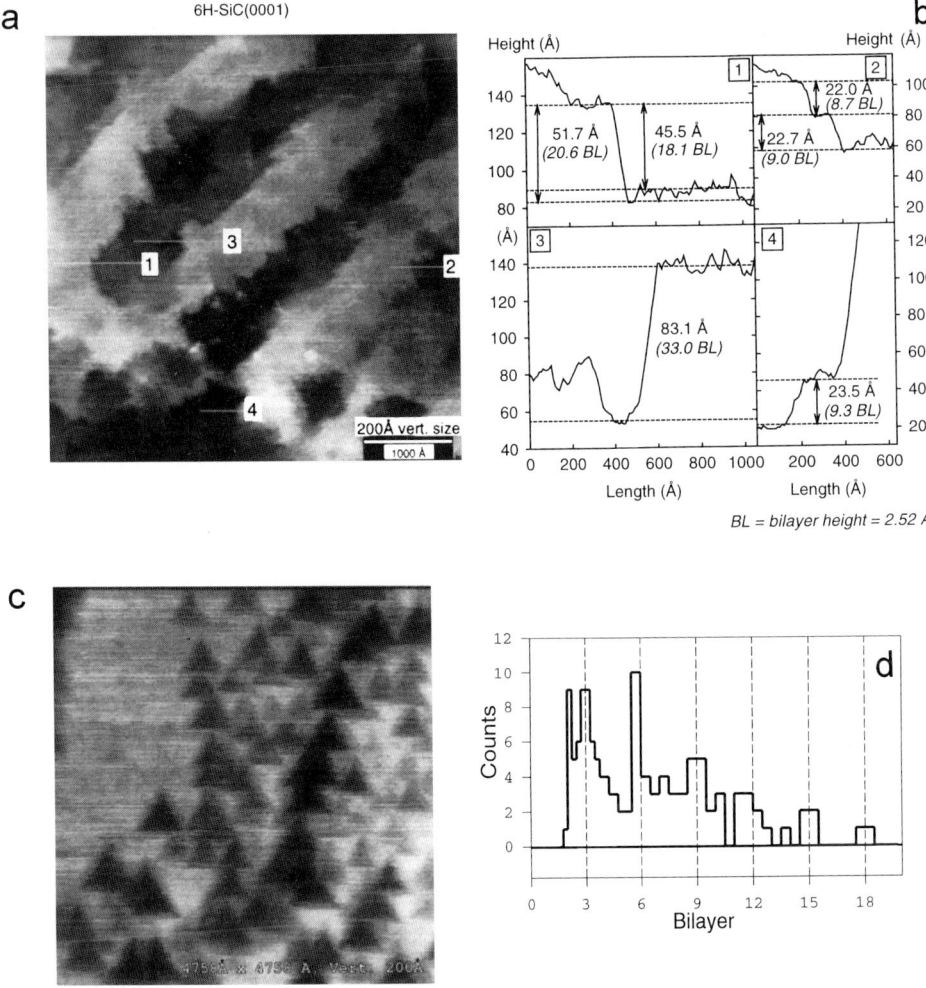

Fig. 9. Morphology of 6H-SiC(0001) samples after chemical preparation. a) 5000 × >5000 Å² STM image in topographic mode obtained from a sample prepared by sacrificial oxidation and HF dip. b) Line scans taken from part a showing step heights of multiples of three bilayer spacings ($n \times 3 \times 2.52$ Å). c) STM image obtained from a sample etched in molten NaOH (topographic mode) showing a large number of etch pits. d) Statistical analysis of step heights within the etch pits (part c) showing maxima at 3, 6, 9... bilayers

could be used to determine step heights as shown in Fig. 9a. Most steps found on the surface have a height corresponding to an odd or even number of three bilayers (Fig. 9b) confirming a single domain type morphology [50, 55]. Of course, which of the domains is present on the surface could only be determined by LEED. A sample which was etched in molten NaOH previous to the chemical preparation showed a large number of etch pits on the surface (see Fig. 9c). Within these etch pits the step heights were statistically evaluated. The step height distribution shown in Fig. 9d is clearly peaked around values corresponding to 3, 6, 9, 12 and 15 bilayer heights. So the etching process also seemed to prefer one of the three possible domains [50, 52]. A preference of triple or multiple triple bilayer height steps was also shown by Owman and Martensson using STM on annealed 6H-SiC(0001) samples [61] and after hydrogen etching [56], and by Kimoto et al. [85 to 87] on CVD samples investigated immediately after the growth process using transmission electron microscopy (TEM). Yet, other groups using STM reported single and double bilayer height steps on annealed samples [64, 67]. So, obviously the morphology of 6H-SiC(0001) surfaces can be differing considerably. However, it seems to be unclear under what circumstances step bunching or a single step morphology is preferred.

Recently, the (1×1) structure of a 6H-SiC(000$\bar{1}$) surface was investigated using LEED, XPS and angle resolved photoelectron spectroscopy (ARPES) [37, 38, 72]. In this orientation only bulk grown material was available. To remove polishing damage from the surface thermal oxidation was used for the LEED experiments, while the sample for the spectroscopic measurements was plasma etched in hydrogen. In both cases a well ordered (1×1) surface was obtained (see Fig. 4b). The preparation of this C-phase of SiC seems very reliable as concluded from the fact that LEED spot intensities acquired on samples from different sources were identical. In the LEED structure analysis of this (000$\bar{1}$) orientation the same domain types and mixtures were tested as for the (0001) samples. Of course, carbon and silicon atoms had to be exchanged in the structural models. The best fit was found for a mixture of all three domains occupying approximately the same surface area. This is shown in Fig. 10 by an R-factor plot similar to that in Fig. 7b. A structural model of the clean surface without adlayers is shown in Fig. 11a which, however, seems unlikely in view of the electronic band structure of the surface. Different from theoretical predictions by Sabisch et al. [80, 88] a dangling bond state in the electronic band gap was not found in the ARPES experiments [72]. Consequently, the LEED analysis was refined by testing several different adlayer models which are displayed in detail in Fig. 11b to d. No good R-factor could be achieved for the linear SiO model shown in Fig. 11c. For hydrogen and oxygen adatoms (part b) and the SiO bilayer model (part d) the R-factor could be slightly reduced, yet, for none of the models far enough to exclude the clean surface model (part a) by means of the R-factor variance. However, the SiO bilayer model seems unlikely as the Si–O bond length is unphysically large in the best-fit geometry. Finally, in core level spectra from the XPS experiments [72] no evidence was present for oxygen bound to

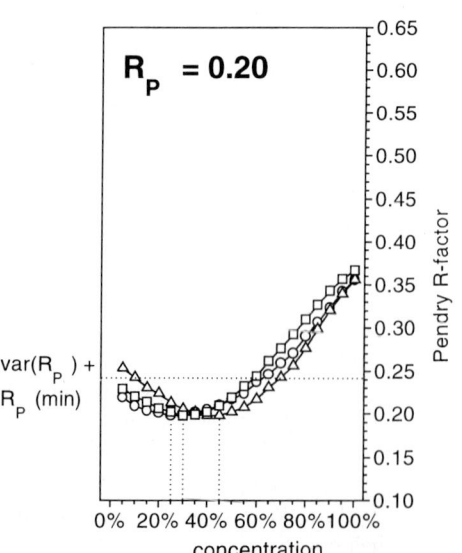

Fig. 10. R-factor plot for the mixing ratio of domains with different surface stacking sequences (see Fig. 2 for terminology) for 6H-SiC(000$\bar{1}$) (○ S1-6H, △ S2-6H, □ S3-6H).

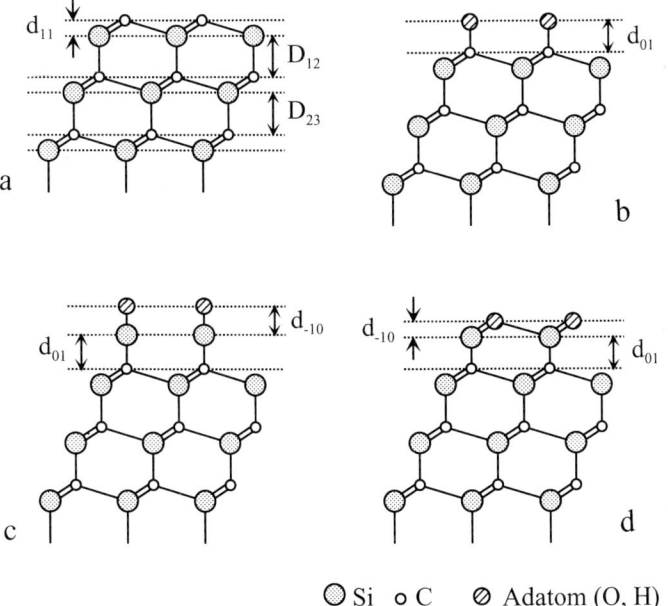

O Si o C ⊘ Adatom (O, H)

Fig. 11. Adatom models tested for 6H-SiC(000$\bar{1}$) including parameter terminology. a) Clean unre-constructed surface terminated by a full bilayer. b) Adatoms in T1 coordination. c) Model with linear Si−O adspecies oriented in surface normal direction. d) Si−O bilayer model continuing the substrate stacking sequence

carbon atoms. That eliminates the oxygen adatom model. So, the only model found compatible with LEED, ARPES and XPS is a hydrogen adatom model as shown in Fig. 11b. In Table 3 the best-fit geometry parameters for the three domains individually are compiled together with layer spacing values averaged according to the domain percentages. These latter values showing less statistical fluctuations than the numbers for single domains indicate that the hydrogen adatoms suppress any layer contraction that would be expected to be considerable on a clean unreconstructed (000$\bar{1}$) surface with dangling bonds [78 to 80, 88].

5.2 Reconstructed phases

The atomic structure of the reconstructed surfaces is still under debate. Many models have been proposed for the different phases. However, a successful crystallographic analysis has not been reported. Kaplan [60] investigated the (3×3) phase on 6H-SiC(0001) and 3C-SiC(111) and found a strong Si enrichment based on AES observations. From a close inspection of surface plasmons using ELS he concluded that the SiC surface should be covered by about two monolayers of silicon and proposed a dimer adatom stacking fault (DAS) model in close analogy to the (7×7) reconstruction of Si(111). The model he suggested consists of a Si bilayer on top of the SiC bulk structure topped by additional Si adatoms. It includes dimers, corner holes and the stacking fault as found in the (7×7)-Si(111) in order to accommodate for the large misfit between the SiC bulk layers and the adlayer consisting purely of silicon. Kaplan's model consists of adatoms on both

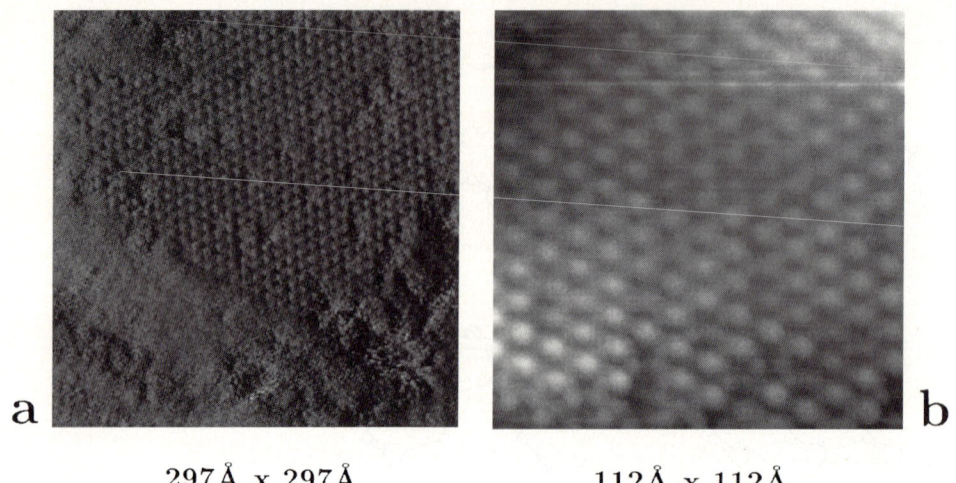

a   297Å x 297Å   112Å x 112Å   **b**

Fig. 12. STM images of a (3×3) phase on 4H-SiC(0001). The images were obtained using a positive tip bias of 1.4 V and a tunneling current of 0.4 nA. The grey scale represents a differentiated corrugation in a); b) shows true z-data after application of a tilt correction

the faulted and unfaulted halves of the (3×3) unit cell. Yet, when the (3×3) phase was investigated by STM with atomic resolution by different groups on 6H-, 4H- and 3C-SiC [65, 68, 89] only one large structure was found in each unit cell as shown in Fig. 12 regardless of the bias applied between tip and sample. Therefore the (3×3) phase most likely consists of only one adatom per unit cell on top of the Si adlayer as suggested by Kulakov et al. [65] in a model derived as modification of the DAS model. For both models a sketch is displayed in Fig. 13 that includes the topmost SiC bulk layer, the Si adlayer and the additional Si adatoms. It should be noted that the adlayer bond lengths chosen for the plot are merely suggestive, indicating a presumable distortion of this layer due to the lattice mismatch between SiC and Si. In connection with the different number of adatoms in the two models the adlayer/adatom configurations are also displaying different 2D symmetries, i.e. sixfold rotational symmetry in the DAS model and threefold rotational symmetry in the single adatom model. The single adatom model does not contain an intrinsic stacking fault because all adatoms are equally coordinated. However, two orientations are feasible for the whole adlayer that differ by a 60° rotation. The two possibilities correspond to either a linearly stacked Si adlayer (shown in Fig. 13) or a stacking fault between the SiC bulk and the Si adlayer (not shown). Li and Tsong [68] suggested an adatom cluster model. However, it seems that this model cannot explain the large amount of Si found on the surface by AES. AES and ELS data are more in line with one of the Si adlayer models from which STM data would clearly favor the single adatom model. However, none of the experiments reviewed here provides evidence on the detailed nature of the Si layer and the site coordination of the adatoms. In fact, it may well be that none of the models is entirely correct.

A similar situation of different model proposals has to be noticed for the $(\sqrt{3} \times \sqrt{3})$-R30° phase where even different preparation conditions have been used. As discussed in Section 4 this phase can be prepared by annealing the Si rich (3×3) or starting from a native (1×1) surface after chemical preparation. There is no agreement between the

a) DAS model

Fig. 13. Structural models proposed for the Si-rich (3×3) phase on hexagonal SiC [60, 65] shown in top and side views

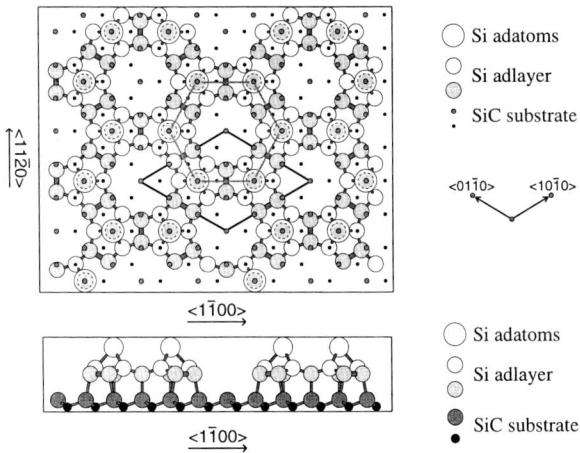

b) Single adatom model

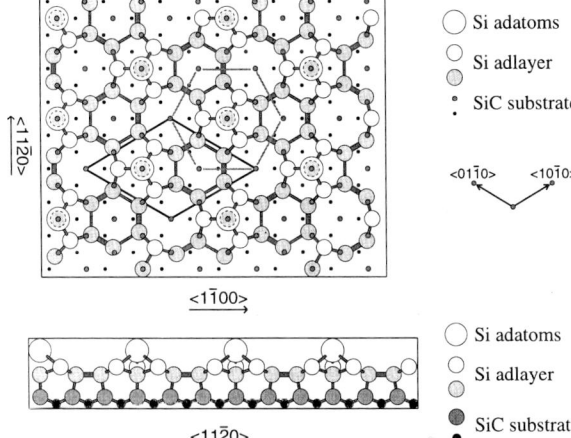

spectroscopic results of different groups whether the surface contains more Si or carbon than a bulk truncated SiC surface. In view of AES measurements Bermudez [71] suggested Si vacancies in the terminating SiC layer, i.e. a carbon rich surface. On the contrary Gunella et al. [90] proposed a Si adlayer of submonolayer density on top of the topmost SiC bilayer on account of XPS and XAES data. Johannson et al. [74, 91 to 93] reported a carbon rich surface. Yet, they need a complex adlayer model in order to interpret core level shifts measured with XPS. STM experiments by two groups revealed a single structure per $(\sqrt{3} \times \sqrt{3})$ unit cell [61, 68] indicating that a single adatom terminates the surface reconstruction. Yet, the two groups presented a different interpretation for site symmetry and chemical nature of the adatom. Similarly, entirely different models ranging from Si adatoms to carbon clusters have been presented resulting

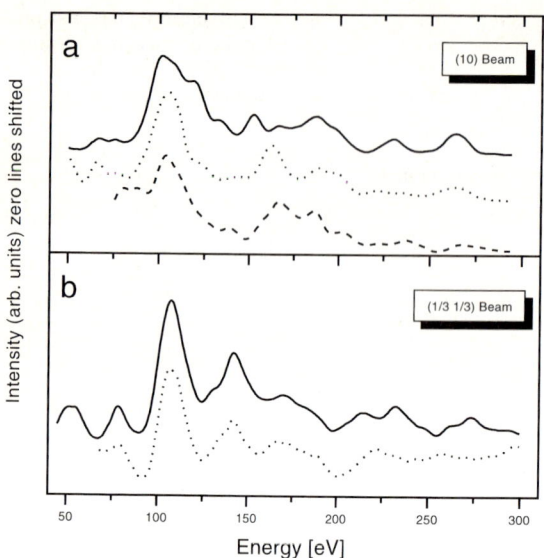

Fig. 14. LEED intensities of the $(\sqrt{3} \times \sqrt{3})$-R30° phase on 4H-SiC(0001) compared for two preparation methods, one starting from a (3×3) phase after Si evaporation (solid lines) and the other from annealing a native (1×1) surface after chemical preparation (dotted lines). a) The integer order spectra are different. The second phase obtained by annealing partly resembles the data obtained from the native sample itself (dashed line). b) The $(1/3, 1/3)$ spectra are very similar

from theoretical investigations [79, 80, 88, 94 to 96]. A possible explanation of the diversity of results would be that the preparation methods used produce different surface structures with a common $(\sqrt{3} \times \sqrt{3})$-R30° periodicity. However, fractional order LEED intensities of structures prepared by either method are very similar as shown in Fig. 14b, indicating that the structures obtained are identical [89]. However, the $(\sqrt{3} \times \sqrt{3})$ phase seems to cover different surface areas in the two cases as judged from the (10) spot which partly resembles the native (1×1) surface when the $(\sqrt{3} \times \sqrt{3})$ is only prepared by annealing, far more than after annealing a (3×3) surface (Fig. 14a). The carbon rich phases, the so called $(6\sqrt{3} \times 6\sqrt{3})$-R30° on the Si side and the (3×3) on the C side of SiC have not been subjected to detailed structural investigations with the exception of STM work by Owman and Martensson on $(6\sqrt{3} \times 6\sqrt{3})$-R30° [62, 63] which, together with the structural models discussed for the $(\sqrt{3} \times \sqrt{3})$ phase is reviewed in a separate article by Martensson et al. in this volume [69].

6. Conclusion

Several ex situ methods have been applied successfully to prepare hexagonal surfaces of SiC to be atomically clean and well ordered. After a final HF dip which is common to all recipes adatoms or admolecules such as oxygen, hydrogen or hydroxyl species are attached to the topmost SiC like bilayers and serve to saturate the dangling bonds. The substrate geometry is rather similar to the SiC bulk structure with layer spacings and bilayer thicknesses deviating less than typically 0.05 Å. While this layer geometry is independent of sample polytype and polarity, the surface morphology seems to depend on both and in addition on the sample pretreatment before the final HF dip. On several nominally silicon terminated samples, i.e. (0001) and (111) orientations, the surface terminating stacking sequence is observed to be favorably linear which leads to a step bunching with step heights of odd or even multiples of half the hexagonal unit cell length along the c-axis. On the opposite polarity this effect is not observed. It remains to identify the reason for this polytype specific morphology on the silicon terminated orientation and to determine the influence of different surface pretreatment steps on the effect.

By annealing the samples in vacuum which has to be accompanied by Si exposure in order to accommodate for the Si depletion due to its higher vapor pressure in comparison to carbon several ordered reconstructions can be prepared. From STM images it seems that at least in the (3×3) and $(\sqrt{3} \times \sqrt{3})$-R30° phases the topmost surface element is a single adatom per unit cell. Yet, neither the nature of this adatom nor the exact atomic geometry underneath has been unambiguously determined up to this point.

Acknowledgements I would like to thank different colleagues for making available manuscripts of their work prior to publication. I am indebted to the Deutsche Forschungsgemeinschaft (DFG) for financial support through SFB 292 and grant Sta315/4.

References

[1] H. MORKOC, S. STRITE, G. B. GAO, M. E. LIN, B. SVERDLOV, and M. BURNS, J. Appl. Phys. **76**, 1363 (1994).
[2] G. L. HARRIS, in: emis Datareviews Series, Vol. 13, INSPEC, The Institution of Electrical Engineers, London (UK) 1995.
[3] R. VERMA and P. KRISHNA, Polymorphism and Polytypism in Crystals, Wiley, New York 1966.
[4] G. PENSL and R. HELBIG, in: Festkörperprobleme/Adv. Solid State Phys. **30**, 133 (1990).
[5] C. CHENG, R. J. NEEDS, V. HEINE, and N. CHURCHER, Europhys. Letters **3**, 475 (1987).
[6] C. CHENG, R. J. NEEDS, and V. HEINE, J. Phys. C **21**, 1049 (1988).
[7] C. CHENG, V. HEINE, and I. L. JONES, J. Phys. CM **2**, 5097 (1990).
[8] C. CHENG, V. HEINE, and R. J. NEEDS, Europhys. Letters **12**, 69 (1990).
[9] P. KÄCKELL, B. WENZIEN, and F. BECHSTEDT, Phys. Rev. B **50**, 10761 (1994).
[10] P. KÄCKELL, B. WENZIEN, and F. BECHSTEDT, Phys. Rev. B **50**, 17037 (1994).
[11] C. H. PARK, B.-H. CHEONG, K.-H. LEE, and K. J. CHANG, Phys. Rev. B **49**, 4485 (1994).
[12] F. BECHSTEDT and P. KÄCKELL, Phys. Rev. Letters **75**, 2180 (1995).
[13] G. ERTL and J. KÜPPERS, Low Energy Electrons and Surface Chemistry, VCH Verlagsgesellschaft, Weinheim 1985.
[14] D. BRIGGS and M. P. SEAH, Practical Surface Analysis by Auger and X-Ray Photoelectron Spectroscopy, Wiley, New York 1983.
[15] J. C. FUGGLE, in: Electron Spectroscopy — Theory, Techniques and Applications, Vol. 4, Ed. C. R. BRUNDLE and A. D. BAKER, Academic Press, London 1981 (p. 85).
[16] N. MARTENSSON and A. NIELSSON, in: Application of Synchrotron Radiation, Springer Ser. Surface Sci., Vol. 35, Ed. W. EBERHARDT, Springer-Verlag, Berlin 1995 (p. 65).
[17] B. FEUERBACHER and B. FITTON, Photoemission and the Electronic Properties of Surfaces, Wiley, Chichester 1978.
[18] W. EBERHARDT, in: Synchrotron Radiation Research: Advances in Surface and Interface Science, Ed. R. Z. BACHRACH, Plenum Press, New York 1992 (p. 139).
[19] H. FROITZHEIM, in: Electron Spectroscopy for Surface Analysis, Ed. H. IBACH, Springer-Verlag, Heidelberg 1977 (p. 205).
[20] H. IBACH and D. L. MILLS, Electron Energy Loss Spectroscopy and Surface Vibrations, Academic Press, London/New York 1982.
[21] J. B. PENDRY, Low Energy Electron Diffraction, Academic Press, London 1974.
[22] M. A. VAN HOVE, W. H. WEINBERG, and C. M. CHAN, Low Energy Electron Diffraction, Springer Ser. Surface Sci., Vol. 6, Springer-Verlag, Berlin 1986.
[23] K. HEINZ, Progr. Surface Sci. **27**, 239 (1988).
[24] K. HEINZ, Rep. Progr. Phys. **58**, 637 (1995).
[25] K. HEINZ and K. MÜLLER, Springer Tracts mod. Phys. **91**, 1 (1982).
[26] K. MÜLLER and K. HEINZ, in: The Structure of Surfaces, Ed. M. A. VAN HOVE and S. Y. TONG, Springer-Verlag, Berlin/Heidelberg 1985 (p. 105).

[27] D. VON GEMÜNDEN, Dissertation, University Erlangen-Nürnberg, 1990.
[28] M. A. VAN HOVE and S. Y. TONG, Surface Crystallography by LEED, Springer-Verlag, Berlin/Heidelberg 1979.
[29] P. J. ROUS, J. B. PENDRY, D. K. SALDIN, K. HEINZ, K. MÜLLER, and N. BICKEL, Phys. Rev. Letters **57**, 2951 (1986).
[30] Y. GAUTHIER, Y. JOLY, R. BAUDOING, and J. RUNDGREN, Phys. Rev. B **31**, 6216 (1985).
[31] Y. GAUTHIER and R. BAUDOING, in: Low Energy Electron Diffraction from Alloy Surfaces, Ed. P. DAWBEN and A. MILLER, CRC Press, Boca Raton 1990 (p. 169).
[32] S. CRAMPIN and P. J. ROUS, Surface Sci. Letters **244**, L137 (1991).
[33] J. B. PENDRY, J. Phys. C **13**, 937 (1980).
[34] M. A. VAN HOVE, W. MORITZ, H. OVER, P. J. ROUS, A. WANDER, A. BARBIERI, N. MATERER, U. STARKE, and G. A. SOMORJAI, Surface Sci. Rep. **19**, 191 (1993).
[35] M. KOTTKE and K. HEINZ, Surface Sci. **376**, 352 (1997).
[36] J. SCHARDT, C. BRAM, S. MÜLLER, U. STARKE, K. HEINZ, and K. MÜLLER, Surface Sci. **337**, 232 (1995).
[37] J. SCHARDT, J. BERNHARDT, U. STARKE, and K. HEINZ, Surface Rev. and Letters, in press.
[38] U. STARKE, J. BERNHARDT, M. FRANKE, J. SCHARDT, and K. HEINZ, Diamond relat. Mater., accepted.
[39] J. SCHARDT, J. BERNHARDT, M. FRANKE, U. STARKE, and K. HEINZ, unpublished.
[40] J. SCHARDT, Diplomarbeit, University Erlangen-Nürnberg, 1994.
[41] H. J. GÜNTHERODT and R. WIESENDANGER (Ed.), Scanning Tunneling Microscopy I, Springer Ser. Surface Sci., Vol. 20, Springer-Verlag, Berlin 1992.
[42] C. J. CHEN, Introduction to Scanning Tunneling Microscopy, Oxford University Press, New York 1993.
[43] J. A. DILLON, JR., R. E. SCHLIER, and H. E. FARNSWORTH, J. Appl. Phys. **30**, 675 (1959).
[44] F. BOZSO, L. MUEHLHOFF, M. TRENARY, W. J. CHOYKE, and J. T. YATES, JR., J. Vacuum Sci. Technol. A **2**, 1271 (1984).
[45] K. SAIKI, H. TANAKA, and S. TANAKA, J. Nuclear Mater. **128/129**, 744 (1984).
[46] M. SCHLÖGEL, Diplomarbeit, University Erlangen-Nürnberg, 1992.
[47] C. BRAM, Diplomarbeit, University Erlangen-Nürnberg, 1994.
[48] A. J. VAN BOMMEL, J. E. CROMBEEN, and A. VAN TOOREN, Surface Sci. **48**, 463 (1975).
[49] S. ADACHI, M. MOHRI, and T. YAMASHINA, Surface Sci. **161**, 479 (1985).
[50] W. HARTNER, Diplomarbeit, University Erlangen-Nürnberg, 1994.
[51] K. BRACK, J. Appl. Phys. **36**, 3560 (1965).
[52] U. STARKE, J. SCHARDT, P.-R. STEINER, W. HARTNER, S. MÜLLER, L. HAMMER, K. HEINZ, and K. MÜLLER, in: Proc. Symp Surface Science 3S' 95, Ed. P. VARGA and F. AUMAYR, Inst. f. Allg. Physik, TU Wien, 1995 (p. 11).
[53] R. KAPLAN and T. M. PARRILL, Surface Sci. Letters **165**, L45 (1986).
[54] L. MUEHLHOFF, M. J. BOZACK, W. J. CHOYKE, and J. T. YATES, J. Appl. Phys. **60**, 2558 (1986).
[55] U. STARKE, C. BRAM, P.-R. STEINER, W. HARTNER, L. HAMMER, K. HEINZ, and K. MÜLLER, Appl. Surface Sci. **89**, 175 (1995).
[56] F. OWMAN, C. HALLIN, P. MARTENSSON, and E. JANZEN, J. Crystal Growth **167**, 391 (1996).
[57] F. OWMAN, in: Linköping Studies in Science and Technology, Dissertations, Vol. 420, Department of Physics and Measurement Technology, Linköping University, 1996.
[58] M. FRANKE, N. NORDELL, A. SCHÖNER, K. ROTTNER, U. STARKE, and K. HEINZ, to be published.
[59] V. M. BERMUDEZ, T. M. PARRILL, and R. KAPLAN, Surface Sci. **173**, 234 (1986).
[60] R. KAPLAN, Surface Sci. **215**, 111 (1989).
[61] F. OWMAN and P. MARTENSSON, Surface Sci. Letters **330**, L639 (1995).
[62] F. OWMAN and P. MARTENSSON, Surface Sci. **369**, 126 (1996).
[63] F. OWMAN and P. MARTENSSON, J. Vacuum Sci. Technol. B **14**, 933 (1996).
[64] C.-S. CHANG, I. S. T. TSONG, Y. C. WANG, and R. F. DAVIS, Surface Sci. **256**, 354 (1991).
[65] M. A. KULAKOV, P. HEUELL, V. F. TSVETKOV, and B. BULLEMER, Surface Sci. **315**, 248 (1994).
[66] Y. MARUMOTO, T. TSUKAMOTO, M. HIRAI, M. KUSAKA, M. IWAMI, T. OZAWA, T. NAGAMURA, and T. NAKATA, Japan. J. Appl. Phys. **34**, 3351 (1995).

[67] S. TANAKA, R. S. KERN, R. F. DAVIS, J. F. WENDELKEN, and J. XU, Surface Sci. **350**, 247 (1996).
[68] L. LI and I. S. T. TSONG, Surface Sci. **351**, 141 (1996).
[69] P. MARTENSSON, F. OWMAN, and L. I. JOHANSSON, phys. stat. sol. (b) **202**, 501 (1997).
[70] S. NAKANISHI, H. TOKUTAKA, S. NISHIMORI, S. KISHIDA, and N. ISHIHARA, Appl. Surface Sci. **41/42**, 44 (1989).
[71] V. M. BERMUDEZ, Appl. Surface Sci. **84**, 45 (1995).
[72] M. HOLLERING, A. ZIEGLER, R. GRAUPNER, B. MATTERN, L. LEY, A. P. J. STAMPFL, J. D. RILEY, R. C. G. LECKEY, J. BERNHARDT, J. SCHARDT, U. STARKE, and K. HEINZ, to be published.
[73] U. STARKE, J. BERNHARDT, J. SCHARDT, N. NORDELL, A. SCHÖNER, and K. HEINZ, to be published.
[74] L. I. JOHANSSON, F. OWMAN, and P. MARTENSSON, Phys. Rev. B **53**, 13793 (1996).
[75] V. V. ELSBERGEN, T. U. KAMPEN, and W. MÖNCH, Surface Sci. **365**, 443 (1996).
[76] S. KARMANN, W. SUTTROP, A. SCHÖNER, M. SCHADT, C. HABERSTROH, F. ENGELBRECHT, R. HELBIG, and G. PENSL, J. Appl. Phys. **72**, 5437 (1992).
[77] N. NORDELL, A. SCHÖNER, and S. G. ANDERSSON, J. Electrochem. Soc. **143**, 2910 (1996).
[78] E. PEARSON, T. TAKAI, T. HALICIOGLU, and W. TILLER, J. Crystal Growth **70**, 33 (1984).
[79] M. SABISCH, P. KRÜGER, A. MAZUR, and J. POLLMANN, Surface Rev. and Letters, in press.
[80] M. SABISCH, P. KRÜGER, and J. POLLMANN, Phys. Rev. B, to be published.
[81] P.-R. STEINER, Diplomarbeit, University Erlangen-Nürnberg, 1994.
[82] J. BERNHARDT, Diplomarbeit, University Erlangen-Nürnberg, 1996.
[83] W. KERN and D. A. PUOTINEN, RCA Rev. **31**, 187 (1970).
[84] H. NIENHAUS, T. U. KAMPEN, and W. MÖNCH, Surface Sci. Letters **324**, L328 (1995).
[85] T. KIMOTO, A. ITOH, and H. MATSUNAMI, Appl. Phys. Letters **66**, 3645 (1995).
[86] T. KIMOTO, A. ITOH, and H. MATSUNAMI, Inst. Phys. Conf. Ser. No. 142, 241 (1996).
[87] T. KIMOTO, A. ITOH, and H. MATSUNAMI, Appl. Phys. Letters, submitted.
[88] M. SABISCH, Dissertation, University Münster, 1996.
[89] U. STARKE and M. FRANKE, unpublished.
[90] R. GUNELLA, J. Y. VEUILLEN, A. BERTHET, and T. A. NGUYEN TAN, Surface Rev. and Letters, in press.
[91] L. I. JOHANSSON, F. OWMAN, and P. MARTENSSON, Surface Sci. Letters **360**, L478 (1996).
[92] L. I. JOHANSSON, F. OWMAN, and P. MARTENSSON, Surface Sci. Letters **360**, L483 (1996).
[93] L. I. JOHANSSON, F. OWMAN, P. MARTENSSON, C. PERSSON, and U. LINDEFELT, Phys. Rev. B **53**, 13803 (1996).
[94] P. BADZIAG, Surface Sci. **337**, 1 (1995).
[95] P. BADZIAG, Surface Sci. **352/354**, 396 (1996).
[96] J. E. NORTHRUP and J. NEUGEBAUER, Phys. Rev. B **52**, R17001 (1995).

phys. stat. sol. (b) **202**, 501 (1997)

Subject classification: 68.35.Bs; 73.20, At; S6

Morphology, Atomic and Electronic Structure of 6H-SiC(0001) Surfaces

P. MÅRTENSSON, F. OWMAN, and L. I. JOHANSSON

Department of Physics and Measurement Technology, Linköping University, S-58183 Linköping, Sweden

(Received January 31, 1997)

Recent findings concerning primarily the $\sqrt{3} \times \sqrt{3}$ and $6\sqrt{3} \times 6\sqrt{3}$ reconstructed surfaces of 6H-SiC(0001) are reviewed. First, the morphology of some different types of 6H-SiC crystals is discussed. The scanning tunneling microscopy (STM) and atomic force microscopy (AFM) results presented show that surfaces with a morphology suitable for surface investigations can be prepared using sublimation- or hydrogen-etching. Then results obtained concerning the atomic and electronic structure for the reconstructed surfaces, prepared using an ex situ method for oxide removal and in situ heating, are presented. For the $\sqrt{3} \times \sqrt{3}$ reconstruction, recent STM and photoelectron spectroscopy (PES) data are discussed in view of available theoretical results. The STM images presented are shown to be consistent with a structural model of Si or C adatoms in threefold symmetric sites. The theoretical results favor Si adatoms in T_4 sites as the optimal configuration for this reconstruction. However, the surface shifted components extracted in studies of the C 1s and Si 2p core levels and the location of a surface state band mapped out in angle resolved experiments cannot be explained using this structural model. At present, there is no structural model that satisfactorily can explain all experimental findings for the $\sqrt{3} \times \sqrt{3}$ reconstruction. A monocrystalline graphite overlayer on top of bulk-terminated or $\sqrt{3} \times \sqrt{3}$-reconstructed SiC has previously been proposed to explain the $6\sqrt{3} \times 6\sqrt{3}$-reconstructed surface. However, STM and PES results are presented that unambiguously show that there is no graphite on the surface when a well developed $6\sqrt{3} \times 6\sqrt{3}$ low-energy electron diffraction (LEED) pattern is observed. The STM images recorded during the gradual development of the $6\sqrt{3} \times 6\sqrt{3}$ surface show growing fractions of pseudo-periodic 6×6 and 5×5 reconstructions. These reconstructed regions dominate on the surface, but small $\sqrt{3} \times \sqrt{3}$-reconstructed regions are still present when a well developed $6\sqrt{3} \times 6\sqrt{3}$ LEED pattern is observed. It is shown that the $6\sqrt{3} \times 6\sqrt{3}$ LEED pattern can be fully explained by scattering from surfaces with a mixture of 6×6, 5×5 and $\sqrt{3} \times \sqrt{3}$ reconstructions. Due to the complexity of the STM data, no structural model is proposed for the 6×6 and 5×5 reconstructions. STM and PES results are presented showing that graphitization of the surface is obtained only after heating at higher temperatures than that required for observing a well developed $6\sqrt{3} \times 6\sqrt{3}$ LEED pattern. The STM images then show that the graphite appears as a monocrystalline overlayer on top of the 6×6 reconstruction and not on bulk-terminated or $\sqrt{3} \times \sqrt{3}$-reconstructed SiC(0001).

1. Introduction

Among the different polytypes of SiC [1] the cubic 3C and hexagonal 6H have been most intensely studied. The 3C-SiC{111} and 6H-SiC{0001} surfaces exhibit many similarities, as expected since the stacking sequences [1] of these two polytypes are closely related. The dimensions of the 3C unit cell in the $\langle 111 \rangle$ direction and the 6H unit cell in the $\langle 0001 \rangle$ direction are 7.55 and 15.18 Å, respectively, and the distance between the Si–C bilayers is 2.52 Å [2]. The nearest-neighbor distance on the hexagonal {111} and {0001} surfaces is 3.08 Å [2]. These surfaces are polar such that the ideal (111) and

(0001) surfaces are terminated by silicon atoms while the $(\bar{1}\bar{1}\bar{1})$ and $(000\bar{1})$ surfaces are terminated by carbon atoms. Real surfaces of SiC, however, exhibit various reconstructions as discussed by Starke [3]. In this article we review recent experimental findings concerning the morphology and the atomic and electronic structure of primarily the $\sqrt{3} \times \sqrt{3}$ and $6\sqrt{3} \times 6\sqrt{3}$ reconstructions of 6H-SiC(0001) prepared by in situ heating.

The fundamental properties of 6H-SiC{0001}, and also 3C-SiC{111} surfaces, have been studied with techniques such as low-energy electron diffraction (LEED) [4 to 9], Auger electron spectroscopy (AES) [4 to 11], scanning tunneling microscopy (STM) [12 to 22], electron energy loss spectroscopy (EELS) [5, 7, 10], and photoelectron spectroscopy (PES) [6, 9 to 11, 23 to 25]. Moreover, the step structure of various on- and off-oriented SiC(0001) surfaces, both of wafer material and of epitaxially grown material, has been studied with STM [7, 16, 26], atomic force microscopy (AFM) [27 to 29], and transmission electron microscopy (TEM) [29 to 32]. Theoretical studies of the reconstructed 6H-SiC{0001} surfaces using semi-empirical cluster calculations [33, 34] and ab initio calculations within the local density functional approximation of the local density theory [35, 36] have been reported. Below, we give only a brief summary of earlier results relevant for the presentation of recent findings.

In studies of clean SiC surfaces there are two important issues concerning the surface preparation. First the scratches from the cutting and polishing process of the material must be sufficiently removed since reasonably large flat terraces are required. A method called sublimation-etching can be used to achieve this, as described below in Section 2.3. Secondly, the native surface oxide and possible contaminants must be removed. This can be accomplished in various ways as reported in the literature. Van Bommel et al. [4] showed that annealing 6H-SiC(0001) samples in vacuum at temperatures above 250 °C[1]) without any preceding chemical treatment resulted in surfaces which exhibited sharp $\sqrt{3} \times \sqrt{3}$ and $6\sqrt{3} \times 6\sqrt{3}$ LEED patterns. Bozso et al. [37] prepared SiC(0001) samples, treated in concentrated HF prior to introduction into the vacuum chamber, by Ar-ion sputtering and high-temperature annealing. The resulting surfaces were clean as determined with X-ray photoelectron spectroscopy (XPS) and AES, but no LEED data were presented. An in situ method for removal of the surface oxide from SiC surfaces has been presented by Kaplan and coworkers [5], who evaporated Ga or Si onto samples heated at temperatures around 850 °C, such that the surface oxide was removed as volatilized Ga_2O and SiO. The resulting surfaces were free from oxide as determined by AES and exhibited a 3×3 LEED pattern. Upon further heating at 950 °C in the absence of a Ga/Si flux, this pattern transformed to a $\sqrt{3} \times \sqrt{3}$ pattern. This method of surface preparation has been utilized in several more recent studies of both 6H- and 3C-SiC surfaces [9, 18, 21, 22, 38].

An ex situ method for oxide removal in aqueous solutions of HF [6] or buffered HF [7, 8] has also been presented. It was shown that such samples load-locked into UHV exhibit sharp 1×1 LEED patterns without any further treatment in situ. Heating these surfaces at temperatures around 950 °C results in the development of a $\sqrt{3} \times \sqrt{3}$ LEED pattern and further heating to around 1150 °C in the development of a $6\sqrt{3} \times 6\sqrt{3}$ LEED pattern (below often labeled $\sqrt{3}$ and $6\sqrt{3}$, respectively). An additional method

---

[1]) The temperatures stated in this work are in general significantly lower than the corresponding temperatures reported in more recent studies. It is possible that this discrepancy is related to differences in the methods used for temperature measurements.

for treating SiC surfaces was recently presented, involving etching by hydrogen at atmospheric pressure in a CVD reactor at temperatures around 1500 °C [39]. This method gives the desired removal of polishing-induced surface damage and results in a surface that does not require further processing before introduction into vacuum and subsequent heat treatments. Most of the STM and PES results presented below have been obtained for sublimation-etched on-axis samples, but the conclusions are general and apply to different types of (0001) oriented 6H-SiC crystals.

2. Morphology of Various Types of 6H-SiC(0001) Crystals

Compared with the high perfection of Si wafers, commercially available SiC wafers have a poor surface morphology with large amounts of damage from the cutting and polishing process. The morphology of some different types of 6H-SiC crystals will be discussed below. The AFM images shown were recorded in air while the STM images were recorded in ultra-high vacuum after heating the samples a few minutes at 950 °C in order to obtain well-developed $\sqrt{3} \times \sqrt{3}$ surfaces for which acquisition of STM images is possible.

2.1 As-polished samples

Fig. 1a shows a large-scan AFM image of an as-polished on-axis 6H-SiC(0001) wafer where the scratches can be seen as streaks running across the surface. As a consequence of the scratches, the step structure on the surface is highly irregular both with respect to height and direction. This is illustrated in Fig. 1b which is a filled-state STM image recorded for an on-axis as-polished sample that has been heated to 950 °C for 5 min such that the surface exhibits a well-developed $\sqrt{3}$ LEED pattern. The inset in Fig. 1b is a large-scan STM image recorded for the same surfce which, when compared with the

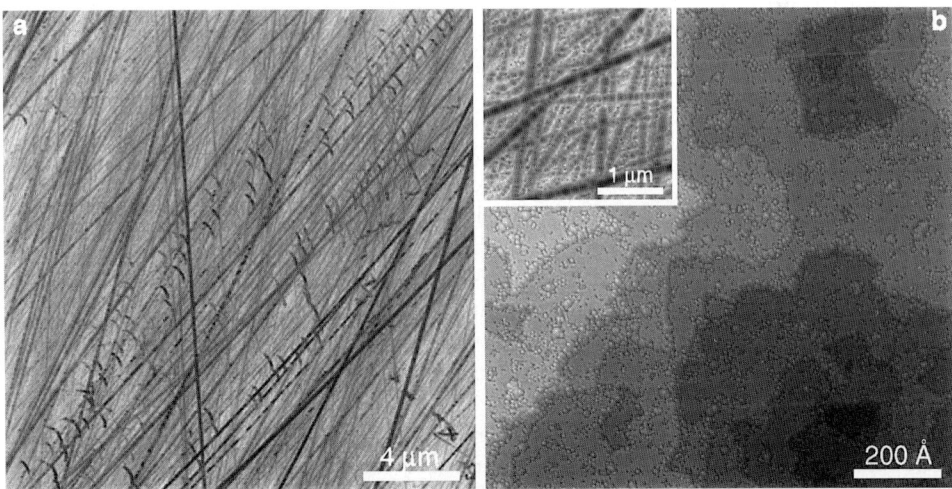

Fig. 1. a) $20 \times 20 \, \mu m^2$ AFM image recorded for an as-polished sample. Black to white corresponds to approximately 200 Å. b) Filled-state STM image of an as-polished sample recorded with a tip voltage of $+2.9$ V at a constant current of 30 pA. The image scale is 1100×1100 Å^2 and black to white corresponds to 30 Å. The inset shows a $3 \times 3 \, \mu m^2$ filled-state STM image recorded with a tip voltage of $+3.0$ V at a constant current of 20 pA. Black to white corresponds to 170 Å

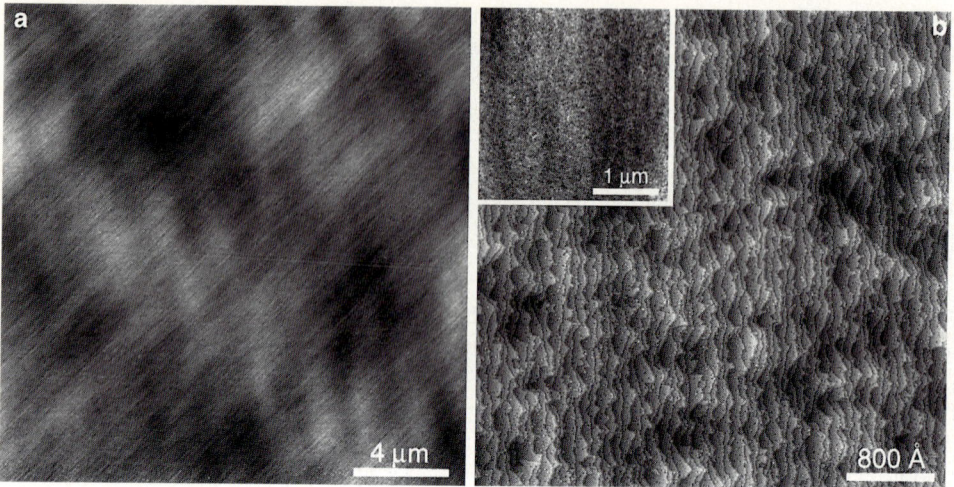

Fig. 2. a) $20 \times 20\,\mu m^2$ AFM image recorded for an as-grown epitaxial layer. Black to white corresponds to approximately 40 Å. b) Filled-state STM image of the same sample as in a) after dry oxidation and the usual chemical treatment, recorded with a tip voltage of $+2.5$ V at a constant current of 20 pA. The image scale is $4300 \times 4300\,Å^2$ and black to white corresponds to 20 Å. The inset shows a $3 \times 3\,\mu m^2$ filled-state STM image recorded with a tip voltage of $+3.0$ V at a constant current of 10 pA. Black to white corresponds to 20 Å

AFM image, shows that the surface morphology is not significantly affected by the heat treatment. Since the step structure of the surface is determined by the scratches rather than the exact orientation of the surface normal, the difference between nominally on-axis and off-axis samples, as seen by STM/AFM, is negligible. We have not observed any improvement of the morphology of polished samples be repeated cycles of dry oxidization at $1250\,°C$ (≈ 1000 Å) and oxide removal using HF.

2.2 Epilayers

Fig. 2a shows a large-scan AFM image of an as-grown off-oriented 6H-SiC(0001) epilayer from Cree Research. The surface is smooth and no traces of scratches can be seen. After ex situ oxide removal the sample was load-locked into the UHV chamber for STM analysis. LEED showed a weak 1×1 pattern with a high background in contrast to the bright and sharp 1×1 patterns normally observed for other types of samples. The epilayer sample was heated at increasingly higher temperatures, but no $\sqrt{3}$ LEED pattern developed. We believe that this could be due to the presence of a highly-disordered or non-stoichiometric outermost SiC layer or to a thick contamination layer. After removing the sample from the UHV chamber, it was dry-oxidized at $1250\,°C$ for about 3 h and subsequently annealed in Ar at $1250\,°C$ for 30 min. This resulted in an oxide with a thickness of the order of 1000 Å (as determined by ellipsometry). After the oxidation, the sample was again wet-chemically treated and load-locked into the vacuum chamber, where it was possible to obtain a well-developed $\sqrt{3}$ LEED pattern after heating at $950\,°C$ for 4 min. Fig. 2b shows filled-state STM images recorded for this sample. The large-scan image in the inset shows that the surface is very smooth without morphological features such as scratches. However, as can be seen in the higher-resolution im-

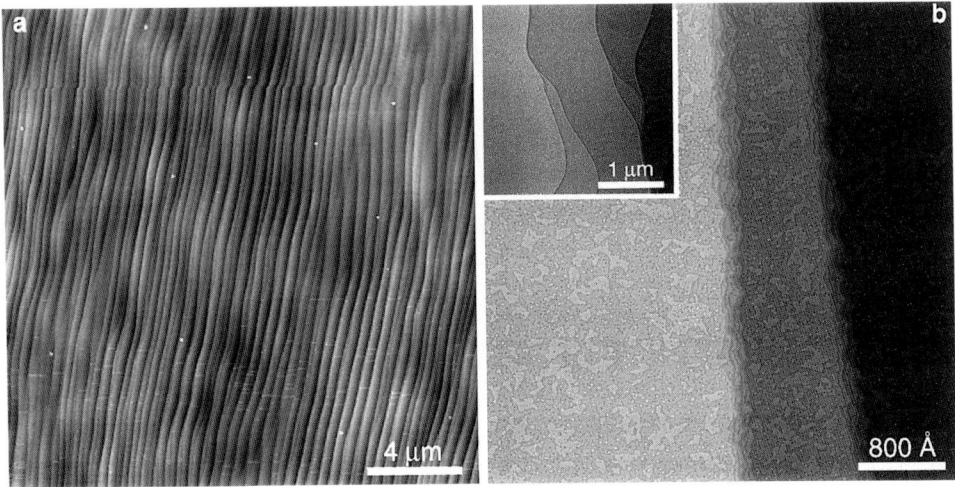

Fig. 3. a) $20 \times 20\ \mu m^2$ AFM image recorded for a sublimation-etched on-axis sample. Black to white corresponds to approximately 100 Å. b) Filled-state STM image of a sublimation-etched on-axis sample recorded with a tip voltage of $+3.0$ V at a constant current of 30 pA. The image scale is $4300 \times 4300\ \text{Å}^2$ and black to white corresponds to 50 Å. The inset shows a $3 \times 3\ \mu m^2$ filled-state STM image recorded with a tip voltage of $+3.0$ V at a constant current of 10 pA. Black to white corresponds to 100 Å

age, the step distribution (determined by the off-orientation) is such that the terraces are very narrow, making this type of sample less suited for atomic-scale STM investigations.

2.3 Sublimation-etched samples

So-called sublimation-etching offers possibilities for obtaining SiC surfaces free from scratches and with a regular step structure. The sublimation-etch is performed at temperatures of the order of 1900 °C in a sublimation-growth reactor with the temperature gradient reversed [40]. Fig. 3a shows a large-scan AFM image of an on-axis 6H-SiC(0001) sample grown with the Lely method. Following cutting and polishing, this sample has been subjected to a sublimation-etch at the Ioffe Institute, St. Petersburg, Russia. Practically no remnants of damage induced by the polishing can be seen and the step structure is well defined. The STM images in Fig. 3b show a sublimation-etched 6H-SiC(0001) surface which has been heated for 3.5 min at 950 °C, resulting in a sharp $\sqrt{3}$ LEED pattern. The large-scan image in the inset shows that the terraces are large, several thousand Å in extent, and are separated by steps which have heights in the range 20 to 40 Å. In the high-resolution STM image, a large amount of small one bilayer high islands can be seen on the terraces.

An off-axis sublimation-etched sample is shown in the large-scan AFM image in Fig. 4a, which exhibits a smooth wavy pattern. The step structure of this off-oriented surface is more difficult to determine than for the on-axis sample shown in Fig. 3 due to the limited resolution in the AFM images. Fig. 4b shows STM images recorded for the same sample after heating at 950°C for 5 min. In the higher-resolution image, it can be seen that the surface consists of regions with few steps and relatively large terraces (as

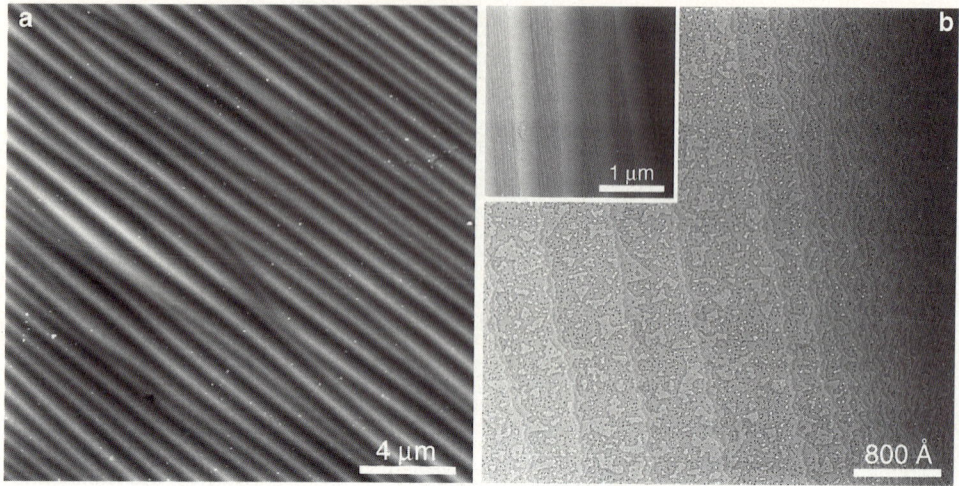

Fig. 4. a) $20 \times 20 \, \mu m^2$ AFM image recorded for a sublimation-etched off-axis sample. Black to white corresponds to approximately 300 Å. b) Filled-state STM image of the same sample as in a) recorded with a tip voltage of +3.0 V at a constant current of 30 pA. The image scale is $4300 \times 4300 \, Å^2$ and black to white corresponds to 180 Å. The inset shows a $3 \times 3 \, \mu m^2$ filled-state STM image recorded with a tip voltage of +3.0 V at a constant current of 20 pA. Black to white corresponds to 1200 Å

seen to the left) separated by heavily-stepped regions (as seen to the right). In the large-scan image in the inset, apparently smooth strips can be seen separated by rougher areas. The rougher areas actually correspond to the parts of the surface having few steps while the smooth strips correspond to the heavily-stepped regions. The observed variation in the step structure across the surface indicates a large tendency for step bunching under the conditions used for the sublimation-etch [27]. The wavy pattern observed in the AFM image does not correctly reflect the morphoplogy of the sample. This is due to the applied background-subtraction method, in which an average parabolic plane has been subtracted from the data.

2.4 Hydrogen-etched samples

A new method for removal of polishing-induced damage from SiC wafers, based on hydrogen etching in a CVD reactor at atmospheric pressure and temperatures around 1500 °C was recently presented [39]. Fig. 5a shows a large-scan AFM image of an on-axis 6H-SiC(0001) surface which has been hydrogen-etched for 30 min at 1550 °C. As can be seen, practically no traces of the polishing process remain on the surface and the step structure is very regular. Fig. 5b shows STM images of a similarly-prepared surface that after the hydrogen etch was quickly load-locked into the UHV system via normal air and then heated for 2 min at 950 °C in order for a well-ordered $\sqrt{3}$ structure to develop. The inset is a large-scan STM image where the regularity of the step structure is clearly visible. All steps separating the main terraces on this surface have a height corresponding to one 6H unit cell (15 Å). In the higher-resolution STM image, the stepped region between two main terraces is shown. Such regions correspond to about 30% of the total area on this surface.

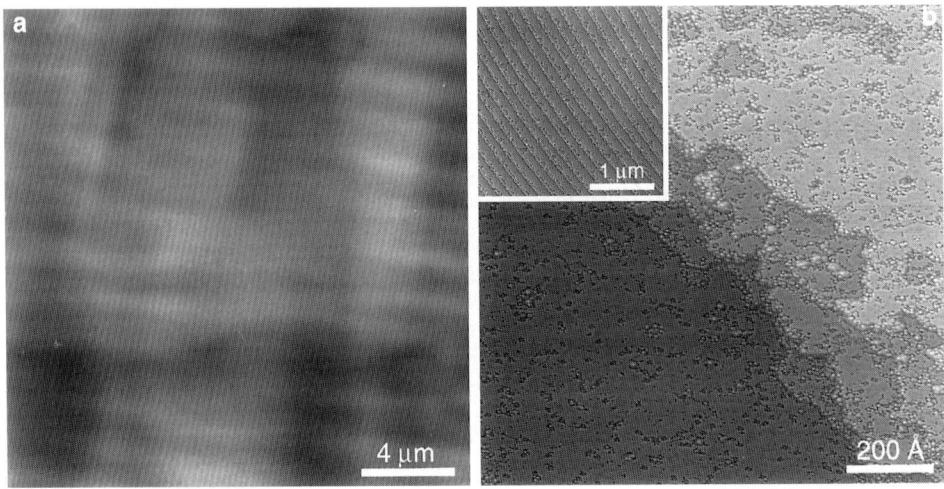

Fig. 5. a) $20 \times 20 \, \mu m^2$ AFM image recorded for a hydrogen-etched on-axis sample. Black to white corresponds to approximately 70 Å. b) Filled-state STM image of a hydrogen-etched on-axis sample recorded with a tip voltage of $+3.0 \, V$ at a constant current of 30 pA. The image scale is $1100 \times 1100 \, \text{Å}^2$ and black to white corresponds to 17 Å. The inset shows a $3 \times 3 \, \mu m^2$ filled-state STM image recorded with a tip voltage of $+3.0 \, V$ at a constant current of 10 pA. Black to white corresponds to 30 Å

3. Temperature Effects

In this section some characteristics of the 1×1, the $\sqrt{3} \times \sqrt{3}$ and the $6\sqrt{3} \times 6\sqrt{3}$ surface structures are presented as observed in LEED, PES and STM studies before and after different heat treatments.

Directly upon load-locking into vacuum the SiC(0001) samples exhibit a sharp and bright 1×1 LEED pattern, as shown in Fig. 6a. The Si 2p and C 1s core-level spectra (Fig. 6b) both consist of a single peak. The Si 2p spectrum is quite broad, suggesting that it consists of at least two components. No clear evidence of oxide-related features can be seen in the Si 2p spectrum, but spectra recorded for the O 1s core level show a strong signal, approximately corresponding to one layer of oxygen atoms on top of the SiC surface. In addition to the oxygen, our PES data show the presence of very small amounts of fluorine which probably are residues from the chemical oxide-removal process. No atomically resolved STM images were possible to record for the as-introduced surfaces due to severely unstable tunneling conditions most likely related to the oxygen present on the surfaces. Based on EELS and LEED experiments, Starke et al. [7] have suggested that the oxygen on the chemically prepared 1×1 surfaces is present in the form of hydroxyl groups saturating the Si dangling bonds.

Heating the samples at temperatures below 850 °C has little effect on the chemical composition of the surfaces which remain 1×1 reconstructed. Oxide formation, as suggested by Starke et al. [7], is however observed by the appearance of a Si 2p component shifted to larger binding energy [23]. Heating above 900 °C results in oxygen removal and the development of a sharp $\sqrt{3}$ LEED pattern that is shown in Fig. 6a. In the C 1s spectrum, a weak component appears on the high binding-energy side of the main peak (see Fig. 6b). In the Si 2p core-level spectrum, more significant changes can be seen as

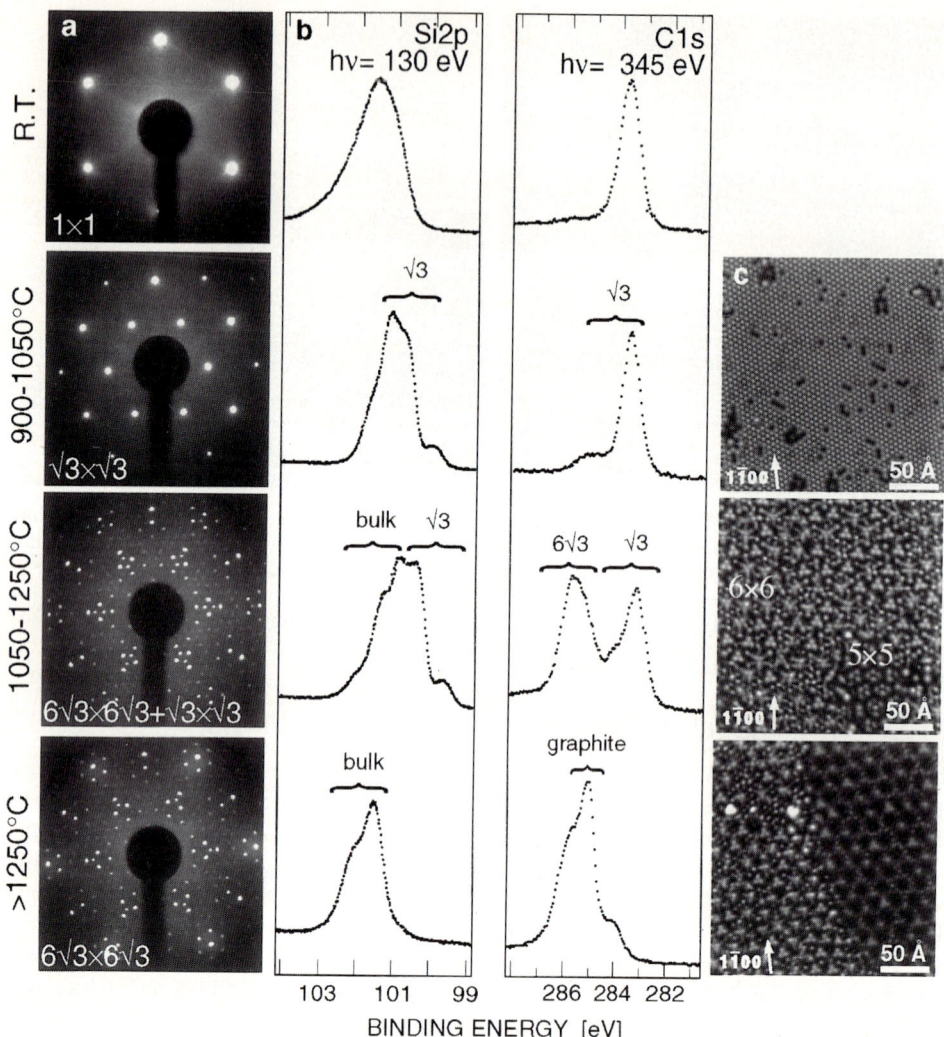

Fig. 6. a) LEED patterns recorded at $E = 80$ eV, b) Si 2p and C 1s core-level spectra recorded for photon energies $h\nu = 130$ eV and 345 eV, respectively, and c) filled-state STM images for SiC(0001) samples heated at different temperatures

surface-shifted components appearing on the low binding-energy side, clearly indicating a modification of the bonding configuration of the surface atoms. In the corresponding filled-state STM image in Fig. 6c, the $\sqrt{3}$ structure can be seen as a hexagonal array of protrusions.

Heating the samples above 1050 °C for a few minutes results in the gradual development of the $6\sqrt{3}$ LEED pattern with clearly visible $\sqrt{3}$ spots [4], as shown in Fig. 6a. In the C 1s spectrum, surface components related to the $6\sqrt{3} \times 6\sqrt{3}$ reconstruction develop at higher binding energies (see Fig. 6b and note that these components do not originate from graphitic carbon). In the Si 2p spectrum, only an additional bulk-related component appears. In the corresponding filled-state STM image in Fig. 6c, two new reconstructions, labeled 5×5 and 6×6, can be observed.

As shown in Fig. 6a, the $\sqrt{3}$ spots initially present in the $6'\sqrt{3}$ LEED pattern gradually become weaker and finally disappear upon heating at the highest temperatures. This is accompanied by the disappearance of the $\sqrt{3}$-reconstructed areas in the STM images as well as by a large increase in the C 1s component corresponding to graphitic carbon (at about 285 eV, see Fig. 6b). At the same time, other C 1s components diminish and the Si 2p spectrum becomes increasingly bulk-like as can be seen in the spectra in Fig. 6b. The corresponding filled-state STM image in Fig. 6c is recorded for a surface heated at 1350 °C for 15 s. The 6×6 and 5×5 reconstructions cover about 90% and 10% of this surface, respectively. To the left in the image, part of a 6×6-reconstructed region can be seen, having the same characteristic features as for the initial 6×6 reconstruction. To the right, the 6×6 structure can be seen with a lower contrast, but it is still possible to distinguish some of the features defining the initial 6×6 reconstruction. The reduced contrast results from the presence of a monocrystalline graphite overlayer as will be shown below.

The changes observed in the valence band region for these three surfaces are illustrated by the spectra shown in Fig. 7. The spectrum from the 1×1 surface is seen to be dominated by contributions from O 2p and O 2s derived states located around 7 and 25 eV, respectively. The valence band spectra recorded from the two reconstructed surfaces look distinctly different and contain no oxygen-related structures. Instead they exhibit several sharp and strong features between 0 and 10 eV and some broader and weaker features at larger energies. Calculated results [24, 41] show that the valence band density of states of SiC can be divided into two subbands. A higher-lying subband, approximately 10 eV wide, composed of hybridized C 2p and Si 3s + 3p states and a lower-lying subband, about 6 eV in width, dominated by contributions from C 2s electrons. A detailed comparison between recorded valence band spectra and calculated partial densities of states was recently published [24]. The important point to notice here is

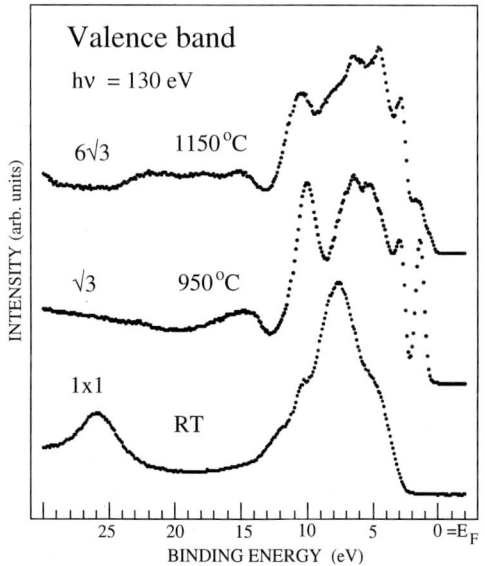

Fig. 7. Valence band spectra of 6H-SiC(0001) recorded using a photon energy of 130 eV from the unreconstructed 1×1 surface and the reconstructed $\sqrt{3} \times \sqrt{3}$ and $6\sqrt{3} \times 6\sqrt{3}$ surfaces (from [23])

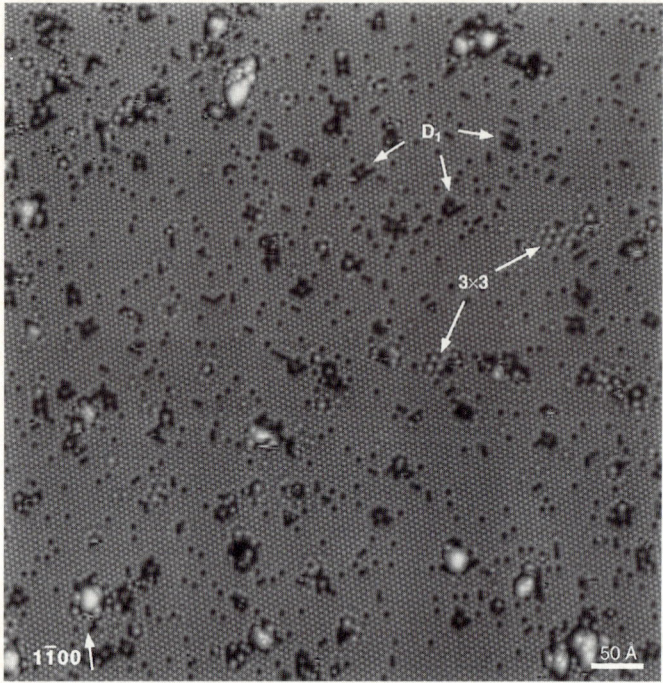

Fig. 8. Filled-state STM image of a $\sqrt{3} \times \sqrt{3}$ surface. The image was recorded with a tip voltage of +2.7 V at a constant current of 60 pA. The image scale is $\approx 630 \times 630$ Å^2 and the atomic corrugation in the $\sqrt{3} \times \sqrt{3}$ areas is approximately 0.15 Å. The arrow indicates one of the six in the bulk equivalent $\langle 1\bar{1}00 \rangle$ directions. Some defects of the D_1 type as well as small regions with the 3×3 reconstruction are marked in the image (from [16])

that the $\sqrt{3}$ surface shows a semiconducting nature, i.e. there are no occupied states close to or at the Fermi level, while the $6\sqrt{3}$ surface appears to be metallic.

4. The $\sqrt{3} \times \sqrt{3}$ Reconstruction

Recent efforts to determine the structure of the SiC(0001) $\sqrt{3} \times \sqrt{3}$ reconstruction are discussed in this section.

4.1 STM results

Fig. 8 shows a high-resolution STM image of the SiC(0001) $\sqrt{3} \times \sqrt{3}$ reconstruction. The surface was prepared by heating a sublimation-etched on-axis 6H-SiC(0001) sample, etched in a buffered HF solution prior to introduction into the vacuum chamber, at 950 °C for 3 min. In the image, a hexagonal array of protrusions is clearly visible together with some amounts of defects. The measured side length of the observed unit cell is 5.4 Å, in good agreement with the expected value of 5.3 Å for the $\sqrt{3}$ reconstruction, and the corrugation in the well-ordered areas is typically 0.15 Å. Similar STM images have been obtained by Li and Tsong [18] for $\sqrt{3}$ surfaces prepared by annealing 3×3-reconstructed surfaces obtained by heating SiC(0001) surfaces under a Si flux.

The most commonly observed defects on the $\sqrt{3}$ surfaces are those which occupy a single $\sqrt{3}$ site, amounting to $\approx 4\%$ of the available $\sqrt{3}$ sites, while the class of more

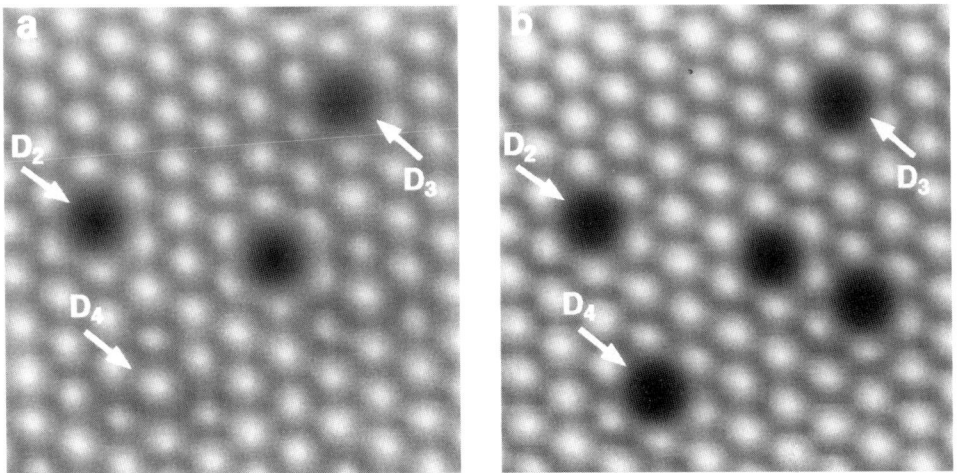

Fig. 9. Simultaneously acquired 50×50 Å^2 STM images recorded with a tip voltage of a) $+3.0$ V (filled states) and b) -2.7 V (empty states) at a constant current of 30 pA. Three different types of defects occupying a single $\sqrt{3}$ site are marked D$_2$, D$_3$, and D$_4$, respectively (from [16] where these defects are discussed in detail)

extended defects denoted D$_1$ in Fig. 8 occurs with a number density corresponding to $\approx 0.5\%$ of the $\sqrt{3}$ sites. In Fig. 8, it is also possible to see small 3×3-reconstructed regions, consisting of only a few sites. In the work by Li and Tsong [18], larger 3×3 and $\sqrt{3} \times \sqrt{3}$-reconstructed regions have been shown to occasionally coexist on surfaces prepared by repeated annealing of SiC(0001) surfaces with and without a Si flux.

To further elucidate the nature of the $\sqrt{3}$ structure, STM images were acquired at various bias voltages at both polarities. In order to be able to acquire empty-state images, we have found it necessary to heat the samples to at least 1000 °C. We believe that this is due to non-ohmic contacts between the sample and the holder on which the sample is mounted using Ta clips. Although the additional heating results in the formation of small 6×6- and 5×5-reconstructed regions, it has no effect on the $\sqrt{3} \times \sqrt{3}$ reconstruction as judged from filled-state STM images. Although there is no significant dependence of the images on the magnitude of the bias voltage within the voltage range used, 2.0 to 3.0 V, a significant polarity dependence can be observed in the appearance of some of the defects in the $\sqrt{3}$ structure as illustrated in Fig. 9. This figure shows a pair of simultaneously recorded images for bias voltages of $+3.0$ and -2.7 V. The positions of the topographic maxima of the $\sqrt{3} \times \sqrt{3}$ structure are identical for both polarities, suggesting that the maxima correspond to positions of atoms at the surface. The appearance of the STM images of the SiC(0001) $\sqrt{3} \times \sqrt{3}$ surface strongly resemble those of the $\sqrt{3} \times \sqrt{3}$ adatom reconstructions observed for a 1/3 monolayer coverage of metals like Al, Ga, In, or Pb on Si(111) [42 to 45]. Thus, a possible configuration of the SiC(0001) $\sqrt{3} \times \sqrt{3}$ surface is a similar type of adatom structure with either Si or C as adatoms on top of a bulk-like Si–C bilayer. As shown in the models in Fig. 10, there are two possible sites for the adatoms in a $\sqrt{3} \times \sqrt{3}$ adatom structure: the fourfold-coordinated site above a second-layer atom (T$_4$) and the threefold-coordinated hollow site (H$_3$). From the STM data alone, it is neither possible to identify which of the elements (Si or C) constitutes the adatoms nor to determine in which of the two sites (T$_4$ or H$_3$)

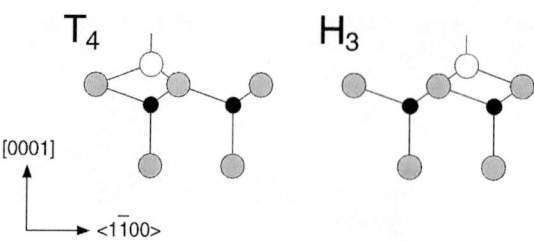

Fig. 10. Model showing two possible sites, T_4 and H_3, for adatoms on a $\sqrt{3} \times \sqrt{3}$ surface

the adatom is located. However, a mixture of Si and C adatoms can most probably be excluded, since all the adatoms (except for the 4% single-site defects) appear identical in the STM images for a given bias voltage. Additionally, registry measurements across boundaries between different domains in the $\sqrt{3}$ structure show that all protrusions are located in equivalent sites so that a mixture of T_4 and H_3 adatoms can be excluded. Another possible configuration for the atomic structure of the SiC(0001) $\sqrt{3} \times \sqrt{3}$ surface would be a structure where the $\sqrt{3}$ sites are occupied with more complex building blocks such as trimers of C or Si, similar to what has been observed on the Si(111) $\sqrt{3} \times \sqrt{3}$-Sb surface [46]. Although no evidence of trimers or other complex structures can be found in the STM images, such configurations cannot be excluded based on the STM data.

In recent STM studies of 6H-SiC(0001) $\sqrt{3} \times \sqrt{3}$ surfaces prepared by heating in the presence of a Si flux, Li et al. [21, 22] have observed STM images quite different from the ones presented above. The STM images that we have recorded only show a polarity dependence in the appearance of the defects, as illustrated in Fig. 9. The STM data of Li et al., however, show a clear contrast reversal with polarity such that the filled-state images show a honeycomb structure with one depression per $\sqrt{3}$ unit cell, while the empty-state images show a hexagonal structure with one protrusion per $\sqrt{3}$ unit cell. It was argued [21, 22] that the observed contrast reversal was inconsistent with an adatom model and different kinds of vacancy models as well as a Si–C alloy model were proposed to account for the observed STM images. In the simplest of the proposed vacancy models, type A, the outermost Si–C bilayer has 1/3 of its Si atoms missing [21]. In an alternative vacancy model, type B, an extra 2/3 of a layer of Si or C atoms is located on top of the outermost Si–C bilayer [21, 22]. In the proposed Si–C alloy model, 2/3 of the surface Si atoms are substituted by C atoms [22]. In [21] it was argued that the type-B vacancy model with an extra 2/3 of a Si layer was more likely in view of the Si-rich conditions during the formation of the $\sqrt{3}$ surface. In [22], on the other hand, the type-B vacancy model with extra C atoms or the Si–C alloy model were judged as more favorable in view of the Si depletion which is likely to occur at elevated temperatures.

Although the models proposed by Li et al. are consistent with their STM images, they seem to be inconsistent with the STM images that we have obtained for the $\sqrt{3}$ surface. Thus, an important question that needs to be addressed concerns the apparent disagreement between the different STM results. We have never observed a honeycomb structure in filled-state images of the $\sqrt{3}$ surface although we have imaged the surface using about ten different samples and thirty different tips, but also the results by Li et al. [47] have been reproduced using several different samples and tips. Since the different STM images have been obtained for $\sqrt{3}$ surfaces prepared using different methods (heating as compared to heating in the presence of a Si flux), one obvious explanation for the discrepancy would be that the different methods result in different surface structures

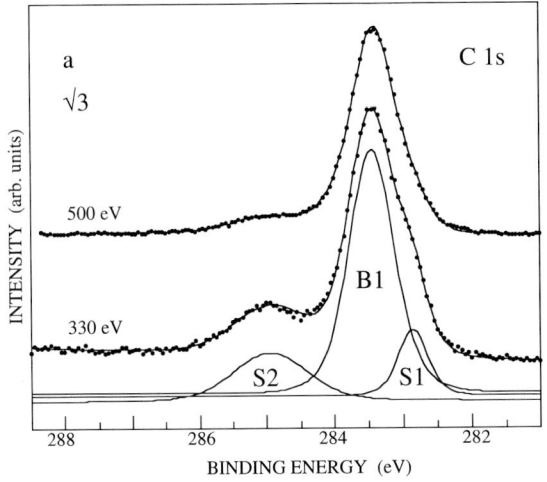

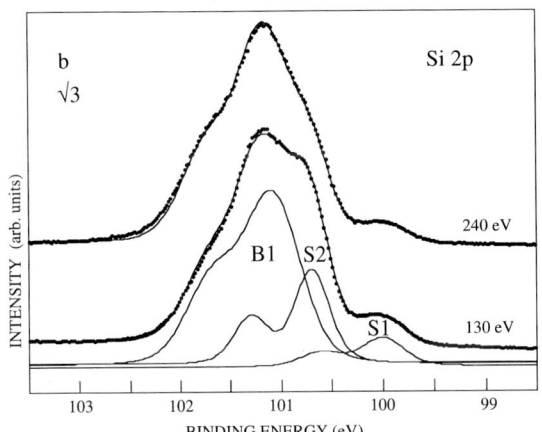

Fig. 11. a) C 1s and b) Si 2p spectra, recorded from the $\sqrt{3} \times \sqrt{3}$ surface using two different photon energies, are shown by dots. The solid curves through the data points show the results obtained using a curve fit procedure and the curves underneath the recorded spectra show the components used (from [23])

which have the same periodicity. As shown in the paper by Starke elsewhere in this volume [3], however, the measured fractional order LEED intensities for $\sqrt{3}$ surfaces prepared using the two methods are very similar, suggesting that the surface reconstructions are identical. Another explanation for the discrepancy between the different experiments could be some reproducible difference in the STM imaging process itself, e.g. due to differences in the tunneling tips. At present, it is not clear what is causing the apparent disagreement between the experiments, and more measurements are needed to clarify this matter.

4.2 Core level PES results

High resolution core level PES studies were conducted [23, 48] in an effort to obtain more information on the structure of the $\sqrt{3}$ surface. The idea was to record core-level spectra at such a high surface sensitivity that the surface shifted components would reveal the type of atoms in the outermost layers. C 1s and Si 2p spectra recorded from the $\sqrt{3}$ surface are shown by the dots in Fig. 11. In both cases the lower spectrum was

recorded at a photon energy giving very high surface sensitivity and the upper at a higher photon energy providing a less surface sensitive spectrum. In order to extract the
shifts and relative intensities of the different components, a curve fitting procedure was
utilized (see [23] for details concerning the fit procedure). The components used are
shown below the recorded spectra and the curves through the data points show the
result of applying the fit procedure. One bulk component (B1) and two surface shifted
components (S1 and S2) are clearly observed in the C 1s as well as in the Si 2p spectrum. Both surface components exhibit negative shifts (i.e. are shifted to lower binding
energy) in the Si 2p spectrum while in the C 1s spectrum the S1 component shows a
negative shift and S2 a positive shift. Several 6H-SiC(0001) samples with different doping concentrations were investigated and they all yielded essentially identical results. In
order to minimize broadening effects the measurements were made at a sample temperature of about 100 K.

Ideally one would like to assign the surface shifted components with specific sites for
the atoms. In a previous investigation of the 3×2 Si-rich 3C-SiC(100) surface [38], two
surface shifted Si 2p components, with fairly similar shifts as the ones shown in Fig. 11b,
were revealed. They were interpreted to originate from the two outermost layers, a
buckled topmost Si layer and a Si layer underneath. The component exhibiting the largest shift was assigned to up-atoms in the buckled topmost layer and the component
showing the smaller shift was interpreted to originate from the other atoms in the topmost layer plus the atoms in the underlying layer. The 3×2 Si-rich 3C-SiC(100) surface
was prepared by exposing the sample to a flux of Si during heating. In our case, the
reconstructed surfaces were prepared by in situ heating of 6H-SiC samples. That there
exist two layers comprised solely of Si atoms on the $\sqrt{3}$ reconstructed surface appears
unlikely since two prominent surface C 1s components, shifted in opposite directions, are
observed. These two carbon components are interpreted to originate from different
atomic layers since no earlier results, experimental or theoretical, have shown that
atoms in a distorted layer can give rise to a difference in surface shifts of about 2.1 eV,
which is the difference obtained between the extracted S1 and S2 components. The
structural model composed of 1/3 layer of Si or C adatoms in threefold symmetric sites
on top of the outermost Si–C bilayer, as suggested by the STM results, can therefore
not explain the two surface shifted components observed in both the Si 2p and C 1s
levels. If the adatoms are assumed to be Si atoms one can account for the two shifted
components observed in the Si 2p spectrum. The S1 component can be assigned to the
Si adatoms and the S2 component to the Si atoms in the topmost bilayer. The problem
encountered is then how to interpret the two shifted components appearing in the C 1s
spectrum. One component can be accounted for, the S1 component which can be interpreted to originate from carbon atoms in the topmost bilayer. The S2 component, appearing at a larger binding energy than the bulk peak, is however difficult to find an
explanation for. Presence of defect areas may be envisioned but the STM results show
that the concentration of defects is too small to account for a surface component of this
magnitude. A similar difficulty in assigning the four surface shifted components is encountered if the adatoms instead are assumed to be C atoms. If the S2 component in the
C 1s spectrum is assigned to the C adatoms and the S1 component to the C atoms in
the topmost bilayer, the problem then concerns the two shifted components in the Si 2p
spectrum. One of these can then be assigned to the Si atoms in the topmost bilayer but
occurrence of a second component with a distinctly different relative strength is difficult

to explain. One possibility to account for these shifted levels would be to assume that regions with different terminations, i.e. both C and Si, exist on the surface, but this possibility is ruled out by the STM results. The above arguments concerning the existence of two surface shifted components in both the Si 2p and C 1s levels also rule out an interpretation in terms of a trimer arrangement [34] of C, or Si, atoms on top of the outermost Si–C bilayer as well as the type-B vacancy model [21, 22] proposed recently.

Other possible arrangements have therefore to be considered. One option is to assume that the topmost bilayer is modified in the $\sqrt{3}$ reconstruction such that 2/3 of the Si atoms are desorbed such that the only Si atoms remaining are the adatoms on top of an arrangement of C atoms. The second bilayer could be assumed to be affected by these rearrangements to such an extent that also these atoms should give rise to surface shifted components. This would explain the observation of two shifted components in both levels and also the appearance of a surface C 1s component shifted to larger binding energy. The observed components could then be assigned as follows: the S1 component in the Si 2p spectrum to the Si adatoms and the S2 component to Si atoms in the second bilayer; the S2 component in the C 1s spectrum to C atoms in a modified ("carburized") topmost bilayer, and the S1 component to C atoms in the second bilayer. This would also be consistent with the extracted relative intensities of the different components. In the Si 2p spectrum the S1 component is considerably smaller than the S2 components while in the C 1s spectrum the magnitudes of the broad S2 and the narrower S1 component are fairly similar. At present we cannot propose a specific arrangement of C atoms in such a modified layer since no theoretical results have been published that predict the core level shift resulting from a certain change in carbon bonding configuration on SiC. The recently proposed [21, 22] Si–C alloy model and also the type-A vacancy model are consistent with the observation of two surface shifted components in the Si 2p and C 1s levels. Theoretical efforts are however needed to provide an explanation for the observed core level shifts.

4.3 Theoretical results

First-principles total energy calculations for different structural configurations of the SiC(0001) $\sqrt{3} \times \sqrt{3}$ reconstruction were recently reported [35, 36, 49]. Calculations were performed for structures with Si or C adatoms and Si or C trimers adsorbed in T_4 or H_3 positions as well as for a Si–vacancy structure (type A). Adatoms of Si in T_4 sites was found to be the optimal configuration. This model produces a $\sqrt{3}$ surface of metallic nature since a half-filled surface state band, derived from dangling bonds on the Si adatoms, is obtained [35, 36]. This is in conflict with the valence band PES results [24], illustrated in Fig. 7, which clearly show a semiconducting nature for the $\sqrt{3}$ surface. The structure located about 1.3 eV below the Fermi level was identified to originate from a surface state. The calculation for the Si-T_4 model [35] showed a half-filled surface state band, labeled Σ_1, centered about 1.2 eV above the valence band maximum (VBM). It was pointed out [35] that the existence of a half-filled dangling bond derived band would seem to be an exception to the electron counting rule, which suggests that reconstructions which give rise to a metallic occupation are energetically unfavorable with respect to at least one reconstruction for which a semiconducting occupation can be obtained. It was suggested that photoemission experiments to probe the location and dispersion of the Σ_1 band would be interesting.

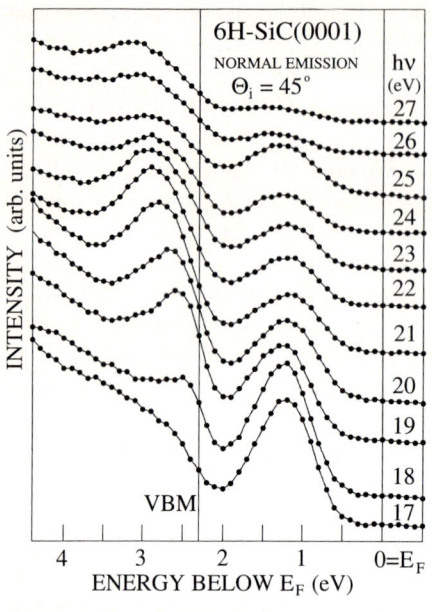

Fig. 12. Normal emission spectra recorded from the 6H-SiC(0001) $\sqrt{3} \times \sqrt{3}$ surface using photon energies between 17 and 27 eV. An incidence angle of $\theta_i = 45°$ was used. The vertical lines indicate the position determined for the Fermi level (E_F) and an estimated position of the valence band maximum (VBM). See text for details (from [25])

4.4 Angle resolved PES results

Photoemission spectra recorded from the $\sqrt{3}$ surface [25] at normal emission, using photon energies between 17 and 27 eV, are shown in Fig. 12. The structure located at about $1.2(\pm 0.1)$ eV below E_F shows no dispersion with photon energy and was found to be more sensitive to surface contamination than the other valence band features. These observations further strengthen the earlier interpretation that the structure originates from a surface state. The dispersing structure located between 2.4 and 3.0 eV below E_F is interpreted to reflect the uppermost valence band. An accurate determination of the VBM cannot be made from the recorded spectra. However, considering the location of the uppermost valence band and that the energy resolution is about 0.2 eV in the measurements, a location of the VBM at $2.3(\pm 0.2)$ eV below E_F can be estimated (see Fig. 12). Earlier results for 3C-SiC(100) [50], with a similar donor-doping concentration, show the Fermi level to be pinned at 2.0 eV above the VBM. In view of that the 6H polytype has a somewhat larger band gap than the 3C polytype (3.02 eV compared to 2.39 eV) [51], our estimated location of the VBM seems reasonable. It deserves to be noticed that this disagrees with the pinning of E_F at 1.2 eV above the VBM determined recently from the difference obtained between ionization energy and thermal work function

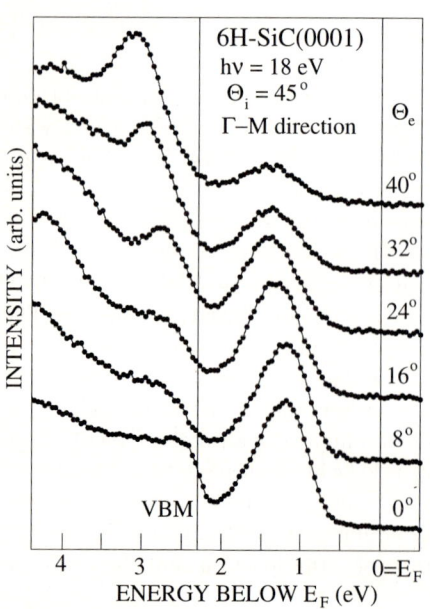

Fig. 13. Spectra recorded from the 6H-SiC(0001) $\sqrt{3} \times \sqrt{3}$ surface at different emission angles θ_e along the Γ–M direction of the 1×1 SBZ. A photon energy of 18 eV and an incidence angle of $\theta_i = 45°$ were used (from [25])

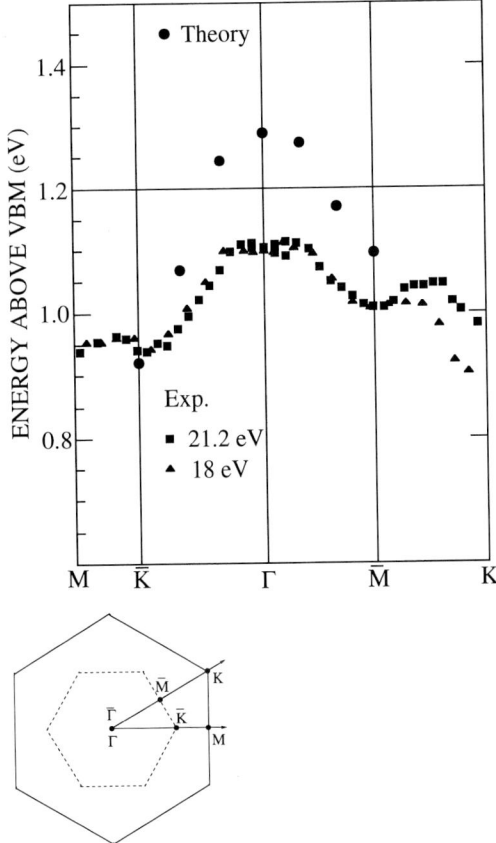

Fig. 14. Experimental and calculated surface state band dispersions. The experimental results were collected using two different photon energies. The 21.2 eV spectra were recorded in angular steps of 2° from Γ and out to the K̄ and M̄ points. The horizontal line at 1.2 eV shows the calculated [35] position of the Fermi level. Experimentally, the Fermi level was determined to be located at 2.3(±0.2) eV above the VBM. The 1 × 1 and √3 × √3 surface Brillouin zones are shown with solid and dashed lines, respectively, in the lower part of the figure (from [25])

of the clean 6H-SiC(0001) surface [11]. The ionization energy was, however, determined assuming that there was no emission from occupied surface states within the band gap while our normal emission spectra show a surface state located about 1.1 eV above the VBM.

Spectra recorded at different emission angles θ_e along the Γ–M direction in the 1 × 1 surface Brillouin zone (SBZ), using a photon energy of 18 eV, are shown in Fig. 13. The surface state exhibits a dispersion as it is seen to move away from E_F with increasing θ_e.

Furthermore, it shows the largest relative strength close to normal emission. Fig. 14 shows the energy location of the surface state versus $k_\parallel$ determined along the Γ–M and Γ–K azimuthal directions of the 1 × 1 SBZ using two different photon energies. Also included in the figure is the calculated energy location [35] of the Σ_1 surface state band in the √3 × √3 SBZ. The experimental and calculated dispersions show the same trends, i.e. the surface state disperses downwards towards the VBM along both the Γ̄–K̄ and Γ̄–M̄ directions. However, the band width obtained experimentally is seen to be about 0.2 eV while the calculated width is 0.35 eV. The line at 1.2 eV above the VBM in the figure illustrates the calculated position of E_F [35]. The experimental results show unambiguously that the surface state has a semiconducting occupation since E_F is found to be located at 2.3(±0.2) eV above the VBM. Experiments carried out for the 4H-SiC(0001) √3 × √3 surface [25] gave very similar results for both the location and dispersion of the surface state band and for the location of the VBM relative to E_F. This is not surprising since the local surface atomic structures are expected to be similar as the difference between the 4H and 6H polytypes is only in the stacking sequence of the Si–C double layers. In contrast to the experiments, a half-filled band, i.e. a metallic surface, was obtained in the calculations made using a single particle theory [35]. It was, however, pointed out that, in view of the narrowness of the band, it is conceivable that correlation effects beyond the scope of a one-electron theory are important. The experimental results suggest this to be the case since it appears that any structural model for the $\sqrt{3}$ reconstruction, involving Si and C atoms, gives an odd number of electrons in each $\sqrt{3} \times \sqrt{3}$ surface unit cell which in a single particle picture results in a metallic surface.

The above discussion shows that both the core level results and the comparison of angle resolved photoemission data with recent theoretical findings indicate that Si adatoms, in T_4 sites, on top of an outermost Si–C bilayer appears to be an inadequate structural model for explaining the experimental findings for the $\sqrt{3} \times \sqrt{3}$ reconstructed surface.

5. The $6\sqrt{3} \times 6\sqrt{3}$ Surface

A structural model consisting of a single monocrystalline layer of graphite on top of the bulk-terminated SiC crystal has been proposed [4] for the $6\sqrt{3} \times 6\sqrt{3}$-reconstructed surface, obtained by heating at temperatures above 1100 °C. The results in recent STM [12, 13, 18, 20] and AES/PES [9] studies of the $6\sqrt{3}$ surface were interpreted to confirm this model. In a theoretical study [35], the $6\sqrt{3}$ surface was proposed to arise from a monocrystalline graphite layer on top of a $\sqrt{3}$ arrangement of Si adatoms rather than on top of a bulk-terminated surface in view of the high stability obtained for Si adatoms in T_4 sites. However, below we present STM and PES results that unambiguously show that there is no graphite on the surface when a well-developed $6\sqrt{3}$ LEED pattern is observed. Carbon enrichment is observed in the surface region but not in the form of graphite. Moreover the STM images of the $6\sqrt{3}$ surface show pseudo-periodic reconstructions with approximate periodicities of 6×6 and 5×5, respectively, in addition to $\sqrt{3} \times \sqrt{3}$-reconstructed regions. A comparison between Fourier transforms of the STM images and the $6\sqrt{3}$ LEED pattern shows that the LEED pattern can be fully explained by scattering from surfaces with a mixture of $\sqrt{3} \times \sqrt{3}$, 5×5 and 6×6 reconstructions. For surfaces heated to higher temperatures than that required for obtaining a well-developed $6\sqrt{3}$ LEED pattern graphitization is observed both in STM and PES.

5.1 STM results

Fig. 15 shows a filled-state STM image of a 6H-Si(0001) surface heated at 1250 °C for 1.5 min. The $6\sqrt{3}$ LEED pattern observed for this surface was similar to the pattern shown in Fig. 6a for temperatures in the range 1150 to 1250 °C, with clearly visible $\sqrt{3}$ spots. In the STM image, a $\sqrt{3}$-reconstructed region can be seen, in accordance with the presence of $\sqrt{3}$ spots in the LEED pattern. In addition to the $\sqrt{3}$-reconstructed region in Fig. 2, two regions with complex reconstructions can be seen, neither of which is strictly periodic.

The reconstruction seen to the left in Fig. 15 consists of a hexagonal array of structural units, labeled A, of different shape, size, and orientation. The hexagonal array is rotated 30° with respect to the $\sqrt{3} \times \sqrt{3}$ lattice, i.e. it is aligned with the 1×1 lattice. Each of the type A structural units consists of between two and seven topographic maxima which appear in a close-packed arrangement within the unit. The separation between neighboring topographic maxima within the units is approximately 6 Å, while the average nearest-neighbor distance between the type A units is approximately 16 Å, i.e., approximately five times the unit cell side length of the 1×1 lattice (3.1 Å). Consequently, we refer to this reconstruction as 5×5.

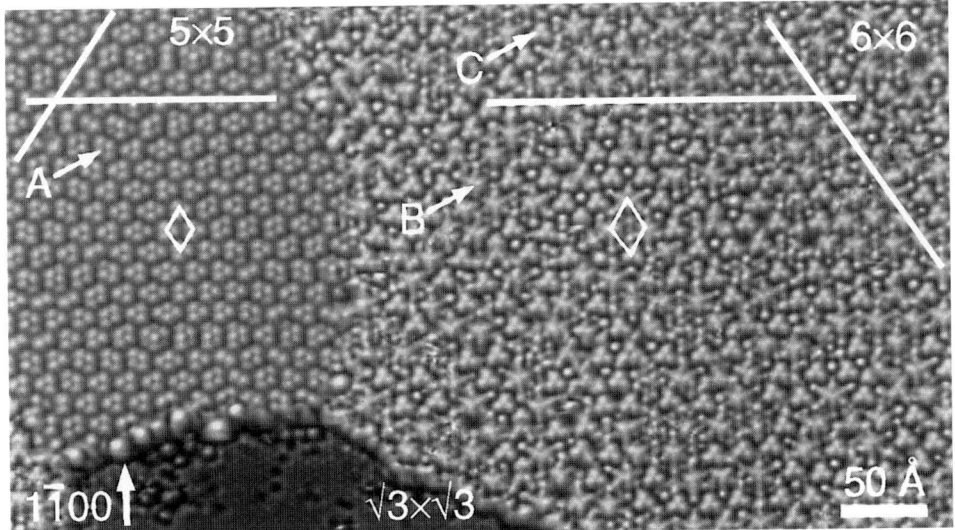

Fig. 15. Filled-state $550 \times 300 \, \text{Å}^2$ image of a surface heated at 1250 °C for 1.5 min recorded with a tip voltage of $+2.5$ eV at a constant current of 60 pA. The approximate unit cells of the 6×6 and 5×5 reconstructions and a few lines along main symmetry directions are shown in the image. The features labeled A to C are discussed in the text (from [19])

The reconstruction on the right-hand side of Fig. 15 is more complex than the 5×5 reconstruction. It can be defined by two types of structural units arranged in a hexagonal array: asterisk-shaped units (labeled B) and trimer-like units (labeled C). The orientation of the hexagonal array is the same as for the 5×5 reconstruction, i.e., aligned with the 1×1 lattice. The approximate unit cell side length is 19 Å, i.e., about 6 times the 1×1 unit cell side length and we refer to this structure as the 6×6 reconstruction.

As can be seen in Fig. 16, the complexity of the 5×5 and 6×6 reconstructions is even more apparent in dual-polarity STM images. While the 5×5 structure appears rather similar when imaging filled and empty states, the 6×6 structure looks significantly different for the two polarities. In addition to the asterisk-shaped type B and trimer-like type C units, the 6×6 structure can be seen to consist of individual topographic maxima labeled D and more complex units labeled E which appear as individual maxima imaging filled states and as irregular-shaped units exhibiting a weak corrugation in a hexagonal close-packed geometry imaging empty states. Further details concerning the STM images obtained for the 5×5 and 6×6 reconstructions can be found in [19], but due to the complexity of the STM data, it has not been possible to propose any models for these reconstructions.

It is important to note that the STM images shown in Fig. 15 and 16 are quite different from images presented by other groups [12, 13, 18, 20] which, in addition to a 6×6 periodicity, clearly show the presence of a monocrystalline graphite overlayer on surfaces exhibiting the $6\sqrt{3} \times 6\sqrt{3}$ LEED pattern. This disagreement can, however, be explained by differences in the annealing temperatures used in the different experiments. Fig. 17 shows a filled-state STM image recorded for a SiC(0001) surface heated at 1350 °C for 15 s, resulting in a $6\sqrt{3} \times 6\sqrt{3}$ LEED pattern with no $\sqrt{3} \times \sqrt{3}$ spots as shown in Fig. 6a for temperatures above 1250 °C. For this surface, no $\sqrt{3} \times \sqrt{3}$-reconstructed re-

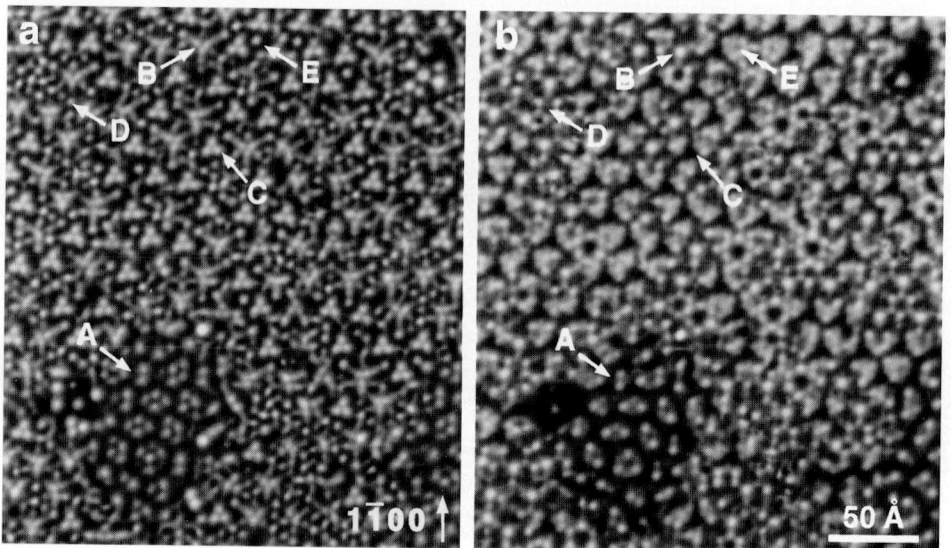

Fig. 16. Simultaneously acquired STM images of a surface heated 1.5 min at 1250 °C. The images were recorded with a tip voltage of a) +2.5 V (filled states) and b) −1.6 V (empty states) at a constant current of 30 pA. The image scale is $\approx 260 \times 290$ Å^2 (from [17])

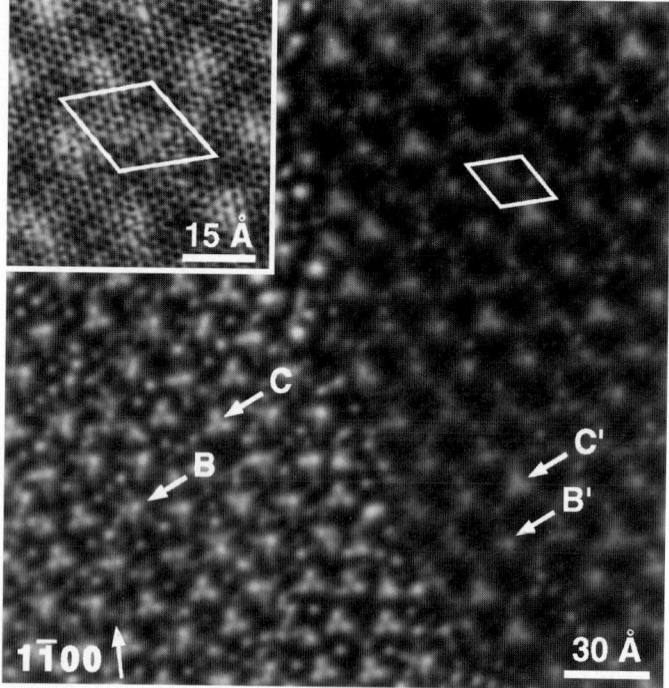

Fig. 17. Filled-state STM image of a surface heated 15 s at 1350 °C recorded with a tip voltage of +2.8 V at a constant current of 60 pA. The image scale is $\approx 240 \times 240$ Å^2. The inset shows a high-resolution filled-state image of a graphite-covered part of the surface, recorded with a tip voltage of +2.6 V at a constant current of 300 pA (from [17])

gions could be seen in STM. The fractions of surface area exhibiting 6×6- and 5×5-reconstructed regions were about 90% and 10%, respectively. In the left-hand side of the image in Fig. 17, part of a 6×6-reconstructed region can be seen with the same characteristic features as described above for the initial 6×6 reconstruction obtained at lower temperatures. On the right-hand side, a region can be seen which exhibits the 6×6 structure with a lower contrast. In this region, the 6×6 structure appears more like a honeycomb structure with the 6×6 unit cell connecting neighboring centers in the honeycomb structure. The inset of Fig. 17 shows a higher magnification image of the modified 6×6 reconstruction. In addition to the 6×6 periodicity, the hexagonal structure of a graphite overlayer is clearly resolved, with a measured unit cell side length of 2.5 Å, in good agreement with the expected value 2.46 Å. In the lower-contrast regions it is still possible to distinguish some of the features defining the initial 6×6 reconstruction. The type C trimers are clearly visible at some locations, labeled C′, whereas the asterisk-shaped features, labeled B above, are only vaguely distinguishable as in the location labeled B′.

We observe the first sign of the graphitization of SiC(0001) surfaces after heating to 1250 °C for a few minutes, resulting in the formation of graphite-covered 6×6 regions on a few percent of the surface area. The amount of graphitization increases with increasing annealing time and temperature. It is important to note that, while the observation of the graphitization is in agreement with previous STM, AES, and XPS results [4, 9, 12, 13, 18, 20], the previous interpretation that the $6\sqrt{3} \times 6\sqrt{3}$ LEED pattern occurs due to the formation of a graphite layer is incorrect. The above STM results, as well as the photoemission results that will be discussed below, clearly show that the $6\sqrt{3} \times 6\sqrt{3}$ pattern is fully developed well before the first signs of graphitization can be seen. Once the graphitization starts, it merely modifies the 6×6 reconstruction that is already present on the surface. The fractional coverage determined for the different reconstructions versus heating temperature and heating time is shown in Table 1.

5.2 Core level PES results

High resolution and surface sensitive C 1s and Si 2p spectra recorded after heating at temperatures between 950 and 1350 °C are shown in Fig. 18. Spectra recorded using higher photon energies, up to 500 and 240 eV, respectively, to determine the contribution from bulk components [23] are not shown. The curves labeled 1 to 5 correspond to

Table 1

The fractional coverage of the different reconstructions as a function of heating temperature and heating time (from [19])

$T(°C)$	time (min)	LEED	$\sqrt{3} \times \sqrt{3}$	5×5	initial 6×6	modified 6×6
950	4.5	$\sqrt{3} \times \sqrt{3}$	100%	—	—	—
1050	1.0	$\sqrt{3} \times \sqrt{3}$	98%	1%	1%	—
1100	2.0	$6\sqrt{3} \times 6\sqrt{3}$ w $\sqrt{3} \times \sqrt{3}$	90%	1%	9%	—
1150	2.0	$6\sqrt{3} \times 6\sqrt{3}$ w $\sqrt{3} \times \sqrt{3}$	72%	1%	26%	—
1200	1.0	$6\sqrt{3} \times 6\sqrt{3}$ w $\sqrt{3} \times \sqrt{3}$	36%	7%	57%	—
1250	0.5	$6\sqrt{3} \times 6\sqrt{3}$ w/o $\sqrt{3} \times \sqrt{3}$	11%	9%	80%	—
1350	0.5	$6\sqrt{3} \times 6\sqrt{3}$ w/o $\sqrt{3} \times \sqrt{3}$	—	8%	45%	47%

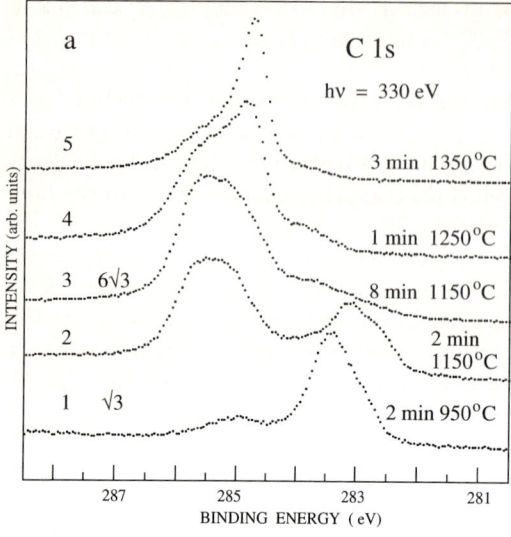

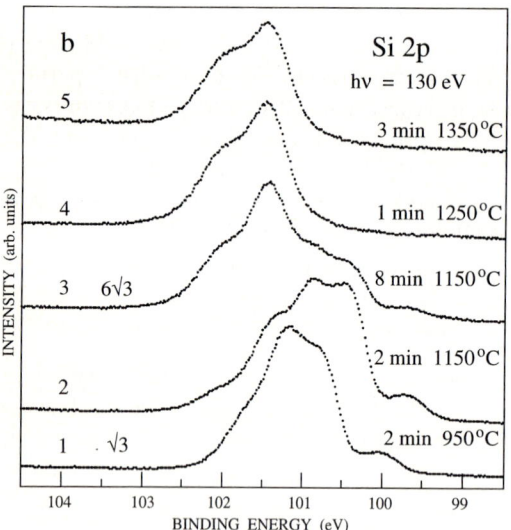

Fig. 18. Normalized core level spectra, a) C 1s at $hv = 330$ eV and b) Si 2p at $hv = 130$ eV, recorded after heating for: (1) 2 min at 950 °C (the $\sqrt{3}$ structure), (2) 2 min at 1150 °C, (3) 8 min at 1150 °C (the 6 $\sqrt{3}$ structure), (4) 1 min at 1250 °C, and (5) 3 min at 1350 °C (from [23])

different heat treatments: (1) 2 min at 950 °C (i.e. the fully developed $\sqrt{3}$ structure), (2) 2 min at 1150 °C, (3) 8 min at 1150 °C (i.e. the well developed 6 $\sqrt{3}$ structure), (4) 1 min at 1250 °C, and (5) 3 min at 1350 °C. The drastic changes observed in both the C 1s and Si 2p spectra show that the structural change from $\sqrt{3}$ to 6 $\sqrt{3}$ is accompanied by quite large changes in the chemical composition in the surface region. The spectra recorded at the two highest temperatures, labeled 4 and 5, show clear evidence of graphitization. In the C 1s spectrum a graphite C 1s peak appears at a binding energy of 284.7 eV, most pronounced at the highest temperature. The Si 2p spectrum is seen to consist essentially of one broad spin−orbit split bulk component, located at about 101.4 eV, and shows no surface related component. The curves labeled 3 correspond to the well-developed 6 $\sqrt{3}$ surface. In this case the Si 2p spectrum exhibits surface related features on the low binding energy side of the strong bulk component while the C 1s spectrum is dominated by a broad surface related structure located at a slightly larger binding energy than the graphite C 1s peak. This C 1s spectrum clearly shows that the 6 $\sqrt{3}$ structure is not related to a graphite surface layer, since the shape of the carbon peak is so different from that of the graphite peak which appears at higher temperatures. The relative strength of the surface-related C 1s peak indicates, however, a carbon enrichment in the surface region. An estimate from extracted surface to bulk intensity ratios, using a layer attenuation model, indicates that more than one layer is affected by the reconstruction [23]. The large width of the surface related C 1s peak moreover shows that it contains more than one component, i.e. that there are carbon atoms occupying inequivalent sites. The curves labeled 2 represent a surface with a mixture of $6\sqrt{3}$ and $\sqrt{3}$ reconstructed areas since weaker $6\sqrt{3}$ superstructure spots were visible in the LEED pattern while the $\sqrt{3}$ spots were still fairly strong. The C 1s spectrum exhibits in this case two dominant features which both show additional structure. The high binding energy feature is interpreted as related to the $6\sqrt{3}$ reconstructed areas and the low binding energy feature to the $\sqrt{3}$-reconstructed areas, although the latter is seen to be shifted compared to the dominant structure observed in the C 1s spectrum for the fully developed $\sqrt{3}$ structure (curve 1). The Si 2p spectrum, curve 2, can be interpreted in a similar fashion if the shoulders visible on the high binding energy side are assigned to contributions from $6\sqrt{3}$-reconstructed areas (the above-assigned bulk peak). The other structures can then be accounted for by shifting the Si 2p peak obtained for the fully reconstructed $\sqrt{3}$ surface (curve 1) to a smaller binding energy [23]. These PES results show unambiguously carbon enrichment in the surface region for the $6\sqrt{3}$-reconstructed surface, but not in the form of graphite. A graphite C 1s peak is only observed after heating to a higher temperature than that required for obtaining a well-developed $6\sqrt{3}$ LEED pattern, in agreement with the STM observations.

5.3 Fourier analysis of STM images

An important issue that remains to be discussed concerns the relationship between the $6\sqrt{3}$ LEED pattern and the real-space structure observed in STM. Van Bommel et al. [4] proposed an explanation for the LEED pattern in terms of multiple scattering from a bulk-terminated SiC surface and a monocrystalline graphite overlayer. Using this model, Tsong and coworkers interpreted the different periodicities observed in their STM images in comparison to the diffraction pattern as the result of the electronic interaction between the graphite monolayer and the bulk-terminated SiC crystal [12, 13]. Recently, they presented a study where the changes observed in STM images of the 6×6 surfaces upon Si evaporation were interpreted as a further confirmation of their model [18]. In the recent theoretical study by Northrup and Neugebauer [35], it was suggested that the $6\sqrt{3}$ reconstruction arises from a graphite monolayer on top of a $\sqrt{3} \times \sqrt{3}$ arrangement of Si adatoms rather than on top of a bulk-terminated surface in view of the high stability calculated for the T_4 Si–adatom structure. The STM and PES data presented above, however, show that these models cannot explain the $6\sqrt{3} \times 6\sqrt{3}$ LEED pattern since this pattern is also observed for surfaces formed by heating in the range 1050 to 1250 °C on which no graphite is present in the C 1s spectra or STM images. For surfaces heated above 1250 °C, a monocrystalline graphite overlayer is observed in the STM images of 6×6-reconstructed regions and a clear graphite peak can be seen in the C 1s spectra while the LEED pattern remains $6\sqrt{3} \times 6\sqrt{3}$. Although these observations seem consistent with the previously proposed model of a bulk- or $\sqrt{3} \times \sqrt{3}$-terminated SiC surface with a monocrystalline graphite overlayer [4, 12, 13, 18, 35], the STM images of the graphite-covered 6×6 surface presented above show features of the initial 6×6 structure under the graphite layer which cannot be explained by either a bulk-terminated or a $\sqrt{3} \times \sqrt{3}$-terminated SiC crystal. Therefore, an alternative explanation of the observed LEED pattern is required.

Information about the relationship between the real-space structure observed in STM and the corresponding diffraction pattern observed in LEED can be obtained by Fourier analysis of the STM images [52]. Fig. 19a shows the power spectrum of the two-dimen-

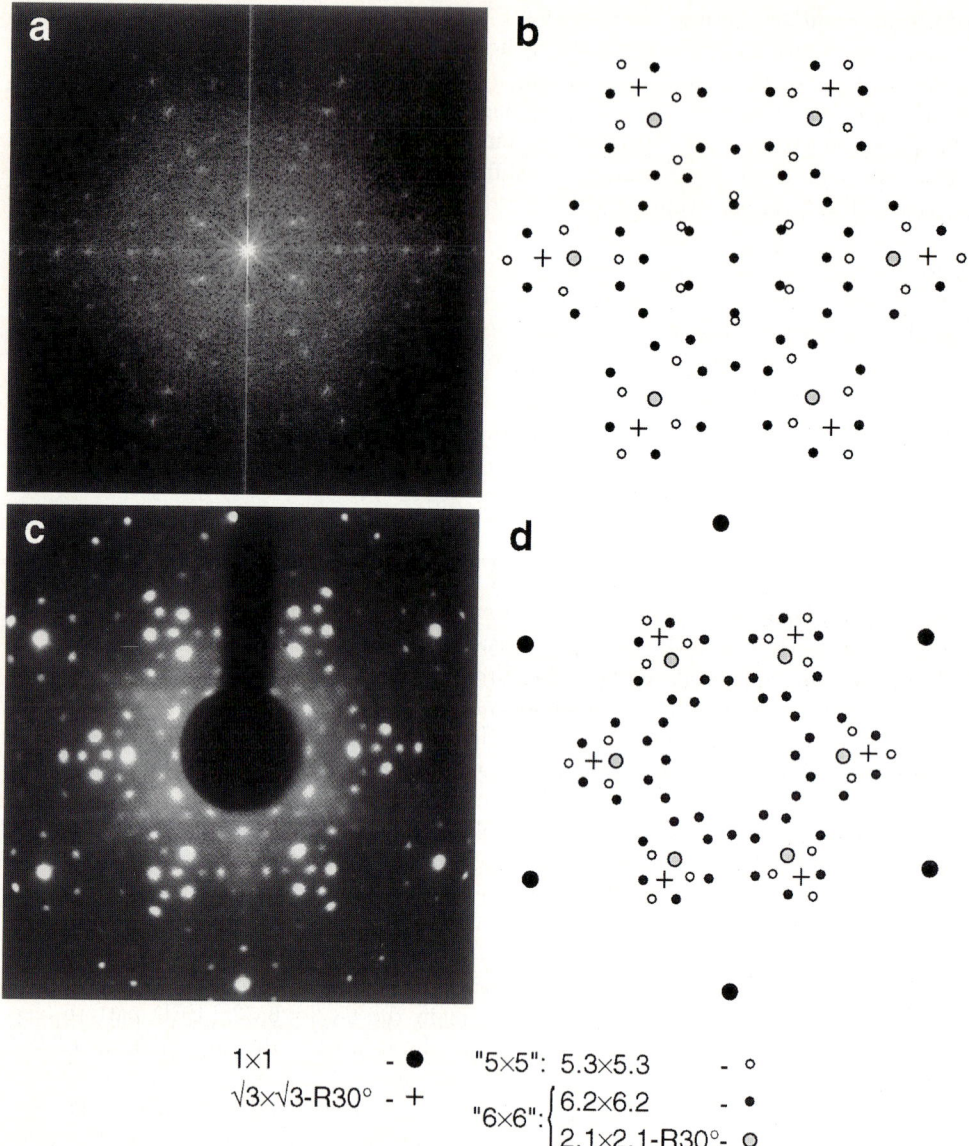

Fig. 19. a) 2D Fourier transform power spectrum of an STM image showing the $\sqrt{3} \times \sqrt{3}$, 6×6, and 5×5 reconstructions and b) its schematic. (c) $6\sqrt{3} \times 6\sqrt{3}$ LEED pattern obtained at an electron energy of 50 eV and d) its schematic (from [19])

sional Fourier transform (2D-FT PS) calculated for a 700×700 Å^2 filled-state STM image acquired in the same region as the image shown in Fig. 15. Thus, the Fourier transform includes components from the $\sqrt{3} \times \sqrt{3}$, 5×5, and 6×6 reconstructions, and the known orientation and periodicity of the $\sqrt{3} \times \sqrt{3}$ reconstruction with respect to the 1×1 unit cell of the SiC(0001) surface can be used as a reference in order to accurately determine the periodicities corresponding to the different peaks observed in the Fourier transform. Furthermore, the origin of the peaks in the Fourier transform can be determined by analyzing the Fourier transforms obtained after masking the image such that only one reconstruction is visible at a time. The results of this analysis are schematically shown in Fig. 19b. The average periodicities of the 6×6 and 5×5 reconstructions are found to be 6.2×6.2 and 5.3×5.3, respectively, referenced to the 1×1 unit cell of the SiC(0001) surface. Additionally, the 2D-FT PS of 6×6-reconstructed regions has components corresponding to an incommensurate hexagonal lattice which, referenced to the 1×1 unit cell, can be denoted 2.1×2.1-R30°. This structure thus has the same orientation with respect to the bulk as the $\sqrt{3} \times \sqrt{3}$ reconstruction, but a unit cell approximately 20% larger than that of the $\sqrt{3} \times \sqrt{3}$ reconstruction. As discussed in more detail in [19], the 2.1×2.1-R30° lattice is formed by the topographic maxima in the close-packed type E unit and the type D maxima which can be seen in Fig. 15. Furthermore, all the different types of structural units which form the 6×6 reconstruction have well-defined positions with respect to the 2.1×2.1-R30° lattice which therefore represents the true underlying periodicity of the 6×6 reconstruction.

By comparing the Fourier transform in Fig. 19a with the corresponding $6\sqrt{3}$ LEED pattern in Fig. 19c it is possible to assign each spot in the LEED pattern to peaks in the 2D-FT PS as schematically shown in Fig. 19d. The $6\sqrt{3}$ LEED pattern can thus be completely described by a combination of reciprocal lattice vectors from the 6×6- and 5×5-reconstructed regions of the surface together with the reciprocal lattice vectors from the $\sqrt{3} \times \sqrt{3}$ reconstruction. When no $\sqrt{3}$-reconstructed areas remain on the surface as determined with STM, the $\sqrt{3}$ spots in the LEED pattern completely disappear and the LEED pattern can be described by the reciprocal lattice vectors of the 6×6 and 5×5 reconstructions. Thus, the commonly used $6\sqrt{3} \times 6\sqrt{3}$ notation for the LEED pattern and reconstruction is incorrect and misleading. It would be more correct to speak about a mixture of 6×6 and 5×5 reconstructions with or without $\sqrt{3} \times \sqrt{3}$ reconstruction depending on annealing temperature.

6. Summary

The STM and AFM results presented in Section 2 showed that SiC(0001) surfaces with a morphology suitable for surface investigations can be prepared using sublimation- or hydrogen-etching. The results presented in Sections 3 to 5 were all obtained for sublimation-etched samples prepared using an ex situ method for oxide removal and in situ heating, which was shown to produce clean and well-ordered surfaces.

Concerning the structure of the $\sqrt{3} \times \sqrt{3}$ reconstruction, the results presented in Section 4 cannot provide a conclusive answer. Our STM images were shown to be consistent with a structural model composed of 1/3 layer of Si or C adatoms in threefold-symmetric sites on top of the outermost Si–C bilayer. Recent theoretical results favored Si adatoms in T_4 sites as the optimal configuration for the $\sqrt{3}$ surface. This adatom model could, however, not explain the existence of the surface shifted components observed in both the C 1s and the Si 2p core level spectra. Two surface shifted C 1s components of fairly similar intensities were observed and interpreted to originate from different atomic layers since they exhibited a difference in surface shifts of 2.1 eV. Therefore a model of 1/3 layer of Si adatoms on top of a modified C layer was proposed as a model which could explain the shifted core level components. The theoretical results for Si adatoms in T_4 sites predicted a half-filled surface state band derived from dangling bonds on the Si adatoms while a surface state band located well below the Fermi level, i.e. of semiconducting nature, was observed in angle resolved photoemission experiments. However, in view of the narrowness of the band it is conceivable that correlation effects can resolve this discrepancy. Until that has been shown to be the case both the core level results and the comparison between angle resolved data and present theoretical results indicate that Si adatoms, in T_4 sites, on top of an outermost Si–C bilayer appears to be an inadequate structural model for explaining recent findings for the $\sqrt{3} \times \sqrt{3}$-reconstructed surface. Further efforts, both theoretical and experimental, are certainly warranted. On the theoretical side, calculations including correlation effects could possibly resolve the issue concerning the semiconducting nature observed for the $\sqrt{3}$ surface and model calculations of surface core level shifts for SiC may provide new insight concerning the core level shifts observed. On the experimental side, other methods for structure determinations such as dynamic LEED and surface X-ray diffraction could possibly give new valuable information concerning the structure of the $\sqrt{3}$ reconstruction.

For the $6\sqrt{3} \times 6\sqrt{3}$ surface the results presented in Section 5 suggested a different structural model than the previously proposed monocrystalline graphite overlayer. During the gradual development of the $6\sqrt{3}$ LEED pattern accomplished by in situ heating at increasing temperatures, the STM images showed growing fractions of pseudo-periodic 6×6 and 5×5 reconstructions. When a well developed $6\sqrt{3}$ LEED pattern was observed the STM images showed that 6×6 and 5×5-reconstructed regions dominated on the surface but that small $\sqrt{3} \times \sqrt{3}$-reconstructed regions were still present. A comparison between Fourier transforms of the STM images and the $6\sqrt{3} \times 6\sqrt{3}$ LEED pattern showed that the LEED pattern can be fully explained by scattering from surfaces with a mixture of 6×6, 5×5 and $\sqrt{3} \times \sqrt{3}$ reconstructions such that the commonly used $6\sqrt{3} \times 6\sqrt{3}$ notation for the LEED pattern and reconstruction is incorrect. Both the STM and PES results showed clearly that there was no graphite on the surface when a well-developed $6\sqrt{3}$ LEED pattern was observed. Carbon enrichment in the surface region was obtained, but not in the form of graphite. A grahite C 1s level could only be detected after heating to a higher temperature than that required for observing a well-developed $6\sqrt{3}$ LEED pattern. The STM images then showed that the 6×6 reconstruction completely dominated the surface coverage and that two apparently different 6×6-reconstructed regions could be observed. In the modified regions a 6×6 structure with lower contrast was observed and images at higher magnification clearly showed the hexagonal structure of a graphite overlayer. The graphite thus appeared on top of a 6×6 reconstruction and not on bulk-terminated or $\sqrt{3}$-terminated SiC as proposed earlier. Due to the complexity of the STM data for the 5×5 and 6×6 reconstructions it was not possible to propose any model for these reconstructions. Further experimental as well as theoretical work concerning structural models for these reconstructions is needed.

References

[1] G. Pensl and R. Helbig, in: Festkörperprobleme/Adv. Solid State Phys. **30**, 133 (1990).
[2] O. Madelung, M. Schulz, and H. Weiss (Ed.), Semiconductors, Physics of Group IV Elements and III–V Compounds, Landolt-Börnstein, Numerical Data and Functional Relationships in Science and Technology, New Series, Group III, Springer-Verlag, Berlin 1982.
[3] U. Starke, phys. stat. sol. (b) **202**, 475 (1997).

[4] A. J. VAN BOMMEL, J. E. CROMBEEN, and A. VAN TOOREN, Surf. Sci. **48**, 463 (1975).
[5] R. KAPLAN, Surf. Sci. **215**, 111 (1989).
[6] S. NAKANISHI, H. TOKUTAKA, K. NISHIMORI, S. KISHIDA, and N. ISHIHARA, Appl. Surf. Sci. **41/42**, 44 (1989).
[7] U. STARKE, C. BRAM, P.-R. STEINER, W. HARTNER, L. HAMMER, K. HEINZ, and K. MÜLLER, Appl. Surf. Sci. **89**, 175 (1995).
[8] J. SCHARDT, C. BRAM, S. MÜLLER, U. STARKE, K. HEINZ, and K. MÜLLER, Surf. Sci. **337**, 232 (1995).
[9] V. V. ELSBERGEN, T. U. KAMPEN, and W. MÖNCH, Surf. Sci. **365**, 443 (1996).
[10] L. MUEHLHOFF, W. J. CHOYKE, M. J. BOZACK, and J. T. YATES, JR.,J. Appl. Phys. **60**, 2842 (1986).
[11] V. V. ELSBERGEN, T. U. KAMPEN, and W. MÖNCH, J. Appl. Phys. **79**, 316 (1996).
[12] C. S. CHANG, I. S. T. TSONG, Y. C. WANG, and R. F. DAVIS, Surf. Sci. **256**, 354 (1991).
[13] M.-H. TSAI, C. S. CHANG, J. D. DOW, and I. S. T. TSONG, Phys. Rev. B **45**, 1327 (1992).
[14] I. S. T. TSONG, J. Amer. Ceram. Soc. **76**, 269 (1993).
[15] M. A. KULAKOV, P. HEUELL, V. F. TSVETKOV, and B. BULLEMER, Surf. Sci. **315**, 248 (1994).
[16] F. OWMAN and P. MÅRTENSSON, Surf. Sci. Lett. **330**, L639 (1995).
[17] F. OWMAN and P. MÅRTENSSON, J. Vacuum Sci. Technol. B **14**, 933 (1996).
[18] L. LI and I. S. T. TSONG, Surf. Sci. **351**, 141 (1996).
[19] F. OWMAN and P. MÅRTENSSON, Surf. Sci. **369**, 126 (1996).
[20] Y. MARUMOTO, T. TSUKAMOTO, M. HIRAI, M. KUSAKA, M. IWAMI, T. OZAWA, T. NAGAMURA, and T. NAKATA, Japan. J. Appl. Phys. **34**, 3351 (1995).
[21] L. LI, Y. HASEGAWA, and T. SAKURAI, J. Appl. Phys. **80**, 2524 (1996).
[22] L. LI, Y. HASEGAWA, I. S. T. TSONG, and T. SAKURAI, J. Physique (1996), in press.
[23] L. I. JOHANSSON, F. OWMAN, and P. MÅRTENSSON, Phys. Rev. B **53**, 13793 (1996).
[24] L. I. JOHANSSON, F. OWMAN, P. MÅRTENSSON, C. PERSSON, and U. LINDEFELT, Phys. Rev. B **53**, 13803 (1996).
[25] L. I. JOHANSSON, F. OWMAN, and P. MÅRTENSSON, Surf. Sci. Lett. **360**, L478 (1996).
[26] F. OWMAN, C. HALLIN, P. MÅRTENSSON, and E. JANZÉN, J. Crystal Growth **167**, 391 (1996).
[27] S. TYC, J. Physique (I) **4**, 617 (1994).
[28] J. A. POWELL, D. J. LARKIN, and P. B. ABELN, J. Electronic Mater. **24**, 295 (1995).
[29] T. KIMOTO, A. ITOH, and H. MATSUNAMI, Appl. Phys. Lett. **66**, 3645 (1995).
[30] L. B. ROWLAND, R. S. KERN, S. TANAKA, and R. F. DAVIS, J. Mater. Res. **8**, 2753 (1993).
[31] S. TANAKA, R. S. KERN, and R. F. DAVIS, Appl. Phys. Lett. **65**, 2851 (1994).
[32] F. R. CHIEN, S. R. NUTT, W. S. YOO, T. KIMOTO, and H. MATSUNAMI, J. Mater. Res. **9**, 940 (1994).
[33] P. BADZIAG, Surf. Sci. **236**, 48 (1990).
[34] P. BADZIAG, Surf. Sci. **337**, 1 (1995).
[35] J. E. NORTHRUP and J. NEUGEBAUER, Phys. Rev. B **52**, R17001 (1995).
[36] M. SABISCH, P. KRÜGER, and J. POLLMAN, Phys. Rev. B (1997), in press.
[37] F. BOZSO, L. MUEHLHOFF, M. TRENARY, W. J. CHOYKE, and J. T. YATES, JR., J. Vacuum Sci. Technol. A **2**, 1271 (1984).
[38] M. L. SHEK, K. E. MIYANO, Q. Y. DONG, T. A. CALCOTT, and D. E. EDERER, J. Vacuum Sci. Technol. A **12**, 1079 (1994).
[39] C. HALLIN, A. S. BAKIN, F. OWMAN, P. MÅRTENSSON, O. KORDINA, and E. JANZÉN, in: Proc. Silicon Carbide and Related Materials, Kyoto (Japan), Inst. Phys. Publ., 1995.
[40] M. M. ANIKIN, A. A. LEBEDEV, S. N. PYATKO, A. M. STRELCHUK, and A. L. SYRKIN, Mater. Sci. Engng. B **11**, 113 (1992).
[41] Y. LI and P. J. LIN-CHUNG, Phys. Rev. B **36**, 1130 (1987).
[42] R. J. HAMERS, Phys. Rev. B **40**, 1657 (1989).
[43] J. NOGAMI, S.-I. PARK, and C. F. QUATE, Surf. Sci. Lett. **203**, L631 (1988).
[44] J. NOGAMI, S.-I. PARK, and C. F. QUATE, Phys. Rev. B **36**, 6221 (1987).
[45] E. GANZ, I.-S. HWANG, F. XIONG, S. K. THEISS, and J. GOLOVCHENKO, Surf. Sci. **257**, 259 (1991).

[46] P. Mårtensson, G. Meyer, N. M. Amer, E. Kaxiras, and K. C. Pandey, Phys. Rev. B **42**, 7230 (1990).
[47] L. Li, C. Tindall, O. Takaoka, Y. Hasegawa, and T. Sakurai, Surf. Sci. (1997), in press.
[48] L. I. Johansson, F. Owman, and P. Mårtensson, Surf. Sci. Lett. **360**, L483 (1996).
[49] P. Käckel, J. Furthmüller, and F. Bechstedt, Proc. 1st Europ. SiC Conf., Heraklion (Crete), 1996.
[50] H. Hoechst, M. Tang, B. C. Johnson, J. M. Meese, G. W. Zajac, and T. H. Fleisch, J. Vacuum Sci. Technol. A **5**, 1640 (1987).
[51] W. H. Backes, P. A. Bobbert, and W. v. Haeringen, Phys. Rev. B **49**, 7564 (1994).
[52] J. E. Demuth, U. K. Koehler, R. J. Hamers, and P. Kaplan, Phys. Rev. Lett. **62**, 641 (1989).

phys. stat. sol. (b) **202**, 529 (1997)

Subject classification: 68.55.Jk; 68.55.Ln; S6

Process-Induced Morphological Defects in Epitaxial CVD Silicon Carbide

J. A. POWELL and D. J. LARKIN

NASA Lewis Research Center, 21000 Brookpark Road, Cleveland, OH 44135, USA

(Received January 31, 1997)

Silicon carbide (SiC) semiconductor technology has been advancing rapidly, but there are numerous crystal growth problems that need to be solved before SiC can reach its full potential. Among these problems is a need for an improvement in the surface morphology of epitaxial films that are grown to produce device structures. Various processes before and during epilayer growth lead to the formation of morphological defects observed in SiC epilayers grown on SiC substrates. In studies of both 6H and 4H-SiC epilayers, atomic force microscopy (AFM) and other techniques have been used to characterize SiC epilayer surface morphology. In addition to the well-known micropipe defect, SiC epilayers contain growth pits, triangular features (primarily) in 4H-SiC, and macro step due to step bunching. In work at NASA Lewis, it has been found that factors contributing to the formation of some morphological defects include: defects in the substrate bulk, defects in the substrate surface caused by cutting and polishing the wafer, the tilt angle of the wafer surface relative to the basal plane, and growth conditions. Some of these findings confirm results of other research groups. This paper presents a review of published and unpublished investigations into processes that are relevant to epitaxial film morphology.

1. Introduction

The recent advances of SiC semiconductor technology have been greatly aided by the commercial availability of SiC wafers of reasonable size and quality. Bulk crystals of various SiC polytypes (e.g. 4H, 6H, and 15R) are now being grown by vapor sublimation processes at various institutions. The 4H and 6H polytypes are available commercially as polished wafers. Currently, both of these polytypes are actively being developed, but recent emphasis has shifted to the 4H polytype because of the larger electron mobility (800 versus 370 cm^2/Vs) and lower donor activation energy (45 versus 90 meV) of 4H-SiC compared to 6H-SiC [1, 2]. These commercial wafers typically contain dislocations at a density greater than 10^4 cm^{-2} and a defect known as micropipes at a density of about 100 cm^{-2} [3, 4]. The micropipes are tubular voids, approximately a micrometer in diameter, that extend along the crystal c-axis (the $\langle 0001 \rangle$ growth direction). The micropipes propagate from the SiC substrate into the subsequently grown epitaxial films [5]. These micropipes can negatively impact devices that are fabricated from the epitaxial films [4].

Chemical vapor deposition (CVD) is currently the method of choice to produce SiC device structures consisting of thin doped epitaxial films. Step-controlled epitaxy has been used to obtain homoepitaxial growth (i.e. film and substrate are the same polytype) on SiC substrates [6]. This epitaxy takes place by the lateral growth of atomic-scale steps that are present on the substrate whose growth surface has been tilted "off-axis" from the (0001) basal plane as shown schematically in Fig. 1. The steps on the

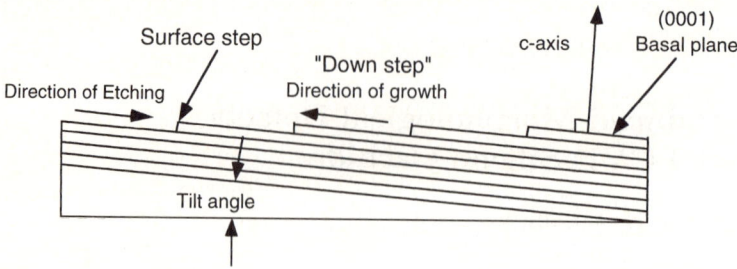

Fig. 1. Cross-sectional schematic diagram illustrating step flow growth on the surface of an "off-axis" SiC substrate

surface contain the polytype stacking sequence of the substrate. In this process, growth occurs at steps rather than at sites of two-dimensional nucleation on atomically-flat terraces between steps or at other unwanted nucleation sites caused by defects or contamination [7]. The polished growth surface of SiC substrates used for device fabrication is typically tilted "off-axis" from the (0001) plane by approximately 3° to 8°. The term "vicinal" (0001) SiC surface is often used for surfaces that are "off-axis" by an angle of up to a few degrees. The precise definition of "vicinal" means "in the neighborhood of", so any angle from zero up to a few degrees is included. The use of the term in the literature is somewhat ambiguous. In this paper, substrates with tilt angles less than 1° will be designated as "on-axis" and substrates with tilt angles greater than 1° will be designated as "off-axis". Also, the "down-step" direction will refer to the direction of the lateral growth of the steps (refer to Fig. 1). The "up-step" direction is the opposite of "down-step".

It has been found that the actual step height of steps on smooth vicinal epitaxially-grown (0001) SiC surface are not equal to 0.25 nm, which is the height of a single double-layer of Si and C atoms that are stacked in the c-direction to form the various polytypes. For example, Tyc [8] observed by atomic force microscopy (AFM) that the average step height on a 3° off-axis, 10 μm thick, 6H epitaxial film grown by Cree was approximately 15 nm high, which is ten times the unit height (i.e. 6×0.25 nm $= 1.5$ nm) of the 6H polytype. Tyc attributed the large steps to a process known as "step bunching" whereby small steps on the surface coalesce into larger steps during the growth process. He also suggested that the large steps could affect the electrical conductivity in different directions parallel to the surface in thin channels of field effect transistors (FETs). Others have shown that step bunching can affect the distribution of dopants that are added during the growth of GaAs and Si device structures [9]. These and other effects of the large surface steps may negatively impact the fabrication of high density SiC integrated circuits.

Commercial fabrication of SiC semiconductor devices is vitally dependent on the achievement of reproducible control over the magnitude and uniformity of epitaxial film thickness and doping. However, the growth rate and dopant incorporation are functions of many growth parameters. The recent NASA-Lewis developed site-competition epitaxy process [10, 11] has enabled reproducible doping of SiC epitaxial films over a much wider range (10^{14} to $>10^{19}$ cm^{-3}) than previously possible (10^{16} to 10^{18} cm^{-3}), and also the ability to produce "ohmic as deposited" electrical contacts. However, growth parameters associated with the use of this process remain to be optimized. For example, this process involves the variation of the Si/C ratio in the gas phase during growth to achieve different doping levels. Since the site-competition epitaxy process has been found to be most effective on the Si-face of (0001)SiC wafers, we have mostly used Si-face wafers in our epitaxial film growth research and to fabricate SiC semiconductor devices.

An essential requirement of commercial SiC CVD processes is that the resulting films be consistently free of defects and morphological features that would negatively impact either the device fabrication processes or the operation of resulting devices. In the case of current SiC epilayers, a variety of defects and features have been observed in addition to the large steps mentioned previously. In this paper, the term "morphological defect" will be used to denote any structural defects (e.g. dislocation, micropipe, etc.) or any unwanted surface features (e.g. pit, hillock, large steps, etc.). This paper will review morphological defects observed in SiC wafers and epilayers and will discuss processes that contribute to the formation of the defects.

2. Characterization of Wafers and Epilayers

The crystal structure and quality of selected wafers and epilayers were determined by low temperature photoluminescence (LTPL) [12] and transmission electron microscopy (TEM) [13] as described previously. Topographical maps of defects in selected wafers and epilayers were produced by Synchrotron White Beam X-Ray Topography (SWBXT) [14]. The surface morphologies of the epilayers in this study were character- ized by Nomarski differential interference contrast (NDIC) optical microscopy and atom- ic force microscopy (AFM) [15]. The latter technique allowed the observation of features down to near atomic scale. The AFMs used were a Park Scientific Instruments Auto- Probe LS and a Digital Instruments Dimension 3000, each with a lateral scan range of up to 100 μm. The vertical noise level of the AFMs were usually less than 0.1 nm which was sufficient to distinguish between atomic step heights that were single or multiples of the unit bilayer SiC step height (0.25 nm) in the c-axis stacking direction.

3. Commercial Wafer Quality

The SiC substrates used in the NASA work were obtained from 6H-SiC and 4H-SiC wafers up to 35 mm in diameter that were produced by two commercial suppliers from sublimation-grown boules. In this article, the vendors will be identified as source A and source B. The wafers were sliced "on-axis" (less than 1° tilt) or "off-axis" (3° to 8° tilt) from the (0001) basal plane in the $\langle 11\bar{2}0 \rangle$ direction with the Si-face side of the wafers polished to form the growth surface. The SiC wafers were usually cut into either four equal-sized pie-shaped pieces, or into smaller 7.5×6 mm^2 pieces to reduce the substrate cost of epilayer growth experiments.

In some wafers, the as-received polished surface was specular but exhibited faint par- allel ridges and random scratches when viewed with NDIC. We believe that the parallel ridges are remnants of the process of slicing wafers from the boule and that the scratch marks are due to the polishing process. The ridges were shallow undulations in the sur- face and the scratches were narrower in width, but both had depths up to about 5 nm (measured by AFM). Also, by adjusting the contrast while observing with NDIC, very small "white dots" against a darker background could be seen. Some of these white dots were revealed by AFM to be small holes in the surface with diameters in the range 0.5

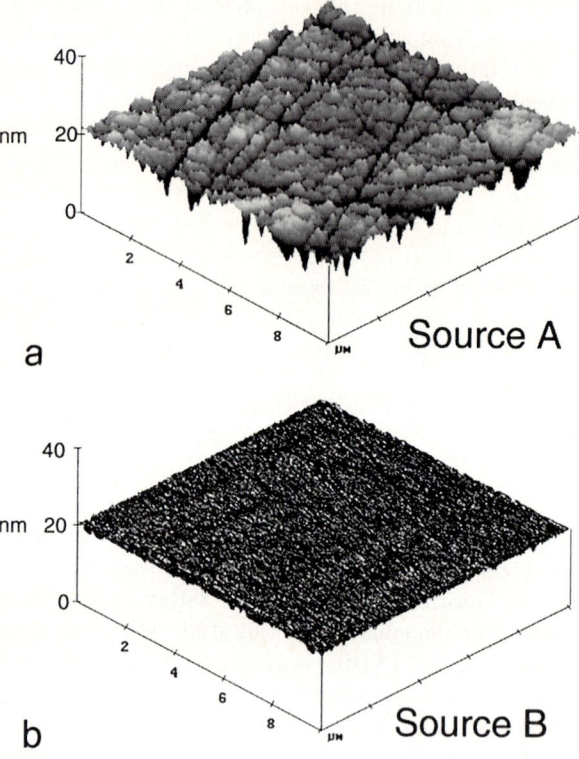

to 1 μm. We did not determine whether these holes as seen by AFM were actually micropipes, but it appeared that the density of the white dots was an order of magnitude higher than the micropipe density of about 100 cm^{-2}. It is believed that these holes are caused by plasma etching by the vendor which was used to reduce polishing damage. Occasional irregular-shaped voids of various sizes intersected the surface. Compared to 6H-SiC, 4H-SiC as-received generally had more of the larger irregularly shaped voids as revealed by optical microscopy. Fig. 2 illustrates the surface of wafers from two commercial sources. Most of the work reported in this paper were carried out on wafers from source A (Fig. 2a) because more wafers were available from this source. Only a few wafers from source B (Fig. 2b) were obtained; this was not sufficient to allow definitive conclusions regarding the impact of the better surface of source B wafers on growth morphology.

4. Typical Epilayer Growth Processes

The epilayer growth experiments were carried out in a CVD system that has been described previously [12, 15, 16]. The horizontal fused silica CVD chamber was water cooled with an inside diameter of 50 mm. All pregrowth etching and epitaxial growth was carried out at atmospheric pressure. The substrates were heated by an rf-heated SiC-coated graphite susceptor, 80 mm long × 30 mm wide × 10 mm thick. The susceptor was held horizontally in the chamber by a fused silica carrier that blocked the flow in the lower half of the chamber forcing the process gases over the top of the susceptor and

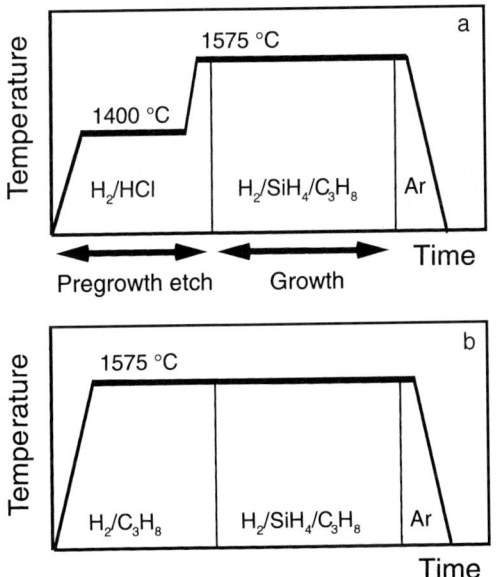

Fig. 3. Two CVD processes used in growth of SiC epilayers on SiC substrates. a) Typical pregrowth etch using H_2/HCl and b) investigated pregrowth etch using H_2/C_3H_8

substrates. Before loading into the CVD chamber, substrates were sequentially cleaned with organic solvents, hot H_2SO_4, scrubbed with liquid detergent, rinsed with $18\ M\Omega$ cm water, and then blown dry with nitrogen.

During etching and growth, the H_2 carrier gas flow was maintained at $3\ l/$ min. Prior to growth, the substrates were typically subjected to a 4 min HCl etch (3% in H_2) at $(1400 \pm 25)\ °C$ in order to remove unintentional contamination and to reduce the surface damage caused by the wafer cutting and polishing process. A pregrowth etch in H_2 was also investigated and will be discussed in Section 8. Epilayer growth was carried out at a fixed temperature in the range 1450 to 1600 °C with SiH_4 (3% in H_2) and C_3H_8 (3% in H_2) as the sources of Si and C. The temperatures given in this paper are those of the substrate; the susceptor surface temperature was determined to be about 50 °C higher than the substrate. The temperature was measured with an automatic optical pyrometer which was calibrated by observing the melting point of a small amount of previously melted Si on both the susceptor and SiC substrate. The actual temperature was known to within 25 °C, but the reproducibility of the temperature control was better than 2 °C. Typically, the SiH_4 concentration in the growth chamber was held at about 200 ppm and the C_3H_8 was varied to produce Si/ C atomic ratios in the gas in the range 0.1 to 0.8. Epilayer growth rates were 3 to 4 μm/ h. Two CVD process schedules that were used in the growth of the epilayers in this study are shown in Fig. 3. If, during CVD, the substrate/susceptor is heated in H_2 above the melting point of silicon (1410 °C), silicon droplets are produced on the SiC substrate. It is believed that the Si droplets contribute to the formation of morphological defects in SiC epilayers. Hence, the CVD process must minimize the time of heating in pure H_2. During the startup portion of the CVD process (Fig. 3a), the presence of HCl inhibits the formation of Si droplets. During cool-down, an inert gas flow, instead of H_2, inhibits the formation of Si droplets. The process shown in Fig. 3b will be discussed in Section 8.

5. Localized Defects (Growth Pits, Triangles)

Epilayers of both 6H and 4H, with thickness in the range 1 to 10 μm, were grown and characterized. Typically, the epilayers appeared specular and free of gross features when examined by eye. When examined by optical microscopy using NDIC illumination, the epilayers exhibited small isolated triangular features and, in the case of 4H, larger trian-

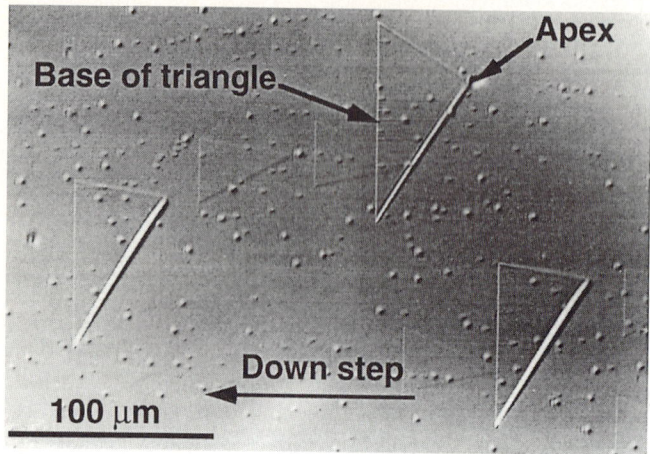

Fig. 4. NDIC optical image of the surface morphology of a 4H-SiC epilayer with isolated morphological defects

gular features as shown in Fig. 4. Both small and large triangular features were always oriented in the same direction, with one corner always pointed in the up-step direction. The distribution of the small features was nonuniform and the density varied widely, often greater than 10^5 cm^{-2}. Often these features were observed to form along scratches present in the as-received substrate. A typical 4H epilayer, grown under the same conditions as a 6H epilayer would exhibit the same small triangule features, but also the larger triangular features with a size extending to a few tens of micrometers. The distribution of the large triangles, which were only observed in the 4H epilayers, was also very nonuniform. They often were observed to be clustered along scratches in the substrate.

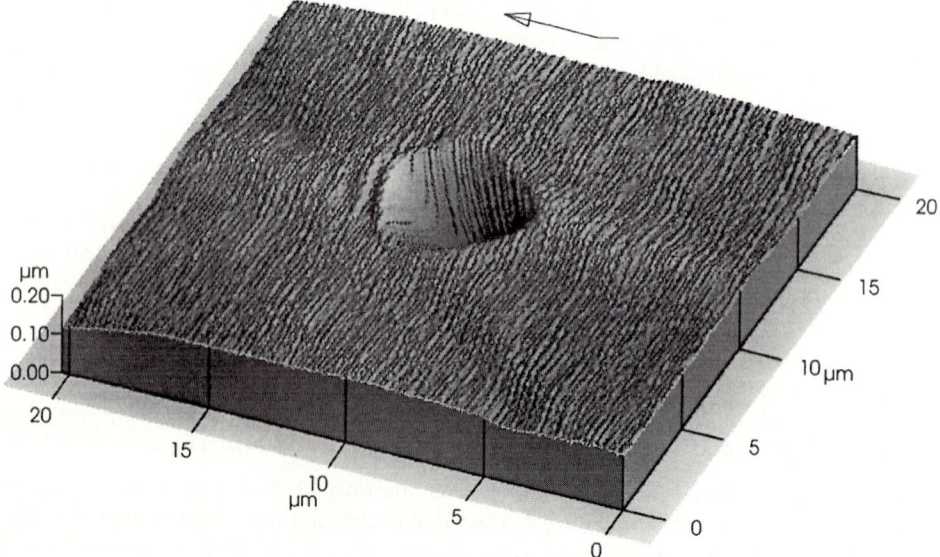

Fig. 5. AFM image of a shallow growth pit produced in a 6H-Si epilayer

Observations of the smaller triangular features with AFM revealed that they were shallow growth pits and that the shape of the pits was somewhat dependent on the specific conditions of the CVD process. An example of the more common of the observed pits is shown in the three-dimensional (3D) AFM image in Fig. 5. The size of the growth pits increased with film thickness. The diameter and depth of the growth pits were approximately 2 μm and 15 nm, respectively, in 1 μm thick epilayers and were larger, about 6 μm and 150 nm, respectively, in 10 μm thick epilayers. The growth pits (sometimes called amphitheaters) are thought to be caused by interruptions in the desired step flow growth.

As seen in Fig. 4, the large triangular features in the 4H epilayers were oriented with an apex at the up-step end and a straight-line base at the down-step end. AFM observations demonstrated that the triangular features were slightly depressed regions. The triangle base was a very straight deep groove and this groove usually extended beyond the corners of the base. The other two sides of the triangle were either straight or curved; these two sites were shallow grooves. In some of the large triangle, the area within the triangle was a depressed facet that was parallel to the (0001) basal plane. The growth on this basal plane area usually was the 3C-SiC polytype.

5.1 Source of growth pits

An investigation [17] was carried out with the purpose of distinguishing which morphological defects were caused by bulk defects in the substrates such as dislocations, micropipes, low-angle grain boundaries, etc. in the original boule crystal and which were caused by surface defects that had been generated by processes involved in cutting, polishing, and preparing the wafer for growth. The following approach was used. First, a typical epilayer (designated as epilayer 1) was grown on some selected commercial 4H and 6H boule-grown substrates (all 3.5° off-axis, Si-face substrates). The epilayer surfaces were characterized with NDIC optical microscopy. Second, epilayer 1 was then polished off to below the original growth surface. The final step of this repolishing used a chemical mechanical polishing (CMP) procedure that will be described in Section 7. A second epilayer (2) was then grown on the same repolished samples. Epilayer 2 was also characterized by NDIC optical microscopy. Specially placed laser-etched markers on the backside of the transparent samples enabled NDIC photographs to be taken of identical locations on epilayers 1 and 2. The purpose of this procedure was to compare the morphology of these two epilayers. We would expect that a) features common to both epilayers would be caused by defects that are present in the original boule crystal, and b) features that were common to only one of the epilayers would be caused by defects at the surface of the wafer (features caused by processes related to preparing the wafer surface for growth).

The results of the repolishing/regrowth experiments are shown in Fig. 6 and 7. NDIC photographs of epilayers 1 and 2 for each of two samples are shown in these figures. Photographs of the same locations on the 4H and 6H boule-derived samples are shown. The "dot-like" features seen in these photographs are the growth pits that we have observed in all of our SiC epilayers. The results can be summarized as follows:

a) In the 4H epilayers (Fig. 6), the growth pit density is much less in epilayer 2 (Fig. 6b). Also, new triangular features appear in epilayer 2, apparently the result of scratches that were produced in the repolishing. There does not appear to be any correlation between any features seen in these two 4H epilayers.

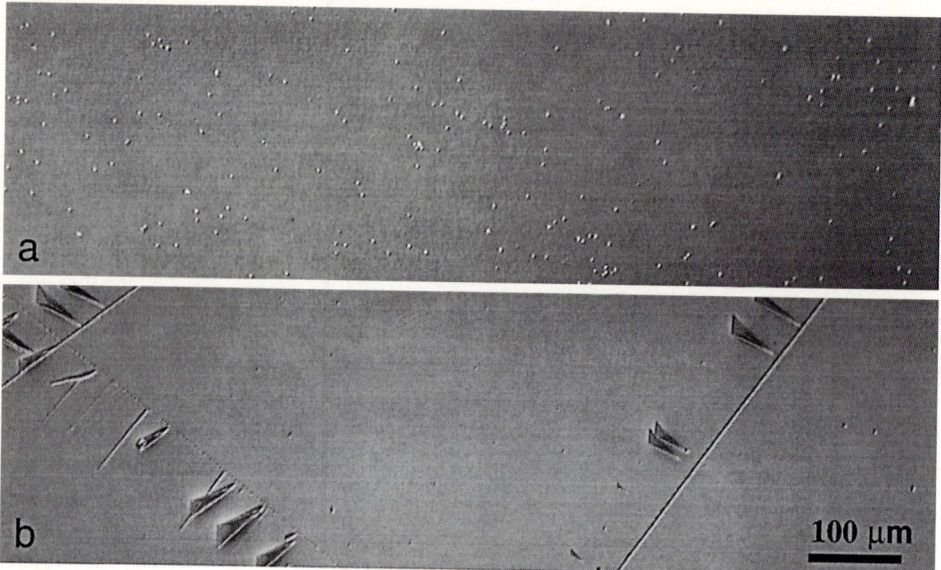

Fig. 6. NDIC optical images of the same area of a 4H-SiC sample with two different epilayers. a) First epilayer, b) second epilayer, after the first was removed. Note triangular features growing from scratches

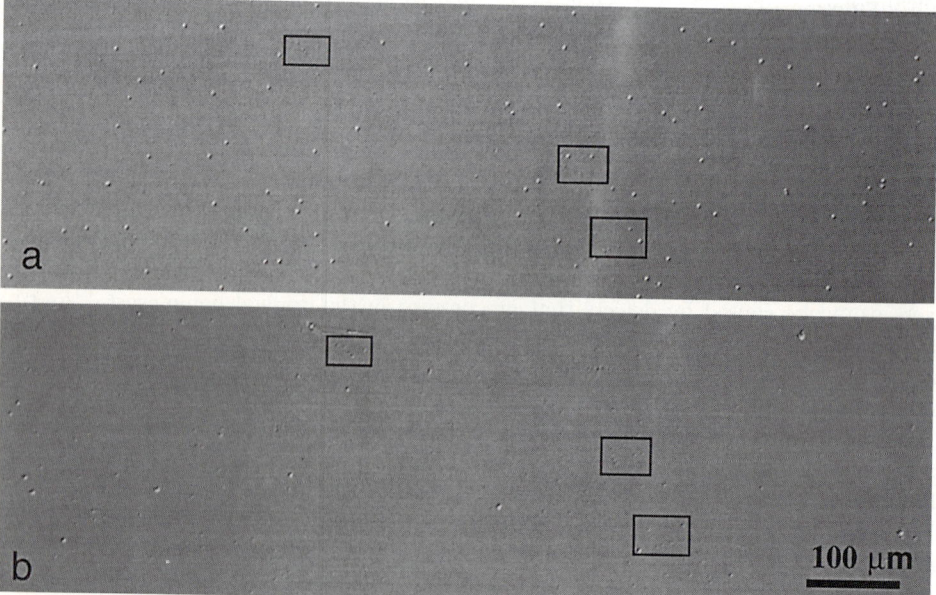

Fig. 7. NDIC optical images of the same area of a 6H-SiC sample with two different epilayers. a) First epilayer, b) second epilayer, after the first was removed. Boxes indicate growth pits common to both epilayers

b) In the 6H epilayers (Fig. 7), there are two groupings of growth pits: large and small. There does not appear to be any correlation between the large pits seen in the two 6H epilayers. Also, the density of large pits is much smaller in epilayer 2; whereas the density of the small pits is about the same. For convenience of comparison, we have put boxes around three groupings of small pits. These small pits are present in both epilayers 1 and 2.

The immediate first conclusion from Figs. 6 and 7 is that all features in the 4H epilayers and the large growth pits in the 6H epilayer are caused by surface defects; that is, from the polishing. The second conclusion is that many of the small pits the 6H epilayer are caused by bulk defects in the substrate.

The above procedure of removing epilayers, repolishing the surface, regrowing epilayers, and making before and after comparisons is very useful in tracking the source of epilayer defects. In the past, comparing epilayers grown on different samples was frustrating because of the variability of the results. The reason, in retrospect, is that poor polishing procedures probably contributed to variable surface quality.

As mentioned in Section 3, micrometer-sized holes, probably due to plasma etching, were observed in as-received wafers. In one experiment, three specific 0.5 μm diameter holes were identified by AFM in an as-received substrate, and then in the subsequent epitaxial film grown on this substrate. These holes were deeper than the detection limits of the AFM probe tip. A shallow growth pit occurred at each of these three holes and the holes were located at the up-step end of each pit. This demonstrates that holes produced by post-polishing etching can play a role in the formation of epilayer growth pits.

Based on the above observations, it appears that growth pits can be caused by different sources. We are not at the point where we can identify the source of each and every pit. For example, we do not know the cause of the specific pit shown in Fig. 7. As stated in the previous section, holes in the surface cause some pits. Polishing damage causes some pits, and bulk defects cause some defects. As we eliminate each known cause, the pit density should decrease. The procedure, described above, whereby a given substrate is reused several times (i.e. repolished, followed by a new epitaxial growth), can be used to determine the source of specific growth pits.

5.2 Effect of tilt angle

Epitaxial growth runs were carried out to confirm the effect of larger tilt angle (8° versus 3.5°) on the morphology of 4H epilayers. Transmission electron microscopy (TEM) was used to study structural defects and to confirm the presence of 3C-SiC in the triangular-shaped features.

Epilayers grown on 4H substrates with tilt angles of 3.5° and 8° in the same growth run are shown in Fig. 8. As can be seen, the triangular features are entirely absent in the epilayer grown on the 8°-tilt substrate. In general, the epilayer morphology of the 8° 4H samples exhibited fewer localized defects than the 3.5° 4H samples. However, the pregrowth etches used to remove scratches were less effective for samples with larger tilt angles; hence, remnants of polishing scratches seen in the epilayer in Fig. 8b is a typical result.

The growth results with the 3.5° and 8° 4H samples do confirm that the triangular features (3C-SiC inclusions) can be largely eliminated by growing on 4H substrates with large tilt angles. It remains to be seen if there will be a price to be paid for using large tilt angles. For example, an electrical anisotropy resulting from large tilt angle could be

Fig. 8. NDIC optical image of 12 μm thick 4H-SiC epilayers grown on 4H-SiC substrates with tilt angles of a) 3.5° and b) 8°

a negative factor in the operation of some devices. One can speculate that perhaps such large tilt angles will not be necessary when the 4H substrates are polished and prepared in a manner that does not leave surface defects which can cause 3C inclusions.

5.3 Effect of polytype (6H versus 4H)

Epilayer of 4H were more susceptible than 6H to 3C inclusions. For example, triangular patterns frequently seen [15] in 4H epilayers, but seldom in 6H, were characterized by oxidation [18] and found to contain surface layers of 3C-SiC within some of the triangular areas. Cross section TEM demonstrated the presence of thin planar 3C sections parallel to the (0001) plane in 4H epilayers containing a high density of the triangles. Also, we attempted to grow homoepitaxial films of 4H and 6H on 0.4° tilt 4H and 6H substrates, respectively, in the same growth runs. In every case, no continuous films of 4H were achieved, only 3C films with a high density of double positioning boundaries (DPBs) [7] and isolated 4H hillocks were obtained. In contrast, continuous 6H films on 6H substrates were achieved in most cases. At present, we do not know why 4H is more susceptible than 6H to 3C inclusions.

5.4 Effect of precursor concentration

The result of a series of three 6H-SiC epilayer growth runs with different silane and propane concentrations, but constant Si/C ratio, is shown in Fig. 9. In these runs, the Si/C ratio was constant at 0.19. The silane concentration for the three runs was Fig. 9a (run 2120): 170 ppm, Fig. 9b (run 2121): 200 ppm, and Fig. 9c (run 2119): 220 ppm.

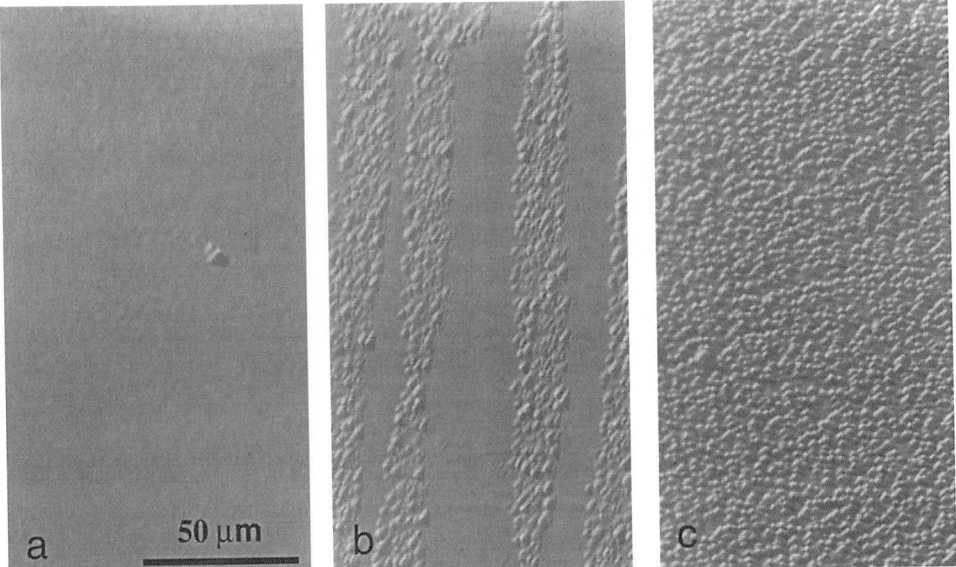

Fig. 9. NDIC optical image of 6H-SiC epilayers grown with an Si/C ratio of 0.19, but with different SiH$_4$ and C$_3$H$_8$ concentrations. Silane concentrations were a) 170, b) 200, c) 220 ppm

Note that the first run in the sequence was Fig. 9c, then Fig. 9a, and then Fig. 9b. This demonstrates that increasing the concentration of the precursors increases the probability of the formation of growth pits. Apparently, the step flow cannot keep up with the higher flux or Si-containing and C-containing species at the surface.

The epilayer results shown in Fig. 9 demonstrate that growth conditions can also be a factor in the formation of growth pits. This is important because, in the application of site-competition epitaxy [11] to dope SiC epilayers, a large range of Si/C ratios (and perhaps large concentrations of either precursor) may be desired.

5.5 Substrate edge effects

We found that the size and position distribution of substrates within the CVD reactor were also factors in determining the growth pit density. For example, when substrates of size 7.5 × 6 mm^2 and about four times larger (i.e. a quarter wafer) were included in the same growth run, the larger substrates consistently yielded epilayers with a lower pit density. When four 7.5 × 6 mm^2 substrates were arranged on the susceptor in a closely-spaced 2 × 2 array, the resulting films tended to have a lower pit density than with a single isolated 7.5 × 6 mm^2 substrate. The resulting growth pit density was typically higher near the edges of an isolated substrate or near the outside edge of a closely-grouped set of substrates. We suspect that contamination from the susceptor plays a role in this edge effect.

6. Step Bunching

As mentioned previously, the thickness of a single SiC double-layer in the c-axis stacking direction in SiC polytypes is 0.25 nm, and the unit heights (c-axis repeat distance) of

Fig. 10. AFM image of large steps on a 6H-SiC epilayer illustrating the nonuniform distribution of steps. Step height: approximately 25 nm

the 6H and 4H polytypes are 1.5 and 1.0 nm, respectively. AFM observations of SiC epilayers confirmed Tyc's observations that steps much larger than these values can occur in epitaxial films grown on SiC substrates. Our observations revealed that step heights and morphologies varied widely for the SiC films studied. Large steps were evident over some millimeter-sized regions, but were absent over other millimeter-sized regions. Step heights ranged from less than 1 nm to greater than 24 nm. For large steps, some had fairly flat terraces and sharp edges, others had very rounded edges. Fig. 10 illustrates the occurrence of a relatively flat area next to periodic arrays of large steps. We have not yet observed any correlation between step height, morphology, and the growth conditions.

6.1 Anisotropic step bunching

Isolated hexagonal hillocks were observed in both 4H and 6H epilayers grown on substrates with tilt angles less than 0.5°. Step bunching on the hillocks was quite different from the typical behavior shown in Fig. 10. Multiple spiral patterns of growth steps, assumed to be caused the by screw dislocations, propagated from the peak of each hillock. An example is illustrated in Fig. 11, which shows two spirals of 0.5 nm high steps propagating from the peak of a hillock grown on a 4H substrate. Each 0.5 nm step is a doublet step of the basic 0.25 nm high SiC bilayer step. The spiral evolves into a hexagonal pattern as shown. Since the total height of the two steps is 2×0.5 nm $= 1.0$ nm, the repeat height (Burgers vector) for this screw dislocation is equal to the repeat height of the 4H polytype. Thus, each 0.5 nm step is one of the two doublet steps of the 4H stacking sequence. As can be seen from Fig. 11, the doublet steps bunch to form a wide step, narrow step combination step with a height of 1.0 nm in the $\langle 1\bar{1}00 \rangle$ directions (e.g. $[10\bar{1}0]$ and $[01\bar{1}0]$). As the growth direction rotates to the next $\langle 1\bar{1}00 \rangle$ direction, the step pattern undergoes a transition at the $[11\bar{2}0]$ direction. At this transition direction, the step heights revert back to a doublet step height, i.e. 0.5 nm. This demonstrates that step bunching is dependent on the growth direction (i.e., it is anisotropic with respect to growth direction).

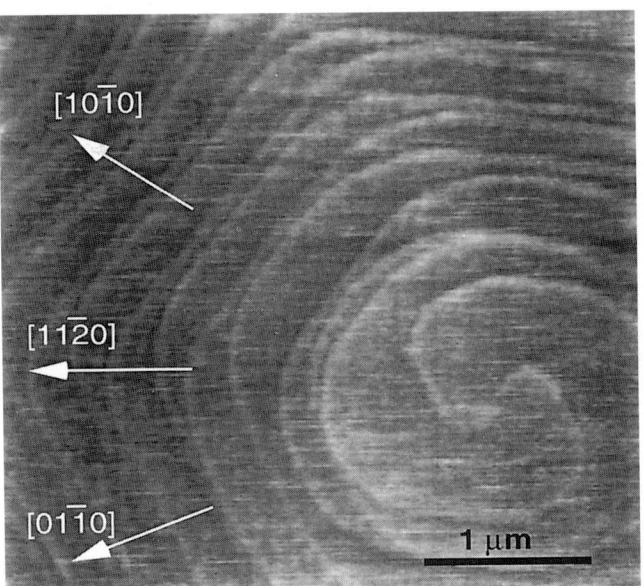

Fig. 11. AFM image of spiral steps propagating from the center of a 4H-SiC hillock

6.2 Model for anisotropic step bunching

The following is a model which we have proposed [19] to explain the anisotropic nature of the step bunching. First, a few aspects of SiC polytypism must be recalled. The 6H structure consists of the stacking of alternate triplet-step layers of the 3C polytype and these are usually designated as ABC and ACB. The difference in structure between these two triplet-step layers is a rotation of 60° about the stacking axis (the c-axis). Similarly, the 4H structure consists of alternate doublet-step layers of the 3C polytype and are designated AB and AC. Sakamoto et al. [20] pointed out that for steps on a Si(111) surface, there are either two dangling bonds per atom or one dangling bond per atom on the step riser depending on the crystallographic direction of the step. Because of threefold symmetry of Si(111), the number of bonds per atom alternates every 60° between one and two as the direction rotates about the surface normal. They suggested that faces (steps) with two dangling bonds per atom grow faster than faces with one dangling bond per atom. We applied these ideas to SiC with the following result. For a given {1$\bar{1}$00} face of 6H-SiC, there is a particular pattern of one and two dangling bonds per atom; if we assume, for example, that the ABC triplet-step layers in a stacking sequence have one dangling bond per atom on this given {1$\bar{1}$00} face, then the other triplet-step layers in the sequence (the ACB layers) on this same {1$\bar{1}$00} face will have two dangling bonds per atom. If one examines the pattern of bonds per atom on adjacent {1$\bar{1}$00} faces, the pattern is reversed: the ABC triplet-step layers have two bonds per atom, and the ACB triplet-step layers have one bond per atom.

We use ideas of the previous section to explain the anisotropic step bunching demonstrated in Figs. 11 and 12 (a special case of a super screw dislocation). We will use the 4H hillock depicted in Fig. 11 to describe the model. If, in the case of Fig. 11, we designate the top (wider) doublet steps in the [10$\bar{1}$0] direction as AB doublet steps and as-

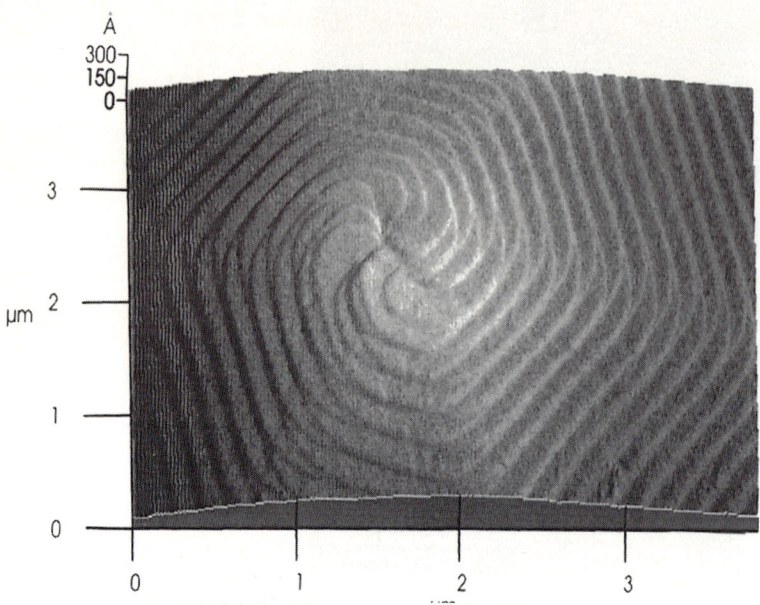

Fig. 12. AFM image of spiral steps propagating from the center of a 6H-SiC hillock. Steps are produced by a super screw dislocation

sume that AB steps have two dangling bonds per atom, they will grow faster in this direction than the AC doublet steps which have only one dangling bond per atom; hence, step bunching of doublet-step pairs will occur as shown. If we examine growth in the adjacent $\langle 1\bar{1}00 \rangle$ direction (e.g. the $[01\bar{1}0]$ direction) the AB steps will now be growing slower because they will have only one dangling bond per atom, and the AC steps will now be growing faster; hence reversed step bunching will occur. In this new direction (i.e. $[01\bar{1}0]$), the faster growing AC steps becomes the top (wider) step, overtaking the slower growing AB steps. The transition between these two states causes a localized "debunching" of the paired doublet steps in the $[11\bar{2}0]$ direction between the two adjacent $\langle 1\bar{1}00 \rangle$ directions. The same process occurs for 6H-SiC triplet steps as shown in Fig. 12.

Calculations were recently carried out by Heuell et al. [21] on the lateral growth of steps on a 6H-SiC vicinal (0001) surface tilted in a $\langle 1\bar{1}00 \rangle$ direction. They assumed (1) single bilayer steps on the surface and (2) each successive group of steps consists of one dangling bond per atom on the leading step edge of three successive steps followed by two bonds per atom on the next successive three steps. The result of the calculations indicated that the single bilayer steps structure would evolve into a structure with a step height of six bilayers. Our results are consistent with these calculations. Also, our model explains the occurrence of 0.75 nm height steps when 1.5 nm high steps turn by 60° as observed by Tyc [22] on a sublimation-grown epilayer on a 6H-SiC Lely crystal.

As can be seen in Figs. 11 and 12, the transition between the two bunched-step orientations in adjacent $\langle 1\bar{1}00 \rangle$ directions occurs over a very small angular change in growth direction in the vicinity of the $\langle 11\bar{2}0 \rangle$ direction between these two $\langle 1\bar{1}00 \rangle$ directions. The instability in step bunching in this transition region may be a factor in the macro scale step bunching that is observed in the SiC epilayers. Any small undulation of the growth surface, or defect on the surface could aggravate the effect of this instability. This may be significant since most SiC epilayer growth is carried out on vicinal (0001) wafers tilted toward a $\langle 11\bar{2}0 \rangle$ direction.

6.3 Super screw dislocation

The unusual hillock shown in Fig. 12 was grown on a 6H substrate. It consists of 16 growth steps propagating from the center of the spiral. The number 16 was determined by following a specific step around one complete revolution and counting the number of steps along the transition region back to the starting point. Each of these 16 steps is a triplet step, 3×0.25 nm = 0.75 nm high. These triplet steps bunch into 8 triplet-step pairs in the $\langle 1\bar{1}00 \rangle$ directions. The bunching forms single large steps rather than wide step, narrow step combination steps as was observed for 4H. Each triplet-step pair has a total height of 2×0.75 nm = 1.5 nm, the repeat height of the 6H polytype. The Burgers vector for this screw dislocation is equal to $8 \times$ the repeat height of the 6H polytype (total height: 12 nm). Screw dislocations with such large Burgers vectors have been called super screw dislocations [23]. The step bunching seen for this hillock is similar to that for the 4H hillock of Fig. 11 except that triplet-step pairs form in the $\langle 1\bar{1}00 \rangle$ directions, instead of doublet-step pairs as was observed for 4H.

Previously, we found that dislocations and micropipes propagate from the SiC substrate into the epilayer [5]. We believe that the observed hillocks are produced from screw dislocation that have their origins in the substrate; however, we do not know the cause of these dislocations. Current thinking is that the strain associated with dislocations with large Burger vectors causes some of these super screw dislocations to become micropipes [4, 24, 25]. The above observations demonstrate that AFM images of epitaxially-grown hillocks can be used to make direct measurements of the Burgers vector of super screw dislocations.

7. Chemical Mechanical Polishing

The results described in Section 5.1 indicated that residual damage caused by cutting and polishing SiC wafers was a major source of growth pits observed in SiC epilayers. In an effort to improve SiC substrate surfaces prior to growth, a collaborative effort between NASA Lewis and Case Western Reserve University (CWRU) was initiated to develop a chemical mechanical polishing process (CMP) for SiC. Based on an idea of P. Pirouz, the use of colloidal silica was successfully applied to SiC [26]. The best surfaces were obtained after colloidal silica polishing under conditions that combined elevated temperatures (approx. 55 °C) with a high slurry alkalinity (pH > 10) and high solute content. Removal rates of about 200 nm/h were achieved. Crosssectional TEM showed no observable sub-surface damage, and AFM showed a significant reduction in roughness compared to commercial diamond-polished wafers. The effectiveness of CMP for a 4H-SiC sample are shown in the "before" and "after" AFM images in Fig. 13 where a scratch-free surface was achieved (Fig. 13b) after 5 h of polishing. If the as-received surface of the substrate was more like that shown in Fig. 2b, then the CMP time required to reach a scratch-free state would have been much less. The CMP process for SiC is still in an early state of development and much optimization of parameters is required. Growth experiments following colloidal silica polishing have yielded a signifi-

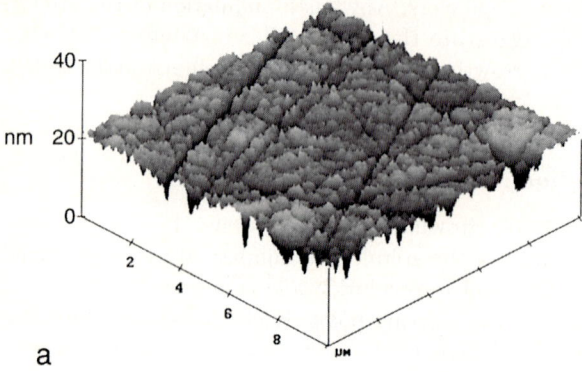

Fig. 13. AFM images of a 4H-SiC wafer, a) as-received surface and b) after chemical mechanical polishing (CMP)

a

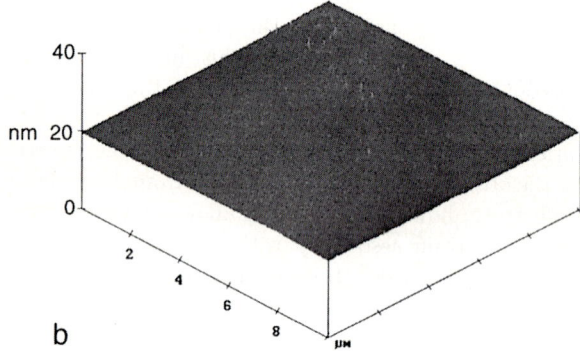

b

cant reduction in the density of growth pits. For example, the final stage of polishing for the samples in Figs. 6 and 7 was accomplished by CMP. As can be seen in Fig. 6b, several scratches are present in this sample. These were most likely caused by chipping of the sample edge during the CMP. This is a problem that has occurred sporadically in our polishing experiments no matter what type of polishing was being used. Edge rounding prior to the polishing has help reduce this problem.

8. Hydrogen Etching

It has been known for many years that H_2 will etch SiC at elevated temperatures [27] and will produce gaseous hydrocarbons and free silicon. In fact, the growth profile described in Section 4 was developed to reduce the time of heating in pure H_2 to a minimize the formation of free Si droplets on the substrate surface. Recently, H_2 etching prior to SiC CVD growth has been investigated by several research groups. Burk and Rowland [28] reported that Si droplets were formed when SiC was exposed to H_2 at temperatures above the melting point of Si (1410 °C). In addition, they found that Si droplet formation was suppressed over the temperature range 1450 to 1520 °C when there was C_3H_8 present in concentrations of 140 to 1200 ppm. There was indication that etching was taking place at a temperature of 1520 °C and C_3H_8 concentrations below

700 ppm. The SiC CVD of Burk et al. and at NASA Lewis were carried out in cold-wall quartz reactors. In contrast, the SiC research group at Linköping University (LU) has carried out SiC CVD and H_2 etching experiments in a hot-wall graphite reactor and have reported [29, 30] that H_2 can be a useful pregrowth etchant for 4H and 6H SiC substrates. Their reactor environment probably produced an overpressure of hydrocarbons. Their results using pure H_2 as a source gas demonstrated a strong dependence on the tilt angle of the substrate. Etching in H_2 carried out at 1550 °C for 30 min was effective at removing polishing scratches from on-axis (0001) SiC samples. The resulting surface of the 6H samples consisted of parallel steps with a height of 1.5 nm (the c-axis repeat distance of the 6H polytype). Under the same etching conditions, scratches were only partially removed from off-axis (3.5°) 4H and 6H SiC samples; the step bunching was less distinct for the off-axis samples. The conditions also produced mesa-like triangular features whose bases were parallel to the steps. The bases of these triangle were much longer (tens of micrometer) than the height of the triangles, producing features that appeared as lines on the surface when viewed with NDIC optical microscopy.

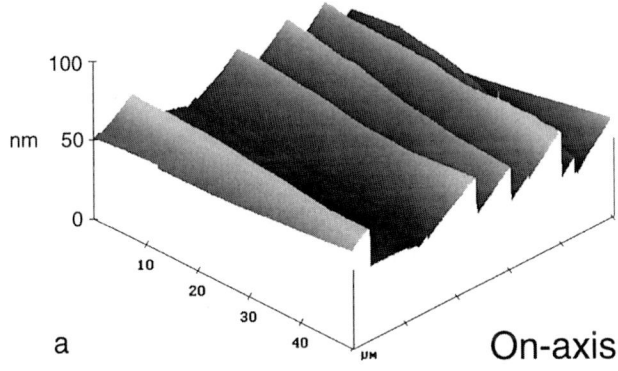

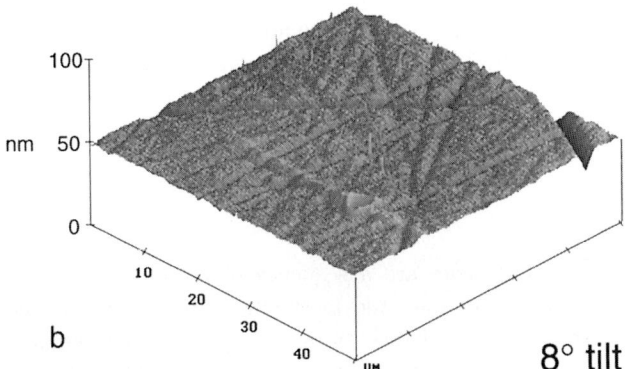

Fig. 14. AFM images of 4H-SiC surfaces after a H_2/C_3H_8 etch at 1575 °C for 30 min. a) On-axis 4H-SiC sample and b) 8° off-axis 4H-SiC sample

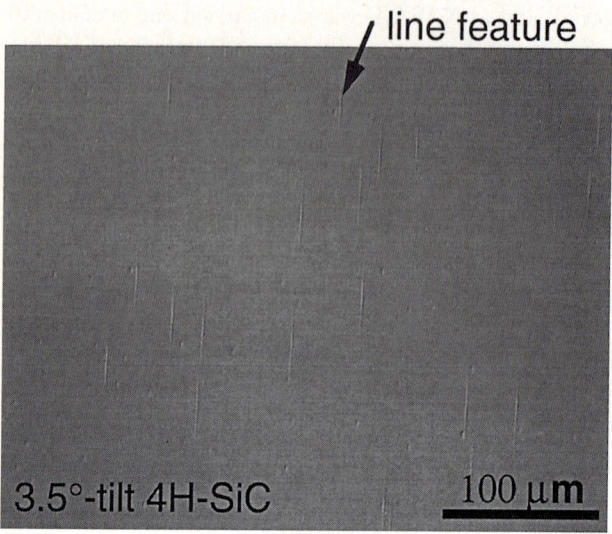

Fig. 15. NDIC optical image of a 3.5° off-axis 4H-SiC surface after a H_2/C_3H_8 etch at 1575 °C for 30 min showing elongated triangular "line" features

An investigation of H_2 etching of 4H and 6H SiC substrates was carried out at NASA Lewis in a cold-wall quartz reactor. In order to suppress Si droplet formation, 100 ppm of C_3H_8 was added to the H_2 carrier gas. Typically, etching was carried out at 1575 °C for 30 min. The NASA results were similar to that of the LU group. For the on-axis samples, the H_2 etching removed polishing scratches and produced steps due to step bunching. While the steps on the on-axis 6H samples were 1.5 nm in height, the steps on the on-axis 4H samples were tens of nm in height as shown in Fig. 14a. This is much larger than the 1 nm height of c-axis repeat distance of the 4H polytype. For the off-axis samples, scratches were only partially removed from 4H and 6H samples. A result for an 8° off-axis 4H sample is shown in Fig. 14b. The "line" features observed by the LU group for 3.5° of-axis 6H samples were also observed in the NASA work as shown in Fig. 15. There was much more of a tendency for the H_2 etching to decorate defects in 3.5° off-axis 4H samples than in the 8° off-axis 4H samples. Growth runs carried out using the schedule shown in Fig. 3b were carried out to compare pregrowth etches of H_2/HCl and H_2/C_3H_8. The typical 4 min, 1400 °C H_2/HCl pregrowth etch was more effective in producing an epilayer free of scratch remnants than a 30 min, 1575 °C H_2/C_3H_8 etch. It appears that H_2/C_3H_8 etching may be useful for on-axis substrates and for decorating defects.

9. Concluding Remarks

From the above sections, it is clear that there are a sequence of processes that can impact the morphology of SiC epilayers. The first (and most important) process is the growth of the bulk crystal since micropipes and dislocations are known to propagate into SiC epilayers. The increasing number of SiC developers who use commercial wafers are at the mercy of the quality of the wafers available from the few commercial SiC bulk crystal growers.

A second set of processes is that of cutting and polishing the wafer. The finding that these processes are having a major impact on morphology was somewhat surprising. It had been assumed that the defects in the bulk crystal were the dominant cause of morphological defects. Wafer users may be forced to repolish wafers to achieve scratch-free and damage-free surfaces. This problem of imperfect wafers surfaces prompted NASA Lewis to develop an in-house polishing capability and to collaborate with others to develop new polishing processes. The lead to the development of a new SiC CMP process. Although in an early state of development, it shows promise of being suitable for the final stage of polishing which will provide a scratch-free surface free of subsurface damage. Our characterization of commercial wafers showed that at least one supplier can produce a nearly scratch-free surface. Hopefully, commercial SiC wafer suppliers will be able to apply this new technology and provide the required polished surfaces that will yield improved SiC epilayers.

Another important process is the pregrowth etch to improve the growth surface, although with improved starting surfaces, this process may not be as important in the future. The H_2/C_3H_8 may prove to be useful in the future in decorating defects in wafers and epilayers.

The final process is the actual epilayer growth procedure. The challenge here will be to grow epilayers with uniform thickness and doping. The growth conditions presented herein cover only a narrow range of possible operating conditions. Much work is still required in order to achieve excellent morphology under the various conditions that will be required to achieve epitaxial films with large area and wide doping ranges.

Acknowledgements Some funding for this work was provided by DARPA (DO No. D149). Structural characterization and repolishing of substrates was carried out at CWRU supported under NASA Grant No. NAG3-1702. The authors would like to thank Andrew Trunek and Luann Keys for their support in carrying out the etching and growth runs and for producing the AFM images.

References

[1] W. J. SCHAFFER, H. S. KONG, G. H. NEGLEY, and J. W. PALMOUR, Silicon Carbide and Related Materials, Proc. 5th Conf. 1–3 Nov. 1993, Washington, DC, USA, Ed. M. G. SPENCER et al., Institut of Physics Publishing, Bristol and Philadelphia 1994 (p. 155).

[2] W. J. SCHAFFER, G. H. NEGLEY, K. G. IRVINE, and J. W. PALMOUR, Mater. Res. Soc. Symp. Proc. **339**, 595 (1994).

[3] K. KOGA, Y. FUJIKAWA, Y. UEDA, and T. YAMAGUCHI, Amorphous and Crystalline Silicon Carbide IV, Vol. 71, Eds. C. Y. YANG, M. M. RAHMAN, and G. L. HARRIS, Springer-Verlag, Berlin/Heidelberg 1992 (p. 96).

[4] P. G. NEUDECK and J. A. POWELL, IEEE Electron Device Lett. **15**, 63 (1994).

[5] J. A. POWELL, D. J. LARKIN, P. G. NEUDECK, J. W. YANG, and P. PIROUZ, see [1] (p. 161).

[6] H. MATSUNAMI, K. SHIBAHARA, N. KURODA, W. YOO, and S. NISHINO, Amorphous and Crystalline Silicon Carbide. Vol. 34, Eds. G. L. HARRIS, C. Y.-W. YANG, Springer-Verlag, Berlin/Heidelberg 1989 (p. 34).

[7] J. A. POWELL, J. B. PETIT, J. H. EDGAR, I. G. JENKINS, L. G. MATUS, J. W. YANG, P. PIROUZ, W. J. CHOYKE, L. CLEMEN, and M. YOGANATHAN, Appl. Phys. Lett. **59**, 333 (1991).

[8] S. TYC, J. Physique I **4**, 617 (1994).

[9] E. BAUSER and H. P. STRUNK, J. Cryst. Growth **69**, 561 (1984).

[10] D. J. LARKIN, P. G. NEUDECK, J. A. POWELL, and L. G. MATUS, see [1] (p. 51).

[11] D. J. LARKIN, P. G. NEUDECK, J. A. POWELL, and L. G. MATUS, Appl. Phys. Lett. **65**, 1659 (1994).

[12] J. A. POWELL, D. J. LARKIN, L. G. MATUS, W. J. CHOYKE, J. L. BRADSHAW, L. HENDER-SON, M. YOGANATHAN, J. YANG, and P. PIROUZ, Appl. Phys. Lett. **56**, 1442 (1990).
[13] J. A. POWELL, P. PIROUZ, and W. J. CHOYKE, in: Semiconductor Interfaces, Microstructures, and Devices: Properties and Applications, Ed. Z. C. FENG, Inst. Phys. Publ., Bristol (UK) 1993 (p. 257).
[14] M. DUDLEY, W. HUANG, S. WANG, J. A. POWELL, P. G. NEUDECK, and C. FAZI, J. Phys. D **28**, A56 (1995).
[15] J. A. POWELL, D. J. LARKIN, and P. B. ABEL, J. Electronic Mater. **24**, 295 (1995).
[16] J. A. POWELL, L. G. MATUS, and M. A. KUCZMARSKI, J. Electrochem. Soc. **134**, 1558 (1987).
[17] J. A. POWELL, D. J. LARKIN, L. ZHOU, and P. PIROUZ, Trans. 3rd Internat. High Temperature Electronics Conf., Vol. 2, Sandia National Laboratories, Albuquerque, (NM); 1996, (p. II).
[18] J. A. POWELL, J. B. PETT, J. H. EDGAR, I. G. JENKINS, L. G. MATUS, W. J. CHOYKE, L. CLEMEN, M. YOGANATHAN, J. W. YANG, and P. PIROUZ, Appl Phys. Lett. **59**, 183 (1991).
[19] J. A. POWELL, D. J. LARKIN, P. B. ABEL, L. ZHOU, and P. PIROUZ, Silicon Carbide and Related Materials 1995. Vo. 142, Eds. S. NAKASHIMA, H. MATSUNAMI, S. YOSHIDA, and H. HARIMA, Inst. Phys. Publ., Bristol (UK) 1996 (p. 77).
[20] K. SAKAMOTO, K. MIKI, and T. SAKAMOTO. J. Cryst. Growth **99**, 50 (1990).
[21] P. HEUELL, M. A. KULAKOV, and B. BULLEMER, See [1], Vol. 137 (p. 253).
[22] S. TYC, See [1] (p. 333).
[23] S. WANG, M. DUDLEY, C. CARTER, JR., D. ASBURY, and C. FAZI, Applications of Synchrotron Radiation Techniques to Materials Science, Eds. D. L. PERRY, N. D. SHINN, R. L. STOCKBAUER, K. L. D'AMICO, and L. J. TERMINELLO, Mater. Res. Soc. Symp. Proc. **307**, 249 (1993).
[24] F. C. FRANK, Acta Cryst. **4**, 497 (1951).
[25] W. SI, M. DUDLEY, R. GLASS, V. TSVETKOV, and C. H. CARTER, JR., J. Electronic Mater. **26**, 128 (1997).
[26] L. ZHOU, V. AUDURIER, P. PIROUZ, and J. A. POWELL, to be published in J. Electrochem. Soc. (1997).
[27] T. L. CHU and R. B. CAMPBELL, J Electrochem. Soc. **112**, 955 (1965).
[28] A. A. BURK, JR. and L. B. ROWLAND, J. Cryst. Growth **167**, 586 (1996).
[29] C. HALLIN, A. S. BAKIN, F. OWMAN, P. MARTENSSON, O. KORDINA, and E. JANZEN, See [19] (p. 613).
[30] F. OWMAN, C. HALLIN, P. MARTENSSON, and E. JANZEN, J. Cryst. Growth **167**, 391 (1996).

phys. stat. sol. (b) **202**, 549 (1997)

Subject classification: 73.40.Ns; 68.35.Bs; 73.20.At; 73.30.+y; S6

Surface Studies on SiC as Related to Contacts

M. J. BOZACK

Surface Science Laboratory, Department of Physics, Auburn University, Auburn, AL 36849, USA

(Received January 31, 1997)

We review the current status of surface studies on SiC as related to development of metal–semiconductor contacts, focusing on ten selected aspects of work through 1996. The ten areas include high resolution microscopy, XPS core-level binding energies versus temperature, surface graphitization, plasmon-loss features, work function and electron affinity, sputtering, etching, surface preparation and cleaning, surface states, and the index of interface behavior for SiC. Special consideration is given to the Ni/SiC ohmic contact system, and we discuss possible future directions in SiC contact surface research.

1. Introduction

In this paper we review the current status of surface studies on SiC as related to development of metal–semiconductor contacts. A literature study of the growth of research papers concerned with SiC surfaces is summarized in Fig. 1. The data was gathered from a search of the INSPEC database, which includes over 7000 scientific journals and conference proceedings. The rapid increase in activity over the last few years attests to the growing interest in SiC technology and its relation to surface science. This review covers selected aspects of surface studies as related to SiC contact technology through 1996.

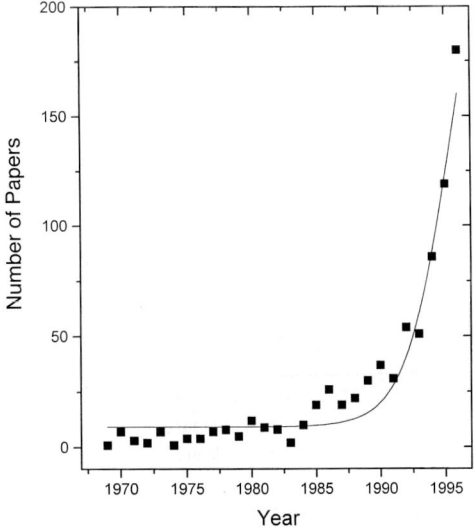

Fig. 1. Rate of scientific publication on SiC surfaces. Data was obtained using a computer search of the INSPEC database

2. Electrical Properties of Metal–Semiconductor Contacts

Currently, the relationship between SiC surface properties and contact performance is poorly understood, despite the fact that the Schottky barrier height in a contact arises because the work functions of the metal and semiconductor are different. Ideally, one can predict the behavior of a metal on a semiconductor if the work functions are known. For example, for an n-type semiconductor, if ϕ_M (the work function for the metal) is greater than ϕ_S (the work function of the semiconductor), the contact between the two is rectifying. On the other hand, if ϕ_M is less than ϕ_S, the contact is ohmic.

The standard equation giving the current density J in terms of the applied voltage V across a Schottky barrier is given by

$$J = A^* T^2 \left[\exp\left(-q\phi_B/kT\right)\right] \left[\exp\left(qV/nkT\right) - 1\right], \tag{1}$$

where A^* is the Richardson constant, T is the temperature, q is the electronic charge, ϕ_B is the Schottky barrier height, k is the Boltzmann constant, and n is the ideality factor, which is unity for the ideal Schottky diode. Small variations in the Schottky barrier height result in substantial changes in contact performance due to the exponential dependence of the current density on the barrier height.

2.1 Schottky model

We review the surface physics of Schottky barrier formation between a metal and a semiconductor in order to introduce the important physical parameters which afffect contact performance. Control of the surface parameters which affect the Schottky barrier guides the development of ohmic and rectifying contacts in semiconductors systems. The classical description of the Schottky barrier formed between a metal and a semiconductor starts with the band diagram shown in Fig. 2.

The work function of the metal, ϕ_M, is the energy needed to remove an electron from the Fermi level of the metal (E_F^M) to an infinite distance from the metal surface (i.e., the vacuum level). The work function of the semiconductor is ϕ_S, but a more useful quantity is the electron affinity χ_S, such that

$$\phi_S = \chi_S + \xi, \tag{2}$$

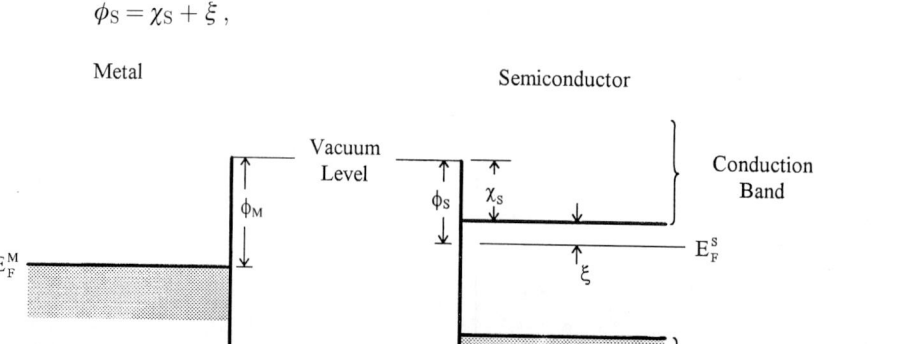

Fig. 2. Energy band diagrams of a metal and a semiconductor separated by a large distance

where ξ is the position of the Fermi level in the semiconductor (E_F^S) relative to the conduction-band edge. We have chosen the flat-band condition for an n-type semiconductor and have assumed no electric field exists within the semiconductor. This implies that the semiconductor terminates at the surface without distortion of the electron energy levels and that no surface states exist. We have also chosen the special case of $\phi_S < \phi_M$.

Bringing the metal and the semiconductor together and assuming the vacuum level is the same in both materials, electrons flow from the semiconductor to the metal, creating positive donor ions, and accumulate at the surface of the metal. The resulting dipole electric field opposes further electron flow, and in equilibrium

$$E_F^M = E_F^S .\tag{3}$$

As shown in Fig. 3, if the semiconductor is uniformly doped, the charge density is uniform to a depth w, called the depletion width, and the field E is linear with distance. If we define the barrier energy ϕ_B as the energy difference between the Fermi level in the metal and the bottom of the conduction band in the semiconductor at the interface, then

$$\text{Schottky Barrier Contact:} \quad \phi_B = \phi_M - \chi_S .\tag{4}$$

This description of a metal–semiconductor contact was first presented by Schottky [1] and is referred to as the *Schottky limit*. Equation (4) states that the barrier ϕ_B is directly proportional to the metal work function ϕ_M. The term barrier is chosen because ϕ_B is the energy necessary for electrons in the metal to penetrate the semiconductor. The barrier for electrons in the semiconductor relative to the metal is V_d, the diffusion potential or the band bending in the semiconductor at equilibrium. The fact that V_d is bias dependent and that ϕ_B is not responsible for the rectifying I–V characteristics of Schottky diodes.

The Schottky model predicts that it is possible to obtain a rectifying diode or an ohmic contact by simply choosing a metal with the appropriate work function. Experimentally, however, it is found that Schottky diodes formed on many of the III–V semiconductors do *not* show this behavior. Thus, the Schottky limit is thus not a complete description of all metal–semiconductor interfaces. Recent evidence for SiC suggests that SiC *approximates* Schottky behavior.

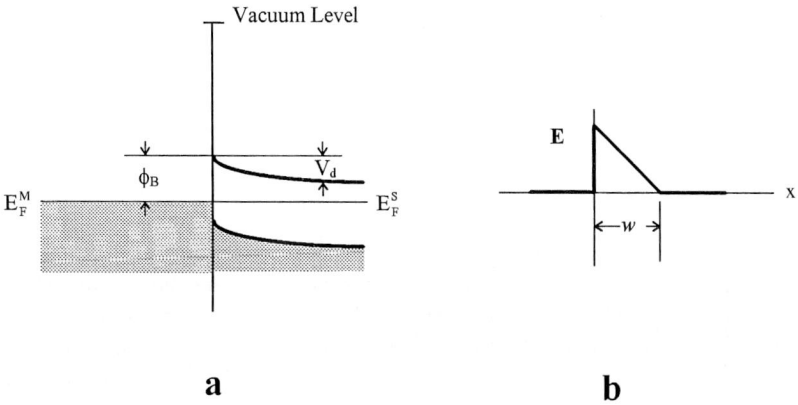

a   **b**

Fig. 3. a) Band diagram of an ideal metal–semiconductor interface, b) electric field distribution at the interface for an n-type semiconductor of uniform doping

2.2 Bardeen model

In 1947 Bardeen [2] proposed that if surface states exist at the metal–semiconductor interface in sufficient numbers, then ϕ_B would be independent of ϕ_M. Surface states are electronic states localized at the surface of the semiconductor crystal produced by the interruption of the perfect periodicity of the crystal lattice. The states can be occupied or empty depending on their position in energy relative to the Fermi level at the surface. We may define a neutral level ϕ_0 as the energy level (measured relative to the valence band) to which the surface states are filled when the surface is neutral. If the states are filled to an energy greater than ϕ_0, the surface possesses a net negative charge and the states are acceptor-like in behavior; if the states are filled to a level below ϕ_0, the surface has a net positive charge and the states behave in a donor-like manner. The parameter ϕ_0 is convenient in describing surface states and its value depends on the particular surface under consideration.

Assume that a thin insulating layer separates the metal from the semiconductor. The layer is so thin enough to be transparent to electron flow yet can withstand a potential difference across it. If the number of surface states is large, then the Fermi level at the surface of the semiconductor will be at ϕ_0. The band bending in the Schottky model was due entirely to the difference between ϕ_S and ϕ_M. However, if an interfacial layer exists and the surface state density is large, the potential difference $\phi_M - \phi_S$ will appear entirely across the interfacial layer since the charge in the surface states will fully accommodate the necessary potential difference. Thus, no change in the charge within the depletion region of the semiconductor is necessary when a metal is brought into contact with the semiconductor. Hence, ϕ_B is independent of ϕ_M, E_F^S at the surface of the semiconductor is the same as in the metal E_F^M, and we obtain

$$\text{Bardeen Barrier Contact:} \quad \phi_B = E_g - \phi_0 . \tag{5}$$

The Fermi level is pinned by the surface states to an energy ϕ_0 above the valence band. This equation is known as the *Bardeen limit*.

In practice, most Schottky barrier heights for semiconductors are predicted more accurately by Equation (5) than by Equation (4) — there is only small dependence on the metal work function. This is true for Si, Ge, and GaAs. For Si, as with Ge, GaAs, and GaP, the quantity $q\phi_0$ is found experimentally to be $\approx \frac{1}{3} E_g$ so that the barrier height $q\phi_B$ from Equation (5) is typically close to $\frac{2}{3}$ of the band gap or roughly 0.75 eV for silicon. The reason for this may be the characteristic high density of states that is common to the diamond lattice and which pins the Fermi level at this energy.

2.3 General case

In general the value of the barrier energy ϕ_B will be somewhere between the Schottky limit and the Bardeen limit. Cowley and Sze [3] were the first to analyze the general case accounting for both surface states and work functions. Consider the metal–semiconductor interface shown in Fig. 4, where the semiconductor is n-type and has permittivity ε_S; the interfacial layer has thickness δ and permittivity ε_i; and the surface states are characterized by the density D_S (per unit area per unit energy) and the neutral level ϕ_0.

It can be shown from their results that ϕ_B^0, the barrier energy with no electric field inside the semiconductor (i.e., the flat-band condition), is given by

$$\phi_B^0 = \gamma(\phi_M - \chi_S) + (1 - \gamma)(E_g - \phi_0), \tag{6}$$

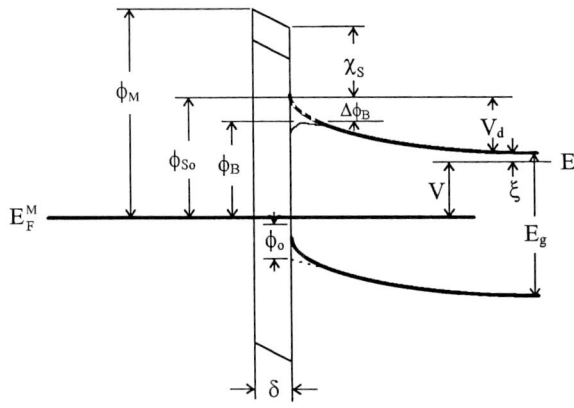

Fig. 4. Band diagram for a nonideal metal–semiconductor interface under forward bias V. An insulating layer of thickness δ exists between the metal and semiconductor and surface states are filled to the level ϕ_0

where

$$\gamma = \frac{\varepsilon_i}{\varepsilon_i + qD_S\delta} . \tag{7}$$

Equations (4) and (5) are limiting cases of this relation. The dimensionless parameter γ varies between zero and unity, depending primarily on the density of surface states. If $q\delta D_S \ll \varepsilon_i$, then $\gamma \approx 1$ and this equation reduces to the Schottky limit. If $q\delta D_S \gg \varepsilon_i$, then $\gamma \ll 1$ and this equation reduces to the Bardeen limit. Assuming $\delta \approx 10$ Å and a value for ε_i typical of common insulating films, D_S would have to exceed about 10^{13} states/cm² eV to pin the Fermi level at ϕ_0. The number of surface atoms on a typical semiconductor is $\approx 10^{15}$ cm⁻², so that a small relative density of surface states can have a significant influence on the Schottky barrier. A density of 10^{13} states/cm² eV corresponds to about one surface state for every 500 atoms of the semiconductor at the metal–semiconductor interface (other contributions to the measured values of ϕ_B such as image-force lowering will not be discussed). Such high surface state concentrations have been measured directly on many semiconductors prepared by chemical etching, and Schottky diodes formed on such surfaces exhibit values of ϕ_B which are independent of the metal as well as the electron affinity of the semiconductor.

Although the linear potential theory of Cowley and Sze has been extensively applied, it is clearly too simple, as it assumes a distribution of surface states which is unlikely to occur universally on all surfaces. Nevertheless, it has been extremely useful. The common procedure is to evaluate D_S from the slope of the linear relationship between ϕ_B and ϕ_M for a range of metals on a given semiconductor. Unfortunately, the values of ϕ_M quoted in the literature often vary over a range up to 0.5 eV. In view of the uncertainties, the application of the linear model must be used with caution.

2.4 Mechanisms of barrier formation

The physical mechanisms determining the barrier to charge carriers at the metal–semiconductor interface in a practical Schottky diode are not well understood. The formation of a practical Schottky diode consists of depositing a metal film onto a semiconductor surface which is not ideal but contaminated by a few monolayers of adsorbed foreign atoms or covered by a thin layer of native oxide. In the case of many III–V semiconductors, the Fermi level E_F^S at the interface is found to be pinned and almost independent of the metal work function. Presently, it is unclear whether the Fermi level is pinned as a result of chemical reactions between the metal and the semiconductor or is due to defects (crystal imperfections, impurities, etc.) at the interface. Further, for many surfaces, it is unknown whether the interface is truly abrupt on the atomic level and whether the simple potential diagrams above provide an accurate representation of a real metal–semiconductor interface.

There are four primary variables which control the Schottky barrier height at metal–semiconductor interfaces: the work function ϕ_M of the metal; the crystalline or amorphous structure at the metal–semiconductor interface; the diffusion of metal atoms across the interface into the semiconductor; and, the outermost electronic configuration of the metal atoms. Several attempts have been made to correlate ϕ_B to known physical or chemical properties of the metal or interfacial film. Although most of the work has been done on silicon and III–V semiconductors, the basic ideas are relevant to SiC contact technology.

Andrews and Phillips [4] plotted the ϕ_B values as a function of the heat of formation (ΔH_f) of the silicides. The result is a straight line defined by ϕ_B (eV) $= 0.83 - 0.18(\Delta H_f)$. According to this relationship, no barrier height should exceed 0.83 eV. The result of PtSi ($\phi_B = 0.88$ eV) is an exception. The result for IrSi ($\phi_B = 0.93$ eV) is similarly too high [5].

The failure to predict ϕ_B values greater than 0.83 eV resulted in a correlation by Ottaviani et al. [6] between ϕ_B and the eutectic temperature. The eutectic closest to the metal side was chosen for silicides that form by metal-atom diffusion, and the eutectic closest to the silicon side was chosen for silicides that form by silicon-atom diffusion. The eutectic temperature was chosen because it relates to the interfacial layer between the metal and silicon. In this correlation both PtSi and IrSi no longer appear as exceptions.

Early models of the metal–semiconductor interface were based on the empirical fact that ϕ_B for some semiconductors obeyed the Schottky limit while others (many of the III–V and elemental semiconductors) followed the Bardeen limit. To fit the wide range of behavior, Kurtin et al. [7] proposed a "linear interface" model where the barrier energy ϕ_B was assumed to be linearly dependent on the work function of the metal. Furthermore, it was more convenient to relate the Schottky barrier ϕ_B to the electronegativity χ_M of the metal rather than to the work function ϕ_M. The electronegativity has been shown [8] to have a linear relationship with the work function – the work function is dependent on the crystal plane, whereas electronegativity is an atomic property. In this case,

$$\phi_B = S\chi_M + C, \tag{8}$$

where S is a dimensionless parameter called the index of interface behavior that measures the sensitivity of ϕ_B to the metal ($S = \partial\phi_B/\partial\chi_M$) and C is a constant. In effect, S gives a quantitative measure of how closely an interfacial system follows Schottky behavior. When $S \approx 1$, the contact displays pure Schottky behavior and is determined by work function differences. When $S \approx 0$, the contact follows Bardeen behavior and is dominated by surface states.

It has been known for years that Fermi level pinning is more common in covalent materials than in ionic materials [9, 10]. Work by Kurtin et al. [11] plotted results [12, 13] of ϕ_B versus χ_M for several metals on Si, GaSe, and SiO$_2$ substrates (χ_M is the electronegativity of the metal). The values of S for Si, GaSe, and SiO$_2$ were ≈ 0.05, 0.6, and 1.0, respectively. The slopes increased with the degree of ionicity of the substrate, as defined by the difference in the electronegativities, $\Delta\chi$, of the two components [14].

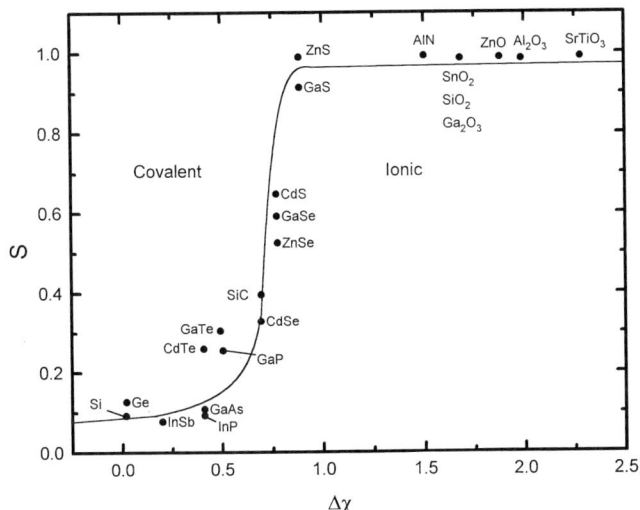

Fig. 5. Index of interface behavior S for several semiconductors versus the electronegativity difference $\Delta\chi$ in elemental components. Data from Kurtin et al. [11]

Kurtin's plot of ϕ_B versus $\Delta\chi$ revealed an important trend. The plot, shown in Fig. 5, shows a region of low S (<0.1) for semiconductors with $\Delta\chi < 0.6$ and a region of high S (>0.9) for semiconductors with $\Delta\chi > 0.8$ connected by an abrupt transition region ($0.6 < \Delta\chi < 0.8$). Carbon and silicon have electronegativities of 2.55 and 1.90, respectively, which corresponds to $\Delta\chi = 0.65$ for SiC. The data indicates that Fermi level pinning dominates Schottky barrier formation in covalent materials, plays a variable role in materials with a mixture of ionic and covalent bonding ($0.6 < \Delta\chi < 0.8$), and does not determine the Schottky barrier height in the more ionic materials. SiC is in the transition between strong Fermi level pinning (surface states) and no Fermi level pinning (no surface states), which is in good agreement with the empirical results for SiC discussed below.

McGill and Mead [15] took $S = 1$ in the linear interfacial model for the Schottky limit and $S = 0$ for the Bardeen limit. By examining $\phi_B - \chi_M$ data for a wide variety of semiconductors, Mead [16] found that S depended strongly on the type of chemical bonding in the semiconductor, with ionic materials exhibiting values of $S \approx 1$ and covalently bonded materials (Ge, Si, GaAs, InP, and SiC) exhibiting values of $S \approx 0.1$. The data on which the linear interface model of the covalent–ionic transition was based have been reexamined as a result of more recent experiments by Schluter [17] and the validity of the parameter S is now in question, although still generally applied. In addition, it is unlikely that the linear interface potential model is always applicable for interfaces formed between metals and a range of atomically clean semiconductor surfaces.

The covalent versus ionic theory of Kurtin et al. has been extended to include the heat of formation, ΔH_f, of the semiconductors. Brillson [18] plotted the S values taken from Mead [19] versus ΔH_f for a variety of semiconductors. He has shown that ϕ_B exhibits a strong correlation with the semiconductor heat of formation ΔH_f. Fig. 6 shows Brillson's [18] plot. It shows the same transition from covalent to ionic as for S versus $\Delta\chi$. Covalent semiconductors have low heats of formation and are less stable against chemical reaction. Both $\Delta\chi_M$ and ΔH_f, which are zero for elemental semiconductors, can be used to predict the approximate ionicity of a compound (the higher the $\Delta\chi$ and ΔH_f,

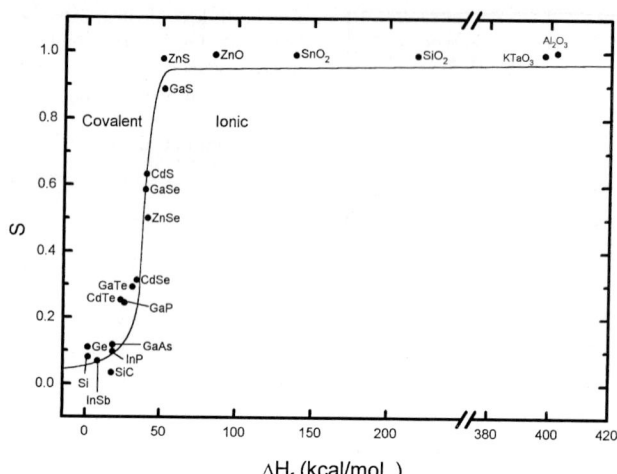

Fig. 6. Index of interface behavior S for several semiconductors versus heat of formation ΔH_f. Data from Brillson [18]

the higher the ionicity). Covalent semiconductors, such as Si, Ge, GaAs, and SiC react more readily with the metal at the metal–semiconductor interface, and lead to a low ϕ_B that depends very little on metal or semiconductor properties such as work function and electronegativity. Ionic semiconductors such as ZnS, ZnO, SnO$_2$ have a high heat of formation and are less likely to react with increasing χ_M. Thus, a well-defined transition exists between interfacial properties corresponding to ionic and covalent semiconductors. The importance of chemical reactivity between the metal and the semiconductor has been underscored by Phillips [20] and later by Andrews and Phillips [4] and Brillson [21].

The role of chemical reactions between the metal and native oxide has been examined by Kowalczyk et al. [22] to determine if reduction of the oxide layer will unpin E_F^S. The experiments have demonstrated that the surface of GaAs covered with approximately 10 Å of As$_2$O$_3$ and Ga$_2$O$_3$ or pure Ga$_2$O$_3$ can be altered by chemical reaction with certain metals during Schottky-barrier formation. For example, the reactive metals Al, Mg, Ti, and Cr were found to reduce the oxides on GaAs and, in some cases, new oxides (Al$_2$O$_3$, Cr$_2$O$_3$) formed during metal deposition at room temperature. However, little change in the position of the Fermi level at the interface occurred as a result of the chemical reactions. Similar experiments are needed on SiC to determine the role of metal–oxide reactions in practical diodes.

Interdiffusion between metal and semiconductor can also affect the Schottky barrier. Brillson et al. [23] have examined the deposition of several metals onto atomically clean InP using X-ray photoemission spectroscopy (XPS) and observed the outdiffusion of P into an overlaying Au film and the outdiffusion of In into Al and Ni films. A later study [24] found that ϕ_B varied with ΔH_f in chemically etched n- and p-type InP Schottky diodes. Further, outdiffusion of In into Au and Ag overlayers was observed but not the type of interdiffusion reported by Brillson on atomically clean InP. The importance of interdiffusion and chemical reactions between a metal and a semiconductor on the Schottky barrier is not yet fully understood.

Woodall and Freeouf [25] proposed an effective work function model to explain the metal–semiconductor interface. Here, the value of ϕ_B is not determined by surface states but is related to the work function of several different metallic-like phases at the metal–

semiconductor interface. The phases result from oxygen contamination or metal–semiconductor reactions which occur during metallization. Since each phase will have its own work function, the barrier height can be written as $\phi_{Bn} = \phi_{eff} - \chi_S$, where ϕ_{eff} is the weighted average of the work functions of the different phases present. Woodall and Freeouf believe that, for the III–V semiconductors, the anion is the primary contributor to ϕ_{eff}. Thus, for case of GaAs Schottky barrier contacts, ϕ_{eff} is the work function of elemental As; not the work function of the metal – GaAs serves as a reference and ϕ_B is independent of the metal. The Fermi level is pinned, not by surface states, but by the presence of excess As at the metal–semiconductor interface.

Murarka [26] plotted silicide barrier heights ϕ_B as a function of the metal position in the periodic table; the value of ϕ_B increases slowly with the number of d-electrons in the metal. The refractory silicides have a ϕ_B which is independent of the number of d-electrons. The value of ϕ_B increases with the number of d-electrons in cases where the outermost electronic orbitals have more than 5 d-electrons. Murarka also noted that metals with higher affinity for oxygen have nearly the same barrier height (about half the silicon band gap energy). On the other hand, metals which have little or no affinity for oxygen have large barrier heights that increase with increasing number of d-electrons.

Mead and Spitzer [27] noted that for covalent semiconductors, the Schottky-barrier energy on n-type semiconductors ϕ_{Bn} is approximately $2E_g/3$, where E_g is the band gap. Such a rule has been found to be approximately correct for GaAs, GaP, and AlAs, but does not predict the barrier energy on InP, InAs, and GaSb. McCaldin et al. [28] further developed this rule using gold as a reference metal to a variety of p-type compound semiconductors, and showed that ϕ_{Bp} varied inversely with the electronegativity of the anion of the semiconductor. This evolved into the "common anion rule" which states that $E_F - E_v$ is determined by the anion of the semiconductor and is fixed and independent of the cation or the metal. Thus, those semiconductors with a common anion should exhibit the same ϕ_{Bp}. The common anion rule implies that the Schottky barrier energy is a property of the bulk semiconductor and not the interface and has correctly predicted the Schottky barrier energies in studies of ternary alloys of the III–V compounds GaAs–InAs and GaP–InP, but has failed in describing the GaAs–AlAs system.

Spicer and coworkers [29] have introduced a defect model to explain Fermi-level pinning at the metal–semiconductor interface. Using XPS, Spicer measured E_F^S on GaAs, InP, and GaSb surfaces during metal deposition under ultrahigh vacuum conditions. Less than a monolayer of metal oxygen perturbed the semiconductor, producing lattice defects at the surface. The defects in turn produce new interface states ("metal-induced gap states" – MIGS) due to the presence of the metal which pin the Fermi level. Since less than 1% surface coverage of oxygen was necessary to pin E_F^S in GaAs, InP, and GaSb, small amounts of oxygen contamination during contact fabrication may be responsible for the observed values of ϕ_B in practical Schottky diodes.

To summarize, there are several physical and chemical mechanisms that affect the Schottky barrier, but currently there is no simple, self-consistent, comprehensive model for the formation of Schottky barriers on semiconductor surfaces. The ability to engineer the metal–semiconductor interface and precisely control ϕ_B prior to or during metal deposition is not yet possible. A standard, useful classification of metal–semiconductor contacts is given in Table 1.

Table 1
Classification of metal–semiconductor contacts

contact type	contacting surface	chemical bonding	contact formed
1	nonmetal is a weakly polarized insulator or semi-conductor	weak (physisorption)	Schottky barrier contact (ϕ_B is nearly proportional to the work function differences between metal and semi-conductor)
2	nonmetal is a highly polarizable semicon-ductor ($\varepsilon_S > 7$)	weak (no interfacial compound formed)	Bardeen barrier contact (ϕ_B determined largely by in-terface states and nearly inde-pendent of the work function difference)
3	nonmetal is highly polarized semi-conductor	strong (interfacial compound formed)	ϕ_B is still influenced by surface states, by these have been largely reduced in density because of the strong chemical bonding
4	nonmetal is a thin oxide	strong	ϕ_B determined largely by properties of the oxide layer

3. Surface Chemistry of Clean 6H-SiC

We examine properties of the free SiC surface relevant to metal–semiconductor con-tacts. Our focus will be on surface studies addressing the behavior of 6H-SiC as a func-tion of temperature. Elsewhere in this volume, Bermudez examines general properties of C-SiC surfaces, while Starke discusses the current, intensive effort to elucidate the atom-ic structure and reconstruction of SiC surfaces. A further article by Ueno considers the oxidation of SiC surfaces and its dependence on orientation.

3.1 High resolution microscopy of 6H-SiC surfaces

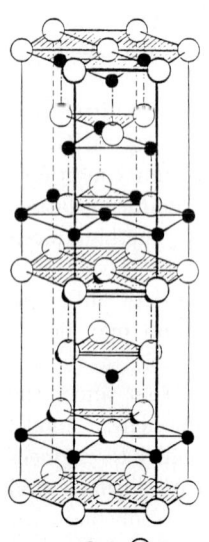

The ideal SiC(000$\bar{1}$) and SiC(0001) polar faces are terminated with a layer of C and Si atoms, respectively (Fig. 7). Producing an ideal SiC surface has proven difficult. High resolution studies using atomic force microscopy (AFM) on "as received" commercial CREE [30] and Epitronics [31] SiC wafers show an extremely rough surface (Fig. 8). Numerous polishing marks are observed, in addition to surface corregations ≈50 to 100 Å in vertical height. It is remarkable that current SiC contacts perform as good as they do give the extreme roughness of the surface.

● C ○ Si   Fig. 7. Crystal structure of the 6H-SiC polytype

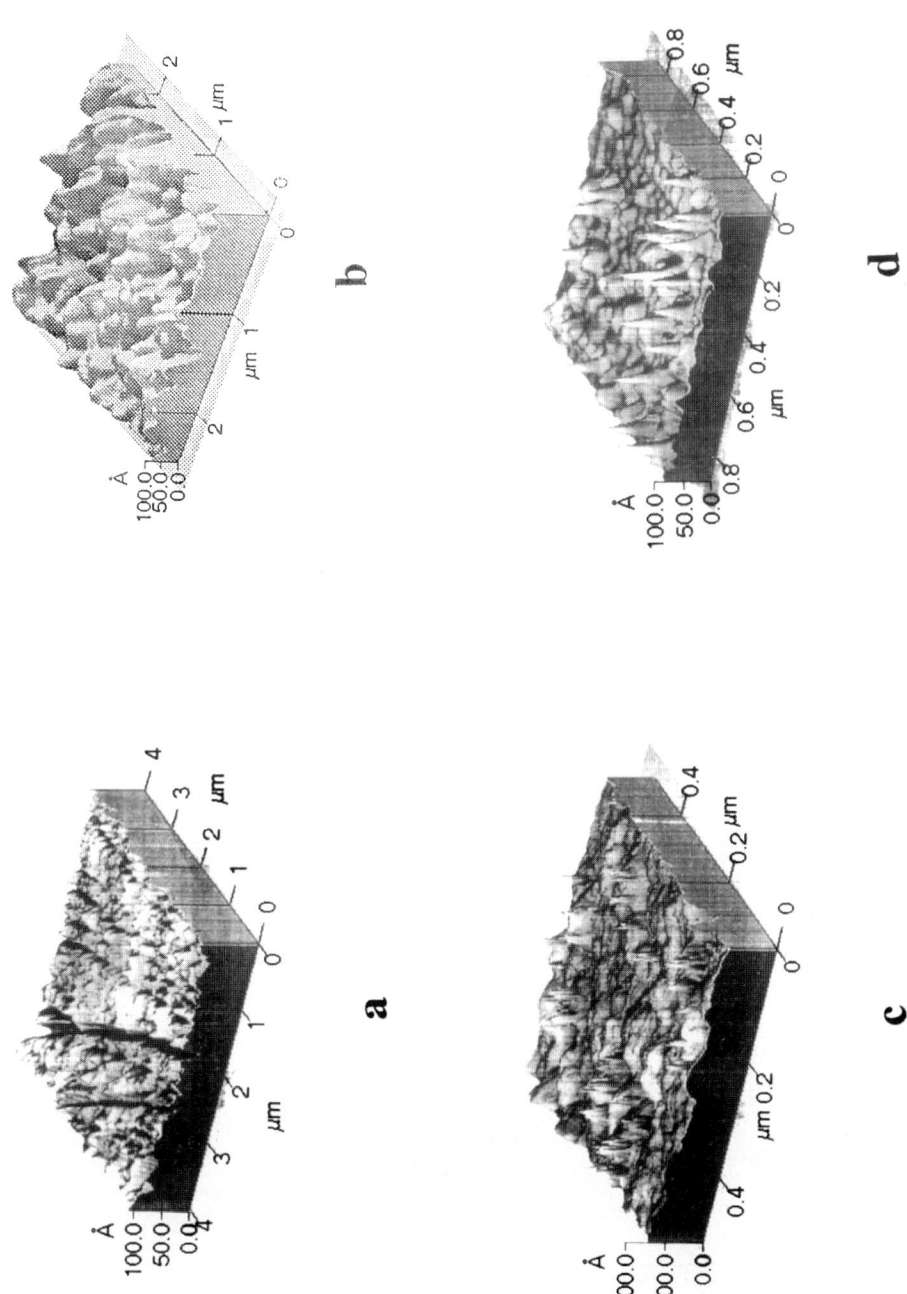

Fig. 8. Atomic force microscopy photographs of commercial "as-received" SiC. a) On-axis CREE 6H-SiC(0001) face, b) on-axis Epitronics 4H-SiC(0001), c) off-axis CREE 6H-SiC(0001)-2°, d) off-axis CREE 6H-SiC(0001)-2° with epitaxial SiC layer

It is frequently useful to verify the polarity of the 6H-SiC surface. There are three methods to distinguish the SiC(000$\bar{1}$)C-terminated surface from the SiC(0001) Si-terminated surface. First, the two polarities have differences in etch characteristics. Brack [32] established the correlation between the polar faces and their etch patterns by X-ray reflection intensity measurements. Small differences in X-ray reflection intensities are theoretically expected and experimentally observed on the opposite crystal faces. Etching SiC with Na_2CO_3 at 1173 K for 5 min yields distinct etch patterns. A hexagonally-pitted etch pattern belongs to the Si-rich surface and a wormy pattern to the C-rich surface. Various etching characteristics of SiC are discussed by Faust [33]. Second, the two polar faces of 6H-SiC are distinguished by differences in gross oxidation of the two faces. The C-rich face oxidizes faster than the Si-rich face [34 to 36] and contains a thicker oxide for equal oxidation times and pressures. Third, the two polar faces can be distinguished by using the thermal annealing properties of the SiC(000$\bar{1}$) and SiC(0001) crystal faces. Above 1300 K, the C face graphitizes at a higher rate than the Si face.

It should be emphasized that Brack's X-ray work is the only absolute determination of polarity. The chemical methods (etching, oxidation, and graphitization) are relative. They establish that the two faces have different properties but do not, by themselves, indicate which is Si- or C-terminated. Such methods must be traceable back to Brack's work via his molten salt etch and are dependent on the validity of his X-ray measurements. Unfortunately, it appears that no one has yet performed ion-scattering spectroscopy (ISS) on α-SiC, which would be a very direct method to determine polarity.

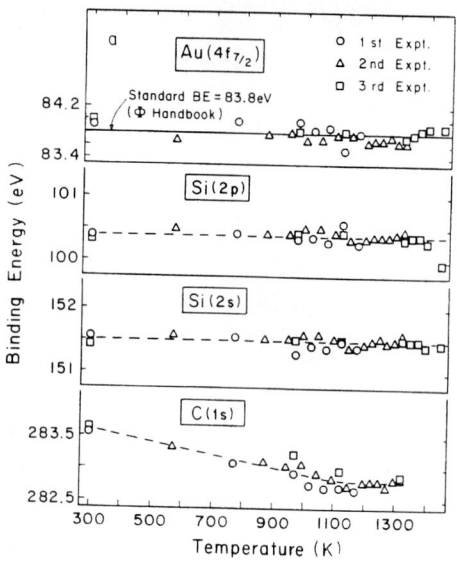

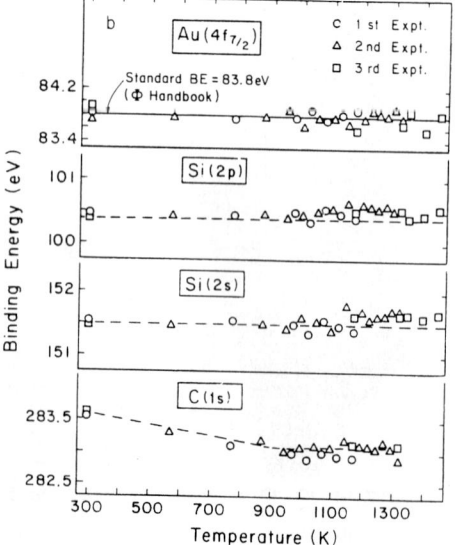

Fig. 9. Core-level binding energies for the a) 6H-SiC(000$\bar{1}$) C-rich and b) 6H-SiC(0001) Si-rich surfaces following thermal annealing. Data from Muehlhoff et al. [39]

3.2 XPS core-level binding energies versus temperature for SiC

Fig. 9 shows core-level binding energy (E_B) measurements for SiC(000$\bar{1}$) and SiC(0001) surfaces as a function of annealing temperature in ultrahigh vacuum (UHV). For comparison, the Au($4f_{7/2}$) reference level measurements are also shown. The Au data have been used to correct the SiC binding energies for small variations in the spectrometer effective work function and specimen charging. Both the Si(2p) and the Si(2s) binding energies are independent of annealing temperature to within ±0.2 eV in the range 300 to 1400 K, whereas the C(1s) binding energy decreases by 0.5 to 1.0 eV. The decrease in E_B for C(1s) is greater for the C-rich SiC(0001) face ($\approx$1.0 eV). This suggests that the chemical environment of the Si remains constant during annealing, while changes occur in the C environment due to the onset of graphitization. The average Si(2p) and C(1s) binding energies agree with recent, careful measurements on CREE material by Bozack [37].

3.3 Surface graphitization versus temperature for SiC

Both SiC(000$\bar{1}$) and SiC(0001) surfaces graphitize with rising temperature due to preferential volatility of silicon. Although different graphitization behavior for the two faces has been known since the late 1970s, careful investigations by Bozso et al. [38] and Muelhoff et al. [39] using surface science techniques have addressed graphitization in SiC as a function of temperature.

Fig. 10 shows the results of XPS measurements made of the C(1s)/Si(2p) integrated intensity ratios for 6H-SiC as a function of annealing temperature in UHV. A clear difference in the thermal behavior of the two SiC faces is observed which is also observed in Auger electron spectroscopy (AES). Four temperature ranges are discernable: 1. a low-temperature region ($<$900 K) exhibits no evidence for C-surface segregation on either SiC face; 2. an intermediate temperature region (900 to 1100 K) involves preferential C-surface segregation for the C-rich SiC(000$\bar{1}$) face, while the Si-rich SiC(0001) face exhibits little or no change; 3. a high-temperature region (1100 to 1300 K) again exhibits stabil-

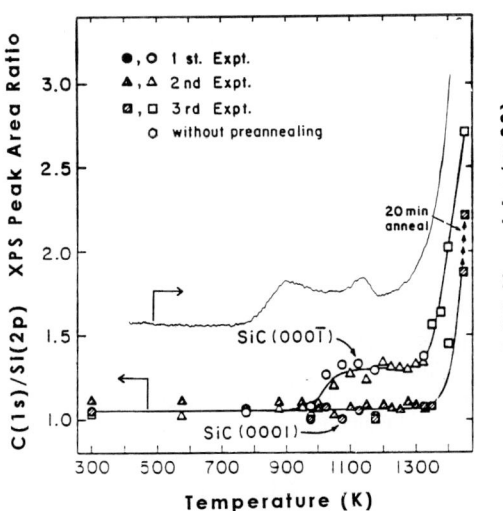

Fig. 10. Relative carbon surface concentration on 6H-SiC(000$\bar{1}$) and 6H-SiC(0001) with annealing temperature as measured by XPS. The right axis shows the volatility of Si from 6H-SiC(0001) over the same temperature interval. Data from Muehlhoff et al. [39]

ity of the C/Si ratio on both SiC faces; 4. a very high-temperature region (1300 to 1450 K) involves massive surface segregation of C to the surface of *both* SiC faces, with the C face exhibiting a higher C-surface enrichment. The C-surface segregation in the temperature range (1300 to 1450 K) is due to preferential volatility of silicon, as verified in mass spectrometer studies of a freely vaporizing SiC crystal. The C segregation in the intermediate temperature ranges is complex. It is probably due to a combination of carbon segregation and surface reconstruction at the high temperatures.

The graphitized layer on SiC is thin. On the assumption that the C segregation observed by XPS and AES is attributed to a surface graphite layer, the thickness of the graphite layer can be estimated from the attenuation of the Si signal. On the C-rich surface, the attenuation of the AES Si(LVV) signal at 1200 K corresponds to a 1.3 Å thick graphite layer, while at 1400 K, it corresponds to a 6 Å thick graphite layer. On the Si-rich surface at 1400 K, the thickness is 9 Å at 1400 K.

Surface graphitization of SiC at elevated temperatures is also observed in XPS C(1s) lineshape studies versus temperature (Fig. 11). The results show that 6H-SiC exhibits

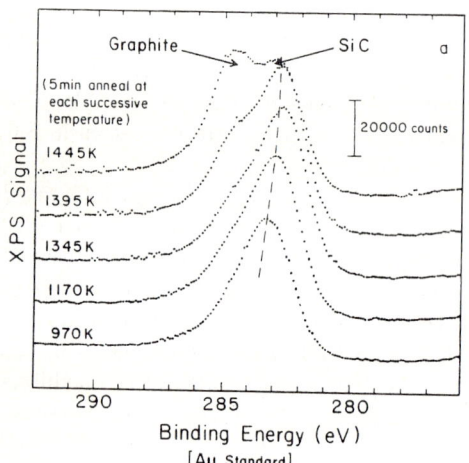

different behavior indicative of different chemistry for the two faces. In the range 900 to 1100 K, the C(1s) binding energy shifts by ≈ 0.3 eV to lower binding energy on the SiC(000$\bar{1}$) C-rich surface but does not shift on the SiC(0001) Si-rich surface. Simultaneously, a low-intensity C(1s) feature at higher binding energy is seen when the C-rich surface is annealed. Annealing the crystal at temperatures above 1100 K increases the intensity of the XPS peak at higher binding energy (284.6 eV) on the C-rich surface. Pate et al. [40] reported the graphite C(1s) XPS binding energy to be 284.7 eV, while Rogers et al. [41] reported it to be 284.6 eV. Based upon this assignment, the C(1s) feature at 284.6 eV is at-

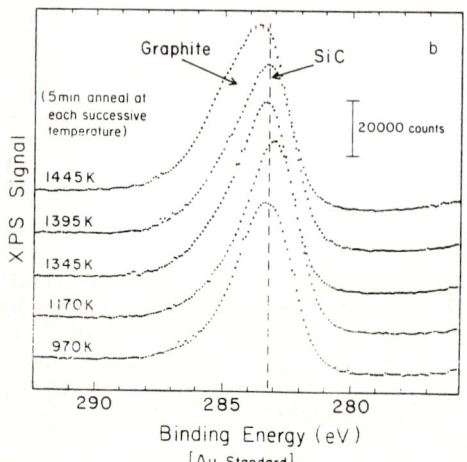

Fig. 11. C(1s) XPS lineshape for a) 6H-SiC(0001) C-rich and b) 6H-SiC(0001) Si-rich surfaces versus annealing temperature. Data from Muehlhoff et al. [39]

tributed to graphite formation on top of the SiC. On the Si-rich surface. the graphite peak is only observed above 1400 K, and its intensity is much less when the two crystal faces are compared after identical annealing conditions.

3.4 Plasmon loss features for SiC

Bulk plasmon excitation [42 to 44] by X-ray and Auger electrons is experimentally observed in the loss structure of surface electron spectroscopies. Bozso et al. [38] and Ghamnia et al. [45] found direct evidence for the presence of a bulk plasmon feature at 22.5 eV in SiC which has proven useful in studies of SiC growth and graphitization. The corresponding graphite loss features occur at 6.5 and 27 eV. Electron energy loss (ELS)

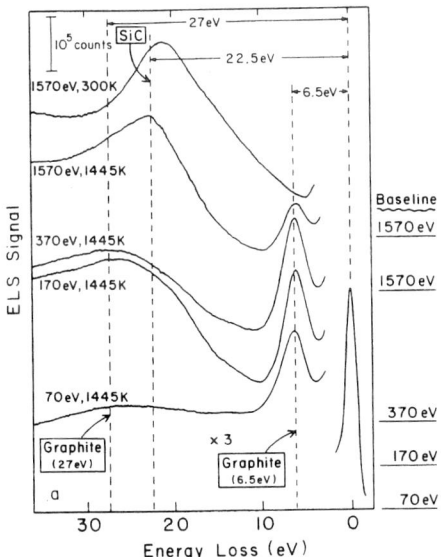

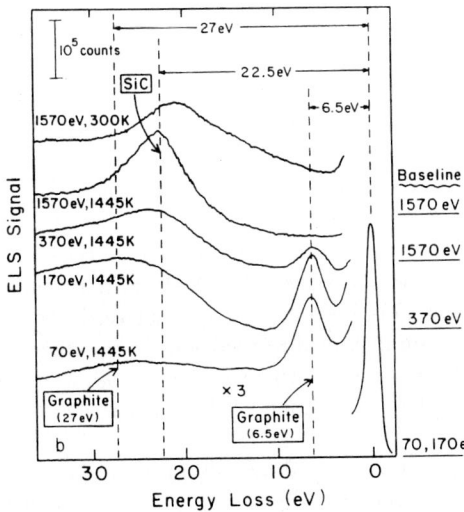

studies of the SiC faces after annealing above ≈ 900 K confirm the presence of surface graphite (Fig. 12). In ELS, an incident electron beam loses energy to excite plasmon oscillations in the solid. ELS spectra of the annealed SiC at 1445 K are shown in Fig. 12 as a function of the primary electron energies 70, 170, 370, and 1570 eV. The higher primary electron energies probe deeper surface layers, and an approximate estimate of the depth profile of different species in the surface region can be made [46].

Two carbon species are present on the annealed SiC surfaces. By comparing the intensities of the 6.5 and 27 eV graphite loss features with the 22.5 eV SiC-loss feature as a function of primary electron energy, we conclude that graphite has segregated to the surface. The 70 eV primary electron energy spectrum probing a thin surface layer indicates no difference between the two SiC faces. The 6.5 and 27 eV graphite loss peaks are the dominant loss features in the spectra. With increasing primary electron energy probing deeper layers, the graphite loss diminishes and the 22.5 eV SiC bulk plasmon loss

Fig. 12. Energy loss (ELS) spectroscopy of a) 6H-SiC(0001) C-rich and b) 6H-SiC(0001) Si-rich surfaces versus primary electron energy. Data from Muehlhoff et al. [39]

dominates the spectrum. The C-rich face shows a thicker graphite layer than the Si-rich face, based on the observation that graphite loss features are observed by ELS with 1570 eV primary electrons on the C-rich side but not on the Si-rich side.

3.5 Work function and electron affinity for SiC

Two properties of a semiconductor which are important in terms of contacts are the work function and electron affinity. The work function (ϕ) is a macroscopic electronic property of a surface defined as the difference in energy between an electron at rest in the vacuum just outside the solid and the most loosely bound electrons inside the solid.

There are several physical factors which contribute to ϕ. First, the value of ϕ depends on the depth (W) of the attractive potential for the conduction electrons inside the solid. This is a bulk property determined by the attraction of a solid's electrons with the lattice of positive ions as a whole. This contribution to W depends on the type and arrangement of positive ions in the lattice, and yields an energy of the order of a few eV.

A surface contribution to W is the strength of a surface electrostatic field, often called the double layer. Because surface atoms are unbalanced (they have matter on one side and not on the other), the electron distribution around them will be asymmetrical with respect to the positive ion cores. Two important effects of this layer is that the work function is sensitive to surface contamination and the exposed crystallographic face. Surface contamination will affect ϕ because it modifies the layer in a way depending upon the affinity of the contaminant for electrons. As the electron affinity of atoms depends on their type, so ϕ will vary according to the type of contaminant. Further, the orientation of the exposed double layer depends on the density of positive ion cores, which in turn varies from one crystal face to another. The contribution of the double layer to ϕ is of the order of a few tenths of an eV. Experimental results for the dependence of ϕ on the crystal face of tungsten range from 4.39 eV for W(111) to 4.69 eV for W(112). The smallest values of ϕ are associated with the least densely packed face. The rearrangement of electron density on the least densely packed face produces a net positive charge on the vacuum side of the surface and a net transfer of electrons to the inside. This lowers the work function.

The work function will also change if the adsorption of atoms or molecules results in the transfer of charge to or from a surface. The adsorbed species may be polarized or ionized by the attractive interaction with the solid surface. If the adsorbed atoms are ionized and transfer electrons into the surface, the work function decreases. Conversely, the formation of adsorbed negative ions increases the work function. More frequently, the adsorbed atoms or molecules are merely polarized by the attractive surface forces and may be viewed as dipoles aligned perpendicular to the surface. If the adsorbed layer is polarized with the negative pole outward, the work function increases due to the lowering of the free electron density by charge transfer and charge localization in the adsorbed layer. Conversely, if the positive pole of the polarized adsorbate is outwards, then the work function will decrease. As long as the number of adsorbed ions (or polarized atoms or molecules) is low enough for interactions between them to be negligible, the change in the work function will be proportional to the number of ions adsorbed. A simple physical explanation follows.

The adsorbed molecules have a dipole moment which modifies the total dipole at the surface and changes the work function. A dipole layer of σ dipoles/area, all identically

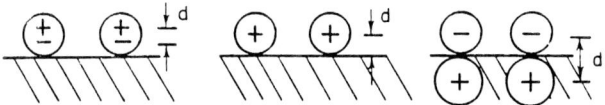

Fig. 13. Generalized diagram of charge separation for different types of adsorbate bonding. Left part: van der Waals, middle part: ionic, right part: covalent

oriented normal to the surface, can be treated in first approximation as a parallel plate capacitor with plate separation d. The plates then have a uniform charge density σe and the product of d and the electric charge e is the effective dipole moment μ_e. The external field of this layer is zero and the change in work function $\Delta\phi$ is equal to the change in surface potential $\Delta\chi$. Thus, $\Delta\phi = \Delta\chi = 4\pi P$, where P is the polarization (dipole moment/area). Since $P = \pm\sigma\mu_e$, then $\Delta\phi = +4\pi\varrho\mu_e$. Writing the coverage $\theta = \sigma/\sigma_0$ (σ_0 is the number of adsorption sites), then

$$\Delta\phi = \pm 4\pi\theta\sigma_0\mu_e \ . \tag{9}$$

In words, this states that the change in work function depends on the average dipole moment per atom and the number of molecules per square centimeter adsorbed on the surface. We note that the sign (+ for negatively charged ions) of μ_e is dependent on the charge q of the adsorbed molecules and the separation d of the two charge layers (i.e., $\mu_e = qd$). The three cases corresponding to van der Waals', ionic, and covalent bonding are shown in Fig. 13. An example of van der Waals' adsorption is inert gas adsorption where little, if any, charge transfer with the substrate occurs and the dipole moment is contained within the adsorbate. Cesium adsorbed on refractory metals is an example of ionic adsorption which involves charge transfer to the substrate. Covalent adsorption involves localized orbital overlap and charge transfer between adsorbate and surface atoms (e.g., oxygen on metals).

It usually turns out that at values of θ close to monolayer coverage, μ_e itself is dependent on the coverage since interaction with the surrounding adsorbate molecules affects the dipole contribution from any given adsorbed molecule. The magnitude of the dipole moment per adsorbed atom varies due to a reduction in μ_e with decreasing lateral spacing (i.e., increasing θ). Although several physical mechanisms may contribute to the coverage dependence of μ_e, the most common explanation is based on the Topping [47] model, which assumes mutual depolarization of adsorbed particles. This leads to $\mu_e(\theta)$ given by

$$\mu_e(\theta) = \mu_{e0}[1 + 9\alpha(\theta\sigma_0)^{3/2}]^{-1} \ , \tag{10}$$

where μ_{e0} is the dipole moment per adsorbed particle at $\theta = 0$ and α is its effective polarizability. Substituting (10) into (9) predicts the well-known maximum or minimum in the $\Delta\phi$ versus θ curve. Only at very low coverages will $\Delta\phi$ be linear in coverage.

Both hydrogen and oxygen adsorption on metal surfaces appears to increase the work function. The chemisorption of π-bonded organic molecules (olefins, for example), decreases the work function. Co, on the other hand, increases the work function of some metals (Fe, Ni, Co, W, Pt) but decreases the work function of others. It is possible to change the work function of metals by ≈ 1.5 eV in either direction. Currently, there is little data on the change in work function with adsorbed species for SiC surfaces.

Pelletier et al. [48] have measured the work function of n- and p-type 6H-SiC as a function of temperature between 300 and 600 K using the Shelton retarding field method. The crystals were grown by the Lely technique, with the n-type material doped with N and the p-type material doped with Al. The surfaces were carefully cleaned to avoid the introduction of extrinsic surface states caused by impurities. However, the normally Si-terminated SiC(0001) surfaces were heated to $T \leq 1230\,°C$, and, since surface graphitization occurs at $\approx 900\,K$ for both polar surfaces of 6H-SiC, the surfaces were largely graphitized.

The difference in work functions between n- and p-type samples (4.75 and 4.85 eV, respectively) was small, and much lower than the Fermi level variation in the gap. Further, the work function for each dopant type remained constant within the temperature range. The results led Pelletier et al. [48] to conclude that the Fermi level is pinned at the surface due to intrinsic surface states. However, if the surfaces were graphite-terminated, it is likely that the surface states were extrinsic in nature.

Using the method of Frese [49], where $\chi_S = \phi_S - 0.5 E_g$, the room temperature electron affinity was calculated by Pelletier to be 3.7 eV. However, using the work function value of 4.80 eV and subtracting 1.5 eV, which is half of the band gap of 6H-SiC, one obtains a value of 3.3 eV. Because the 3.3 eV value follows from a measured work function value, it is frequently quoted as the electron affinity of 6H-SiC. An error in χ_S simply shifts all of the predicted Schottky barrier heights by the same amount and does not affect the *slope* of ϕ_B versus ϕ_M. While the room temperature work function was not measured for 3C-SiC, the room temperature electron affinity was reported to be 4.0 eV. Using the method above, this corresponds to a work function of 5.2 eV for β-SiC.

Kennou [50] has studied the adsorption of Er on 6H-SiC(0001) and concludes that deposition of $>15\,\text{Å}$ of Er at room temperature gives a Schottky barrier height of 1.40 eV. The work function value of clean 6H-SiC (4.5 eV) is reduced to $\approx 2.9\,eV$ after 5 Å of Er deposition beyond which the Schottky barrier begins to develop. Upon annealing to a temperature $>650\,K$, the Schottky barrier height increases and reaches 1.80 eV at 900 K due to an interfacial reaction which leads to silicide formation at the interface.

Van Elsbergen et al. [51] have investigated the adsorption of Cs on 6H-SiC(0001) and find that the Fermi levels is pinned at 1.2 eV above the valence band maximum. At 130 K, Cs grows layer by layer. The films become metallic after the deposition of the first Cs layer. For submonolayer coverages, Cs-induced surface donors form at 2.96 eV above the valence band maximum. They are due to covalent Cs–Si bonds. The barrier height of Cs/6H-SiC Schottky contacts was found to be 0.57 eV with n-type and 2.28 eV with p-type doped specimens. The results support the notion that metal-induced gap states determine the barrier height.

3.6 Sputtering of SiC surfaces

Ion bombardment is a useful technique for modifying, cleaning, and sputtering surfaces. Because the surface structure and composition is of great importance for epitaxial growth, polytype control [52, 53] and the quality of semiconductor devices, the effect of Ar^+ sputtering of SiC surfaces is important.

Ion bombardment affects the crystal surface in different ways. The first effects is removal of material by sputtering. Sputtering yields of SiC under Ar^+ bombardment and for selected sputtering angles have been determined in several studies [54 to 56]. Pezoldt et al. [57] determined SiC sputter yields for Ar^+ as a function of ion energy, shown in Fig. 14. There is an expected, continuous rise in the sputtering yield with increasing ion

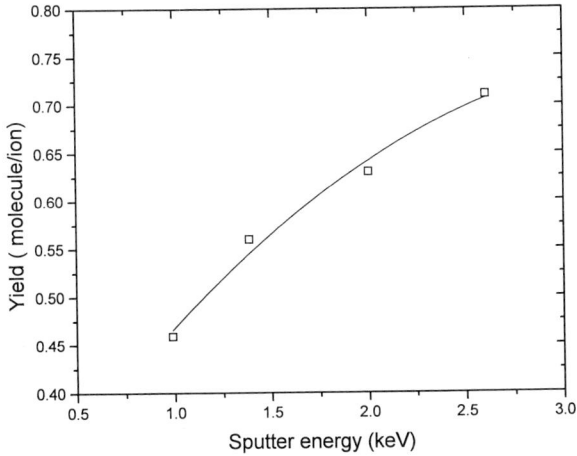

Fig. 14. Sputter yield versus Ar$^+$ ion energy for 6H-SiC. The sputter yield was determined by measuring the step height of mesa structures with a microinterferometer. Data from Pezoldt et al. [57]

energy. Within experimental error ($\pm 10\%$) of the experiments to date, there is no observed dependence on either the polarity or the polytype of the SiC surface. It is likely that ion bombardment disorders the surface region and the structural properties relate more to amorphous SiC than to the original ordered monocrystalline polytype substrate. This has been confirmed by Pezoldt et al. who observed by reflection high energy electron diffraction (RHEED) that, after sputtering at room temperature, a diffuse diffraction pattern was observed on SiC which corresponds to an amorphous surface. The composition near the surface can also be altered by diffusion below the surface owing to mobile defects and atoms created by ion impact (ion beam mixing). The effects result in an altered distribution of the two constituents near the surface compared with the original distribution. The energy loss of the sputtering ion can lead to structural changes from crystalline to amorphous.

The temperature dependence of the sputtering yield at 2 keV Ar$^+$ for 6H-SiC in the temperature range 20 to 1000 °C is shown in Fig. 15. A drop in sputtering yield is observed between 400 and 700 °C, which may be attributed to morphology or phase

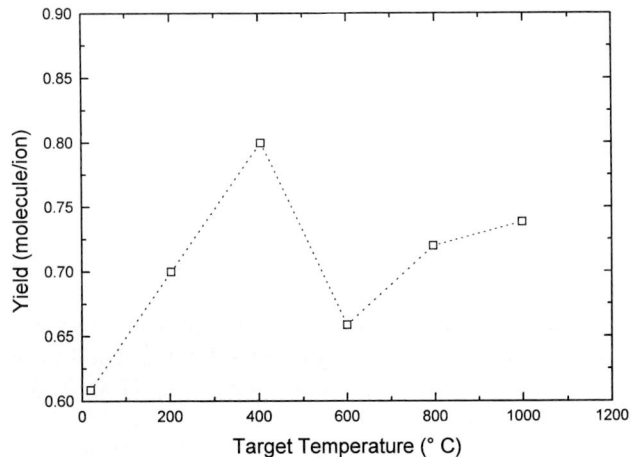

Fig. 15. Sputter yield versus surface temperature for 6H-SiC at an Ar$^+$ ion energy of 2 keV. Data from Pezoldt et al. [57]

changes due to surface recrystallization [58]. RHEED investigations after Ar^+ sputtering show that at room temperature, SiC has a thin amorphous layer on the top of the surface. At a substrate temperature of 200 °C, a partial phase transition of the type $6H \rightarrow 3C$ occurs, whereas at 400 °C a partial phase transition of the type $6H \rightarrow 15R$ occurs. At high temperatures, no change in the polytype structure is observed.

A second effect of ion bombardment is preferential sputtering, where one surface component is sputtered at a faster rate than others, leaving the surface enriched with the component with the smaller sputtering yield. There is currently conflicting evidence that SiC suffers preferential sputtering during low energy Ar^+ bombardment. In several studies [59 to 63, 38] it was pointed out that preferential sputtering does not occur on SiC surfaces under Ar^+ bombardment, but sputtering led to the same surface composition at the polar surfaces. For example, in studies by Muehlhoff et al. [62], the SiC surfaces were sputter cleaned at 300 K without subsequent annealing and both basal plane faces were compared by AES and XPS. A low surface concentration (~ 2 at% in the sampling depth of XPS) of embedded Ar is present and can be removed only by heat treatment above 1273 K. Within the experimental error ($\pm 5\%$ XPS; $\pm 10\%$ AES), the relative XPS and AES intensity ratios for C and Si were the same after sputtering both the SiC($000\bar{1}$) C-rich and SiC(0001) Si-rich faces. The XPS and AES peak energies on both crystal faces are identical for the sputtered SiC surface. Sputtering appears to restore stoichiometry and preserve short-range order of SiC with tetrahedral coordination, as the AES C(KLL) line shape observed is characteristic of SiC. Sputtering also appears to yield the same Si/C stoichiometry on both C-rich and Si-rich SiC surfaces as judged by XPS and AES intensity ratios.

In contrast, Dillon et al. [64] reported excess C on SiC surfaces cleaned by Ar^+ ion bombardment followed by 1273 K annealing. Later, Morgen et al. [65] showed that thin films of SiC_x ($x \approx 1$) grown by sputter deposition on a Si(100) substrate, heated to 1100 K, and subsequently sputtered with 1 keV Ne^+ ions were found to be homogeneous in the bulk and strongly C-enriched in a thin surface layer. Finally, Pezoldt et al. [57] using AES and computer simulation of Ar^+ sputtering as a function of ion energy at room temperature showed changes in surface composition during sputtering due to preferential sputtering of silicon. Thus SiC suffered a slight enrichment of C in the surface layer which depended on the energy of the impacting ions.

3.7 Etching of SiC surfaces

A clean, fast, reproducible, and reliable etch process is essential for fabrication of SiC devices. At the present time, reactive ion etching (RIE) processes are being refined for the various polytypes of SiC. Most research efforts indicate slow etch rates without oxygen. Yih and Steckl [66] have obtained etch rates on 6H-SiC of ≈ 32 Å/min in pure CHF_3. After adding a mixture of H_2 and O_2 to the CHF_3 source gas, the etch rate increased to 263 Å/min. To avoid the use of H_2, Yih and Steckl [67] later tried fluorinated plasmas combining NF_3, CF_4, and SF_6 in varying fractions.

Casady et al. [68] have devised a RIE process for 6H- and 4H-SiC utilizing a source gas of pure NF_3. The process yields a relatively high etch rate of ≈ 1500 Å/min and is free of etch residues. It utilizes a self-induced DC bias ranging from 25 to 50 V, a forward RF power of 275 W (1.7 W/cm^2), chamber pressure of 225 mT, and a NF_3 flow rate between 95 and 110 sccm. A peak etch rate of ≈ 2000 Å/min was obtained at

375 W, but surface cleanliness and smoothness were compromised at the higher power. The etch rate ranged between a minimum of ≈ 1200 Å/min at 75 mT pressure, to a peak of ≈ 1800 Å/min at 325 mT.

3.8 SiC surface preparation and cleaning

We review methods to prepare a clean SiC surface. The condition of the surface before metal deposition is crucial for contact performance. Surface roughness, reconstruction, and native oxide layers can dramatically affect the Schottky barrier height. Preparation of a chemically-clean, atomically-ordered SiC surface is nontrivial and a major source of irreproducibility in contact performance.

Van Bommel et al. [69] first investigated the surface structure of crystalline 6H-SiC as a function of temperature. By using low energy electron diffraction (LEED), they found that, at 250 °C, the Si-terminated SiC(0001) surface exhibited a $(\sqrt{3} \times \sqrt{3})$ LEED pattern while the C-terminated SiC(0001) surface exhibited a (1×1) pattern. Between 800 and 1000 °C, a $(6\sqrt{3} \times 6\sqrt{3})$-R 30° pattern developed on the (0001) face. At 1500 °C, the (0001) surface exhibited a (1×1) graphite phase. Correspondingly, the SiC(0001) face showed (2×2) or (3×3) patterns after 800 and 1000 °C. After 1500 °C, graphite rings appeared on SiC(0001), indicative of polycrystalline graphite. AES studies after annealing at 800 °C showed carbidic C(KVV) peaks. At 1000 °C, the AES C/Si ratio increased, and C(KVV) changed from carbidic to graphitic, [70], indicating the formation of graphite.

Van Bommel et al. attributed the LEED results to surface reconstruction. The $(6\sqrt{3} \times 6\sqrt{3})$-R 30° pattern observed at 250 °C was due to a SiC surface with an overlayer of crystalline graphite which was rotated 30° with respect to the underlying SiC substrate. The indication is that graphite begins to form on both 6H-SiC surfaces at temperatures as low as 800 °C. The presumed mechanism consists of three collapsing C layers to form one graphite layer due to preferential evaporation of Si.

Workers have since experimented with methods to clean the SiC surface without forming a graphite surface layer. Kaplan and Parrill [71] heated β-SiC(100) and 6H-SiC (0001) and SiC(0001) samples at successively higher temperatures in the presence of Ga flux to volatilize the SiC surface oxide. The oxide was reduced at temperatures as low as 850 °C, but complete removal was very slow at temperatures below 1050 °C. The temperature necessary to completely remove oxygen also increased the C/Si ratio. After annealing at 1185 °C for 3 min, the C/Si ratio increased dramatically and the surface formed graphite.

Standard chemical cleaning procedures do little to remove the native oxide on SiC. After most chemical cleaning procedures, a SiC surface contains ≈ 10 at% oxygen. Waldrop and Grant [72] have calculated the amount of native O on SiC to be ≈ 0.75 monolayer. To eliminate the native oxide, Kaplan [73] heated SiC in the presence of an elemental Si flux, which removed the surface oxide as volatilized SiO. Faint satellite spots appeared with a small amount of oxide still present. After heat treatment at 850, 950, and 1000 °C, the respective LEED patterns were (3×3); $(\sqrt{3} \times \sqrt{3})$-R 30°; and (1×1). The Si surface concentration decreased with temperature. Electron energy loss spectroscopy (EELS) revealed the beginning of graphite segregation at 950 °C. The (3×3) phase was attributed by Kaplan to coincide with an adsorbed Si bilayer with a high vacancy concentration. Annealing at 950 °C depleted the surface Si and caused the surface reconstruction to change to a $(\sqrt{3} \times \sqrt{3})$-R 30° phase. Annealing at $>1000\,°C$ resulted in further depletion of surface Si and the development of a (1×1) pattern. By comparison, the work of van Bommel et al. exhibited a direct change from $(\sqrt{3} \times \sqrt{3})$-R 30° to a graphitic state without first exhibiting a (1×1) phase. As the Si depletion increased, the amount of graphitic C increased. The appearance of the (1×1) phase was attributed to the formation of defects resulting from the elimination of the Si adlayer.

Bozso et al. [38] investigated the SiC(0001) Si-terminated surface using XPS and AES. The crystal was cleaned by a combination of flow energy Ar$^+$ sputtering and high temperature annealing. Low angle (80° incidence angle) sputtering was employed to minimize implantation of both Ar and residual gas atoms to a depth which prevented desorption during annealing. After 30 min of low angle sputtering, small concentrations of Ar, F, and O remained at the surface. The F and O was removed by suddenly heating the crystal to 1000 °C for 2 min. The Ar which remained was eventually eliminated after further low angle sputtering and heating cycles. An atomically-clean surface was achieved and preferential removal of Si or C was not observed.

Andreev et al. [74] attempted to clean oxidized 6H-SiC(0001) and 6H-SiC(0001) by electron bombardment rather than ion bombardment, using 640 eV electrons at a flux of $(2 \text{ to } 5) \times 10^{17}\,e/(cm^2\,s)$. The sample temperature was between 700 and 1000 °C. Electrons rather than ions were employed because of their lower tendency to damage the surface. AES analysis showed an uncontaminated, stoichiometric surface. However, graphitization was observed to occur when the procedure was continued after the oxide was removed. Although evidence was not given, the surface was reported to have been undamaged. This cleaning procedure may be useful if a low energy source of electrons is available and the crystal temperature carefully controlled.

Porter et al. [75] have employed a chemical and thermal cleaning process for SiC which is easily implemented in manufacturing environments. A thermally-oxidized 6H-SiC(0001) surface is etched in a 10% HF aqueous solution. The resulting surface contains hydrocarbons evidenced by a small peak component on the high binding energy side of the C1s XPS peak. After heating the samples in a vacuum ($\approx 10^{-7}$ Pa) at 700 °C for 15 min, the hydrocarbons were eliminated, while residual O and F were still present. Porter et al. also used XPS to determine the amount of band bending before and after deposition of various metal contacts. After the cleaning steps an upward band bending of a few tenths of an eV was calculated, indicating the presence of surface states.

In our laboratory, we routinely employ a cleaning procedure for SiC surfaces which exploits the graphitization of SiC at elevated temperatures. The specimen is annealed in UHV at high temperature, allowing the surface to graphitize, and then the thin graphitized layer is sputtered away using a standard Ar$^+$ sputter ion gun. The procedure is simple, results in an atomically clean SiC surface with no oxygen, preserves the proper Si/C ratio at the surface, and minimizes the amount of implanted argon. No ex-situ chemical cleaning procedures are required. The three steps are: 1. anneal the crystal at 1300 K for 10 min, allowing the surface to graphitize; 2. reduce the crystal temperature to 700 K (below the graphitization temperature) and sputter the surface using 5 keV Ar$^+$ for 5 min, 3 keV Ar$^+$ for 10 min, and 1 keV Ar$^+$ for 10 min; 3. anneal the crystal at 700 K for 10 min (minimizes sputter damage and desorbs implanted Ar). Based on the SiO$_2$ sputtering rate of our Ar$^+$ ion gun, the sputtering step removes ≈ 100Å of the surface. Possible surface damage due to the sputtering process has not been investigated in depth. The resulting state of the surface should be further investigated with LEED and

STM before this preparation technique is employed for metal contants. If surface damage has been eliminated by heating, this technique may be a viable solution to the surface cleaning problem.

3.9 Surface states on SiC

Surface states are extra allowed states for electrons that are present at the semiconductor surface but not within the bulk. They are traveling wave solutions of the Schrödinger equation where electron waves travel parallel to the surface but not into the bulk. If charge is situated in the surface states it will result in electrostatic fields penetrating into the solid and create a varying electrostatic potential on passing from the surface to the bulk. The varying potential distorts the band structure of the bulk solid and results in band bending.

Band bending arises because an energy barrier exists at a cleaved semiconductor surface. An electron from the interior of the semiconductor must overcome this energy barrier to come to the surface. To see how the energy barrier is created, consider the semiconductor immediately after it is cleaved. At this moment there will be a deficit of electrons at the surface because the two surfaces are created by cleavage and on average each surface contains half the electrons originally in the cleavage plane. As a result of the electron deficit at the cleaved semiconductor surface, electrons will diffuse from the interior of the semiconductor to the surface where they become captured by the surface states. The charge in the surface states sets up an electric field that repels electrons diffusing from the bulk to the surface. In the equilibrium conditions, a potential barrier exists at the surface which repels electrons that approach the semiconductor surface.

Surface states arise from a number of sources. They are associated not only with the sudden termination of potential at a perfect clean bulk exposed plane but also with changes in surface potential due to relaxation, reconstruction, structural imperfections, or adsorbed impurities. It is well known that surface states occur on many clean semiconductor surfaces and are associated with the termination of the periodic bulk lattice potential and to the unsaturated or "dangling" bonds at the surface. The surface states affect the electrical properties of a surface by acting as a source (or a sink) of electrons.

Consider the clean surface states associated with atoms of the host lattice. There will be extra states on the surface because electrons in the surface region are bonded only from the side directed toward the bulk, and the characteristic energies for electrons at such sites will differ from the characteristic energies in the bulk. Surface states of this type are called Tamm or Shockley ones. The density of the Shockley-Tamm states in a particular semiconductor is of the order of the density of atoms at the surface, or roughly $N_0^{2/3}(\text{cm}^{-2})$ where N_0 is the bulk atomic density (atoms cm^{-3}). For Si and SiC, $N_0 \approx 5 \times 10^{22}$ cm^{-3} and the density of the Tamm-Shockley states is about 10^{15} cm^{-2}. The energetic distribution of these states is not well established, although studies for many III–V materials indicate their density to be peaked roughly one-third of the forbidden-gap energy above the valence band. Bonded contaminant atoms at the surface or surface crystal defects are sources for other types of surface states, often called interface states. The detailed origin of the interface states and the way they relate to the microscopic structure of the interface region is currently being worked out.

It is difficult to make quantitative estimates of band shapes due to charge localized in surface states. If the free carrier density in the material is very high, then compensation of the charge at the surface by flow of carriers out of the bulk is effective (the conduction band electrons screen the surface charge). Thus, metals tend to have negligible band bending. In semiconductors and insulators, however, complete compensation is not possible because of the much lower free-carrier densities, and the existence of occupied surface states results in surface potentials of the order of mV or V, with band bending effects extending some microns into the solid.

The presence of surface states modifies the simple Schottky contact theory. If the semiconductor surface states are neutral or if they change their charge state when the contact is formed, then the charge configuration obtained by applying Schottky theory is not valid. Any electronic states clustered near the Fermi level will cause the Fermi level to be pinned when the state density becomes large because slight changes in Fermi energy would result in very sizeable charge transfer. A filled surface state band leads to Fermi level pinning in p-type materials, whereas an empty band causes pinning in n-type materials.

The surface state population on SiC is currently the subject of active investigation. Evidence for the absence of surface states on SiC has been reported by Parrill and Bermudez [76] who determined both surface and bulk valence band photoemission spectra for SiC. The bulk valence band was studied using high energy photons (MgKα, 1253.6 eV, ≈7 monolayer attenuation length) while the surface valence band was investigated using low energy photons (ZrMζ, 151.4 eV, ≈2.5 monolayer attenuation length). Thin platelets of α-SiC of unspecified polytype and undetermined termination were cleaned in acetone, methanol, concentrated HF, and methanol followed by annealing at 850 to 950 °C in a Si flux. This procedure removed the native oxide but the surfaces contained ≈0.8 monolayers of excess silicon as measured by AES. The excess Si was removed by annealing for 1 to 2 min at temperatures ranging from 960 to 1160 °C. After annealing, the surfaces showed a (1×1) LEED pattern. Subsequent heating at ≈1160 °C yielded a (3×3) reconstruction and no graphite formation according to AES. The photoemission results showed no states within the optical band gap.

On the other hand, work by Verenchikova et al. [77] indicates that there is some pinning due to intrinsic C vacancies, which vary in concentration with polytype. The experimentally-determined barrier heights of Cr metal on several polytypes compared favorably with the theoretically predicted behavior of C surface vacancies for several hexagonal and rhombohedaral polytypes of SiC. As the concentration of vacancies increased, the barrier height decreased. Presumably, the carbon vacancies form an energy level in the upper half of the band gap and a high concentration of carbon vacancies causes a shift of E_F towards E_C.

Apparently, certain adatom termination (e.g, Si or O) on SiC results in a low or absence of surface states within the band gap. In the case of a submonolayer coverage of oxygen, there appears to be a sufficient density of surface states to reduce the index of interface behavior from 1.0 but which is insufficient to completely pin the Fermi level. This will be discussed in more detail in the next section. Syrkin et al. [78, 79] compare experimental and theoretical Schottky barrier heights for several metals (Al, Au, Mo, Cr) on 6H-SiC and estimate the dependence of the barrier height on surface state density, pinning, metal work function, and the concentration of uncompensated donors. The average surface energy level for 6H-SiC(0001) was found to be 0.3 E_b at room temperature (E_b is the band gap) with a corresponding surface state density of ≈10^3 cm^{-2} eV^{-1}. Davidov and Tikhonov et al. [80] use the model of adatom-induced surface states to explain the formation of the Schottky barrier of 6H-SiC. The positions of local levels induced by group I and II metal atoms adsorbed on the surface are calculated in the strong-coupling approximation. The barrier height is predicted to depend on the ionicity of the semiconductor.

3.10 Index of interface behaviour for SiC

We review several studies to determine whether contacts to SiC follow Schottky behavior, indicated by a near-unity index of interface behavior. Mead and Spitzer [81] report barrier heights for metals on SiC and other semiconductors nearly independent of work function. For semiconductors where both vacuum cleavage and chemical preparation could be employed, the barrier heights showed no appreciable dependence on the method of surface preparation.

Hagen [82] also found ϕ_B on terraced α-SiC surfaces to be independent of ϕ_M. The barrier height on n-type samples was 1.45 eV for Au, Ag, and Al despite the large variation in work function between the metals (Au: 5.28 eV; Ag: 4.31 eV; Al: 4.20 eV). In addition, ϕ_B was the same for both 6-H and 15R-SiC polytypes. The results indicated that the Fermi level is fixed by surface states at the middle of the band gap, which agreed with Mead's conclusion that surface states are present in predominantly covalent semiconductors such as SiC. In related work, Wu and Campbell [83] studied Au contacts on the 6H-SiC(0001) C-terminated surface and measured a barrier height of 1.45 eV, in agreement with Hagen for Au on the SiC(0001) Si-terminated surface.

Kosyachenko et al. [84] have reported that sputter-deposited Pt, Au, Ag, Ni, Cr, and Al produce rectifying contacts on 6H-SiC. Depending on the metal and the surface preparation, the band bending at the semiconductor surface ranged from 0.6 to 2.5 eV. Porter and coworkers [85 to 88] studied room temperature, UHV deposited Ti films on epitaxial n-type 6H-SiC(0001). The contacts were rectifying. Surfaces were cleaned in a solution of ethanol/hydrofluoric acid/deionized water (10:1:1) following by a 700 °C thermal anneal in UHV. Titanium was chosen because both Ti ($a = 2.95$ Å, $c = 4.68$ Å) and 6H-SiC ($a = 3.08$ Å, $c = 15.11$ Å) are hexagonal and have a −4% lattice mismatch between close-packed planes. XPS studies of unannealed, thin (≈10 Å) Ti films showed Ti–C bonding at the Ti/SiC interface. Thicker (1000 Å) films had leakage currents as low as 9×10^{-8} A cm^{-2} at −10 V, with ideality factors between 1.01 and 1.09. Annealing for 1 h at 700 °C resulted in a thin, crystalline TiC layer at the SiC interface and Ti$_5$Si$_3$ in the remainder of the reacted film. Barrier heights before and after annealing were 0.88 and 1.04 eV, respectively, attributable to the chemical reaction betwen Ti and SiC.

Waldrop et al. [89] have examined the barrier heights of room temperature, UHV deposited Pd, Au, Ag, Tb, Er, Mn, Al, and Mg on epitaxial n-type 6H-SiC. XPS analysis prior to metal deposition showed the presence of ≈1/2 monolayer of O on the SiC(0001) C-terminated face and ≈3/4 monolayer of O on the SiC(000$\bar{1}$) Si-terminated face. The resulting ϕ_B values vary over a 1.3 eV range. With the exclusion of the rare earth elements, there was a general increase in ϕ_B with ϕ_M. Porter and Davis [90] have plotted the trend in Fig. 16 where the Schottky barrier heights (SBHs) calculated from the XPS, I–V, and C–V techniques are graphed versus the work functions of the metals and compared to the theoretical SBHs according to the Schottky limit. The slope of the lines is the index of interface behavior S, which varies from 0.47 to 0.63. It is evident that S depends on the techniques used to measure the SBHs. That $0 < S < 1$ gives evidence for partial pinning of the Fermi level in 6H-SiC.

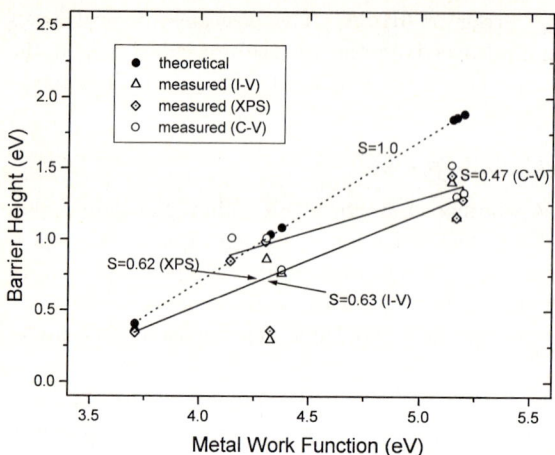

Fig. 16. Experimental and theoretical barrier heights on n-type 6H-SiC(0001) versus work function of the metal contacts. The experimental points are from Waldrop et al. [89, 91]. S is the slope of best linear fit through each set of data points

Waldrop and Grant [91] later measured the barrier heights of Ni, Ti, and Al contacts to n-type 6H-Si(0001) and SiC(0001), which are also plotted in Fig. 16. XPS analysis indicated the formation of TiC and TiSi$_x$ in the unannealed Ti/SiC interface region. No chemical reactions were detected on the Si face before or after annealing Ni/SiC at 400 °C. Porter et al. [92 to 94] have also studied the SBHs of a large number of metals on SiC (Ti, Pt, Hf, Co, and Ni). The relationship of ϕ_B to ϕ_M is shown in Fig. 17. The slopes S vary from 0.12 for the data points calculated from I–V measurements to 0.40 and 0.41 for those calculated from C–V and XPS measurements, respectively. The S values are somewhat less than those derived from Waldrop's data. It is clear that both sets of results show a significant, positive correlation between SBH and metal work function and a susceptibility to measurement technique.

Earlier we saw that the contacts studied by Mead and Spitzer [81] and Hagen [95] gave barrier heights for metals on SiC nearly independent of work function. The fact that the Fermi level is completely pinned in these studies as opposed to more recent studies in which partial pinning is evident may be due to different surface preparation and the quality of the substrate material. Currently, the dependence of barrier height on

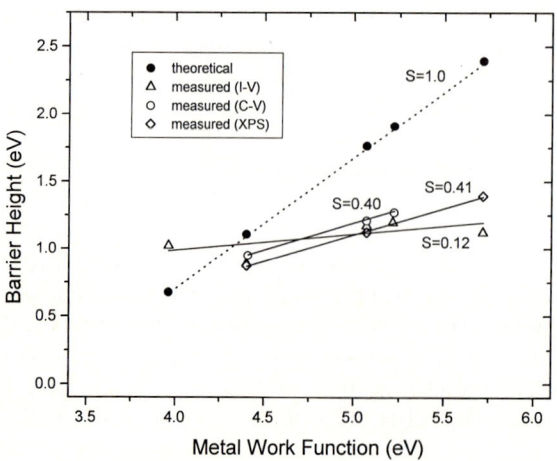

Fig. 17. Experimental and theoretical barrier heights on n-type 6H-SiC(0001) versus work function of the metal contacts. The experimental points are from Porter et al. [92 to 94]. S is the slope of best linear fit through each set of data points

surface preparation is poorly understood for SiC. Further, SiC has proven difficult to grow and only in the last few years has crystalline SiC material of a single polytype been available.

Porter and Davis [96] provide a useful tabular summary of Schottky barrier heights for rectifying contacts on both n-type and p-type 6H-SiC. Most of the metals on the SiC(0001) Si-terminated face had SBHs between approximately 0.8 and 1.25 eV. On the SiC(0001) C-terminated face, most of the SBHs were between 1.0 and 1.6 eV. The limited range is presumably related to the fact that most transition metals have work functions within about 0.8 eV between each other.

4. The Ni/SiC Ohmic Contact

Nickel has been the most widely used metal for ohmic contacts to n-type SiC. By annealing the contact at temperatures >900 °C, room temperature contact resistances less than $9 \times 10^{-6}\ \Omega\,cm^2$ ((7 to 9) $\times 10^{18}\ cm^{-3}$ carrier concentration) have been achieved. We discuss the thin film growth dynamics vs. temperature for Ni/SiC. Elsewhere in this volume, Crofton et al. [97] discuss the electrical and device parameters of this contact system. Useful information about ohmic contacts on SiC is found in the EMIS DataReviews [98].

Ohdomari et al. [99] studied a 2200 Å thick Ni film on 6H-SiC annealed at 870 K using ion resonance scattering and X-ray diffraction (XRD). They reported Ni_2Si formation with 25 at% C uniformly distributed in the film. There was no evidence of Ni carbide formation from XRD. Pai et al. [100] found that a 1000 Å thick Ni film on SiC, after prolonged annealing at 970 K, reacted to form Ni_2Si, while Slijkerman et al. [101] observed a similar reaction sequence on β-SiC. No mixing between SiC and a thin (≈15 Å) film of Ni occurs at room temperature, but at 570 K, Ni begins to react with Si and forms a disordered film with a composition close to Ni_2Si. The Ni_2Si phase is stable up to ≈770 K, but C segregates to the surface to form a layer of graphite. A related study by Levit et al. [102] considers the interaction of $Ni_{90}Ti_{10}$ on 6H-SiC.

Work in our laboratory [103] using AES and Rutherford backscattering (RBS) has shown that Ni_2Si is formed on 6H-SiC during 3 min anneals at ≈770 K in vacuum of ≈10^{-4} Pa. The resulting contact is rectifying; an ohmic contact is formed during annealing to ≈1220 K. The Ni_2Si film and the electrical performance are unchanged after 300 h at 923 K, demonstrating that Ni/SiC is stable *over long periods of time at high temperatures*. As the operating temperature increases, the contact resistance *decreases*. Marinova and Krastev [104] have also observed a decrease in the Ni/SiC contact resistance with temperature.

The product phases which form when a metal is deposited on SiC are very different from reaction of the same metals with silicon. Although SiC readily reacts with most metals, the reaction temperatures are typically greater than those for silicon. It is likely that the chemical reaction between SiC and certain metals is limited by the dissociation of SiC because of the strong Si–C bond. This is seen by the fact that metal-rich silicides are formed first (e.g., between SiC and Ni [105 to 107], Co [105, 108, 109], and Pd [110, 111]), followed in most cases by formation of less metal-rich silicides. The metals Ti [112 to 116], Mo [117], Fe [118, 119], Nb [120], and W [121] formed both carbides and silicides after annealing. It is evident that the product phases depend on temperature, film thickness, and time.

5. Future Directions in SiC Contact Surface Research

Although work in SiC surface physics is increasing rapidly at the present time, the field is still in its infancy and a number of fundamental problems remain.

– Contact technology on SiC has been hampered by the cost and quality of commercial substrates. A typical commercial wafer of SiC contains polishing marks, scratches, and bears little resemblance to an ideally flat surface. It is remarkable how good SiC contacts perform given the degree of surface roughness. Advances in growth, cutting, and polishing technology are needed to reduce the observed irreproducibility in contact performance. The characterization of polish-related surface damage on 6H-SiC surfaces has been studied by Qian et al. [122] who report that the extent of subsurface damage during polishing can be minimized by optimum conditions of abrasive particle size, vertical load, and rotation speed. Ducke et al. [123] report a two-step process for preparation of highly perfect 6H-SiC surfaces with microroughness <10 Å. In the first step, etching in H_2/O_2 RF plasma removes the damaged region of the surface and subsurface. The second step involves thermal oxidation in O_2/Ar gas at 1200 °C and subsequent removal of the grown oxide.

– The reaction dynamics of SiC with metals to form silicides and/or carbides has been studied in detail for only few metal systems. The high temperature stability of the chemical phases formed when a metal interacts with SiC is potentially a serious problem for long-term operation of SiC devices at high temperature. It will be necessary to develop diffusion barriers which work at elevated temperatures for contact metallizations.

– There is little information in the literature on the long-term, high temperature performance of SiC contact. For a contact system to be commercially viable, it must endure both high temperatures and temperature cycling. In our laboratory, we have several systems to test contacts for extended operation at high temperature and routinely examine the I–V characteristics of contacts *during* high temperature operation. More studies should be undertaken which address the lifetime of SiC contact systems.

– The industry is currently undergoing a shift from 6H-SiC to 4H-SiC due to its wider band gap (3.3 eV) and higher electron mobility (500 cm^2/Vs), which give 4H-SiC advantages in power density, temperature range, and speed. Currently, there are no surface studies of the 4H polytype. Issues such as surface structure, reconstruction, graphitization, oxidation, work function, Fermi level pinning, surface states, and variations in surface chemistry with polar face will become important as contact studies on 4H-SiC proceed.

– There is good evidence for the presence of surface states on 6H-SiC. If the metal–semiconductor interface is relatively clean, however, the density of surface states is not sufficiently high to completely pin the Fermi level. The partial pinning has been shown to be associated with the percent ionicity of SiC, in agreement with the results of Kurtin. It should be possible to modify the work function and surface state population of the SiC surface to optimize contact performance. Currently, little information exists on the clean surface work functions of the various polytypes and orientations of SiC or on the changes in work function with adsorbed metal and gas atoms. The ability to alter the interfacial work function and the density of surface states on SiC would enhance our ability to engineer optimum SiC contacts.

– The role of incorporated impurities on SiC contact performance is currently unknown. Contacts to SiC are currently fabricated using conventional thin film techniques under moderately low pressures (10^{-4} to 10^{-5} Pa), and fabrication procedures unavoidably introduce contaminant species such as H_2, C, N_2, and O_2 which are believed to adversely affect the electrical and physical characteristics of the contacts through surface state and work function changes. The influence of residual gas impurities on contact performance may be studied by examining comparative thin films deposited under ultra-high vacuum conditions.

Acknowledgements This work was supported by the Center for the Commercial Development of Space: Space Power and Advanced Electronics, located at Auburn University, with funds from NASA Cooperative Agreement NCC3-511, Auburn University, and the Center's industrial partners. Thanks are also due to Mr. K. Bryant who helped with the literature research. The author gratefully acknowledges Prof. W. J. Choyke of the University of Pittsburgh for continued encouragement and support.

References

[1] W. SCHOTTKY, Naturwissenschaften **26**, 843 (1938).
[2] J. BARDEEN, Phys. Rev. **71**, 717 (1947).
[3] A. M. COWLEY and S. M. SZE, J. Appl. Phys. **36**, 3212 (1965).
[4] J. M. ANDREWS and J. C. PHILLIPS, Phys. Rev. Lett. **35**, 56 (1975).
[5] I. OHDOMARI, T. S. KUAN, and K. N. TU, J. Appl. Phys. **50**, 7020 (1979).
[6] G. OTTAVIANI, K. N. TU, and J. W. MAYER, Phys. Rev. Lett. **44**, 284 (1980).
[7] S. KURTIN, T. C. McGILL, and C. A. MEAD, Phys. Rev. Lett. **22**, 3212 (1969).
[8] W. GORDY and W. J. O. THOMAS, J. Chem. Phys. **24**, 439 (1956).
[9] C. A. MEAD, Appl. Phys. Lett. **6**, 103 (1965).
[10] C. A. MEAD, Solid State Electronics **9**, 1023 (1966).
[11] S. KURTIN, T. C. McGILL, and C. A. MEAD, Phys. Rev. Lett. **22**, 1433 (1969)
[12] B. E. DEAL, E. H. SNOW, and C. A. MEAD, J. Phys. Chem. Solids **27**, 1873 (1966).
[13] M. J TURNER and E. H. RHODERICK, Solid State Eletconics **11**, 291 (1968).
[14] L. PAULING, The Nature of the Chemical Bond, 3rd ed., Cornell University Press, Ithaca, NY, 1967.
[15] T. C. McGILL and C. A. MEAD, J. Vacuum Sci. Technol. **11**, 122 (1974).
[16] C. A. MEAD, Solid State Electronics **9**, 1023 (1966).
[17] M. SCHLUTER, J. Vacuum Sci. Technol. **15**, 1374 (1978).
[18] L. J. BRILLSON, Phys. Rev. B **18**, 2431 (1978).
[19] C. A. MEAD, Solid State Electronics **9**, 1023 (1966).
[20] J. C. PHILLIPS, J. Vacuum Sci. Technol. **11**, 947 (1974).
[21] L. J. BRILLSON, Phys. Rev. Lett. **40**, 260 (1978).
[22] S. P. KOWALCZYK, J. R. WALDROP, and R. W. GRANT, Appl. Phys. Lett. **38**, 167 (1981); J. Vacuum Sci. Technol. **19**, 611 (1981).
[23] L. J. BRILLSON, C. F. BRUCKER, A. D. KATNANI, N. G. STOFFEL, DANIELS, and G. MARGARITONDO, Physics and Chemistry of Semiconductor Interfaces Conference, Asilomar, Calif., 1982.
[24] E. HOKELEK and G. Y. ROBINSON, Appl. Phys. Lett. **40**, 426 (1982).
[25] J. M. WOODALL and J. L. FREEOUF, J. Vacuum Sci. Technol. **19**, 794 (1981); Appl. Phys. Lett. **39**, 727 (1981).
[26] S. P. MURARKA, J. Vacuum Sci. Technol. **17**, 775 (1980).
[27] C. A. MEAD and W. G. SPITZER, Phys. Rev. **134**, A713 (1964).
[28] J. O. McCALDIN, T. C. McGILL, and C. A. MEAD, Phys. Rev. Lett. **36**, 56 (1976); J. Vacuum Sci. Technol. **13**, 802 (1976).
[29] W. E. SPICER, I. LINDAU, P. SKEATH, C. Y. SU, and P. CHYE, Phys. Rev. Lett. **44**, 420 (1980); J. Vacuum Sci. Technol. **17**, 1019 (1980).
[30] CREE Research, Inc., 2810 Meridian Parkway, Durham, NC 27713, USA.
[31] Epitronics, 7 Commerce Drive, Danbury, CT 06810, USA.

[32] K. BRACK, J. Appl. Phys. **36**, 3560 (1965).
[33] J. W. FAUST, JR., in: Silicon Carbide, A High Temperature Semiconductor, Eds. J. R. O'CONNOR and J. SMILTENS, Pergamon Press, New York 1960 (p. 403).
[34] A. SUZUKI, H. I. MATSUNAMI, and T. TANAKA, J. Electrochem. Soc. **125**, 1897 (1978).
[35] R. C. A. HARRIS and R. L. CALL, in: Silicon Carbide 1973, Eds. R. C. MARSHALL, J. W. FAUST, JR., and C. E. RYAN, University of South Carolina, Columbia 1974 (p. 329).
[36] L. MUEHLHOFF, W. J. CHOYKE, M. J. BOZACK, and J. T. YATES, JR., J. Appl. Phys. **60**, 2558 (1986).
[37] M. J. BOZACK, Surf. Sci. Spectra **3**, 82, 86 (1995).
[38] F. BOZSO, L. MUEHLHOFF, M. TRENARY, W. J. CHOYKE, and J. T. YATES, JR., J. Vacuum Sci. Technol. A **2**, 1271 (1984).
[39] L. MUEHLHOFF, W. J. CHOYKE, M. J. BOZACK, and J. T. YATES, JR., J. Appl. Phys. **60**, 2842 (1986).
[40] B. B. PATE, M. OSHIMA, J. A. SILBERMAN, G. ROSSI, I. LINDAU, and W. E. SPICER, J. Vacuum Sci. Technol. A **2**, 957 (1984).
[41] J. W. ROGERS, J. E. HOUSTON, R. R. RYE, F. H. HUTSON, and D. E. RAMAKER, J. Vacuum Sci. Technol. A **4**, 1601 (1986).
[42] J. W. GADZUK, J. Electron Spectroscopy Related Phenomena **11**, 355 (1977).
[43] W. J. PARDEE, G. D. MAHAN, D. E. EASTMAN, and R. A. POLLAK, Phys. Rev. B **11**, 3615 (1975).
[44] A. M. BRADSHAW and W. WYROBISCH, J. Electron Spectroscopy Related Phenomena **7**, 45 (1975).
[45] M. GHAMNIA, C. JARDIN, D. KADRI, and M. BOUSLAMA, Vacuum **47**, 141 (1996).
[46] F. BOZSO, J. T. YATES, JR., W. J. CHOYKE, and L. MUEHLHOFF, J. Appl. Phys. **57**, 2771 (1985).
[47] J. TOPPING, Proc. Roy. Soc. (London) **A114**, 67 (1927).
[48] J. PELLETIER, D. GERVAIS, and C. POMOT, J. Appl. Phys. **55**, 994 (1984).
[49] K. W. FRESE, JR., J. Vacuum Sci. Technol. **16**, 1042 (1979).
[50] S. KENNOU, J. Appl. Phys. **78**, 587 (1995).
[51] V. VAN ELSBERGEN, T. U. KAMPEN, and W. MUNCH, J. Appl. Phys. **79**, 316 (1996).
[52] J. A. POWELL, J. B. PETIT, J. H. EDGAR, I. G. JENKINS, L. G. MATUS, J. W. YANG, P. PIROUZ, W. J. CHOYKE, L. CLEMEN, and M. YOGANATHAN, Appl. Phys. Lett. **59**, 333 (1991).
[53] J. PEZOLDT, A. A. KALNIN, and W. D. SAVELYEV, Nuclear Instrum. and Methods B **65**, 361 (1992).
[54] J. COMAS and C. B. COOPER, J. Appl. Phys. **37**, 2820 (1966).
[55] M. MOHRI, K. WATANABE, K. YAMASHITA, H. DOI, and K. HAYAKAWA, J. Nuclear Mater. **75**, 309 (1978).
[56] E. E. VIOLIN, Izv. Leningr. Elektrotekh. Int. **250**, 77 (1979).
[57] J. PEZOLDT, B. STOTTKO, G. KUPRIS, and G. ECKE, Mater. Sci. Engng. B **29**, 94 (1995).
[58] J. BOHDANSKY, H. LINDNER, E. HECHT, A. P. MARTINELLI, and J. ROTH, Nuclear Instrum. and Methods B **18**, 509 (1987).
[59] M. MOHRI, K. WATANABE, and T. YAMASHINA, J. Nuclear Mater. **75**, 7 (1978).
[60] T. YAMASHINA, M. MOHRI, K. WATANABE, H. DOI, and K. HAYAKAWA, J. Nuclear Mater. **76/77**, 202 (1978).
[61] W.-Y. LEE, J. Appl. Phys. **51**, 3365 (1980).
[62] L. MUEHLHOFF, W. J. CHOYKE, M. J. BOZACK, and J. T. YATES, JR., J. Appl. Phys. **60**, 2842 (1986).
[63] B. JORGENSEN and P. MORGEN, J. Vacuum Sci. Technol. **4**, 1701 (1986).
[64] J. A. DILLON, R. E. SCHLIER, and H. E. FARNSWORTH, J. Appl. Phys. **30**, 675 (1959).
[65] P. MORGEN, K. L. SEAWARD, and T. W. BARBEE, J. Vacuum Sci. Technol. A **3**, 2108 (1985).
[66] P. H. YIH and A. J. STECKL, Proc. 5th Internat. Conf. SiC and Related Materials, 1993 (p. 317).
[67] P. H. YIH and A. J. STECKL, J. Electrochem. Soc. **140**, 1813 (1993).
[68] J. CASADY, E. D. LUCKOWSKI, M. J. BOZACK, D. SHERIDAN, R. W. JOHNSON, and J. R. WILLIAMS, Inst. Phys. Conf. Ser. No. 142, 625 (1996).

[69] A. J. van Bommel, J. E. Crombeen, and A. van Tooren, Surf. Sci. **48**, 463 (1975).
[70] T. W. Hass, J. T. Grant, and G. J. Dooley, Jr., J. Appl. Phys. **43**, 1853 (1972).
[71] R. Kaplan and T. M. Parrill, Surf. Sci. **165**, L45 (1986).
[72] J. R. Waldrop and R. W. Grant, Appl. Phys. Lett. **62**, 2685 (1993).
[73] R. Kaplan, Surf. Sci. **215**, 111 (1989).
[74] A. N. Andreev, M. M. Anikin, A. L. Syrkin, and V. E. Chelnokov, Semicond. **28**, 577, 377 (1994).
[75] L. M. Porter, J. S. Bow, R. C. Glass, M. J. Kim, R. W. Carpenter, and R. F. Davis, J. Mater. Res. **10**, 668 (1995).
[76] T. M. Parrill and V. M. Bermudez, Solid State Commun. **63**, 231 (1987).
[77] R. G. Verenchikova, V. I. Sankin, and E. I. Radovanova, Soviet Phys. – Semicond. **17**, 1123 (1983).
[78] A. L. Syrkin, A. N. Andreev, A. A. Lebedev, M. G. Rastegaeva, and V. E. Chelnokov, Mater. Sci. Engng. B **29**, 198 (1994).
[79] A. L. Syrkin, A. N. Andreev, A. A. Lebedev, M. G. Rastegaeva, and V. E. Chelnokov, J. Appl. Phys. **78**, 5511 (1995).
[80] S. Yu. Davidov and S. K. Tikhonov, Phys. Solid State **37**, 1514 (1995).
[81] C. A. Mead and W. G. Spitzer, Phys. Rev. **134**, A713 (1964).
[82] S. H. Hagen, J. Appl. Phys. **39**, 1458 (1968).
[83] S. Y. Wu and R. B. Campbell, Solid State Commun. **17**, 683 (1974).
[84] L. A. Kosyachenko, E. F. Kukhto, and V. M. Sklyarchuk, Zh. Prikladnoi Spektroskopii **41**, 615 (1984).
[85] L. M. Porter, J. S. Bow, R. C. Glass, M. J. Kim, R. W. Carpenter, and R. F. Davis, J. Mater. Res. **10**, 668 (1995).
[86] L. M. Spellman, R. C. Glass, R. F. Davis, T. P. Humphreys, H. Jeon, R. J. Nemanich, S. Chevacharoenkul, and N. R. Parikh, Mater Res. Soc. Symp. Proc. **221**, 99 (1991).
[87] L. M. Spellmann, R. C. Glass, R. F. Davis, T. P. Humphreys, R. J. Nemanich, K. Das, and S. Chevacharoenkul, in: Springer Proc. Phys., Vol. 71, Eds. C. Y. Yang, M. M. Rahman, and G. L. Harris, 1992 (p. 417).
[88] J. S. Bow, L. M. Porter, M. J. Kim, R. W. Carpenter, and R. F. Davis, Ultramicroscopy **52**, 289 (1993).
[89] J. R. Waldrop, R. W. Grant, Y. C. Wang, and R. F. Davis, J. Appl. Phys. **72**, 4757 (1992).
[90] L. M. Porter and R. F. Davis, Mater Sci. Engng. B **34**, 83 (1995).
[91] J. R. Waldrop and R. W. Grant, Appl. Phys. Lett. **62**, 2685 (1993).
[92] L. M. Porter, R. C. Glass, R. F. Davis, J. S. Bow, M. J. Kim, and R. W. Carpenter, Mater. Res. Soc. Symp. Proc. **282**, 471 (1993).
[93] L. M. Porter, J. S. Bow, M. J. Kim, R. W. Carpenter, and R. F. Davis, J. Mater. Res. **10**, 26 (1995).
[94] L. M. Porter, J. S. Bow, M. J. Kim, R. W. Carpenter, and R. F. Davis, J. Mater. Res. (1995).
[95] S. H. Hagen, J. Appl. Phys. **39**, 1458 (1968).
[96] L. M. Porter and R. F. Davis, Mater. Sci. Engng. B **34**, 83 (1995).
[97] J. Crofton, P. G. McMullin, J. R. Williams, and M. J. Bozack, Trans. 2nd High Temperature Electronics Confer., Charlotte (NC) 1994, IOP Publ. Ltd., London 1994.
[98] EMIS DataReviews, Vol. 13, Ed. G. L. Harris, INSPEC, London 1995.
[99] I. Ohdomari, S. Sha, H. Aochi, T. Chikyow, and S. Suzuki, J. Appl. Phys. **62**, 3747 (1987).
[100] C. S. Pai, C. M. Hanson, and S. S. Lau, J. Appl. Phys. **57**, 618 (1985).
[101] W. F. J. Slijkerman, A. E. M. J. Fischer, J. F. van der Veen, I. Ohdomari, S. Yoshida, and S. Misawa, J. Appl. Phys. **66**, 666 (1989).
[102] M. Levit, I. Grimberg, and B.-Z. Weiss, J. Appl. Phys. **80**, 167 (1996).
[103] J. Crofton, P. G. McMullin, J. R. Williams, and M. J. Bozack, J. Appl. Phys. **77**, 1317 (1995).
[104] T. Marinova and V. Krastev, Appl. Surf. Sci. **99**, 119 (1996).
[105] M. Nathan and J. S. Ahearn, J. Appl. Phys. **70**, 811 (1991).

[106] C. S. PAI, C. M. HANSON, and S. S. LAU, J. Appl. Phys. **57**, 618 (1985).
[107] H. HOCHST, D. W. NILES, G. W. ZAJAC, T. H. GLEISCH, B. C. JOHNSON, and J. M. MEESE, J. Vacuum Sci. Technol. B **6**, 1320 (1988).
[108] L. M. PORTER, J. S. BOW, M. J. KIM, R. W. CARPENTER, and R. F. DAVIS, J. Mater. Res. **10**, 26 (1995).
[109] N. LUNDBERG, C.-M. ZETTERLING, and M. OSTLING, Appl. Surf. Sci. **73**, 316 (1993).
[110] C. S. PAI, C. M. HANSON, and S. S. LAU, J. Appl. Phys. **57**, 618 (1985).
[111] V. M. BERMUDEZ, Appl. Surf. Sci. **17**, 12 (1983).
[112] L. M. PORTER, J. S. BOW, R. C. GLASS, M. J. KIM, R. W. CARPENTER, and R. F. DAVIS, J. Mater. Res. **10**, 668 (1995).
[113] M. BACKHAUS-RICOULT, in: Metal–Ceramic Interfaces, Vol. 4, Acta Scripta Metallurgica, Proc. Series, Eds. M. RUHLE, A. G. EVANS, M. F. ASHBY, and J. P. HIRTH, Pergamon Press, New York 1990 (p. 79).
[114] C. G. RHODES and R. A. SPRULING, in: Recent Advances in Composites in the United States and Japan, Vol. 864, ASTM Spec. Tech. Publ., Eds. J. R. VINSON and M. TAYA, Amer. Soc. for Testing and Materials, Philadelphia (PA) 1985 (p. 585).
[115] M. BACKHAUS-RICOULT, Ber. Bunsenges. Phys. Chem. **93**, 1277 (1989).
[116] I. GOTMAN, E. Y. GUTMANAS, and P. MOGILEVSKY, J. Mater. Res. **8**, 2725 (1993).
[117] S. HARA, K. SUZUKI, A. FURUYA, Y. MATSUI, T. UENO, and I. OHDOMARI, J. Appl. Phys. **29**, L394 (1990).
[118] R. KAPLAN, P. H. KLEIN, and A. ADDAMIANO, J. Appl. Phys. **58**, 321 (1985).
[119] K. M. GEIB, C. W. WILMSEN, J. E. MAHAN, and M. C. BOST, J. Appl. Phys. **61**, 5299 (1987).
[120] D. L. YANEY and A. JOSHI, J. Mater. Res. **5**, 2197 (1990).
[121] L. BAUD, C. JAUSSAUD, R. MADAR, C. BERNARD, J. S. CHEN, and M. A. NICOLET, Mater. Sci. Engng. B **29**, 126 (1995).
[122] W. QIAN, M. SKOWRONSKI, G. AUGUSTINE, R. C. GLASS, I. McD. HOBGOOD, and R. H. HOPKINS, J. Electrochem. Soc. **142**, 4290 (1995).
[123] E. DUCKE, R. KRIEGEL, A. FISSEL, U. KAISER, B. SCHROTER, P. MULLER, and W. RICHTER, in: Silicon Carbide and Related Materials 1995, Proc. 6th Internat. Confer., Kyoto (Japan) 1995, IOP Publ. Ltd., London 1995 (p. 18).

phys. stat. sol. (b) **202**, 581 (1997)

Subject classification: 73.40.Cg; 73.40.Ns; S6

The Physics of Ohmic Contacts to SiC

J. CROFTON (a), L. M. PORTER[1]) (b), and J. R. WILLIAMS (c)

(a) Department of Physics and Engineering Physics, Murray State University, Murray, KY 42071, USA

(b) Department of Materials Science and Engineering, North Carolina State University, Raleigh, NC 27695-7907, USA

(c) Department of Physics, Auburn University, Auburn, AL 36849, USA

(Received January 31, 1997)

The specific contact resistance of an ohmic contact will be discussed including ways to calculate and measure this parameter. Ohmic contacts to n- and p-type hexagonal SiC will then be detailed. Low resistance n-type ohmic contacts are predominately fabricated by annealing a refractory metal, thereby forming a silicide with a lowered Schottky barrier height at the metal–SiC interface. P-type contacts on the other hand generally use Al or Al alloys which upon annealing enable Al to diffuse into the SiC thus resulting in ohmic properties. Aluminium alloys however suffer from many problems which will be discussed. Other novel contacting schemes to p-type SiC will also be reviewed.

1. Introduction

Silicon carbide, SiC, has great promise for uses in high power and high temperature applications. One of the key technology issues, however, that still must be addressed before SiC can realize its great potential is in the area of ohmic contacts. Ohmic contacts are electrical connections between a metal and a semiconductor which have linear current–voltage (I–V) characteristics. Metal–semiconductor combinations generally upon preparation are not ohmic but instead are rectifying. These rectifying contacts are often referred to as Schottky contacts due to the Schottky barrier at the metal semiconductor interface responsible for rectification. There are many factors which can affect the Schottky barrier height (SBH); one of the more important is the work function difference between the metal and the semiconductor (see Fig. 1) which gives rise to an energy barrier that carriers must surmount in order to pass from one material to the other [1]. These Schottky contacts have asymmetric I–V characteristics similar to a p–n junction diode.

Ohmic contacts may be considered a limiting case of the more general class of Schottky contacts. A metal–semiconductor combination that upon preparation has Schottky characteristics can be converted into an ohmic contact with certain processing steps that modify the Schottky barrier. Although there are metal–semiconductor combinations which are ohmic as prepared, these too have a Schottky barrier at the metal–semiconductor interface which is either too low or too thin to produce an asymmetric I–V characteristic.

---

[1]) Present address: Department of Materials Science and Engineering, Carnegie Mellon University, Pittsburgh, PA 15213, USA.

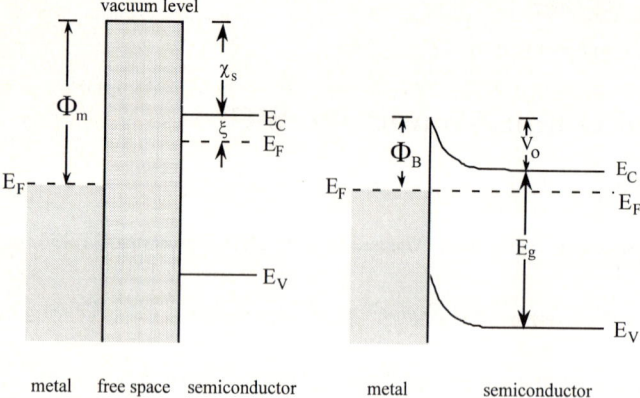

Fig. 1. Energy band diagrams depicting electron energy levels of a typical metal and n-type wide bandgap semiconductor before and after contact, respectively. E_F is the Fermi level, Φ_m the metal work function, E_C the conduction band minimum, E_V the valence band maximum, E_g the band gap, χ_s the electron affinity, and Φ_B the Schottky barrier height

Contact resistance is a term that describes the electrical resistance present at the interface between the metal and semiconductor in an ohmic contact. The contact resistance associated with an ohmic metal–semiconductor interface is the total electrical resistance of the interface and depends on the area or geometry of the contact. The interface itself, however, has certain properties which influence the contact resistance, R_c, independent of the contact geometry. This interfacial property is known as the specific contact resistance, r_c, and should in principal be independent of the area of the metal–semiconductor interface. The difference in contact resistance and specific contact resistance is analogous to the difference in the resistance R of a standard carbon resistor and the resistivity ϱ of the carbon from which the resistor is made; the resistance is equal to the resistivity times a geometric factor, specifically the length l divided by the cross sectional area A or $R = \varrho l/A$. Similarly the contact resistance of a metal–semiconductor interface is equal to the specific contact resistance divided by the cross-sectional area of the metal–semiconductor interface or $R_c = r_c/A$; this of course assumes the entire contact area takes part in the conduction process which is not always true. The units of contact resistance are Ω while specific contact resistance has units of Ω times length squared. Generally the unit of length used is cm thus specific contact resistance is given in Ω cm^2. The specific contact resistance is related to the current density J (units of A/cm^2) by

$$r_c = \lim_{V \to 0} \left(\frac{dJ}{dV}\right)^{-1}. \tag{1}$$

2. Methods of Determining Specific Contact Resistance

2.1 Theoretical calculations

Bethe's thermionic emission theory accounted for the rectification in Schottky barriers made on lightly doped semiconductors by assuming the carriers had to possess enough thermal energy to surmount the Schottky barrier in order to pass from one material to the other [2]. In the years that followed, it was realized that in many cases the difference

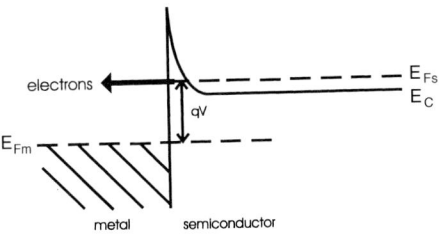

field emission

Fig. 2. Energy band diagram depicting field emission, thermionic field emission, and thermionic emission for an n-type semiconductor. Field emission occurs when the Schottky barrier is sufficiently thin to allow tunneling at the semiconductor Fermi energy, E_{Fs}. When the barrier is not sufficiently thin due to insufficient doping, the electrons tunnel at an energy greater than the semiconductor Fermi energy. For lightly doped semiconductors, the barrier is too thick at all energies therefore no tunneling can occur. For this case, carriers must have sufficient energy to pass over the Schottky barrier; this is known as thermionic emission

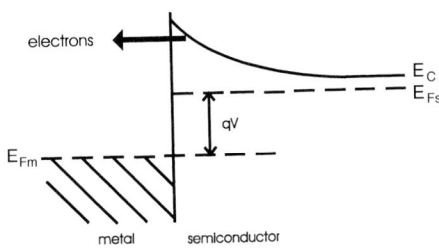

thermionic field emission

between a Schottky and an ohmic contact was nothing more than the doping of the semiconductor immediately beneath the metal. A metal deposited on a heavily doped semiconductor will have the same barrier height as the same metal deposited on a more lightly doped sample of the same semiconductor; the barrier, however, will be much thinner and therefore affords carriers, holes or electrons, the chance to quantum mechanically tunnel through the barrier in the contact made on the more heavily doped sample (see Fig. 2). Ohmic contacts which are made on heavily doped semiconductors are therefore dependent on carriers tunneling through the Schottky energy barrier as opposed to traveling over the barrier.

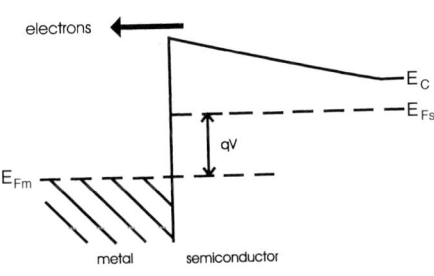

thermionic emission

Many of the early investigations on tunneling were performed by Stratton in the 1960's [3]. Based on theoretical considerations, Stratton described a process in which carriers tunneled from the Fermi level in a degenerately doped semiconductor into a metal. This process is known as field emission and explains much of the data observed in both ohmic contacts and in more heavily doped Schottky diodes. Several years later, Padovani and Stratton [4] described theoretically a conduction process between the two extremes of thermionic emission and field emission. This process, known as thermionic field emission, involves the carriers with sufficient thermal energy to tunnel at an energy above the Fermi level where the barrier is thin but without sufficient thermal energy to surmount the barrier as in the case of thermionic emission. Thermionic field emission generally takes place in Schottky diodes made on moderately doped semiconductors. The relationship between temperature and doping which determines the dominant conduction mechanism is given by the constant

E_{00} (in V) given by

$$E_{00} = \frac{h}{4\pi} \left(\frac{N}{m\varepsilon}\right)^{1/2}, \tag{2}$$

where h is Planck's constant, N the semiconductor doping, m the effective mass, and ε the semiconductor dielectric constant. The relationship between temperature and doping which determines which conduction mechanism is dominant is given by

$$
\begin{aligned}
&\text{field emission for} &kT &\ll qE_{00}, \\
&\text{thermionic field emission for} &kT &\approx qE_{00}, \\
&\text{thermionic emission for} &kT &\gg qE_{00}.
\end{aligned}
\tag{3}
$$

The constant E_{00} also appears in the expressions for the field and thermionic field emission current density equations. The field and thermionic field emission equations are complicated and sometimes difficult to use analytically, thus numerical methods to calculate current densities and specific contact resistances have recently been used to model ohmic contacts formed on SiC [5 to 7].

Because it strongly affects both current density and contact resistance, one must be able to both control and determine the SBH at the metal–SiC interface. An ideal Schottky barrier on n-type material would have a barrier height Φ_B given by $\Phi_B = \Phi_m - \chi_s$ where Φ_m is the metal work function and χ_s is the semiconductor electron affinity (see Fig. 1). However, most metal–semiconductor interfaces, including SiC–metal interfaces, deviate from this due to effects such as surface states and metal-induced gap states [8], therefore this theoretical relation is seldom satisfied for "real" SiC contacts. The barrier height is usually experimentally determined using one of three methods: 1. capacitance–voltage; 2. current–voltage or 3. a photoresponse technique. Barrier heights determined using current–voltage methods include an image-force lowering term. Once the barrier height is known, a variety of methods and assumptions can be used to calculate the actual shape of the energy barrier. The depletion approximation is often used to calculate the form of the potential barrier since it is easy to use and gives reasonably accurate results [1]. This approximation assumes the depletion region present at the metal–semiconductor interface is devoid of any charge carriers, thus the only charges present are the ionized donor or acceptor atoms. This assumption results in a parabolic barrier that is mathematically tractable. Both field and thermionic field emission theories assume a parabolic barrier in deriving expressions for the current density.

The tunneling probability through a particular energy barrier is also associated with specific contact resistance and is usually calculated using the Wentzel-Kramers-Brillouin (WKB) approximation [9]. If a parabolic energy barrier is assumed, a WKB calculation of the tunneling probability as a function of energy, $T(E)$, can be expressed in closed form; unfortunately, the closed form expression for $T(E)$ does not lend itself to a simple evaluation of the current density integral. The general theories of field and thermionic field emission deal with this problem by expanding the tunneling probability in a Taylor series as a function of energy about the Fermi level of the source material in the case of field emission or about some energy greater than the Fermi level in the case of thermionic field emission. The resulting expressions for the current density for both forward and reverse bias are complicated and will not be repeated here. Nevertheless, several simplifying assumptions can be made for the calculation of specific contact resistance based on these theories. For ohmic contacts with high doping concentrations, the predominant mode of conduction is field emission. Taking the expresssion for the forward field emission current density and applying equation (1), one can show that the specific contact resistance r_c is proportional to the exponential of the barrier height divided by the square root of the doping or

$$r_c \propto \exp\left(\frac{\Phi_B}{\sqrt{N}}\right). \tag{4}$$

This relationship is very useful for predicting trends in contact resistance as a function of barrier height and the semiconductor doping. More accurate calculations require a knowledge of the semiconductor Fermi level, tunneling mass, and other physical properties, and in these cases, a numerical evaluation of the current density and contact resistance is preferred [10, 11]. If a numerical approach is used, the closed form expressions for the tunneling probability can be used to numerically integrate the expressions for current density. If all that is required is a specific contact resistance, then $T(E)$ may be used to directly evaluate the contact resistance by

$$\frac{1}{r_c} = \frac{4\pi m q^2}{h^3} \int_0^\infty \frac{T(E)}{1 + \exp\left[\dfrac{E - E_F}{kT}\right]} \, dE. \tag{5}$$

Fig. 3 shows calculated and measured specific contact resistances versus doping for n- and p-type 6H-SiC. The calculated values were determined using Equation (5). The appropriate masses were taken from the literature [12, 13] so the tunneling mass could be calculated while the barrier height was adjusted until the calculated values gave the best agreement with the experimental values [6, 7]. The specific contact resistance measurements as a function of doping for p-type SiC are seen to be approximately an order of magnitude greater than those on n-type material.

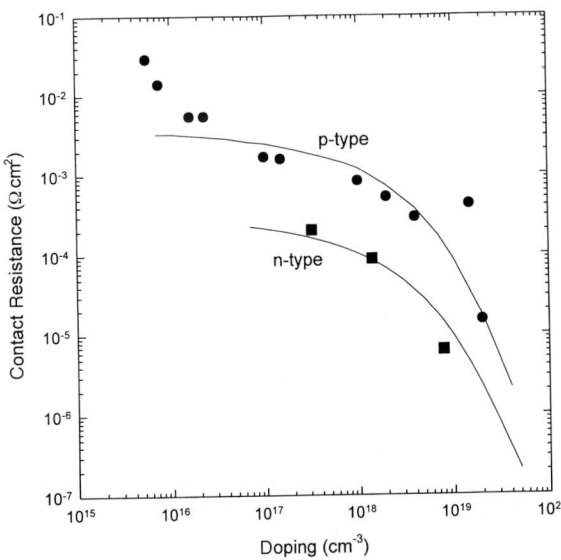

Fig. 3. Specific contact resistance vs. doping for n- and p-type 6H-SiC. The n- and p-type data are taken from [6] and [7], respectively. The solid lines were calculated assuming barrier heights of 0.37 and 0.35 eV for the p- and n-type data, respectively

2.2 *Experimental techniques*

Table 1 is a summary of the most widely used experimental methods to measure specific contact resistance. The method of Cox and Strack [14] which is one of the oldest techniques was first proposed in 1967. The test structure consists of varying diameter dots which are patterned on the top surface of the semiconducting layer with a large broad area ohmic contact on the semiconductor's back surface. The measurement is made vertically through the substrate allowing for a vertical measurement of the specific contact resistance as opposed to a lateral measurement. This is often useful where the contacts to be studied will find application in vertical devices such as static induction transistors (SITs). This method has not been widely used for contact resistance measurements on SiC, therefore it will not be discussed in detail here.

Two other well-known methods to measure contact resistance also shown in Table 1 are the method of Kuphal [15] and the Kelvin technique [16]. The method of Kuphal, also known as a four-point method, uses four co-linear equal area dots to determine the contact resistance. The technique is a lateral measurement but is not very accurate since it does not account for current crowding at the contact edges, and often does not accurately account for the spreading resistance in the semiconductor. The Kelvin technique

Table 1

Summary of methods for experimentally determining specific contact resistance. The most widely used method for SiC contact work has been the Linear Transmission Line Method

method	test structure	advantage	disadvantage	ref.
Cox and Strack	varying diameter dots on top surface with large contact on back of substrate	easy to fabricate; gives vertical measurement of contact resistance	difficult to measure small values of contact resistance	[14]
Kuphal	4 co-linear dots with the same diameter	easy to fabricate	inaccurate, does not account for current crowding at contact edge	[15]
Kelvin	diffused path and metal layer in a four-point structure	capable of measuring very small contact resistance	test structure requires diffused layer	[16]
linear TLM	equal area rectangular contacts with varying spaces between contacts on an isolated mesa	easy to extract moderate to small values of lateral contact resistance	test structure requires two masking levels	[17, 18]
circular TLM	varying radii concentric circles on epitaxial or implanted material	capable of extracting moderate to small values of lateral contact resistance with only one masking level	requires extremely low metal sheet resistance and the analysis of modified Bessel functions	[19]

is very accurate and useful for measuring small values of specific contact resistance; however, the Kelvin test structure requires a diffused or implanted layer, therefore it has not been used often for SiC contact work. The most widely used methods to determine specific contact resistance on SiC have been the Transmission Line Methods (TLM) which are lateral techniques.

A technique known as the Linear Transmission Line Method was proposed in 1964 by Shockley [17] to empirically determine specific contact resistance. This method was later refined by Berger [18] in 1972 and has become the method of choice for specific contact resistance measurements in general, not just to SiC. This technique gives information on the semiconductor sheet resistance as well as the specific contact resistance. Another form of the TLM method known as the Circular Transmission Line Method was proposed by Reeves in 1980 [19]. This model gives the same information as the linear TLM but the contact structures are easier to fabricate. One major disadvantage of the circular TLM is that it requires very low metal sheet resistances in order to ensure accuracy. Often, low metal sheet resistances do not exist either because the metal layers are thin or else the resistivity is high. In these cases, additional measurements and calculations are necessary to ensure accuracy [20] or the metal sheet resistance must be lowered through the use of highly conductive overlays. Another disadvantage of the circular TLM is the complexity of the modified Bessel function equations which must be analyzed using this method. The most widely used method for measuring contact resistances on SiC is the linear TLM method.

The linear TLM method requires a thin isolated semiconducting layer upon which a number of rectangular ohmic contacts are fabricated. The rectangular contacts are identical in size therefore they should possess equal contact resistances (in Ω) provided the specific contact resistance (in $\Omega \, cm^2$) does not vary along the length of the TLM mesa. Plotting the total resistance between pairs of contacts as a function of spacing between the adjacent contacts should result in a straight line with a slope proportional to the semiconductor sheet resistance and with a vertical intercept equal to twice the total resistance of one of the contacts. From the knowledge of the total resistance of the contact, R_c, the semiconductor sheet resistance under the contact, R_{sk}, and the length d and the width w of the contact, the specific contact resistance r_c can be found from a solution of

$$R_c = \frac{(r_c R_{sk})^{1/2}}{w} \coth\left[d \left(\frac{R_{sk}}{r_c} \right)^{1/2} \right]. \tag{6}$$

The total resistance of the contact R_c and the specific contact resistance r_c are not related by $R_c = r_c/A$ because for planar structures, current crowding at the edge of the contact occurs. The effective length of the contact is equal to the transfer length L_t which is given by

$$L_t = \sqrt{\frac{r_c}{R_{sk}}}. \tag{7}$$

The effective area of the contact is therefore $L_t w$. In equations (6) and (7), R_{sk} is the semiconductor sheet resistance directly under the ohmic contact which generally is equal to the bulk semiconductor sheet resistance R_{sh} for non-alloyed contacts. The TLM methods allow for the fact that the sheet resistance under the metal contact may be modified by the ohmic contact fabrication process. The linear TLM method will also yield information on the uniformity or reproducibility of the contact resistance [21, 28].

The TLM method requires accurate knowledge of the spacings between the contacts which typically vary between a few and a few tens of microns. Errors on the order of a few tenths of a μm can mean errors in extracted contact resistances of factors of two or greater. Accurate measurements therefore require very precise reproducible linewidths from run to run or necessitate measurements of the contact spacings be made after processing. Finally, the TLM method requires the metal contact to be an equipotential surface. This requires the metal which forms the ohmic contact to have a very low sheet resistance, typically less than 1% of the underlying semiconductor sheet resistance for typical linear geometries where contacts are no more than a few hundred μm in width. This requirement is even more stringent for the circular TLM method, since the spatial extent of a circular contact can be many hundreds of μm. Measurements using the circular TLM method usually require a highly conductive overlay such as Al or Au be deposited over the contact material. The measured contact resistances shown in Fig. 3 for p-type SiC were made using circular TLM patterns which necessitated the overlay of approximately 2 μm of Al for accurate measurements.

3. Ohmic Contacts to n-Type α-SiC

3.1 Introduction to n-type contacts

The discussion in this section will be limited to the 4H and 6H hexagonal (α) polytypes of SiC. These polytypes are by far the most widely used for SiC device fabrication, although a limited number of hetero-devices have also been fabricated using cubic (β) SiC. For a summary of contact work to n-type β-SiC, see the review articles by Porter and Davis [22] or by Harris et al. [23].

Ohmic contacts to n-type α-SiC have been studied extensively over the last several years. During this time, the quality of SiC bulk and epitaxial material has steadily improved, and heavily doped n-type layers have become more readily available as the result of improved techniques for doping during epitaxial growth and for doping by ion implantation/activation. As a result of this progress, measured specific contact resistances on n-type α-SiC have decreased to levels which are difficult to measure. Specific contact resistances in the 10^{-6} Ω cm^2 range on heavily doped material ($\approx 10^{19}$ cm^{-3}) are now common. Much of the contact work has been performed on the 6H-SiC which was the first high-quality α-polytype widely available to researchers. The 6H polytype is currently less expensive than 4H material; however, 4H-SiC is now the material of choice among device designers due to its superior mobility characteristics. Studies have shown that the mechanisms responsible for ohmic contact formation on 6H material occur on 4H material as well (e.g., metal silicide formation), and it is generally believed that n-type contact results on α-SiC are relatively polytype insensitive.

Attempts to model n-type ohmic contacts have been generally successful [5, 6, 68]. Results show that the specific contact resistance increases with increasing barrier height and that the rate of decrease for the specific contact resistance as a function of doping increases dramatically for heavily doped materials. Specific contact resistances have been measured and compared to calculated results for several different contacts; however, barrier heights used to achieve good agreement with measured results have been very

Table 2

Ohmic contacts on n-type α(6H or mixed)-SiC. Multi-layered contacts are designated with slashes to separate the distinct layers; layers at the surface to the interface with SiC proceed from left to right. The surface preparations which consisted of at least a surface oxide etching step and a hydrocarbon removal step by heating in high or ultra-high vacuum were rated as 'very good'. Those which consisted of at least a chemical clean and a surface oxide etching step but lacked the capability for heating in high vacuum were rated as 'good'. This rating system is based on analysis of the SiC surface by XPS. Reprinted from [22] with kind permission from Elsevier S.A., P.O. Box 564, 1001 Lausanne, Switzerland

metalliza-tion	deposition method	annealing condition	r_c (Ω cm^2)	SiC carr. conc. (cm^{-3})	SiC surface	method of r_c meas.	origin of SiC	surface preparation	ref.
Cr	melting	melting (≥ 2130 °C)	N.R	N.R.	N.R.	–	N.R.	N.R.	[63]
Ni	e-beam evap.	1000 °C/20 s	1.7×10^{-4}	4.5×10^{17}	(0001)	TLM	seeded sub.; CVD epi	N.R.	[64]
TiN	ion-assisted e-beam evap.	600 °C/30 min	4×10^{-2}	$\approx 1 \times 10^{18}$	(0001)	TLM	seeded sub.	very good	[65, 66]
TiW	sputtering	O$_2$ plasma + 600 °C/5 min	7.8×10^{-4}	4.7×10^{18}	(0001)	circular TLM	seeded sub.; CVD epi	O$_2$ plasma	[67]
Ni	e-beam evap.	950 °C/5 min	mid 10^{-2}	4.7×10^{18}	(0001)	four-point	seeded sub.; CVD epi	good	[67]
Ni–Cr (60/40 wt%)	sputtering	950 °C/5 min	1.8×10^{-3}	4.7×10^{18}	(0001)	circular TLM	seeded sub.; CVD epi	O$_2$ plasma	[67]
W	thermal evap.	1200 to 1600 °C	5×10^{-3} to 1×10^{-4}; 1×10^{-2} to 5×10^{-4}	3×10^{18} to 1×10^{19}; 1×10^{17} to 1×10^{19}	(0001); (000$\bar{1}$)	four-point	Lely	varied	[34]
TiW	sputtering	750 °C/5 min	$\approx 8 \times 10^{-4}$	7 to 8×10^{18}	(0001)	circular TLM	seeded sub.; CVD epi	good	[5]
Ti	thermal evap.	none	1×10^{-2} to $<2 \times 10^{-5}$	2×10^{18} to 1×10^{20}	(0001)	circular TLM	seeded sub.	good	[68]
Mo	sputtering	none	$\approx 1 \times 10^{-4}$	$>1 \times 10^{19}$	(0001)	four-point and TLM	seeded sub.; CVD epi	good	[45]
Ta	sputtering	none	$\approx 1 \times 10^{-4}$	$>1 \times 10^{19}$	(0001)	four-point and TLM	seeded sub.; CVD epi	good	[45]

Table 2 (continued)

metalliza-tion	deposition method	annealing condition	r_c (Ω cm²)	SiC carr. conc. (cm⁻³)	SiC surface	method of r_c meas.	origin of SiC	surface preparation	ref.
Ni/3C-SiC	resistive evap.	1000 °C/30 s	$<1.7 \times 10^{-5}$ $<6 \times 10^{-5}$	1 to 2×10^{18}	(0001) (000$\bar{1}$)	Cox and Strack	Lely	very good	[43]
Ni	e-beam evap.	950 °C/2 min	$<5 \times 10^{-6}$	7 to 9×10^{18}	(0001) (000$\bar{1}$)	TLM	seeded sub.; CVD epi	good	[69]
Ni	sputtering	1050 °C/5 min	10^{-3} to 10^{-4}	9.8×10^{17}	(0001)	TLM	seeded sub.	good	[70]
W/Ti/Ni	sputtering	1050 °C/5 min	10^{-3} to 10^{-4}	9.8×10^{17}	(0001)	TLM	seeded sub.	good	[70]
Ni	thermal evap.	1000 °C/5 min	1×10^{-6}	4.5×10^{20}	(000$\bar{1}$)	contact area	LPE	good	[44]
Ti–Al	thermal evap.	1000 °C/5 min	$<1 \times 10^{-3}$	4.5×10^{20}	(000$\bar{1}$)	contact area	LPE	good	[44]
TiC	CVD; 1260 °C etched at 1300 °C for 15 min in H₂		1.3×10^{-5}	4×10^{19}	(0001)	TLM	CVD epi	N.R.	[28]

N.R. not reported, sub. sublimation, epi epilayer

low [5, 6, 68]. For Ni_2Si, which is perhaps the most widely used ohmic contact to n-SiC, agreement was achieved by assuming a barrier height of 0.35 eV (n-type curve in Fig. 3, see also [6]). Specific contact resistances for non-reacted TiW ohmic contacts formed on n-type SiC following 600 °C anneals have also been measured as a function of temperature and compared with calculated values [5]. Best agreement was achieved by assuming a barrier height of 0.49 eV. These barrier heights for both Ni_2Si and TiW are relatively low compared to as-deposited barrier heights (≈ 1 eV) for Ni and TiW measured using capacitance–voltage or current–voltage techniques. The post-deposition annealing process lowers the as-deposited barrier height thus forming the ohmic contact.

Table 2 summarizes ohmic contact data reported on n-type α-SiC. Contacts have been deposited on SiC with a wide range of doping concentrations and processed under a variety of conditions. Reported specific contact resistances vary between 1×10^{-6} and 1×10^{-2} Ω cm^2. Several additional observations may also be made: (1) the lowest reported contact resistances have been measured on the most heavily doped material; (2) Ni is the metal most widely used for the fabrication of n-type ohmic contacts and (3) with proper processing, Ni makes a relatively good ohmic contact to moderately doped ($\approx 10^{18}$ cm^{-3}) SiC.

3.2 Nickel contacts

Uemoto [44] reported a specific contact resistance of 1×10^{-6} Ω cm^2 following 5 min/ 1000 °C anneals of Ni deposited on heavily doped 6H material (4.5×10^{20} cm^{-3}) grown using LPE techniques. This doping concentration is approximately 10 times the highest concentration normally achieved using CVD techniques, and, as might be expected, the

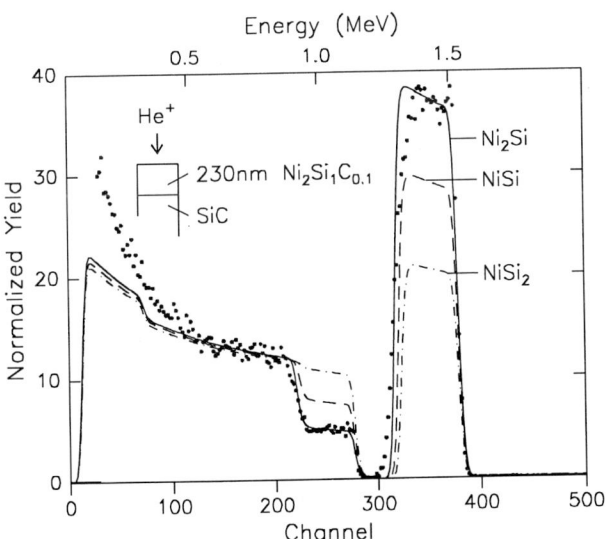

Fig. 4. RBS spectrum for an annealed Ni/6H-SiC sample. The silicide layer thickness for the NiSi (dashed line) and $NiSi_2$ (dash-dotted line) simulations have been slightly reduced in order to keep the simulations distinct. The Ni_2Si silicide phase is stable after annealing at 500 °C and $\approx 10^{-4}$ Pa for more than 300 h

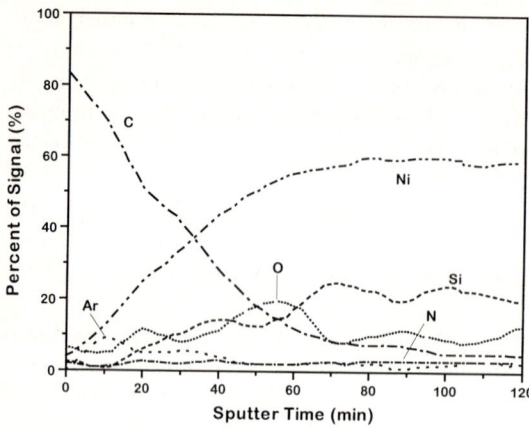

Fig. 5. Auger spectra for a Ni_2Si thin film on SiC. The C concentration is significantly higher at the free surface of the silicide layer compared to the concentration deeper into the layer

measured value of the specific contact resistance is the lowest value listed in Table 2. The 5 min/1000 °C anneal likely forms a nickel silicide ohmic contact; however, the authors have not reported the results of any physical analysis.

The Ni/6H-SiC system has been characterized electrically and physically by Crofton et al. [24]. Specific contact resistances of $<5 \times 10^{-6}\,\Omega\,cm^2$ were measured following 2 min/950 °C vacuum anneals of Ni layers deposited on epilayers with doping concentrations between 7 and $9 \times 10^{18}\,cm^{-3}$. Decreasing specific contact resistance was observed for increasing operating temperature, and at 500 °C, measured values were, on average, lower by a factor of about 5 compared to room temperature results. Physical characterization using Rutherford backscattering spectrometry (RBS) and Auger electron spectroscopy (AES) showed that during the anneal cycle, Ni and SiC react to form Ni_2Si [24] (see the RBS spectrum shown in Fig. 4). Nickel silicide formation has also been observed by Liu et al. [25]. The formation of the nickel silicide, however, is not sufficient to ensure ohmic behavior. For anneal times of a few minutes, silicide formation begins at temperatures as low as 600 °C, whereas ohmic behavior is not observed at temperatures below about 900 °C for these short anneal times. The movement of C away from the interface during formation of the silicide may be responsible for this behavior. Fig. 5 shows that C migrates away from the interface at higher annealing temperatures and accumulates on the free silicide surface, resulting in a clean SiC/Ni_2Si interface [26]. Note however that, although formation of a silicide and the removal of excess C from the Ni_2Si/SiC interface accompanies the onset of ohmic behavior, other mechanisms such as the formation of vacancies during the chemical reaction between Ni and SiC may also contribute to ohmic behavior. This reasoning is evidenced by the observation that unannealed, sputter-deposited nickel silicide layers do not exhibit ohmic behavior on heavily doped SiC epitaxial material. Liu et al. [25] have included Cr with Ni in a Ni/W/Cr composite contact in an effort to improve contact stability while maintaining a low specific contact resistance. Chromium reacts with C and Ni reacts with Si to form two stable compounds, Cr_3C_2 and Ni_2Si, both of which were observed by the authors using X-ray diffraction. The specific contact resistances reported are quite similar to those reported using pure Ni, and preliminary data on temperature stability did not indicate enhanced performance compared to purely Ni-based contacts.

The choice of metal for the formation of silicide ohmic contacts to n-type material is not limited to Ni. Other metals including Co, Hf, and Ta have been shown to form silicides with physical and electrical properties similar to those of nickel silicide [27]. However, as mentioned previously, Ni has been used more than any other metal for the formation of silicide ohmic contacts. Nickel silicide contacts to moderate and heavily doped SiC (see Table 2) are characterized by low specific contact resistances which are necessary for device applications. These contacts also exhibit excellent electrical and physical stability for long-term operation at temperatures of 500 °C and below.

3.3 Other n-type contacts

Alok et al. [68] have used Ti to form ohmic contacts to n^+ ($\approx 10^{20}$ cm^{-3}) 6H layers formed by nitrogen ion implantation into bulk material followed by a post-implant activation anneal at 1250 °C. Specific contact resistances as low as 2×10^{-5} Ω cm^2 were measured using the circular transmission line method (CTLM). This measurement technique has been analyzed in some detail by the authors who also reported results for theoretical calculations which show that the specific contact resistance decreases with increasing doping concentration and increases with increasing barrier height. Details of the theoretical calculations were not presented.

Other novel ohmic contacts to n-type SiC have also been developed. One such contact is TiC which has been investigated by Chaddha et al. [28]. Specific contact resistances as low as 1.3×10^{-5} Ω cm^2 were measured for material of doping concentration 4×10^{19} cm^{-3}. These contacts were formed by growing an epitaxial TiC layer at 1260 °C on heavily doped 6H-SiC. Crystalline TiC has a small lattice mismatch with 6H-SiC, so that the resulting TiC/SiC interface presumably has electrical properties which are quite different compared to those typically observed for metals which are sputter-deposited or evaporated onto SiC. Because of the thermodynamic stability of TiC with SiC [29], no silicides should form at the interface, so that the reaction products differ substantially from those in a standard Ni-based contact.

3.4 Packaging issues for nickel contacts

For device applications, ohmic contacts very often must be wire bonded for connection to a die package. Nickel silicide layers which are formed following short, high temperature anneals (≈ 1000 °C for 2 min) exhibit low specific contact resistances and excellent physical stability for temperatures up to 500 °C; however, for wire bonding applications with Au cap layers, pure nickel silicide contacts exhibit two problems which must be overcome. Gold cap layers often do not adhere well to the silicide contact layers and for high temperature applications, an intermediate diffusion barrier layer is required to prevent interdiffusion between the nickel silicide contact layer and the Au cap layer. These problems are eliminated when the ohmic contacts are formed using NiCr rather than pure Ni [30]. Fig. 6 shows Auger depth profiles for nickel silicide layers formed using 80/20 wt% NiCr ($\approx 3.5/1$ at%). Small concentrations of Ni, Cr, and Si are present near the surface of the thin sample together with a substantial amount of C, while Ni and Cr appear in the thick sample only after several minutes of ion sputtering. Results for the different NiCr layer thicknesses indicate that the surface accumulation process for C is source limited rather than diffusion rate limited. Success rates for direct wire bonding to

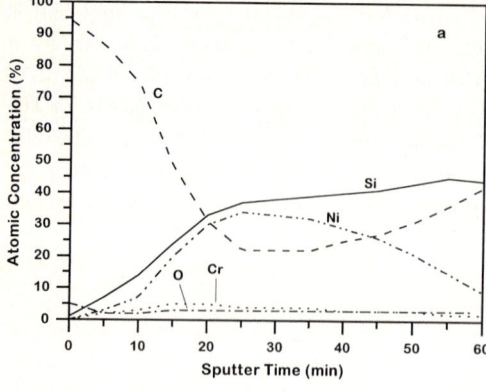

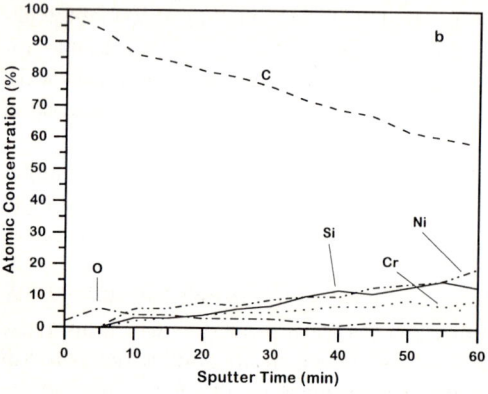

Fig. 6. Auger depth profiles for Ni$_2$Si layers on 6H-SiC formed by annealing; a) 50 nm and b) 200 nm NiCr layers for 3 min at 1100 °C

Au cap layers ($\approx$200 nm) sputter-deposited over the contact layers are very nearly 100% for thin NiCr layers compared to approximately 20% for thick layers. These poor results for the thick layers may be attributed to excessive carbon buildup on the free surfaces of the silicide layers.

Rutherford backscattering analysis (not shown) was performed to investigate the effect of long-term 300 °C anneals of Au/NiCr and Au/Ni samples. The Au cap layers were sputter-deposited following short, high temperature anneals used to produced ohmic contacts for both samples. For the Ni sample, interdiffusion between the Au and the nickel silicide contact layer is apparent after 990 h while the Au and NiCr contact layers show essentially no mixing after 2500 h. The Cr concentration in the NiCr contact layer acts as a barrier to suppress interdiffusion.

3.5 Summary

In summary, models based on standard transport mechanisms which incorporate relatively low barrier heights ($\approx$0.3 to 0.5 eV) predict specific contact resistances for n-SiC which are in good agreement with experimental results. Measured values reported to date range between 10^{-2} and 10^{-6} Ω cm^2 with the lower value requiring an n-type doping concentration of $\geq 10^{19}$ cm^{-3}. This concentration can be achieved either by doping during epitaxial growth or by ion implantation/activation. Nickel is widely used for ohmic contacts to n-SiC, and contact formation proceeds normally through high temperature short-term anneals ($\approx$1000 °C/1 to 5 min) which form a metal-rich (Ni$_2$Si) silicide layer. Excessive carbon can accumulate at the free silicide surface during annealing, and care must be taken during subsequent processing steps such as contact wire bonding. However, nickel silicide ohmic contacts to n-SiC provide both the low specific contact resistance and the excellent electrical and physical stability required for long-term high temperature and high power device applications.

4. Ohmic Contacts to p-type α-SiC

4.1 Introduction to p-type contacts

This discussion will be limited to hexagonal p-type SiC; for a review of contacts to p-type β-SiC, see either Porter and Davis [22] or Harris et al. [23]. P-type materials with large bandgaps, such as SiC, typically form high Schottky barrier heights (SBHs) at their interfaces with metals. This phenomenon can be understood by considering Fig. 7 which shows the typical energy band relationship between a metal and a p-type wide bandgap semiconductor. The large bandgap of SiC (3.0 eV) combined with its electron affinity (3.3 eV) [31] corresponds with the position of the valence band being >6 eV away from the vacuum level. Because the work functions of most metals are between 4 and 5.5 eV, a notably high energy difference exists between the conducting carriers in the two materials. Thus, a large SBH is formed at the interface.

As shown in the chronological listing in Table 3, the earliest reported ohmic contacts to p-type SiC contained either Al or B [32] and were annealed at temperatures above 1700 °C. An enhanced p-type concentration at the SiC surface was thought to result from recrystallization from solution rather than from diffusion of Al into the SiC. Other ohmic contacts comprising Cu–Ti and Al–Si eutectics [33], W [32], and W/AuW/W/Al [34] resulted in extensive reaction depths that would be unacceptable in the fabrication and operation of current devices.

4.2 Aluminum and aluminum alloy contacts

When tunneling dominates the current transport as occurs for sufficiently high doping concentrations and a finite barrier, the specific contact resistance varies as given by Equation (4) (see also [13]). Reducing the width of the depletion (Schottky barrier) region via high doping concentrations at the surface, rather than reducing the height of the Schottky barrier, has been the method used nearly exclusively to form ohmic contacts to p-type SiC. Conventionally, contacts containing Al [7, 35 to 39] (also see Table 3) are annealed at high temperatures (>800 °C), which likely results in diffusion of

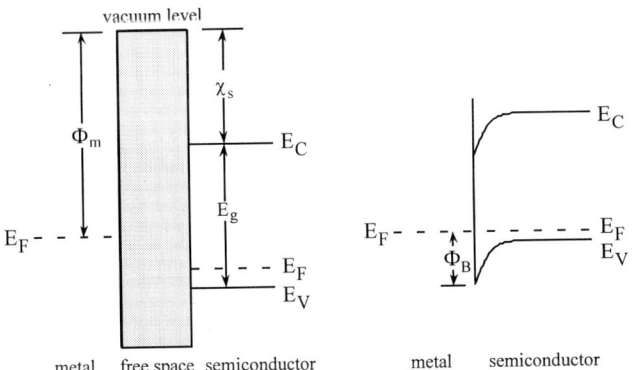

Fig. 7. Energy band diagrams depicting electron energy levels of a typical metal and p-type wide bandgap semiconductor before and after contact, respectively. E_F is the Fermi level, Φ_m the metal work function, E_C the conduction band minimum, E_V the valence band maximum, E_g the bandgap, χ_s the electron affinity, and Φ_B the Schottky barrier height

Table 3

Ohmic contacts to p-type α(6H or mixed)–SiC. Reprinted from [22] with kind permission from Elsevier S.A., P.O. Box 564, 1001 Lausanne, Switzerland

metalliza-tion	deposition method	annealing condition	r_c ($\Omega\,cm^2$)	SiC carr. conc. (cm^{-3})	SiC surface	method of r_c meas.	origin of SiC	surface prepara-tion	ref.
Al–Si (≈1:1)	melting	1700 °C	N.R.	N.R.	(0001)	–	Lely	N.R.	[32]
Si–B (a few %B)	melting	1700 to 2000 °C	N.R.	N.R.	(0001)	–	Lely	N.R.	[32]
W	melting	1900 °C	N.R.	N.R.	(0001)	–	Lely	N.R.	[32]
Cu–Ti (71:29 at%)	melting	>880 °C	N.R.	N.R.	N.R.	–	Carborundum Co.	good	[33]
Al–Si (88.7:11.3 at%)	melting	900 to 1000 °C	N.R.	N.R.	N.R.	–	Carborundum Co.	good	[33]
Al–Ti	N.R.	950 °C/5 min	N.R.	N.R.	(0001)	–	seeded sub.; LPE epi	good	[35]
Al	e-beam evap.	700 °C/10 min	1.7×10^{-3}	1.8×10^{18}	(0001)	TLM	seeded sub.; CVD epi	good	[67]
W/AuW/W/Al	sputtering	1800 °C/120 s	$(2 \text{ to } 5) \times 10^{-4}$	N.R.	(0001 and 000$\bar{1}$)	four-point	CVD epi Lely	varied	[34]
Al–Ti	sputtering	1000 °C/5 min	2.9×10^{-2} to 1.5×10^{-5}	5×10^{15} to 2×10^{19}	(0001)	circular TLM	seeded sub.; CVD epi	good	[7]
Mo	sputtering	none	2×10^{-4}	$>1 \times 10^{19}$	(0001)	four-point and TLM	seeded sub.; CVD epi	good	[45]
Ta	sputtering	none	7×10^{-4}	$>1 \times 10^{19}$	(0001)	four-point and TLM	seeded sub.; CVD epi	good	[45]
Ti	sputtering	none	3×10^{-4}	$>1 \times 10^{19}$	(0001)	four-point and TLM	seeded sub.; CVD epi	good	[45]
Al	sputtering	800 °C/10 min	10^{-2} to 10^{-3}	8×10^{18}	(0001)	TLM	seeded sub.	good	[70]
W/Pt/Al	sputtering	800 to 850 °C/10 min	10^{-2} to 10^{-3}	8×10^{18}	(0001)	TLM	seeded sub.	good	[70]

Contact	Deposition	Annealing	ρ_c ($\Omega\,cm^2$)	Doping	Orientation	Method	Substrate/epi	Quality	Ref.
Ti/Al/ 3C-SiC	e-beam evap./ LPCVD	950 °C/2 min	$(2\ \text{to}\ 3) \times 10^{-5}$	N.R.	(0001)	TLM	Lely sub.; LPE epi	very good	[43]
Pt	e-beam evap.	450 to 850 °C/ 20 min	N.R.	$>1 \times 10^{18}$	N.R.	–	seeded sub.; CVD epi	very good	[71]
Ti/Al	thermal evap.	1000 °C/5 min	N.R.	N.R.	$(000\bar{1})$	–	Acheson sub.; LPE epi	good	[44]
Ti/Al	sputtering	500 °C/20 min (1650 °C/30 min postimplant)	5.6×10^{-4}	Al implant dose: $1 \times 10^{15}\ cm^{-2}$	N.R.	four-point	seeded sub.; CVD epi	N.R.	[38]
Ti/Al	sputtering	1050 °C/20 min	1.2×10^{-2}	1.4×10^{18}	(0001)	four-point	seeded sub.; CVD epi	N.R.	[37]
Al/Ti/Al	thermal evap.	650 to 800 °C	$(1\ \text{to}\ 2) \times 10^{-4}$	9×10^{20}	N.R.	TLM	seeded sub.; CVD epi	N.R.	[39]
Al/Ti/Al	thermal evap.	800 to 950 °C	1×10^{-3} to 2×10^{-4}	$8 \times 10^{17}, 1 \times 10^{20}$	N.R.	TLM	seeded sub.; CVD epi	N.R.	[39]
Si/Co	e-beam evap.	500 °C/5 h 900 °C/2 h	$<4 \times 10^{-6}$	2×10^{19}	(0001)	TLM	seeded sub.; CVD epi	good	[47]

N.R. not reported, sub. sublimation, epi epilayer.

Al into the SiC, and presumably yields an enhanced p-type doping concentration near the surface. A higher doping concentration corresponds to a narrower depletion region through which holes can quantum mechanically tunnel. It has been assumed that the Al heavily dopes the surface of the SiC leading to enhanced tunneling as described above. However, recent work [21] on Al–Ti contacts including surface studies of etched Al–Ti contact layers has revealed many pits at the contact surface of sufficient size and density to suggest that the diffused Al may not actually dope the SiC surface, but may instead lead to enhanced field emission by the creation of many hemispherical intrusions down into the SiC surface similar in nature to those observed by Braslau [40] in Au–Ge contacts to GaAs. This remains to be demonstrated conclusively, however.

Certain properties of Al have complicated the process of making ohmic contacts and hindered the necessary progress in reducing contact resistance and improving thermal stability. The extremely high thermodynamic driving force for oxidation of Al [41] places strict requirements on annealing ambients and passivating layers. The insulating oxide layer that can form when Al is annealed interferes with electrical measurements and the corresponding calculations of specific contact resistance [39]. In addition, the reaction of Al with O opposes the diffusion and presumed activation of Al in SiC. Because of the low diffusion coefficient of Al in SiC [42], high annealing temperatures are required which in turn enhance the tendency for oxidation. Metals which have lower driving forces for oxidation such as Ti have been combined in metallization schemes with Al; [7, 35, 37 to 39, 43, 44], however, it would be informative to investigate metals which oxidize more readily than Al such as Mg in combination with Al. This type of metallization scheme should prevent the oxidation of Al, thereby liberating it to react with or diffuse into the SiC.

Another serious problem associated with Al based compounds is the tendency for the Al to volatilize during the anneal. Recent work [21] has shown that during anneals of a 90/10 wt% Al–Ti alloy at 1000 °C, sufficient amounts of Al can be lost to the annealing environment to drive the Al–Ti sheet resistance into the hundreds or even thousands of $\Omega/\square$ range thus rendering the metallic layer essentially non-conducting. This problem was shown to be dependent on the initial layer thickness and could be avoided by depositing a thick (>2500 Å) layer of Al–Ti provided the annealing time was no more than

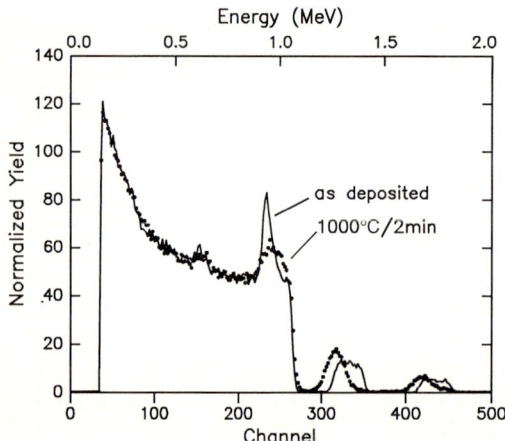

Fig. 8. Rutherford backscattering data for Al–Ti on SiC before and after annealing. The RBS data indicate a segregation of Al and Ti during the anneal with the Al moving toward the metal surface

several minutes. The loss of Al was also shown to result from the segregation of Al and Ti in the alloy during the anneal (see Fig. 8).

The negative effects that Al oxidation and or volatilization can have on an ohmic contact can be seen in the results of Nordell et al. [39] (also see Table 3). Specific contact resistances in the range of $(1 \text{ to } 2) \times 10^{-4} \, \Omega \, cm^2$ were obtained for Al–Ti contacts that were annealed between 600 and 800 °C to material that was doped $9 \times 10^{20} \, cm^{-3}$. Increasing the anneal temperature from 600 to 800 °C lowered the contact resistance by a factor of two. Al–Ti contacts on material doped $1 \times 10^{20} \, cm^{-3}$ were not ohmic until annealing at 800 °C which gave contact resistances of approximately $1 \times 10^{-3} \, \Omega \, cm^2$, however further annealing did not improve the contact resistance. This was attributed to a resistive film that was produced on the metal contact which either masked any improvement in the contact resistance or prevented any lowering in contact resistance from occurring. Comparison of these results with the experimental (Table 3) and theoretical (Fig. 3) values demonstrates the dependence of the contact resistance of Al–Ti alloys on composition, annealing conditions, and the resultant oxidation characteristics. It is also clear from these results that both semiconductor surface doping and high anneal temperatures play a crucial role in the formation of low specific contact resistance contacts using Al–Ti.

4.3 Other non-aluminum based contacts

The need for using Al based contacts can be reduced by heavily doping the SiC surface by in-situ doping or ion implantation. As-deposited Mo, Ta, and Ti contacts on p^+ SiC grown by CVD yielded ohmic contacts [45] with calculated contact resistances in the $10^{-4} \, \Omega \, cm^2$ range. A patent by Glass et al. [46] combines the processes of increasing the doping concentration and decreasing SBH to create ohmic contacts to p-type SiC. Platinum contacts on p^+ SiC were ohmic in both the as-deposited and annealed (to 850 °C) conditions. Platinum's beneficial properties include its high work function, high melting point and oxidation resistance. However, because its work function (5.65 eV) was not sufficient to eliminate the SBH, ion implantation of Al and post-implant annealing at 1500 °C were performed prior to metallization.

Very recently, a novel contact assembly reportedly yielded specific contact resistances below $4 \times 10^{-6} \, \Omega \, cm^2$ on p^+ $(2 \times 10^{19} \, cm^{-3})$ 6H-SiC [47]. This value is half an order of magnitude lower than the lowest contact resistance reported for p-type SiC. The contacts consisted of a 1600 Å layer of Si deposited on a 500 Å layer of Co which was annealed at 500 °C for 5 h followed by 900 °C for 2 h. Rutherford backscattering spectrometry (RBS) data indicated the formation of $CoSi_2$ with a smooth interface and the absence of a graphite phase. The authors incorporated Si to prevent the formation of a C-rich phase as found in previous investigations of annealed Co/SiC structures [48 to 53]. While the high contact resistances reported for annealed Co contacts on p-type SiC [47, 53] were claimed to be associated with the formation of a residual C phase [47], the reason(s) for the ohmicity of the $CoSi_2$ contacts are difficult to surmise.

Recent results [21, 54] using metals such as Ti, Cr, and alloys such as CrB_2 suggest that the formation of the ohmic contact has less to do with the choice of metal or alloy and more to do with an unidentified reaction that takes place during annealing of these materials. Deposition of pure Ti or CrB_2 on heavily doped $(1.3 \times 10^{19} \, cm^{-3})$ p-type material yielded contact resistances in the low $10^{-5} \, \Omega \, cm^2$ range after annealing for approximately a minute at temperatures near 800 °C for Ti or for 30 min at 1100 °C for

CrB$_2$. RBS analysis indicates that neither Ti nor Cr substantially reacts with the SiC during the anneals. Boron evaporates from the boride layer during the anneal, possibly in the form of B$_2$O$_3$ after reacting with oxygen incorporated in the boride layer during sputter deposition.

These results are seen to be consistent with those reported by Lundberg and Ostling [47] when considering the difference in doping of the p-type SiC material used. A plot of contact resistance versus doping for Al–Ti p-type ohmic contacts by Crofton et al. [7] (see also Fig. 3) shows a very strong doping dependence for the contact resistance at high doping concentrations. The curve suggests almost an order of magnitude decrease in contact resistance when the doping is increased from 1×10^{19} to 2×10^{19}, thus a value in the low 10^{-5} Ω cm^2 range on material doped 1.3×10^{19} is consistent with a contact resistance in the mid 10^{-6} Ω cm^2 range on material doped 2×10^{19} cm^{-3}, assuming that the doping dependence of contact resistance for Al–Ti contacts is typical of all p-type ohmic contacts.

4.4 Ion implantation for p-type contacts

The variety of materials and processes that yield ohmic contacts to p-type SiC however have not been shown to yield acceptable contact resistances on moderately or lightly doped material. To date, Al–Ti is the only known metal alloy that yields good ohmic contacts on moderately doped material, thus emphasizing the importance of solving the problems with the Al–Ti alloy as well as being able to accurately control the p-type doping in contact regions of device structures.

While ion implantation is a technique which can be used for selective doping for ohmic contact formation, further research on ion implantation of p-type dopants is needed to yield the high doping concentrations necessary for reducing contact resistances. Many problems are associated with the implantation of high fluences of Al$^+$ and B$^+$ in SiC. Amorphous layers, which are created above a critical dose (Al$^+$: 1×10^{15} cm^{-2}; B$^+$: 5×10^{15} cm^{-2} at room temperature) [55] could not be recrystallized by annealing at temperatures to 1700 °C [56]. Diffusion of the implanted species occurs during annealing or implanting at temperatures above 1500 °C, resulting in spreading of the profile and depletion of the implanted species by evaporation from the surface (observed for temperatures above 1700 °C [56 to 58]). However, high temperatures are necessary to reverse the lattice damage and activate the dopants. One method used to reduce the lattice damage created during the implantation process involves heating the samples during implantation [59, 60]. Several authors report better activation of dopants after implanting at 800 to 850 °C [61] and/or annealing at temperatures in the 1500 to 1700 °C range [56, 61].

4.5 Summary

Low resistance ohmic contacts to p-type SiC can be fabricated using Al–Ti alloys on heavily doped SiC layers. Both the doping and the Al–Ti anneal are crucial to achieving low specific contact resistances. Many problems are associated with Al–Ti, therefore other contacts which appear to rely on tunneling such as CoSi$_2$ and Ti are currently being studied. Since extremely heavy doping is required for non-Al based contacts, typically greater than 1×10^{19} cm^{-3}, implant technology is being developed to enable extremely high surface dopings to be achieved.

5. Future Work

N-type SiC ohmic contacts have developed to the point where specific contact resistances on heavily doped material are now in the $10^{-6}\,\Omega\,cm^2$ range. This is due to the fact that high quality heavily doped material is now available and that processes involving silicide formation using metals such as Ni appear to lead to lowered Schottky barrier heights at the metal–semiconductor interface. Implant technology has not generally been required to meet most device specifications for n-type contacts. N-type contacts which have been aged at high temperatures for long times and then characterized at room temperature indicate good high temperature stability. However, it is important to assess the stability of these contacts both at temperature and under bias; this has yet to be done. Finally, more suitable metallization schemes involving alloys such as NiCr are being developed for wire bonding and packaging applications in order to take fullest advantage of the robust physical and electrical properties of SiC for n-type devices.

P-type ohmic contacts are not as well developed or understood as n-type contacts. The very large Schottky barrier height that exists at the metal–p-type SiC interface has led to the need for extremely heavy surface doping since sufficient barrier lowering to enable ohmic contact formation has not been achieved. Enhanced doping for the formation of p-type ohmic contacts has been achieved either during epitaxial growth, by ion implantation, or it is generally believed, by contact processes using Al and Al based alloys, although currently there is no definitive work that shows that contact anneals of Al alloys lead to electrically active Al doping of p-type SiC. Work is currently underway to definitely determine whether diffusion of Al into p-type SiC results in Al atoms occupying electrically active sites thus yielding enhanced surface doping or whether the process of ohmic contact formation with Al involves some other mechanism such as enhanced field emission. Oxidation and volatilization problems for Al are also being studied while work is underway to find alternative metallization schemes which do not include Al.

Finally, consideration of the thermodynamic properties of contacts, both n- and p-type should be continued. Recent work by De Lucca and Mohney [62] shows that a careful study of the multi-component phase equilibria is necessary for designing contacts with the desired thermodynamic stability. Ternary isotherms for Co–Si–C, Ni–Si–C and other systems complement the electrical work that has been done recently and will help lead to a fuller understanding of ohmic contacts to both n- and p-type SiC.

Acknowledgements The authors would like to acknowledge the Kentucky Space Grant Consortium, the Center for Commercial Development of Space Power and Advanced Electronics, located at Auburn University, with funds from NASA Cooperative Agreement NCC3-511, Auburn University, and the Center's industrial partners. The authors also wish to thank Jim Barnett, Louis Beyer, Mike Bozack, Eric Luckowski, Tamara Isaacs-Smith, Nelson Oder, and Suzanne Mohney for help with obtaining TLM, RBS and Auger data.

References

[1] E. H. RHODERICK and R. H. WILLIAMS, Metal Semiconductor Contacts, 2nd ed., Clarendon Press, Oxford 1988.
[2] H. A. BETHE, M.I.T. Rad. Lab. Rep. 1942.
[3] R. STRATTON, Phys. Rev. **126**, 2002 (1962).
[4] F. A. PADOVANI and R. STRATTON, Solid State Electronics **9**, 695 (1966).

[5]   J. CROFTON, J. R. WILLIAMS, M. J. BOZACK, and P. A. BARNES, Inst. Phys. Conf. Ser.
      No. 137, 719 (1994).
[6]   J. CROFTON, E. D. LUCKOWSKI, J. R. WILLIAMS, T. ISAACS-SMITH, M. J. BOZACK, and
      R. SIERGIEJ, Inst. Phys. Conf. Ser. No. 142, 569 (1996).
[7]   J. CROFTON, P. A. BARNES, J. R. WILLIAMS, and J. A. EDMOND, Appl. Phys. Lett. **62**, 384
      (1993).
[8]   W. MONCH, Barrier Heights of 3C- and 6H-SiC Schottky Contacts: Explanation by the MIGS
      and Electronegative Model, in: Control of Semiconductor Interfaces, Eds. I. OHDOMARI,
      M. OSHIMA, and A. HIRAKI, Elsevier Publ. Co., Amsterdam/New York 1994 (pp. 169 to 174).
[9]   A. T. FROMHOLD, Quantum Mechanics for Applied Physics and Engineering, Academic Press,
      New York 1981.
[10] J. CROFTON and P. A. BARNES, J. Appl. Phys. **69**, 7660 (1991).
[11] J. CROFTON, P. A. BARNES, and M. J. BOZACK, Amer. J. Phys. **60**, 499 (1992).
[12] N. T. SON, O. KORDINA, A. O. KONSTANTINOV, W. M. CHEN, E. SORMAN, B. MONEMAR,
      and E. JANZEN, Appl. Phys. Lett. **65**, 3209 (1994).
[13] S. M. SZE, Physics of Semiconductor Devices, 2nd ed. Wiley, New York 1981.
[14] R. H. COX and H. STRACK, Solid State Electronics **10**, 1213 (1967).
[15] E. KUPHAL, Solid State Electronics **24**, 69 (1981).
[16] S. J. PROCTOR and L. W. LINHOLM, IEEE Electron Device Lett. **3**, 10, 294 (1982).
[17] W. SHOCKLEY, Air Force Avionics Tab. Tech. Rep. No. AL TDR 64-207, Appendix B, Wright
      Patterson Air Force Base, Ohio 1964.
[18] H. H. BERGER, Solid State Electronics **15**, 145 (1972).
[19] G. K. REEVES, Solid State Electronics **23**, 487 (1980).
[20] J. S. CHEN, A. BACHLI, M. A. NICOLET, L. BAUD, C. JAUSSAUD, and R. MADAR, Mater. Sci.
      Engng. B **29**, 185 (1995).
[21] J. CROFTON, L. BEYER, J. R. WILLIAMS, E. D. LUCKOWSKI, S. E. MOHNEY, and J. M. DE-
      LUCCA, accepted for publication in Solid State Electronics.
[22] L. M. PORTER and R. F. DAVIS, Mater. Sci. Engng. B **34**, 83 (1995).
[23] G. L. HARRIS, G. KELNER, and M. SHUR, EMIS Datarev. Ser. No. 13, 231 (1995).
[24] J. CROFTON, P. G. McMULLIN, J. R. WILLIAMS, and M. J. BOZACK, J. Appl. Phys. **77**, 1317
      (1995).
[25] S. LIU, K. REINHARDT, J. SCOFIELD, and C. SEVERT, Workshop High Temperature Power
      Electronics for Vehicles, Fort Monmouth, N.J., USA, April 1995. (Sponsored by Army Re-
      search Lab.).
[26] J. R. WILLIAMS, M. J. BOZACK, T. ISACCS-SMITH, E. E. LUCKOWSKI, C. MEADOWS,
      J. CROFTON, and P. G. McMULLIN, AIP Conf. Proc. **325**, 135 (1995).
[27] J. CROFTON and J. R. WILLIAMS, unpublished.
[28] A. K. CHADDHA, J. D. PARSONS, and G. B. KRUAVAL, Appl. Phys. Lett. **66**, 760 (1995).
[29] M. BACKHAUS-RICOULT, in: Metal–Ceramic Interfaces, Acta-Scripta Metall. Proc., Eds.
      M. F. ASHBY, and J. P. HIRTH, Pergamon Press, New York 1990 (pp. 79 to 92).
[30] E. D. LUCKOWSKI, J. R. WILLIAMS, M. J. BOZACK, T. ISACCS-SMITH, and J. CROFTON, Ma-
      ter. Res. Soc. Symp. Proc. **423**, 119 (1996).
[31] J. PELLETIER, D. GERVAIS, and C. POMOT, J. Appl. Phys. **55**, 994 (1984).
[32] R. N. HALL, J. Appl. Phys. **29**, 914 (1958).
[33] J. S. SHIER, J. Appl. Phys. **41**, 771 (1970).
[34] M. M. ANIKIN, M. G. RASTEGAEVA, A. L. SYRKIN, and I. V. CHUIKO, Amorphous and Crys-
      talline Silicon Carbide III, Vol. 56, Eds. G. L. HARRIS, M. G. SPENCER, and C. Y. YANG,
      Springer-Verlag, Berlin 1992 (pp. 183 to 189).
[35] T. NAKATA, K. KOGA, Y. MATSUSHITA, Y. UEDA, and T. NIINA, Single Crystal Growth of
      6H-SiC by a Vacuum Sublimation Method, and Blue LEDs, Springer-Verlag, Berlin 1989
      (pp. 26 to 34).
[36] A. SUZUKI, Y. FUJII, H. SAITO, Y. TAJIMA, and K. FURUKAWA, J. Cryst. Growth **115**, 623
      (1991).
[37] O. NENNEWITZ, L. SPIESS, and V. BRETERNITZ, Appl. Surf. Sci. **91**, 347 (1995).
[38] L. SPIESS, O. NENNEWITZ, and J. PETZOLDT, Inst. Phys. Conf. Ser. No. 142, 585 (1996).
[39] N. NORDELL, S. SAVAGE, and A. SCHÖNER, Inst. Phys. Conf. Ser. No. 142, 573 (1996).
[40] N. BRASLAU, J. Vacuum Sci. Technol. **19**, 803 (1981).

[41] M. W. Chase, Jr., C. A. Davies, J. R. Downey, Jr., D. J. Frurip, R. A. McDonald, and N. A. Syverud, JANAF Thermochemical Tables, Vol. 14, The American Chemical Society and the National Institute of Physics for the National Bureau of Standards, Midland, (MI) 1985.

[42] E. N. Mokhov, Y. A. Vodakov, and G. A. Lomakina, Soviet Phys. – Solid State 11, 415 (1969).

[43] V. A. Dmitriev, K. Irvine, and M. Spencer, Appl. Phys. Lett. 64, 318 (1994).

[44] T. Uemoto, Jpn. J. Appl. Phys. 34, L7 (1995).

[45] J. B. Petit, P. G. Neudeck, C. S. Salupo, D. J. Larkin, and J. A. Powell, Inst. Phys. Conf. Ser. No. 137, 679 (1994).

[46] R. C. Glass, J. W. Palmour, R. F. Davis, and L. M. Porter, North Carolina State University, Raleigh, NC; Cree Research, Inc., Durham (NC) 1994.

[47] N. Lundberg and M. Ostling, Solid State Electronics 39, 1559 (1996).

[48] T. C. Chou, A. Joshi, and J. Wadsworth, J. Vacuum Sci. Technol. A 9, 1525 (1991).

[49] T. C. Chou, A. Joshi, and J. Wadsworth, J. Mater. Res. 6, 796 (1991).

[50] M. Nathan and J. S. Ahearn, J. Appl. Phys. 70, 811 (1991).

[51] N. Lundberg, C.-M. Zetterling, and M. Ostling, Appl. Surf. Sci. 73, 316 (1993).

[52] N. Lundberg and M. Ostling, Appl. Phys. Lett. 63, 3069 (1993).

[53] L. M. Porter, J. S. Bow, M. J. Kim, R. W. Carpenter, and R. F. Davis, J. Mater. Res. 10, 26 (1995).

[54] T. N. Oder, J. R. Williams, S. E. Mohney, and J. Crofton, submitted to J. Electronic Mater.

[55] T. Kimoto, A. Itoh, H. Matsunami, T. Nakata, and W. Watnabe, J. Electronic Mater. 25, 879 (1996).

[56] G. Pensl, V. V. Afanasev, M. Bassler, M. Schadt, T. Troffer, J. Heindl, H. P. Strunk, M. Maier, and W. J. Choyke, Inst. Phys. Conf. Ser. No. 142, 275 (1996).

[57] A. V. Suvorov, D. A. Plotkin, V. N. Makarov, and V. N. Svetlov, Mater. Res. Soc. Symp. Proc. 279, 415 (1993).

[58] A. V. Suvorov, I. O. Usov, V. V. Sokolov, and A. A. Suvorov, Mater. Res. Soc. Symp. Proc. 396, 239 (1996).

[59] J. A. Edmond, K. Das, and R. F. Davis, J. Appl. Phys. 63, 922 (1988).

[60] J. A. Edmond, J. Ryu, J. T. Glass, and R. F. Davis, J. Electrochem. Soc. 135, 359 (1988).

[61] J. A. Freitas, Jr., J. Gardner, and M. V. Rao, Inst. Phys. Conf. Ser. No. 142, 529 (1996).

[62] J. M. DeLucca and S. E. Mohney, Mater. Res. Soc. Symp. Proc. 423, 137 (1996).

[63] A. Addamiano, Semiconductor Crystals of Silicon Carbide with Improved Chromium-Containing Electrical Contacts, US Patent No. 3510733 (1970).

[64] G. Kelner, S. Binari, M. Shur, and J. W. Palmour, Electronics Lett. 27, 1038 (1991).

[65] R. C. Glass, L. M. Spellman, and R. F. Davis, Appl. Phys. Lett. 59, 2868 (1991).

[66] R. C. Glass, L. M. Spellman, S. Tanaka, and R. F. Davis, J. Vacuum Sci. Technol. A 10, 1625 (1992).

[67] J. Crofton, J. M. Ferrero, P. A. Barnes, J. R. Williams, M. J. Bozack, C. C. Tin, C. D. Ellis, J. A. Spitznagel, and P. G. McMullin, Amorphous and Crystalline Silicon Carbide IV, Eds. C. Y. Yang, M. M. Rahman, and G. L. Harris, Springer-Verlag, Berlin 1992 (p. 176).

[68] D. Alok, B. J. Baliga, and P. K. McLarty, IEDM Technical Digest, IEDM 1993 (p. 691 to 694).

[69] J. Crofton, P. G. McMullin, J. R. Williams, and M. J. Bozack, Trans. 2nd High Temperature Electronics Conf., Charlotte (NC) 1994 (p. XIII-15).

[70] S. Liu, S. R. Smith, S. Adams, C. Severt, and J. Leonard, ibid. (p. XIII-9).

[71] R. C. Glass, J. W. Palmour, R. F. Davis, and L. M. Porter, Method of Forming Ohmic Contacts to p-Type Wide Bandgap Semiconductors and Resulting Ohmic Contact Structure, US Patent No. 5323022 (1994).

phys. stat. sol. (b) **202**, 605 (1997)

Subject classification: 61.80.Jh; 68.35.Bs; 73.40.Ns; S6

A Review of SiC Reactive Ion Etching in Fluorinated Plasmas

P. H. Yih (a), V. Saxena (b), and A. J. Steckl[1]) (b)

(a) Bell Laboratories, Lucent Technologies, Orlando, FL 32819, USA

(b) Nanoelectronics Laboratory, Department of Electrical and Computer Engineering/ Computer Science, University of Cincinnati, Cincinnati, OH 45221-0030, USA

(Received January 31, 1997)

Research and development in semiconducting silicon carbide (SiC) technology has produced significant progress in the past five years in many areas: material (bulk and thin film) growth, device fabrication, and applications. A major factor in this rapid growth has been the development of SiC bulk crystals and the availability of crystalline substrates. Current leading applications for SiC devices include high power and high temperature devices and light emitting diodes. Due to the strong bonding between Si and C (Si–C = 1.34 × Si–Si), wet chemical etching can only be performed at high temperature. Therefore, plasma-based ("dry") etching plays the crucial role of patterning SiC for the fabrication of various electronic devices. In the past several years, reactive ion etching (RIE) of SiC polytypes (3C and 6H) has been investigated in fluorinated gases (primarily CHF_3, $CBrF_3$, CF_4, SF_6, and NF_3), usually mixed with oxygen and occasionally with other additives or in a mixture of fluorinated gases. In this paper, a review of SiC RIE is presented. The primary emphasis is on etching of the 3C and 6H polytypes, but some results on RIE of the 4H polytype are included. The paper covers the basic etching mechanisms, provides typical etching properties in selected plasma conditions, discusses the effects of changes in various etching parameters, such as plasma pressure, density and power, etching time, etc. The etching of features of sizes varying from sub-μm to tens of μm's is addressed. Finally, optimum etching conditions and trade-offs are considered for various device configurations.

1. Introduction

In the early 1960's, at the beginning of modern silicon carbide (SiC) development, SiC research was conducted only on small pieces (<1 cm) produced by the Acheson process [1], which is currently used for the sand paper industry. In the recent years, larger area 6H- and 4H-SiC substrates (reaching ≈5 cm diameter) obtained from boules grown by the modified Lely sublimation method have become commercially available [2]. The availability of these SiC substrates has led to greatly increased research and development in all aspects of SiC. High-quality cubic and hexagonal SiC epitaxial layers have been successfully grown on the crystalline SiC substrates. For useful SiC device production, large area and defect-free [3] substrates are required. At the same time, the growth of 3C-SiC on Si substrates, has continued to improve [4 to 7]. This heteroepitaxial approach has the ultimate goal to provide truly large area and low cost SiC substrate materials, as well as the potential integration of SiC high voltage technology with Si

---

[1]) The author to whom correspondence on this paper should be addressed.
   e-mail: a.steckl@uc.edu

microelectronics. This general material availability has opened the way to significant progress in the fabrication of various SiC electronic devices for applications requiring high-temperature, high-speed, and high-power operation [8 to 13]. Promising electrical performance for MESFET's [14, 15], MOSFET's [16, 17], thyristors [18, 19], non-volatile memory devices [20, 21], heterojunction bipolar transistors (HBT) [22 to 27], charge-coupled devices (CCD) [28], and high breakdown voltage Schottky [29 to 40] and p–n junction diodes [41 to 45] have been reported. Various etching characteristics are required for the patterning of these devices. Reviews [46, 47] of higher power semiconductor electronics show SiC as the best candidate material for the near future smart power technology. This is mainly due to its outstanding material properties such as high thermal conductivity, high electric field breakdown, high electron saturation velocity, and relatively better developed device fabricated processes.

The paramount reason for using plasma-based etching of Si or III–V semiconductor materials is to take advantage of the relative anisotropy of plasma etching in order to precisely control the line-width. This becomes extremely important when the device feature is at the sub-µm scale (<1 µm). For SiC, another important reason to employ plasma etching is the chemical stability of SiC which makes "wet" etching of device structure very difficult. Indeed, wet etching of SiC has to be done either at elevated temperature ($>600\,^\circ$C) in alkaline solutions [48] or with photoelectrochemical etching at room temperature [49 to 51]. It is important to note that line-width control is very difficult in the wet etching of SiC process under such high temperature or photo-assisted chemical etching. This explains why plasma-assisted ("dry") etching plays a crucial role in the fabrication of various of SiC devices, for both large and small dimensions.

Reactive ion etching (RIE) of SiC in fluorinated plasmas has been developed to the point where it is now widely employed in both the research and development environment and commercial product fabrication [2]. In our research laboratory at the University of Cincinnati, we have worked for the past ten years on the RIE of SiC for electronic device fabrication. We have investigated aspects of SiC RIE in fluorine-based plasmas, leading to a greater understanding of dry etching issues of SiC materials. In the early stage of these investigations, amorphous [52] and polycrystalline SiC [53] were employed. More recently, 3C-, 4H- and 6H-SiC RIE have been reported in a variety of fluorinated gases (CHF_3, $CBrF_3$, CF_4, SF_6, and NF_3) usually in combination with oxygen. SiC RIE in fluorinated plasmas has been shown to produce useful etch rates (100 to 1000 Å/min) and a high degree of etching anisotropy leading to the patterning of sub-µm features. One difficult aspect of SiC RIE has been the formation of residues (which lead to a rougher surface) after longer term etching under many conditions. This is (in part) because the commercial RIE systems are designed to accommodate multiple large Si wafers rather than the much smaller SiC substrates currently available (thus the area of the electrode is usually much larger than that of the SiC samples being etched). Residue formation can be a serious problem for subsequent processes, such as metal contact (ohmic or Schottky) formation. Several techniques which have now been successfully developed to prevent residue formation are reviewed in the latter section of this paper. Very limited information has been reported [54] on SiC etching in chlorine-based plasmas, which has the potential for using non-metallic etch mask materials (e.g. SiO_2) and obtaining residue-free etching. SiC etching in fluorine-based high density electron cyclotron resonance (ECR) plasma was reported recently [55, 56] with promising results for high etch rate, anisotropic profile, and residue-free (smooth) surface topography.

In this paper, a review of the current understanding and practice of reactive ion etching of SiC is presented. We concentrate on the fluorine-based RIE of 6H-SiC, the most widely used polytype. However, some results in the plasma-assisted etching of 3C and 4H polytypes are also discussed. The topics include: a) basic plasma etching mechanisms in fluorinated-oxygen mixtures; b) etching characeristics — etch rate (ER), etch aspect ratio (EAR), etch rate ratio (ERR or selectivity) to Si, SiO$_2$ — and their dependence on plasma parameters (power, pressure, and density); c) techniques for obtaining smooth (residue-free) etched surfaces; d) techniques for etching sub-μm features; e) an introduction to various high density plasma sources; f) a discussion of suggested etching conditions and trade-offs in light of various device configurations. Finally, we discuss the recent progress on high density plasma etching of SiC.

2. Basic Plasma Etching Mechanisms

In this section a brief overview is given of the basic mechanisms at work during reactive ion etching (RIE) of SiC at room temperature. Both physical and chemical processes participate in the overall removal of Si and C atoms of SiC. Basically, all chemical etching processes probably consist of three sequential steps [57, 58]: 1. adsorption of the etching species; 2. product formation; 3. product desorption. During a plasma discharge a variety of species are produced [59]: charged particles (ions and electrons), photons, and neutrals (radicals). Plasma etching of materials can proceed via a combination of physical and chemical mechanisms. The dominant mechanism is determined by the volatility of the reaction by-products and the energy of the ionized species. In practice, this translates into choices regarding the feed gas (inert or reactive), the plasma pressure and the choice of connecting the sample biasing electrode to the RF power or to ground. These conditions result in plasma etching processes which can be grouped into four categories: a) *sputtering* — purely physical removal of the material by energetic ions of the gas molecules; b) *chemical plasma etching* — neutral radicals formed in the plasma react with the substrate material to produce volatile species ; c) *ion-enhanced chemical etching* — energetic ions damage the etch surface, enhancing its reactivity; d) *inhibitor-controlled chemical etching* — ion bombardment removes inhibitor layers from surfaces orthogonal to the ion flux, allowing chemical etching to proceed. Reactive ion etching, which encompasses the last two categories, operates at relatively low pressures (from a few mTorr to hundreds of mTorr) with the sample placed on the cathode, thus resulting in the production of fairly energetic ions along with the formation of reactive radicals. Plasma-based etching in the RIE mode generally allows for the most useful trade-offs between etch rates and anisotropy.

The overall etch rate is given by the combination for the material removing mechanisms outlined above:

$$R = R_{\text{SPUTTER}} + R_{\text{NEUT}} + R_{\text{IEN}} + R_{\text{ICN}} , \tag{1}$$

where R_{SPUTTER} is the ion sputter-removal rate, R_{NEUT} is the chemical etching performed by neutral radicals, R_{IEN} is the ion-enhanced neutral chemical etching, and R_{ICN} is the inhibitor-controlled neutral etching. To understand the effect of the arrival rate of ions and neutrals on the overall etch rate, Eq. (1) can be expanded [60 to 62] as follows:

$$R = F_{\text{I}} \varphi_{\text{S}} + F_{\text{N}} (1 - \alpha - \beta) \, \varphi_{\text{N}} + F_{\text{N}} \alpha \varphi_{\text{N}}^* + F_{\text{N}} \beta \varphi_{\text{N}}^{**} , \tag{2}$$

where F_I is the ion flux (ions/cm² s), F_N is the flux of neutral particles, φ_S is the sputtering efficiency (cm³/ion), φ_N is the chemical etch rate efficiency of neutral species (cm³/neutral), φ_N^* and φ_N^{**} are the chemical etch rate efficiencies of neutral species on the fraction (α) of the surface which has been ion bombarded ("sensitized") and on the surface fraction (β) covered by an etch inhibitor. The sensitized surface fraction obviously increases with the ion flux. The ion flux also (negatively) affects the inhibitor covered fraction, thus enhancing the chemical (neutral) etch rate. Basically, this model simply explains the relationship between the two major plasma species (neutral radicals and ions) involved in the plasma-assisted ("dry") etching process.

For SiC etching in mixtures of fluorinated gases and oxygen, the most likely chemical reactions associated with the removal of Si and C atoms are given in Eq. (3) to (5) [53], leading to the combined chemical reaction shown in Eq. (6) for the removal of SiC molecules. However, it is assumed that actual molecular removal is unlikely. To some extent, we do not consider the other reaction compounds, such as COF_2, in the following discussion:

$$Si + mF \rightarrow SiF_m \qquad (m = 1 \text{ to } 4), \tag{3}$$

$$C + mF \rightarrow CF_m, \tag{4}$$

$$C + nO \rightarrow CO_n \qquad (n = 1 \text{ to } 2), \tag{5}$$

$$SiC + mF + nO \rightarrow SiF_m + CO_n + CF_m. \tag{6}$$

In silicon RIE, the presence of oxygen in fluorinated plasma produces important effects, due to several mechanisms [63]. *First,* [O] atoms react with unsaturated fluoride species generating reactive F atoms, while simultaneously depleting these polymer-forming species. *Second,* when increasing O_2 is added to the feed gas, sufficient [O] chemisorbs on the silicon surface making it more "oxide-like", thus reducing the available Si sites for etching. *Third,* if the oxygen additive is introduced as a replacement for the fluorinated gas in order to keep the total flow rate constant, dilution effects reduce the etch rate. The effect of oxygen additive can be readily seen in the etching of Si in $CBrF_3/O_2$, CF_4/O_2, and SF_6/O_2 gas mixtures [53]. The highest etch rate is obtained when a relatively low oxygen ($\approx$5 to 20%) percentage is added to the fluorinated gas to enhance the generation of etchant species and to deplete polymer species. Beyond this point, as the oxygen percentage continues to increase, the etch rate monotonically decreases. A model was proposed for the etching of Si [64, 65]. When Si is exposed to [F] atoms, it acquires a fluorinated skin (fluorine atom adsorption) which extends a few monolayers below the surface. While most [F] atoms involved in the etching process react with the surface, a small fraction attack underlying Si–Si bonds (3.38 eV), liberating SiF_m molecules. In Si etching, the [F/C] ratio, where [C] originates in the gas phase (etchant), was proposed [58] as a determinant of the etching dependence on the process pressure, input power, and additives. A high [F/C] ratio in the gas produces a high etch rate, while at a sufficiently low [F/C] ratio a polymer film could be deposited on the surface during the process, resulting in a negative etch rate or a taper (width of top < bottom) etching profile. Polymer formation could be even more severe when H_2 is present in the plasma. This is a typical example of competition between etching and deposition that determines the etching profile [66].

Turning now to SiC RIE in fluorinated plasmas, the etching process contains both similarities and differences from the Si case. For example, highly anisotropic etching profiles are obtained (>10:1) during the residue-free etching process which employs a graphite sheet covering the powered electrode [67]. In this case, an [F/C] ratio much smaller than unity is expected, since copious levels of [C] are produced from several sources: the SiC itself, the etchant gas (in the case of fluorocarbons), and most importantly from the graphite sheet. This indicates that the [F/C] ratio model used successfully for Si etching must be modified for SiC etching. For example, an anisotropic etching profile was obtained for SiC in SF_6/O_2 mixtures, which is not the case of normally undercut etching profile for Si [68, 69]. The reason is due to SiC itself providing carbon, which enhances the polymer formation, preventing the side wall from being etched. However, there were no fluorinated gases we have investigated which produced an undercut profile during SiC etching.

In addition to the indirect roles of oxygen in the gas phase reaction, oxygen also participates by directly removing C atoms in SiC through the reaction given in Eq. (5). Carbon can be etched in either a pure fluorine-containing plasma (Eq. (4)) or a pure O_2 (Eq. (5)) plasma. As reported by several research groups [52, 53, 67, 70], a thin carbon-rich layer is formed on the etched surface. This indicates that C is not removed sufficiently fast from the etched surface through the reaction of either carbon–fluorine [C–F] or carbon–oxygen [C–O] reactions. At low (or zero) % O_2, it has been suggested [70] that carbon is preferentially removed through the formation of CF_m (Eq. (4)) rather than C–O_n (Eq. (5)), whereas at high $O_2\%$ removal of carbon is dominated by the [C–O] reaction. However, generally the SiC etch rate decreases as the O_2 percentage increases. This indicates that the [C–O] reaction through Eq. (5) may not be as efficient as the [C–F] reaction through Eq. (4). This explains why one obtains the highest 3C-SiC etch rate under low oxygen percentage conditions, which produce high fluorine intensity in CF_4/O_2, NF_3/O_2, and SF_6/O_2 mixtures (Fig. 1 to 3), and for higher oxygen percentage (60 to 80%) in CHF_3/O_2.

Along with the purely chemical plasma etching process, the effect of the energetic ion flux needs to be considered. This primarily consists of damage or breaking of the surface Si–C bond (4.52 eV), which enhances the chemical reaction efficiency, and removal of non-volatile surface species, which enables the chemical reaction to proceed. The latter includes providing sufficient energy to break the strong C–C bonding (6.27 eV) that could exist in the C-rich layer. This combination of effects have led to a two regime model for the effect of dc bias on the etch rate of polycrystalline SiC [53]: a) at low dc bias conditions, the low energy (and effectiveness) of the ion flux is the dominant mechanism; b) at sufficiently high values of the dc bias, the ion energy is high enough to no longer limit the process and the etch rate is determined by the removal efficiency of the chemical reaction.

3. Experimental Conditions

In order to draw meaningful conclusions regarding effects of various etching parameters we have mostly utilized a common set of etching conditions in all experiments. This consists of an rf power of 200 W (0.4 W/cm^2), a pressure of 20 mTorr, a total flow rate of 20 sccm, and etch times of 5, 20, and/or 30 min. In the following sections, we refer to these experimental values as the "standard" etching condition. A few exceptions to the standard condition are made: a) for the use of low flow rate (5 sccm) H_2 additive in which case the plasma pressure was increased by 5 mTorr to minimize the reactant-limited effect; b) for evaluating mixtures of NF_3/SF_6 where a flow rate as high as 35 sccm was utilized. Several other research groups also reported on the RIE etching of SiC in a variety of conditions. Usually, they operated at a much higher etching pressure, ranging as high as 200 mTorr, or in a microwave plasma. The process pressure plays a critical role in determining the value of the dc bias (ion energy), the etchant species lifetime (reciprocal to pressure), macro- and micro-uniformity, and the level of polymer formation. Unfortunately, in many cases, there is not sufficient information on the etch rate for a meaningful comparison. However, relevant data from other groups is discussed and referenced in separate sections (Sections 4.3, 5.3.2 and 6).

To quantitatively measure the relative intensity of the primary plasma species, a computer controlled optical emission spectrometer (OES) monitored the photoemission from the plasma during the process. The dc bias and plasma species intensity were recorded after striking the plasma for 1 min. Noble gas actinometry [71] was regularly employed, using 0.6 sccm (3% of total gas flow rate) of Ar added to the gas feed, to obtain the relative concentration of various species, such as [H], [F], [Ar], [O], and [Br]. Plasma photoemission was monitored at wavelengths of 486 nm for hydrogen [H], 703 nm for [F], 750 nm for argon [Ar], 777 nm for oxygen [O], and 336 nm for bromine [Br].

Several categories of SiC material were utilized for our etching experiments: a) *3C-SiC*: 6 to 8 μm unintentionally doped n-type layers grown on p-type Si(100) by conventional chemical vapor deposition [4]; b) *6H-SiC*: primarily n-type (0001) Si-face substrates [2] with doping concentration of 10^{17} cm^{-3}; in one set of experiments: n$^+$ substrates with doping concentration of 10^{19} cm^{-3}, and p-type substrates with doping concentration of 10^{18} cm^{-3}; c) *4H-SiC*: n-type (0001) Si-face substrates [2] with doping concentration of 9.9×10^{18} cm^{-3}. Sputter deposited aluminum patterned by lift-off process was employed as the etch mask. The Al metal mask was removed prior to measuring the etch step height (and associated etch rate) and surface analysis. The step height was measured with a surface profilometer (Dektak). Generally, the sample size was kept small enough so that it does not significantly influence the nature of the discharge during the process. This is extremely important for the comparison of experiments with the same nominal etching condition.

4. Etching in Fluorinated Gas Mixtures

As discussed in Section 2, the RIE of SiC is performed in a primary fluorinated gas to which secondary gases (oxygen and hydrogen) may be added to control or enhance the process. In the work reviewed here, the primary gas consists of one of the following gases: CF$_4$, SF$_6$, NF$_3$, CBrF$_3$, and CHF$_3$. Mixtures of certain fluorinated gases were also investigated. The reasons for choosing these gases are given below along with their etching characteristics. Other characteristics frequently investigated include the etching anisotropy (through the line edge profile) and the density of residues in the etch field. The most important characteristic to be measured is, of course, the etch rate. In addition, plasma characteristics, such as the self-induced bias and the density of the main plasma species (such as [F], [O], [H]), are normally measured in order to understand the mechanisms determining the process. Finally, the process parameter most commonly varied is the level of the secondary gas.

4.1 Fluorinated-oxygen mixtures

In this section, we review the important characteristics of 3C-SiC etching in mixtures of fluorinated gases and oxygen and compare 3C- and 6H-SiC RIE. In all experiments described below, the standard etching conditions mentioned before, with an etch time of 5 min were employed. The SiC samples were unintentionally-doped n-type 3C-SiC grown on p-type Si [4], with the exception of RIE in $CBrF_3/O_2$ mixture where high temperature annealed polycrystalline 3C-SiC sputter deposited on Si was employed [53]. Si and SiO_2 (thermally grown) samples were also etched simultaneously with 3C-SiC samples, to provide the etch rate ratios (ERR) of SiC/Si and SiC/SiO_2 required for fabrication of SiC/Si heterojunction devices, such as HBT. The 6H-SiC samples were n-type Si-face (0001) commercially available [2] substrates. The 3C-SiC etching characteristics (etch rate, ERR and relative density of main etching species) are shown in Figs. 1 to 5 for CF_4, SF_6, NF_3, CHF_3, and $CBrF_3$ as a function of $O_2\%$. Fig. 6a shows the comparison of the etch rates obtained using each of these gases on the same plot. The self-induced dc bias resulting in the various etching conditions is shown in Fig. 7. Some of the main conclusions regarding 3C-SiC RIE are: a) a relatively high etch rate can be obtained in SF_6/O_2 and NF_3/O_2 mixtures; b) a high SiC/Si ERR is produced in low fluorinated gases, such as CHF_3/O_2 and $CBrF_3/O_2$ mixtures; c) the dc bias is not a portable parameter; d) highly anisotropic etching profile can be obtained in all fluorinated oxygen plasmas except CHF_3/O_2 mixtures. Details of the etch aspect ratio for these cases were reported elsewhere [53] and will not be repeated here.

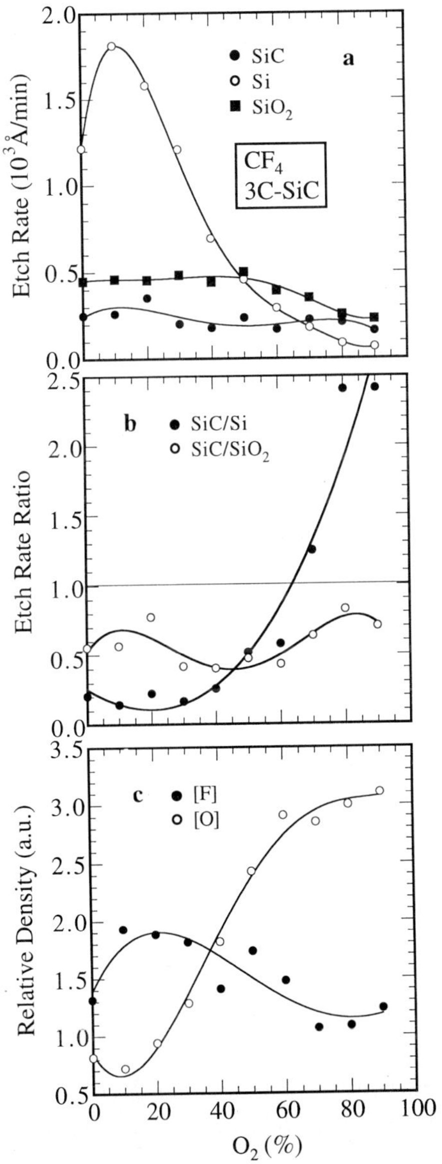

4.1.1 3C-SiC etching in CF_4/O_2 [52, 67, 70, 72 to 76]

CF_4 was one of the first halogens used for plasma-assisted etching [58, 61] and its characteristics in etching Si are well-known [63]. The SiC, Si, and SiO_2 etch rates in CF_4/O_2 mixtures and related information are given in Fig. 1. The typical effect of the O_2 percentage

Fig. 1. 3C-SiC RIE in CF_4 as a function of O_2 plasma under the standard conditions a) SiC, Si, and SiO_2 etch rate, b) SiC/Si and SiC/SiO_2 etch rate ratio under the standard conditions, c) relative [F] and [O] densities [73] (a.u. means arb. units)

on the Si etch rate is observed, with the etch rate experiencing a sharp maximum ($\approx 1.8 \times 10^3$ Å/min) in the vicinity of the peak in the [F] concentration in the plasma ($\approx 10\%$ O$_2$) and then decreasing rapidly with increasing O$_2$ percentage. The SiC etch rate exhibited a much less pronounced peak value of 350 Å/min at 20% O$_2$. No effect of O$_2$ percentage on the SiO$_2$ etch rate (≈ 450 to 500 Å/min) was observed in the 0 to 50% range. At higher O$_2$ levels, the SiO$_2$ etch rate slowly decreases and reaches a low value of ≈ 210 Å/min at 90% O$_2$. ERRs of SiC/Si and SiC/SiO$_2$ are shown in Fig. 1b. An ERR > 1 for SiC/Si is obtained at high (70 to 90%) O$_2$ percentage, where its effect is more pronounced on the Si etch rate than on the SiC etch rate. No ERR > 1 for SiC/SiO$_2$ was found in the entire range of O$_2$ percentage. As shown in Fig. 1c, the [F] and [O] plasma emission intensities were monitored as the main etching species, thus providing a measure of their relative density in the plasma. As expected, the [F] intensity experiences a maximum at a relatively low level of O$_2$ in the gas stream, whereas the [O] intensity increases monotonically with the O$_2$ percentage. A cross-over point between the [F] and [O] intensities is observed at $\approx 40\%$ O$_2$. As noted by other workers in the field [53], the pattern of the Si etch rate with O$_2$ percentage shows at first a strong increase due to additional [F] liberated by oxygen, followed by eventual reduction due to the dilution effect. On the other hand, no such strong relationship was found between the SiC etch rate and the [F] and/or [O] emission intensities.

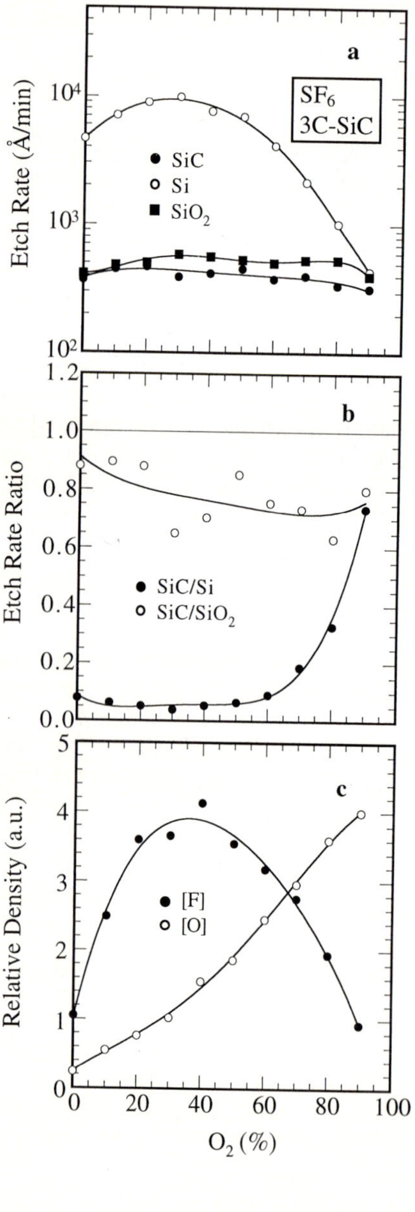

4.1.2 3C-SiC etching in SF$_6$/O$_2$ [53, 73, 75 to 78]

The abundance of [F] in the SF$_6$ plasma produces a high Si etch rate, which is particularly favorable for applications requiring etching of thick layers such as micromachining [68] and power device patterning [69]. Re-

Fig. 2. 3C-SiC RIE in SF$_6$ as a function of O$_2$ plasma under the standard conditions a) SiC, Si, and SiO$_2$ etch rate, b) SiC/Si and SiC/SiO$_2$ etch rate ratio, c) relative [F] and [O] densities [3]

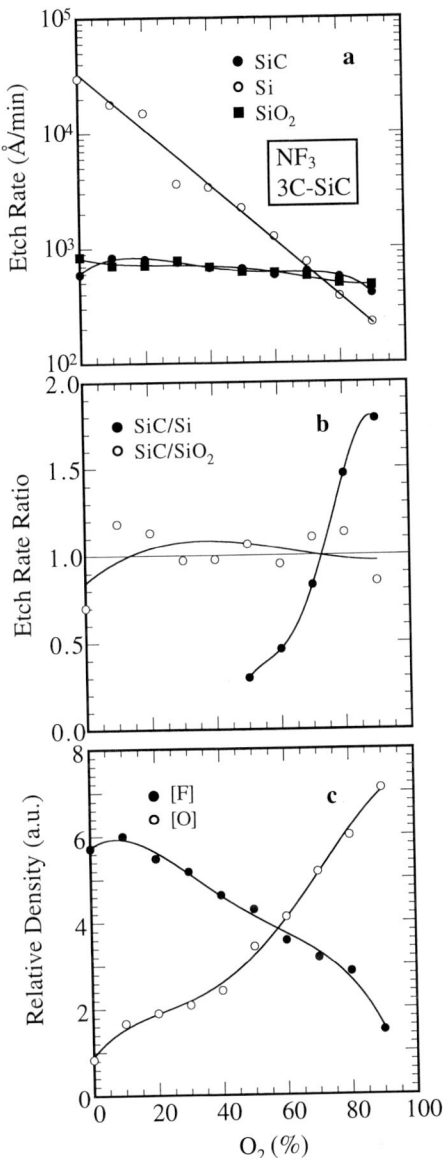

Fig. 3. 3C-SiC RIE in NF_3 as a function of O_2 plasma under the standard conditions a) SiC, Si, and SiO_2 etch rate, b) SiC/Si and SiC/SiO_2 etch rate ratio, c) relative [F] and [O] densities [73]

garding SiC etching, SF_6 does not provide a source of C in the gas phase, unlike CF_4. Results obtained in etching of SiC, Si, and SiO_2 in SF_6/O_2 mixtures are shown in Fig. 2. The Si etch rate follows the same trend as the [F] concentration as a function of O_2%, reaching a maximum of 9×10^3 Å/min at 30% O_2. By comparison, the SiC and SiO_2 etch rates are nearly constant with O_2 percentage. The highest SiC etch rate (≈ 450 Å/min) is obtained at 20% O_2. ERRs of SiC/Si and SiC/SiO_2 are shown in Fig. 2b. No etch selectivity greater than unity was found for either SiC/Si or SiC/SiO_2 within the full range of O_2 percentage. SF_6/O_2 is the only gas mixture reported in this paper which shows no ERR > 1 for either set of materials. As shown in Fig. 2c, a cross-over point of [F] and [O] emission intensities is present at $\approx 70\%$ O_2. No simple relationship was found between either [F] or [O] emission intensity and the SiC etch rate.

4.1.3 3C-SiC etching in NF_3/O_2 [67, 73, 75, 78]

NF_3 is perhaps the most copious producer of [F] species in the plasma, hence resulting in very high Si etch rates [73]. With regard to SiC etching, NF_3 like SF_6 is not a source of C, but unlike SF_6 it produces only volatile by-products. Etching results for SiC, Si, and SiO_2 in NF_3/O_2 mixtures are given in Fig. 3. The Si ER is highest (≈ 3 μm/min) in pure NF_3 and drops exponentially with the addition of O_2. In sharp contrast, the SiC and SiO_2 ERs experience neither a tremendous etch rate in pure NF_3 nor a rapid decrease with O_2 dilution. Nearly equal ERs of SiC and SiO_2 were obtained in the full range of O_2 percentage. The highest SiC etch rate of ≈ 830 Å/min was obtained at 10% O_2. Etch selectivities of SiC/Si and SiC/SiO_2 shown in Fig. 3b indicate that a SiC/Si ERR greater than unity is obtained at high O_2 percentage (>70%). The [F] and [O] plasma emission intensities in NF_3/O_2 mixtures are shown in Fig. 3c. Among the gas mixtures investigated, NF_3/O_2 plasmas produce the

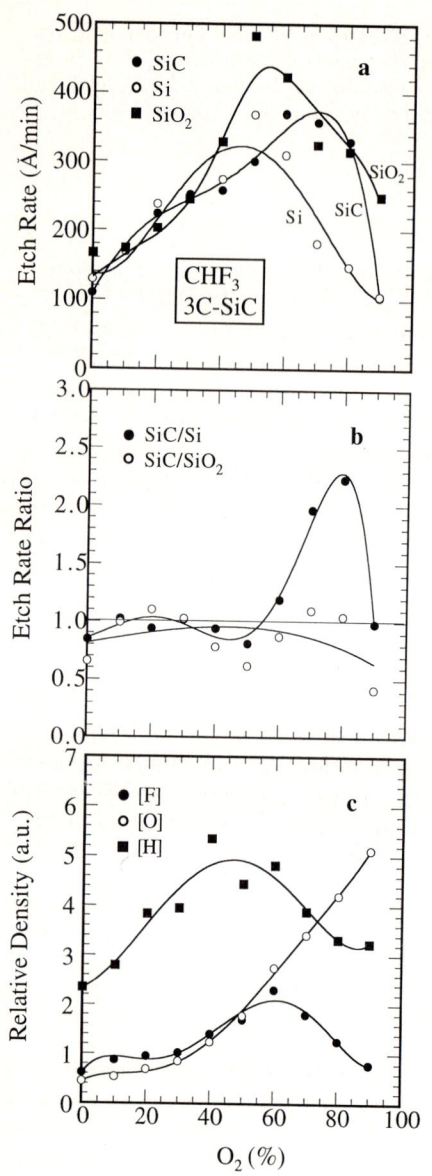

Fig. 4. 3C-SiC RIE in CHF_3 as a function of O_2 plasma a) SiC, Si, and SiO_2 etch rate, b) SiC/Si and SiC/SiO_2 etch rate ratio, c) relative [F], [O], and [H] densities [73]

highest [F] emission levels. Highly anisotropic SiC etching profile in NF_3/O_2 mixture was reported in detail elsewhere [73]. The [F] emission intensity decreases linearly as a function of O_2 percentage so that O_2 plays a role of diluent. A cross-over point of [F] and [O] emission intensity was obtained at $\approx 60\%$ O_2.

4.1.4 3C-SiC etching in CHF_3/O_2 [53, 73, 75]

While not an obvious choice, CHF_3 is an interesting gas for etching SiC. By comparison to CF_4, in CHF_3 one of the F atoms is replaced with H. This significantly reduces the [F] density since hydrogen also acts as a scavenger for [F]. In turn, as shown in Fig. 4, this greatly decreases the Si etch rate at low O_2 percentage, but does not significantly affect the SiC etch rate. At the same time the presence of [H] in the plasma has the important positive effect of reducing the density of residues present in the etch field (see discussion in Section 5). The highest SiC etch rate of ≈ 370 Å/min was obtained at 60% O_2 plasma, while the highest Si and SiO_2 etch rates of 370 and 480 Å/min were both obtained at 50% O_2. Beyond the 50% O_2 level, the effect of increasing O_2 percentage is to rapidly decrease the Si etch rate, while the SiC etch rate still increases up to the 60% O_2 level and then decreases slowly until the 80% level is reached. A SiC/Si ERR > 1 was obtained in the range of 60 to 80% oxygen, with the highest ERR of ≈ 2 being obtained at 80% O_2. Interestingly, the etch rates of SiC and SiO_2 exhibit similar trends with oxygen percentage, resulting in a SiC/SiO_2 ERR of 0.5 to 1.0 over the entire range of CHF_3/O_2 mixtures. Since the effect of the dc bias (see Fig. 6) in increasing the etch rate is probably stronger for Si than SiC, the fact that the SiC/Si ERR increases with O_2 in the 60 to 80% range indicates the dominance of the C–O chemical reaction. The [F], [O], and [H] plasma emission intensities are shown in Fig. 4c. It is interesting to note that the same cross-over point of 50% O_2 was obtained for the [F] and [O] relative concentrations and for the SiC and Si etch rates (Fig. 4b). Of

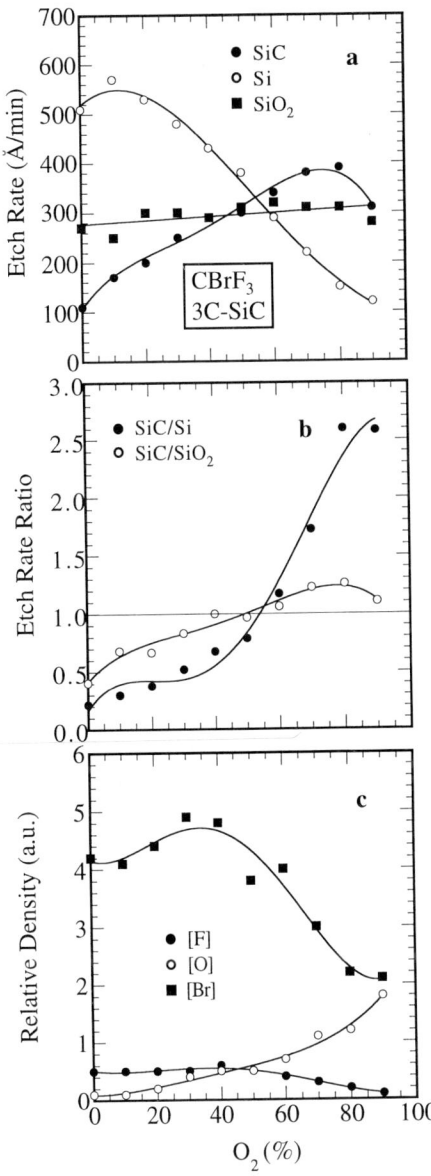

Fig. 5. Polycrystalline 3C-SiC RIE in CBrF$_3$ as a function of O$_2$ plasma a) SiC, Si, and SiO$_2$ etch rate, b) SiC/Si and SiC/SiO$_2$ etch rate ratio, c) relative [F], [O], and [Br] densities [53]

course, at high enough O$_2$ percentage the [F] level decreases to the point where it becomes the rate limiting step in the etching process and the SiC etch rate drops off very quickly. Therefore, multiple effects, involving the [F], [O], and [H] intensities (and the dc bias level) combine to control the reactive ion etching process of SiC.

4.1.5 3C-SiC etching in CBrF$_3$/O$_2$ [53]

The last etching gas to be discussed, CBrF$_3$, is the only one investigated which combines two different halogen atoms in a single molecule. By comparison to CF$_4$, in CBrF$_3$ one of the F atoms is replaced with Br. The results of etching polycrystalline 3C-SiC in CBrF$_3$/O$_2$ plasma are shown in Fig. 5. The Si ER pattern as a function of O$_2$ percentage is similar to that observed for CF$_4$ (see Fig. 1) but with much lower ER values. For example, the peak CF$_4$ ER of $\approx 1.8 \times 10^3$ Å/min which occurs at 10% O$_2$ is three times higher than the peak ER of ≈ 580 Å/min obtained in CBrF$_3$ at the same O$_2$ percentage. This clearly shows the lower Si removal efficiency of [Br] as compared to [F]. At higher levels of O$_2$, where the [F] level is depressed, the Si ERs in CF$_4$ and in CBrF$_3$ are very similar. The SiC etch rate is observed to increase nearly linearly with O$_2$ percentage up to 80% where it reaches the highest etch rate of ≈ 380 Å/min. Interestingly, this SiC etch pattern and ER levels are similar to those in CHF$_3$/O$_2$ plasma, where a H atom is found in place of the Br, and different from the pattern obtained in CF$_4$, where a F atom replaces the Br. In Si processing technology, CBrF$_3$/CF$_4$ mixtures are employed for the selective etching of poly-silicon over SiO$_2$, whereas CHF$_3$/CF$_4$ mixtures are utilized for the selective SiO$_2$ via etching. As shown in Fig. 5b, the etch rate cross-over point of SiC and Si occurs at $\approx 60\%$ O$_2$, and the highest SiC/Si ERR of ≈ 2 is observed at the usual high O$_2$ percentage (80 to 90%). The [F], [O], and [Br] plasma emission intensities are shown in Fig. 5c. The cross-over point between [F] and [O] occurs at $\approx 50\%$ O$_2$, which is approximately

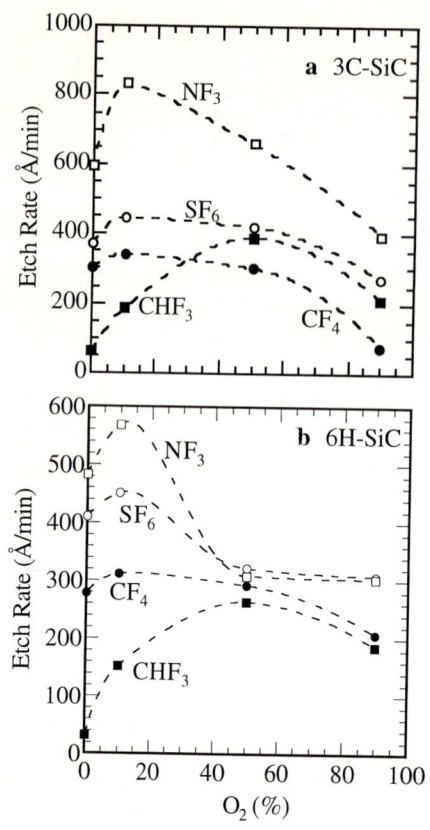

Fig. 6. Comparison of SiC etch rates in fluorinated/oxygen mixtures under the standard conditions for different polytypes. a) 3C-SiC, b) 6H-SiC

the same as for CHF_3/O_2 mixtures. It is interesting to point out that as the [F] level is relatively low and constant up to $\approx 50\%$ O_2, the linear increase in SiC etch rate with O_2 percentage in this range must be due to the increasing effect of C–O reactions.

4.1.6 Comparison of 3C- and 6H-SiC RIE

In general, as seen in Fig. 6, RIE of 3C-SiC grown on Si substrates results in a higher etch rate than that of 6H-SiC substrates. Based on current reported 3C-SiC material quality, a higher defect density of 3C-SiC heteroepitaxially grown on Si could in part enhance the charge transfer rate (chemical reaction). Unfortunately, there are no reports on the etching of 3C-SiC grown on 6H-SiC substrates. An in-depth etching comparison between 3C-SiC grown on Si and 6H-SiC substrate is therefore difficult to make based on the limited current data.

4.1.7 3C-SiC effect of dc bias [53, 73]

The self-induced dc bias generated by plasmas using the fluorinated gases discussed above are shown in Fig. 7 as a function of O_2 percentage. Interestingly, the fluorinated gases which use more strongly electronegative species (e.g. NF_3 and SF_6) develop generally lower dc bias levels and higher etch rates than the other fluorinated oxygen plas-

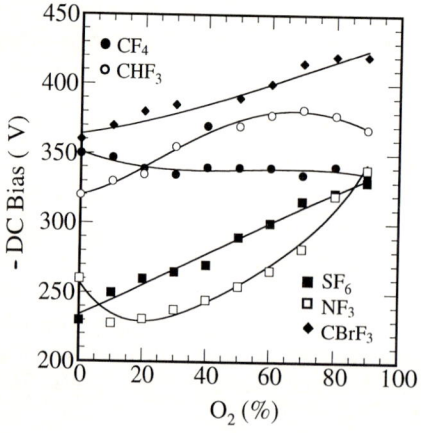

Fig. 7. The self-induced bias produced in the fluorinated plasmas of Fig. 1 to 5

mas. This would tend to indicate that the "cross-over" bias level separating the dc bias and reaction-rate limited SiC etching regimes [53] is a function of the etching gas used. The dc bias increases with O_2 percentage much more rapidly in NF_3/O_2 and SF_6/O_2, than for the other mixtures. In contrast, in CF_4/O_2 mixtures the self-induced bias remains nearly constant with O_2 percentage, while in CHF_3/O_2 and $CBrF_3/O_2$ mixtures the dc bias increases slowly as a function of O_2 percentage.

4.1.8 3C-SiC etching in remote NF_3/O_2 [79]

A remote plasma apparatus is frequently used for plasma-enhanced chemical vapor deposition (PECVD). In this case, a plasma produced remotely from the substrate is used to dissociate gas molecules in order to reduce the growth temperature. Pure NF_3 gas is currently used as a cleaning gas in PECVD to remove deposits from the reactor walls. The reason is the low polymer production of NF_3 plasma [80] and high etch rates of Si, SiO_2, and Si_3N_4 [81]. Interestingly, etching of 6H-SiC in a remote NF_3/O_2 mixture plasma hot-wall PECVD system has been reported with highly isotropic profile [79]. Although the etching experiment was performed at elevated temperature (330 °C), only a relatively low etch rate of 220 Å/min was obtained. In this case, thermal energy was the major energy input to the chemical etching process of SiC. The ion energy was limited by the plasma potential in which the sample was floating. This set of experimental conditions and results indicate that ion bombardment plays an important role in determining the SiC etching profile [53] and etch rate (through the breaking of the strong Si–C bonds). This confirms that both chemical and physical processes are needed in the etching of SiC which is in agreement with our previous discussion in Section 2 of this paper.

4.1.9 3C-SiC effect of crystallinity [72]

The effect of crystallinity on the SiC etch rate was explored by Padiyath et al. [72]. As shown in Fig. 8, they have found that etching of hydrogenated amorphous SiC (a-SiC), polycrystalline (poly-SiC), and crystalline 3C-SiC (c-SiC) in CF_4O_2 plasma at high dc bias (>300 V) results in a decreasing etch rate with increasing level of crystallinity. This etch rate pattern is consistent with the trend in the number of broken Si–C bonds in

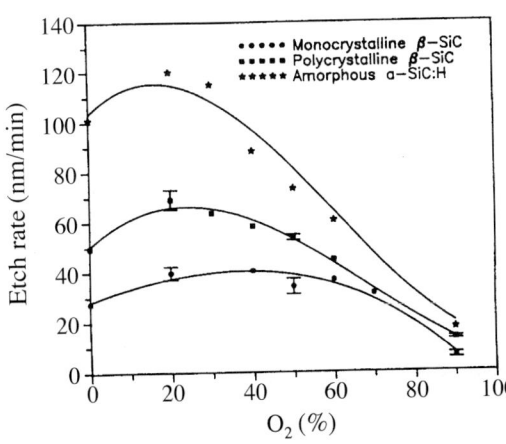

Fig. 8. Etch rates of 3C-SiC, poly-SiC, and a-SiC:H in CF_4 as a function of O_2 plasma [72]

these materials (c-SiC < poly-SiC < a-SiC), and hence the charge transfer rate during the chemical reaction process. The effect of crystallinity (or lack of it) on the etch rate is consistent with the effect of the dc bias, which increases the etch rate by partially damaging the SiC surface.

4.2 Mixtures of two fluorinated gases

In this section, we review the reactive ion etching of SiC in plasmas containing various combinations of two fluorinated gases. Since the fluorinated gases we have primarily investigated (CF_4, SF_6, NF_3, and CHF_3) each produce different etch rates, it was considered of general interest to investigate the effects of dual gas mixtures. The first set of mixtures consists of CHF_3 as the primary gas and CF_4, NF_3 and SF_6 as the secondary gases. Pure CHF_3 produces a residue-free etch surface but has a relatively low etch rate, while the reverse is the case for the other three gases. We were, therefore, interested in determining the residue-free parameter space obtainable with a dual fluorinated gas mixture. To that end, in this set of experiments data was obtained for the RIE of 3C- and 6H-SiC samples placed on the bare Al electrodes, which normally produces residues in the etch surface. The second set of experiments explored conditions for increasing the SiC etch rate while preventing residues through the use of a graphite cover on the Al electrodes. For these experiments, mixtures of NF_3 and SF_6 were used to etch 6H- and 4H-SiC samples.

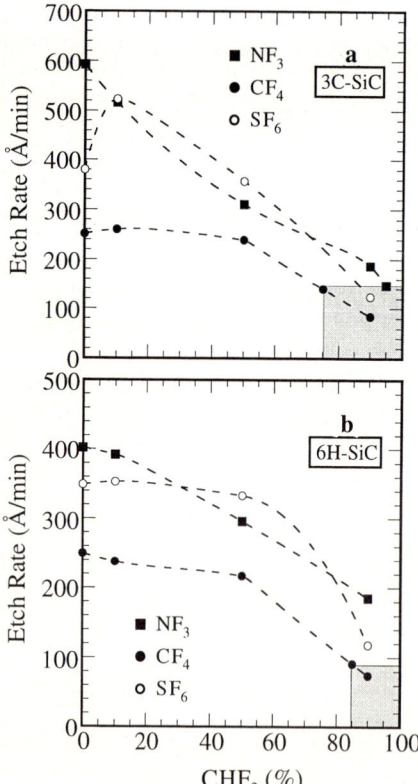

4.2.1 CHF₃ plus other fluorinated gases [82]

In this section we discuss the etching of SiC in dual fluorinated mixtures of CF_4/CHF_3, NF_3/CHF_3, and SF_6/CHF_3. As in the cases involving a mixture of a single fluorinated gas and oxygen (see Section 4.1), the experiments with dual fluorinated mixtures are performed with the standard etching conditions, and etch times of 5 and 30 min. Etch rates of 3C- and 6H-SiC in NF_3, CF_4, and SF_6 as a function of CHF_3 percentage are shown in Fig. 9a and b, respectively. The corresponding self-induced dc bias is shown in Fig. 10. The shadowed rectangular areas contain experimental conditions for which residue-free etching was obtained. A detailed

Fig. 9. Etch rates of a) 3C- and b) 6H-SiC in NF_3, CF_4, and SF_6 as a function of CHF_3 mixture plasmas. The shadowed rectangular area indicates a residue-free condition [82]

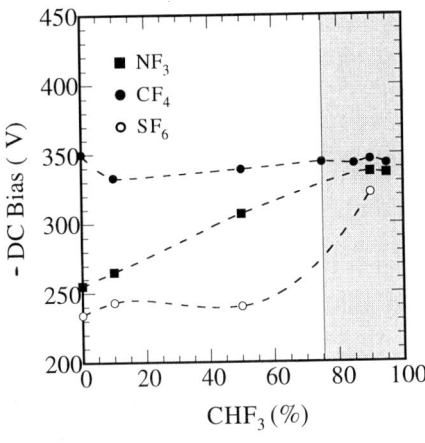

Fig. 10. Self-induced dc bias in fluorinated mixture plasmas [82]

discussion of the residue-free etching aspect is presented in the next section (Section 5) of this paper. Generally, 3C-SiC has a higher etch rate than 6H-SiC. In part, defect density in 3C-SiC [83] could be an important factor resulting in a higher etch rate. In Fig. 9, the experimental results (data points) were primarily obtained at 0, 10, 50, and 90% of CHF_3 in the total gas mixture. Additional data points were obtained for the residue-free etching study. Due to lack of full range of data points from 0 to 90%, dashed lines were used for the data point curve fitting for both Figs. 9 and Fig. 10. Without the O_2 additive effects, the [F] emission intensity changes almost linearly as a function of CHF_3.

Currently, etching of Si in fluorinated gas mixtures is frequently employed in the fabrication of Si integrated circuits. A variety of fluorinated gas mixtures, such as NF_3/CHF_3 [84], SF_6/CHF_3 [68], and CF_4/CHF_3 [66] have been reported for different applications. Usually, CHF_3 is used as a primary gas [66] for increasing selective etching of SiO_2 over Si, side-wall protection for anisotropic etching, resolving corrosion issues, and obtaining a residue-free (smooth) surface [68]. The CF_4/CHF_3 mixture has been particularly effective for the SiO_2 contact via and profile control etching. Under these near polymer formation conditions (without O_2 present in the main stream gas), care must be taken against contamination from particulate and excess polymerization. This could occur due to hydrogen-containing fluorinated gases (specially CHF_3) which produce more polymer than hydrogen-free gases (CF_4 and SF_6).

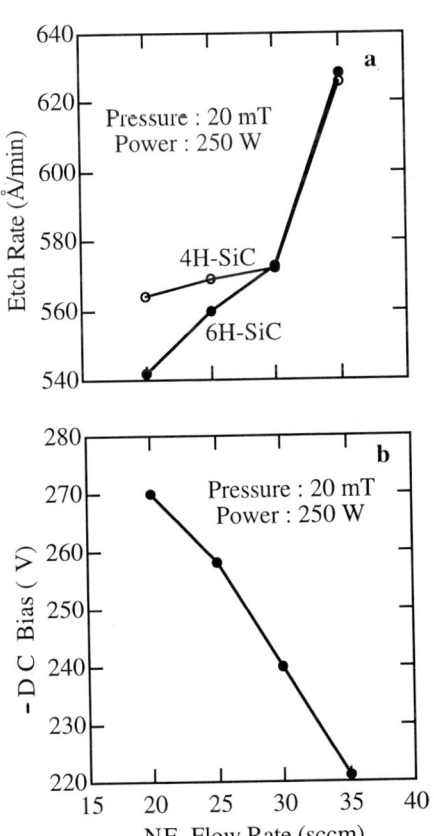

Fig. 11. Etch rate of 6H- and 4H-SiC vs. flow rate of NF_3

The etch rate of 3C-SiC generally decreases with increasing CHF_3 percentage in the dual gas mixtures. As shown in Fig. 9a, the most pronounced decrease occurs for the NF_3/CHF_3 and SF_6/CHF_3 mixtures. For example, the etch rate reduces from a high of 592 Å/min in pure NF_3 to 120 Å/min in the $NF_3/90\%$ CHF_3 mixture. In the case of CF_4/CHF_3 plasma, the etch rate was constant at low CHF_3 percentage, and much reduced compared to equivalent NF_3/CHF_3 and SF_6/CHF_3 mixtures. CF_4/CHF_3 mixtures containing more than 50% CHF_3 result in a reduction in etch rate similar to that of the other two mixtures.

The etch rates of 6H-SiC in the same dual fluorinated mixtures are shown in Fig. 9b. The highest etch rates for 6H-SiC were all obtained in pure NF_3, CF_4, and SF_6 gases and the etch rates in the dual mixtures generally decrease as the CHF_3 percentage increases. The etch rate of 6H-SiC in NF_3/CHF_3 mixtures decreases almost linearly with increasing CHF_3 percentage. Similar trends in etch rates were obtained for both of CF_4/CHF_3 and SF_6/CHF_3 mixtures.

Considering the dc bias and etch rate, it is interesting to note although both etch rates of 3C- and 6H-SiC fall in conditions of $SF_6/50\%$ $CHF_3 > NF_3/50\%$ $CHF_3 > CF_4/50\%$ CHF_3, the corresponding dc biases are in the reverse order of $SF_6/50\%$ $CHF_3 < NF_3/50\%$ $CHF_3 < CF_4/50\%$ CHF_3. This indicates a higher dc bias results in a lower etch rate, which means the etch rate does not have a simple linear relationship with the applied dc bias. However, a certain level of bias is required to physically or chemically etch SiC. To obtain a high etch rate, a combination of physical and chemical processes is generally required.

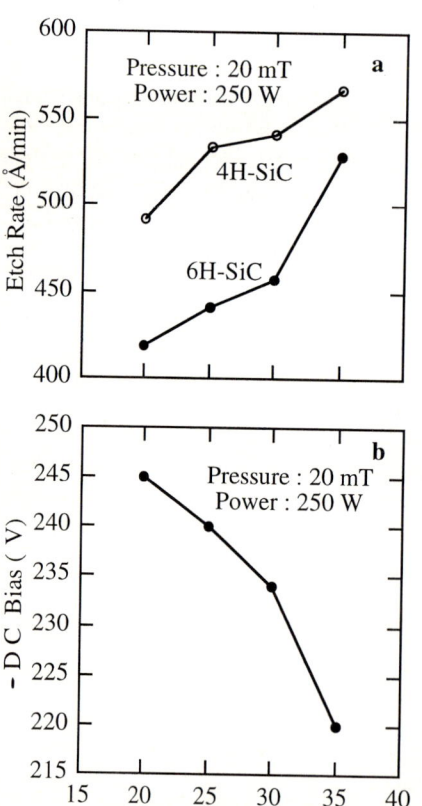

4.2.2 NF₃ plus SF₆ mixtures

4H- and 6H-SiC samples cut from single-crystal wafers were etched in mixtures of NF_3 and SF_6. The objective of investigating these dual fluorinated mixtures was to achieve higher etch rates, but without producing residues in the process. Therefore, we utilized a graphite cover between the Al electrode and the samples during RIE. Furthermore, we modified the standard etching parameters that was given in Section 3 (and followed in our previously reported results) by increasing the rf power from 200 to 250 W and investi-

Fig. 12. Etch rate of 6H- and 4H-SiC vs. flow rate of SF_6

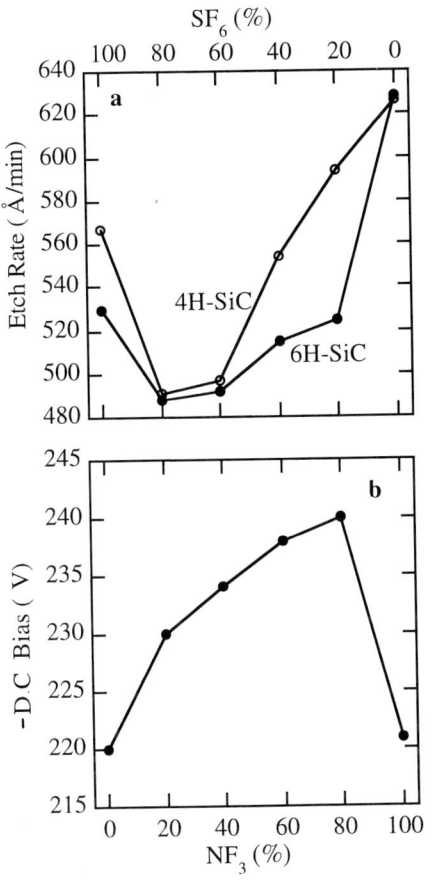

Fig. 13. Etch rate of 6H- and 4H-SiC in NF_3/SF_6 mixture plasmas

gating the effect of increasing the flow rate from 20 to 35 sccm (while keeping the pressure fixed at 20 mTorr). With these modifications, we found several conditions for which the resulting etch rate was greater than 500 Å/min, which is reasonably high for SiC RIE.

Figs. 11 and 12 show the etch rates of 4H- and 6H-SiC as a function of the flow rates of NF_3 and SF_6, respectively. The overall etch rate increases with the gas flow rate, despite the accompanying reduction in the residence time of the chemically active species. This shows that the critical flow rate is still not reached at 35 sccm. No significant difference in the etch rates for the two SiC polytypes was observed for NF_3, while for SF_6 the 4H polytype showed a somewhat higher etch rate (567 Å/min at 35 sccm) than the 6H polytype (529 Å/min at 35 sccm). The highest etch rate exceeded 600 Å/min in NF_3 plasma, and 500 Å/min in SF_6 plasma for both polytypes.

Fig. 13 shows the etch rates measured for the two SiC polytypes in different mixtures of NF_3 and SF_6 with a total flow rate of 35 sccm. The same general etch rate trend is

Fig. 14. SEM photograph of 6H-SiC (with 300 nm of Al mask) etched in pure NF_3 for 25 min under the following conditions: pressure: 20 mTorr, flow rate: 35 sccm, rf power: 250 W

Table 1

Summary of results reported on RIE of primarily hexagonal SiC polytypes

polytypes etched	source gas(es)	process type	typical process conditions: pressure, power, dc bias, flow rates	etch rate (Å/min)	ref.
3C	CF_4/O_2	plasma/ RIE (rf)	180 to 200 mT, 0.8 W/cm^2, 67% O_2, 33% CF_4	60 to 260	Dohmae et al. [85]
4H, 6H	SF_6	RIE (rf)	20 mT, 250 W, -220 to -250 V, 20 sccm 35 sccm	490, 420 570, 530	this work
6H	SF_6/O_2 NF_3/O_2	RIE (rf)	20 mT, 200 W, -220 to -250 V, $SF_6 : O_2 = 18 : 2$ (sccm) $NF_3 : O_2 = 18 : 2$ (sccm)	450 570	this work
6H	SF_6/O_2	RIE (rf)	50 mT, 200 W, -250 V, $SF_6 : O_2 = 5 : 5$ (sccm)	360	Kothandara- man et al. [86]
6H	SF_6/O_2, CF_4/O_2 with N_2 additive	RIE (rf)	190 mT, 300 W, $CF_4 : O_2 : N_2$ $= 40 : 15 : 10$ (sccm) $SF_6 : O_2 : N_2$ $= 40 : 2 : 0$ (sccm)	2200 3000	Wolf and Helbig [76]
4H, 6H	NF_3	RIE (rf)	20 mT, 250 W, -220 to -250 V, 20 sccm 35 sccm	565, 540 630	this work
4H, 6H	NF_3	RIE (rf)	225 mT, 275 W, -25 to -50 V, 95 to 110 sccm	1500	Casady et al. [78]
6H	$Cl_2/SiCl_4/O_2$ and Ar/N_2	RIE (rf)	190 mT, 300 W, $Cl_2 : SiCl_4 : O_2 : N_2$ $= 40 : 20 : 8 : 10$ (sccm) $Cl_2 : SiCl_4 : O_2 : Ar$ $= 40 : 20 : 0 : 10$ (sccm)	1600 1900	Niemann et al. [54]
3C, 6H	SF_6/O_2	ECR (µwave)	1 mT, 1200 W, -20 to -110 V $SF_6 : O_2 = 4 : 0$ to 8 (sccm) $SF_6 : O_2 = 4 : 0$ to 6 (sccm)	1000 to 2700	Lanois et al. [87]
4H, 6H	CF_4/O_2	ECR (µwave)	1 mT, 650 W, -100 V, $CF_4 : O_2 = 41.5 : 8.5$ (sccm)	700	Flemish and Xie [88]

observed for both polytypes. A broad and shallow minimum in the etch rate occurs at 20 to 40% NF_3, while the highest etch rates are obtained for the pure gases. The fact that the etch rate of 4H-SiC is very close to that of 6H-SiC under most conditions enables us to transfer most RIE processes developed from one polytype to the other.

An example of RIE using pure NF_3 to etch a trench deeper than 1 µm is shown in Fig. 14. The etched surface and the sidewalls are quite smooth and without any residues (due to the use of the graphite cover). These process parameters could therefore provide favorable conditions for etching deep trenches in SiC for applications such as power device fabrication and micromachining.

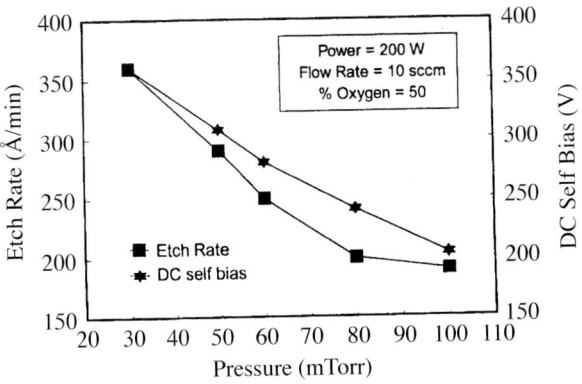

Fig. 15. Effect of variation in chamber pressure on etch rate and dc self bias [86]

4.3 Review of other results on SiC reactive ion etching

In this section, we review some of the results reported by other groups on primarily hexagonal SiC reactive ion etching. Table 1 summarizes the etching conditions and resulting characteristics. For completeness, selected results from our own work are also included. The table also contains information on microwave-based plasma etching of SiC, which is discussed in more detail in Section 6, and chlorinated gas plasmas, which is discussed in Section 5.3.2.

Plasma etching of different polytypes of SiC have been reported by other researchers. Dohmae et al. [85] reported the first plasma etching results on the 3C-SiC polytype. They used the CF_4/O_2 mixture at 180 to 200 mTorr to obtain plasma etch rate of 6 to 26 nm/min.

More recently, Kothandaraman et al. [86] have reported the effect of gas flow rate, chamber pressure, and rf power on the etch rate of 6H-SiC in SF_6/O_2 mixtures. Fig. 15

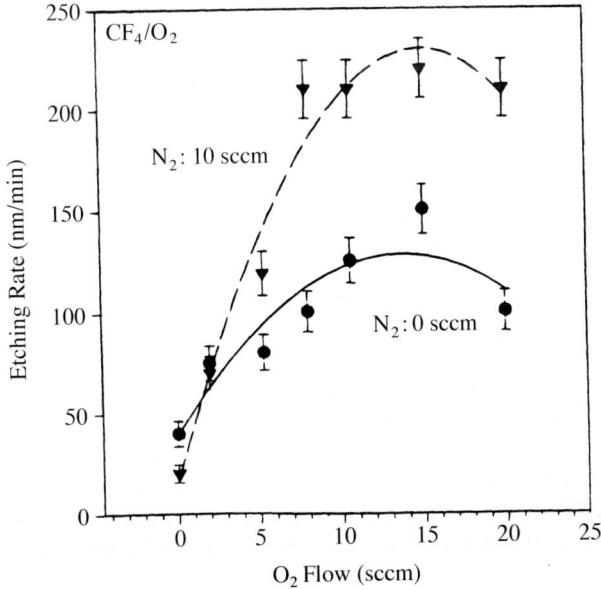

Fig. 16. RIE in CF_4 as a function of O_2 gas flow with and without N_2 additive [76]

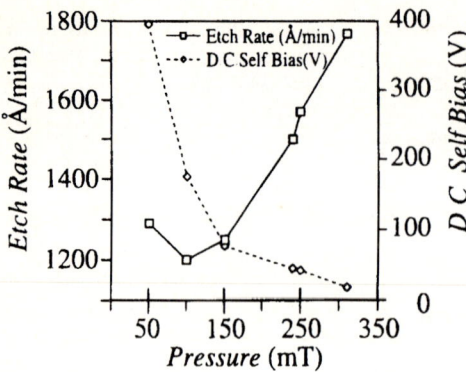

Fig. 17. Etch rate of 6H-SiC and self-induced dc bias as a function of chamber pressure with constant rf power of 275 W (1.7 W/cm^2) [78]

shows the variation of the etch rate as a function of pressure for a SF$_6$/50% O$_2$ mixture, an rf power of 200 W, and a total flow rate of 10 sccm. A peak etch rate of 360 Å/min was obtained at 30 mTorr pressure, which was the lowest pressure evaluated. Wolf and Helbig [76] have reported the effect of nitrogen additive to CF$_4$/O$_2$ and SF$_6$/O$_2$ plasmas on the etch rate of 6H-SiC at relatively higher pressure (190 mTorr). As seen in Fig. 16, the etch rate at this pressure is enhanced by a factor of 1.5 to 2 through the addition of N$_2$ in CF$_4$/O$_2$ plasmas. In SF$_6$/O$_2$, the etch rate is higher than for CF$_4$/O$_2$ and actually decreases with N$_2$ addition due to the formation of complex molecules. Casady et al. [78] have reported RIE of both 6H- and 4H-SiC polytypes in pure NF$_3$ source gas at a very high flow rate of 100 sccm. As shown in Fig. 17, the etch rates obtained exceed 1200 Å/min, and increase almost monotonically as the pressure is increased from 75 to 325 mTorr. They report an optimum pressure of $\approx$225 mTorr for best etch rate and surface cleanliness.

5. Techniques for Obtaining Residue-Free Etching

Interestingly, SiC etching in most of the fluorinated oxygen plasmas results in residue formation (surface roughness) [89] in the etch field except in some of the CHF$_3$ plasma conditions for all SiC polytypes [73, 75]. It has been shown using Auger surface analysis [67, 89] that residues were formed due to contamination from the cathode material through the micromasking effect. Several issues need to be addressed regarding how the residues are formed. As discussed before, the electrode area in the commercially available RIE systems is much larger than that of the SiC samples utilized in the etching experiments. This results in simultaneous etching (sputtering) of the metal electrode, leading to residue formation. Residue-free etching of SiC using graphite or kapton sheet to cover the Al elecrode has reaffirmed the micromasking effect [67, 73, 75].

To obtain residue-free etching of SiC, we have previously reported the use of fluorinated oxygen mixtures such as CHF$_3$/O$_2$, CF$_4$/O$_2$, SF$_6$/O$_2$, and NF$_3$/O$_2$ with H$_2$ additive [73, 75], and of mixtures of two fluorinated gases such as CF$_4$/CHF$_3$ and NF$_3$/CHF$_3$ [82]. In this paper, we review several approaches to prevent residue formation in the etch field including (i) adding pure H$_2$ gas into fluorinated oxygen plasmas; (ii) utilizing H-containing gas as one of the etching gases, such as CHF$_3$; (iii) employing high density plasma sources — sample bias control and control of the location of the sample which is separated from plasma generation region; (iv) covering the metal exposed to the plasma by using teflon, graphite, and kapton. The trade-off between residue formation and other etching parameters is presented.

We have previously demonstrated the etch rate and anisotropic profile of SiC in a variety of fluorinated oxygen and fluorinated mixture plasmas, such as CHF_3/O_3, $CBrF_3/O_2$, CF_4/O_2, SF_6/O_2, NF_3/O_2, CF_4/CHF_3, SF_6/CHF_3, and NF_3/CHF_3. Considering the surface morphology after long term etching, rough surfaces (residues) were obtained after etching in most of these gas mixtures. The density and physical shape of residues were determined by the plasma chemistry [73, 75]. Interestingly, similar rough surfaces were reported for Si etching in pure CF_4 plasma [90, 91]. The mechanism proposed was again metal contamination from the electrode. The rough etched surface usually shows as black when observed directly under room light. This is due to the scattering of light from the surface microstructure. These areas have been called "black silicon". For n-type 6H-SiC, we observe dark non-transparent green areas with residues in the etch field. To remove the electrode metal contamination, H_2 additive in pure CF_4 plasma has been reported in Si RIE etching [90]. H_2 could possibly be seen as an aluminum scavenger ("gettering effect"). This indicates that the H_2 additive process could not only be applied to just Si and SiC, but possibly to any semiconductor material in general [92]. In addition to using H_2 to obtain a residue-free etched surface, it has been used as an additive in fluorinated plasmas for the selective etching of SiO_2 and Si_3N_4 over Si [66]. One of the side effects of the H_2 additive in the plasma is that the H radicals are implanted into the Si substrate, and deactivate the dopants [93, 94]. Implant depths of 400 Å into the Si substrate were reported [95] resulting in the possible reduction of the surface doping concentration. This result has been confirmed [56] by fabricating Schottky diode to measure the surface doping concentration under various etching conditions.

To obtain a residue-free etching of 3C- and 6H-SiC, one can (i) etch in fluorinated oxygen mixture with H_2 additive, such as $CHF_3/O_2/H_2$, $CF_4/O_2/H_2$, $SF_6/O_2/H_2$, and $NF_3/O_2/H_2$ [73, 75]; (ii) etch in CF_4/CHF_3 and NF_3/CHF_3 fluorinated mixtures [84]; (iii) perform etching with the electrode covered with a non-metallic sheet [67, 73, 75]. These processes for residue-free etching can easily be utilized by choosing the right etchants or by covering the electrode with graphite, polymide, or quartz sheet. Possible mechanisms have been proposed including (i) gettering effect − gas phase reaction of H_2 and Al clusters to form aluminum hydride volatile compounds (AlH_x); and (ii) etching enhancement of carbon-rich surface. A material with low etch rate and sputter yield could be the best candidate. Care must be taken to avoid excess polymerization or particle generation in this case. The greatest advantage of using H_2 additive for residue-free etching is that the process is independent of the reactor material, and is highly reproducible and portable. However, H_2 addition also results in a reduction of the etch rate due to gas phase reaction of fluorine and hydrogen. The presence of excess H_2 in the plasma also reduces the etch aspect ratio. This is because the hydrogen-containing fluorinated gases produce more polymer than hydrogen-free gases [66]. A trade-off between residue-free etching with H_2 process and other etching requirement is often needed.

5.1 H_4 additive in fluorinated oxygen mixtures

5.1.1 Etching of 3C-SiC [73]

Etching of 3C-SiC in fluorinated oxygen mixture plasmas has been reviewed in detail in the previous section. Adding pure H_2 in the fluorinated oxygen plasmas can prevent residue formation. Due to the gas phase reaction between fluorine and hydrogen, various levels of H_2 are required under a variety of etching conditions. The 3C-SiC etch rate in

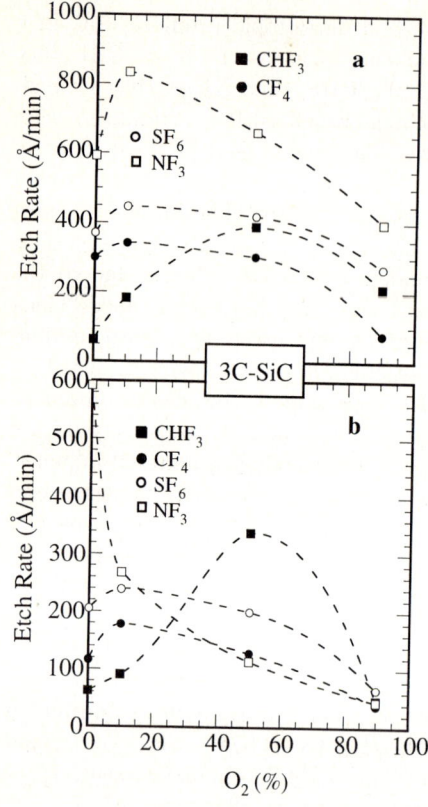

Fig. 18. Etch rates of 3C-SiC a) without and b) with H_2 additive in CHF_3, CF_4, SF_6 and NF_3 as a function of O_2 plasma [73]

CHF_3, CF_4, SF_6, and NF_3 as a function of O_2 plasma without and with H_2 additive is shown in Fig. 18a and b, respectively. The corresponding minimum H_2 flow rate of Fig. 18b is shown in Fig. 19. As shown in Fig. 18, limited data points were measured at 0, 10, 50, and 90% of O_2 so that the dashed line was employed for the curve fitting. Generally, the etch rate decreases as the H_2 percentage increases. This is the reason why finding the minimum H_2 flow rate is so important. Furthermore, the minimum H_2 flow rate required for residue-free etching in fluorinated oxygen gas mixtures decreases as the O_2 percentage in the mixture increases. This is due to increased consumption as a result of the gas phase reaction. An exception is the CHF_3/O_2 plasma. This may be because CHF_3 is by itself a H_2-containing gas. Etching in pure NF_3 plasma with few spikes was obtained only for 3C-SiC with high etch rate and anisotropic profile. As shown in Fig. 19, 16 sccm H_2 was used as the highest flow rate in all of the fluorinated oxygen mixture plasmas we have investigated. As the required H_2 flow rate is greater than 16 sccm, no further experiments were performed to find the minimum H_2 flow rate. As shown in Fig. 18b, the highest residue-free etch rate of ≈ 338 Å/min was obtained in $CHF_3/50\%$ O_2 (20 sccm) with H_2 (2 sccm) additive plasma. The process pressure was changed from 20 to 25 mTorr.

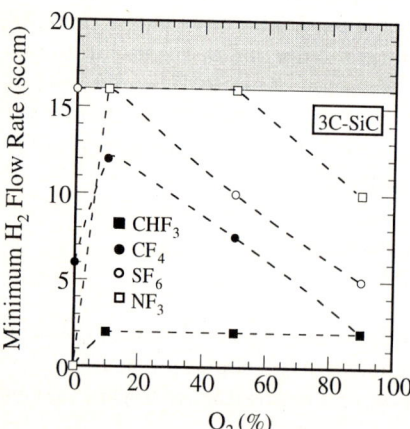

Fig. 19. Minimum H_2 flow rate of residue-free etching condition for 6H-SiC in fluorinated oxygen plasmas [73]

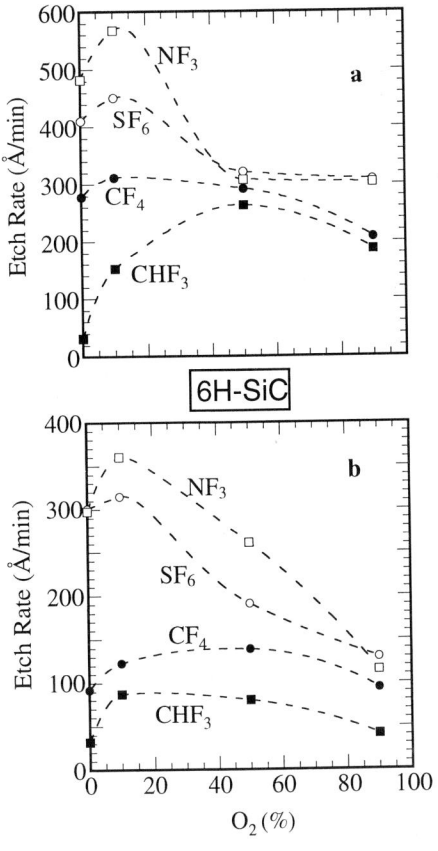

Fig. 20. Etch rates of 6H-SiC in CHF_3/50% O_2 a) without and b) with H_2 additive in CHF_3, CF_4, SF_6 and NF_3 as a function of O_2 plasma [73]

5.1.2 Etching of 6H-SiC [75]

Etching of 6H-SiC in CHF_3, CF_4, SF_6 and NF_3 as a function of O_2 plasma without and with H_2 additive is shown in Fig. 20a and b, respectively. The corresponding H_2 flow rate of Fig. 20b is shown in Fig. 21. As shown in Fig. 20, limited data points were measured at 0, 10, 50, and 90% of O_2 so that the dashed line was employed for curve fitting. As for 3C-SiC, the etch rate generally decreases as the H_2 percentage increases. The highest etch rates were obtained by using the minimum H_2 flow rate needed for residue-free etching. As shown in the shadowed rectangular region of Fig. 21, an upper limit of 16 sccm was set for the H_2 flow rate. As shown in Fig. 21, for the NF_3/O_2 and SF_6/O_2 mixtures, residue-free etching condition was obtained only at 90% O_2 for both mixtures. The etch rate obtained was ≈ 150 Å/min. At low O_2 percentage, the minimum H_2 requirement for residue-free etching increases significantly. Except for the CF_4/10% O_2 composition, residue-free etching was obtained in both CF_4/O_2 and CHF_3/O_2 mixtures with <16 sccm of H_2 additive. Comparing the H_2 requirement to 3C-SiC (Fig. 19), 6H-SiC needs a higher level of H_2 to prevent the residue formation. The reason is still unknown based on the current

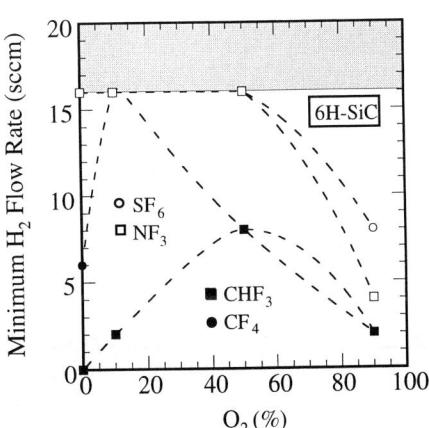

Fig. 21. Minimum H_2 flow rate of residue-free etching condition for 3C-SiC in fluorinated oxygen plasmas [73]

experimental results. As shown in Fig. 20b, the highest residue-free etch rate of ≈ 150 Å/ min was obtained in CF$_4$/50% O$_2$ + H$_2$ (28%) mixture plasma. The process pressure was changed from 20 to 30 mTorr.

5.2 Fluorinated mixtures

CHF$_3$ is the only fluorinated plasma in which residue-free etching can be obtained for both 3C- and 6H-SiC polytypes without any gas additive and without the need for covering the Al electrode. Unfortunately, the etch rate in pure CHF$_3$ plasma is much lower compared to the etch rates in the plasmas of other fluorinated gases investigated i.e. NF$_3$, CF$_4$, and SF$_6$. Hence the mixtures of these gases with CHF$_3$ as the primary gas have been employed for the etching of 3C- and 6H-SiC to get a higher etch rate while maintaining residue-free etching. The etch rate and etched surface morphology of 3C- and 6H-SiC in NF$_3$, CH$_4$, and SF$_6$ as a function of the percentage of CHF$_3$ in the mixture are shown in Fig. 9a and b, respectively [82]. The corresponding dc bias is shown in Fig. 10. Etch rate in fluorinated mixture plasmas has been discussed previously (Section 4.2) and elsewhere [82]. Residue-free etching of 3C- and 6H-SiC in fluorinated mixture plasmas has been obtained in the shadowed rectangular region of Fig. 9. As shown in Fig. 9a, residue-free etching of 3C-SiC was obtained only in CF$_4$/CHF$_3$ and NF$_3$/CHF$_3$ mixture plasmas. At CHF$_3$ percentages higher than 75%, residue-free etching can be obtained in CF$_4$/CHF$_3$ mixtures. In this paper, we have expressed this as CF$_4$/ >75% CHF$_3$. With this nomenclature, NF$_3$/ >95% CHF$_3$ is required to obtain residue-free etching for the 3C-SiC. In addition, even though the mixture of SF$_6$/90% CHF$_3$ is located in the residue-free etching region in terms of etch rate, no residue-free etching was found in the mixture of SF$_6$/CHF$_3$.

For 6H-SiC, as shown in Fig. 9b, CF$_4$/CHF$_3$ is the only fluorinated mixture plasma in which we can achieve residue-free etching. CF$_4$/ >85% CHF$_3$ is required for this condition. Comparing to 3C-SiC (Fig. 9a), etching of 6H-SiC (Fig. 9b) needs a higher level of CHF$_3$ to achieve residue-free conditions. Interestingly, this result is similar to that for the H$_2$ additive in fluorinated oxygen mixture plasmas i.e. a higher level of H$_2$ is needed for 6H-SiC to obtain residue-free etching. In NF$_3$/CHF$_3$ plasma, even at NF$_3$/ >95% CHF$_3$, residues in the etch field were obtained in 6H-SiC, whereas residue-free etching was obtained in 3C-SiC for these compositions. This difference in the etch results for

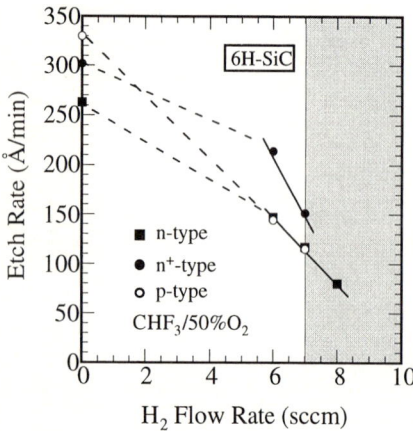

Fig. 22. Etch rate 6H-SiC of various doping in CHF$_3$/50% O$_2$ as a function of H$_2$ flow rate [75]

3C- and 6H-SiC for the same gas composition can be used to identify the polytype by employing RIE. Similar to 3C-SiC, residue-free etching could not be obtained in SF_6/CHF_3 mixtures for 6H-SiC as well.

To study the residue-free etching process on various doping levels and dopant type, n-, n^+-, and p-type 6H-SiC samples were etched in $CHF_3/50\%$ O_2, and the minimum hydrogen additive required for each was evaluated. The doping level of n-, n^+-, and p-type 6H-SiC are about 10^{17}, 10^{19}, and 10^{18} cm^{-3}, respectively. The etch rate with different doping level and dopant types as a function of H_2 flow rate are shown in Fig. 22. Similar etch rates were obtained for the n- and p-type samples. The n^+-type 6H-SiC samples, however, had a somewhat higher etch rate. Similar levels of H_2 (7 to 8 sccm) were required to obtain residue-free etching for samples with different doping level and dopant type. This result indicates that the doping level and the dopant type is insignificant for the residue-free etching process utilizing H_2 additive.

5.3 Other techniques for residue-free etching

5.3.1 Non-metallic electrodes

Metal contamination (micromasking) from the electrode or surrounding reactor wall is the main source for contamination resulting in residues in the etch field. The easy way is to cover with non-metallic material such as SiO_2, Si, graphite, kapton, and teflon. A low etch rate and sputter yield material can be the best candidate. Notably, loading effect from the covering material becomes inevitable during etching. In Si technology, the most common material used to cover the electrode/reactor walls is teflon. Since most systems are designed for etching Si wafers, teflon could be used to cover the entire metal surface so that one can easily minimize the metal contamination during the etching process. However, for SiC, we have reported [73, 75] that it is sufficient to cover the powered electrode (sample electrode) by a graphite or a kapton sheet and that we can leave the upper electrode and the reactor walls intact (without covering). Since the material is in form of a thin sheet, it does not affect the distance between the electrodes. One should be aware that the use of a covering material could change the discharge conditions by leading to polymer formation (if the cover material is a polymer), and by altering the distance between the electrodes (if the sheet is too thick). However, utilizing the right material for the etch mask, and avoiding metal contamination from the etching reactor can prevent the formation of residues in the etch field.

5.3.2 Non-fluorinated plasmas

RIE of 6H-SiC in chlorine-based plasmas ($Cl_2/SiCl_4/O_2/Ar/N_2$) has been reported by Niemann et al. [54]. Interestingly, non-metallic CVD SiO_2 was employed as the etch mask and the SiC sample was affixed on the oxidized Si wafer to facilitate the transferring system. Although the size (1″) of SiC wafer is small, the Si wafer can fully avoid the contamination from the metallic electrode. No residues were observed in this gas mixture. A higher etch rate of $(1 \text{ to } 2) \times 10^3$ Å/min was obtained in this gas mixture by varying the flow of O_2 and choosing Ar or N_2 inert gas independently. These experiments show promising results for SiC etching in Cl-based plasma with a high etch rate and non-metallic etch mask. However, higher processing costs usually accompany the use of chlorinated plasma-assisted etching. Further investigation of SiC etching in chlorine based plasmas are required to compare the etch rate and etched surface morphology to those obtained in fluorinated plasmas.

6. Etching in High Density Plasmas

As we progress into the Si Ultra-Large-Scale Integrated (ULSI) circuit era, Si technology is currently employing etching at lower pressures (<10 mTorr) and use of higher density (10^{10} to 10^{12} cm^{-3}) plasmas. Such plasmas have relatively lower plasma potential and higher ionization efficiencies [66], and it is possible to have independent control over the ion energy and the flux to the wafer. They are able to provide sufficient plasma densities for etching feature sizes below $1\,\mu$m and for aspect ratios (depth/width) much larger than one. Etching in such plasmas results in a higher etch rate (and thus a higher throughput) and a highly anisotropic etch profile (desirable for sub-μm etching). The process utilizes lower pressures which reduces the scattering of ions and increases directionality, thus giving more control over the etch profile and a better etch uniformity. This is extremely important when device size shrinks to $<1\,\mu$m. The process also results in reduced surface damage (and hence lower leakage current), and allows the use of pure etchant gases without the need for additives to avoid residues. These advantages make it desirable to perform dry etching of SiC in high density plasmas. With etch rates higher than 1000 Å/min, reactors utilizing such plasmas would be able to fulfill the nominal mass-production requirement at a throughput of ≈ 10 wafers/h when larger area SiC wafers would be used for device fabrication. Investigation of etching of SiC in high density plasmas could indeed be an active area for research.

Many widely used high density plasma systems employ electron cyclotron resonance (ECR), and in such systems, the samples to be etched are placed remotely from the plasma generation region. Recently, plasma etching of 6H-SiC has been reported [55] in $\approx$ CF$_4$/20% O$_2$ plasmas with relatively lower dc bias (<200 V). A higher etch rate and a residue-free ("smooth") surface was obtained without any gas additive. The absence of residues in this process indicates that there is negligible sputtering from the etch mask and the electrode materials. Interestingly, indium tin oxide (ITO) has been employed as an etch mask material rather than aluminum (Al). Although a higher etch rate was obtained in ECR etching, a higher input power (>500 W) was utilized to achieve this goal. The results reported from ECR etching in CF$_4$/O$_2$ plasma [55] are shown in Fig. 23. To understand the advantages of SiC ECR etching over RIE, it is important to consider the discharge mechanism and conditions of the ECR plasma. ECR plasma can

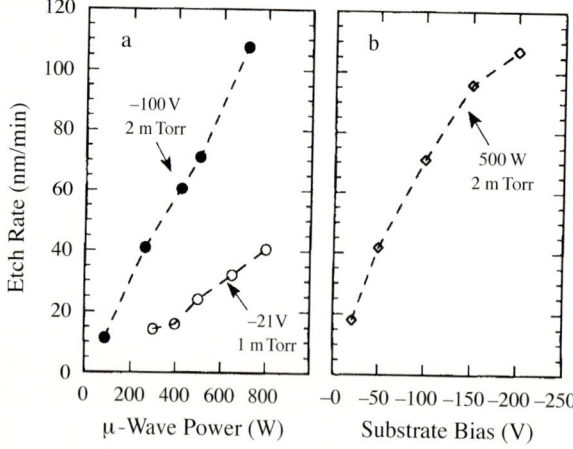

Fig. 23. Etch rates of 6H-SiC in $\approx$ CF$_4$/20% O$_2$ ECR plasma as a function of a) input μwave power and b) applied rf bias power [55]

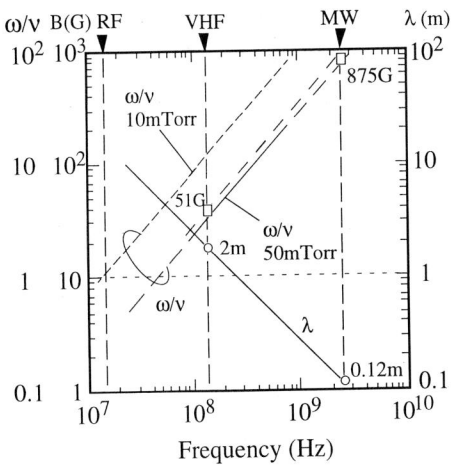

Fig. 24. Conditions of electron cyclotron resonance (ECR) plasma as a function of frequency [92]

be produced either by microwaves (µwave) or by VHF waves with different levels of magnetic flux density [96]. ECR plasma parameters such as excitation frequency and magnetic field as a function of frequency are shown in Fig. 24. The condition of $\omega/\nu \gg 1$ should be satisfied, where $\omega(2\pi f)$ and ν are the angular frequency and electron collision frequency, respectively. The magnetic field corresponding to µwave and VHF waves are 875 and 51 G, respectively. The value of ω/ν also indicates the intensity of the plasma. Under certain pressure (<5 mTorr), a plasma density higher than that in RIE can be obtained from charge particle resonance.

In addition to ECR plasmas, there are many other high density remote plasma sources commercially available, such as the Transformer/Inductive Coupled Plasma (TCP/ICP) [97], Helicon (H) Plasma [98 to 101], and the Helical Resonance (HR) Plasma [102, 103]. For comparison, the plasma density of various plasma sources as a function of pressure is shown in Fig. 25. The plasma density decreases as the pressure increases. Generally, the RIE plasma density is in the range of 10^9 to 10^{11} cm^{-3}. The plasma density in an RIE reactor can be enhanced by confining the charge particles with a magnetic field. This process is referred to as magnetically enhanced reactive ion etching (MERIE). However, in the presence of the magnetic field, plasma uniformity in such a process becomes a key issue. To have a generic idea of the various plasma sources, system schematic diagrams of RIE, PE, ECR, TCP/ICP, H, and HR plasma are shown in Fig. 26. RIE (Fig. 26a) and PE (Fig. 26b) are both capacitively coupled

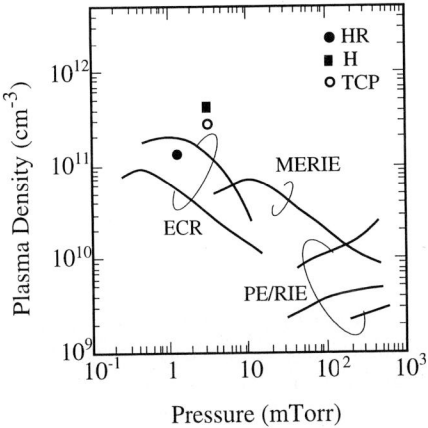

Fig. 25. Plasma densities as a function of pressure in various plasma sources

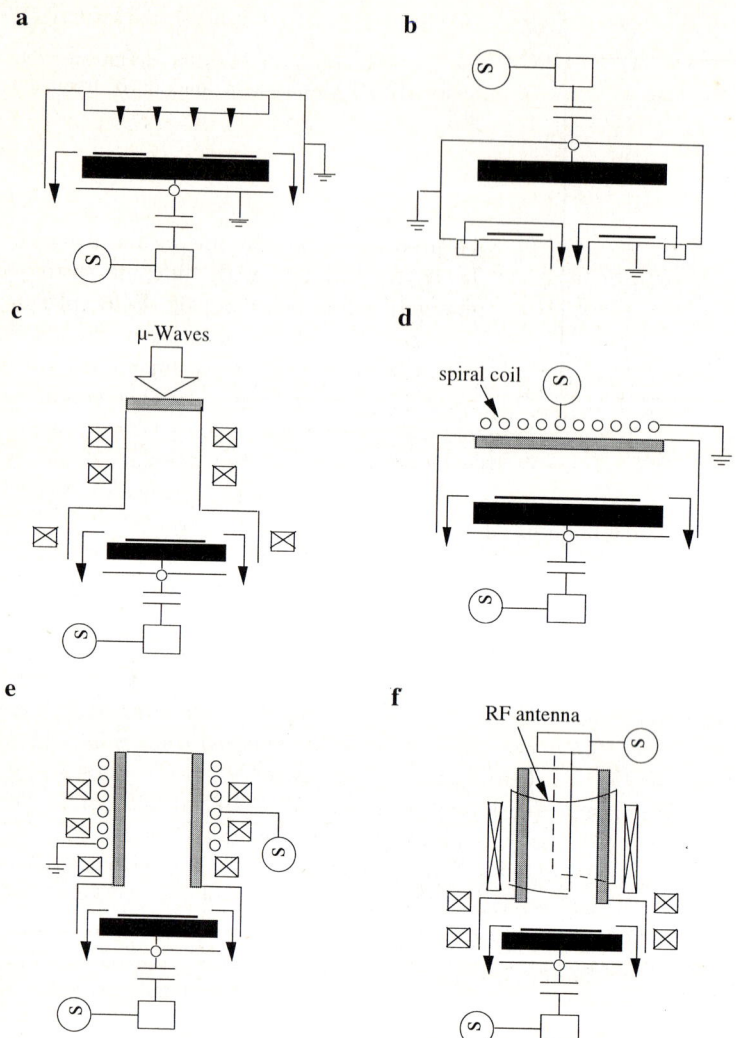

Fig. 26. System configuration of remote plasma sources: a) reactive ion etching (RIE), b) plasma etching (PE), c) electron cyclotron resonance plasma (ECR), d) transformer coupled plasma (TCP), e) helical resonance plasma (HR), f) helicon plasma (H)

plasmas (low plasma density and high plasma potential). For RIE, the area of the bottom electrode (wafer electrode) is smaller than that of upper electrode so that a higher sheath bias can be generated across the wafer. The wafer to be etched is placed on the electrode that has the higher bias and hence ion bombardment can participate in the overall etching process.

The discharge mechanisms of TCP/ICP [104], H [105, 106], and HR [107] plasma have been reported elsewhere. A detailed discussion on these plasmas is beyond the scope of this paper. However, a brief discussion is given next. The transformer coupled plasma (Fig. 26d) is generated by the application of rf power to a non-resonant, induc-

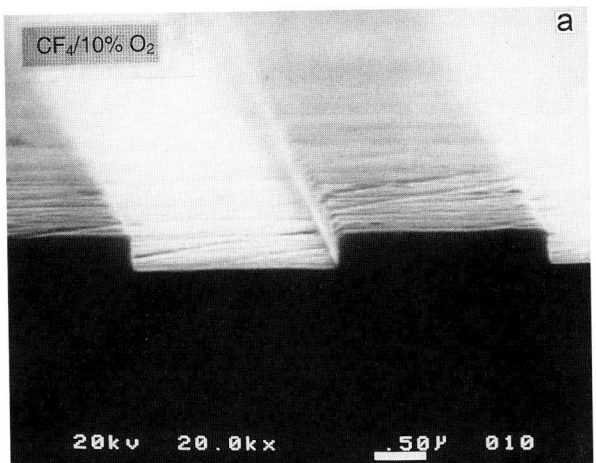

CF$_4$/10% O$_2$

a

20kv 20.0kx .50μ 010

Fig. 27. Helicon plasma etching of
6H-SiC at 10 mTorr in a) CF$_4$/
10% O$_2$ and b) SF$_6$/10% O$_2$

SF$_6$/10% O$_2$

b

20kv 20.0kx .50μ 009

tive coil, resulting in the breakdown of the process gas within or near the coil by the induced rf electric field. Power is mainly transferred from the electric field to the plasma electrons by collision dissipation (ohmic heating). Inductive sources have several advantages over other sources, including simplicity in concept, no need for magnetic field (resulting in better plasma uniformity), non-resonant operation, and cable-ready rf power (no waveguide needed). The helical resonator plasma (Fig. 26e) is created by either capacitive or inductive coupling. Careful hardware design can force the operation in the desired inductive mode (high density and low plasma potential). Helicon plasma (Fig. 26f) is excited by an rf-driven antenna that is coupled with the transverse mode structure across an insulating chamber wall. The input energy is absorbed by the electrons in the plasma through collision or through Landau damping.[2]) Currently, high

---

[2]) Landau damping is a process by which a wave transfers energy to electron having velocity near the phase velocity $v_{\mathrm{ph}} = \omega/k_z$ of the wave.

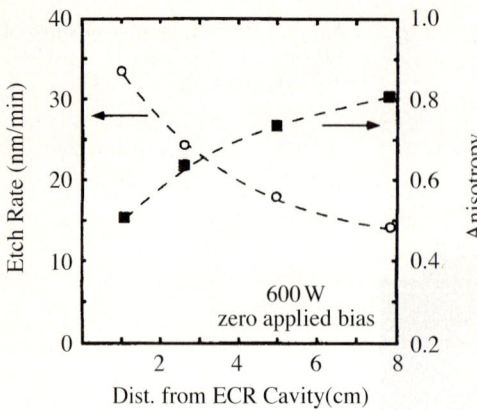

Fig. 28. Effect of sample proximity in ECR system on SiC etch rate and anisotropy of etching [88]

density plasma sources are being widely utilized in the fabrication of Si ULSI circuits [108 to 111].

We have explored the etching of SiC in the helicon plasma. In these experiments, we performed the etching of n-type 6H-SiC substrate in $CF_4/10\%\ O_2$ and $SF_6/10\%\ O_2$ gas mixtures with an input rf power (helicon source) of 1000 W, an applied rf bias power of 100 W, a process pressure of 10 mTorr, and a total gas flow rate of 44 sccm. Al was employed as the etch mask. Using kapton tape, the SiC sample was affixed to a large area glass substrate (Corning 7049) to facilitate the automatic transferring system into the etching reactor. The sample is located right underneath the plasma generation zone and the area is much smaller than the glass plate. We can assume that there was no metal contamination* from the electrode which was covered by the glass plate. Under this condition, the only possible contamination source is the etch mask. The etching profiles with etch mask in place in $CF_4/10\%\ O_2$ and $SF_6/10\%\ O_2$ plasmas are shown in Fig. 27a and b, respectively. As shown in Fig. 27, highly anisotropic etching profiles were obtained from both mixture gases. The result obtained in SF_6/O_2 reaffirmed that SiC does enhance the polymer formation during etching to prevent the side wall from being etched (no undercut). In these experiments, the etched area was $\approx 10\%$ of the

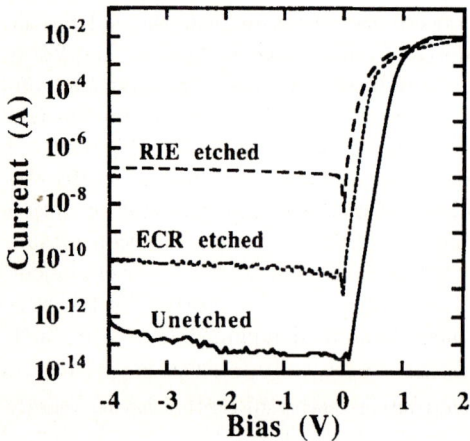

Fig. 29. Current–voltage characteristics of Pd Schottky diodes on RIE etched with addition of H_2, ECR etched, and unetched SiC samples [56]

total sample electrode area exposed to the plasma. The plasma was generated inside a quartz tube and the SiC sample was located on top of the glass plate. Interestingly, residues were found fully covering the etch field in both mixture gases. This is in contrast to ECR plasma etching in which no residues were reported. The possible explanation for residues in the etch field is etch mask contamination. These results indicate that etching in high density plasma does not always result in residue-free etching. Optimization of process conditions including etch mask material, etching condition, and system configuration is required to achieve a high etch rate and residue-free etching.

ECR plasma etching using microwave plasma has been reported by Lanois et al. [87] and Flemish and Xie [88]. Lanois et al. achieved an etch rate of 100 to 270 nm/min for both 3C and 6H polytypes. Flemish and Xie reported effective control of the etch profile and morphology of 4H- and 6H-SiC polytypes using CF_4/O_2 mixtures. Fig. 28 shows the effect of sample proximity to the ECR source on the etch rate and anisotropy. The anisotropy can be quantified as $A = (d - l)/d$, where d and l are the vertical and lateral etch depths, respectively. In addition, ECR etching has been reported [56] to produce significantly reduced damage on the SiC surface. Fig. 29 shows the current–voltage characteristics of Pd Schottky diodes formed on the unetched SiC surface, as well as on RIE etched and ECR etched surfaces. The ECR etched surface results in a much reduced leakage current compared with the RIE etched case.

7. Etching of Sub-μm Features

Sub-μm reactive ion etching is one of the critical processes for device fabrication when the device size shrinks to $<1\,\mu m$. In the etching of Si sub-μm features, highly anisotropic etching profile is necessary. Microloading and microscopic etching uniformity are the two key issues for the etching of sub-μm features [112]. Microloading effect is dependent on the microstructure density in the etched area and the gas depletion near the surface. Microscopic etching uniformity refers to achieving similar etch depths and profiles throughout the surface of the wafer. Due to reasons such as gas depletion and plasma non-uniformity, a wide open etched area often exhibits a higher etch rate than that observed in the smaller areas (microloading). Ion scattering results in an unexpected etching profile depending on the process pressure and the etchant gas. For SiC, basic digital integrated circuit on a single wafer is still under development [113]. Applying the same principles as for Si, a high degree of etching profile control is required for the etching of sub-μm SiC. To achieve this condition, gases resulting in higher etch rates such as NF_3 and SF_6 are advantageous. For example, we have previously reported [73] etching of 3C-SiC in pure NF_3 and $NF_3/10\%\ O_2$ mixture with etch rates in excess of 600 Å/min, and a highly anisotropic profile ratio (vertical-to-lateral etch distance) of 10:1 to 15:1 (equivalent to $A = 0.9$ to 0.93). It is important also to consider the issue of etch residues during sub-μm etching under these conditions because metal contact is usually the subsequent process.

8. Etching Conditions and Trade-Offs

In the previous sections, RIE etching of SiC has been shown to produce usable etch rates (100 to 1000 Å/min) and a high degree of etching anisotropy [53, 73, 75]. In most cases, the etch rate increases proportionately with the input rf power [53]. To obtain

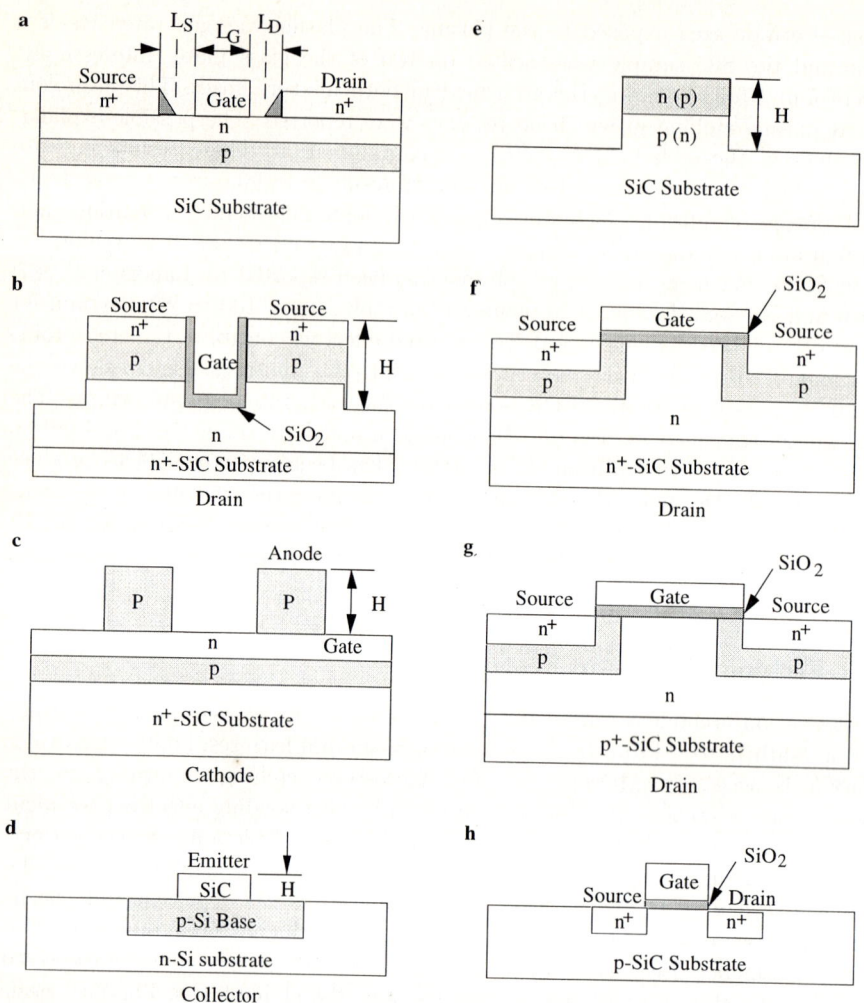

Fig. 30. Device structures: a) MESFET, b) UMOS, c) thyristor, d) HBT, e) diode, f) DMOS, g) IGT, and h) MOS [111]. H indicates the etch step height

residue-free etching of SiC, one can utilize either H_2 additive in fluorinated oxygen gas mixtures or H_2 containing gas in fluorinated gas mixtures. Both these approaches suffer from a reduced etch rate and anisotropy. For example, in the fluorinated mixture, one could obtain an etch rate of only $\approx 150\,\text{Å/min}$ and an etch aspect ratio (depth/width) of ≈ 3 to 4, equivalent to anisotropy $A = 0.66$ to 0.75. From a device fabrication point of view, higher etch rates and etch aspect ratios are desirable for etching of trenches for power devices and for device isolation. For example, as shown in the cross-sectional view of a MESFET in Fig. 30a, the surface morphology of gate trench etching becomes extremely important for the subsequent gate metal deposition. In a currently published MESFET device structure [14], a $0.15\,\mu\text{m}$ n^+ SiC layer has been used for forming the metal–semiconductor contact. The length of the gate (L_G), and spacing gate–source (L_S) and

Table 2
Summary of etching results and issues on RIE of 3C-SiC

issues	source gas(es)	$\approx$ etch rate (Å/min)	selectivity SiC/Si	etch aspect ratio	residue-free [a])
high selectivity	$CHF_3/80\% O_2$	350	2		N
	$CBrF_3/80\% O_2$	380[b])	2		N
	$CF_4/80\% O_2$	200	2		N
high etch rate	NF_3	600		> 10	spikes
	$NF_3/10\% O_2$	830		> 10	spikes
	$SF_6/20\% O_2$	450		> 10	N
	$CF_4/20\% O_2$	350		7 to 8	N
	$CHF_3/60$ to $80\% O_2$	350 to 370		3 to 4	N
	$CBrF_3/70$ to $80\% O_2$	350 to 400[b])			
etch aspect ratio	NF_3	600		15	spikes
	$NF_3/10\% O_2$	830		> 10	spikes
	$SF_6/20\% O_2$	450		> 10	N
residue-free non-metallic electrode	one can use all of the conditions given above				Y
H_2 additive (Fig. 10b and 11)	$NF_3/ > 50\% > O_2$	40 to 150		3 to 4	Y
	$SF_6/ > 10\% O_2$	50 to 200		3 to 4	Y
	$CF_4/0$ to $90\% O_2$	40 to 180	2	3 to 4	Y
	$CHF_3/0$ to $90\% O_2$	50 to 350	2	3 to 4	Y
fluorinated mixtures	$CF_4/ > 75\% CHF_3$	100 to 150		3 to 4	Y
	$NF_3/ > 95\% CHF_3$	150			Y

[a]) Y: Yes; N: No.
[b]) Poly-SiC.

gate–drain (L_D) are given as 0.7, 0.3, and 0.8 µm, respectively. We have shown earlier that average etch steps of the order of ≈ 1.5 µm can be achieved in fluorinated mixture plasmas [82]. With an expected aspect ratio of ≈ 3, the lateral etch resulting while etching through the 0.15 µm thick n^+ layer is only 0.05. The n^+ source and drain islands can therefore be readily patterned under residue-free conditions.

In addition to the high speed MESFET device, several other devices [114], such as UMOS, thyristor, heterojunction bipolar transistor (HBT), p–n diode, double diffusion MOS (DMOS), insulated gate transitor (IGT), and MOS capacitors are reported with promising electrical performance. The corresponding device structures are shown in Fig. 30b to h, respectively. For example, UMOS, HBT, and thyristor strictly require residue-free etching for subsequent metal contact process. For MOS devices, the critical issue is to improve the interface quality between SiO_2 and SiC. However, a high etch rate is required to minimize the cost, i.e. increase the throughput. The various etching requirements are dependent on the specific device structure. High etch rates and highly anisotropic profiles are usually necessary for patterning of most of the devices. In the past, we have reported on the reactive ion etching of SiC by discussing the key issues such as selectivity, surface morphology, etch aspect ratio, and etch rate. These key etching issues and etching results for various processes are outlined in Table 2 for 3C-SiC.

Information on the etching of hexagonal (4H and 6H) SiC is summarized in Table 1. Often a trade-off between a variety of requirements is needed.

Interestingly, the residues observed on the etched regions in SiC are similar to those found on Si and attributed to electrode metal contamination [89] when it was etched in pure CF_4 plasma. With H_2 additive in the CF_4 plasma, the residues on the Si etch field were removed [73, 75]. This indicates that the explanation of surface roughening mechanism by attribution to the "micromasking effect" is correct. Since residue-free etching was obtained in Si etching with H_2 additive as well, the mechanism of "etching enhancement of carbon-rich surface" that we proposed for SiC etching in RIE may in fact not be a dominant process parameter during the etching of SiC. Si etching results indicate that gas phase reaction of volatile compounds plays an extremely important role in removing the metal contamination. This mechanism [90] during the Si residue-free etching results in a lower etch rate with the CF_4/H_2 mixture. For etching SiC, however, one can easily find a residue free process resulting in etch rates that are comparable to that resulting in residues. For example, in Fig. 9a, within the shadowed rectangular residue-free region, one finds similar etch rate with $(SF_6/90\% CHF_3)$ and without $(CF_5/>75\% CHF_3$ and $NF_3/>95\% CHF_3)$ residues. However, several side-effects may be encountered. For example when H_2 is used as an additive to obtain a residue-free process, it may be implanted into the etched surface resulting in deactivated dopants (thus lowering the surface doping concentration). H_2 may also lead to increased polymer generation, it often affects the etch rate, and results in safety issues (particularly when used eith O_2) during the process.

9. Current Issues and Future Directions

Nearly all plasma processes create damage to the semiconductor materials. Non-uniform plasma may develop lateral currents which result in gate dielectric damage (charging damage) [115 to 118] in the fabrication of Si CMOS ULSI device. The damage itself could also be a part of the overall etching mechanism, such as damage induced chemical reaction (see Section 2). Not only do the charge particles (ions) create this damage, but also high energy photons may produce thin damage layers. For example, when high energy electrons are swept away from the plasma zone and collide with the side walls of the chamber, they may create potentially damaging high energy X-ray [119, 120] and ultraviolet [121] photons. As shown in Fig. 29 [56], plasma etching damage of SiC has been observed by measuring the leakage current in SiC Schottky diodes: the higher the level of damage, the higher is the leakage current. RIE results in higher damage because of the high self-induced bias. It has no control over this bias which is almost fixed for a given system and input rf power. Unlike RIE, high density plasma sources (such as ECR plasma) usually have low plasma potential (<50 V) so that ion bombardment energy can be controlled at a minimum level of plasma potential. The use of high density plasma sources could therefore minimize the problem of damage due to ion bombardment.

From a production point of view, Si ULSI technology is transferring from a multi-wafer process to a single-wafer process. As the wafer size becomes larger, etching uniformity will become more and more critical in determining the overall etching performance. To increase macro-and micro-etching uniformity [123], substrate temperature and process pressure are the two major factors. To ensure uniform etching of small size samples of SiC in a multi-wafer RIE system, one should control the pressure to reduce ion scattering and maintain laminar etchant gas flow. Various high density plasma sources are better candidates than RIE, which is operated in a lower process pressure and plasma potential.

Highly anisotropic etching profiles are extremely important for patterning most device structures. As the Si device size shrinks, high aspect ratio of the etching profile becomes critical. For SiC, highly anisotropic etching profiles have been obtained in most of the fluorinated oxygen mixtures we investigated, except mixtures with CHF_3. The reasons are directly related to the polymer generation, where the SiC provides the carbon to protect the side wall from being etched. In mixtures of CHF_3 and H_2, short-range pro-grade etching profiles were obtained, because of additional polymer deposition in the etched area. On the other hand, a residue-free etching process is obtained with CHF_3/H_2 without the need for covering the metallic electrode. Another approach is the use of a non-metallic material cover to prevent metal contamination during RIE. In Si etching, teflon and quartz are the two most popular materials used to cover the metal chamber wall. However, a trade-off between the etch rate, the etching profile, the material design for a given reactor chamber wall, and the type of plasma source utilized is required.

10. Summary

In this paper, we have reviewed recent literature on reactive ion etching of SiC in a variety of fluorinated oxygen mixture gases such as CHF_3/O_2, $CBrF_3/O_2$, CF_4/O_2, SF_6/O_2, NF_3/O_2 and fluorinated mixture gases such as CF_4/CHF_3, SF_6/CHF_3, and NF_3/CHF_3 and SF_6/NF_3. The following items have been discussed: 1. basic etching mechanisms; 2. etch rate, etch aspect ratio (depth/width), and ERR; 3. techniques for obtaining residue-free etching (smooth etched surface); 4. techniques for obtaining sub-μm feature etching; 5. recent progress of high density plasma etching of SiC; 6. the limitations of SiC reactive ion etching.

Acknowledgements   The authors would like to acknowledge the support of this work at the University of Cincinnati by the Edison Materials Technology Center, NASA Lewis Research Center, and Wright Patterson Air Force Base. One of us (A.J.S.) would like to acknowledge the early support and encouragement in this direction by Dr. David Nelson.

References

[1] W. F. KNIPPENBERG, Philips Res. Rep. **18**, 161 (1963).
[2] Cree Research, Inc., Durham, NC 27713, USA.
[3] P. G. NEUDECK and J. A. POWELL, IEEE Electron Device Lett. **15**, 63 (1994).
[4] J. A. POWELL, L. G. MATUS, and M. A. KUCZMARSKI, J. Electrochem. Soc. **134**, 46 (1987).
[5] A. J. STECKL and J. P. LI, IEEE Trans. Electron Devices **39**, 2672 (1988).
[6] A. J. STECKL, C. YUAN, J. P. LI, and M. J. LOBODA, Appl. Phys. Lett. **63**, 3347 (1993).
[7] C. YUAN, A. J. STECKL, and M. J. LOBODA, Appl. Phys. Lett. **64**, 3000 (1994).
[8] J. A. POWELL and L. G. MATUS, Springer Proc. Phys. **34**, 2 (1989).
[9] J. A. POWELL, P. G. NEUDECK, L. G. MATUS, and J. B. PETIT, Mater. Res. Soc. Symp. Proc. **242**, 495 (1991).
[10] R. J. TREW, J.-B. YAN, and P. M. Mock, Proc. IEEE **79**, 598 (1991).
[11] R. F. DAVIS, G. KELNER, M. SHUR, J. W. PALMOUR, and J. A. EDMOND, Proc. IEEE **79**, 677 (1991).

[12] H. Matsunami, Springer Proc. Phys. **71**, 3 (1992).

[13] M. Bhatnagar and B. J. Baliga, IEEE Trans. Electron Devices **40**, 645 (1993).

[14] C. E. Weitzel, J. W. Palmour, C. H. Carter, Jr., and K. Nordquist, IEEE Electron Device Lett. **15**, 406 (1994).

[15] J. W. Palmour, C. E. Weitzel, K. Nordquist, and C. H. Carter, Jr., Inst. Phys. Conf. Ser. No. 137, Chap. 6, 495 (1993).

[16] J. W. Palmour, H. S. Kong, and R. F. Davis, J. Appl. Phys. **64**, 2168 (1988).

[17] D. M. Brown, E. Downey, and M. Ghezzo, Solid State Electronics **39**, 1531 (1996).

[18] V. A. Dimitriev, M. E. Levinshtein, S. N. Vainshtein, and V. E. Chelnokov, Electronics Lett. **24**, 1031 (1988).

[19] J. W. Palmour, J. A. Edmond, H. S. Kong, and C. H. Carter, Jr., Inst. Phys. Conf. Ser. No. 137, Chap. 6, 499 (1993).

[20] W. Xie, J. A. Cooper, Jr., M. R. Melloch, J. W. Palmour, and C. H. Carter, Jr., IEEE Electron Device Lett. **15**, 212 (1994).

[21] C. T. Gardner, J. A. Cooper, Jr., M. R. Melloch, J. W. Palmour, and C. H. Carter, Jr., Appl. Phys. Lett. **61**, 1185 (1992).

[22] T. Sugii, T. Ito, Y. Furumura, M. Doki, F. Mieno, and M. Maeda, J. Electrochem. Soc. **134**, 2545 (1987).

[23] T. Sugii, T. Aoyama, and T. Ito, J. Electrochem. Soc. **136**, 3111 (1989).

[24] T. Sugii, T. Ito, Y. Furumura, M. Doki, F. Mieno, and M. Maeda, IEEE Electron Device Lett. **9**, 87 (1988).

[25] T. Sugii, T. Yamazaki, and T. Ito, IEEE Trans. Electron Devices **37**, 2331 (1990).

[26] S. Nishino, K. Yamazaki, H. Tanaka, and J. Saraie, Springer Proc. Phys. **71**, 411 (1992).

[27] M. Kondo, T. Shiba, Y. Tamaki, and T. Nakamura, J. Electrochem. Soc. **143**, 1949 (1996).

[28] S. T. Sheppard, M. R. Melloch, and J. A. Cooper, IEEE Electron Device Lett. **17**, 4 (1996).

[29] S. Y. Wu and R. B. Campbell, Solid State Electronics **17**, 683 (1974).

[30] S. Yoshida, K. Sasaki, E. Sakuma, S. Misawa, and S. Gonda, Appl. Phys. Lett. **46**, 766 (1985).

[31] S. Yoshida, H. Daimon, M. Yamanaka, E. Sakuma, S. Misawa, and K. Endo, J. Appl. Phys. **60**, 2989 (1986).

[32] N. A. Papanicolaou, A. Christou, and L. Gipe, J. Appl. Phys. **65**, 3526 (1989).

[33] K. Das, H. S. Kong, J. B. Petit, J. W. Bumbarner, and R. F. Davis, J. Electrochem. Soc. **137**, 1958 (1990).

[34] J. R. Waldrop and R. W. Grant, Appl. Phys. Lett. **56**, 557 (1990).

[35] M. Bhatnagar, P. K, McLarty, and B. J. Baliga, IEEE Electron Device Lett. **13**, 501 (1992).

[36] K. Ueno, T. Urushidani, K. Hashimoto, and Y. Seki, IEEE Electron Device Lett. **16**, 331 (1995).

[37] N. Lundberg, M. Östling, P. Tägtström, and U. Jansson, J. Electrochem. Soc. **143**, 1662 (1996).

[38] T. Urushidani, S. Kobayashi, T. Kimoto, and H. Matsunami, Inst. Phys. Conf. Ser. No. 137, Chap. 6, 471 (1993).

[39] A. J. Steckl, J. N. Su, P. H. Yih, C. Yuan, and J. P. Li, Inst. Phys. Conf. Ser. No. 137, Chap. 6, 653 (1993).

[40] A. J. Steckl and J. N. Su, IEEE Electr. Dev. Meet. Tech. Digest 93CH3361-3, 695 (1993).

[41] J. E. Edmond, K. Das, and R. F. Davis, J. Appl. Phys. **63**, 923 (1988).

[42] L. G. Matus, J. A. Powell, and C. S. Salupo, Appl. Phys. Lett. **59**, 1770 (1991).

[43] M. Ghezzo, D. M. Brown, E. Downey, J. Kretchmer, W. Hennessy, D. L. Polla, and H. Bakhru, IEEE Electron Device Lett. **13**, 639 (1992).

[44] M. Ghezzo, D. M. Brown, E. Downey, J. Kretchmer, and J. J. Kopanski, Appl. Phys. Lett. **63**, 1206 (1993).

[45] S. Yaguchi, T. Kimoto, N. Ohyama, and H. Matsunami, Jpn. J. Appl. Phys. **34**, 3036 (1995).

[46] H. Morkoç, S. Strite, G. B. Gao, M. E. Lin, and B. Sverdlov, and M. Burns, J. Appl. Phys. **76**, 1363 (1994).

[47] B. J. BALIGA, IEEE Spectrum **32**, 34 (1995).
[48] J. W. FAUST, JR., in: The Etching of SiC, Eds. J. R. O'CONNOR and J. SMILTENS, Pergamon Press, London/Oxford 1960.
[49] J. S. SHOR and R. M. OSGOOD, and A. D. KUTZ, Appl. Phys. Lett. **60**, 1001 (1992).
[50] J. S. SHOR and A. D. KUTZ, J. Electrochem. Soc. **141**, 778 (1994).
[51] Y. HIBI, Y. ENOMOTO, K. KIKUCHI, and N. SHIKATA, Appl. Phys. Lett. **66**, 817 (1995).
[52] J. SUGIURA, W. J. LU, K. C. CADIEN, and A. J. STECKL, J. Vacuum Sci. Technol. B **4**, 349 (1986).
[53] W.-S. PAN and A. J. STECKL, J. Electrochem. Soc. **137**, 212 (1990).
[54] E. NIEMANN, A. BOOS, and D. LEIDICH, Inst. Phys. Conf. Ser. No. 137, Chap. 7, 695 (1993).
[55] J. R. FLEMISH, K. XIE, and J. H. ZHAO, Appl. Phys. Lett. **64**, 2315 (1994).
[56] K. XIE, J. R. FLEMISH, J. H. ZHAO, W. R. BUCHWALD, and L. CASAS, Appl. Phys. Lett. **67**, 368 (1995).
[57] H. F. WINTERS, J. Appl. Phys. **49**, 5165 (1978).
[58] J. W. COBURN and H. F. WINTERS, J. Vacuum Sci. Technol. **16**, 391 (1979).
[59] H. F. WINTERS, J. W. COBURN, and T. J. CHUANG, J. Vacuum Sci. Technol. B **1**, 469 (1983).
[60] D. L. FLAMM and G. K. HERB, in: Plasma Etching: An Introduction, Eds. D. M. MANOS and D. L. FLAMM, Chap. 1, Academic Press, New York 1989.
[61] J. W. COBURN and E. KAY, IBM J. Res. Develop. **23**, 33 (1979).
[62] T. M. MAYER and R. A. BARKER, J. Vacuum Sci. Technol. **21**, 757 (1982).
[63] D. L. FLAMM, V. M. DONELLY, and J. A. MUCHA, J. Appl. Phys. **52**, 3633 (1981).
[64] C. J. MOGAB, A. C. ADAMS, and D. L. FLAMM, J. Appl. Phys. **49**, 3796 (1978).
[65] D. L. FLAMM, see [60] (Chap. 2).
[66] G. W. GRYNKEWICH and J. N. HELBERTIN, in: Handbook of Multilevel Metallization for Integrated Circuits, Eds. S. R. WILSON, C. J. TRACY, and J. L. FREEMAN, JR., Chap. 7, Noyes Publication, New Jersey 1993.
[67] J. W. PALMOUR, R. F. DAVIS, P. ASTELL-BURT, and P. BLACKBOROW, Mater. Res. Soc. Symp. Proc. **76**, 185 (1987).
[68] R. LEGTENBERG, H. JANSEN, M. DE BOER, and M. ELWENSPOEK, J. Electrochem. Soc. **142**, 2020 (1995).
[69] T. SYAN, B. J. BALIGA, and R. W. HAMAKER, J. Electrochem. Soc. **138**, 3076 (1991).
[70] J. W. PALMOUR, R. F. DAVIS, T. W. WALLETT, and K. B. BHASIN, J. Vacuum Sci. Technol. A **4**, 590 (1986).
[71] J. W. COBURN and M. CHEN, J. Appl. Phys. **51**, 3134 (1980).
[72] R. PADIYATH, R. L. WRIGHT, M. I. CHAUDHRY, and S. V. BABU, Appl. Phys. Lett. **58**, 1053 (1991).
[73] P. H. YIH and A. J. STECKL, J. Electrochem. Soc. **140**, 1813 (1993).
[74] J. WU, J. D. PARSONS, and D. R. EVANS, J. Electrochem. Soc. **142**, 669 (1995).
[75] P. H. YIH and A. J. STECKL, J. Electrochem. Soc. **142**, 312 (1995).
[76] R. WOLF and R. HELBIG, J. Electrochem. Soc. **143**, 1037 (1996).
[77] G. KELNER, S. C. BINARI, and P. H. KLEIN, J. Electrochem. Soc. **134**, 253 (1987).
[78] J. B. CASADY, E. D. LUCKOWSKI, M. BOZACK, D. SHERIDAN, R. W. JOHNSON, and J. R. WILLIAMS, J. Electrochem. Soc. **143**, 1750 (1996).
[79] B. P. LUTHER, J. RUZYLLO, and D. L. MILLER, Appl. Phys. Lett. **63**, 171 (1993).
[80] N. J. IANNO, K. E. GREENBERG, and J. T. VERDEYEN, J. Electrochem. Soc. **128**, 2174 (1981).
[81] D. R. SPARKS, J. Electrochem. Soc. **139**, 1736 (1992).
[82] P. H. YIH and A. J. STECKL, J. Electrochem. Soc. **142**, 2853 (1995).
[83] P. G. NEUDECK, D. J. LARKIN, J. E. STARR, J. A. POWELL, C. S. SALUPO, and L. G. MATUS, IEEE Electron Device Lett. **14**, 136 (1993).
[84] H. PARK, K. KWON, and J. LEE, J. App. Phys. **76**, 4596 (1994).
[85] S. DOHMAE, K. SHIBAHARA, S. NISHINO and H. MATSUNAMI, Japan. J. Appl. Phys. **24**, L873 (1985).
[86] M. KOTHANDARAMAN, D. ALOK, and B. J. BALIGA, J. Electronic Mater. **25**, 875 (1996).
[87] F. LANOIS, P. LASSAGNE, D. PLANSON, and M. L. LOCATELLI, Appl. Phys. Lett. **69**, 236 (1996).

[88] J. R. Flemish and K. Xie, J. Electrochem. Soc. **143**, 2620 (1996).
[89] A. J. Steckl and P. H. Yih, Appl. Phys. Lett. **60**, 1966 (1992).
[90] L. M. Ephrath and R. S. Bennett, J. Electrochem. Soc. **129**, 1822 (1982).
[91] M. Valente and G. Queirolo, J. Electrochem. Soc. **131**, 1132 (1984).
[92] J. Kaindl, S. Sotier, and G. Franz, J. Electrochem. Soc. **142**, 2418 (1995).
[93] F. Gendron, L. M. Porter, C. Porte, and E. Bringuier, Appl. Phys. Lett. **67**, 1253 (1995).
[94] S. J. Fonash, J. Electrochem. Soc. **137**, 3885 (1990).
[95] G. S. Orehrlein, R. M. Tromp, Y. H. Lee, and E. J. Petrillo, Appl. Phys. Lett. **45**, 420 (1984).
[96] S. Oda, J. Noda, and M. Matsumura, Jpn. J. Appl. Phys. **28**, L1860 (1989).
[97] J. B. Carter, J. P. Holland, E. Peltzer, B. Richardson, E. Bogle, H. T. Nguyen, Y. Melaku, D. Gates, and M. Ben-Dor, J. Vacuum Sci. Technol. A **11**, 1301 (1993).
[98] R. W. Boswell and R. K. Porteous, Appl. Phys. Lett. **50**, 1130 (1987).
[99] A. J. Perry and R. W. Boswell, Appl. Phys. Lett. **55**, 148 (1989).
[100] R. W. Boswell, A. J. Perry, and M. Emami, J. Vacuum Sci. Technol. A **7**, 3345 (1989).
[101] A. J. Perry, D. Vender, and R. W. Boswell, J. Vacuum Sci. Technol. B **9**, 310 (1991).
[102] J. M. Cook, D. E. Ibbotson, and D. L. Flamm, J. Vacuum Sci. Technol. B **8**, 1 (1990).
[103] J. M. Cook, D. E. Ibbotson, P. D. Foo, and D. L. Flamm, J. Vacuum Sci. Technol. A **8**, 1820 (1990).
[104] Lam Research Corporation, Technical Note TN-003 (1992).
[105] G. Chevalier and F. F. Chen, J. Vacuum Sci. Technol. A **10**, 1389 (1992).
[106] R. W. Boswell, Plasma Physics and Controlled Fusion **26**, 1147 (1984).
[107] D. L. Flamm, D. E. Ibbotson, and W. L. Johnson, US Patent No. 4,918,031 (1990).
[108] I. Tepermeister, N. Blayo, F. P. Klemens, D. E. Ibbotson, R. A. Gottscho, J. T. C. Lee, and H. H. Sawin, J. Vacuum Sci. Technol. B **12**, 2310 (1994).
[109] I. Tepermeister, D. E. Ibbotson, and J. T. Lee, and H. H. Sawin, J. Vacuum Sci. Technol. B **12**, 2322 (1994).
[110] G. W. Gibson, Jr., H. H. Sawin, I. Tepermeister, D. E. Ibbotson, and J. T. C. Lee, J. Vacuum Sci. Technol. B **12**, 2333 (1994).
[111] N. Blayo, I. Tepermeister, J. L. Benton, G. S. Higashi, T. Boone, A. Onuoha, F. P. Klemens, D. E. Ibbotson, J. T. C. Lee, and H. H. Sawin, J. Vacuum Sci. Technol. B **12**, 1340 (1994).
[112] R. A. Gottscho, C. W. Jurgensen, and D. J. Vitkavage, J. Vacuum Sci. Technol. B **2133**, 10 (1992).
[113] P. G. Neudeck, J. Electronic Mater. **24**, 283 (1995).
[114] B. J. Baliga, in: Modern Power Devices, John Wiley & Sons, New York 1987.
[115] D. J. DiMaria, L. M. Ephrath, and D. R. Young, J. Appl. Phys. **50**, 4015 (1979).
[116] L. M. Ephrath and D. J. DiMaria, Solid State Technol. 182 (April 1981).
[117] C. T. Gabriel and J. P. McVittie, Solid State Technol. 81 (June 1992).
[118] C. T. Gabriel and Y. Melaku, J. Vacuum Sci. Technol. B **12**, 454 (1994).
[119] T. J. Castagna, J. L. Shohet, K. A. Ashitani, and N. Hershkowitz, J. Vacuum Sci. Technol. A **10**, 1325 (1992).
[120] T. J. Castagna, J. L. Shohet, D. D. Denton, and N. Hershkowitz, Appl. Phys. Lett. **60**, 2856 (1992).
[121] D. A. Buchanan and G. Fortuño-Wiltshire, J. Vacuum Sci. Technol. A **9**, 804 (1991).
[122] K. P. Giapis, G. R. Scheller, W. S. Hobson, R. A. Gottscho, and Y. H. Lee, Appl. Phys. Lett. **57**, 983 (1990).

Contents of Volume II

Characterization of SiC

Processing

Devices